NOW EXTRA HELP IN
IS ONLY A PHONE CA

GW00363741

SOFT WARE

SAS Institute
Powerful—handles unlimited data sets. Easy to learn. Completely menu-driven. Combines classical statistics with today's most interactive graphics! Available for Windows & Macintosh.

Stata Corporation
Inexpensive! Powerful. Easy-to-use. Graphically superior. Contains a 250 page user's guide. Windows, Macintosh, and DOS versions.

Strategy Plus, Inc.
Analytically powerful, easy-to-use, intuitive format. Menu-Driven. MiniGuide available (sold seperately). DOS version.

HAND BOOKS

Middleton
Clear. Simple. Demonstrates Microsoft Excel's basic and advanced data analysis features—from simulations to graphics to database management.

Ryan and Joiner
Updated, now covers Releases 7-10! Ideal for hands-on sessions in the computer lab.

Dilorio
Straightforward. Practical. Self-directed. Contains everything needed to get started with SAS.

George and Mallery
Ideal statistics guide for the social sciences. Provides step-by-step instruction for learning SPSS.

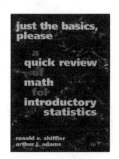

▼

To Order Call Your
Local Bookstore

A COURSE IN BUSINESS STATISTICS
FOURTH EDITION

WILLIAM MENDENHALL
University of Florida, Emeritus

ROBERT J. BEAVER
University of California, Riverside

BARBARA M. BEAVER
University of California, Riverside

DUXBURY PRESS
An Imprint of Wadsworth Publishing Company

I(T)P® An International Thomson Publishing Company

Belmont • Albany • Bonn • Boston • Cincinnati • Detroit • London • Madrid • Melbourne
Mexico City • New York • Paris • San Francisco • Singapore • Tokyo • Toronto • Washington

Editor	CURT HINRICHS
Assistant Editor	JENNIFER BURGER
Editorial Assistants	JANIS BROWN, CYNTHIA MAZOW
Marketing Manager	JOANNE TERHAAR
Advertising Project Manager	JOSEPH JODAR
Production Editor	SANDRA CRAIG
Print Buyer	BARBARA BRITTON
Permissions Editor	PEGGY MEEHAN
Copy Editor	MARY ROYBAL
Text Design	CLOYCE WALL
Cover Design	JANET BOLLOW
Cover Illustration	RON LOWERY, THE STOCK MARKET
Screen Captures	LAUREL TECHNICAL SERVICES
Illustrations, Composition	INTERACTIVE COMPOSITION CORPORATION
Printer	R. R. DONNELLEY & SONS/CRAWFORDSVILLE

Printed on acid-free recycled paper.

For more information, contact Duxbury Press at Wadsworth Publishing Company:

Wadsworth Publishing Company
10 Davis Drive
Belmont, California 94002, USA

International Thomson Publishing Europe
Berkshire House 168-173
High Holborn
London, WC1V 7AA, England

Thomas Nelson Australia
102 Dodds Street
South Melbourne 3205
Victoria, Australia

Nelson Canada
1120 Birchmount Road
Scarborough, Ontario
Canada M1K 5G4

International Thomson Editores
Campos Eliseos 385, Piso 7
Col. Polanco 11560 México D.F. México

International Thomson Publishing GmbH
Königswinterer Strasse 418
53227 Bonn, Germany

International Thomson Publishing Asia
221 Henderson Road
#05-10 Henderson Building
Singapore 0315

International Thomson Publishing Japan
Hirakawacho Kyowa Building, 3F
2-2-1 Hirakawacho
Chiyoda-ku, Tokyo 102, Japan

Library of Congress Cataloging-in-Publication Data

Mendenhall, William.
    A course in business statistics / William Mendenhall, Robert J.
Beaver, Barbara M. Beaver. — 4th ed.
      p. cm.
    Includes index.
    ISBN 0-534-26508-1 (hardcover)
    1. Commercial statistics. I. Beaver, Robert J.   II. Beaver,
Barbara M.   III. Title.
HF1017.M458   1996
519.5—dc20
                                  95-34227

A COURSE IN BUSINESS STATISTICS

BRIEF CONTENTS

CONTENTS

CHAPTER 10   QUALITY CONTROL   421

A Course in Business Statistics is intended for a one- or two-term business statistics course with an algebra prerequisite. The Fourth Edition retains the emphasis and flavor of earlier versions while using more computer graphics from several statistical software packages to present text material. The goal of the chapter presentations is to make students aware of the important role of statistics in their daily lives, as reported in the media, and its role in making informed business decisions. This is accomplished by using a case study taken from the current media to introduce the major topic of each chapter. In addition, timely exercises reinforce chapter topics.

The content of this revision reflects the conclusions of the conference on business statistics originally held at the Graduate School of the University of Chicago in June of 1986[†] that courses in business statistics are most effective when:

- Motivation is stimulated by real examples.

- Applications are emphasized rather than probability theory.

- Topics such as sampling, experimental design, quality control, time series analysis, and report writing receive emphasis.

ORGANIZATION AND COVERAGE

The sequence of chapters and the material within chapters reflect the pedagogy followed by the authors. We believe that Chapters 1–8 should be covered in the order presented. Chapter 10, Quality Control, and Chapter 15, The Chi-Square Goodness-of-Fit Test, can be covered before Chapter 9, The Analysis of Variance. However, coverage of Chapter 9 should precede the study of Chapters 11 and 12, which deal with regression and correlation using the printouts from statistical packages, most of which use an analysis of variance in testing for significant regression. The remaining chapters can be covered in any order.

REVISION HIGHLIGHTS

Modifications in this edition reflect the changing role of business statistics and our goal of keeping the text current in its presentation of topics and applications for which statistical techniques are necessary tools. Specific changes include the following:

- Appendix III contains thirteen additinoal case studies that can be integrated into various chapters. Their data appear on the data disk that accompanies this text.

- In addition, two chapter case studies have been updated and nine have been replaced based on timely new situations from current literature.

[†] George Easton, Harry V. Roberts, and George C. Tiao, "Preliminary Report on Making Statistics More Effective in Schools of Business: Overview and Summary," University of Chicago, December 1986.

- The order of Chapters 1–8 is the same as that of the Third Edition. However, Chapter 10 is now Quality Control, while The Chi-Square Goodness-of-Fit Test has become Chapter 15, following Sampling Methods.

- Two new sections in Chapter 2 deal with graphical methods of data presentation. The demonstration that the sum of squared deviations can be calculated using the sum and the sum of squares of each observation is now integrated into the section in which the sample variance is introduced. A new section on bivariate data now appears in Chapter 2. Examples involve bivariate data in which one variable is qualitative and the other is quantitative, and the case in which two variables are quantitative. Scatterplots are used for graphical description, and Pearson's coefficient of correlation is introduced as a measure of linear association between two quantitative variables. Population percentiles remain, but examples of sample percentiles have been deleted.

- In Chapters 3, 4, and 5, which cover probability, discrete probability distributions, and continuous probability distribution, including the normal, many small but significant changes have clarified and shortened the text.

- Chapter 6, Sampling Distributions, describes a direct application of the sampling distribution of the sample mean in the context of quality control using an x-bar chart. This topic introduces the variation in sample means that can be observed when a process is in control, and what variation in a series of sample means would lead an experimenter to determine that a process is not in control.

- In Chapter 7, the sections on large sample point estimation and confidence interval estimators have been combined and the discussion streamlined.

- In addition to the ideas of statistical process control, the topic of process capability has been added to Chapter 10 to improve the presentation of quality management.

- Chapter 11, Simple Linear Regression, has been heavily revised to emphasize examining and interpreting the output of regression programs. Pertinent formulas are introduced in Section 11.6, allowing instructors and students to verify or obtain the results of regression analysis or both.

- The former Chapter 17, Decision Analysis, has been removed from the text but is available from the publisher.

- Exercise sets have been revised. Application problems are based on data published in 1990 or more recently (the few exceptions illustrate an exceptional technique or application). Problems in the multiple regression and time series chapters have been updated. The many new replacement exercises are from a variety of fields.

- The tables in Appendix II have been rearranged to reflect their order of introduction in the text. The table of values of $e^{-\mu}$ and that of the arcsin transformation have been deleted because both of these quantities can be easily found using standard statistical software or scientific calculators.

THE ROLE OF THE COMPUTER

The Minitab and Execustat statistical packages are integrated throughout the text, the Statistical Analysis Systems (SAS) package is also used in regression analysis, and Microsoft Excel, a commonly used spreadsheet package with statistical applications, appears in several chapters. The disk that accompanies each new copy of the book includes all data sets that contain ten or more entries. Data are formatted in ASCII, Microsoft Excel, Execustat, Minitab, SAS, and StataQuest.

Appendix I describes the contents of the data disk, including five large data sets that can be used throughout the course. Data sets A, salaries for male and female college professors for institutions having faculty in each of five ranks; B, the yield characteristics of broccoli; and E, an updated time series consisting of the total manufacturer's inventories, given as book values in billions of dollars recorded at the end of the period unadjusted for seasonal variation, from the Third Edition, are augmented by two new data sets: C, the record of three variables recorded for 604 market money funds for the period ending July 13, 1994; and D, the 100 best small companies (*Business Week,* May 23, 1994).

ACKNOWLEDGMENTS

We would like to recognize the assistance of the reviewers who aided this revision: Sung K. Ahn, Washington State University; Bruce Bowerman, Miami University; Paul W. Guy, California State University, Chico; Robert Hannum, University of Denver; Shu-ping Hodgson, Central Michigan University; Jacqueline F. Hoell, Virginia Polytechnic Institute; David W. Pentico, Duquesne University; and Sue B. Schou, Idaho State University.

We also want to thank the members of the Duxbury Press staff who helped produce this edition: our editor, Curt Hinrichs; Jennifer Burger, assistant editor; Janis Brown and Cynthia Mazow, editorial assistants; Mary Roybal, copy editor; and Sandra Craig, production editor.

A COURSE IN BUSINESS STATISTICS, FOURTH EDITION
BY WILLIAM MENDENHALL, ROBERT J. BEAVER, AND BARBARA M. BEAVER

Imagine, if you will, landing that ideal job you spent your whole college career working toward. Then one morning your boss walks into your office and hands you a stack of data. "Give me an analysis of this by tomorrow morning," she commands.

Growing numbers of employers now expect new college-educated employees to understand data. *A Course in Business Statistics, Fourth Edition* is an introduction to the concepts and methods businesspeople need to know. This book explains, quantitatively and qualitatively, the way many modern organizations use statistics. The examples and case studies it contains pertain to business administration and its related subfields.

To get the most out of this book, you should familiarize yourself with its many learning features. Some of them are graphical, some are based on computer simulation, and others are anecdotal. On the following pages, you will find examples of some features found in this book and suggestions on how to make the best use of them.

CASE STUDIES

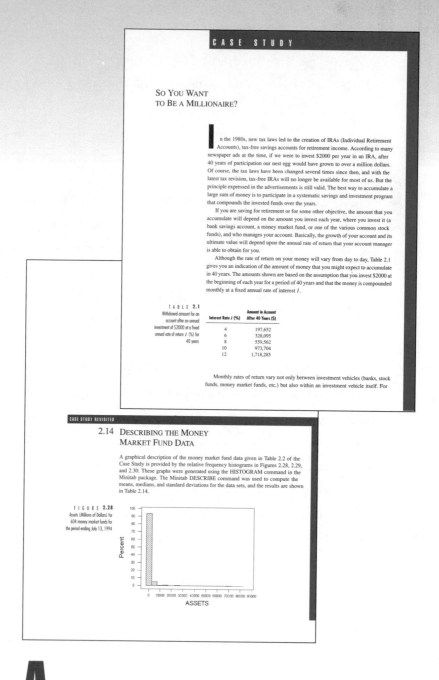

CASE STUDY

SO YOU WANT TO BE A MILLIONAIRE?

In the 1980s, new tax laws led to the creation of IRAs (Individual Retirement Accounts), tax-free savings accounts for retirement income. According to many newspaper ads at the time, if we were to invest $2000 per year in an IRA, after 40 years of participation our nest egg would have grown to over a million dollars. Of course, the tax laws have been changed several times since then, and with the latest tax revision, tax-free IRAs will no longer be available for most of us. But the principle expressed in the advertisements is still valid. The best way to accumulate a large sum of money is to participate in a systematic savings and investment program that compounds the invested funds over the years.

If you are saving for retirement or for some other objective, the amount that you accumulate will depend on the amount you invest each year, where you invest it (a bank savings account, a money market fund, or one of the various common stock funds), and who manages your account. Basically, the growth of your account and its ultimate value will depend upon the annual rate of return that your account manager is able to obtain for you.

Although the rate of return on your money will vary from day to day, Table 2.1 gives you an indication of the amount of money that you might expect to accumulate in 40 years. The amounts shown are based on the assumption that you invest $2000 at the beginning of each year for a period of 40 years and that the money is compounded monthly at a fixed annual rate of interest I.

TABLE 2.1
Withdrawal amount for an account after an annual investment of $2000 at a fixed annual rate of return I (%) for 40 years

Interest Rate I (%)	Amount in Account After 40 Years ($)
4	197,652
6	328,095
8	559,562
10	973,704
12	1,718,285

Monthly rates of return vary not only between investment vehicles (banks, stock funds, money market funds, etc.) but also within an investment vehicle itself. For

CASE STUDY REVISITED

2.14 DESCRIBING THE MONEY MARKET FUND DATA

A graphical description of the money market fund data given in Table 2.2 of the Case Study is provided by the relative frequency histograms in Figures 2.28, 2.29, and 2.30. These graphs were generated using the HISTOGRAM command in the Minitab package. The Minitab DESCRIBE command was used to compute the means, medians, and standard deviations for the data sets, and the results are shown in Table 2.14.

FIGURE 2.28
Assets (Millions of Dollars) for 604 money market funds for the period ending July 13, 1994

At the beginning of every chapter you will find a **Case Study,** which presents a problem related to the main chapter topic. The **Case Study Revisited** at the end of the chapter demonstrates how the statistical subject discussed in the chapter can be applied to resolve the problem. Case Studies are taken from the current media.

hroughout this book, you will find references to commonly used statistical software packages, including **Minitab, Execustat, SAS,** and spreadsheets such as **Microsoft Excel.** These programs run on most computers. They help to illustrate important concepts, and—as you will come to learn—they handle most of the calculations performed when analyzing large or small groups of data. A disk at the back of this book contains all of the data sets with ten or more entries.

FIGURE **2.8**
Minitab histogram for the dividend yield data

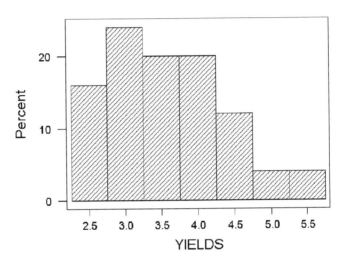

FIGURE **5.4**
Execustat printout showing normal probability distributions with differing values of μ and σ

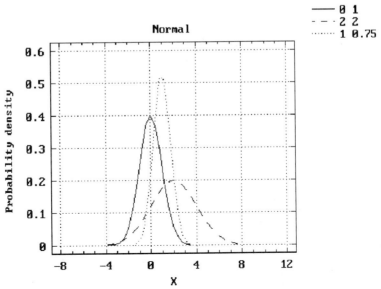

72 Chapter 2 Describing Sets of Data

EXAMPLE 2.20 Consider a sample of $n = 10$ measurements:

$$3, 2, 0, 15, 2, 3, 4, 0, 1, 3$$

You can see at a glance that the measurement $x = 15$ appears to be an outlier. Calculate the z-score for this observation, and state your conclusions.

Solution For the sample, we have the following calculations:

$$\sum_{i=1}^{10} x_i = 33 \quad \text{and} \quad \sum_{i=1}^{10} x_i^2 = 277$$

Then

$$\bar{x} = \frac{\sum_{i=1}^{10} x_i}{n} = \frac{33}{10} = 3.3$$

$$s^2 = \frac{\sum_{i=1}^{n} x_i^2 - \frac{\left(\sum_{i=1}^{n} x_i\right)^2}{n}}{n-1}$$

$$= \frac{277 - \frac{(33)^2}{10}}{9} = \frac{168.1}{9} = 18.6778$$

$$s = 4.32$$

Using these quantities to calculate the z-score for the suspect outlier $x = 15$, we find

$$z\text{-score} = \frac{x - \bar{x}}{s} = \frac{15 - 3.3}{4.32} = 2.71$$

Thus the measurement $x = 15$ lies 2.71 sample standard deviations away from the sample mean $\bar{x} = 3.3$. Since this z-score exceeds 2, we identify $x = 15$ as a possible outlier. We should examine our sampling procedure to see whether evidence exists to indicate that $x = 15$ is a faulty observation. •

Many of the numerical descriptive measures that we have discussed are easily found using the command DESCRIBE in the Minitab package. As part of its output, this program produces the values of the mean, the standard deviation, the median, the largest (**MAX**) and the smallest (**MIN**) of the observations, and the lower and upper quartiles, together with the values of some other statistics that we have not discussed. The **trimmed mean** (given as **TRMEAN**) is the mean of the middle 90% of the measurements after deleting the smallest 5% and the largest 5% of the observations.

Tips on Problem Solving

1. Be careful about rounding numbers. Use a calculator with built-in statistical functions if possible. If you calculate s^2 and s by hand, carry your calculations to at least six significant figures.

2. After you have calculated the standard deviation s for a set of data, compare its value with the range of the data. The Empirical Rule tells you that approximately 95% of the data should fall into the interval $\bar{x} \pm 2s$;

2.11 Graphical and Numerical Techniques for Bivariate Data 67

Most data analysts begin any data-based investigation by examining plots of the variables involved; if the relationship between two variables is of interest, bivariate plots are also explored in conjunction with numerical measures of location, dispersion, and correlation. Graphs and numerical descriptive measures are only the first of many statistical tools you will soon have at your disposal.

Exercises

Basic Techniques

2.46 A set of bivariate data consists of measurements on two variables, x and y, as follows:

$$(3, 6), (5, 8), (2, 6), (1, 4), (4, 7), (4, 6)$$

a. Draw a scatterplot to describe the data.
b. Does there appear to be a relationship between x and y? If so, how would you describe it?

2.47 Consider the set of bivariate data shown below:

x	1	2	3	4	5	6
y	5.6	4.6	4.5	3.7	3.2	2.7

a. Draw a scatterplot to describe the data.
b. Does there appear to be a relationship between x and y? If so, how would you describe it?
c. Calculate the correlation coefficient, r. Does the value of r substantiate your conclusions in part (b)? Explain.

2.48 The value of a quantitative variable is measured once a year for a ten-year period. The data are shown below.

Year	Measurement	Year	Measurement
1	61.5	6	58.2
2	62.3	7	57.5
3	60.7	8	57.5
4	59.8	9	56.1
5	58.0	10	56.0

a. Draw a scatterplot to describe the variable as it changes over time.
b. Describe the measurements using the graph constructed in part (a).

Applications

2.49 The accompanying data represent the average seller's asking price, the average buyer's bid, and the average closing bid for each of ten types of used computer equipment ("Used Computer Prices," 1992).

68 Chapter 2 Describing Sets of Data

Data for Exercise 2.49

Machine	Average Seller's Asking Price	Average Buyer's Bid	Average Closing Bid
20MB PC XT	$400	$200	$300
20MB PC AT	700	400	575
IBM XT 089	450	200	325
IBM AT 339	700	350	600
20MB IBM PS/2 30	950	500	725
20MB IBM PS/2 50	1050	700	875
60MB IBM PS/2 70	2000	1600	1725
20MB Compaq SLT	1200	700	875
Toshiba 1600	1000	700	900
Toshiba 1200HB	1150	800	975

2.50 A Minitab software package was used to find the correlation coefficient, r, for the three pairs of variables in Exercise 2.49. Find the correlation coefficients on the printout. Interpret the values of r in terms of the practical problem of buying a used computer.

```
MTB > Correlation 'ASK-PRC' 'BYR-BID' 'CLOS-BID'

          ASK-PRC  BYR-BID
BYR-BID    0.985
CLOS-BID   0.991    0.994
```

2.51 The tax burden borne by citizens of the United States varies substantially from state to state, and even from city to city. The accompanying table gives an estimate of the state and local taxes paid in two different cities by a family of four with an annual income of $100,000 in 1992 (Tannenwald, 1994).

	City	
Tax	Boston	Kansas City
Income tax	$4142	$3285
Property tax	1734	930
Sales tax	1213	1583
Auto tax	987	1105

Draw a stacked bar... describe the...

O ne of the best ways to learn important concepts is to practice them. *A Course in Business Statistics* contains dozens of **Examples** on essential statistics ideas, explained step by step.

You will also find **Exercises** designed to help test your understanding of the material you just covered in the text and in class.

Tips on Problem Solving are an added feature in this text. They augment your instructor's lessons and reading assignments with tips on selecting the appropriate method for a problem.

Because business students have diverse career interests, this book contains application exercises drawn from many fields. Specially designed icons indicate the area of business each question addresses. Here is a key to the application icons found in the book.

	Agriculture		*Investment*
	Auditing/tax		*Law*
	Biology		*Marketing/advertising*
	Business		*Medicine*
	Chemistry		*Physics*
	Computer oriented		*Political science*
	Economics		*Polling/surveys*
	Education		*Psychology*
	Engineering/technical		*Quality/production*
	Environmental studies		*Real estate*
	Flash cards		*Sociology*
	General		*Sports*
	Geology		*Table usage*
			Transportation

Y ou will find numerous graphical displays in this book. Graphs and tables illustrate concepts that are more easily described by a "picture" than with words.

Definitions of key terms and concepts are called out in the margin and high-lighted with color. These are useful references you can use during reading assignments, while solving class or textbook problems, or when preparing for exams.

TABLE **2.7**

Tabulation of relative frequencies for a histogram

Class	Class Boundaries	Tally	Class Frequency	Class Relative Frequency
1	2.25–2.75	IIII	4	4/25 = .16
2	2.75–3.25	̶H̶H̶ I	6	6/25 = .24
3	3.25–3.75	̶H̶H̶	5	5/25 = .20
4	3.75–4.25	̶H̶H̶	5	5/25 = .20
5	4.25–4.75	III	3	3/25 = .12
6	4.75–5.25	I	1	1/25 = .04
7	5.25–5.75	I	1	1/25 = .04
		Total	$n = 25$	1.00

FIGURE **2.19**

Histogram for the dividend yield data (Figure 2.7) with intervals $\bar{x} \pm s, \bar{x} \pm 2s,$ and $\bar{x} \pm 3s$ superimposed

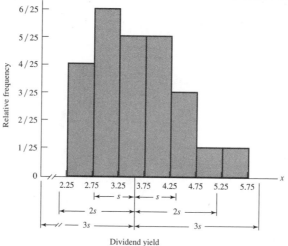

DEFINITION •

The **sample z-score** corresponding to an observation x is a measure of relative standing defined by

$$z\text{-score} = \frac{x - \bar{x}}{s} \quad \blacksquare$$

A   COURSE   IN   BUSINESS   STATISTICS

WHAT IS STATISTICS?

About This Chapter

The purpose of this chapter is to identify the nature of statistics, its objective, and how it plays an important role in the sciences, in industry, and, ultimately, in our daily lives.

Contents

LOOKING AHEAD TO THE CENSUS IN THE YEAR 2000

The Census Bureau is considering wholesale changes in the way it will collect data in the decennial census in the year 2000. This falls on the heels of soaring costs and the seemingly flawed results of the 1990 census, which some believe to be the least accurate in decades. Steven A. Holmes, writing for the *New York Times* News Service (*The Press-Enterprise,* Riverside, California, May 16, 1994), indicates that census officials will use sophisticated estimates based upon sample surveys to supplement data obtained by actual counting. In its plans to scrap the long-form survey that has been used every ten years to gather information ranging from household income to the number of bedrooms and bathrooms per household, the Census Bureau plans to conduct monthly surveys throughout the decade in order to produce timelier information. The bureau is also considering steps to increase the percentage of households that return the form—a record low of 65% in 1990.

How can the government be certain that these sample surveys will yield accurate results? When will sampling provide results that are more accurate than an actual enumeration? These and other questions regarding the role of sampling in the field of statistics will be addressed in Section 1.5.

1.1 Illustrative Statistical Problems

What is statistics? How does it function? How does it help solve certain practical problems? Rather than attempt a definition at this point, let us examine several problems that might come to the attention of the statistician. From these examples we can then select the essential elements of a statistical problem.

To predict the outcome of a national election, pollsters interview a predetermined number of people throughout the country and record their preferences. On the basis of this information a prediction is made. Similar problems are encountered in market research (What fraction of potential buyers prefer automobile brand A?); in sociology (What fraction of rural homes have electricity?); and in industry (What fraction of items purchased, or produced, are defective?).

An auditor wants to determine the inventory of a large hospital. To count the number and value of each expendable and nonexpendable item in stock would not only be very costly but would also, because of the size of the task, be subject to error. To reduce the cost and obtain a reliable estimate of value, the auditor selects a sample of items from the list of the hospital's supplies and equipment, carefully counts the number of each item on hand, and records its value. One minus the ratio of the total value of this sample of items to the total value shown in the hospital's records provides an estimate of the shrinkage due to theft, failure to record use of items, and so on. This shrinkage rate can then be applied to the total value of inventory shown in the hospital's records, thereby obtaining an estimate of the actual value of current inventory. How accurate is this estimate? How far might we expect the estimate to deviate from the actual value of the hospital's inventory?

The yield (production) of a chemical plant is dependent upon factors such as reaction time, reaction temperature, or purity of the raw material. By observing these factors and the yield over a period of time, we can construct a prediction equation relating yield to the observed factors. As another example, the economist wants to develop prediction equations that will be useful in forecasting growth or some other measure of economic health as a function of variables such as percentage of unemployed workers, inflation rate, gross national product, or trade deficit. Similarly, a manager may want to predict the sales of a product as a function of advertising expenditure, number of salespersons employed, or various other variables that may be related to the company's sales.

How do we find a good prediction equation? If the equation is used to predict yield, the prediction will rarely equal the true yield; that is, the prediction will almost always be in error. Can we place a limit on the prediction error? Which factors are the most important in predicting yield?

In addition to being involved in prediction, statistics is concerned with decision making based on observed data. Consider the problem of determining the effectiveness of a new flu vaccine. For simplification, let us assume that ten people have received the new flu vaccine and are observed over a winter season. Of these ten, eight survive the winter without acquiring the flu. Is the vaccine effective?

Two different teaching techniques are used to present a subject to two groups of students of comparable ability. At the end of the instructional period a measure of achievement is obtained for each group. On the basis of this information we ask: Do the data present sufficient evidence to indicate that one method produces, on the average, higher student achievement?

Consider the inspection of items purchased for a manufacturing plant. On the basis of such an inspection each lot of incoming goods must be either accepted or rejected and returned to the supplier. The inspection might involve drawing a sample of ten items from each lot and recording the number of defectives. The decision to accept or reject the lot could then be based on the number of defective items observed.

A company manufacturing complex electronic equipment produces some systems that function properly but also some that, for unknown reasons, do not. What makes good systems good and bad systems bad? To answer this question, we might make certain internal measurements on a system to find important factors that differentiate between an acceptable and an unacceptable product. From a sample of good and bad systems, data could then be collected that might shed light on the fundamental design or on production variables that affect system quality.

1.2 The Population and the Sample

The examples we have cited vary in nature and complexity, but each involves prediction or decision making. In addition, each of these examples involves sampling. A specified number of items (objects or bits of information)—a **sample**—is drawn from a much larger body of data, which we call the **population.** The pollster draws a sample of opinions (those interviewed) from the statistical population, which is the set of opinions corresponding to all the eligible voters in the country. In predicting the fraction of potential buyers who prefer a tartar-control toothpaste, we assume that those interviewed yield a representative sample of the population of all potential toothpaste buyers. The sample for the flu vaccine experiment consists of observations made on the ten individuals receiving the vaccine. The sample is presumably representative of data pertinent to a much larger body of people—the population—who could have received the vaccine.

Which is of primary interest, the sample or the population? In all of the examples given above, we are primarily interested in the population. We cannot interview all the people in the United States; therefore, we must predict their behavior on the basis of information we obtain from a representative sample. Similarly, it is practically impossible to give all possible users a flu vaccine. The manufacturers of the drug are interested in its effectiveness in preventing the flu in the purchasing public (the population). They can predict this effectiveness from information extracted from the sample. Therefore, the sample may be of immediate interest, but we are primarily interested in describing the population from which the sample is drawn.

DEFINITION ■

> The **population** is the set representing all measurements of interest to the investigator. ■

DEFINITION ■

> A **sample** is a subset of measurements selected from the population of interest. ■

Most people give the word *sample* two meanings. They refer to it as the set of objects on which measurements are to be taken, or they use it to refer to the measurements themselves. A similar double use could be made of the word *population.*

For example, we read in the newspaper that a Gallup Poll was based on a sample of 1823 people. In this use of the word *sample,* the objects selected are obviously people. Presumably, each person is interviewed on a particular question and that person's response represents a single item of data. The collection of responses obtained represents a sample of data.

In a study of sample survey methods we must distinguish between the objects measured and the measurements themselves. To experimenters, the objects measured are called **experimental units.** The sample survey statistician calls these objects **elements of the sample.**

To avoid a proliferation of terms, we will use the word *sample* in its everyday meaning. Most of the time, we will be referring to the set of measurements made on the experimental units (elements of the sample). If occasionally we use the term to refer to a collection of experimental units, the context of the discussion will clarify the meaning.

EXAMPLE   1.1   A fast-food company wishes to know how much money Americans, say age 16 and over, will spend on fast foods during the first week of June. Describe the population of data of interest to the company. Explain how the company might acquire the information it desires.

Solution   The information that the fast-food company wishes to acquire is associated with a population of measurements—one measurement for a week's expenditure on fast foods for each of the many millions of Americans in the age group 16 and over. These measurements will vary from person to person.

Some persons will purchase no fast foods during the week; others will purchase varying amounts. To acquire information on this vast population of measurements, the fast-food company likely will hire a marketing research organization to sample the population. From the sample measurements the marketing organization will estimate the average amount spent by all Americans age 16 and over, will estimate the proportion of all Americans who buy fast foods, and will answer other similar questions posed by the fast-food company. Thus the marketing organization will use information contained in the sample to infer the nature of the population of fast-food expenditures during the first week of June.   ▪

1.3   The Essential Elements of a Statistical Problem

You can see from the preceding discussion that statistics is concerned with describing a data set called a population. In some rare instances, such as the United States census, the population will be stored in a computer, and the statistical problem is to describe and extract information from a large mass of data. It is equally important to organize and summarize sample data. The branch of statistics concerned with these two types of problems is called **descriptive statistics.** Usually, however, the population is unavailable. Either observing and recording every single member of the population is too costly (as in the case of the hospital audit), or the population is conceptual—that is, the set of measurements (such as the daily yields of a chemical plant over the next two years) exists in our minds but is not actually available. Using a sample from a population, we attempt to deduce the population's nature by using the branch of statistics known as **inferential statistics.** Each example described in Section 1.1 represents a problem in inferential statistics. All have the same objective.

The Objective of Inferential Statistics

The objective of inferential statistics is to make inferences (predictions, decisions) about a population based on information contained in a sample.

How will we achieve this objective? We will find that every statistical problem contains five elements. The first and foremost of these is a clear **specification of the question** to be answered and **identification of the population** of data related to it.

The second element of a statistical problem is the decision about how the sample will be selected. This element, called the **design of the experiment** or the **sampling procedure,** is important because obtaining data costs money and time. In fact, it is not unusual for an experiment or a statistical survey to cost $50,000 to $500,000, and the costs of many biological or technological experiments can run into the millions of dollars. What do these experiments and surveys produce? Numbers on a sheet of paper or, in brief, information. Therefore, planning the experiment is important. Including too many observations in the sample is often costly and wasteful; including too few is also unsatisfactory. Most importantly, you will learn that the method used to collect the sample will often affect the amount of information per observation. A good sampling design can sometimes reduce the costs of data collection to one-tenth or as little as one-hundredth of the cost of another sampling design.

The third element of a statistical problem involves the **analysis of the sample data.** No matter how much information the data contain about the practical question, you must use an appropriate method of data analysis to extract the desired information from the data.

The fourth element of a statistical problem is the use of the sample data to make an **inference** about the population. As you will learn, many different procedures can be used to make an estimate or decision about some characteristic of a population or to predict the value of some member of the population. For example, two different methods may be available to estimate consumer response to an advertising campaign, but one procedure may be much more accurate than another. Therefore, you will want to use the best inference-making procedure when you use sample data to make an estimate or decision about a population, or a prediction about some member of a population.

The final element of a statistical problem identifies what is perhaps the most important contribution of statistics to inference making. It answers the question, how good is the inference? To illustrate, suppose you manage a small manufacturing concern. You arrange for an agency to conduct a statistical survey for you, and it estimates that your company's product will gain 34% of the market this year. How much faith can you place in this estimate? You will quickly realize that you are lacking some important information. Of what value is the estimate without a measure of its reliability? Is the estimate accurate to within 1%, 5%, or 20%? Is it reliable enough to be used in setting production goals? Statistical estimation, decision making, and prediction procedures enable you to calculate a **measure of goodness** for every inference. Consequently, in a practical inference-making situation, every inference should be accompanied by a measure that tells you how much faith you can place in the inference.

To summarize, a statistical problem involves the following:

1   A clear definition of the objective of the experiment and the pertinent population.

2   The design of the experiment or sampling procedure.

3   The collection and analysis of data.

4   The procedure for making inferences about the population on the basis of sample information.

5   The provision of a measure of goodness (reliability) for the inference.

Note that the steps in the solution of a statistical problem are sequential; that is, you must identify the population of interest and plan how you will collect the data before you can collect and analyze it. And all these operations must precede the ultimate goal, which is making inferences about the population based on information contained in the sample. These steps—carefully identifying the pertinent population and designing the experiment or sampling procedure—often are omitted. The experimenter may select the sample from the wrong population or may plan the data collection in a manner that intuitively seems reasonable or logical but that may be an extremely poor plan from a statistical point of view. The resulting data may be difficult or impossible to analyze or may contain little or no pertinent information, or, inadvertently, the sample may not be representative of the population of interest. Thus every experimenter either should be knowledgeable in the statistical design of experiments and/or sample surveys or should consult an applied statistician for the appropriate design *before* the data are collected.

1.4   The Role of Statistics in Inference Making

What is the role of the statistician, given our description of statistical problems? People have been making observations and collecting data for centuries. Furthermore, they have been using the data as a basis for prediction and decision making completely unaided by statistics. What, then, do statisticians and statistics have to offer?

Statistics is an area of science concerned with the extraction of information from numerical data and its use in making inferences about a population from which the data are obtained. The statistician quantifies information, studies various designs and sampling procedures, and searches for the procedure that yields a specified amount of information in a given situation at a minimum cost. Therefore, one major contribution of statisticians is the design of experiments and surveys that greatly reduce the cost and size of studies of larger populations. The second major contribution is in the inference making itself. The statistician studies various inferential procedures and looks for the best predictor or decision-making process for a given situation. More importantly, the statistician provides information concerning the goodness of an inferential procedure. When we predict something, we would like to know something about the error in our prediction. If we make a decision, we want to know the chance that our decision is correct. Our built-in individual prediction and decision-making systems do not provide immediate answers to these important questions. They can be evaluated only by observation over a long period of time. In contrast, statistical procedures do provide answers to these questions. Thus statistics enables us to make inferences from sample data and to evaluate the reliability of those inferences; this information will help us make informed business decisions.

CASE STUDY REVISITED

1.5 MORE ON THE CENSUS IN THE YEAR 2000

We now examine some of the questions posed in the Case Study. Will total enumeration (a census) produce more accurate information about the American population than that obtained by survey sampling? Let us examine how errors and possible undercounts might occur in a census or total enumeration. One serious problem in obtaining information is nonresponse—the failure of households to return their census forms for one or more reasons. Nonresponse contributes to undercounts across the gamut of areas included within the survey form. Undercounts affect the amounts that states and cities receive in federal dollars, which in turn causes problems in public planning for programs ranging from highway construction and repair to school lunch programs.

Even if nonresponse were not a problem, errors occur in transferring and recording information gleaned from the millions of census long-forms. Also, there will always be some respondents who intentionally report incorrect information. In light of these and other problems, a census in and of itself may not obtain totally accurate information.

How will sampling provide results that are more accurate than an actual enumeration? When an appropriate sample representing a fixed percentage of the total population is selected, statistical techniques allow us to produce estimates of important quantities, such as average household income or the proportion of American households that currently have health insurance coverage, within a margin of error that can be made as small as we desire by simply taking a large enough sample.

This is not to say that all results of the 2000 census will be based upon sampling but rather that sampling can be used as a supplement to census information. For example, rather than trying to obtain information from all nonresponse households by secondary visits, a sample of the nonresponse households would be selected for several visits until the information was obtained. The results obtained for these sampled households would then be extrapolated to all nonresponse households. According to Steven Holmes, the cost of the last census was "$2.6 billion over a ten-year cycle, double the cost of the 1980 enumeration. The Census Bureau predicts that if it does not change its methods, the cost for the 2000 census will more than double." In view of cost and other considerations, it is likely that sampling will play a large part in the 2000 census unless Congress decides to limit changes or to keep the status quo.

1.6 Summary

Statistics is an area of science concerned with the design of experiments or sampling procedures, the analysis of data, and the making of inferences about a population from information contained in a sample. Statistics is concerned with developing and using procedures for design, analysis, and inference making that will provide the best inference at a minimum cost. In addition, statistics is concerned with providing a quantitative measure of the goodness of the inference-making procedure.

A careful identification of the target population and the design of the sampling procedure are often omitted, but they are essential steps in drawing inferences from experimental data. Poorly designed sampling procedures will often produce data that are of little or no value (although this may not be obvious to the experimenter). After you make an inference, examine it carefully; be sure you acquire a measure of its reliability.

1.7 A Note to the Reader

We have stated the objective of statistics and have attempted to give you some idea of the types of business problems for which statistics can be of assistance. The remainder of this text is devoted to the development of the basic concepts of statistical methodology. In other words, we want to explain how statistical techniques actually work and why.

Statistics is a very heavy user of applied mathematics. Most of the fundamental rules (called theorems in mathematics) are developed and based on a knowledge of the calculus or higher mathematics. Inasmuch as this text is meant to be an introductory one, we omit proofs except where they can be easily derived. Where concepts or theorems can be shown to be intuitively reasonable, we attempt to give a logical explanation. Hence, we attempt to convince you with the aid of examples and intuitive arguments rather than with rigorous mathematical derivations.

You should refer back to this chapter occasionally and review the objective of statistics and the elements of a statistical problem. Each of the following chapters should, in some way, answer the questions posed here. Each is essential to completing the overall picture of statistics.

Exercises

Understanding the Concepts

1.1   In order to control the cost of inventory, the manager of a hardware store wants to limit the number of each item in stock and, at the same time, to avoid turning customers away due to insufficient inventory. To accomplish this goal, the manager wants to know the monthly demand for each item.

   a   Describe the population of interest to the manager for a single item.

   b   Can the manager actually acquire this population so that it can be studied? Explain.

c   Explain how you might learn something about the characteristics of the population without actually having the population in hand.

1.2   Suppose that you are a potential buyer of 100 acres of land. How much is it worth? If you were to ask 100 different land appraisers, you might receive 100 different answers. Thus the "value of the land" might be characterized by a population of measurements.

a   Describe the population.

b   Does this population actually exist, or is it conceptual? Explain.

c   Does the population associated with experienced appraisers differ from the population associated with buyers, sellers, or the public at large? Explain.

d   If you plan to base your offering price on the advice of experienced appraisers, would you employ one appraiser or would you seek the advice of several? Explain.

1.3   An emerging new market in the automotive industry is the market for electric cars. These cars are currently in the production stage, and one particular make, the Renaissance Tropica, was scheduled to go on sale in emission-conscious California by the end of 1994 (Parrish, 1994). The manufacturer has provided numerous facts about the Tropica's performance, including a top speed of 62 miles per hour, a range of 44.7 miles before recharging, and a time of 8.6 seconds to accelerate from 0 to 60 mph. How did the manufacturer arrive at these figures?

a   Describe the population of interest to the experimenter for any of the variables mentioned above.

b   Do you think that the figures given in the article were based on population measurements or on a sample taken from the population?

1.4   Refer to Exercise 1.3. The manufacturer of the Tropica plans to sell the car in California at a cost of under $15,000. Suppose that the manufacturer wants to determine whether it is profitable to market the car and needs to know what percentage of Californians would be interested in buying the Tropica.

a   Describe the population of interest to the experimenter.

b   How would you obtain a sample from the population in part (a)?

c   Would a sample of 200 residents of Los Angeles County be representative of all residents of California with respect to the marketability of this car? Explain.

1.5   A study reported by the *Associated Press* ("Ozone Damage," 1990) indicates that the number of pine trees in Yosemite National Park that have been damaged by ozone, a colorless and odorless gas produced by automobile and industrial emissions, has increased dramatically. The study reports in 1990 that 29.7% of Yosemite's pines showed mottling due to ozone damage, compared to 5.7% in 1985.

a   Describe the population of interest to the experimenter.

b   Do you think that the percentages given in the article are the actual population percentages, or are they estimated from a sample drawn from the population?

c   If you wanted to take a sample of trees from Yosemite National Park, describe some methods you might use.

1.6   Unless you are a close follower of corporate common stock, you may choose to invest some of your future savings in shares of one or more mutual stock funds. The growth in the value of a fund's shares over a period of time will depend upon general economic conditions and the skill of the fund manager in choosing the fund's investments.

a   Suppose that you wish to consider the percentage return of mutual stock funds over the past year. Does this population exist, or is it conceptual?

b   Suppose that you wished to consider the percentage return of mutual stock funds over the next 12 months. Does this population exist, or is it conceptual?

1.7   Marketing experts conducted a study to investigate the effect of background music on the behavior of supermarket shoppers. One objective of the study was to determine whether or not shoppers preferred to hear music playing in the background while they shopped. In this particular study, 300 shoppers at a Los Angeles area supermarket were surveyed.

a Describe the population and the sample associated with this survey.

b Would it be possible to examine the entire population if you wanted to? Explain.

c Will the percentage of patrons in the sample equal the percentage of patrons in the population who prefer background music while they shop? Explain.

1.8 Many valuation criteria, including P/E ratios, dividend yields, and price-to-book ratios, can be used by an investor to find "cheap" stocks. In an attempt to find "cheap" stocks based on a combination of these criteria, a computer program scanned the stocks of 8000 public companies (Palmer, 1990). To qualify for the final list, companies had to have a P/E ratio below 14.8, a price-to-book ratio of less than 2.45, a yield of more than 3.4%, and a return on assets higher than 7.4%. Furthermore, companies with declining earnings and large long-term debts, with low prices per share, or with low market capitalization were eliminated. A final list of 34 companies with "cheap" stocks was provided.

a Describe the population and sample of interest.

b Is the sample a representative sample from the population of interest? Why or why not?

CHAPTER 2

DESCRIBING SETS OF DATA

About This Chapter

Sometimes, the data that we have collected represent a sample selected from a population. Other times (such as a national census), the data may represent the entire population. In either case, we need to be able to describe the data set. The objective of this chapter is to present two types of methods for describing data sets: (1) graphical descriptive methods and (2) numerical descriptive methods. Graphical descriptive methods describe the data by using charts and graphs. Numerical descriptive methods use numbers to help us construct a mental picture of the data.

Contents

So You Want to Be a Millionaire?

I n the 1980s, new tax laws led to the creation of IRAs (Individual Retirement Accounts), tax-free savings accounts for retirement income. According to many newspaper ads at the time, if we were to invest $2000 per year in an IRA, after 40 years of participation our nest egg would have grown to over a million dollars. Of course, the tax laws have been changed several times since then, and with the latest tax revision, tax-free IRAs will no longer be available for most of us. But the principle expressed in the advertisements is still valid. The best way to accumulate a large sum of money is to participate in a systematic savings and investment program that compounds the invested funds over the years.

If you are saving for retirement or for some other objective, the amount that you accumulate will depend on the amount you invest each year, where you invest it (a bank savings account, a money market fund, or one of the various common stock funds), and who manages your account. Basically, the growth of your account and its ultimate value will depend upon the annual rate of return that your account manager is able to obtain for you.

Although the rate of return on your money will vary from day to day, Table 2.1 gives you an indication of the amount of money that you might expect to accumulate in 40 years. The amounts shown are based on the assumption that you invest $2000 at the beginning of each year for a period of 40 years and that the money is compounded monthly at a fixed annual rate of interest I.

TABLE 2.1
Withdrawal amount for an account after an annual investment of $2000 at a fixed annual rate of return I (%) for 40 years

Interest Rate I (%)	Amount in Account After 40 Years ($)
4	197,652
6	328,095
8	559,562
10	973,704
12	1,718,285

Monthly rates of return vary not only between investment vehicles (banks, stock funds, money market funds, etc.) but also within an investment vehicle itself. For example, consider a money market fund, one that invests in short-term commercial notes. Since the notes will be negotiated at different times and will vary in length of time to maturity, the rate of return for the total fund at any one point in time will depend upon the fund manager's skill in placing the loans. If the interest rate rises in the future, it pays to hold notes with a small average time to maturity. If it drops, it pays to hold notes with a large average time to maturity.

The characteristics of money market funds as a vehicle for investment are shown in the data of Table 2.2. Table 2.2 presents the asset size (in millions of dollars), the average maturity (in days) for notes, and the average 7-day yields (%) for the period ending July 13, 1994, for 604 large money market funds available to investors. An examination of Table 2.2 clearly indicates a statistical problem. While it is possible to obtain a general feeling about the asset size, the average maturity time, and the average rate of return by examining the data, it is difficult to obtain a clear picture of the characteristics of these data sets by simply scanning the table. This problem motivates the topic of Chapter 2. In this chapter we examine methods for describing data sets. Then in Section 2.14, we apply these techniques to the money market fund data and see how this descriptive information is relevant to our prospects of becoming millionaires.

2.1 Variables and Data

Our primary objective in Chapter 2 will be to present some basic techniques in **descriptive statistics**—the branch of statistics concerned with describing sets of measurements, both **samples** and **populations.** After we have collected a set of measurements, how can we display this set in a clear, understandable, and readable form? First, we must be able to define what is meant by measurements or data and to categorize the types of data we are likely to encounter in real life. We begin by introducing some definitions, some new terms in the statistical language that you need to know.

DEFINITION ■

> A **variable** is a characteristic that changes or varies over time, or a characteristic that varies across different individuals or objects under consideration at a particular point in time. ■

For example, stock price is a variable that changes over time within a single stock; it also varies from stock to stock at a given point in time. Political affiliation, ethnic origin, income, age, and number of offspring are all variables—characteristics that vary depending on the individual chosen.

T A B L E **2.2**   Data on 604 money market funds for the period ending July 13, 1994[†]

Fund	Mat.	Yld.	Assets	Fund	Mat.	Yld.	Assets	Fund	Mat.	Yld.	Assets	Fund	Mat.	Yld.	Assets
AALMny	60	2.91	67	CardGvt	3	3.42	397	Fid FDIT	37	3.96	2090	HanvGov	33	3.87	915
AIM MM C	23	3.46	340	Carnegie	23	3.27	18	FidGvRes	45	3.78	1137	HanvUSTr	42	3.81	1271
AIM MMA	23	3.46	122	CascdCs	21	3.29	100	FidGvPr	47	3.95	159	HanvTreas	41	3.56	952
AVESTA Tr	54	3.82	23	CshActMM	32	3.50	147	FidUS Tr	28	3.84	170	Harbor	68	3.83	67
AccUSGov	28	3.72	19	CshAcctGv	23	3.50	36	FidRetGov	41	3.92	1635	HrtgCsh	32	3.71	964
ActAsGv	41	3.55	495	CashEqv	36	3.65	3408	FidRetMM	41	4.12	2871	HiMrkDv	36	3.51	304
ActAsMny	54	3.89	4392	CshEqGv	27	3.51	1564	FidSpGov	44	3.87	761	HiMrkUS	28	3.43	188
AetnaAdvs	38	4.23	31	CashMgt	21	3.71	2799	FidSpMM	47	4.08	70100	HiMrkUST	33	3.27	201
Aetna Sel	38	4.23	150	CshTrGv	31	3.39	442	FidSpUSTr	70	3.81	1552	HmestdDly	48	3.63	32
AlexBwn	49	3.90	1296	CshTrPr	35	3.46	800	59WallStTreas	56	3.57	143	HorznPr	10	4.04	2812
AlxBTr	46	3.60	551	CshTrTreas	29	3.21	400	59WallStMM	56	3.67	591	HorznTr	45	4.15	2137
AlgerMM	42	4.18	148	CshTrIl	29	3.38	235	FnclRsv	54	3.96	387	Hummer	30	3.59	146
AliaCpRs	51	3.31	240	CentnGv	35	3.85	640	FstAmerInstit	32	4.04	1082	IAATrMM	27	3.89	37
AliaGvR	49	3.22	2089	Centen	42	3.81	2670	FtBostin	25	4.16	150	IAIMnyMktFd	27	3.63	22
AlliMny	52	3.27	1788	ChchCsh	29	3.80	191	First Muni	38	2.32	79	IDS CshM	30	3.55	1174
AmAAdTrl	1	4.03	78	ColonIMMB	24	2.53	60	FstOmahaGv	56	3.55	78	IDS PLA	21	4.04	29
AmAAdMMI	30	4.25	1280	ColonIMM A	24	3.49	94	FirPrTreasTr	54	3.77	97	IMGLiq	36	3.52	138
AmCRes	17	3.47	438	ColDln f	46	3.77	652	First USGv	51	4.27	71	IndCaGv	41	3.37	246
AmPerCsh	35	3.82	172	ComBkrUSGv	27	4.39	121	FtBost	47	4.29	2149	IndCaMM	41	3.36	313
AmPerTrs	1	3.46	137	ComSens	17	3.34	56	FtInvCs	20	3.70	120	IndOnPr	19	3.75	286
AmSouth Pr	44	3.74	578	CompCs	52	3.93	439	FtPraGv	23	4.05	157	IndOnUS	28	3.60	247
AmSouth US	39	3.51	294	CompUS	48	3.66	279	FtPraMM	34	3.62	146	InfnAlGv	32	3.87	73
AmbMMF	32	3.84	323	CmpCshMM A	40	3.35	121	First USTrs	47	3.81	76	InfnCCR Inst	61	4.07	63
AmbTreas F	46	3.60	208	CnstgCs	13	3.72	124	CalvFtGv	31	3.58	255	InfCCR	61	3.41	333
AmbTreasI	46	3.45	73	CnstgUS	28	3.44	307	FlexFd	51	4.12	160	InstCsh	31	4.49	417
AmbMMI	32	3.69	517	CG CapGov	82	3.68	194	Fortis	20	3.52	109	InstFd	21	3.30	11
Amcore Gv	15	3.67	115	CoreFd	33	3.93	518	ForumDATrs	23	3.74	29	InstGov	53	4.17	195
AmAAdMMM	30	4.00	49	CoreTr a	35	3.72	432	Founders	42	3.63	241	InvCshGv	26	4.00	125
ArchUSTr	44	3.33	2	CortldGn	33	3.21	903	FountSGv	41	3.89	123	InvCshR	37	3.96	779
ArchFd	40	3.66	43	CortldUS	56	3.22	212	FountSCP	28	3.87	221	InvGvtMF	29	3.62	75
ArkMnyMkt	40	4.24	205	CowenStdResv	24	3.72	699	FountSTO	31	3.78	330	Invesco	11	4.13	131
ArkUSGovt	22	4.08	430	CrestCshTr	41	3.48	411	FrnklFT	41	4.07	216	IvyMny	40	3.10	25
ArkUSTrsy	56	3.67	142	CrestUST Tr	49	3.42	311	FrnkLate	4	4.16	53	JPM InstP	34	4.15	305
AMF St Lq	13	3.77	49	DG InvUSGv	54	3.61	166	FrkUSTr	44	4.03	191	JPMInstlTrsy	37	4.08	86
AutCsh	36	3.94	1032	DailCsh I	43	3.81	3338	FrklFTGS	15	4.03	215	JanMS Gov	41	3.37	246
AutGvt	32	3.65	2500	DlyPasp	32	3.31	2076	FrkFdl b	12	3.36	210	JanMS MM	41	3.36	313
AuGvSvc	33	3.73	407	DWitrLq	54	3.70	8637	FreeCsh	39	3.67	1098	JHanCshM	52	3.55	217
AutTreasC	37	3.51	172	DWitrUS	47	3.19	784	FreeGv	53	3.47	272	KemperGvt	25	3.83	717
BB&T UST Tr	43	3.46	73	DelaCR f	33	3.33	738	FremntMM	44	4.06	118	KemperM	35	4.02	4182
BNY Hmltn	36	4.02	357	DelaTr	73	3.02	21	FrkMny	48	3.60	1131	KeyLqd	13	3.46	453
BT InstCash	17	4.19	874	DryBasic	59	4.26	1575	FstAminstGv	35	3.99	436	KidPeCsh	39	3.71	1789
BT InstCshRv	17	4.24	786	DryBasGov	46	4.05	248	FstAmerInv	24	3.66	58	KidPeGv	45	3.43	303
BT Inst Trsy	19	3.78	156	Dry100 US	63	3.34	1635	FirPrTreasIn	54	3.37	27	KidPePr	38	3.56	813
BT InvCash	17	3.69	163	DryGvt	42	3.18	480	FirstUnMMI	44	3.86	92	LdmkUSTrs	56	3.32	209
BT InvTrsy	19	3.28	610	DryInG	46	3.28	104	FirstUnTMMT	43	4.00	192	LdmkCashRs	73	3.80	431
BT Pyramid	17	4.07	839	DryInst	46	3.55	332	FirstUnTMMI	43	3.70	451	LdmkPrmLq	73	4.10	207
Babson	40	3.42	43	DryfLA	63	3.43	4851	FdTrMny	74	3.99	84	LdmkInstLq	73	4.30	208
BartCsR	38	3.65	75	DryMM	52	3.44	173	GEMnyMktC	31	3.88	31	LandlUSTr	56	3.77	115
BayFdMM	37	3.90	144	DryWld	59	3.28	2713	GOC UST	32	3.75	133	LaurlUSTOII	51	3.91	32
BayFdMMInv	37	3.65	55	EatVCsh	34	3.56	101	GT Mny	34	3.17	321	LaurCashInv	31	3.51	199
BedfdGv	51	3.43	176	ElfunMM	32	4.03	84	Gab OC DP	38	3.82	157	LaurGvtInv	49	3.53	49
BdfdMM	43	3.45	721	EnterMM	24	3.43	24	Gab OC UST	36	3.59	143	LaurUSGI	59	4.11	472
BenchmrkCA Mu	37	2.00	91	EvgrnM	51	4.09	281	GabelliUST	48	3.79	205	LaurPRI	38	3.98	681
BenchmrkDivAst	32	4.08	2955	ExcelsrMM	73	4.31	585	GalaxyGv	31	3.75	563	LaurPrTst	52	3.83	115
BenchmrkGovSel	41	4.15	409	ExcelsrTM	56	3.80	59	GalxyMM	29	3.59	836	LaurUSTI	55	4.01	497
BenchmrkGovt	19	3.92	782	FBL	25	2.07	18	GalaxTr	48	3.60	460	LaurUSTTst	39	3.77	190
BenhGvAg	46	3.86	500	FFB Csh	46	3.76	544	GnGvSec	44	3.72	524	LdmkPrUST	56	3.57	226
BenhPrime	68	4.51	484	FFB US Gvt	36	3.64	189	GnMMkt	55	3.47	592	LegMUS	28	3.47	205
BiltPrCshlst	39	4.27	775	FFB US Tr	30	3.70	616	GSILFd	44	4.13	1619	LegMCR f	21	3.58	797
BiltTrea	39	3.89	74	FFBLexicn	21	3.81	117	GSILGv	42	3.83	1394	Lehman100GvA	11	4.11	9
BILTmm	44	3.93	142	FidUSTI	65	4.09	1059	GSILMM	36	4.08	876	Lehman100TrA	68	4.08	65
BiltMMIv	44	3.63	39	FMBCons	12	3.62	3	GSILTrsOblg	44	3.83	942	LehmanGvtOblA	15	4.19	98
WmBlrRdy	40	3.74	481	FMBInst	12	3.62	60	GSILPO	31	3.99	2426	LehmanPrmeVal	41	4.31	1175
BlnchGv	72	3.43	237	FdShtUS	33	3.91	939	GSILTrsinst	43	4.01	472	LehmanPrime	23	4.30	2160
BradGovObl	51	3.43	39	FedMstr	36	3.93	808	GlnmndGvCash	25	4.17	351	LehmanTreaslA	16	4.12	362
Bradfd	38	3.49	716	FidInDom	28	4.34	885	GoldenOK-PC	28	3.87	87	Lexingt a	27	3.30	104
BullBDir	69	3.43	74	FidInGov	50	4.18	3096	GvInvTr	15	3.13	75	LibtyUS	31	3.27	766
CRTGovt	43	3.58	919	FidInMM	33	4.37	4001	GrdMcDnldGvRv	43	3.45	993	Liq Ins Gv	34	4.00	68
CRTMMkt	35	3.64	200	FidInTr	29	4.14	1336	GrtHallGv	37	3.49	56	LiqInt	31	4.04	272
CapCash	24	4.00	2	FidInTrIl	31	4.06	3875	GrtHallPr	45	3.56	1002	LiqCapital	26	3.56	287
CapPre II	1	3.56	281	FidCRMM	44	3.56	671	Griffin	54	4.40	37	LiqCshTr f	6	4.23	449
CapPrsv	42	3.72	2803	FidCRGv	29	3.34	331	GrdCsFd	22	3.87	383	LrdAbCR	24	3.33	156
CapitolGovB	74	3.76	143	FidCsRes	43	3.90	13299	GrdCsMg	22	3.56	47	LuthrnBr	32	3.35	273
CapitolMMB	26	3.69	13	FidDMM	39	3.81	1561	HTInsgtCs	39	3.77	428	MASCashRes	18	4.04	33
CapitolTreas	25	3.94	283	FidDUS	29	3.63	2100	HTInsgtGv	45	3.73	243	MFSGovMonA	23	3.26	42
CapitolMM	26	3.74	69	FidDom	24	4.09	374	HanvCsh	64	3.86	659	MFSMonMkA	34	3.51	476

[†]The quotations, collected by the National Association of Securities Dealers Inc., represent the average of annualized yields and dollar-weighted portfolio maturities ending Wednesday, July 13, 1994. Yields don't include capital gains or losses.

T A B L E **2.2** (continued)

Fund	Mat.	Avg. 7 Day Yld.	Assets	Fund	Mat.	Avg. 7 Day Yld.	Assets	Fund	Mat.	Avg. 7 Day Yld.	Assets	Fund	Mat.	Avg. 7 Day Yld.	Assets
MIMLIC a	36	3.45	25	PFAMCo MM	28	3.97	7	RemTreasTr	27	3.42	101	StepstnInv	38	3.47	83
MainSty	48	3.87	184	PNCGovtS	49	3.94	298	RemGovtTr	21	3.79	156	StepstnTrInst	39	3.74	148
MngdCsh	34	3.89	346	PNCMoneyS	58	3.84	539	RenaisGvt	1	3.60	46	StepstnTrInv	39	3.49	29
MngdGv	50	3.82	68	PNCMoneyl	58	4.14	470	RenaisMM	31	3.71	322	Strong	33	3.89	456
Map Gvt	37	3.51	72	PNCMunil	52	2.25	29	ResrveFd Gvt	15	3.38	750	StrongUST	30	3.78	97
MarinCsh	42	4.01	203	PcfCptlCshAs	25	3.80	388	ReserveFd	41	3.36	1388	SumtCsh	33	3.62	135
MarineGv	47	3.88	148	PcfCptlUSTrs	41	3.53	87	RetirGv	27	3.15	261	SunAmMMA	36	3.54	302
MarinerUS	45	3.46	135	PcHrzGvHor	29	4.13	531	RimcoTrs	72	3.67	93	TCW MM	37	3.92	83
MrkTwGovtT	20	3.43	148	PacAmerMM	47	3.68	514	RIMCOPrm	65	3.85	392	TFI GovtMMIS	16	3.77	44
MarquisTrl	34	3.72	463	PacAmerUST	41	3.82	690	RiverUSGv	26	3.72	144	TRowPRF	42	3.65	3741
MarquisTrR	34	3.63	92	PcHrzGvPH	29	3.81	531	RiversdeCap	34	3.23	137	TRowUST	42	3.48	672
Marshall Tr	37	4.09	1085	PcHrzPr	10	3.72	2812	RdSqMM	19	3.80	842	TempltnM	67	3.54	223
ML CBAMon	37	3.67	1281	PcHrzPrVal	18	4.29	779	RdSqUS	18	3.71	399	ThomMB	19	3.67	82
ML CMAGv	44	3.61	3278	PcHrzPValH	18	4.29	779	RshFGI	41	3.45	550	ThmNtl	22	3.77	709
ML CMAMn	40	3.79	26917	PcHrzTr	45	3.83	2137	Rshmre	29	3.65	23	TowerUSTreas	30	3.56	42
MerLyGv	37	4.10	1520	PcHrzTO PH	55	3.45	362	RydexUSGv	1	3.23	113	TowerCsh	34	3.43	172
MerLITr	28	3.67	285	PacificGv	45	3.65	99	SBSF MM	73	3.72	15	TransamCs	29	3.78	382
MerLyIn	44	4.18	4437	Pacifica	50	3.88	156	SEICshTrea	21	4.06	64	TransamUS	33	3.94	100
MerLyRdy	42	3.71	6532	PWMnyMkA	29	3.48	36	SEI CsPrB	27	3.88	11	TrstShtGv	34	3.84	1431
MerLyRet	44	3.80	7360	PWMnyMkB	29	3.00	43	SEI CsPrC	27	3.67	1	TCU MMP	39	4.11	263
ML CMATr	35	3.42	1099	PWMnyMkD	29	3.02	44	SEI CsFd	21	3.44	11	TrGvCsh	34	3.94	952
MerLyUSA	35	3.38	570	PW Cash	40	3.77	3527	SEI CshGvII	27	4.03	701	TrstUSTrOb	33	3.76	3527
ML US Tr	30	3.26	59	PW RMA	52	3.70	4413	SEI CsMM	33	4.35	219	TwCntPrCapR	17	3.97	45
MetLfSt E	33	3.71	155	PW RM US	50	3.48	899	SEI CsPr	27	4.18	2252	TwCntCs	18	3.68	1358
MdIncTrGv	63	3.17	81	PW Retr	41	3.58	2484	SEI LqGv	37	3.80	259	231 Prime	33	4.25	1437
MdInInst	58	3.90	49	ParagTr	36	3.81	290	SEI LqPr	26	3.97	1037	231 Trsy	26	4.03	120
MonarchCash	27	4.13	53	ParkTrinvA	36	3.57	58	SEI LqTr	31	3.73	1344	UMB Fed	45	3.67	168
MonarchCU	27	4.41	34	ParkPrinvA	43	3.64	106	SEI CsTrlIA	52	4.01	324	UMB Prim	39	3.76	191
MonarchGvt	22	4.16	71	ParkPrinst	43	3.74	572	SEI CsTrlIB	52	3.69	24	US TreSec	18	3.10	163
MonarchGU	22	4.42	95	ParkTrinst	36	3.67	71	SITMMkt	21	3.89	15	US Tr Am	39	3.37	198
MonarchTrsyInst	4	3.87	37	ParkUS C	40	3.68	193	STIPrQuTr	43	3.84	614	US Treasry	49	3.62	20
MonMMgt f	36	3.21	128	ParkUSInvA	40	3.58	186	STIUSGvTr	35	3.63	247	USAA Mutl	52	4.10	980
MonMkTrst	35	3.92	561	PeachtGov	55	4.06	57	STIUSGvIv	35	3.51	37	USAA Treas	53	3.71	38
MonitorMMT	40	3.97	327	Peacht Pr	52	4.13	58	Safeco f	40	3.56	167	UST Gvt	14	3.64	647
MonitorUST	40	3.79	248	PerfrmTrCsr	43	4.02	1	SalomonUST	41	3.68	28	UST Mny	14	3.91	800
MonitrMMI	40	3.87	22	PerfrmTrInst	43	4.27	172	SchbValAdv	50	4.14	1904	UST Treas	64	3.77	262
MontGovRes	53	3.74	200	Phonix	54	3.38	206	Schwablntl	45	4.03	67	USTrCshIns	38	3.81	274
MS Mony	48	3.92	688	PierpontMM	34	3.92	2044	SchwbRetir	43	3.78	23	UtdCshM	61	3.51	317
MunderC	27	3.84	162	PierpontTrsy	37	3.88	100	SchwbGv	40	3.57	1994	UtdSvsGvt	22	3.93	626
NatnsPrTrA	67	4.25	2722	PillarPrObA	39	3.65	133	SchbMM	52	3.73	1055	VIMMP	41	4.30	554
NCC Govtl	33	3.95	759	PillarUST A	42	3.42	384	Schb UST	51	3.49	559	ValLin	37	3.74	356
NCC GovtR	33	3.85	6	PionrCs	46	3.68	174	ScudCshln	34	3.75	1503	VnEckUS a	88	4.09	49
NCC MMR	30	3.95	73	PionUS	51	3.71	29	Scud UST	61	3.55	384	VanKmpMM	19	3.28	27
NCC MMI	30	4.05	925	Piprlnst	59	4.19	35	SecurityCsh	32	3.39	50	VangAdmUST	41	4.01	1076
NYL Inst	36	3.91	71	PiprMM	47	3.46	1239	SelectGv	17	3.61	45	VangFdl	41	4.01	2052
NatnsTrinvA	48	3.65	79	PiprUS	53	3.52	206	SeligCsh Gvt	24	3.29	17	VangPr f	42	4.11	13361
NatnsPrinvA	67	3.80	523	PortgeGv	62	3.49	62	SeligCshPrA	35	3.59	170	VangUST	40	3.84	1983
NatnsGvMMTrA	51	4.08	393	PortInstMn	37	4.04	717	SentinelUST	48	3.26	73	VictoryPt	39	4.28	462
NatnsTreTrA	48	4.02	2530	PortMM	41	3.88	150	SevnSea f	49	3.94	3219	VictoryUSTrs	33	3.71	145
NatwMM	43	3.75	451	PortUS	36	3.80	196	SvnSeaGv	66	3.87	240	VisnMM	36	4.05	275
NeubCsh	40	3.67	313	PortUSFed	44	3.46	40	1784InstUSTr	36	4.02	180	VisnTr	45	3.82	225
NeubGvt	44	3.26	204	PrimeCsh	24	3.77	52	ShawPrmMMIn	39	3.68	103	VistaFedInst	60	4.28	78
NewEngCMTUSG	16	3.47	63	PrinPresCs	38	3.53	52	ShmtPrTr	39	3.93	443	VistaGloPS	79	4.07	494
NewEngCMT	33	3.48	696	PrincorCash	36	3.68	292	ShtTrmincoA	50	3.36	680	VistaGloVS	79	3.92	200
Nicholas	26	3.98	124	ProspectHill	38	4.00	52	ShTrinUS	32	3.34	449	VistaGlol	79	4.25	328
Northern	23	3.99	369	PIFTrsTr	52	4.01	1325	SierraGlbA	65	3.72	53	VistaGovPS	28	3.80	549
NorthernMun	46	1.97	466	PIFTpCs	31	4.28	2381	SierraGovA	30	3.34	31	VistaGov VS	28	3.55	335
NorthernUSGv	37	3.95	111	PIFTFd	33	4.05	1088	SigntSIMM	72	3.88	135	VistaGovtInt	28	4.08	195
NorwestCash	46	3.76	1390	PIFFed	40	4.15	1133	SigntSITr	54	3.72	394	VistaPrInst	25	4.38	65
NorwestMunill	46	2.21	183	PIFFedTr	41	4.28	361	SmBarCash	48	3.76	2955	VistaPrimeP	25	4.20	161
NorwestRdy	38	3.42	166	PIFTmp	50	4.17	5609	SmBarGvt	44	3.54	625	Voyager	33	3.41	34
NorwestTr	50	3.58	554	PruCdGvt	37	3.49	326	SmBarRetir	45	3.63	1119	WPG GovMM	22	3.53	217
NorwestUS	47	3.70	1131	PruCdMny	37	3.86	2551	SmBarShrDDiv	49	3.50	15186	WarburgPCR	50	3.91	272
OLDE MM	54	3.38	190	PruGvt	39	3.35	677	SmBarShrGov	44	3.50	2972	WestcrCs	11	3.81	252
OLDE PrPl	50	4.44	262	PruInMM	41	3.98	356	SoTrVlcnTres	43	3.75	261	WestcrGv	38	3.62	73
OLDE Prem	56	3.79	82	PruMart	36	3.79	6849	SocietyPrOb	65	3.96	781	WestcrPr	37	4.06	258
OneGrGovt	45	4.20	670	PruSMM	31	3.88	463	SocietyUSGv	15	3.67	426	WestcrTr	35	4.05	369
OneGrPrA	52	4.08	1566	PruTr	42	3.45	278	Stagecoach	41	3.80	869	WoodTreas	48	3.78	627
OneGrPr B	52	3.83	77	PruInstMM	38	3.83	45	StagecoachInt	43	3.99	115	WoodGv	76	3.82	368
OneGrUSTScA	36	3.82	1329	PutDDiv	38	3.94	1323	StarPrmeObl	49	3.41	77	WoodMM	62	3.89	1510
OneGrUSTScB	36	3.57	56	QltyCsh	49	3.46	97	StarTreas	47	3.48	312	WorkAsets	45	3.37	104
OneGrTresOn	48	3.98	228	QuestCshGov	36	3.39	122	StarbGovT	50	3.44	172	ZweigGvt	37	3.59	87
OppMoney	40	3.84	836	QuestCshPr	37	3.51	1545	StarbMMT	56	3.66	139				
Oppenh Csh	38	3.37	94	RNC Liq	66	3.87	38	SteinroeCRs	70	3.83	551				
OvidExTrs	46	3.43	193	RegisDSI	35	3.75	118	SteinroeGvt	38	3.43	111				
OvidExMM	44	3.76	365	RemTaxTr	22	3.93	483	StepstnInst	38	3.67	489				

Source: Data from "Money Market Summary," *Wall Street Journal,* July 14, 1994, p. C27.

In the introduction, we defined an **experimental unit** as the object on which a measurement is taken. Equivalently, we could define an experimental unit as the object on which a variable is measured. When a variable is actually measured on a set of experimental units, a set of measurements or **data** results.

DEFINITION ▪ | An **experimental unit** is the individual or object on which a variable is measured. A single **measurement** or data value results when a variable is actually measured on an experimental unit. ▪

If a measurement is generated for every experimental unit in the entire collection, the resulting data set constitutes the **population** of interest. Any smaller subset of measurements is a **sample.**

EXAMPLE **2.1**   A set of five employees is selected from the employees at a large firm, and the following measurements are recorded. Discuss the variables measured for these five employees.

Employee	Performance Score (0–20)	Sex	Number of Years of Service	Job Classification	Salary ($ thousand)
1	18	F	12	Sales	35
2	15	F	9	Managerial	55
3	10	M	2	Clerical	23
4	19	M	15	Managerial	58
5	15	F	13	Sales	36

Solution   There are several **variables** in this example. The **experimental unit** on which each variable is measured is a particular employee in the firm. Five variables are measured for each employee: performance score, sex, years of service, job classification, and salary. Each of these characteristics varies from employee to employee. If we consider the performance scores of all employees at this firm to be the population of interest, the five performance scores represent a **sample** from this population. If the performance score of each employee of the firm had been measured, we would have generated the entire **population** of measurements for this variable.

The second variable measured on the employees is *sex,* which can fall into one of two categories—male or female. It is not a numerically valued variable, and hence it is somewhat different from *performance score.* The population, if it could be enumerated, would consist of a set of Ms and Fs, one for each employee at the firm. Similarly, the fourth variable, *job classification,* generates nonnumerical data, with one category for each job classification at the firm. The third and fifth variables, *number of years employed* and *salary,* are numerically valued, generating a set of numbers rather than a set of categories.

Although we have discussed each variable individually, remember that we have measured each of these five variables on five experimental units—the five employees.

Therefore, in this example, an observation on one individual consists of five measurements. For example, the observation taken on employee 2 produces the following measurement:

$$(15, \ F, \ 9, \ managerial, \ 55) \quad \blacksquare$$

You can see that there is a difference between a *single* variable measured on a single experimental unit and *multiple* variables measured on a single experimental unit. If a single variable is measured, the resulting data are called **univariate data.** If two variables are measured on a single experimental unit (such as sex and salary), the resulting data are called **bivariate data.** If more than two variables are measured, as in Example 2.1, the data are called **multivariate data.**

2.2 Types of Variables

Example 2.1 demonstrated that measuring variables produces data that might be either numerical or nonnumerical. Variables that give rise to nonnumerical data in which observations are categorized according to similarities or differences in kind are called **qualitative variables.** Political affiliation, occupation, marital status, and high school attended are examples of qualitative variables, as are the variables "sex" and "job classification" in Example 2.1. Variables used to measure an attribute that produce numerical observations are said to be **quantitative variables.** The Dow-Jones Industrial Average, the prime interest rate, the number of nonregistered taxicabs in a city, and the daily power usage for an industrial plant are examples of quantitative variables, which give rise to quantitative data.

DEFINITION ■ **Quantitative variables** give rise to numerical observations that represent an amount or quantity. **Qualitative variables** give rise to nonnumerical observations that can be categorized. ■

Quantitative variables, which are often represented by the letter x, can be further categorized according to the range of numerical values that a measurement can assume. Variables such as the number of family members in families in Arizona, the number of new car sales at the Riverfront Auto Mall, and the number of defective tires returned to a manufacturer for replacement take values corresponding to some subset of the counting integers 0, 1, 2, ... Specifically, these variables can take on a countable number of values and are called **discrete variables.** The name *discrete* reflects the fact that there are discrete gaps between the possible values that the data can assume. On the other hand, measurements on variables such as height, weight, time, distance, or volume can assume values corresponding to all of the points on a line interval. Variables of this type are called **continuous variables.** A third value can always be found between any two values of a continuous variable.

DEFINITION   ▪

> A **continuous variable** is one that can assume all of the infinitely many values corresponding to a line interval. A **discrete variable** can assume only a countable number of values.   ▪

EXAMPLE   **2.2**   Identify each of the following variables as *qualitative* or *quantitative.*

   a   The most frequent use of your microwave (reheating, defrosting, warming, other) during December 1995.

   b   The number of consumers refusing to answer a telephone survey.

   c   The type of cable service delivered to residences (standard cable, premium cable or antenna only) in Atlanta.

   d   The completion time for a particular task performed by a computer software program.

   e   The number of stocks on the New York Stock Exchange showing a gain from March 1, 1995, to July 1, 1995.

Solution   Variables (a) and (c) are both qualitative variables, since only an attribute is measured on each experimental unit. The categories for these two variables are shown in parentheses. The other three variables are quantitative. The number of consumers is a *discrete* variable; it could take on any of the values 0, 1, 2, ..., with a maximum value depending on the number of consumers called. Similarly, the number of stocks showing a gain could take any of the values 0, 1, 2, ..., with a maximum value depending on the number of stocks on the NYSE. Variable (d), completion time for a particular task, is the only *continuous* variable in the list. The completion time could be 121 seconds, 121.25 seconds, or any value in between any two values listed.   ▪

Why should we be concerned about the different kinds of variables and the data that they generate? The techniques used for summarizing and describing data sets depend upon the kind of data collected. Qualitative data are usually summarized by determining the number or proportion of observations in each of several categories. The results are then presented using tables and graphs. Graphical presentations differ somewhat for discrete and continuous quantitative variables, but in general they focus on graphs in which the number of observations in a class or category is plotted against classes or categories. For each set of data you encounter, the trick will be to determine what type of data is involved and how you can present it in a way that is clear and understandable to your audience (see Figure 2.1).

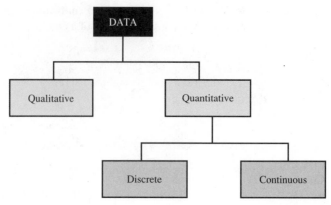

2.3 Statistical Tables and Graphs

When data are collected, the measurements on a variable need to be organized before any summary graphical presentations can be made. An important tool that you can use for data summary is a *statistical table*. These tables vary for each situation, but their primary purpose is to consolidate and summarize the data set so that it is more understandable to the reader. The statistical table can then be used to create charts and graphs that display the data pictorially.

Statistical tables for *qualitative data* consist of a list of the categories being measured along with either the frequency of occurrence of each category, the proportion of the observations in each category, or a quantity or amount measured for each category. These categories must be chosen so that an observation will belong to one and only one category and so that each observation has a category to which it can be assigned. If categories are defined using these guidelines, there will be no ambiguity about which category an observation belongs to.

If we were interested in categorizing meat products according to the type of meat, we might use the categories listed in Table 2.3. Categories describing the ranks of college faculty are given in Table 2.4. The "other" category is included in both cases to allow for the possibility that an observation cannot be assigned to one of the five categories specified.

T A B L E **2.3**
Type of meat

Beef	Pork
Chicken	Turkey
Seafood	Other

T A B L E **2.4**
Rank of college faculty

Professor	Instructor
Associate Professor	Lecturer
Assistant Professor	Other

Once the observations have been categorized, the data summary is presented as a *statistical table,* displaying an amount or quality in each category, or a *frequency distribution,* displaying the frequency or number of measurements in each category. The data can then be displayed graphically in a *pie chart,* in which the "pieces of the pie" represent the proportion of the total falling in each category, or a *bar graph,* in which the height of the bars represents the quantity or frequency of each category.

E X A M P L E   **2.3**   In a survey concerning the health of the U.S. economy, 400 economists were asked to rate the health of the U.S. economy. Their responses are summarized in Table 2.5. Construct a pie chart and a bar graph for this set of data.

T A B L E **2.5**
U.S. economic ratings by 400 economists

Rating	Frequency	Percentage
A	35	9
B	260	65
C	93	23
D	12	3
Total	400	100

Solution   In constructing a pie chart, we assign a sector of a circle to each category so that the angle of a sector is in proportion to the percentage of the measurements in that category. Since a circle contains $360°$, the number of degrees of the circle corresponding to each category is found using the formula (Percentage/100) $\times 360°$.

Category	Angle
A	$.09 \times 360° = 32.4°$
B	$.65 \times 360° = 234.0°$
C	$.23 \times 360° = 82.8°$
D	$.03 \times 360° = 10.8°$
Total	$360°$

Figure 2.2 shows the pie chart for Example 2.3, which has been constructed using the table above. While pie charts use percentages to determine the relative sizes of the pie slices, bar graphs plot frequency or proportion in a category against the categories. A bar graph for these data is shown in Figure 2.3. The visual impact of these two graphs is somewhat different. The pie chart displays the relationship of the parts to the whole, while the bar graph emphasizes the actual quantity or frequency for each category.   ▪

FIGURE **2.2**
Pie chart for Example 2.3

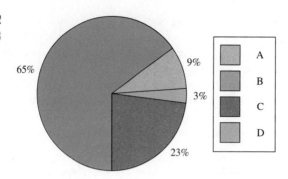

FIGURE **2.3**
Bar graph for Example 2.3

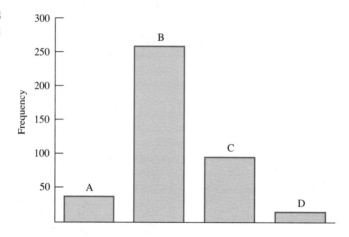

The *line graph* is used to display the change in a variable over time. In a line graph, the time intervals or categories (day, month, year, etc.) are located on one axis (usually the horizontal axis) and the variable to be charted is located on the other. (This type of graph is called a *time series,* the subject of Chapter 13.)

EXAMPLE **2.4** The yearly profit for a small manufacturing firm is given in the following statistical table for the years 1989–93. Construct a line graph for the data.

Year	Profit ($ million)
1991	1.5
1992	1.9
1993	2.2
1994	2.3
1995	2.5

FIGURE **2.4**
Line graph for Example 2.4

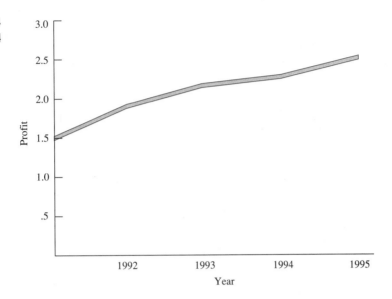

Solution   The categories, time intervals in years, are located on the horizontal axis, and the profit is plotted using the vertical axis. Note the increase in profit over time in Figure 2.4.   ▪

Guidelines for Constructing Pie Charts, Bar Graphs, and Line Graphs

- Use a small number of categories if possible. Oftentimes, too many categories will produce a visual dilution of important information.
- For bar graphs, label frequencies, proportions, or amounts along one axis and categories along the other. Leave a space between each bar to emphasize the distinctiveness of each category.
- For line graphs, label units of time along the horizontal axis and the measured variable along the vertical axis.

Exercises

Basic Techniques

2.1   Identify the following variables as quantitative or qualitative.
  a   The amount of time it takes to perform a simple assembly-line task.
  b   The number of new employees hired by a company in 1994.
  c   The rating of the CEO of a large corporation (excellent, good, fair, poor).
  d   The state in which a person lives.

2.2   Identify the following quantitative variables as discrete or continuous.

   a   The appraised value of a house, in dollars.

   b   The length of time until the fan belt of a car needs replacement.

   c   The number of miles traveled by a salesperson in a given month.

   d   The number of accounts held by a bank at a certain point in time.

   e   The length of time a customer must wait at a bank's cashier counter.

2.3   A group of 50 people are grouped into four categories—A, B, C, and D—and the number of people who fall into each category is shown below.

Category	Frequency
A	11
B	14
C	20
D	5

   a   Construct a pie chart to describe the data.

   b   Construct a bar graph to describe the data.

   c   Does the shape of the bar graph in part (b) change depending on the order of presentation of the four categories? Is the order of presentation important? Explain.

2.4   A stockbroker monitored the performance of one particular stock by recording its closing price for five days. The results are shown below. Create a line graph to describe the data. Do you think the stock is becoming more or less profitable?

Day	1	2	3	4	5
Price ($)	45	43	46	32	25

Applications

2.5   The price of health care in the United States has increased dramatically in the last twenty years and has become a major political issue in Washington, D.C. The consumer price indexes for medical services in three different categories are shown below for the years 1970, 1982, and 1992.

Year	Total Services	Physician Fees	Hospital Room Charges
1970	32	37	24
1982	93	93	90
1992	191	181	209

Source: Adapted from *Statistical Abstract of the United States, 1993,* p. 114.

   a   Use a line graph to describe the consumer price index for total services over time.

   b   Use line graphs to describe the other two consumer price indexes shown in the table.

   c   What conclusions can you draw from these graphs?

2.6   Pay raises for city and state employees have been almost nonexistent in the state of California during the last five years. In a recent *Los Angeles Times* poll, citizens of Los Angeles were asked if they approved or disapproved of the pay raise being offered to the Los Angeles Police Department, with the following results:

Response	Percent
Approve	74
Disapprove	20
Don't know	6

Source: Los Angeles Times,
June 28, 1994.

Construct a pie chart and a bar graph to describe these data. Which of the two presentations is more effective?

2.4 Relative Frequency Distributions

The statistical tables and graphs described in Section 2.3 are useful when the data have been categorized according to one or more *qualitative* variables. In this situation, pie charts, bar graphs, and line graphs allow us to describe the frequencies or amounts falling into each category. These methods will not always be appropriate, however, when the data are *quantitative*.

Quantitative variables give rise to numerical observations that represent an amount or a quantity. As we saw in Section 2.2, if the variable can take on only a finite or countable number of values, it is a *discrete* variable, while a variable that can assume an infinite number of values corresponding to all the points on a line interval is called a *continuous variable*. A quantitative data set, either discrete or continuous, consists of a set of numerical values that cannot be easily separated into qualitative categories. How could we graphically describe this type of data?

The simplest graphical method available for describing quantitative data is the **scatterplot.** For a small set of measurements—for example, the set 2, 3, 5, 8, and 6—the measurements can simply be plotted as points on a horizontal line. Such a scatterplot is shown in Figure 2.5. However, as the number of measurements increases, the scatterplot becomes uninformative, as shown in Figure 2.6 (based on a set of 120 measurements).

A better way to make sense of quantitative data is to use a type of graph called a **relative frequency histogram.** The data presented in Table 2.6 represent the dividend yields for 25 bank common stocks. The dividend yield for a stock is the percentage of the stock's price represented by its dividend. Although these yields are recorded only to one-decimal-place accuracy, the data are modeled by the continuous variable "dividend yield," which can take on any positive value. By examining the table, you can quickly see that the smallest and largest yields are 2.3 and 5.3, respectively, but how are the remaining 23 yields distributed? To answer this question, we divide the interval into an arbitrary number of subintervals or **classes** of equal length. As a rule of thumb, the number of classes should range from 5 to 20; the more data available, the more classes you need. The classes must be chosen so that each measurement can fall in *one and only one* class. *These classes are similar to the categories used for qualitative data.* Once the classes are formed, the measurements are categorized according to the class into which they fall, and a graph resembling a bar graph is drawn. This graph is called a **relative frequency histogram.**

F I G U R E **2.5**

Scatterplot for a small set of data

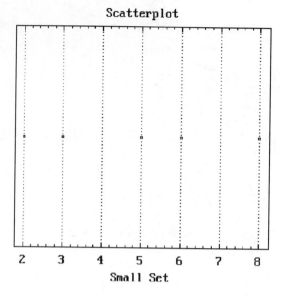

F I G U R E **2.6**

Scatterplot for a large set of data

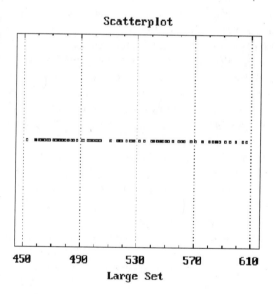

T A B L E **2.6**

The dividend yields (%) for 25 bank common stocks

3.1	4.2	2.3	3.3	2.8
5.3	3.5	3.1	2.6	3.3
4.7	3.7	3.0	2.6	4.0
3.8	4.4	3.2	3.2	3.8
5.1	3.7	2.3	4.3	3.9

DEFINITION ∎

> The **relative frequency histogram** for a quantitative data set is a bar graph in which the height of the bar represents the proportion or relative frequency of occurrence for a particular class or subinterval of the variable being measured. The classes or subintervals are plotted along the horizontal axis. ∎

For the dividend yields in Table 2.6, we choose to use seven intervals of equal length. Since the length of the total yield span is $(5.3 - 2.3) = 3.0$, a convenient choice of interval length is $(3 \div 7) = .43$, rounded off to .5. Rather than beginning the first interval at the lowest value, 2.3, we choose a starting value of 2.25 and form subintervals from 2.25 to 2.75, 2.75 to 3.25, and so on. By choosing 2.25 as the starting value, we make it impossible for any measurement to fall on the class boundaries and eliminate any ambiguity regarding the disposition of a particular measurement.

The 25 measurements are now categorized according to the class into which they fall, as shown in Table 2.7. The seven classes, along with a tally of the number of measurements falling into each class, are given in the second and third columns of the table. The fourth column gives the *class frequency,* the number of measurements falling into a particular class. The last column of the table presents the fraction or proportion of the total number of measurements falling into each class. We call this proportion the **class relative frequency.** If we let n represent the total number of measurements—in our example, $n = 25$—then the relative frequency for a particular class is calculated as

$$\text{Relative frequency} = \frac{\text{Frequency}}{n}$$

TABLE **2.7**
Tabulation of relative frequencies for a histogram

Class	Class Boundaries	Tally	Class Frequency	Class Relative Frequency
1	2.25–2.75	IIII	4	$4/25 = .16$
2	2.75–3.25	IHI I	6	$6/25 = .24$
3	3.25–3.75	IHI	5	$5/25 = .20$
4	3.75–4.25	IHI	5	$5/25 = .20$
5	4.25–4.75	III	3	$3/25 = .12$
6	4.75–5.25	I	1	$1/25 = .04$
7	5.25–5.75	I	1	$1/25 = .04$
		Total	$n = 25$	1.00

To construct the relative frequency histogram, plot the class boundaries along the horizontal axis. Draw a bar over each class interval, with height to the relative frequency for that class. The relative frequency histogram for the dividend yield data, Figure 2.7, shows at a glance how the yields are distributed over the interval 2.25 to 5.75.

FIGURE **2.7**
Relative frequency histogram

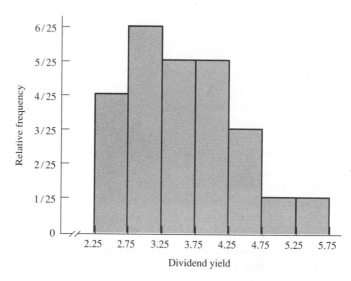

In recent years, computers have become readily available to many students, providing them with an invaluable tool. In the study of statistics, even the beginning student can use packaged programs to perform statistical analyses with a high degree of speed and accuracy. Some of the more common program packages (called *statistical software*) are Minitab, Statistical Analysis Systems (SAS), Statistical Packages for the Social Sciences (SPSS), Biomedical Packages (BMDP), as well as Execustat, Microsoft Excel, and other packages for personal computers.

Minitab is a package that is fast and easy to implement interactively while working at a terminal, even though its individual programs are quite powerful. We will show you how to implement specific programs in the Minitab package as we encounter those techniques in the text. Although other packages may also be introduced to implement statistical procedures presented in the text, Minitab will be the primary statistical package that we will use. Minitab programs are implemented using instructions called **commands.** These commands operate on data stored in one or more columns, designated C1, C2, and so on.

The Minitab package can be used to create a histogram using the command HISTOGRAM, followed by the column number, say C1, in which the data are stored. The program automatically chooses the number of classes, the class width, and class midpoints unless instructed by subcommands to do otherwise. If a data value falls on a class boundary, it is assigned to the adjacent class with the *larger* midpoint. The Minitab printout of a histogram for the dividend yields of the 25 bank common stocks is given in Figure 2.8; Figure 2.9 is the equivalent histogram generated by Execustat. The graphs are almost identical to the one shown in Figure 2.7, except for the scale of the vertical axis.

F I G U R E **2.8**
Minitab histogram for the
dividend yield data

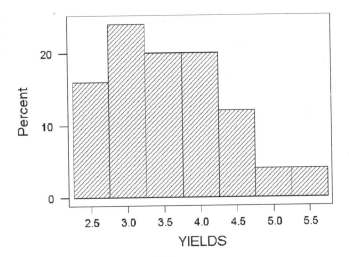

F I G U R E **2.9**
Execustat histogram for the
dividend yield data

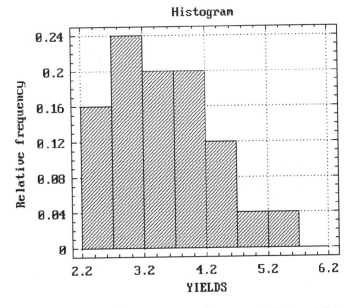

We can use the relative frequency histogram in Figure 2.7 to answer some interesting questions. What fraction of the stocks had dividend yields greater than 4.25%? Checking the relative frequency histogram, we see that the fraction involves all classes to the right of 4.25. Using Table 2.7, we see that five stocks possessed dividend yields greater than 4.25. Hence, the fraction is 5/25, or 20%. Suppose that we were to write each dividend yield on a piece of paper, place the 25 slips of paper in a hat, mix them, and then draw one piece of paper from the hat. What is the chance that this paper will contain a dividend yield greater than 4.25? Because 5 of the 25 slips are marked with numbers greater than 4.25, we say that we have 5 chances out of 25. Or we might say that the probability is .2. (You have undoubtedly encountered the word *probability* in ordinary conversation, and we defer definition and discussion of its significance until Chapter 3.)

What fraction of the dividend yields for *all* bank stocks were greater than 4.25? If we possessed the relative frequency histogram for the population, we could give the exact answer to this question by calculating the fraction of total area lying to the right of 4.25. Unfortunately, since we do not have such a histogram, we are forced to make an *inference*. We must estimate the true population fraction, basing our estimate upon information contained in the sample. Our estimate would likely be .2 or 20%. Suppose that we want to state the chance or probability that a bank stock from the population will have a dividend yield greater than 4.25. Without knowledge of the population relative frequency histogram, we infer that the population histogram is similar to the sample histogram and that approximately .2 of the measurements in the population will be greater than 4.25. Naturally, this estimate is subject to error. We will examine the magnitude of this estimation error in Chapter 7.

The relative frequency histogram is often called a **relative frequency distribution** because it shows the manner in which the data are distributed along the horizontal axis of the graph. **The rectangles constructed above each class are subject to two interpretations. They represent the fraction of the observations in a given class. Also, if a measurement is drawn from the data, a particular class relative frequency is also the chance or probability that the measurement will belong to that class.** The most significant feature of the sample frequency histogram is that it provides information on the population frequency histogram that describes the population. We would expect the two frequency histograms, sample and population, to be similar. Such is the case. The degree of resemblance will increase as more and more data are added to the sample. If the sample were enlarged to include the entire population, the sample and population would be synonymous, and the histograms would be identical.

Guidelines for Constructing a Relative Frequency Distribution

- *Determine the number of classes.* It is usually best to have from 5 to 20 classes. The larger the amount of data, the more classes can be used.

- *Determine the class width.* As a rule of thumb, divide the difference between the largest and smallest measurements by the number of classes desired and round up to a convenient figure for class width. All the classes should be of equal width.

- *If the data are discrete,* classes can be chosen to correspond to the values taken by the data. In this case, if the number of values taken by the data is small, each value can serve as a class.

- *Locate the class boundaries.* The lowest class must include the smallest measurement. Then add the remaining classes. Class boundaries should be chosen so that it will be impossible for a measurement to fall on a boundary.

- *Construct a table* containing the classes along with their frequencies and relative frequencies.

> ▪ *Construct the relative frequency histogram,* plotting the class intervals on the horizontal axis and drawing bars with heights corresponding to the appropriate relative frequencies.

Exercises

Basic Techniques

2.7   Consider the following set of data:

3	4	4	3	6	5
7	5	3	5	4	2
5	4	6	4	3	5
2	7	5	6	2	6
3	4	5	1	8	4

a   Construct a relative frequency histogram for the data. Use classes starting at .5 with a class width of 1—that is, .5 to 1.5, 1.5 to 2.5, and so forth.

b   What fraction of the measurements is less than 4.5?

c   What fraction of the measurements falls between 3.5 and 5.5?

2.8   Consider the following set of data:

3.1	4.9	2.8	3.6	2.5
4.5	3.5	3.7	4.1	4.9
2.9	2.1	3.5	4.0	3.7
2.7	4.0	4.4	3.7	4.2
3.8	6.2	2.5	2.9	2.8
5.1	1.8	5.6	2.2	3.4
2.5	3.6	5.1	4.8	1.6
3.6	6.1	4.7	3.9	3.9
4.3	5.7	3.7	4.6	4.0
5.6	4.9	4.2	3.1	3.9

a   Suppose that you wish to construct a relative frequency histogram for the data. Approximately how many class intervals should you use?

b   Suppose that you decide to use classes starting at 1.55 with a class width of .5—that is, 1.55 to 2.05, 2.05 to 2.55, and so on. Construct the relative frequency histogram for the data.

c   What fraction of the measurements is less than 5.05?

d   What fraction of the measurements is larger than 3.55?

2.9   A certain discrete variable can take only the values 0, 1, or 2. A set of 20 measurements on this variable is shown below:

1	2	1	0	2
2	1	1	0	0
2	2	1	1	0
0	1	2	1	1

a   Construct a relative frequency histogram for the data.

b   What proportion of the measurements is greater than 1?

c   What proportion of the measurements is less than 2?

d   If a measurement is selected at random from the 20 measurements shown, what is the probability that it is a 2?

Applications

2.10   In the table that follows, the U.S. Census Bureau details the per capita debt and the per capita taxes for each of the 50 states in the fiscal year 1992.

State	Debt per Capita	Taxes per Capita	State	Debt per Capita	Taxes per Capita
AL	998	1019	MT	2266	1153
AK	8418	2730	NE	1092	1176
AZ	743	1259	NV	1457	1369
AR	809	1145	NH	3882	770
CA	1225	1495	NJ	2540	1643
CO	857	1018	NM	1015	1415
CT	3644	1846	NY	3083	1661
DE	5140	1944	NC	558	1316
FL	911	1068	ND	1615	1186
GA	662	1076	OH	1106	1099
HI	4040	2335	OK	1138	1206
ID	1210	1303	OR	2114	1113
IL	1611	1157	PA	1079	1354
IN	913	1143	RI	5125	1270
IA	669	1280	SC	1300	1092
KS	192	1110	SD	2657	794
KY	1762	1353	TN	558	900
LA	2331	991	TX	453	964
ME	2135	1347	UT	1187	1096
MD	1698	1324	VT	2706	1339
MA	4002	1651	VA	1160	1101
MI	1097	1195	WA	1400	1648
MN	924	1662	WV	1431	1297
MS	621	954	WI	1457	1380
MO	1213	988	WY	1920	1386

Source: Data from U.S. Department of Commerce, Bureau of the Census, *The World Almanac and Book of Facts,* 1994 ed., p. 105.

a   Construct a relative frequency histogram to describe the data on debt per capita for the 50 states.

b   Construct a relative frequency histogram to describe the data on taxes per capita for the 50 states.

c   Compare the shapes of the relative frequency histograms in (a) and (b). Are they similar or different?

2.11   In order to decide on the number of service counters needed for stores built in the future, a supermarket chain wanted to obtain information on the length of time (in minutes) required to service customers. To obtain information on the distribution of customer service times, a sample of 1000 customers' service times was recorded. Sixty of these are shown in the accompanying table.

3.6	1.9	2.1	.3	.8	.2
1.0	1.4	1.8	1.6	1.1	1.8
.3	1.1	.5	1.2	.6	1.1
.8	1.7	1.4	.2	1.3	3.1
.4	2.3	1.8	4.5	.9	.7
.6	2.8	2.5	1.1	.4	1.2
.4	1.3	.8	1.3	1.1	1.2
.8	1.0	.9	.7	3.1	1.7
1.1	2.2	1.6	1.9	5.2	.5
1.8	.3	1.1	.6	.7	.6

a   Construct a relative frequency histogram for the data.

b   What fraction of the service times are less than or equal to one minute?

2.12   The number of home loan applications granted during a particular month was recorded for a sample of $n = 50$ commercial banks and/or lending institutions. These data are as follows.

2	4	2	32	9
9	2	6	3	1
14	9	16	7	8
19	6	4	4	2
4	18	0	6	13
7	2	8	0	1
14	1	2	2	18
8	24	1	8	5
1	3	11	18	26
3	12	23	5	4

a   Construct a relative frequency histogram for the data.

b   Determine the fraction of the 50 commercial banks and/or lending institutions that granted 10 or fewer home loans during this particular month.

2.13   The new electric sports car Tropica, described in Exercise 1.3, has been clocked by the California Air Resources Board to have a range of 44.7 miles (Parrish, 1994). In an independent test, 30 trials produced the following driving ranges:

45.4	45.9	45.3	42.9	45.8
44.6	44.7	46.4	45.7	45.9
44.1	47.1	46.4	44.3	47.3
46.8	44.2	45.5	45.0	46.4
43.5	46.6	44.5	44.8	44.3
47.0	44.6	43.3	46.1	43.8

Construct a relative frequency histogram to describe these data. Does the reported range of 44.7 miles seem reasonable in light of this histogram? Explain.

2.5 Stem and Leaf Displays

The stem and leaf display is another method for describing a set of data. It is just one of a number of techniques in a newly emerging area of statistics called Exploratory Data Analysis (EDA). This area of statistics provides simple techniques, often pictorial, that allow an investigator to examine the data prior to formal analysis. The stem and leaf display provides a picture of the data that is similar to a histogram; however, the actual values are used in the display. Although the display has a graphical appearance, it also provides a tabulation of the data values.

To demonstrate the steps in constructing a stem and leaf display, we will use the following data, which give the resale price (in thousands of dollars) of 25 condominium units in a condominium complex.

68.9	100.5	82.5	105.1	90.4
83.7	75.1	85.1	80.6	83.4
77.0	93.5	96.4	91.0	75.6
81.0	84.8	89.3	92.0	69.5
70.0	82.8	85.7	78.1	79.4

In creating a stem and leaf display, we divide each observation into two parts: the **stem** and the **leaf.** For these data we could divide each observation at the decimal point so that the portion to the left of the decimal place is the stem and the portion to the right is the leaf. For example, for the observation 68.9 the stem and leaf are

Stem	Leaf
68	.9

Or we could place the point of division between the units and tens places so that the stem and leaf are

Stem	Leaf
6	8.9

Since the stems serve the same function as the class boundaries in a histogram, we would select stem values so that, in general, each stem has one or more leaves. To simplify presentations, leaves are usually given as a rounded single digit. Hence using the second designation for the observation 68.9 the stem and leaf are

Stem	Leaf
6	9

To create a stem and leaf display, list the stem values in a column; then systematically tabulate the leading digit of each leaf to the right of the appropriate stem.

F I G U R E **2.10**
Stem and leaf display of
condominium resale prices

```
 6| 9
 7| 7 0 5 8 6 0 9
 8| 4 1 5 3 3 5 9 6 1 3
 9| 4 6 1 2 0
10| 1 5
```

The completed stem and leaf display for our example is given in Figure 2.10. In this display, each stem value represents $10,000 and each leaf unit represents $1000.

A stem and leaf display can be implemented using the Minitab command STEM AND LEAF[†] followed by the column number that contains the data of interest. The Minitab stem and leaf display for these data is given in Figure 2.11. By turning the stem and leaf display on its side, you can see that we have a frequency distribution with slightly more information about the values of the observations in each class.

F I G U R E **2.11**
Minitab stem and leaf printout of
the condominium resale prices

```
MTB > STEM C1;
SUBC> INCR 10.

Stem-and-leaf of CONDO      N = 25
Leaf Unit = 1.0

    1       6 9
    8       7 0056789
  (10)      8 1133345569
    7       9 01246
    2      10 15
```

There are times when a data set such as the dividend yields in Table 2.6 contains too few stems. In cases such as these, stems are used twice, so that the leaves 0, 1, 2, 3, and 4 are associated with the first use of the stem and the digits 5, 6, 7, 8, and 9 are associated with the second use of the stem. In effect the new class boundaries are 2.0 to 2.4 inclusive, and 2.5 to 2.9 inclusive. The Minitab stem and leaf display of the 25 dividend yields in Table 2.6 is given in Figure 2.12. In those instances when the values of a few observations lie far from the others, they are listed below the display as HI or LO.

To summarize, a stem and leaf display is easy to construct, and the display creates the same sort of figure produced by a relative frequency distribution. In addition, it permits the viewer to reconstruct the data set and also to identify ordered observations (such as the fifth largest).

[†]Only the first four letters of a Minitab command are necessary to implement the command. For example, STEM is sufficient to implement the command STEM AND LEAF. However, for the sake of clarity, we will always give the complete command.

FIGURE **2.12**
Stem and leaf display of the
dividend yields of Table 2.6

```
MTB > STEM AND LEAF C1

Stem-and-leaf of YIELDS      N  = 25
Leaf Unit = 0.10


       2       2 33
       5       2 668
      12       3 0112233
      (6)      3 577889
       7       4 0234
       3       4 7
       2       5 13
```

Exercises

Basic Techniques

2.14 Construct a stem and leaf display for the data given in Exercise 2.8 using the leading digit as the stem. The data are reproduced below for your convenience.

3.1	4.9	2.8	3.6	2.5
4.5	3.5	3.7	4.1	4.9
2.9	2.1	3.5	4.0	3.7
2.7	4.0	4.4	3.7	4.2
3.8	6.2	2.5	2.9	2.8
5.1	1.8	5.6	2.2	3.4
2.5	3.6	5.1	4.8	1.6
3.6	6.1	4.7	3.9	3.9
4.3	5.7	3.7	4.6	4.0
5.6	4.9	4.2	3.1	3.9

a Compare the stem and leaf display with the relative frequency histogram that you constructed in Exercise 2.8.

b Find the sixth smallest measurement when the data are ranked from smallest to largest.

2.15 Refer to Exercise 2.14. Construct a stem and leaf display for the data using each leading digit twice. Has this technique improved the presentation of the data?

2.16 Consider the following set of data:

652	648	658	662
653	654	670	671
674	666	679	679
653	652	667	671
655	677	646	678
682	678	669	650

a What would be the most appropriate choice of stem and leaf in this case?

b Construct a stem and leaf display using the results of part (a).

c Describe the shape of the distribution of measurements. Are any of the measurements unusually high or low? Use the stem and leaf display to find the largest and smallest measurements.

Applications

2.17   Construct a stem and leaf display for the per capita debt data of Exercise 2.10. Compare the stem and leaf display with the relative frequency histogram of Exercise 2.10. Do the two graphical descriptions of the data seem to convey the same information?

2.18   Construct a stem and leaf display for the supermarket service times of Exercise 2.11. Compare the stem and leaf display with the relative frequency histogram of Exercise 2.11. Do the two graphical descriptions of the data seem to convey the same information?

2.19   Refer to Exercise 2.12. Construct a stem and leaf display for the number of home loan applications data. Compare the stem and leaf display with the relative frequency histogram of Exercise 2.12. Do the two graphical descriptions of the data seem to convey the same information?

2.20   Refer to Exercise 2.19. The Minitab command STEM was used to produce the stem and leaf display below for the home loan data. Explain the choice of stem and leaf used in the Minitab analysis. Does this display differ from the display you constructed in Exercise 2.19?

```
MTB > Stem-and-Leaf C1.

Stem-and-leaf of C1       N  = 50
Leaf Unit = 1.0

     8      0 00111111
    18      0 2222222333
    24      0 444455
    (5)     0 66677
    21      0 8888999
    14      1 1
    13      1 23
    11      1 44
     9      1 6
     8      1 8889
     4      2
     4      2 3
     3      2 4
     2      2 6
     1      2
     1      3
     1      3 2

MTB >
```

2.21   The accompanying table gives the amounts of sales (in billions of dollars) for the 50 U.S. industrial corporations with the largest sales in 1992. Construct a stem and leaf display to describe the sales for the 50 corporations.

Table for Exercise 2.21

Company	Sales	Company	Sales
General Motors	132.8	Digital Equipment	14.0
Exxon	103.5	Minnesota Mining	13.9
Ford Motor	100.8	Johnson & Johnson	13.8
IBM	65.1	Tenneco	13.6
General Electric	62.2	International Paper	13.6
Mobil	57.4	Motorola	13.3
Philip Morris	50.2	Sara Lee	13.3
Du Pont	37.6	Coca Cola	13.2
Chevron	37.5	Westinghouse	12.1
Texaco	37.1	Allied-Signals	12.1
Chrysler	36.9	Phillips Petroleum	12.0
Boeing	30.2	Goodyear Tire	11.9
Procter and Gamble	29.9	Georgia Pacific	11.8
Amoco	25.5	Bristol-Meyers-Squib	11.8
Pepsico	22.1	Anheuser-Busch	11.4
United Technologies	22.0	IPB	11.1
Shell Oil	21.7	Rockwell International	11.0
Conagra	21.2	Caterpillar	10.2
Eastman Kodak	20.6	Lockheed	10.1
Dow Chemical	19.2	Coastal	10.1
Xerox	18.3	Merck	9.8
Atlantic Richfield	18.1	Alcoa	9.6
McDonald Douglas	17.5	Archer Daniels	9.3
Hewlett-Packard	16.4	Weyerhaeuser	9.3
USX	16.2	Unilever U.S.	9.2

Source: Data from *Fortune*, *The World Almanac and Book of Facts*, 1994 ed., p. 111.

2.6 Numerical Methods for Describing a Set of Data

Graphical methods are extremely useful in presenting data and in conveying a rapid, general description of collected data. This supports, in many respects, the saying that a picture is worth a thousand words. There are, however, limitations to the use of graphical techniques for describing and analyzing data. For instance, suppose that we want to discuss our data before a group of people and have no overhead projector available! We would be forced to use other descriptive measures that would convey to the listeners a mental picture of the histogram. A second and not so obvious limitation of the histogram and other graphical techniques is that they are difficult to use for purposes of statistical inference. Presumably, we use the sample histogram to make inferences about the shape and position of the population histogram, which describes the population and is unknown to us. Our inference is based upon the correct assumption that some degree of similarity will exist between the two histograms, but we are then faced with the problem of measuring the degree of similarity. We know when two figures are identical, but this situation will not likely occur in practice. If they were identical, we could say "They are the same!" But, if they are different, it is difficult to describe the "degree of difference."

The limitations of the graphical method of describing data can be overcome by the use of **numerical descriptive measures.** Numerical descriptive measures for a population are called **parameters.** The corresponding numerical descriptive measures calculated from a sample are called **statistics.** Thus we would like to use the sample data to calculate a set of numbers, statistics, that will convey a good mental picture of the sample relative frequency distribution and that will be useful in making inferences about the population relative frequency distribution.

DEFINITION   ■
> Numerical descriptive measures computed from population measurements are called **parameters.**   ■

DEFINITION   ■
> Numerical descriptive measures computed from sample measurements are called **statistics.**   ■

2.7 Measures of Central Tendency

In constructing a mental picture of the frequency distribution for a set of measurements on a quantitative variable, x, we would likely envision a histogram similar to that shown in Figure 2.7 for the data on bank stock dividend yields. One of the first descriptive measures of interest is a **measure of central tendency,** which is a measure, such as an average, that locates the center of the distribution. We note that the dividend yields ranged from a low of 2.3 to a high of 5.3, with the center of the histogram located in the vicinity of 3.6. Let us now consider some definite rules for locating the center of a distribution of data.

One of the most common and useful measures of central tendency is the arithmetic average of a set of measurements. This value is also often referred to as the **arithmetic mean,** or simply the **mean,** of a set of measurements. Since we will want to distinguish between the mean of a sample and the mean of a population, we will use the symbol $\bar{x}$ (x bar) to represent the sample mean and μ (lowercase Greek letter mu) to represent the mean of the population.

DEFINITION   ■
> The **arithmetic mean** of a set of measurements is equal to the sum of the measurements divided by the number of measurements.   ■

The procedures for calculating a sample mean and many other statistics are conveniently expressed as formulas. Consequently, we will need a symbol to represent the process of summation. If we denote the n quantities that are to be summed as $x_1, x_2, \ldots, x_n$, then their sum is denoted by the symbol

$$\sum_{i=1}^{n} x_i$$

The Greek capital letter sigma (Σ) is an instruction *to add*. The quantity x_i to the right of Σ is the typical element to be summed. The notations $i = 1$ below and n above the letter Σ indicate that i is the variable of summation and begins with the value 1, increases in increments of 1, and ends with the value n. For example,

$$\sum_{i=1}^{3} x_i = x_1 + x_2 + x_3$$

Using this notation, we can express the formulas for the sample mean and population mean as follows:

Formulas for Calculating the Mean

Sample mean: $\bar{x} = \dfrac{\sum_{i=1}^{n} x_i}{n}$

Population mean: $\mu = \dfrac{\sum_{i=1}^{N} x_i}{N}$

E X A M P L E   **2.5**   Find the mean of the set of measurements 2, 9, 11, 5, 6.

Solution

$$\bar{x} = \frac{\sum_{i=1}^{n} x_i}{n} = \frac{2 + 9 + 11 + 5 + 6}{5} = 6.6 \quad \blacksquare$$

Even more important than locating the center of a set of sample measurements, $\bar{x}$ will be used as an estimator (predictor) of the value of the unknown population mean μ. For example, the mean of the data listed in Table 2.6 is equal to

$$\bar{x} = \frac{\sum_{i=1}^{n} x_i}{n} = \frac{89.2}{25} = 3.568$$

Note that this value falls approximately in the center of the set of measurements. The mean of the entire population of dividend yields, μ, is unknown to us; but if we were to estimate its value, our estimate of μ would be 3.568.

A second measure of central tendency is the **median.**

DEFINITION ▪

> The **median** m of a set of n measurements x_1, x_2, x_3, ..., x_n is the value of x that falls in the middle when the measurements are ranked in order from the smallest to the largest. ▪

If the measurements in a data set are ranked from the smallest to the largest, the median will be the value of x that falls in the middle. If the number n of measurements is odd, this number will be the measurement with rank equal to $(n + 1)/2$. If the number of measurements is even, the median is chosen as the value of x halfway between the two middle measurements—that is, halfway between the measurement ranked $n/2$ and the one ranked $(n/2) + 1$. The rule for calculating the median is given in the following box.

Rule for Calculating a Median

Rank the n measurements from the smallest to the largest.

1 If n is odd, the median m is the measurement with rank $(n + 1)/2$.

2 If n is even, the median m is the value of x that is halfway between the measurement with rank $n/2$ and the one with rank $(n/2) + 1$.

EXAMPLE **2.6** Find the median for the following set of five measurements.

$$9, \ 2, \ 7, \ 11, \ 14$$

Solution We first rank the $n = 5$ measurements from the smallest to the largest, $2, 7, 9, 11, 14$. Then, since $n = 5$ is odd, we choose 9 as the median. This value is the measurement with rank $(n + 1)/2 = (5 + 1)/2 = 3$. ▪

EXAMPLE **2.7** Find the median for the following set of measurements.

$$9, \ 2, \ 7, \ 11, \ 14, \ 6$$

Solution Since $n = 6$ is even, we rank the measurements as $2, 6, 7, 9, 11, 14$ and choose the median halfway between the two middle measurements, 7 and 9. Therefore, the median is equal to 8. ▪

Although both the mean and the median are good measures for the center of a distribution of measurements, the median is less sensitive to extreme values. For example, if the distribution is symmetric about its mean—that is, the left and right halves of the distribution are mirror images—then the mean and the median are equal [see Figure 2.13(a)]. If a distribution is not symmetric and has extreme observations lying in the right tail of the distribution, the distribution is said to be *skewed to the right* [see Figure 2.13(b)]. Since the large extreme values in the upper tail of the distribution increase the sum of the measurements, the mean shifts to the right. The median is not affected by these extreme values, because the numerical values of the measurements are not used in its computation. Finally, if a distribution is *skewed to the left,* the mean shifts to the left.

F I G U R E **2.13**

Relative frequency distributions showing the effect of extreme values on the mean and median

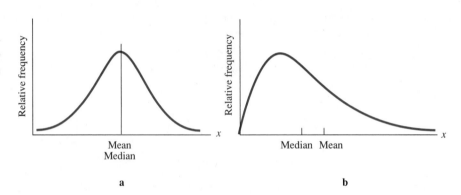

Another measure of central tendency is the **mode,** which is defined as the observation that occurs most often in a data set.

D E F I N I T I O N   ▪

The **mode** of a set of n measurements x_1, x_2, x_3, $\ldots$, x_n is the value of x occurring with the greatest frequency.   ▪

When the measurements are grouped in a relative frequency histogram, the class with the largest relative frequency is called the **modal class,** and the modal class midpoint is taken as the value of the mode.

E X A M P L E   **2.8**   Given the sample measurements

$$5, \ 5, \ 7, \ 7, \ 7, \ 10, \ 15$$

the value 7 occurs three times, the value 5 occurs twice, and the values 10 and 15 each occur once. Therefore the mode of the sample measurements is 7.   ▪

For symmetric distributions, the mean, median, and mode are all equal. In distributions skewed to the right, the mode lies to the left of the median and mean. See Figures 2.13(a) and 2.13(b). If the distribution is skewed left, the positions of these three measures are reversed, and the mode lies to the right of the mean and the median.

It is possible for a distribution of measurements to have more than one mode. For example, the distribution of salaries for a large group of employees might produce a *bimodal distribution,* possibly reflecting a mixture of measurements taken on blue-collar and white-collar employees.

Exercises

Basic Techniques

2.22   Consider $n = 5$ measurements, 0, 5, 1, 1, 3.

a   Draw a scatterplot for the data. [*Hint:* If two measurements are the same, place one dot above the other.] Guess the approximate "center."

b   Find the mean, the median, and the mode.

c   Locate the three measures found in (b) on the scatterplot in part (a). Based on the relative positions of the mean and median, would you say that the measurements are symmetric or skewed?

2.23   Consider $n = 8$ measurements, 3, 1, 5, 6, 4, 4, 3, 5.

a   Find $\bar{x}$.

b   Find m.

c   Based on the results of parts (a) and (b), are the measurements skewed or symmetric? Draw a scatterplot to confirm your answer.

2.24   Given $n = 10$ measurements, 3, 5, 4, 6, 10, 5, 6, 9, 2, 8, find the following.

a $\bar{x}$

b m

c   the mode

Applications

2.25   Many computer buyers have discovered that they can save a considerable amount of money by purchasing a personal computer from a mail-order company—an average of $900 by their estimates ("Who's Tops," 1992). The satisfaction ratings (on a scale of 1 to 9) for seven such companies, based on a survey of 4000 buyers, are shown below.

Company	Rating	Company	Rating
CompuAdd	7.5	Insight	7.8
Dell	7.9	Northgate	7.7
FastMicro	7.4	Zeos	8.0
Gateway	8.2		

a What is the average satisfaction rating for these seven companies?

b What is the median of the satisfaction ratings?

c If you were a computer buyer, would you be interested in the average satisfaction rating? If not, what measure would you be interested in? Explain.

2.26 The earnings per share for the second quarter of 1994 for a sample of 20 companies are given below:

$ .72	.56	.21	.54	.32
1.28	.10	1.64	.29	.33
.29	.73	.29	.33	.43
.56	.89	.84	.62	.44

Source: Data from the *Press-Enterprise*, Riverside, Calif., July 20, 1994.

a Do you think that the distribution of earnings per share is symmetric or skewed?

b Calculate the mean, the median, and the mode for these measurements.

c Draw a relative frequency histogram for the data set. Locate the mean, the median, and the mode along the horizontal axis. Is your answer to part (a) correct?

2.27 *PC World* provides an excellent source for computer users who want to upgrade their current operating systems or buy new ones. A recent issue of *PC World* ("Top 10," 1994) listed the top ten Windows accelerators, along with an overall value rating and the estimated street price, as given in the following table:

Accelerator	Overall Value Rating	Estimated Street Price
Diamond Stealth	87	$249
Number Nine	86	275
Genoa Phantom	85	245
Hercules Dynamite Pro	82	210
miroCrystal8S	82	195
Orchid Kelvin	75	275
Hercules Graphite	73	335
Matrox MGA	73	475
Hercules Dynamite Power	72	237
Paradise Ports o' Call	72	235

a What is the average overall value rating for these ten products?

b What is the average estimated street price?

c If you were going to purchase an accelerator, would these averages be important to you? Explain.

2.28 In an article titled "You Aren't Paranoid If You Think Someone Eyes Your Every Move," the *Wall Street Journal* (March 19, 1985) notes that big business collects detailed statistics on your behavior. Jockey International knows how many pairs of undershorts you own; Frito-Lay, Inc., knows which you eat first—the broken pretzels in a pack or the whole ones; and, to get to specifics, Coca-Cola knows that you put 3.2 ice cubes in a glass. Have you ever put 3.2 ice cubes in a glass? What did the *Wall Street Journal* article mean by that statement?

2.29   The table in Exercise 2.10 gives the per capita debt for each of the 50 states in fiscal year 1992.

    a   Find the average per capita debt for the 50 states.

    b   Find the median per capita debt for the 50 states and compare it with the mean calculated in part (a).

    c   Based on your comparison in part (b), would you conclude that the distribution of per capita debts is skewed? Explain.

2.30   Unit pricing has become an industry-wide standard in the grocery business. The consumer's job is to weigh the quality of the product against the price per unit in an attempt to determine the "best buy." The following measurements are the price per liner recorded for ten different brands of 13-gallon tall kitchen trash-can liners (*Consumer Reports,* February 1994).

$$
\begin{array}{ccccc}
10 & 9 & 13 & 8 & 9 \\
10 & 10 & 6 & 5 & 11 \\
\end{array}
$$

    a   Find the average price per liner.

    b   Find the median price per liner.

    c   If you were writing a report to describe these data, which measure of central tendency would you use? Explain.

2.8   Measures of Variability

Once we have located the center of a distribution of data, the next step is to provide a measure of the **variability,** or **dispersion,** of the data. Consider the two distributions shown in Figure 2.14. Both distributions are located with a center at $x = 4$, but there is a vast difference in the variability of the measurements about the mean for the two distributions. The measurements in Figure 2.14(a) vary approximately from 3 to 5; in Figure 2.14(b) the measurements vary from 0 to 8.

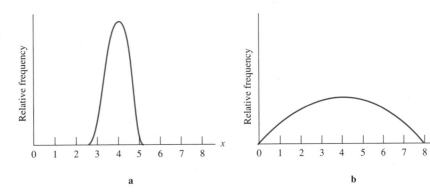

FIGURE **2.14**
Variability or dispersion of data

Variation is a very important characteristic of data. For example, if we are manufacturing bolts, excessive variation in the bolt diameter would imply a high percentage of defective product. On the other hand, if we are using an examination to discriminate between good and poor accountants, we would be most unhappy if the examination always produced test grades with little variation, since this would make discrimination very difficult.

In addition to the practical importance of variation in data, a measure of this characteristic is necessary to the construction of a mental image of the frequency distribution. We will discuss only a few of the many measures of variation.

The simplest measure of variation is the **range.**

DEFINITION ■
> The **range** of a set of n measurements x_1, x_2, x_3, ..., x_n is defined as the difference between the largest and smallest measurements. ■

The dividend yield data varied from 2.3 to 5.3. Hence, the range is $(5.3 - 2.3) = 3.0$. The range is easy to calculate, easy to interpret, and quite adequate as a measure of variation for small sets of data. But for large data sets the range is not an adequate measure of variability. For example, the two relative frequency distributions in Figure 2.15 have the same range but have very different shapes and variability.

FIGURE **2.15**
Distributions with equal range and unequal variablility

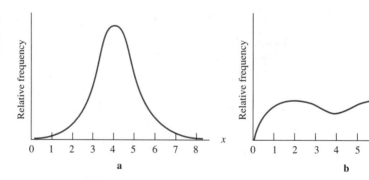

Can we find a measure of variability that is more sensitive than the range? Consider, as an example, the sample measurements 5, 7, 1, 2, 4, displayed as a scatterplot in Figure 2.16. The mean of these five measurements is

$$\bar{x} = \frac{\displaystyle\sum_{i=1}^{n} x_i}{n} = \frac{19}{5} = 3.8$$

as indicated on the scatterplot.

We can now view variability in terms of distance between each dot (measurement) and the mean $\bar{x}$. If the distances are large, we can say that the data are more variable than if the distances are small. More explicitly, we define the **deviation** of a measurement from its mean to be the quantity $(x_i - \bar{x})$. Measurements to the right of the mean produce positive deviations, and those to the left produce negative deviations. The values of x and the deviations for our example are shown in the first and second columns of Table 2.8.

FIGURE **2.16**
Scatterplot

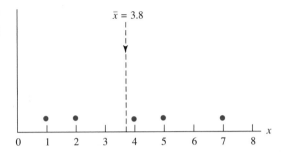

$\bar{x} = 3.8$

TABLE **2.8**
Computations involving sample deviations

x_i	$(x_i - \bar{x})$	$(x_i - \bar{x})^2$	$|x_i - \bar{x}|$
5	1.2	1.44	1.2
7	3.2	10.24	3.2
1	−2.8	7.84	2.8
2	−1.8	3.24	1.8
4	.2	.04	.2
19	0.0	22.80	9.2

If we agree that deviations contain information on variation, our next step is to construct a measure of variation based on the deviations about the mean. As a first possibility, we might choose the mean of the deviations. Unfortunately, the mean will not work, because some of the deviations are positive, some are negative, and the sum is always zero (unless round-off errors have been introduced into the calculations). Note that the deviations in the second column of Table 2.8 sum to zero.

There are two ways to avoid this problem. Why not calculate the average of the absolute values of the deviations? This measure is called the **mean absolute deviation** (MAD).

DEFINITION ▪

The **mean absolute deviation** of a set of n measurements x_1, x_2, ..., x_n is the average of the absolute values of the deviations about the sample mean and is given by

$$\text{MAD} = \frac{\sum\limits_{i=1}^{n} |x_i - \bar{x}|}{n} \quad ▪$$

The absolute deviations for our set of $n = 5$ observations together with their sum are given in Table 2.8. Therefore,

$$\text{MAD} = \frac{\sum\limits_{i=1}^{n} |x_i - \bar{x}|}{n} = \frac{9.2}{5} = 1.84$$

Although MAD is sometimes used as a measure of variability for a set of data, it is mainly used in assessing prediction accuracy. This topic will be revisited in Chapter 13, where we discuss time series and index numbers.

The second way to use the deviations is to work with the sum of squares of the deviations. Using the sum of squared deviations, we calculate a single measure called the **variance** of a set of measurements. To distinguish between the variance of a *sample* and the variance of a *population,* we use the symbol s^2 to represent the sample variance and σ^2 (Greek lowercase sigma) to represent the population variance. *This measure will be relatively large for highly variable data and relatively small for less variable data.*

DEFINITION ▪

> The **variance of a population** of N measurements x_1, x_2, ..., x_N is defined to be the average of the squares of the deviations of the measurements about their mean μ. The population variance is given by the formula
>
> $$\sigma^2 = \frac{\sum_{i=1}^{N}(x_i - \mu)^2}{N} \quad \blacksquare$$

Most often, you will not have all the population measurements available but will need to calculate the *variance of a sample* of n measurements.

DEFINITION ▪

> The **variance of a sample** of n measurements x_1, x_2, ..., x_n is defined to be the sum of the squared deviations of the measurements about their mean $\bar{x}$ divided by $(n-1)$. The sample variance is denoted by s^2 and is given by the formula
>
> $$s^2 = \frac{\sum_{i=1}^{n}(x_i - \bar{x})^2}{n-1} \quad \blacksquare$$

For example, we can calculate the variance for the set of $n = 5$ sample measurements presented in Table 2.8. The square of the deviation of each measurement is recorded in the third column of Table 2.8. Adding, we obtain

$$\sum_{i=1}^{5}(x_i - \bar{x})^2 = 22.80$$

The sample variance is

$$s^2 = \frac{\sum_{i=1}^{n}(x_i - \bar{x})^2}{n-1} = \frac{22.80}{4} = 5.70$$

The variance is measured in terms of the square of the original units of measurement. If the original measurements are in inches, the variance is expressed in square inches. Taking the square root of the variance, we obtain the **standard deviation,** which returns the measure of variability to the original units of measurement.

D E F I N I T I O N ▪ The **standard deviation** of a set of measurements is equal to the positive square root of the variance. ▪

Notation

n: number of measurements in the sample N: number of measurements in the
s^2: sample variance   population
$s = \sqrt{s^2}$: sample standard deviation σ^2: population variance
                                                                            $\sigma = \sqrt{\sigma^2}$: population standard deviation

For the set of $n = 5$ sample measurements in Table 2.8, the sample variance was $s^2 = 5.70$, so the sample standard deviation is $s = \sqrt{s^2} = \sqrt{5.70} = 2.39$. The more variable the data set is, the larger the value of s will be.

For the small set of measurements we used, the calculation of the variance was not too difficult. However, for a larger set, the calculations can become very tedious. Most calculators with statistical capabilities have built-in programs that will calculate $\bar{x}$ and s or μ and σ, so that your computational work will be minimized. The sample or population mean key is usually marked with $\bar{x}$. The sample standard deviation key is usually marked with s or σ_{n-1} and the population standard deviation key with σ or σ_N. In using any calculator with these built-in function keys, be sure you know which calculation is being carried out by each key!

If you need to calculate s^2 and s by hand, it is much easier to use the alternative computing formula given below. This computational form is sometimes called the shortcut method for calculating s^2.

Computing Formula for s^2

$$s^2 = \frac{\sum_{i=1}^{n} x_i^2 - \frac{\left(\sum_{i=1}^{n} x_i\right)^2}{n}}{n - 1}$$

where

$$\sum_{i=1}^{n} x_i^2 = \text{the sum of the squares of the individual observations}$$

$$\left(\sum_{i=1}^{n} x_i\right)^2 = \text{the square of the sum of the individual observations}$$

The *sample standard deviation, s,* is the positive square root of s^2.

E X A M P L E   **2.9**   Calculate the variance and standard deviation for the five measurements in Table 2.8, which are given as 5, 7, 1, 2, and 4. Use the computing formula for s^2 and compare your results with those obtained using the original definition of s^2.

T A B L E **2.9**
Table for simplified calculation of s^2 and s

x_i	x_i^2
5	25
7	49
1	1
2	4
4	16
19	95

Solution   The entries in Table 2.9 are the individual measurements, x_i, and their squares, x_i^2, together with their sums. Using the computing formula for s^2, we have

$$s^2 = \frac{\sum_{i=1}^{n} x_i^2 - \frac{\left(\sum_{i=1}^{n} x_i\right)^2}{n}}{n-1}$$

$$= \frac{95 - \frac{(19)^2}{5}}{4} = \frac{22.80}{4} = 5.70$$

and $s = \sqrt{s^2} = \sqrt{5.70} = 2.39$, as before. ∎

E X A M P L E   **2.10**   Calculate the sample variance and standard deviation for the $n = 25$ yields in Table 2.6.

Solution   Using a calculator with built-in statistical functions, you can verify the following:

$$\sum_{i=1}^{n} x_i = 89.2$$

$$\sum_{i=1}^{n} x_i^2 = 333.82$$

Using the computing formula

$$s^2 = \frac{\sum_{i=1}^{n}(x_i - \bar{x})^2}{n-1} = \frac{\sum_{i=1}^{n} x_i^2 - \frac{\left(\sum_{i=1}^{n} x_i\right)^2}{n}}{n-1}$$

$$= \frac{333.82 - \frac{(89.2)^2}{25}}{24} = \frac{15.5544}{24} = .6481$$

and $s = \sqrt{s^2} = \sqrt{.6481} = .81$ ▪

You may wonder why we divide by $n - 1$ rather than n when we compute the sample variance. The sample mean $\bar{x}$ is used as an estimator of the population mean because it provides a good estimate of μ. If we want to use the sample variance as an estimator of the population variance of σ^2, the sample variance s^2 with $n - 1$ in the denominator provides better estimates of σ^2 than would an estimator calculated with n in the denominator. **For this reason, we will always divide by $n - 1$ when computing the sample variance s^2 and the sample standard deviation s.**

At this point, you have learned how to compute the variance and standard deviation of a set of measurements. Remember the following points:

- The larger the value of s^2 or s, the greater the variability of the data set.

- If s^2 or s is equal to zero, all the measurements must have the same value.

- The standard deviation s is computed in order to have a measure of variability measured in the same units as the observations.

This information allows us to compare several sets of data with respect to their locations and their variability. How can we use these measures to say something more specific about a single set of data? The theorem and rule presented in the next section will help us answer this question.

2.9   On the Practical Significance of the Standard Deviation

We now introduce a useful theorem developed by the Russian mathematician Tchebysheff. Proof of the theorem is not difficult, but we are more interested in its application than its proof.

Tchebysheff's Theorem

Given a number k greater than or equal to 1 and a set of n measurements $x_1, x_2, \ldots, x_n$, at least $[1 - (1/k^2)]$ of the measurements will lie within k standard deviations of their mean.

Tchebysheff's Theorem applies to any set of measurements and can be used to describe either a sample or a population. We will use the notation appropriate

for populations, but you should realize that we could just as easiiy use the mean and the standard deviation for the sample.

The idea involved in Tchebysheff's Theorem is illustrated in Figure 2.17. An interval is constructed by measuring a distance $k\sigma$ on either side of the mean μ. Note that the theorem is true for any number we choose for k as long as it is greater than or equal to 1. Then, computing the fraction $[1 - (1/k^2)]$, we see that Tchebysheff's Theorem states that at least that fraction of the total number n measurements lies in the constructed interval.

FIGURE **2.17**
Illustrating Tchebysheff's
Theorem

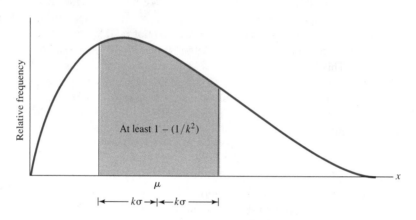

Let us choose a few numerical values for k and compute $[1 - (1/k^2)]$ (see Table 2.10). When $k = 1$, the theorem states that at least $1 - [1/(1)^2] = 0$ of the measurements lie in the interval from $(\mu - \sigma)$ to $(\mu + \sigma)$, a most unhelpful and uninformative result. However, when $k = 2$, we observe that at least $1 - [1/(2)^2] = 3/4$ of the measurements lie in the interval from $(\mu - 2\sigma)$ to $(\mu + 2\sigma)$. At least 8/9 of the measurements lie within three standard deviations of the mean—that is, in the interval from $(\mu - 3\sigma)$ to $(\mu + 3\sigma)$. Although $k = 2$ and $k = 3$ are very useful in practice, k need not be an integer. For example, the fraction of measurements falling within $k = 2.5$ standard deviations of the mean is at least $1 - [1/(2.5)^2] = .84$.

TABLE **2.10**
Illustrative values of
$[1 - (1/k^2)]$

k	$1 - (1/k^2)$
1	0
2	3/4
3	8/9

EXAMPLE **2.11**   The mean and variance of a sample of $n = 25$ measurements are 75 and 100, respectively. Use Tchebysheff's Theorem to describe the distribution of measurements.

Solution We are given $\bar{x} = 75$ and $s^2 = 100$. The standard deviation is $s = \sqrt{100} = 10$. The distribution of measurements is centered about $\bar{x} = 75$, and Tchebysheff's Theorem states the following:

- *At least* 3/4 of the 25 measurements lie in the interval $\bar{x} \pm 2s = 75 \pm 2(10)$, that is, 55 to 95.
- *At least* 8/9 of the 25 measurements lie in the interval $\bar{x} \pm 3s = 75 \pm 3(10)$, that is, 45 to 105. ∎

Since Tchebysheff's Theorem applies to *any* distribution, it is very conservative. This is why we emphasize the "at least $1 - (1/k^2)$" in this theorem. We can state a rule that describes accurately the variability of a particular bell-shaped distribution and describes reasonably well the variability of other mound-shaped distributions of data. The frequent occurrence of mound-shaped and bell-shaped distributions of data in nature—hence, the applicability of our rule—leads us to call it the Empirical Rule.

Empirical Rule

Given a distribution of measurements that is approximately bell-shaped (see Figure 2.18), the interval

$(\mu \pm \sigma)$ contains approximately 68% of the measurements;

$(\mu \pm 2\sigma)$ contains approximately 95% of the measurements;

$(\mu \pm 3\sigma)$ contains all or almost all of the measurements.

The bell-shaped distribution shown in Figure 2.18 is commonly known as the **normal distribution** and will be discussed in detail in Chapter 5. The Empirical Rule applies exactly to data that possess a normal distribution, but it also provides an excellent description of variation for many other types of data.

FIGURE **2.18**
Normal (bell-shaped) distribution

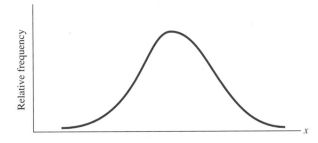

E X A M P L E   **2.12**   A time study is conducted to determine the length of time necessary to perform a specified operation in a manufacturing plant. The length of time necessary to complete the operation is measured for each of $n = 40$ workers. The mean and standard deviation are found to be 12.8 and 1.7, respectively. Describe the sample data by using the Empirical Rule.

Solution   To describe the data, we calculate the intervals

$$(\bar{x} \pm s) = 12.8 \pm 1.7 \qquad \text{or} \qquad 11.1 \text{ to } 14.5$$
$$(\bar{x} \pm 2s) = 12.8 \pm 2(1.7) \qquad \text{or} \qquad 9.4 \text{ to } 16.2$$
$$(\bar{x} \pm 3s) = 12.8 \pm 3(1.7) \qquad \text{or} \qquad 7.7 \text{ to } 17.9$$

According to the Empirical Rule, we expect approximately 68% of the measurements to fall into the interval from 11.1 to 14.5, approximately 95% to fall into the interval from 9.4 to 16.2, and all or almost all to fall into the interval from 7.7 to 17.9.

If we doubt that the distribution of measurements is mound-shaped, or wish for some other reason to be conservative, we can apply Tchebysheff's Theorem and be absolutely certain of our statements. Tchebysheff's Theorem tells us that at least $3/4 = 75\%$ of the measurements fall into the interval from 9.4 to 16.2 and at least $8/9 = 89\%$ into the interval from 7.7 to 17.9.   ▪

How well does the Empirical Rule apply to the dividend yield data of Table 2.6? We have shown that the mean and standard deviation for the $n = 25$ measurements are $\bar{x} = 3.57$ and $s = .81$. The appropriate intervals are then calculated, and the number of measurements falling in each interval are recorded. The results are shown in Table 2.11, with k in the first column and the interval $(\bar{x} \pm ks)$ in the second column, using $\bar{x} = 3.57$ and $s = .81$. The frequency or number of measurements falling in each interval is given in the third column, and the relative frequency is given in the fourth column. Note that the observed relative frequencies agree with Tchebysheff's Theorem and are reasonably close to the relative frequencies specified in the Empirical Rule.

T A B L E   **2.11**
Frequency of measurements lying within k standard deviations of the mean for the data in Table 2.6

k	Interval $\bar{x} \pm ks$	Frequency in Interval	Relative Frequency
1	2.76–4.38	17	.68
2	1.95–5.19	24	.96
3	1.14–6.00	25	1.00

Another way to see how well the Empirical Rule and Tchebysheff's Theorem apply to the dividend yield data is to mark off the intervals $(\bar{x} \pm s)$, $(\bar{x} \pm 2s)$, and

$(\overline{x} \pm 3s)$ on the relative frequency histogram for the data, as shown in Figure 2.19. Now recall that the area under the histogram over an interval is proportional to the number of measurements falling in the interval; visually observe the proportion of the area above the interval $(\overline{x} \pm s)$. You will see that this proportion is near the .68 specified by the Empirical Rule. Similarly, you will note that almost all of the area lies within the interval $\overline{x} \pm 2s$. Clearly, both the Empirical Rule and Tchebysheff's Theorem, using $\overline{x}$ and s, provide a good description for the dividend yield data.

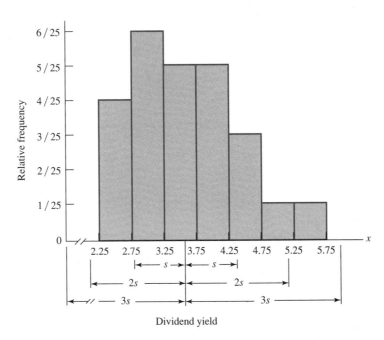

F I G U R E **2.19**
Histogram for the dividend yield data (Figure 2.7) with intervals $\overline{x} \pm s, \overline{x} \pm 2s$, and $\overline{x} \pm 3s$ superimposed

To conclude, note that Tchebysheff's Theorem is a fact that can be proved mathematically, and it applies to any set of data. It gives a *lower* bound to the fraction of measurements to be found in an interval $(\overline{x} \pm ks)$, where k is some number greater than or equal to 1. In contrast, the Empirical Rule is an arbitrary statement about the behavior of data—a rule of thumb. Although the percentages contained in the rule come from the area under the normal curve (Figure 2.18), the same percentages hold approximately for distributions with varying shapes as long as they tend to be roughly mound-shaped (the data tend to pile up near the center of the distribution). We have shown this rule to be true for the dividend yield data. If you need further convincing, calculate $\overline{x}$ and s for a set of data of your choosing and check the fraction of measurements falling in the intervals $(\overline{x} \pm s)$, $(\overline{x} \pm 2s)$, and $(\overline{x} \pm 3s)$. We think you will find that the observed relative frequencies are reasonably close to the values specified in the Empirical Rule.

The mean absolute deviation and the standard deviation are often referred to as *measures of absolute variation* because they retain the units in which $x_1, x_2, \ldots, x_n$ were measured.

A commonly used *measure of relative variation* is the coefficient of variation, in which the standard deviation is expressed as a percentage of the mean.

DEFINITION ▪

> The **coefficient of variation** is a measure of relative variation defined as
>
> $$CV = \left(\frac{s}{\bar{x}}\right) 100\% \quad ▪$$

The coefficient of variation can be used to compare the variation of two sets of data having different means, or measured in different units (for example, miles and kilometers). For the business statistician, the coefficient of variation is used as a measure of risk for a stock or possibly a portfolio. The smaller the value of the coefficient of variation, the more stable the price of the stock, and vice versa.

EXAMPLE 2.13 Find the coefficient of variation for the 25 dividend yields given in Table 2.6. How does this value compare with the coefficient of variation of the resale prices of the 25 condominiums in Section 2.5?

Solution The mean and standard deviation of the 25 dividend yields are

$$\bar{x} = 3.57 \quad \text{and} \quad s = .81$$

Therefore the coefficient of variation is

$$CV = \left(\frac{.81}{3.57}\right) 100\% = 22.7\%$$

The mean and standard deviation of the condominium resale values are

$$\bar{x} = 84.06 \quad \text{and} \quad s = 9.27$$

Therefore the coefficient of variation for these data is

$$CV = \left(\frac{9.27}{84.06}\right) 100\% = 11.03\%$$

Our conclusion is that the absolute variation of the dividend yields is approximately twice that for the condominium resale values. ▪

2.10 A Check on the Calculation of s

Tchebysheff's Theorem and the Empirical Rule can be used to detect gross errors in the calculation of s. We know that at least 3/4 or, in the case of a mound-shaped distribution, approximately 95% of a set of measurements will lie within two standard deviations of their mean. Consequently, most of the sample measurements will lie in the interval ($\bar{x} \pm 2s$), and the range will approximately equal $4s$. This is, of course, a very rough approximation, but from it we can acquire a useful check that will detect large errors in the calculation of s.

Letting R equal the range,

$$R \approx 4s$$

Then s is approximately equal to $R/4$; that is,

$$s \approx \frac{R}{4}$$

The computed value of s using the shortcut formula should be of roughly the same order as the approximation.

EXAMPLE **2.14** Use the range approximation to check the calculation of s for Table 2.6.

Solution The range of the 25 measurements is

$$R = 5.3 - 2.3 = 3.0$$

Then

$$s \approx \frac{R}{4} = \frac{3}{4} = .75$$

This result is of the same order as the calculated value, $s = .81$. ∎

Note that the range approximation is not intended to provide an accurate value for s. Rather, its purpose is to detect gross errors in calculating, such as the failure to divide the sum of squares of deviations by ($n - 1$) or the failure to take the square root of s^2. Both errors yield solutions that are many times larger than the range approximation of s.

EXAMPLE **2.15** Use the range approximation to determine an approximate value for the standard deviation for the data in Table 2.8.

Solution The range is $R = 7 - 1 = 6$. Then

$$s \approx \frac{R}{4} = \frac{6}{4} = 1.5$$

We have shown that $s = 2.39$ for the data in Table 2.8. The approximation is of the same order as the actual value of s. ∎

The range for a sample of n measurements will depend on the sample size n. The larger the value of n, the more likely you will observe extremely large or small values of x. The range for large samples (say, $n = 50$ or more observations) may be as large as $6s$, whereas the range for small samples (say, $n = 5$ or less) may be as small as or smaller than $2.5s$.

The range approximation for s can be improved if it is known that the sample is drawn from a bell-shaped distribution of data. Thus the calculated s should not differ substantially from the range divided by the appropriate ratio given in the following table:

Number of measurements	5	10	25
Expected ratio of range to s	2.5	3	4

Tips on Problem Solving

1   Be careful about rounding numbers. Use a calculator with built-in statistical functions if possible. If you calculate s^2 and s by hand, carry your calculations to at least six significant figures.

2   After you have calculated the standard deviation s for a set of data, compare its value with the range of the data. The Empirical Rule tells you that approximately 95% of the data should fall into the interval $\bar{x} \pm 2s$; that is, a very approximate value for the range will be $4s$. Consequently, a very rough rule of thumb is that

$$s \approx \frac{\text{Range}}{4}$$

This crude check will help you to detect large errors—for example, failure to divide the sum of squares of deviations by $(n - 1)$ or failure to take the square root of s^2.

Exercises

Basic Techniques

2.31   Given $n = 8$ measurements, 4, 1, 3, 1, 3, 1, 2, 2, calculate the following.

   a $\bar{x}$

   b s^2 using the formula given by the definition in Section 2.8

   c s^2 and s using the computing formula. Compare the results with those found in part (b).

2.32   Given $n = 5$ measurements, 2, 1, 1, 3, 5, calculate the following.

   a $\bar{x}$

   b s^2 and s using the computing formula

   c   a range estimate of s (Section 2.10) as a rough check on your calculations in part (b)

2.33   Given $n = 6$ measurements, 3, 0, 2, 2, 1, 4, find the following.

    a $\bar{x}$

    b   MAD

    c s

2.34   Given $n = 5$ measurements, 2, 4, 0, 3, 1, find the following.

    a $\bar{x}$

    b s

    c   coefficient of variation

2.35   Suppose that you want to create a mental picture of the relative frequency histogram for a large data set consisting of 1000 observations and you know that the mean and standard deviation of the data set are equal to 36 and 3, respectively.

    a   If you are fairly certain that the relative frequency distribution of the data is mound-shaped, how might you describe the relative frequency distribution? [*Hint:* Use the Empirical Rule.]

    b   If you had no prior information concerning the shape of the relative frequency distribution, what could you say about the relative frequency distribution?

Applications

2.36   The average weekly earnings for production workers has increased from approximately \$235 in 1980 to \$364 in 1992 (*World Almanac & Book of Facts,* 1994, p. 136) Suppose that you owned a company and you were concerned with the distribution of weekly wages in the production industry. If the standard deviation of the distribution is \$26, within what limits would most weekly 1992 wages fall?

2.37   Refer to Exercise 2.36 and visualize the distribution of weekly earnings in the production industry.

    a   Do you think that the distribution of weekly earnings is symmetric about the mean, or is it skewed? Explain.

    b   If the distribution of weekly earnings is approximately mound-shaped, what proportion of construction workers will have wages below \$312?

    c   If you planned to pay your production workers an average of \$260 per week, could you be accused of underpaying your workers? Explain.

2.38   An inspector from the Food and Drug Administration (FDA) wished to determine the actual average weight of the contents of all boxes of a particular cereal labeled "16 ounces." A sample of $n = 30$ boxes of this cereal was selected, and the contents of each was weighed. The average content weight for the $n = 30$ boxes was found to be 15.92 ounces; the standard deviation was .04 ounce.

    a   Using Tchebysheff's Theorem, what would you expect the distribution of these 30 weights to be like?

    b   Using the Empirical Rule, what would you expect the distribution to be like? (Would you expect the Empirical Rule to be suitable for describing these data?)

    c   Suppose the inspector had weighed the contents of only $n = 4$ boxes of cereal and obtained the weights 15.84, 16.00, 15.92, and 15.84. Would the Empirical Rule be suitable for describing the $n = 4$ measurements? Why or why not?

2.39   The length of time required for a cashier to check out food items at a supermarket that employs automated checkers was recorded for $n = 10$ customers. The times, in seconds, were 15, 62, 53, 11, 38, 75, 112, 40, 22, 57.

    a   Scan the data and use the range approximation to find an approximate value for s. Use this value to check your calculations in part (b).

    b   Calculate the sample mean $\bar{x}$ and the standard deviation s. Compare your answers with the answer obtained in part (a).

2.40 The number of television viewing hours per household and the prime viewing times are two factors that affect television advertising income. A random sample of 25 households in a particular viewing area produced the following estimates of viewing hours per household.

3.0	6.0	7.5	15.0	12.0
6.5	8.0	4.0	5.5	6.0
5.0	12.0	1.0	3.5	3.0
7.5	5.0	10.0	8.0	3.5
9.0	2.0	6.5	1.0	5.0

a Scan the data and use the range approximation to find an approximate value for s. Use this value to check your calculations in part (b).

b Calculate the sample mean $\bar{x}$ and the sample standard deviation s. Compare s with the approximate value obtained in part (a).

c Find the percentage of the viewing hours per household that fall in the interval $\bar{x} \pm 2s$. Compare your answer with the corresponding percentage given by the Empirical Rule.

2.41 "Cruising" is a procedure used by timber owners when determining the amount of saleable lumber in a tract of land. To estimate the amount of lumber in a particular tract of timber, an owner "cruises" randomly selected 50 × 50-foot squares and counts the number of trees with diameters exceeding 12 inches. Suppose seventy 50 × 50 squares were "cruised" and the number of trees with diameters in excess of 12 inches were counted for each. The data follow.

7	8	7	10	4	8	6
9	6	4	9	10	9	8
3	9	5	9	9	8	7
10	2	7	4	8	5	10
9	6	8	8	8	7	8
6	11	9	11	7	7	11
10	8	8	5	9	9	8
8	9	10	7	7	7	5
8	7	9	9	6	8	9
5	8	8	7	9	13	8

a Construct a relative frequency histogram to describe the data.

b Calculate the sample mean $\bar{x}$ as an estimate of μ, the mean number of trees with diameters exceeding 12 inches for all 50 × 50-foot squares in the tract.

c Calculate s for the data. Construct the intervals $(\bar{x} \pm s)$, $(\bar{x} \pm 2s)$, and $(\bar{x} \pm 3s)$. Find the percentage of squares falling in each of the three intervals, and compare your answers with the corresponding percentages given by the Empirical Rule and Tchebysheff's Theorem.

2.42 Refer to Exercise 2.25, in which the satisfaction ratings of seven mail-order computer companies were given as follows:

$$7.5, \ 7.9, \ 7.4, \ 8.2, \ 7.8, \ 7.7, \ 8.0$$

a Calculate the sample variance and standard deviation for these data.

b Calculate the mean absolute deviation.

c Would it be appropriate to use either Tchebysheff's Theorem or the Empirical Rule to describe these data? Explain.

2.43   Refer to the 60 supermarket service times given in Exercise 2.11.

    a   Look at the data and use the range to calculate a rough approximation to the value of s.

    b   Calculate $\bar{x}$ and s.

    c   Calculate the intervals $(\bar{x} \pm s)$, $(\bar{x} \pm 2s)$, and $(\bar{x} \pm 3s)$.

    d   Count the number of service times falling in each interval. Do the fractions falling in these intervals agree with Tchebysheff's Theorem? the Empirical Rule?

2.44   Examine the data on the total sales (in billions of dollars) for the 50 industrial corporations given in Exercise 2.21. The data are reproduced below in a different form.

132.8	103.5	100.8	65.1	62.2	57.4	50.2	37.6	37.5	37.1
36.9	30.2	29.9	25.5	22.1	22.0	21.7	21.2	20.6	19.2
18.3	18.1	17.5	16.4	16.2	14.0	13.9	13.8	13.6	13.6
13.3	13.3	13.2	12.1	12.1	12.0	11.9	11.8	11.8	11.4
11.1	11.0	10.2	10.1	10.1	9.8	9.6	9.3	9.3	9.2

Source: Data from *Fortune, The World Almanac and Book of Facts,* 1994 ed., p. 111.

    a   Use the range to calculate an approximate value for s.

    b   Calculate $\bar{x}$ and s.

    c   Find the proportions of the total sales that fall in the intervals $(\bar{x} \pm s)$, $(\bar{x} \pm 2s)$, and $(\bar{x} \pm 3s)$. Do these proportions agree with Tchebysheff's Theorem? the Empirical Rule?

2.45   Suppose that some measurements occur more than once and that the data $x_1,\ x_2,\ \ldots,\ x_k$ are arranged in a frequency table as shown below.

Observations	Frequency f_i
x_1	f_1
x_2	f_2
$\vdots$	$\vdots$
x_k	f_k
	n

Then

$$\bar{x} = \frac{\sum\limits_{i=1}^{k} x_i f_i}{n} \qquad \text{where } n = \sum_{i=1}^{k} f_i$$

and

$$s^2 = \frac{\sum\limits_{i=1}^{k} x_i^2 f_i - \dfrac{\left(\sum\limits_{i=1}^{k} x_i f_i\right)^2}{n}}{n-1}$$

Although these formulas for grouped data are primarily of value when you have a large number of measurements, demonstrate their use for the sample 1, 0, 0, 1, 3, 1, 3, 2, 3, 0, 0, 1, 1, 3, 2.

a   Calculate $\bar{x}$ and s^2 directly using the formulas for ungrouped data.

b   The frequency table for the $n = 15$ measurements follows.

x_i	f_i
0	4
1	5
2	2
3	4
$n = 15$	

Calculate $\bar{x}$ and s^2 using the formulas for grouped data. Compare the results with your answers to part (a).

2.11   Graphical and Numerical Techniques for Bivariate Data

When experiments are run or when surveys are conducted, the researcher is often interested in two or more variables that can be measured during the investigation. For example, in a survey of policyholders conducted by an auto insurance company, the number of vehicles owned and the number of drivers per household are both important pieces of information for the company. In a survey of households in a certain city, the amount spent on groceries per week and the number of people in the household are two quantitative variables of interest to an economist. In the same survey, the selling price of a residential property and the size of the living areas in square feet are quantitative variables of interest to both a potential seller and a potential buyer in the city. It would not be difficult to think of other variables that would also be of interest to the investigator in these situations.

When two variables are measured on a single experimental unit, the resulting data are called **bivariate data.** How should we display the data? If we want to study each variable separately, the techniques described in earlier sections will be sufficient. However, if we wish to determine any inherent relationships between the two variables, other techniques must be used.

Comparative graphs are graphical techniques used for bivariate data when at least one of the two variables measured is *qualitative*. The following example demonstrates two graphical techniques that can be used in this situation.

EXAMPLE   **2.16**   According to a report by Kenneth Eskey of the Howard News Service (Eskey, 1992), "women are choosing medical careers in record numbers and could fill one of every two seats in the nation's medical schools if the trend continues." The medical school enrollment (in thousands) for men and women over the years 1986–92 is given in Table 2.12.

a   Use two line graphs plotted on the same axes showing the number of men and women enrolled in medical schools over the period given.

b Using these same data, compare the number of men and women enrolled in medical schools using a stacked bar graph.

T A B L E **2.12**
Medical school enrollment

Year	Men	Women	Total
1986	10.5	5.6	16.1
1987	10.2	5.8	16.0
1988	10.1	5.9	16.0
1989	9.8	6.0	15.8
1990	9.8	6.2	16.0
1991	9.8	6.4	16.2
1992	9.5	6.8	16.3

Solution **a** The line graphs must both use the same horizontal and vertical axes. With this in mind, the two lines are plotted in Figure 2.20. Although the two lines are not identical, the number of women enrolled in medical schools appears to be approaching the 1-to-1 ratio that more nearly reflects their numbers in the population at large.

F I G U R E **2.20**
Comparative line graph for
Example 2.16

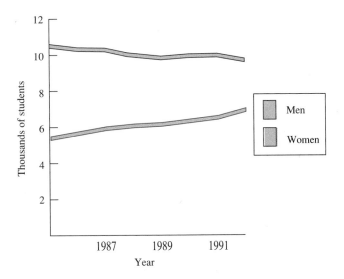

b In preparing a stacked bar graph, we draw a rectangle for each time category showing the total number enrolled and indicate with color or some other marking the proportion of each bar that represents the number of women enrolled in medical schools. The stacked bar graph is shown in Figure 2.21. This presentation may be more effective than that in part (a) in that the total number of persons enrolled in medical schools has remained fairly constant, and the proportion of each bar representing women enrolled is approaching 50%. ▪

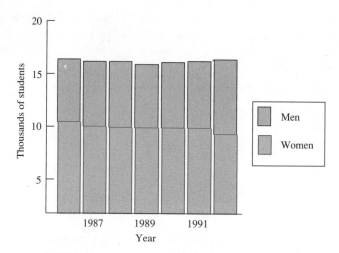

When both variables are *quantitative,* a generalization of the univariate scatterplot can be used to display any apparent relationship between them. A **bivariate scatterplot** of the two variables, usually designated as x and y, will reflect how x varies with y. The distribution of the points in this plot may take one of many different forms, and, based on what we see in this scatterplot, we might postulate one or more possible relationships between x and y. No relationship between x and y would be indicated by a general scattering of the points with no apparent pattern in the plot. The simplest relationship between x and y would be one in which the values of y increase or decrease linearly with the values of x, that is, in which the plotted values of the (x, y) pairs appear to lie on a straight line or perhaps appear to scatter about an underlying line. In the next example, we explore how the price of a home varies as a function of the number of square feet of living space in the home.

E X A M P L E   **2.17**   Table 2.13 gives the size of the living area in square feet, denoted by x, and the selling price, denoted by y, of 12 residential properties in thousands of dollars. Plot the data and comment on any relationship that seems apparent from the data.

Residence	x (sq. ft.)	y ($ thousand)
1	1360	78.5
2	1940	175.7
3	1750	139.5
4	1550	129.8
5	1790	95.6
6	1750	110.3
7	2230	260.5
8	1600	105.2
9	1450	88.6
10	1870	165.7
11	2210	225.3
12	1480	68.8

FIGURE **2.22**

Scatterplot of x versus y for Example 2.17

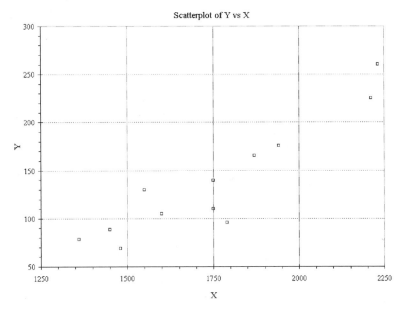

Solution Let the values of x correspond to the horizontal axis and the values of y correspond to the vertical axis. The resulting plot, given in Figure 2.22, indicates that the values of y appear to increase linearly as x increases. ∎

We could describe each variable, x and y, individually using descriptive measures such as the means ($\bar{x}$ and $\bar{y}$) or the standard deviations (s_x and s_y). However, these measures do not describe the relationship between x and y for a particular residence, that is, how the size of the living space affects the selling price of the home. A simple measure that serves this purpose is called the **correlation coefficient,** denoted by r, and is defined as

$$r = \frac{s_{xy}}{s_x s_y}$$

The quantities s_x and s_y are the standard deviations for the two variables, which can be calculated using the shortcut formula in Section 2.8, or using a calculator that has statistical capabilities. The quantity s_{xy} is similar to a standard deviation and is calculated as

$$s_{xy} = \frac{\sum\limits_{i=1}^{n} x_i y_i - \dfrac{\left(\sum\limits_{i=1}^{n} x_i\right)\left(\sum\limits_{i=1}^{n} y_i\right)}{n}}{n-1}$$

Most computer software packages will compute the correlation coefficient when given the appropriate commands. In the Minitab package, the data are stored in two columns, C1 and C2, and the command CORRELATION C1 C2 will generate the correlation coefficient, r.

However you calculate the correlation coefficient, it can be shown that the value of r always lies between -1 and 1. When r is positive, x increases when y increases, and vice versa. When r is negative, x decreases when y increases, or when x increases y decreases. When r takes the value 1 or -1, all the points lie exactly on a straight line. If $r = 0$, then there is no apparent linear relationship between the two variables. The closer the value of r to 1 or -1, the stronger the linear relationship between the two variables.

E X A M P L E   **2.18**   Find the correlation coefficient for the number of square feet of living area and the selling price of a home for the data in Example 2.17.

Solution   Three quantities are needed to calculate the correlation coefficient. The standard deviations of the x and y variables are found using a calculator with a statistical function. You can verify that $s_x = 281.4842$ and $s_y = 59.7592$. Finally,

$$s_{xy} = \frac{\sum x_i y_i - \frac{\left(\sum x_i\right)\left(\sum y_i\right)}{n}}{n-1}$$

$$= \frac{3,044,383 - \frac{(20,980)(1643.5)}{12}}{11} = 15,545.19697$$

Then

$$r = \frac{s_{xy}}{s_x s_y} = \frac{15,545.19697}{(281.4842)(59.7592)} = .9241$$

This value of r is fairly close to 1, indicating that the linear relationship between these two variables is very strong. Additional information about the correlation coefficient and its role in analyzing linear relationships, along with alternative calculation formulas, can be found in Chapter 11. ▪

As you continue to work through the exercises in this chapter, you will become more experienced in recognizing different types of data and in determining the most appropriate graphical or numerical method. Remember that the type of graph you use is not as important as the interpretation that accompanies it. The important characteristics to look for in graphs are the following:

- location or center of the data
- shape of the distribution of data
- unusual observations in the data set
- relationships between variables

Most data analysts begin any data-based investigation by examining plots of the variables involved; if the relationship between two variables is of interest, bivariate plots are also explored in conjunction with numerical measures of location, dispersion, and correlation. Graphs and numerical descriptive measures are only the first of many statistical tools you will soon have at your disposal.

Exercises

Basic Techniques

2.46   A set of bivariate data consists of measurements on two variables, x and y, as follows:

$$(3, 6), (5, 8), (2, 6), (1, 4), (4, 7), (4, 6)$$

a   Draw a scatterplot to describe the data.

b   Does there appear to be a relationship between x and y? If so, how would you describe it?

2.47   Consider the set of bivariate data shown below:

x	1	2	3	4	5	6
y	5.6	4.6	4.5	3.7	3.2	2.7

a   Draw a scatterplot to describe the data.

b   Does there appear to be a relationship between x and y? If so, how would you describe it?

c   Calculate the correlation coefficient, r. Does the value of r substantiate your conclusions in part (b)? Explain.

2.48   The value of a quantitative variable is measured once a year for a ten-year period. The data are shown below.

Year	Measurement	Year	Measurement
1	61.5	6	58.2
2	62.3	7	57.5
3	60.7	8	57.5
4	59.8	9	56.1
5	58.0	10	56.0

a   Draw a scatterplot to describe the variable as it changes over time.

b   Describe the measurements using the graph constructed in part (a).

Applications

2.49   The accompanying data represent the average seller's asking price, the average buyer's bid, and the average closing bid for each of ten types of used computer equipment ("Used Computer Prices," 1992).

a   Draw a scatterplot relating average seller's asking price to average closing bid.

b   Draw a scatterplot relating average buyer's bid to average closing bid.

c   Compare the two scatterplots. Describe the data using the graphs from parts (a) and (b). What relationships appear to exist among the three variables?

Data for Exercise 2.49

Machine	Average Seller's Asking Price	Average Buyer's Bid	Average Closing Bid
20MB PC XT	$400	$200	$300
20MB PC AT	700	400	575
IBM XT 089	450	200	325
IBM AT 339	700	350	600
20MB IBM PS/2 30	950	500	725
20MB IBM PS/2 50	1050	700	875
60MB IBM PS/2 70	2000	1600	1725
20MB Compaq SLT	1200	700	875
Toshiba 1600	1000	700	900
Toshiba 1200HB	1150	800	975

2.50 A Minitab software package was used to find the correlation coefficient, r, for the three pairs of variables in Exercise 2.49. Find the correlation coefficients on the printout. Interpret the values of r in terms of the practical problem of buying a used computer.

```
MTB > Correlation 'ASK-PRC' 'BYR-BID' 'CLOS-BID'

          ASK-PRC  BYR-BID
BYR-BID    0.985
CLOS-BID   0.991    0.994
```

2.51 The tax burden borne by citizens of the United States varies substantially from state to state, and even from city to city. The accompanying table gives an estimate of the state and local taxes paid in two different cities by a family of four with an annual income of $100,000 in 1992 (Tannenwald, 1994).

Tax	City	
	Boston	Kansas City
Income tax	$4142	$3285
Property tax	1734	930
Sales tax	1213	1583
Auto tax	987	1105

Draw a stacked bar graph to describe the estimated tax burden for the families in these two cities. What conclusions can you draw from this graph?

2.52 As the United States becomes more aware of environmental problems, many cities are instituting curbside recycling programs in an attempt to conserve space in local landfills. These investments in recycling programs are more prevalent in some regions of the country than in others. The table below shows the number of curbside recycling programs and the number of landfills in each of seven regions of the United States (*EPA Journal,* July/August, 1992).

Region	Curbside Recycling Programs	Landfills
West	569	1374
Rocky Mountain	44	661
Midwest	108	1402
Great Lakes	1148	531
South	402	1007
Mid-Atlantic	1379	334
New England	305	503

a   Draw a scatterplot relating the number of curbside recycling programs to the number of landfills for these seven regions.

b   Describe the relationship between these two variables using the scatterplot from part (a).

c   Can you explain why there is such a large difference in the pattern of these two variables from region to region?

2.12 Measures of Relative Standing

Sometimes we want to know the position of an observation relative to others in a set of data. For example, if you took a placement examination and scored 640, you might like to know the percentage of participants who scored lower than 640. Such a **measure of relative standing** of an observation within a data set is called a **percentile**.

DEFINITION ▪

> Let x_1, x_2, . . . , x_n be a set of n measurements arranged in increasing order. The pth **percentile** is the value of x such that at most p percent of the measurements are less than that value of x and at most $(100 - p)$ percent are greater. ▪

EXAMPLE **2.19**   Before being accepted into the master of business administration (MBA) program at a university, you have been notified that your score of 610 on the Verbal Graduate Record Examination placed you at the 60th percentile in the distribution of scores. Where does your score of 610 stand in relation to the scores of others who took the examination?

Solution   Scoring at the 60th percentile means that 60% of the other examination scores were lower than your score of 610 and 40% were higher. ▪

Viewed graphically, a particular percentile, say the 60th percentile, is a point on the x axis located so that 60% of the area under the relative frequency histogram for the data lies to the left of the 60th percentile (see Figure 2.23) and 40% of the area lies to the right. Thus by our definition, the median of a set of data is the 50th percentile, because half of the measurements in a data set are smaller than the median and half are larger.

FIGURE **2.23**
The 60th percentile shown on the relative frequency histogram for a data set

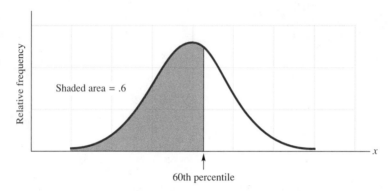

The 25th and 75th percentiles, called the **lower and upper quartiles,** along with the median (the 50th percentile) locate points that divide the data into four sets of equal size. Twenty-five percent of the measurements will be less than the lower (first) quartile, 50% will be less than the median (the second quartile), and 75% of the measurements will be less than the upper (third) quartile. Thus the median and the lower and upper quartiles are located at points on the x axis such that the area under the relative frequency histogram for the data is partitioned into four equal areas, as shown in Figure 2.24. You can see (Figure 2.24) that 1/4 of the area lies to the left of the lower quartile and 3/4 to the right. The upper quartile is the value of x such that 3/4 of the area lies to the left and 1/4 to the right.

FIGURE **2.24**
Location of quartiles

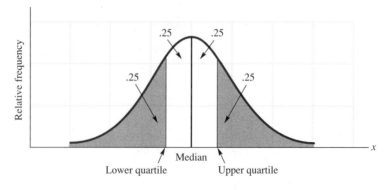

The *z*-**score** is yet another measure of relative standing; it uses both the mean and the standard deviation of the data set.

DEFINITION ∎
> The **sample *z*-score** corresponding to an observation x is a measure of relative standing defined by
>
> $$z\text{-score} = \frac{x - \bar{x}}{s} \quad \blacksquare$$

A *z*-score measures the number of standard deviations between an observation and the mean of the set. Suppose we know that the mean and standard deviation of a set of test scores based on a total of 100 points are $\bar{x} = 74$ and $s = 8$. The *z*-score for your test grade of 92 is calculated to be

$$z\text{-score} = \frac{x - \bar{x}}{s} = \frac{92 - 74}{8} = 2.25$$

Therefore your score lies 2.25 standard deviations above the mean; that is, $92 = 74 + 2.25(8)$.

By themselves *z*-scores merely indicate how many standard deviations a score is above or below the mean. However, if the *z*-score is used in conjunction with Tchebysheff's Theorem, then some conservative statements about the relative standing of an observation can be made. Further, if the data are mound-shaped, the Empirical Rule can be used to make stronger statements about the relative standing of an observation in terms of its *z*-score. Since at least 75%, and more likely 95%, of the observations in a data set lie within two standard deviations of the mean, *z*-scores between -2 and $+2$ are highly likely, and hence not unusual. However, at least 8/9, and more likely all, of the observations lie within three standard deviations of the mean. Therefore *z*-scores between 2 and 3 in absolute value are much less likely to occur, while *z*-scores exceeding 3 in absolute value are very unlikely to occur and should be carefully examined. A test grade whose *z*-score exceeds 3 is outstanding, while a stock whose price-earnings ratio (the price of a stock divided by its annual per-share earnings) has a *z*-score of -3 would be considered an investment with good earning potential.

Extremely large or extremely small *z*-scores raise questions about the validity of an observation. It may be that the observation is simply quite large or quite small relative to the others. However, the observation may have been incorrectly recorded, or for one reason or another it may not belong to the population that we wished to sample. Observations with extremely large or small *z*-scores are often referred to as **outliers** because they lie far from the center of the data set. Observations lying between two and three standard deviations above or below the mean are possible outliers, while those lying more than three standard deviations above or below the mean are considered to be definite outliers.

E X A M P L E   **2.20**   Consider a sample of $n = 10$ measurements:

$$3, \ 2, \ 0, \ 15, \ 2, \ 3, \ 4, \ 0, \ 1, \ 3$$

You can see at a glance that the measurement $x = 15$ appears to be an outlier. Calculate the z-score for this observation, and state your conclusions.

Solution   For the sample, we have the following calculations:

$$\sum_{i=1}^{10} x_i = 33 \qquad \text{and} \qquad \sum_{i=1}^{10} x_i^2 = 277$$

Then

$$\bar{x} = \frac{\sum_{i=1}^{10} x_i}{n} = \frac{33}{10} = 3.3$$

$$s^2 = \frac{\sum_{i=1}^{n} x_i^2 - \frac{\left(\sum_{i=1}^{n} x_i\right)^2}{n}}{n - 1}$$

$$= \frac{277 - \frac{(33)^2}{10}}{9} = \frac{168.1}{9} = 18.6778$$

$$s = 4.32$$

Using these quantities to calculate the z-score for the suspect outlier $x = 15$, we find

$$z\text{-score} = \frac{x - \bar{x}}{s} = \frac{15 - 3.3}{4.32} = 2.71$$

Thus the measurement $x = 15$ lies 2.71 sample standard deviations away from the sample mean $\bar{x} = 3.3$. Since this z-score exceeds 2, we identify $x = 15$ as a possible outlier. We should examine our sampling procedure to see whether evidence exists to indicate that $x = 15$ is a faulty observation.   ▪

Many of the numerical descriptive measures that we have discussed are easily found using the command DESCRIBE in the Minitab package. As part of its output, this program produces the values of the mean, the standard deviation, the median, the largest (**MAX**) and the smallest (**MIN**) of the observations, and the lower and upper quartiles, together with the values of some other statistics that we have not discussed. The **trimmed mean** (given as **TRMEAN**) is the mean of the middle 90% of the measurements after deleting the smallest 5% and the largest 5% of the observations.

Unlike the ordinary arithmetic mean, the trimmed mean is unaffected by the largest and the smallest values in a data set.

The DESCRIBE command produced the printout given in Figure 2.25 summarizing the dividend yields of Example 2.10 (data given in Table 2.6) and the observations in Example 2.20. You may wish to compare the values of the statistics calculated in those examples with the values given in Figure 2.25.

FIGURE 2.25
Minitab printout using the command DESCRIBE for the data of Example 2.10 (C1) and the data of Example 2.20 (C2)

```
MTB > Describe C1 C2.

            N      MEAN    MEDIAN    TRMEAN    STDEV   SEMEAN
C1         25     3.568     3.500     3.548    0.805    0.161
C2         10      3.30      2.50      2.25     4.32     1.37

           MIN       MAX        Q1        Q3
C1       2.300     5.300     3.050     4.100
C2        0.00     15.00      0.75      3.25
```

Exercises

Basic Techniques

2.53 Use the following set of data:

$$3, \ 9, \ 6, \ 5, \ 5, \ 4, \ 7, \ 6, \ 8, \ 2, \ 6, \ 7, \ 3$$

a Calculate $\bar{x}$ and s.

b Calculate the z-score for the smallest and largest observations. Are either of these observations unusually large or small?

2.54 Find the z-score for the largest observation in the following data set:

$$19, \ 12, \ 16, \ 0, \ 14, \ 9, \ 6, \ 1, \ 12, \ 13, \ 10, \ 19, \ 7, \ 5, \ 8$$

2.55 If you scored in the 90th percentile on a graduate school placement test, how would your score stand in relation to others taking the test?

Applications

2.56 Refer to the per capita debt data in Exercise 2.10.

a Find the mean and standard deviation of the per capita debts.

b Consult Exercise 2.10 to find the per capita debt in your state in 1992. Use a z-score to describe how the per capita debt in your state compared with corresponding debts in other states.

2.57 An article in *American Demographics* (Kirchner, R., and Thomas, R., "New Markets for Health Insurance," December 1990, p. 40) provides some interesting statistics on the population of Americans in 1988 who did not have some type of health insurance. In many cases, these uninsured Americans had at least adequate resources to pay for this insurance. According to the article, "nearly 40 percent of the uninsured had incomes of $20,000 or more; 22 percent had incomes of $30,000 or more; and 13 percent, or over 4 million, lived in households with incomes of $40,000 or more." Identify any percentiles that can be determined from this information.

2.58 According to *Consumer Reports* (March 1994), the average price of a Sony SLV-700HF stereo VCR is $410, with a standard deviation of $14. If you purchase this type of VCR for $430, calculate the z-score for your purchase price. Is this price unusually high?

2.13 Box Plots

The **box plot** is another technique used in exploratory data analysis (EDA). It can be used to describe not only the behavior of the measurements in the middle of the distribution but also their behavior at the ends or tails of the distribution. Values that lie very far from the middle of the distribution in either direction are called **outliers.** An outlier may result from transposing digits when recording a measurement, from incorrectly reading an instrument dial, from a malfunctioning piece of equipment, and so on. Even when there are no recording or observational errors, a data set may contain one or more valid measurements that, for one reason or another, differ markedly from the others in the set. These outliers can cause a marked distortion in the values of commonly used numerical measures such as $\bar{x}$ and s. In fact, outliers may themselves contain important information not shared with the other measurements in the set. Therefore, isolating outliers, if they are present, is an important step in any preliminary analysis of a data set. The box plot is designed expressly for this purpose.

A box plot is constructed by using the median and two other measures, known as **hinges.** The median divides the ordered data set into two halves; the hinges are the values in the middle of each half of the data. Hinges are very similar to quartiles and, in effect, serve the same purpose. The actual difference in value between a hinge and a quartile is quite small and decreases as the number of measurements increases. Nevertheless, we retain the distinction between these two measures and use hinges in constructing a box plot. *However, we can think of a hinge as playing the role of a quartile.* Hinges are calculated using the following procedure.

Calculating Hinges

1 Calculate the position of the median, $(n + 1)/2$, and drop the fraction $1/2$ if there is one. This quantity is called $d(M)$ and measures the "depth" of the median from each end of the ordered measurements.

2 Calculate the position of the hinges as

$$\frac{d(M) + 1}{2}$$

The hinges are the values in position $[d(M) + 1]/2$ as measured from each end of the ordered data set.

For example, for a data set with $n = 10$, the position of the median is $(10 + 1)/2 = 5.5$, and $d(M) = 5$. The position of the hinges is then

$$\frac{d(M) + 1}{2} = \frac{5 + 1}{2} = 3$$

The hinges are those values in the third position, as measured from each end of the ten ordered measurements. If the position of the hinges ends in 1/2, a hinge will be the average of the two adjacent values in the ordered data set.

The dispersion of the measurements is now measured in terms of the difference between the hinges, called the **H-spread,** which is approximately equal to $(Q_U - Q_L)$, the **interquartile range.** A data value will be identified as an outlier depending on its relative position with respect to boundary points, called inner and outer fences. The **inner fences** are defined as follows:

Lower inner fence = Lower hinge − 1.5(H-spread)

Upper inner fence = Upper hinge + 1.5(H-spread)

The **outer fences** are defined as follows:

Lower outer fence = Lower hinge − 3(H-spread)

Upper outer fence = Upper hinge + 3(H-spread)

The data values in each tail closest to, but still inside, the inner fences are called the **adjacent values.** Values lying between an inner fence and its neighboring outer fence are termed "outside" and are considered to be mild outliers. Values outside the outer fences are termed "far outside" and are considered to be extreme outliers. A box plot combines all this information in a pictorial display as follows.

Constructing a Box Plot

1 Calculate the median and the upper and lower hinges, and locate them on a horizontal line representing the scale of measurement.

2 Draw a box whose ends are the upper and lower hinges. Draw a line through the box at the value of the median.

3 Calculate the **H-spread** = upper hinge − lower hinge. Then calculate the **inner and outer fences.** Determine mild and extreme outliers, and locate them on the box plot.

4 Locate the **adjacent values** on the box plot. Draw a dashed line from the box to the corresponding adjacent values.

E X A M P L E   **2.21**   Construct a box plot for the data in Example 2.20.

Solution   The $n = 10$ measurements in the sample, ranked from smallest to largest, are

$$0, \ 0, \ 1, \ 2, \ 2, \ 3, \ 3, \ 3, \ 4, \ 15$$

The position of the median is $(n + 1)/2 = 11/2 = 5.5$. Hence, the median is $(2 + 3)/2 = 2.5$, and the depth of the median is

$$d(M) = 5$$

The position of the hinges, then, is

$$\frac{d(M) + 1}{2} = \frac{5 + 1}{2} = 3$$

and the hinges are 1 and 3, the values in the third position as measured from each end of the ordered data set. The H-spread is $3 - 1 = 2$, and the outer and inner fences are calculated as follows:

$$\text{Upper inner fence} = 3 + 1.5(2) = 6$$
$$\text{Lower inner fence} = 1 - 1.5(2) = -2$$
$$\text{Upper outer fence} = 3 + 3(2) = 9$$
$$\text{Lower outer fence} = 1 - 3(2) = -5$$

The value $x = 15$ lies outside the outer fences and hence is considered an extreme outlier.

To construct a box plot for the data, we draw a box whose ends are the upper and lower hinges, as shown in Figure 2.26. A line is drawn through the box at the value of the median. The outlier, $x = 15$, is plotted using a capital O. The adjacent values, lying just inside the inner fences, are 0 and 4 and are connected to the box with a dashed line.

F I G U R E **2.26**
Box plot for the data in
Example 2.21

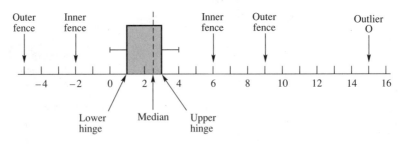

F I G U R E **2.27**
Minitab printout of the box plot
for the data in Example 2.21

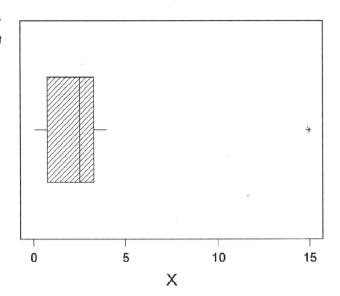

The box plot emphasizes the fact that the outlier lies far from the central 50% of the measurements lying between the hinges. The box plot also indicates that these data are positively skewed (skewed to the right), since the median is not equally spaced between the two hinges but, rather, lies closer to the upper hinge.

For these same data, the command BOXPLOT in the Minitab package produced the box plot in Figure 2.27. Notice that, except for scale, the box plots are identical and lead to the same conclusions concerning the outlier, $x = 15$.   ■

Exercises

Basic Techniques

2.59   Construct a box plot for the following data and identify any outliers.

$$25, \ 22, \ 26, \ 23, \ 27, \ 26, \ 28, \ 18, \ 25, \ 24, \ 12$$

2.60   Construct a box plot for the following data and identify any outliers.

$$3, \ 9, \ 10, \ 2, \ 6, \ 7, \ 5, \ 8, \ 6, \ 6, \ 4, \ 9, \ 22$$

Applications

2.61 The data on total sales (in billions of dollars) for the 50 industrial corporations given in Exercise 2.21 are reproduced below. Construct a box plot for the data and identify any outliers.

132.8	103.5	100.8	65.1	62.2	57.4	50.2	37.6	37.5	37.1
36.9	30.2	29.9	25.5	22.1	22.0	21.7	21.2	20.6	19.2
18.3	18.1	17.5	16.4	16.2	14.0	13.9	13.8	13.6	13.6
13.3	13.3	13.2	12.1	12.1	12.0	11.9	11.8	11.8	11.4
11.1	11.0	10.2	10.1	10.1	9.8	9.6	9.3	9.3	9.2

Source: Data from *Fortune, The World Almanac and Book of Facts,* 1994 ed., p. 111.

2.62 In Exercise 1.8, a computer program was used to scan the stocks of 8000 public companies (Palmer, 1990) in order to find "cheap" stocks, based on a number of valuation criteria. The final list of 34 companies had prices per share as shown in the following table.

$14.8	6.5	10.1	7.6	27.9
17.1	10.1	69.0	10.0	19.3
12.0	53.3	22.0	15.8	10.5
30.6	15.8	30.3	34.3	21.6
19.6	8.9	13.0	25.4	34.1
8.9	25.3	42.1	33.9	79.0
21.1	12.1	29.0	6.5	

a Construct a box plot for the data.

b Do any of the 34 companies appear to be outliers—that is, do they have a price that is inconsistent with the other prices?

2.63 A survey of auto quality by J. D. Power and Associates is closely watched within the auto industry. The survey's Customer Satisfaction Index ranks cars according to the quality of car and dealer service within the first year ("Japan Carmakers," 1994). The scores for 1994 are shown in the table, along with a box plot generated by Minitab. Use the box plot to describe the distribution of the scores.

Carmaker	Score	Carmaker	Score	Carmaker	Score
Lexus	176	Buick	141	Nissan	132
Infinity	171	Mercedes	141	Pontiac	127
Saturn	155	Saab	140	Dodge	125
Acura	150	Subaru	139	Mitsubishi	125
Audi	148	BMW	138	Ford	120
Toyota	146	Oldsmobile	137	Volkswagen	120
Honda	145	**Industry Avg.**	135	Mazda	119
Lincoln	144	Mercury	133	Geo	118
Volvo	144	Plymouth	133	Eagle	116
Cadillac	143	Chevrolet	132	Hyundai	110
Jaguar	143	Chrysler	132	Suzuki	108

Minitab printout for Exercise 2.63

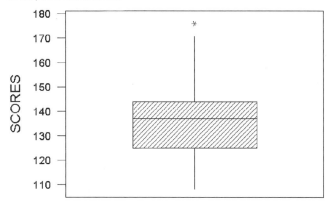

2.14 DESCRIBING THE MONEY MARKET FUND DATA

A graphical description of the money market fund data given in Table 2.2 of the Case Study is provided by the relative frequency histograms in Figures 2.28, 2.29, and 2.30. These graphs were generated using the HISTOGRAM command in the Minitab package. The Minitab DESCRIBE command was used to compute the means, medians, and standard deviations for the data sets, and the results are shown in Table 2.14.

F I G U R E **2.28**
Assets (Millions of Dollars) for 604 money market funds for the period ending July 13, 1994

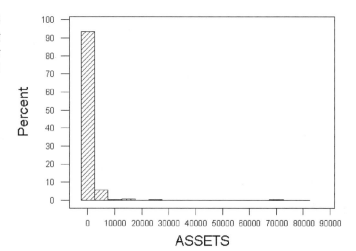

F I G U R E **2.29**
Average maturity (Days) for
604 money market funds for
the period ending July 13, 1994

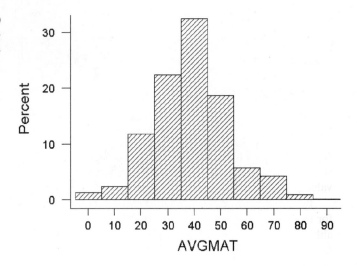

F I G U R E **2.29**
Average maturity (Days) for
604 money market funds for
the period ending July 13, 1994

F I G U R E **2.30**
Seven-day average yields for
604 money market funds for
the period ending July 13, 1994

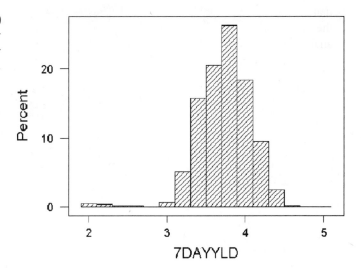

T A B L E **2.14**
Minitab DESCRIBE output for the
data in Table 2.2

	N	MEAN	MEDIAN	TRMEAN	STDEV	SEMEAN
AVGMAT	604	38.560	38.000	38.289	14.224	0.579
7DAYYLD	604	3.7324	3.7500	3.7435	0.3337	0.0136
ASSETS	604	843	261	483	3312	135

	MIN	MAX	Q1	Q3
AVGMAT	1.000	88.000	29.000	47.000
7DAYYLD	1.9700	4.5100	3.5125	3.9500
ASSETS	1	70100	98	694

Figure 2.28 shows the relative frequency histogram for the asset sizes of the 604 funds. Notice that this distribution is strongly skewed to the right, a situation in which the median is a better measure of central tendency than the mean. We can see from the histogram that most of the funds had assets near the median, $261 million. In contrast, the mean $\bar{x} = \$843$ million is quite large because its value is inflated by asset values for a few large funds (one fund, FidSpMM, contained assets of $70,100 million!). More than 75% of the funds' asset values were less than the mean, since the upper quartile, $Q_U = \$694$ million, is less than $\bar{x} = \$843$ million. The fact that the mean is much larger than the median indicates that the distribution is highly skewed to the right. Consequently, the median gives a better measure of the asset value of a typical money market fund.

Figure 2.29 gives the distribution of the average number of days to maturity for the notes held by the funds. This distribution is slightly skewed to the right, although the mean ($\bar{x} = 38.56$ days) and the median ($m = 38$ days) are nearly equal. Surprisingly, the average number of days for notes held to maturity by some of the funds was very small (the smallest was 1 day). The largest average number of days that notes were held to maturity was 88 days. Most of the observations fell within the interval ($\bar{x} \pm 2s$), or 11 to 67 days. The upper quartile computed in the Minitab analysis was 47 days.

The relative frequency distribution of the 7-day yields is shown in Figure 2.30. You can see that the mean and the median of the distribution of yields are nearly equal, indicating a symmetric distribution. The variation in the fund yields is not very large ($2s \approx .7\%$), although a few funds performed very poorly, deviating more than 1.7 percentage points below the mean yield.

How well do the means and standard deviations describe the three distributions of money market fund data? Notice in Table 2.15 that, despite the skewed nature of the relative frequency distributions, the percentages of observations in the interval ($\bar{x} \pm 2s$) are very close to the 95% stated by the Empirical Rule. Thus Table 2.15 provides further evidence that the Empirical Rule is a good rule of thumb for describing a unimodal set of data in terms of its mean and standard deviation.

T A B L E **2.15**
The percentage of observations in the interval $\bar{x} \pm 2s$

Data Type	Percentage
Asset value	99.0
Maturity time	94.4
7-day average yield	97.2

What relevance does the information provided by the relative frequency histograms in Figures 2.28–2.30 have to your becoming a millionaire? Table 2.1 shows that a 2% change in yield on your invested money can make a substantial difference in the amount of money that you will accumulate over a 40-year period. Consequently, you will want to seek investment instruments that, given comparable safety, produce the highest yields. Figures 2.28–2.30 provide useful information for someone seeking to accumulate money for retirement via a money market account. They show the mean rate of return (unadjusted for income taxes) that you might expect to obtain using a money market fund as the vehicle for investment; you can compare this value with the current mean yields for other types of investments. They also show the substantial variation in yields within money market funds and thereby emphasize the need to be selective in acquiring a good money manager.

2.15 Summary

Methods for describing sets of measurements fall into one of two categories, graphical methods and numerical methods. The relative frequency histogram is an extremely useful graphical method for characterizing a set of measurements. Numerical descriptive measures are numbers that attempt to create a mental image of the frequency histogram (or frequency distribution). We have restricted the discussion to measures of central tendency and variation, the most useful of which are the mean and the standard deviation. Although the mean possesses intuitive descriptive significance, the standard deviation is significant only when used in conjunction with Tchebysheff's Theorem and the Empirical Rule. The objective of sampling is the description of (making inferences about) the population from which the sample was obtained. This objective is accomplished by using the sample mean $\bar{x}$ and the quantity s^2 as estimators of the population mean μ and variance σ^2. When the data consist of pairs of observations, a bivariate plot is used to visually assess how x varies with y, while the correlation coefficient is used to determine the strength of the linear relationship between x and y. Other measures such as percentiles or z-scores are used to determine the relative standing of an observation in a population or in a sample. Boxplots are visual data summaries that are useful in detecting outliers.

The descriptive methods and the numerical measures presented constitute only a small number of those that could have been discussed. In addition, many special computational techniques usually found in elementary texts have been omitted. This omission was necessitated by the limited time available in any elementary course. Also, the common use of calculators and computers has minimized the importance of special computational formulas. But more importantly, the inclusion of such techniques would tend to detract from and obscure the main objective of modern statistics and this text: statistical inference.

Supplementary Exercises

2.64   Identify the following variables as qualitative or quantitative.

a   the ethnic origin of a job candidate

b   the score (0–100) on a training examination

c   the fast-food establishment preferred by a consumer (McDonalds, Burger King, or Carl's Jr.)

d   the mercury concentration in a sample of tuna

2.65 Identify the following quantitative variables as discrete or continuous.

    a the number of new clients acquired by a law firm in a month

    b the shelf life of a particular drug

    c the weight of a railway carload of wheat

    d the velocity of a pitched baseball

    e the number of fatal accidents at a manufacturing facility in a year

2.66 Would you expect the distributions of the following variables to be symmetric or skewed? Explain.

    a the size in dollars of nonsecured loans

    b the size in dollars of secured loans

    c the price of an 8-ounce can of peas

    d the number of broken taco shells in a package of 100 shells

2.67 Conduct the following experiment: toss 10 coins and record x, the number of heads observed. Repeat this process $n = 50$ times, thus providing 50 values of x. Construct a relative frequency histogram for these measurements.

2.68 Refer to Exercise 2.67.

    a Calculate $\bar{x}$ and s.

    b Calculate the intervals $(\bar{x} \pm s)$, $(\bar{x} \pm 2s)$, and $(\bar{x} \pm 3s)$.

    c Find the fractions of observations in the intervals. Do these fractions agree with Tchebysheff's Theorem? the Empirical Rule?

2.69 Recall that the range approximation for s can be improved if it is known that the sample is drawn from a bell-shaped distribution of data. Thus the calculated s should not differ substantially from the range divided by the appropriate ratio given in the following table:

Number of measurements	5	10	25
Expected ratio of range to s	2.5	3	4

Consider the $n = 5$ measurements 0, 2, 2, 4, 1.

    a Approximate s by using the accompanying table.

    b Calculate s, and compare it with your approximation.

2.70 Consider the set of $n = 6$ measurements 5, 7, 2, 1, 3, 0.

    a Calculate the mean, the median, and the mode.

    b Use the range approximation to estimate s. Use the table in Exercise 2.69 to approximate s.

    c Calculate the MAD, s^2, and s. Compare the value of s with the approximations in part (b). Which approximation is better?

    d Calculate the coefficient of variation (CV).

2.71 Each week the net amount of electric energy distributed by electric utilities across the country is compiled by the Edison Electric Institute. In one particular week the electric output (in thousands of kilowatt-hours) for ten midsize cities reads as follows: 49, 70, 54, 67, 59, 40, 61, 69, 71, 52.

    a Observe the data and guess the value of s by use of the range approximation.

    b Calculate $\bar{x}$ and s, and compare s with the range approximation of part (a).

2.72 The price-to-earnings (P/E) ratio of a stock is the ratio of a stock's most recent price per share to the stock's earnings per share (averaged over a 12-month period), multiplied by 100. The accompanying table gives the P/E ratios for 45 randomly selected securities traded on the New

York Stock Exchange (NYSE) on Thursday, July 21, 1994. Construct a stem and leaf display for the data. Use the display to describe the shape of the distribution of P/E ratios.

24	9	13	98	17
41	23	22	29	25
20	22	21	18	15
62	22	13	41	18
13	16	75	11	32
9	16	9	16	6
54	16	6	20	10
37	9	4	8	22
21	31	1	16	19

Source: The Press-Enterprise,
Riverside, Calif., July 22, 1994.

2.73 Refer to Exercise 2.72. Construct a box plot for the data. Do any of the P/E ratios appear to be outliers?

2.74 Refer to Exercise 2.72. The Minitab command DESCRIBE used for the data produced the accompanying output. Use the Minitab output to calculate the intervals $(\bar{x} \pm s)$, $(\bar{x} \pm 2s)$, and $(\bar{x} \pm 3s)$. Find the proportion of P/E ratios in the three intervals, and compare them with the proportions specified by Tchebysheff's Theorem and the Empirical Rule.

```
MTB > Describe 'PE-RATIO'

                 N      MEAN    MEDIAN   TRMEAN    STDEV   SEMEAN
PE-RATIO        45     22.89     18.00    20.78    18.66     2.78

               MIN       MAX        Q1       Q3
PE-RATIO      1.00     98.00     12.00    24.50
```

2.75 The accompanying measurements represent the grade-point average of 25 college freshmen marketing majors.

2.6	1.8	2.6	3.7	1.9
2.1	2.7	3.0	2.4	2.3
3.1	2.6	2.6	2.5	2.7
2.7	2.9	3.4	1.9	2.3
3.3	2.2	3.5	3.0	2.5

Construct a relative frequency histogram for these data.

2.76 Refer to Exercise 2.75.

a Calculate $\bar{x}$ and s.

b Calculate the intervals $(\bar{x} \pm s)$, $(\bar{x} \pm 2s)$, and $(\bar{x} \pm 3s)$.

c Find the fractions of observations in the intervals. Do these fractions agree with Tchebysheff's Theorem? the Empirical Rule?

2.77 In a study of unit pricing (Exercise 2.30), the price per liner was recorded for ten different brands of 13-gallon tall kitchen trash-can liners. A similar sample was also taken for ten different brands of 30-gallon trash-can liners. The results are shown below:

13-gallon	10	9	13	8	9	10	10	6	5	11
30-gallon	28	21	28	26	17	12	15	9	20	18

Source: Data from *Consumer Reports,* February 1994, p. 110.

a Find the mean and standard deviation for each of the two sets of data.

b Find the coefficient of variation for each of the two sets.

c Use the results of part (b) to compare the relative variability of the unit prices for the two types of liners. Which unit price is more variable?

 2.78 Consider a population consisting of the numbers of professors employed in the business departments at large four-year universities. Suppose that the number of business professors per college has an average of $\mu = 25$ and a standard deviation of $\sigma = 5$.

a Use Tchebysheff's Theorem to make a statement about the percentage of colleges that employ between 15 and 35 business professors.

b Assume that the population is normally distributed. What fraction of the colleges has more than 30 business professors?

 2.79 One method of generating additional revenue for a periodical is to allow more full-page advertisements in each issue. The following data represent the number of full-page advertisements in randomly selected issues of a weekly magazine.

12	10	16	7	18
13	14	20	9	23
8	13	14	6	19
6	11	15	10	16

Computing s for these data, the magazine's production manager obtained a value of 8.90. Use the range approximation for s as a check on this computation. Do these estimated and computed values for s appear to differ excessively?

 2.80 First introduced in the United States in 1960, industrial robots (e.g., mechanical arms, automated assemblers, computerized welders) have come into much wider use over the past decade. Initially, a typical assembly-line robot cost (on the average) $4.20 per hour, just slightly higher than the average factory worker's hourly wage. Today factory workers are paid between $15 and $20 per hour, but robots can still be operated for less than $5 per hour. Suppose the operating costs per hour were recorded for 100 firms that employ industrial robots. The mean and variance of the sample were found to be $4.86 and $2.50($)^2$. Calculate $(\bar{x} \pm s)$, $(\bar{x} \pm 2s)$, and $(\bar{x} \pm 3s)$, and state the approximate fraction of measurements we would expect to fall in these intervals according to the Empirical Rule.

 2.81 The mean duration of television commercials on a given network is 75 seconds, and the standard deviation is 20 seconds. Assume that duration times are approximately normally distributed.

a Approximately what fraction of commercials will last less than 35 seconds?

b Approximately what fraction of commercials will last longer than 55 seconds?

2.82 A random sample of 100 large law firms was surveyed to determine the prevalence of computerized office equipment in the industry. The number of pieces of computerized equipment (word processors, data terminals, microprocessors, minicomputers, etc.) in use at each firm was recorded. Sixty-nine law firms were found to utilize no computerized equipment, 17 employed one piece, and so on. The following table is a frequency tabulation of the data.

Number of computerized pieces, x	0	1	2	3	4	5	6	7	8
Number of law firms, f	69	17	6	3	1	2	1	0	1

a Construct a relative frequency histogram for x, the number of pieces of computerized office equipment per firm.

b Calculate $\bar{x}$ and s for the sample. [*Hint:* Use the formulas given in Exercise 2.45.]

c What fraction of the equipment counts falls within two standard deviations of the mean? three? Do the results agree with Tchebysheff's Theorem? the Empirical Rule?

2.83 In 30 years, Wal-Mart Stores, Inc., has grown into a large retail empire, taking in more than $1 billion per week. The following table shows Wal-Mart's approximate sales and number of employees over a ten-year period of time.

Year	Sales ($ billion)	Number of Stores	Number of Employees (thousands)
1983	4	551	46
1988	17	1114	183
1993	56	1900	470

Source: Data adapted from *The Press-Enterprise*, August 16, 1993.

a What variables have been measured in this study? Are the variables quantitative or qualitative? discrete or continuous?

b Draw separate line graphs to describe the sales, number of stores, and number of employees over the ten-year period. Interpret the resulting graphs.

c Draw a scatterplot to describe the relationship between sales and number of stores for the three years given. Repeat the procedure for the other two pairs (sales vs. employees and employees vs. stores).

d Look at the results of part (c). How are the three variables in this study interrelated?

2.84 The Sony Corporation is developing a new video machine called the PlayStation in an attempt to break into the video game market, currently dominated by Sega and Nintendo. The new machine will require game CDs, which can be produced for $3 but will cost the consumer between $50 and $100 ("Sony Has Some Very Scary Monsters," 1994). Suppose that the average price of a game CD is $69, with a standard deviation of $8.

a If nothing more is known about the distribution of the CD prices, what can you say about the proportion of game CDs with prices between $53 and $85?

b What can you say about the proportion of game CDs with prices exceeding $85?

2.85 Refer to Exercise 2.84. Suppose that the prices of the game CDs are approximately mound-shaped, with a mean of $69 and a standard deviation of $8.

a What proportion of the game CDs will have prices between $61 and $85?

b What proportion of the game CDs will cost less than $61?

2.86 The rapid increase in house prices during the 1980s has made home ownership more difficult for prospective first-time buyers. Most first-time buyers have to save for several years and even seek a gift or loan from family members to secure the required down payment (Englelhardt, 1994). The following table gives the median income of first-time home buyers in several U.S. cities, along with the median purchase price and the average number of years required to save for the purchase.

City	Purchase Price ($ thousand)	Median Income ($ thousand)	Years to Save
San Francisco	211.1	61.8	3.0
New York	148.1	60.2	4.2
Washington	130.0	66.7	2.7
Seattle	114.3	51.3	2.3
Philadelphia	103.6	47.2	2.8
Atlanta	91.4	50.8	2.5
Cleveland	70.1	41.0	2.4
Phoenix	76.8	47.8	1.4
Detroit	75.3	52.8	2.6

a What types of variables have been measured in this survey? Are the data univariate or multivariate?

b Use any appropriate graphical or numerical descriptive measures to describe these data. Summarize the results of your description in a short report.

2.87 Employment opportunities in Canada provide much reason for optimism. In a survey of 2072 Canadian companies, 22% planned to expand their work forces in 1994, with only 2.8% planning to contract their work forces. The accompanying data show the percent of companies in various sectors whose work forces plan to expand in 1994.

Sector	Percent Expanding
Manufacturing	29.3
Transportation	26.0
Mining	25.0
Wholesaling	24.2
Services	24.1
Construction	18.8
Agriculture	17.2
Retailing	17.1
Finance	8.3

Source: Report on Business Magazine,
September 1993.

a Use an appropriate graphical technique to describe the data.

b Find the average percentage of companies planning to expand their work forces for these nine sectors.

c Does the average percentage calculated in part (b) equal the overall percentage, 22%, given in the magazine? Should they be equal?

Exercises Using the Data Disk

2.88   Refer to data set A: Average Salaries of Instructional Staff Faculty, on the data disk.

   a   For colleges numbered 1 through 30, construct a histogram of average salaries for males in each of the four ranks: professor, associate professor, assistant professor, and lecturer.

   b   For these same colleges, construct a histogram of average salaries for females in each of these four ranks. How do the salary histograms for males compare with those for females?

   c   Find the mean, median, and standard deviation for average salaries paid to males and females for each of the four faculty ranks at colleges 1 through 30. Does Tchebysheff's Theorem apply to these eight data sets? Does the Empirical Rule apply?

2.89   Refer to data set B: Broccoli Data, on the data disk.

   a   For the observations 1–39, construct histograms using fresh-weight, market-weight, and head diameter for variety 0 and for variety 1. Comment on the similarities and/or the differences of the histograms when comparing the variables fresh-weight, market-weight, and head diameter for each variety.

   b   For these same observations, find the mean, median, and standard deviation of the fresh-weight, market-weight, and head diameter for each variety. Are there any apparent differences in these summaries of fresh-weight, market-weight, and head diameter for each variety?

2.90   Refer to data set C on the data disk.

   a   Select the first 10% of the observations for average maturity in days, seven-day average yield, and effective seven-day average yield. Construct a histogram using each of the three variables. How would you describe these histograms? What can be said about symmetry? any apparent outliers? If you think that a histogram appears to have one or more outliers, construct a box plot for that variable.

   b   Find the mean, median, and standard deviation for all three variables. Use the mean and median to comment about possible skewness. Which of the three variables exhibits the least variation?

CHAPTER 3

PROBABILITY AND DISCRETE PROBABILITY DISTRIBUTIONS

About This Chapter

The objective of this chapter is to lay the foundation for statistical inference. We will introduce some basic concepts of probability, and then we will present some models for calculating the probabilities of sample outcomes.

Contents

WILL BABY BOOMERS DUMP DEPARTMENT STORES?

I s most of your shopping done in department stores, or do you spend your shopping dollars at a select number of specialty shops? Research shows that middle-aged people are department stores' best customers. The managers of department stores are hoping that their slow sales will step up as the baby-boom generation moves into middle age. In his article, "Will Baby Boomers Dump Department Stores?" (Schwartz, 1990), the author looks at this question in light of a study by Marvin J. Rothenberg, Inc., a consulting firm headquartered in Fair Lawn, New Jersey. The researchers at Rothenberg interviewed 60,000 regular customers of their client department stores between 1988 and 1990. The stores ranged in size from small to large and were located across the states "from New York in the East to California in the West, and from Wisconsin in the North to Louisiana in the South." According to the report by Rothenberg, "department stores capture a larger share of the retail dollars spent by their older customers than they do from their younger customers."

However, as the baby-boom generation moves into middle age, the concern is that their numbers may not automatically translate into increased sales for department stores. Although older people spend a larger proportion of their retail dollars in department stores, they spend less than their younger counterparts. The distribution of the share of retail spending by department store customers (based on the 60,000 customers sampled) is summarized in Table 3.1.

T A B L E **3.1**
Retail Spending

Age Segment	Share of Retail Spending	Share of Sales	Amount Spent Monthly per Customer
18–24	.22	.15	$ 50
25–34	.24	.27	108
35–44	.26	.25	126
45–54	.28	.16	116
55–64	.28	.12	89
65 and older	.39	.05	53

Source: Marvin J. Rothenberg, Inc., Fair Lawn, N.J.

In this chapter you will be introduced to the role of probability in statistics. In this vein, you will find that polls and studies generate random variables, so named because their values vary from sample to sample, and very often they vary and change over time. Sections 3.1 to 3.5 are concerned with a short but systematic introduction to the basic concepts of probability. Section 3.6 deals with random variables and their probability distributions. We will continue the discussion of the department store share of retail spending and the baby-boom generation in Section 3.7.

3.1   The Role of Probability in Statistics

Probability and statistics are related in an important way. **Probability is the tool that allows the statistician to use sample information to make inferences about or to describe the population from which the sample was drawn.** We can illustrate this relationship with a simple example.

Consider a balanced die with its familiar six faces. When the die is tossed, any of the six sides has an equal chance of being the upper face. If we were to toss this die over and over again, we would generate a population of numbers, in which the upper face, x, would be 1, 2, 3, 4, 5, or 6. What would this population look like? Although it would be infinitely large, we can still say that it would consist of an equal number of 1s, 2s, ..., 6s. Now let us toss the die once and observe the value of x. This is equivalent to taking a sample of size $n = 1$ from the population. What is the probability that $x = 2$? Knowing the structure of the population, we know that each of the six values of x has an equal chance of occurring, so the probability that $x = 2$ is 1/6. This is a simple application of *probability*. When the population is known, we can calculate the probability of observing a particular sample.

Suppose now that the population is *not* known. That is, we do not know whether the die is balanced or not. The population still consists of 1s, 2s, 3s, ..., 6s, but we do not know in what proportion they occur! In an attempt to find out, we toss the die $n = 10$ times and record the upper face, x, after each toss. This is equivalent to taking a sample of size $n = 10$ from the population. Suppose that, on each toss, the upper face turns out to be $x = 1$. What would you infer about the population? Would you believe that the die is balanced? Probably not, because, if the die were balanced, the chance of observing a "1" ten times in a row would be very small. Either we have observed this very unlikely event, or the die is unbalanced. We would likely be inclined toward the latter conclusion, because it is the more *probable* of the two choices.

This example shows the relationship between probability and statistics. When the population is known, probability is used to describe the likelihood of various sample outcomes. When the population is *unknown* and we have only a sample, we have the statistical problem of trying to make inferences about the unknown population. Probability is the tool we use to make these inferences. **Thus probability reasons from the population to the sample, whereas statistics acts in reverse, moving from the sample to the population.**

As we explain the language of probability, we will assume that the population is known and will calculate the probability of drawing various samples. In doing so, we are really choosing a **model** for a physical situation, because the actual composition of a population is rarely known in practice. Thus the probabilist models a physical situation (the population) with probability much as the sculptor models with clay. In the sections that follow, we use simple examples to help you grasp the concept of probability. Practical applications follow these simple examples.

3.2 The Probability of an Event

Data are obtained either by observing uncontrolled events in nature or by observing controlled situations in a laboratory. We will use the term **experiment** to describe either method of data collection.

DEFINITION •

An **experiment** is the process by which an observation (or measurement) is obtained. •

Note that the observation need not produce a numerical value. Here are some typical examples of experiments.

1 Recording the daily production of a manufacturing plant.
2 Recording the exchange rate between the dollar and the British pound.
3 Interviewing a consumer to determine product preference among a group of ten types of automobiles.
4 Inspecting a light bulb to determine whether it is a defective or an acceptable product.
5 Tossing a coin and observing the face that appears.

Experiments may result in one or more outcomes, which are called **events** and are denoted by capital letters.

DEFINITION •

An **event** is the outcome of an experiment. •

EXAMPLE **3.1** Experiment: Toss a die and observe the number appearing on the upper face. Some events would be as follows:

$$\text{event } A : \quad \text{observe an odd number}$$
$$\text{event } B : \quad \text{observe a number less than 4}$$
$$\text{event } E_1 : \quad \text{observe a 1}$$
$$\text{event } E_2 : \quad \text{observe a 2}$$

event E_3 :   observe a 3
event E_4 :   observe a 4
event E_5 :   observe a 5
event E_6 :   observe a 6   ∎

In Example 3.1, there is a distinct difference between events A and B and events E_1, E_2, E_3, E_4, E_5, and E_6. Event A will occur if event E_1, E_3, or E_5 occurs—that is, if we observe a 1, 3, or 5. Thus A could be decomposed into a collection of simpler events, namely, E_1, E_3, and E_5. Likewise, event B will occur if E_1, E_2, or E_3 occurs and could be viewed as a collection of smaller or simpler events. In contrast, it is impossible to decompose events E_1, E_2, E_3, ..., E_6. These events are called **simple events.**

DEFINITION   ∎   An event that cannot be decomposed is called a **simple event.** Simple events will be denoted by the symbol E with a subscript.   ∎

The events E_1, E_2, ..., E_6 represent a complete listing of all simple events associated with the experiment in Example 3.1. **An experiment will result in one and only one of the simple events.** For instance, if a die is tossed, we will observe a 1, 2, 3, 4, 5, or 6, but we cannot possibly observe more than one of the simple events at the same time. Hence, a list of simple events provides a breakdown of all possible indecomposable outcomes of the experiment.

DEFINITION   ∎   The **sample space** S is the set of all possible outcomes of an experiment.   ∎

Finally, we can define the outcomes of an experiment in terms of the associated simple events.

DEFINITION   ∎   An **event** is a collection of one or more simple events.   ∎

The **probability of an event** A is a measure of our belief that an experiment will result in event A. In order to attach meaning to this concept, we note that populations of observations are generated by repeating an experiment over and over again a very large number of times. If in this large number N of repetitions of the experiment event A is observed n times, then we view the probability of event A as

$$P(A) = \frac{n}{N}$$

In fact, $P(A)$ is the limiting value of the fraction n/N as N becomes infinitely large. This practical interpretation of the meaning of probability—an interpretation held by most laypersons—is called the **relative frequency concept of probability.**

Needless to say, it would be very time-consuming to repeat an experiment a very large number of times. For this reason, we will use alternative methods for calculating probabilities that conform to this relative frequency concept. For example, the relative frequency definition of probability implies that $P(A)$ must be a fraction lying between 0 and 1, inclusive, with $P(A) = 0$ if the event A never occurs and $P(A) = 1$ if A always occurs. Therefore, the closer $P(A)$ is to 1, the more likely it is that A will occur.

Some events A and B possess the unique property that when one event occurs the other cannot occur (and vice versa). Events that possess this property are said to be **mutually exclusive.** For example, suppose that an experiment is to observe next month's action by the Federal Reserve Board on the prime interest rate. Define the following events:

A :   the prime rate is raised

B :   the prime rate is lowered

C :   the prime rate is unchanged

Then events A and B are mutually exclusive events because a single experiment cannot result in both A and B. If the next declaration of prime rate policy results in A, then B cannot also occur. If the prime rate was raised, it could not have been lowered at the same time. Events A, B, and C are also said to be mutually exclusive events. If any one of the group is observed, neither of the remaining two events could have occurred.

DEFINITION ▪

> Two events A and B are said to be **mutually exclusive** if when A occurs, B cannot occur (and vice versa). ▪

Simple events are mutually exclusive, and therefore the probabilities associated with simple events satisfy the following conditions.

Requirements for Simple-Event Probabilities

1   Each probability must lie between 0 and 1, inclusive.

2   The sum of the probabilities for all simple events in S equals 1.

When it is possible to write down the simple events associated with an experiment and assess their respective probabilities, we can find the probability of an event A by summing the probabilities for the simple events contained in the event A.

DEFINITION ▪

> The **probability of an event** A is equal to the sum of the probabilities of the simple events contained in A.   ▪

EXAMPLE **3.2**   Using the simple event approach, calculate the probability of observing exactly one head in a toss of two coins.

Solution   Construct the sample space and let H represent a head and T represent a tail. The outcomes associated with this experiment can be displayed using a **tree diagram.** In a tree diagram, each successive level of branching corresponds to the next step required to generate the possible outcomes of the experiment. The tree diagram is shown in Figure 3.1. The simple events that result are given in the last column of Figure 3.1 and in Table 3.2.

TABLE **3.2**
Sample space for Example 3.2

Event	First Coin	Second Coin	$P(E_i)$
E_1	H	H	1/4
E_2	H	T	1/4
E_3	T	H	1/4
E_4	T	T	1/4

FIGURE **3.1**
Tree diagram for Example 3.2

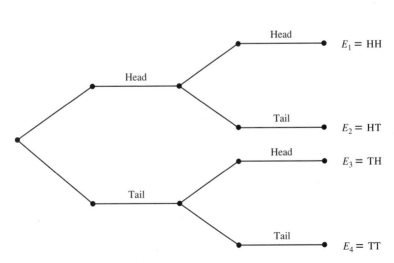

The sample space S for the experiment contains the four simple events listed in Table 3.2. Since a reasonable assumption is that any one of the simple events is as likely to occur as any other, we assign a probability of 1/4 to each. We are interested in

event A :   observe exactly one head

Event A will occur if and only if E_2 or E_3 occurs. Therefore,

$$P(A) = P(E_2) + P(E_3)$$
$$= \frac{1}{4} + \frac{1}{4}$$
$$= \frac{1}{2} \quad \blacksquare$$

EXAMPLE 3.3   Refer to our observation of the Federal Reserve Board's action on the prime rate, and define the events A, B, and C as follows:

A :   the prime rate is raised

B :   the prime rate is lowered

C :   the prime rate is unchanged

Suppose that we know that $P(A) = .2$ and $P(B) = .3$. Find $P(C)$.

Solution   Examining A, B, and C, we can see that the experiment can result in one and only one of these three mutually exclusive events. Therefore, A, B, and C are simple events that represent the sample space S for the experiment, and

$$P(A) + P(B) + P(C) = 1$$

Substituting $P(A)$ and $P(B)$ into this equation yields

$$.2 + .3 + P(C) = 1$$

Solving for $P(C)$, we obtain

$$P(C) = 1 - .3 - .2 = .5 \quad \blacksquare$$

Tips on Problem Solving

Calculating the probability of an event: the simple-event approach

- Use the following steps for calculating the probability of an event by summing the probabilities of the simple events:

  1 Define the experiment.

  2 Identify a typical simple event. List the simple events associated with the experiment, and test each one to make certain that it cannot be decomposed. This defines the sample space S.

  3 Assign reasonable probabilities to the simple events in S, making certain that each probability is between 0 and 1 and that the sum of the simple-event probabilities equals 1.

  4 Define the event of interest A as a specific collection of simple events. (A simple event is in A if A occurs when the simple event occurs. Test *all* simple events in S to locate those in A.)

  5 Find $P(A)$ by summing the probabilities of simple events in A.

- When the simple events are equiprobable, obtain the sum of the probabilities of the simple events in A (step 5) by counting the points in A and multiplying by the probability per simple event.

- Calculating the probability of an event by using the five-step procedure described above is systematic and will lead to the correct solution if all steps are followed correctly. Major sources of error include the following:

  1 Failure to define the experiment clearly (step 1).

  2 Failure to specify simple events (step 2).

  3 Failure to list all the simple events (step 2).

  4 Failure to assign valid probabilities to the simple events (step 3).

Exercises

Basic Techniques

3.1 An experiment involves tossing a single die. Specify the simple events in the following events.

A :   observe a 2

B :   observe an odd number

C :   observe a number less than 4

D :   observe both A and B

E :   observe either A or B or both

F :   observe both A and C

Calculate the probabilities of the events A through F by summing the probabilities of the appropriate simple events.

3.2   An experiment can result in one of four simple events, with

$$P(E_1) = .1 \qquad P(E_2) = .15 \qquad P(E_3) = .6 \qquad P(E_4) = .15$$

Explain why this assignment of probabilities to the simple events is or is not valid.

3.3   An experiment can result in one of five simple events, E_1, E_2, E_3, E_4, and E_5. If $P(E_1) = P(E_2) = P(E_3) = P(E_4) = .15$, find $P(E_5)$.

3.4   A sample space contains five simple events, E_1, E_2, E_3, E_4, and E_5. If $P(E_3) = .4$, $P(E_4) = 2P(E_5)$, and $P(E_1) = P(E_2) = .15$, find the probabilities of E_4 and E_5.

3.5   A sample space contains ten simple events, E_1, E_2, ..., E_{10}. If $P(E_1) = 3P(E_2) = .45$, with the remaining simple events being equiprobable, find the probabilities of these remaining simple events.

3.6   The game of roulette uses a wheel containing 38 pockets. Thirty-six pockets are numbered 1, 2, ..., 36, and the remaining two are marked 0 and 00. A wheel is spun and one of these pockets is identified as the "winner." Assume that the observance of any one pocket is just as likely as any other.

a   Identify the simple events in a single spin of the roulette wheel.

b   Assign probabilities to the simple events.

c   Let A be the event that you observe either a 0 or a 00. List the simple events in the event A and find $P(A)$.

d   Suppose that you were to place bets on the numbers 1 through 18. What is the probability that one of your numbers will be the winner?

Applications

3.7   An oil-prospecting firm hits oil or gas on 10% of its drillings. If the firm drills two wells, the four possible simple events and three of their associated probabilities are as shown in the accompanying table.

Simple Event	Outcome of 1st Drilling	Outcome of 2nd Drilling	Probability
1	Hit (oil or gas)	Hit (oil or gas)	.01
2	Hit	Miss	?
3	Miss	Hit	.09
4	Miss	Miss	.81

a   Construct a tree diagram to represent this experiment.

b   Find the probability that the company will hit oil or gas on the first drilling and miss on the second.

c   Find the probability that the company will hit oil or gas on at least one of the two drillings.

3.8   A marketing survey for a large department store classified the store's customers according to whether they were male or female and according to their residence, suburban or city. The proportions of customers falling in the four categories are shown in the following *probability table*. Each entry in the table represents a possible outcome of the experiment (a simple event).

	Sex	
Residence	**Male**	**Female**
Suburban	.17	.67
City	.04	.12

Suppose a single adult is selected from among this group of customers. Find the following probabilities.

a that the customer resides in the suburbs

b that the customer is a female who lives in the city

c that the customer is a male

  3.9 Two stock market analysts are each asked to forecast whether the Dow-Jones stock average will gain 100 points or more, lose 100 points or more, or change less than 100 points by the end of the next 12 months. Thus the experiment consists of observing the pair of forecasts produced by the two market analysts. Suppose that each of the analysts is as likely to select any one of the three choices as he or she is likely to select any other.

a List the simple events in the sample space S. Use a tree diagram.

b Let A be the event that at least one of the analysts forecasts a rise in the Dow-Jones average of 100 points or more. Find the simple events in A.

c Let B be the event that both analysts agree in their forecasts. Find the simple events in B.

d Assign probabilities to the simple events in S, and find $P(A)$.

e Find $P(B)$.

  3.10 A food company plans to conduct an experiment to compare its brand of tea with that of two competitors. In actual practice a number of tea tasters would be employed. For this example we will assume that only one tea taster tastes each of the three brands of tea, which are unmarked except for identifying symbols, A, B, and C.

a Define the experiment.

b List the simple events in S.

c If the taster had no ability to distinguish a difference in taste between teas, what is the probability that the taster will rank tea type A as best? as the least desirable?

3.11 Four union men, two from a minority group, are assigned to four distinctly different one-man jobs.

a Define the experiment.

b List the simple events in S.

c If the assignment to the jobs is unbiased—that is, if any one ordering of assignments is as probable as any other—what is the probability that the two men from the minority group are assigned to the two least desirable jobs?

3.3 Event Composition and Event Relations

Frequently we are interested in experimental outcomes formed by some composition of two or more events. Often, events can be formed by **unions** or **intersections** of other events, or by some combination of the two.

DEFINITION ■

> The **intersection** of events A and B, denoted by AB, is the event that both A and B occur.[†] ■

DEFINITION ■

> The **union** of events A and B, denoted by $A \cup B$, is the event that A or B or both occur. ■

EXAMPLE 3.4   Refer to the experiment in Example 3.2, where two coins are tossed, and define

event A :   at least one head

event B :   at least one tail

Define events A, B, AB, and $A \cup B$ as collections of simple events, and find their probabilities.

Solution   Recall that the simple events for this experiment were

E_1 :   HH (head on first coin, head on second)

E_2 :   HT

E_3 :   TH

E_4 :   TT

The occurrence of simple events E_1, E_2, or E_3 implies at least one head and hence defines event A, and $P(A) = \frac{1}{4} + \frac{1}{4} + \frac{1}{4} = \frac{3}{4}$. The other events are defined similarly.

event B : E_2, E_3, E_4,	and	$P(B) = \frac{3}{4}$
event AB : E_2, E_3,	and	$P(AB) = \frac{2}{4} = \frac{1}{2}$
event $A \cup B$: E_1, E_2, E_3, E_4	and	$P(A \cup B) = \frac{4}{4} = 1$

Notice that $A \cup B$ comprises the entire sample space and hence is certain to occur. Thus $P(A \cup B) = 1$. ■

[†]Some authors use the symbol $A \cap B$.

When the two events A and B are mutually exclusive, it means that when A occurs, B cannot, and vice versa. Mutually exclusive events are also referred to as **disjoint events.** When A and B are mutually exclusive,

1 $P(AB) = 0$

2 $P(A \cup B) = P(A) + P(B)$

That is, if $P(A)$ and $P(B)$ are known, we do not need to enumerate the simple events comprising $A \cup B$ and sum their respective probabilities; rather, we can simply sum $P(A)$ and $P(B)$.

E X A M P L E **3.5** A summary of the activity (total dollar amount of checks written) of a bank's $1000-minimum-balance personal checking accounts is given in the table that follows.

Activity	Under $1000	$1000–2999	$3000–4999	$5000–7999	$8000 and over
Percentage	17%	43%	28%	9%	3%

A single $1000-minimum-balance checking account is selected at random from those at this bank.

a Find the probabilities of the following events:

  $A :$ checking activity is less than $3000
  $B :$ checking activity is between $1000 and $4999
  $C :$ checking activity exceeds $4999

b Find $P(B \cup C)$.

c Are events A, B, and C mutually exclusive?

Solution The simple events in the experiment are given in the table, with $E_1 = $ (under $1000), $E_2 = $ ($1000–2999), and so on.

a Then

$$P(A) = P(E_1) + P(E_2) = .17 + .43 = .60$$
$$P(B) = P(E_2) + P(E_3) = .43 + .28 = .71$$
$$P(C) = P(E_4) + P(E_5) = .09 + .03 = .12$$

b The event $B \cup C$ consists of the simple events E_2, E_3, E_4, and E_5, so

$$P(B \cup C) = .43 + .28 + .09 + .03 = .83$$

c A and C are mutually exclusive, as are B and C. However, A and B contain the common simple event E_2 and hence are not mutually exclusive. Therefore, A, B, and C are not mutually exclusive. ∎

Complementation is another event relationship that often simplifies probability calculations.

DEFINITION ▪

> The **complement** of an event A, denoted by $\overline{A}$, consists of all the simple events in the sample space that are not in A.   ▪

The complement of A is the event that A does not occur. Therefore, A and $\overline{A}$ are mutually exclusive, and $A \cup \overline{A}$ comprises the total sample space. It follows that $P(A) + P(\overline{A}) = 1$ and

$$P(A) = 1 - P(\overline{A})$$

For example, if the event A is the occurrence of at least one head in the toss of two fair coins (see Example 3.4), then the event $\overline{A}$ is the occurrence of no heads in the toss, and

$$P(A) = 1 - P(\overline{A}) = 1 - \frac{1}{4} = \frac{3}{4}$$

3.4 Conditional Probability and Independent Events

Two events are often related in such a way that the probability of the occurrence of one event depends upon whether the other event has or has not occurred. For example, suppose that you are a money trader who trades dollars against the British pound, and the experiment consists in observing whether the dollar is up (has increased in value) against the pound. Let A be the event that the dollar is up against the pound on the English currency market before our market opens at 9:00 A.M., and let B be the event that the dollar is up on the American market after it opens. Events A and B are certainly related, because the two markets will likely move in the same direction most (but not necessarily all) days. Therefore, the probability $P(B)$ that the dollar will be up against the pound on the U.S. market is not the same as the probability that B will occur given that you know that the dollar is already up (event A) on the British market.

For example, suppose that the dollar is up against the pound 60% of all days on the British market, and 50% of all days it is up on both the U.S. and British markets (i.e., $P(A) = .6$ and $P(AB) = .5$). Then if you already know that the dollar is up against the pound on the British market, the probability that it will be up on the U.S. market is 5/6. This fraction, $P(AB)/P(A)$, is called the **conditional probability** of B given that event A has occurred, and is denoted by the symbol

$$P(B|A)$$

The vertical bar in the expression $P(B|A)$ is read "given," and the events appearing to the right of the bar are the events that have occurred.

DEFINITION ▪

> The **conditional probability** of B given that event A has occurred is
>
> $$P(B|A) = \frac{P(AB)}{P(A)}, \quad \text{for } P(A) > 0$$
>
> and the conditional probability of A given that event B has occurred is
>
> $$P(A|B) = \frac{P(AB)}{P(B)}, \quad \text{for } P(B) > 0 \quad ▪$$

EXAMPLE **3.6** Calculate $P(A|B)$ for the die-tossing experiment described in Example 3.1, where

$$A: \quad \text{observe an odd number}$$
$$B: \quad \text{observe a number less than 4}$$

Solution Given that event B (a number less than 4) has occurred, we have observed a 1, 2, or 3, all of which occur with equal frequency. Of these three simple events, exactly two (1 and 3) result in event A (an odd number). Hence,

$$P(A|B) = \frac{2}{3}$$

Or we could obtain $P(A|B)$ by substituting into the equation

$$P(A|B) = \frac{P(AB)}{P(B)} = \frac{1/3}{1/2} = \frac{2}{3}$$

Note that $P(A|B) = 2/3$ and $P(A) = 1/2$, indicating that A and B are dependent on each other. ▪

DEFINITION ▪

> Two events A and B are said to be **independent** if and only if either
>
> $$P(A|B) = P(A)$$
>
> or
>
> $$P(B|A) = P(B)$$
>
> Otherwise, the events are said to be **dependent.** ▪

Translating this definition into words: Two events are **independent** if the occurrence or nonoccurrence of one of the events does not change the probability of the occurrence of the other event. If $P(A|B) = P(A)$, then $P(B|A)$ will also equal $P(B)$. Similarly, if $P(A|B)$ and $P(A)$ are unequal, then $P(B|A)$ and $P(B)$ will also be unequal.

E X A M P L E   **3.7**   Refer to the die-tossing experiment in Example 3.1, where

A :   observe an odd number

B :   observe a number less than 4

Are events A and B mutually exclusive? Are they complementary? Are they independent?

Solution   In the terms of simple events, we can write

$$\text{event } A : \quad E_1, E_3, E_5$$
$$\text{event } B : \quad E_1, E_2, E_3$$

Event AB is the set of simple events in both A and B. Since AB includes events E_1 and E_3, A and B are not mutually exclusive. They are not complementary because B is not the set of all outcomes in S that were not in A. The test for independence lies in the definition; that is, we will check to see if $P(A|B) = P(A)$. From Example 3.6, $P(A|B) = 2/3$. Then, since $P(A) = 1/2$, $P(A|B) \neq P(A)$ and by definition events A and B are dependent.   ∎

An alternative approach to solving probability problems is based on event relations and two probability rules, which we will state without proof. The first rule is called the **additive rule of probability,** and it deals with unions of events.

Additive Rule of Probability

Given two events A and B, the probability of their union $A \cup B$ is equal to

$$P(A \cup B) = P(A) + P(B) - P(AB)$$

If A and B are mutually exclusive, then $P(AB) = 0$ and

$$P(A \cup B) = P(A) + P(B)$$

Notice that the sum $P(A) + P(B)$ double-counts the simple events that are common to both A and B; subtracting $P(AB)$ gives the correct result.

The second rule of probability is called the **multiplicative rule of probability** and follows from the definition of conditional probability.

> **Multiplicative Rule of Probability**
>
> The probability that both of two events A and B occur is
>
> $$P(AB) = P(A)P(B|A) = P(B)P(A|B)$$
>
> If A and B are independent events,
>
> $$P(AB) = P(A)P(B)$$
>
> Similarly, if A, B, and C are mutually independent events, then the probability that A, B, and C will occur is
>
> $$P(ABC) = P(A)P(B)P(C)$$

EXAMPLE   **3.8**   Suppose that you randomly select three common stocks, A, B, and C, from among the 500 stocks used to calculate Standard & Poor's (S&P) 500-stock average. What is the probability that the yearly gain for all three stocks will exceed the gain of the S&P average? Let A, B, and C represent the events that stocks A, B, and C individually outperform the S&P average, and assume that $P(A) = P(B) = P(C) = 1/2$.

Solution   The event that all three stock choices will beat the S&P average is the intersection of the events A, B, and C. We do not know the conditional probabilities of A, B, and C, so we cannot use the definition of conditional probability to test for independence. Rather, we must rely on our intuition. Since we selected the three stocks in an unrelated manner, it seems unlikely that the selection of one stock from among the 500 would greatly affect the selection of another. For this reason we will declare the events to be independent and calculate

$$P(ABC) = P(A)P(B)P(C)$$
$$= \left(\frac{1}{2}\right)\left(\frac{1}{2}\right)\left(\frac{1}{2}\right) = \frac{1}{8}$$

Thus we calculate the probability of selecting three stocks that will beat the S&P average to be 1/8.   ∎

A **probability table** for the events A and B is a two-way table whose four entries are the four intersection probabilities $P(AB)$, $P(A\overline{B})$, $P(\overline{A}B)$, and $P(\overline{A}\,\overline{B})$ and whose marginal row and column sums correspond to the unconditional probabilities

$P(A)$, $P(\overline{A})$, $P(B)$, and $P(\overline{B})$ as in Table 3.3. There are several interesting points that are quite obvious from a probability table. One is that an event A can be expressed as consisting of two parts—AB, those simple events in A that are also in B, and $A\overline{B}$, those simple events in A that are not in B—so that

$$P(A) = P(AB \cup A\overline{B}) = P(AB) + P(A\overline{B})$$

Another is that $P(A|B)$, the conditional probability of A given B, represents the portion of the simple events in B that are also in A, so that $P(A|B) = P(AB)/P(B)$ is merely a rescaled probability on a reduced sample space equivalent to B.

TABLE 3.3

	B	$\overline{B}$	Totals
A	$P(AB)$	$P(A\overline{B})$	$P(A)$
$\overline{A}$	$P(\overline{A}B)$	$P(\overline{A}\,\overline{B})$	$P(\overline{A})$
Totals	$P(B)$	$P(\overline{B})$	1

EXAMPLE 3.9   A mail order merchandiser of clothing sells two product lines, one relatively expensive, the other inexpensive. A survey of orders produced the relative frequencies of orders by product line and the sex of the customer. In Table 3.4, A is the event that the customer is female and B is the event that the order is from product line 1. Suppose that a customer who has placed an order is selected at random.

a   Find the probability that the customer is female.

b   Find the probability that the order is from product line 1.

c   Find the probability that the order is from product line 1 and the customer is female.

d   Find $P(A \cup B)$, the probability that the customer is female or the order is from product line 1, or both.

e   Find $P(B|A)$.

f   Show that A and B are dependent and that $P(AB) = P(A)P(B|A)$.

TABLE 3.4

	Product Line		
Sex	$1(B)$	$2(\overline{B})$	Totals
Female (A)	.516	.205	.721
Male ($\overline{A}$)	.132	.147	.279
Totals	.648	.352	1.000

Solution $\overline{A}$ is the event that the customer is a male, and $\overline{B}$ is the event that the order is from product line 2.

a $P(A) = .721$ is the first row total.

b $P(B) = .648$ is the first column total.

c $P(AB)$, the probability that the customer is female and the order is from product line 1, is .516, the entry in the first row and first column.

d   Using the entries in Table 3.4,

$$P(A \cup B) = P(A) + P(B) - P(AB)$$
$$= .721 + .648 - .516 = .853$$

e   To find $P(B|A)$, we use the entries in the first row to find

$$P(B|A) = \frac{P(AB)}{P(A)} = \frac{.516}{.721} = .716$$

f   Since $P(B|A) \neq P(B)$ (see parts (b) and (e)), A and B are dependent. Also,

$$P(AB) = .516 \quad \text{directly from the table}$$

while

$$P(B|A)P(A) = (.716)(.721) = .516$$

so $P(AB) = P(A)P(B|A)$.   ▪

E X A M P L E   3.10   A company found that 85% of the persons selected for its trainee program completed the course. Of these, 60% became productive salespersons, compared with only 10% of those trainees who did not complete the trainee program.

a   What is the probability that a person who enters the trainee program becomes a productive salesperson?

b   If a salesperson who entered the trainee program is deemed to be productive, what is the probability that this person completed the trainee program?

Solution   Define the following events:

A :   the trainee completes the trainee program

B :   the trainee becomes a productive salesperson

Use a tree diagram as in Figure 3.2 to display the appropriate probabilities. The second-step probabilities given in parentheses are conditional, given the first step.

a   A person entering the trainee program either will or will not complete the program. Thus,

$$P(B) = P(AB) + P(\overline{A}B) = .5100 + .0150 = .5250$$

F I G U R E **3.2**   Tree diagram for Example 3.10

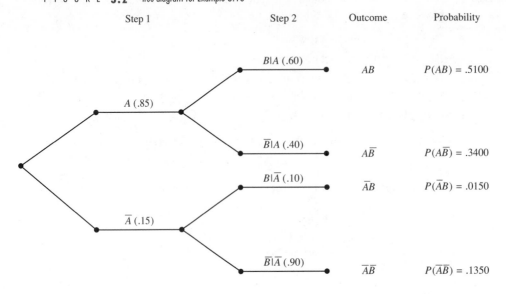

| | Step 1 | Step 2 | Outcome | Probability |

Step 1   Step 2   Outcome   Probability

$B|A$ (.60) AB $P(AB) = .5100$

A (.85)

$\bar{B}|A$ (.40) $A\bar{B}$ $P(A\bar{B}) = .3400$

$B|\bar{A}$ (.10) $\bar{A}B$ $P(\bar{A}B) = .0150$

$\bar{A}$ (.15)

$\bar{B}|\bar{A}$ (.90) $\bar{A}\bar{B}$ $P(\bar{A}\bar{B}) = .1350$

b In order to find $P(A|B)$, we need $P(AB) = .5100$ and $P(B) = .5250$ from part (a). Then

$$P(A|B) = \frac{P(AB)}{P(B)} = \frac{.5100}{.5250} = .9714 \quad \blacksquare$$

Tips on Problem Solving

Calculating the probability of an event: event-composition approach

- Use the following steps to calculate the probability of an event using the event-composition approach:

1 Define the experiment.

2 Clearly visualize the nature of the simple events. Identify a few to clarify your thinking.

3 When possible, write an equation expressing the event of interest (say, A) as a composition of two or more events using either or both of the two forms of composition (unions and intersections). Note that this equates point sets. Make certain that the event implied by the composition and event A represent the same set of simple events.

4 Apply the Additive and Multiplicative Laws of Probability to step 3 and find $P(A)$.

- Be careful with step 3. You can often form many compositions that will be equivalent to event A. The trick is to form a composition in which all the probabilities appearing in step 4 will be known. Visualize the results of step 4 for any composition and select the one for which the component probabilities are known.

- Always write down letters to represent events described in an exercise. Then write down the probabilities that are given, and assign them to events. Identify the probability that is requested in the exercise. This can help you arrive at the appropriate event composition.

Exercises

Basic Techniques

3.12 An experiment can result in one of five equally likely simple events. Events A, B, and C include the following simple events:

$$A: \quad E_1, E_3 \qquad B: \quad E_1, E_2, E_4, E_5 \qquad C: \quad E_3, E_4$$

List the simple events in the following events, and find their respective probabilities.

a S		**b** A	**c** B
d C		**e** $\overline{A}$	**f** $\overline{B}$
g AB		**h** AC	**i** $A\mid B$
j $A \cup B$		**k** $A \cup C$	**l** $A\mid C$

3.13 Refer to Exercise 3.12. Are events A and B mutually exclusive? Independent? Explain.

3.14 An experiment generates a sample space containing eight simple events E_1, E_2, ..., E_8, with $P(E_1) = .1$, $P(E_2) = P(E_3) = P(E_4) = .05$, $P(E_5) = .3$, $P(E_6) = .2$, $P(E_7) = .1$, and $P(E_8) = .15$. The events A and B are

$$A: \quad E_1, E_4, E_6 \qquad B: \quad E_3, E_4, E_5, E_6, E_7$$

a Find $P(A)$.

b Find $P(B)$.

c Find $P(AB)$.

d Find $P(A \cup B)$.

e Are events A and B mutually exclusive? Explain.

f Find $P(A\mid B)$.

g Find $P(B\mid A)$.

h Are A and B independent events? Explain.

3.15 An experiment can result in event A with probability $P(A) = .2$ and event B with probability $P(B) = .6$. The probability of the intersection of A and B is $P(AB) = .15$.

a Construct a probability table for this experiment.

b Use the probability table in part (a) to find $P(\overline{A}B)$, $P(A\mid\overline{B})$, and $P(A \cup B)$.

c Are A and B independent events? Explain.

Applications

3.16 New orders for a company's products vary in dollar size according to the following probability distribution.

Size of sale ($)	0–1000	1001–2000	2001–3000	3001–4000	4001–5000
Probability	.10	.35	.25	.20	.10

a Find the probability that a new order will exceed $2000.

b Find the probability that a new order will be $2000 or less, given that the order exceeds $1000.

c Find the probability that a new order will be larger than $3000, given that the sale exceeds $2000.

3.17 Advertising researchers are always interested to know whether consumers find their advertisements believable, especially when the advertisements contain celebrity endorsements, hidden-camera interviews, or claims that products are "new and improved." The data in the following table represent a small portion of a survey conducted by the Roper Organization. The table indicates the proportion of adults who find celebrity-endorsement ads believable, listed according to the adult's level of educational attainment.

	Less Than High School	High School Graduate	Some College	College Graduate
Believable	.27	.27	.25	.18

Source: American Demographics, December 1990, p. 14.

Suppose that a single adult is randomly selected from among those listed in the survey.

a If the proportion of college graduates in the surveyed group is .24, what is the probability that the chosen adult will be a college graduate who does not believe the ad?

b If the chosen adult has had some college, what is the probability that she does not believe the ad?

c If the proportion of adults in the surveyed group who have never been to college is .4, what is the probability that the chosen adult will not have gone to college and believes the ad?

[*Hint:* Entries in the table are the conditional probabilities of believing the ad, given the adult's level of educational attainment.]

3.18 A recent survey of American businesses indicated that 27% maintain a parental-leave program for parents of newborn children. Of these, one-third offer some kind of salary continuation during the leave, and three-quarters continue to subsidize health care (*Journal of Accountancy,* November 1990, p. 23).

a What is the probability that a randomly chosen business offers parental leave with some form of salary continuation?

b What is the probability that a randomly chosen business offers parental leave without continuing to subsidize health-care costs?

3.19 In a survey involving 100 cars, each vehicle was classified according to whether or not it had antilock brakes and whether or not it had been involved in an accident in the past year.Suppose that one of these cars is randomly selected for inspection.

a What is the probability that the car has been involved in an accident in the past year?

b What is the probability that the car has not been in an accident and has antilock brakes?

c Given that the car has been involved in an accident in the past year, what is the probability that it has antilock brakes?

Table for Exercise 3.19

	Antilock Brakes	No Antilock Brakes
Accident	.03	.12
No Accident	.40	.45

Source: Data adapted from *Consumers' Research*, March 1994.

3.20   An article in *Consumers' Research* ("Antilock Brakes," 1994) states that 43% of all 1993 model cars came equipped with antilock brakes. Suppose three 1993 model cars are selected at random for inspection.

a   Given that the first car chosen has antilock brakes, what is the probability that the second car chosen has antilock brakes?

b   Are the two events described in part (a) independent? Explain.

c   What is the probability that all three cars chosen have antilock brakes?

d   What is the probability that only one of the three cars chosen has antilock brakes?

3.21   A survey was conducted to assess the effect of the information superhighway on businesses in the United States. Based on this survey of senior marketing executives, 40% said that they would use the information superhighway to interact directly with customers, 36% said they would not, and 24% didn't know.

a   If a senior marketing executive is chosen at random, what is the probability that she would use the information superhighway to interact directly with customers?

b   If two senior marketing executives are randomly chosen, what is the probability that only one would use the information superhighway to directly interact with customers?

3.22   An article in *Consumer Reports* ("Ratings: Interior Latex Paints," 1994) ranked 35 brands of interior latex paint using the qualitative classification shown below.

Rating	Number of Brands
Excellent	2
Very good	21
Good	11
Fair	1
Poor	0

Suppose that a consumer selects one of the 35 brands at random.

a   What is the probability that the consumer selects an "excellent" brand?

b   What is the probability that the consumer selects a brand that is rated at least "good"?

c   What is the probability that the consumer selects a brand that is *not* rated "very good" or "excellent"?

d   If the consumer selects two *different* brands to compare, what is the probability that both of the brands were rated "very good"?

3.23   A sprinkler system used on commercial buildings is designed so that each sprinkler can be activated by two independent devices. The sprinkler will function if either of the two devices (or both) is activated. The reliability of the first device (the probability that it is activated when a particular temperature is reached) is .91, while the reliability of the second device is .95. What is the probability that the sprinkler will function properly when the particular temperature is reached?

3.5 Bayes' Rule and Conditional Probability (Optional)

Conditional probabilities allow us to update probabilities by using information as it becomes available. For example, we may know a priori, or before the fact, that all of the items used in a production process are provided by three suppliers, and that supplier S_1 provides 20% of these items, supplier S_2 provides 30%, and supplier S_3 provides the remaining 50%. Therefore, if no other information is available and an item is randomly selected from the production process, the probability that the item came from supplier S_1 is .20, from supplier S_2 is .30, and from supplier S_3 is .50.

From past performance, the percentages of defective items supplied by these three suppliers are .05, .02, and .01, respectively. If an item randomly selected from the production process is found to be defective, what is the probability that the item was supplied by S_1? by S_2? by S_3? That is, how does this information change the *prior* probabilities that the item came from one of the three suppliers?

If D is the event that a defective item was observed, then any one of the three suppliers may have provided the item. We wish to update the *prior* probabilities, $P(S_i)$, using conditional probabilities in the form $P(S_i|D)$. The relevant information is given in the tree diagram in Figure 3.3. The second-step probabilities are conditional, given the first step. For example, the probability of a defective item, given supplier S_1, is $P(D|S_1) = .05$.

Suppose we wish to find the *posterior* probabilities that have been updated with the sample information. For example, to find $P(S_1|D)$, we have

$$P(S_1|D) = \frac{P(S_1 D)}{P(D)}$$

From the tree diagram in Figure 3.3, $P(D)$ is the sum of the three outcomes resulting in D. Hence,

$$
\begin{aligned}
P(D) &= P(S_1 D) + P(S_2 D) + P(S_3 D) \\
&= P(S_1)P(D|S_1) + P(S_2)P(D|S_2) + P(S_3)P(D|S_3) \\
&= .010 + .006 + .005 = .021
\end{aligned}
$$

Since $P(S_1 D) = P(S_1)P(D|S_1) = .20(.05) = .010$, then

$$P(S_1|D) = \frac{.010}{.021} = .476$$

Furthermore, the two remaining posterior probabilities are

$$P(S_2|D) = \frac{.006}{.021} = .286 \qquad \text{and} \qquad P(S_3|D) = \frac{.005}{.021} = .238$$

By use of the *prior* probabilities, an item was most likely to have been supplied by supplier S_3. By use of the *posterior* probabilities, the sampled item was most likely to have come from supplier S_1.

The method for calculating the posterior probabilities is generally known as *Bayes' Rule,* named for the English mathematician Thomas Bayes.

F I G U R E **3.3**   Tree diagram for text example

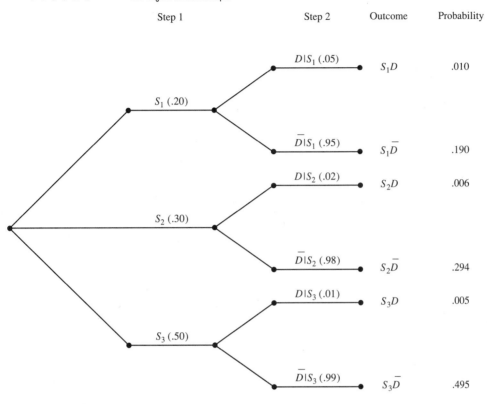

Bayes' Rule

Let S_1, S_2, ..., S_k represent the k mutually exclusive, only possible states of nature with prior probabilities $P(S_1)$, $P(S_2)$, ..., $P(S_k)$. If an event A occurs, the posterior probability of S_i given A is the conditional probability

$$P(S_i|A) = \frac{P(S_i)P(A|S_i)}{\sum_{j=1}^{k} P(S_j)P(A|S_j)}$$

for $i = 1, 2, ..., k$.

E X A M P L E   **3.11**   A department store is considering adopting a new credit management policy in an attempt to reduce the number of credit customers that default on their payments. The credit manager has suggested that in the future credit should be discontinued to any customer who has twice been a week or more late with his monthly installment payment. She supports her claim by noting that past credit records show that 90% of all those defaulting on their payments were late with at least two monthly payments.

Suppose from our own investigation we have found that 2% of all credit customers actually default on their payments and that 45% of those who have not defaulted have had at least two late monthly payments. Find the probability that a customer with two or more late payments will actually default on his payments, and, in light of this probability, criticize the credit manager's credit plan.

Solution   Let the events L and D be defined as follows:

event L :   a credit customer is two or more weeks late with at least two monthly payments

event D :   a credit customer defaults on his payments

and let $\overline{D}$ denote the complement of event D. We seek the conditional probability

$$P(D|L) = \frac{P(DL)}{P(L)} = \frac{P(L|D)P(D)}{P(L|D)P(D) + P(L|\overline{D})P(\overline{D})}$$

From the information given in the problem description, we find that

$$P(D|L) = \frac{(.90)(.02)}{(.90)(.02) + (.45)(.98)} = \frac{.0180}{.0180 + .4410} = .0392$$

Therefore, if the credit manager's plan is adopted, the probability is only about .04—or about 1 in 25—that a customer who loses his credit privileges would actually have defaulted on his payments. Unless management would consider it worthwhile to detect one prospective defaulter at the expense of losing 24 good credit customers, the credit manager's plan would be a poor business policy.   ∎

Exercises

Basic Techniques

3.24   Suppose that three mutually exclusive states of nature, S_1, S_2, S_3, can exist with probabilities

$$P(S_1) = .4 \qquad P(S_2) = .5 \qquad P(S_3) = .1$$

and that given these states of nature, an event I can occur with probabilities

$$P(I|S_1) = .1 \qquad P(I|S_2) = .3 \qquad P(I|S_3) = .2$$

Given that event I is observed, find $P(S_1|I)$, $P(S_2|I)$, and $P(S_3|I)$.

3.25 Suppose that four mutually exclusive states of nature, S_1, S_2, S_3, and S_4, can exist with probabilities

$$P(S_1) = .1 \qquad P(S_2) = .4 \qquad P(S_3) = .3 \qquad P(S_4) = .2$$

and that given these states of nature, an event I can occur with probabilities

$$P(I|S_1) = .6 \qquad P(I|S_2) = .2 \qquad P(I|S_3) = .2 \qquad P(I|S_4) = .5$$

Given that event I is observed, find $P(S_1|I)$, $P(S_2|I)$, $P(S_3|I)$, and $P(S_4|I)$.

Applications

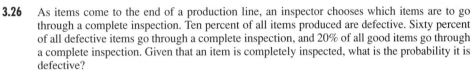

3.26 As items come to the end of a production line, an inspector chooses which items are to go through a complete inspection. Ten percent of all items produced are defective. Sixty percent of all defective items go through a complete inspection, and 20% of all good items go through a complete inspection. Given that an item is completely inspected, what is the probability it is defective?

3.27 With the recent threats of terrorism, airports are increasingly concerned about weapons detection at boarding gates. The detection rates for weapons carried on the person or in their carry-on luggage must be extremely high. In a particular city, airport A handles 50% of all airline traffic, while airports B and C handle 30% and 20%, respectively. The detection rates at the three airports are .99, .95, and .80, respectively. If a passenger at one of the airports is found to be carrying a weapon through the boarding gate, what is the probability that the passenger is using airport A? airport C?

3.28 Suppose that 5% of all people filing the long income tax form seek deductions that they know to be illegal and an additional 2% will incorrectly list deductions because of lack of knowledge of the income tax regulations. Of the 5% guilty of cheating, 80% will deny knowledge of the error if confronted by an investigator. If a filer of the long form is confronted with an unwarranted deduction and he denies knowledge of the error, what is the probability that he is guilty?

3.6 Discrete Random Variables and Their Probability Distributions

In Chapter 2, we defined a *variable* as a characteristic that changes or varies over time and/or for different individuals or objects under consideration. *Quantitative variables* give rise to numerical measurements, while *qualitative variables* give rise to categorical measurements. However, even qualitative data can give rise to numerical measurements, if we count the number of individuals or elements in each of the defined categories.

If we denote the variable being measured by x, then the value that x assumes changes or varies, depending on the particular outcome of the experiment. For example, suppose we let x be the daily production in a manufacturing plant. The variable x can take on a number of different values depending on the day on which we measure x. Similarly, the number of sales closed per day by a salesperson is a quantitative variable x whose value depends on the day on which the measurement is taken. If the value of a variable *x depends on the random outcome of an experiment,* we refer to the variable *x* as a **random variable.**

DEFINITION ▪

A variable *x* is a **random variable** if the value that it assumes, corresponding to the outcome of an experiment, is a chance or random event. ▪

Observing the number of defects on a randomly selected piece of furniture, selecting a college applicant at random and observing the person's SAT score, and measuring the number of telephone calls received by a crisis intervention hotline during a randomly selected time period give rise to random numerical events.

Quantitative random variables are classified as either *discrete* or *continuous,* according to the values that *x* may assume. The distinction between discrete and continuous random variables is important, since different probability models are required for each type of variable. We will focus our attention on discrete random variables in the remainder of this chapter. Continuous random variables are the subject of Chapter 5.

DEFINITION ▪

The **probability distribution** for a discrete random variable is a formula, table, or graph that provides $p(x)$, the probability associated with each of the values of *x*. ▪

The events associated with different values of *x* cannot overlap, because one and only one value of *x* is assigned to each simple event; hence, the values of *x* represent mutually exclusive numerical events. Summing $p(x)$ over all values of *x* equals the sum of the probabilities of all simple events and hence equals 1.

Requirements for a Discrete Probability Distribution

1 $0 \leq p(x) \leq 1$

2 $\displaystyle\sum_{\text{all } x} p(x) = 1$

EXAMPLE **3.12** Consider an experiment that consists of tossing two coins, and let *x* equal the number of heads observed. Find the probability distribution for *x*.

Solution The simple events for this experiment with their respective probabilities are as follows:

Simple Event	Coin 1	Coin 2	$P(E_i)$	x
E_1	H	H	1/4	2
E_2	H	T	1/4	1
E_3	T	H	1/4	1
E_4	T	T	1/4	0

Because E_1 is associated with the simple event "observe a head on coin 1 and a head on coin 2," we assign it the value $x = 2$. Similarly, we assign $x = 1$ to event E_2, and so on. The probability of each value of x can be calculated by adding the probabilities of the simple events in that numerical event. The numerical event $x = 0$ contains one simple event, E_4; $x = 1$ contains two simple events, E_2 and E_3; and $x = 2$ contains one event, E_1. The values of x with respective probabilities are given in Table 3.5. Observe that

$$\sum_{x=0}^{2} p(x) = 1 \quad \blacksquare$$

T A B L E **3.5**

Probability distribution for x (x = number of heads)

x	Simple Events in x	$p(x)$
0	E_4	1/4
1	E_2, E_3	1/2
2	E_1	1/4
	$\sum_{x=0}^{2} p(x) =$	1

The probability distribution in Table 3.5 can be presented graphically in the form of the relative frequency histogram (see Section 2.4).[†] The histogram for the random variable x would contain three classes, corresponding to $x = 0$, $x = 1$, and $x = 2$. Since $p(0) = 1/4$, the theoretical relative frequency for $x = 0$ is 1/4; $p(1) = 1/2$, and hence the theoretical relative frequency for $x = 1$ is 1/2. The histogram is given in Figure 3.4.

You have probably noticed the similarity between the probability distribution for a discrete random variable and the relative frequency histograms discussed in Chapter 2. The difference is that the relative frequency histograms are constructed for a *sample* of n measurements drawn from the population, while the probability histogram is constructed as a model for the *entire population* of measurements. Just

---

[†]The probability distribution in Table 3.5 can also be presented using a formula, which is given in Section 4.2.

F I G U R E **3.4**
Probability histogram showing
$p(x)$ for Example 3.12

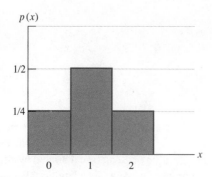

as we calculated the mean and standard deviation for a sample of n measurements to measure the location and the variability of the relative frequency distribution, we can also calculate a mean and standard deviation to describe the probability distribution for a random variable x. The population mean, which measures the average value of x in the population, is also called the **expected value** of the random variable x.

The method for calculating the population mean or expected value of a random variable can be more easily understood by considering an example. Let x equal the number of heads observed in the toss of two coins. For convenience, $p(x)$ is given as

x	0	1	2
$p(x)$	1/4	1/2	1/4

Suppose the experiment is repeated a large number of times—say, $n = 4{,}000{,}000$ times. Intuitively, we would expect to observe approximately 1 million zeros, 2 million ones, and 1 million twos. Then the average value of x would equal

$$\frac{\text{Sum of measurements}}{n} = \frac{1{,}000{,}000(0) + 2{,}000{,}000(1) + 1{,}000{,}000(2)}{4{,}000{,}000}$$

$$= \frac{1{,}000{,}000(0)}{4{,}000{,}000} + \frac{2{,}000{,}000(1)}{4{,}000{,}000} + \frac{1{,}000{,}000(2)}{4{,}000{,}000}$$

$$= (1/4)(0) + (1/2)(1) + (1/4)(2)$$

Note that the first term in this sum is equal to $(0)p(0)$, the second is equal to $(1)p(1)$, and the third is equal to $(2)p(2)$. The average value of x, then, is

$$\sum_{x=0}^{2} xp(x) = 1$$

This result is not an accident, and it provides some intuitive justification for the definition of the expected value of a discrete random variable x.

DEFINITION ▪

> Let x be a discrete random variable with probability distribution $p(x)$. The mean or **expected value of x** is given as
>
> $$\mu = E(x) = \sum_{x} x p(x)$$
>
> where the elements are summed over all values of the random variable x. ▪

EXAMPLE   3.13   In a lottery conducted to benefit a local charity, 10,000 tickets are to be sold at $5 each. The prize is a $12,000 automobile. If you purchase two tickets, what is your expected gain?

Solution   Your gain x may take one of two values. You will either lose $10 (i.e., your gain will be −$10) or win $11,990, with probabilities .9998 and .0002, respectively. The probability distribution for the gain x is as follows:

x	$p(x)$
−$10	.9998
$11,990	.0002

The expected gain will be

$$E(x) = \sum_{x} x p(x)$$
$$= (-\$10)(.9998) + (\$11{,}990)(.0002)$$
$$= -\$7.60$$

Recall that the expected value of x is the average of the theoretical population that would result if the lottery were repeated an infinitely large number of times. If it were, your average or expected gain per lottery would be a loss of $7.60. ▪

EXAMPLE   3.14   Determine the yearly premium for a $1000 insurance policy covering an event that, over a long period of time, has occurred at the rate of 2 times in 100. Let x equal the yearly financial gain to the insurance company resulting from the sale of the policy, and let C equal the unknown yearly premium. We will calculate the value of C such that the expected gain, $E(x)$, will equal zero. Then C is the premium required to break even. To this figure the company will add administrative costs and profit.

Solution   The first step in the solution is to determine the values that the gain x can take and then to determine $p(x)$. If the event does not occur during the year, the insurance company will gain the premium of $x = C$ dollars. If the event does occur, the gain will be negative. That is, the company will lose $1000 less the premium of C dollars already collected. Then $x = -(1000 - C)$ dollars. The probabilities associated with these two values of x are 98/100 and 2/100, respectively. The probability distribution for the gain is as follows:

$x =$ **gain**	$p(x)$
C	$\dfrac{98}{100}$
$-(1{,}000 - C)$	$\dfrac{2}{100}$

Since we want the insurance premium C such that, in the long run (for many similar policies), the mean gain will equal zero, we will set the expected value of x equal to zero and solve for C. Then

$$E(x) = \sum_x x p(x)$$

$$= C\left(\frac{98}{100}\right) + [-(1000 - C)]\left(\frac{2}{100}\right) = 0$$

or

$$\frac{98}{100}C + \left(\frac{2}{100}\right)C - 20 = 0$$

Solving this equation for C, we obtain

$$C = \$20$$

Thus if the insurance company were to charge a yearly premium of \$20, the average gain calculated for a large number of similar policies would equal zero. The actual premium would equal \$20 plus administrative costs and profit.   ∎

In Chapter 2, we defined the population variance σ^2 to be the average of the squares of the deviations of the measurements from their mean. Because taking an expectation is equivalent to "averaging," we define the **variance** and the **standard deviation** as follows.

DEFINITION ▪

Let x be a discrete random variable with probability distribution $p(x)$ and expected value $E(x) = \mu$. The **variance of x** is

$$\sigma^2 = E[(x - \mu)^2] = \sum_x (x - \mu)^2 p(x)$$

where the summation is over all values of the random variable x.[†]   ∎

DEFINITION   ∎   The **standard deviation** σ of a random variable x is equal to the square root of its variance.   ∎

EXAMPLE   **3.15**   Let x be a random variable with the probability distribution given in the following table:

x	0	1	2	3	4	5
$p(x)$	.10	.40	.20	.15	.10	.05

Find μ, σ^2, and σ. Graph $p(x)$ and locate the interval $\mu \pm 2\sigma$ on the graph. What is the probability that x will fall in the interval $\mu \pm 2\sigma$?

Solution

$$\mu = E(x) = \sum_{x=0}^{5} xp(x)$$

$$= (0)(.10) + (1)(.40) + \cdots + (4)(.10) + (5)(.05)$$

$$= 1.90$$

$$\sigma^2 = E[x - \mu)^2] = \sum_{x=0}^{5}(x - \mu)^2 p(x)$$

$$= (0 - 1.9)^2(.10) + (1 - 1.9)^2(.40) + \cdots + (5 - 1.9)^2(.05)$$

$$= 1.79$$

and

$$\sigma = \sqrt{\sigma^2} = \sqrt{1.79} = 1.34$$

The interval $(\mu \pm 2\sigma)$ is $[1.90 \pm (2)(1.34)]$, or $-.78$ to 4.58.

The graph of $p(x)$ and the interval $\mu \pm 2\sigma$ are shown in Figure 3.5. You can see that $x = 0, 1, 2, 3, 4$ fall in the interval. Therefore,

$$P(\mu - 2\sigma < x < \mu + 2\sigma) = p(0) + p(1) + \cdots + p(4)$$

$$= .10 + .40 + .20 + .15 + .10$$

$$= .95   ∎$$

[†]It can be shown (proof omitted) that

$$\sigma^2 = \sum_{x}(x - \mu)^2 p(x) = \sum_{x} x^2 p(x) - \mu^2$$

This result is analogous to the shortcut formula for s^2 given in Chapter 2.

F I G U R E **3.5**
The probability histogram for
$p(x)$ for Example 3.15

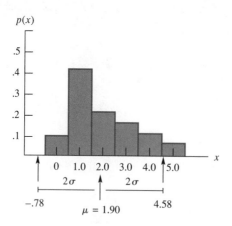

Tips on Problem Solving

1   To find the expected value of a discrete random variable x, construct a table containing three columns, the first for x and the second for $p(x)$. Then multiply each x value by its corresponding probability and enter the results in the third column. The sum of this third column, the sum of $xp(x)$, will give you the expected value of x.

2   To find σ^2 for a discrete random variable x, start with a table containing four columns, the first for x, the second for $(x - \mu)^2$, the third for $p(x)$, and the fourth for the cross products, $(x - \mu)^2 p(x)$. First, calculate the value of $(x - \mu)^2$ for each value of x and enter the results in column 2. Then obtain the cross product $(x - \mu)^2 p(x)$ for each value of x, and enter the results in column 4. The sum of column 4 will give you the value of σ^2.

Exercises

Basic Techniques

3.29   A random variable x possesses the following probability distribution:

x	0	1	2	3	4	5
$p(x)$	.1	.3	.4	.1	?	.05

   a   Find $p(4)$.

   b   Construct a probability histogram to describe $p(x)$.

3.30   Refer to Exercise 3.29.

   a   What is the probability that $x = 2$? at least 2?

   b   What is the probability that x is no greater than 3?

   c   Find $\mu = E(x)$ and σ^2.

3.31   Which of the following distributions do not represent a probability distribution? Explain.

a

x	$p(x)$
1.5	.60
2.0	.25
6.5	.15

b

x	$p(x)$
0	−1
1	2
2	5
3	4

c

x	$p(x)$
1	.2
2	.4
3	.2

3.32   Let x be a discrete random variable with the probability distribution given in the following table:

x	1	2	3	4	5	6	7
$p(x)$	.05	.2	.35	.2	.1	.05	.05

   a   Find μ, σ^2, and σ.

   b   Construct a probability histogram for $p(x)$.

   c   Locate the interval $(\mu \pm 2\sigma)$ on the x axis of the histogram. What is the probability that x will fall in this interval?

   d   If you were to select a very large number of values of x from the population, would most fall in the interval $\mu \pm 2\sigma$? Explain.

Applications

 3.33   A $50,000 diamond is insured to its total value by paying a premium of D dollars. If the probability of theft in a given year is estimated to be .01, what premium should the insurance company charge if it wants the expected gain to equal $1000?

 3.34   The maximum patent life for a new drug is 17 years. Subtracting the length of time required for testing and approval of the drug by the Food and Drug Administration, you obtain the actual patent life of the drug—that is, the length of time that a company has to recover research and development costs and make a profit. Suppose that the distribution of the length of patent life for new drugs is as shown in the following table:

Years, x	3	4	5	6	7	8	9	10	11	12	13
$p(x)$	.03	.05	.07	.10	.14	.20	.18	.12	.07	.03	.01

   a   Find the expected number of years of patent life for a new drug.

   b   Find the standard deviation of x.

   c   Find the probability that x falls in the interval $(\mu \pm 2\sigma)$.

 3.35   When asked whether members of Congress should be limited to 12 years in office, 61% of Americans questioned agree with a 12-year congressional term limit (Galifianakis, 1994).

Suppose that three people were selected at random and asked whether they agree with a 12-year congressional term limit. Let x be the number who agree with a 12-year congressional term limit.

a   Find the probability distribution for x.

b   Construct a probability histogram for x.

c   What is the probability that at least two people agree with a 12-year congressional term limit?

d   Find the mean and standard deviation of x.

 3.36   A manufacturing representative is considering the option of taking out an insurance policy to cover possible losses incurred by marketing a new product. If the product is a complete failure, the representative feels that a loss of $80,000 would be incurred; if it is only moderately successful, a loss of $25,000 would be incurred. Insurance actuaries have determined from market surveys and other available information that the probabilities that the product will be a failure or only moderately successful are .01 and .05, respectively. Assuming that the manufacturing representative would be willing to ignore all other possible losses, what premium should the insurance company charge for the policy in order to break even?

 3.37   A manufacturing company ships its product in two different sizes of truck trailers, an $8 \times 10 \times 30$ and an $8 \times 10 \times 40$. If 30% of its shipments are made by using the 30-foot trailer and 70% by using the 40-foot trailer, find the mean volume shipped per trailerload (assume that the trailers are always full).

CASE STUDY REVISITED

3.7   MORE ON BABY BOOMERS AND DEPARTMENT STORES

The study by Rothenberg, Inc., given in the case study at the beginning of this chapter has several implications for department store managers who wish to step up sales as the baby-boom generation moves into middle age. There is no guarantee that these people will change their shopping habits and suddenly prefer department stores as they grow older. Since younger-aged groups actually spend more, it is important to cater to those in the 25–54 age segment who spend the greatest actual dollars.

In addition to the amounts spent for each age segment, Rothenberg reported that the average department store customer spends $100 a month at department stores. How is that figure calculated using the information we are given? First we see that the share of retail sales actually represents a probability distribution for the age segments of the population. In other words, the probability that a retail customer is between 18 and 24 years of age is .15, while the probability that a customer is 65 years or older is .05. Notice that all these probabilities sum to 1.00. In addition, for each age group, we are given the amount spent monthly in department store sales per customer. If we define x to be the amount spent per customer for a given age group, we have a probability distribution for x given by the following table:

x	$p(x)$
$ 50	.15
108	.27
126	.25
116	.16
89	.12
53	.05

Using this probability distribution for x, we can calculate $E(x)$, the average department store retail spending per customer. Hence,

$$E(x) = \sum xp(x) = 50(.15) + 108(.27) + \cdots + 53(.05)$$
$$= \$100.05$$

a value that agrees with that reported by Rothenberg.

3.8 Remarks

As we proceed through the next several chapters, we will study probability distributions for discrete and continuous random variables that act as models for various business applications. Means and standard deviations of random variables comprise the basic summary information upon which inferences about underlying population parameters are based.

Supplementary Exercises

3.38   A jar contains two black and two white balls. Suppose the balls are thoroughly mixed, and then two are selected from the jar, one at a time without replacement; that is, the first ball is not put back in the jar before drawing the second ball.

  a   List the simple events for the experiment and assign appropriate probabilities to each.

  b   What is the probability that the first ball is black?

  c   What is the probability that the second ball is black, given that the first ball is white?

  d   What is the probability that one black and one white ball are selected?

3.39   Refer to Exercise 3.38 and let x be the number of white balls selected.

  a   Assign the appropriate value of x to each simple event.

  b   Calculate the values of $p(x)$ and display them in tabular form. Show that $\sum_{x=0}^{2} p(x) = 1$.

  c   Construct a probability histogram for $p(x)$.

3.40   Refer to Exercises 3.38 and 3.39. Simulate the experiment by actually drawing two balls (coins, etc.) from a jar that contains two "black" and two "white" balls. Repeat the drawing process 100 times, each time recording the value of x that was observed. Construct a relative frequency histogram for the 100 observed values of x, and compare your result to the probability histogram in Exercise 3.39(c).

3.41   Refer to Exercise 3.38. From the jar containing two black and two white balls, the two balls are now selected *with replacement;* that is, the first ball is returned to the jar before the second ball is selected.

a   List the simple events for the experiment and assign appropriate probabilities to each.

b   What is the probability that the second ball is black?

c   What is the probability that the second ball is black, given that the first ball is white?

d   Calculate the probability distribution for x, the number of white balls in the selection. Compare the results to those of Exercise 3.39(b).

3.42   Moreno Valley, California, a bedroom community lying east of Los Angeles, has experienced rapid growth in the last decade, mainly because of the affordable housing and the 1-hour commute time to many parts of Los Angeles. The age distribution of its residents 18 years or older was reported as follows:

Age	Proportion
18–24 years	.15
25–34 years	.28
35–44 years	.21
45–54 years	.13
55–64 years	.11
65 + years	.12

Source: Press-Enterprise, Riverside, Calif.
(Oct. 12, 1992).

Suppose four persons 18 years or older are randomly selected from this population.

a   What is the probability that all four will be under 25 years of age?

b   What is the probability that all four will be 65 years or older?

c   What is the probability that exactly one person will be under 25 and the remaining three will be 25 years or older?

3.43   Americans are becoming more and more aware of the threat to our environment due to increasing amounts of solid waste that are dumped into our landfills. In a survey conducted by *Better Homes and Gardens* (Arthur, 1990), 30% of shoppers say that individuals have the primary responsibility for solving the garbage problem; 18% say that it is the manufacturers' primary responsibility; and 18% say that it is the government's responsibility. Suppose that two shoppers are randomly selected and asked to assess whose primary responsibility the garbage problem is.

a   What is the probability that both shoppers feel that it is the government's responsibility?

b   What is the probability that one shopper blames the manufacturers and one blames individual consumers?

c   What is the probability that neither shopper blames the manufacturers?

3.44   Identify the following as discrete or continuous random variables.

a   The number of bank failures in a given year.

b   The floor-space area in a new office building.

c   The number of people waiting for treatment at a hospital emergency room.

d   The total points scored in a football game.

e   The number of claims received by an insurance company during a day.

3.45 The probability distribution for a discrete random variable x is as shown in the following table:

x	0	1	2	3	4	5
$p(x)$	.05	.3	.3	.2	.1	.05

    a Find $E(x)$.

    b Find σ^2.

    c Sketch $p(x)$ and locate the interval $(\mu \pm 2\sigma)$ on the graph.

    d Find the probability that x falls in the interval $(\mu \pm 2\sigma)$.

3.46 Records show that 30% of all patients admitted to a medical clinic fail to pay their bills and eventually the bills are forgiven. Suppose that $n = 4$ new patients represent a random selection from the large set of prospective patients served by the clinic. Find the following probabilities.

    a That all the patients' bills will eventually have to be forgiven.

    b That one bill will have to be forgiven.

    c That none will have to be forgiven.

3.47 Refer to Exercise 3.46 and let x equal the number of patients in the sample of $n = 4$ whose bills will have to be forgiven. Construct a probability histogram for $p(x)$.

3.48 A random variable x can assume five values, 0, 1, 2, 3, and 4. A portion of the probability distribution follows.

x	0	1	2	3	4
$p(x)$	.1	.3	.3	?	.1

    a Find $p(3)$.

    b Construct a probability histogram for $p(x)$.

    c Simulate the experiment by marking ten poker chips (or coins), one with a 0, three with a 1, three with a 2, and so on. Mix the chips thoroughly, draw one, and record the observed value of x. Repeat the process 100 times. Construct a relative frequency histogram for the 100 values of x, and compare your result to the probability histogram in part (b).

3.49 An experiment consists of tossing a single die and observing the number of dots shown on the upper face. The events A, B, and C are defined as follows:

$$A: \quad \text{observe a number less than 4}$$
$$B: \quad \text{observe a number less than or equal to 2}$$
$$C: \quad \text{observe a number greater than 3}$$

List the simple events in the following events, and find their respective probabilities.

    a S **b** A **c** B **d** C **e** AB

    f AC **g** BC **h** $A \cup B$ **i** $A \cup C$ **j** $B \cup C$

3.50 Refer to Exercise 3.49.

    a Are events A and B mutually exclusive? A and C? B and C?

    b Find $P(A \cup B)$ using the additive rule of probability. Compare the result with the probability calculated in Exercise 3.49(h).

    c Use the multiplicative rule of probability to find $P(A|B)$ and $P(A|C)$.

    d Are A and B independent events? A and C? Explain.

3.51 Refer to Exercise 3.49.

a Find the probability distribution of x, the number of dots on the upper face.

b Construct a probability histogram for $p(x)$.

c Calculate μ, σ^2, and σ.

d What is the probability that the number observed in a single toss of a die falls in the interval $(\mu \pm 2\sigma)$?

e The probability distribution for the number x observed in the toss of a single die is not mound-shaped—that is, $p(x) = 1/6$. Despite this fact, how does your answer to part (d) agree with the Empirical Rule?

3.52 The number N of residential homes that a fire company can serve depends on the distance r (in city blocks) that a fire engine can cover in a specified (fixed) period of time. If we assume that N is proportional to the area of a circle r blocks from the firehouse, then

$$N = C\pi r^2$$

where C is a constant, $\pi = 3.1416\ldots$, and r, a random variable, is the number of blocks that a fire engine can move in the specified time interval. For a particular fire company, $C = 8$, the probability distribution for r is as shown in the accompanying table, and $p(r) = 0$ for $r \leq 20$ and $r \geq 27$. Find the expected value of N, the number of homes that the fire department can serve.

r	21	22	23	24	25	26
$p(r)$	.05	.20	.30	.25	.15	.05

3.53 Accident records collected by an automobile insurance company give the following information. The probability that an insured driver has an automobile accident is .15; if an accident has occurred, the damage to the vehicle amounts to 20% of its market value with probability .80, 60% of its market value with probability .12, and a total loss with probability .08. What premium should the company charge on a $15,000 car so that the expected gain by the company is zero? Administrative costs will be added to this amount.

3.54 Experience has shown a shipping company that the cost of delivering a small package within 24 hours is $14.80. The company charges $15.50 for shipment but guarantees to refund the charge if delivery is not made within 24 hours. If the company fails to deliver only 2% of its packages within the 24-hour period, what is the expected gain per package?

3.55 According to *Consumer Reports* ("Getting Things Fixed," 1994), products with housings that are welded or permanently sealed—such as electric housewares and personal-care products—are difficult to service, and over half end up in the trash. What happens to broken items? The following table gives information on the disposition of broken shavers and telephone answering machines.

Disposition	Shavers	Answering Machines
Repaired	.34	.31
Thrown away	.34	.30
Kept, but unused	.19	.20
Sold or given away	.13	.19

a What is the probability that a broken shaver is not repaired?

b What is the probability that a shaver is thrown away, given that it was not repaired?

c What is the probability that a broken telephone answering machine is not repaired but is kept unused?

d Given that a broken answering machine was not thrown away, what is the probability that it was repaired?

e What is the probability that only one of a broken shaver and a broken answering machine is repaired?

 3.56 If you wanted to buy another personal computer (PC), would you buy it from the same company? According to Anita Amirrezvani ("Help Is on the Way," 1994), 88% of those customers who purchased a PC from Micron Computer of Nampa, Idaho, would buy from them again because of the company's "can-do" attitude to resolving problems. Suppose four customers who purchased a PC from Micron were ready to purchase another PC.

a What is the probability that all four would purchase another Micron PC?

b What is the probability that none would purchase another Micron PC?

c What is the probability that exactly one would purchase another Micron PC?

3.57 A recent report regarding the ratings of 24 refrigerator brands/models by price ("Ratings: Refrigerators," 1994) was used to create the following table:

	Price Range			
Rating	$600–799	$800–999	$1000–1199	$1200–1399
Excellent	1	3	0	0
Very good	4	5	4	5
Good	1	1	0	0

Suppose one of these 24 refrigerators is chosen at random.

a What is the probability that the refrigerator chosen has been rated as excellent?

b What is the probability that the refrigerator costs between $600 and $999?

c What is the probability that the refrigerator is not rated as excellent?

d What is the probability that the refrigerator is in the $1200–1399 price range and is rated very good?

e What is the probability that the refrigerator is rated excellent, given that the refrigerator is in the $800–999 price range?

f What is the probability that the refrigerator is rated excellent, given that it is in the $1000–1199 price range?

USEFUL DISCRETE PROBABILITY DISTRIBUTIONS

About This Chapter

In this chapter, we introduce several discrete probability distributions that serve as models for measurements made under observational or experimental conditions that arise in the areas of marketing, economics, or general business. The focal point of the chapter is the application of these distributions as models in business and economics.

Contents

DRIVING—A RIGHT
OR A PRIVILEGE?

Americans' abiding love for the automobile has integrated it into the very fabric of our lives. There are few if any days when an American car owner fails to get behind the wheel to drive to work, to carpool children to and from activities, to run errands, to shop, or simply to drive for the pleasure of it. Nonetheless, Americans, both adults and teenagers, feel that a driving license is not a right but a privilege, according to Frank Newport and Leslie McAneny (1993), who surveyed 1003 adults in June and 803 teenagers in September of 1993. They found that 70% of adults polled supported a mandatory physical exam every three years for persons over 65 years of age and that 56% of the teenagers polled supported state laws that would deny driver's licenses to anyone under 21 years of age who drops out of high school. Their report states that the percentages reported for the adults differ from the actual percentage in the population at large by no more than 3 percentage points and that the percentages reported for teenagers differ from the actual percentage in the population at large by no more than 4 percentage points.

How can we be certain that the reported percentages are as accurate as claimed? When polls are conducted using a yes or no response, the obvious questions for a student of statistics to ask are "What statistical model is appropriate in situations such as this?" and, secondly, "How can we use these models to assess the reliability of inferences based on responses to yes or no questions?"

These and other questions relating to inferences will be addressed in Sections 4.2 and 4.3 when we introduce the binomial and Poisson distributions. Remember that inferences are based on means and standard deviations, whose values are determined using sample information. We will return to the question "How reliable is an estimate of the proportion of drivers who favor mandatory physical exams for senior citizens?"

4.1 Introduction

In Chapter 3, we found that random variables defined over a finite or countably infinite number of simple events are called **discrete random variables.** Examples of discrete random variables abound in business and economics, but three discrete probability distributions serve as **models** for a large number of these applications.

These three distributions are **the binomial, the Poisson, and the hypergeometric probability distributions.** In this chapter, we study these distributions and discuss their development as logical models for discrete processes observed in different physical settings.

4.2 The Binomial Probability Distribution

One of the most elementary, useful, and interesting discrete random variables, the binomial random variable, is associated with the coin-tossing experiment described in Examples 3.2 and 3.12. As an illustration, consider a sample survey conducted to determine market acceptance of a new product. Each person interviewed is similar to the toss of a coin, since a person's acceptance of a product is similar to observing a head, and that person's rejection of the product is similar to observing a tail. The difference is that the probability of acceptance for a new product is usually not 1/2.

Similar polls are conducted in the social sciences, industry, and marketing. The sociologist is interested in the proportion of Hispanic citizens registered to vote; the producer of printed circuit boards is interested in the number of boards with at least one defect; an environmental group is interested in the proportion of families actively involved in the recycling of aluminum cans. Although dissimilar in some respects, the experiments described here will often exhibit, to a reasonable degree of approximation, the characteristics of a **binomial experiment.**

DEFINITION ▪

A **binomial experiment** is one that possesses the following properties:

1 The experiment consists of n identical trials.

2 Each trial results in one of two outcomes. For lack of a better nomenclature, we will call the one outcome a success, S, and the other a failure, F.[†]

3 The probability of success on a single trial is equal to p and remains the same from trial to trial. The probability of a failure is equal to $(1 - p) = q$.

4 The trials are independent.

5 We are interested in x, the number of successes observed during the n trials. ▪

EXAMPLE 4.1 Suppose that there are approximately 1 million adults in a certain sales region who are potential buyers for a new product and that an unknown proportion p would purchase the product if it was offered for sale. A sample of 1000 adults will be chosen in such

---

[†]Although it is traditional to call the two possible outcomes of a trial "success" and "failure," they could have been called "head" and "tail," "red" and "white," or any other pair of words. Consequently, the outcome called a success need not be viewed as a success in the ordinary usage of the word.

a way that each of the 1 million in the sales region has an equal chance of being chosen. Each adult in the sample will be asked whether he or she would purchase the product if it was offered for sale. (The ultimate objective of this survey is to estimate the unknown proportion p, a problem that we will learn how to solve in Chapter 7.) Is this a binomial experiment?

Solution   To decide whether this is a binomial experiment, we must see if the sampling satisfies the five characteristics described in the definition.

1   The sampling consists of $n = 1000$ identical trials. One trial represents the selection of a single adult from the 1 million adults in the sales region.

2   Each trial will result in one of two outcomes. A person will state either that she would buy the new product or that she would not. These two outcomes could be associated with the "success" and "failure" of a binomial experiment.

3   The probability of a success will equal the proportion of the 1 million adults who would buy the new product. For example, if 500,000 of the 1 million adults in the region would buy the product, then the probability that the first adult selected would buy the product is $p = .5$. For all practical purposes, this probability will remain the same from trial to trial even though adults selected in the earlier trials are not replaced as the sampling continues.

4   For all practical purposes, the probability of a success on any one trial will be unaffected by the outcome on any of the others (it will remain very close to p).

5   We are interested in the number x of adults in the sample of 1000 who would buy the product.

Because the survey satisfies the five characteristics reasonably well, for all practical purposes it (like many other opinion polls) can be viewed as a binomial experiment.   ∎

EXAMPLE   **4.2**   A purchaser, who has received a shipment containing 20 personal computers, wishes to sample three of the PCs to see whether they are in working order before he unloads the shipment. The 3 nearest PCs are removed for testing and, afterward, are declared either defective or nondefective. Unknown to the purchaser, 2 of the 20 PCs are defective. Is this a binomial experiment?

Solution   As in Example 4.1, we check the sample procedure against the characteristics of a binomial experiment.

1   The experiment consists of $n = 3$ identical trials. Each trial represents the selection and testing of one PC from the total of 20.

2   Each trial results in one of two outcomes. Either a PC is defective (call this a "success") or it is not (a "failure").

3   Suppose that the PCs were randomly loaded into a cargo container, so that any one of the 20 PCs could have been placed near the door. Then the unconditional probability of drawing a defective PC on a given trial will be 2/20.

4   The condition of independence between trials is *not* satisfied because the probability of drawing a defective PC on the second and third trials will be dependent on the outcome of the first trial. For example, if the first trial results in a defective PC, then there is only one defective PC left of the remaining 19 in the shipment. Therefore, the conditional probability of success on trial 2, given a success on trial 1, is 1/19. This result differs from the unconditional probability of a success on the second trial (which is 2/20). Therefore, the trials are dependent and the sampling does not represent a binomial experiment.   ▪

Example 4.2 illustrates an important point. If the sample size n is large relative to the population size N, then the probability of success p will not remain constant from trial to trial. Hence, the trial outcomes will be dependent, and the resulting experiment will not be a binomial experiment. **As a rule of thumb, if $n/N \geq .05$, the resulting experiment will not be binomial.**

The probability distribution for a simple binomial random variable (the number of heads in the tosses of two coins) was derived in Example 3.12. The probability distribution for a binomial experiment consisting of n tosses is derived in exactly the same way, but the procedure is much more complex when the number n of trials is large. We will omit this derivation and simply present the **binomial probability distribution** and its mean, variance, and standard deviation, as shown in the following display.

Binomial Probability Distribution

$$p(x) = C_x^n p^x q^{n-x} = \frac{n!}{x!(n-x)!} p^x q^{n-x}$$

x, the number of successes in n trials, may take values 0, 1, 2, ..., n; p is the probability of success on a single trial; and C_x^n is defined as

$$C_x^n = \frac{n!}{x!(n-x)!}$$

where $n! = n(n-1)(n-2)\cdots(2)(1)$, and $0! \equiv 1$.

Mean: $\mu = np$

Variance: $\sigma^2 = npq$

Standard deviation: $\sigma = \sqrt{npq}$

The factorial notation $n!$ is used to represent the product $n(n-1)(n-2)\cdots(3)(2)(1)$. For example, $5! = (5)(4)(3)(2)(1) = 120$, and 0! is defined to be equal to 1. The notation C_x^n **is shorthand for $n!/[x!(n-x)!]$,** an expression that appears in the formula for the binomial probability distribution.

Graphs of three binomial probability distributions are shown in Figure 4.1: the first for $n = 10$, $p = .1$; the second for $n = 10$, $p = .5$; and the third for $n = 10$, $p = .9$. The height of the bar for any particular value of x is calculated using the binomial formula, $p(x)$, given above. Notice that when $p = .5$, the distribution is symmetric; if p is small, the distribution is skewed to the right; if p is large, the distribution is skewed to the left.

FIGURE 4.1
Binomial probability distributions

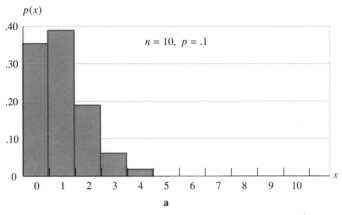

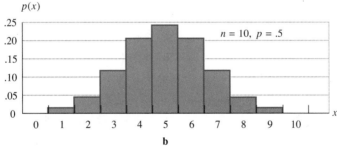

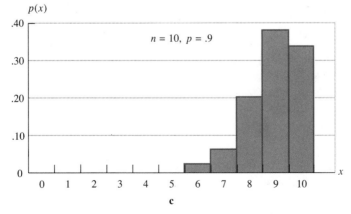

EXAMPLE 4.3 Studies have shown that one in every five apartment dwellers moves within a given year. Suppose that four apartment dwellers are interviewed.

a What is the probability that exactly two have moved within the past year?

b What is the probability that at least two have moved within the past year?

c What is the probability that all four have moved within the past year?

Solution a Define the random variable x to be the number of the four apartment dwellers who have moved within the past year. Assuming that the four apartment dwellers have been independently chosen and are not related, then p remains constant from person to person and x is a binomial random variable with $n = 4$ and $p = .2$. Therefore,

$$p(x) = C_x^4 (.2)^x (.8)^{4-x}$$
$$p(2) = C_2^4 (.2)^2 (.8)^{4-2}$$
$$= \frac{4!}{2!2!}(.04)(.64) = \frac{4(3)(2)(1)}{2(1)(2)(1)}(.04)(.64)$$
$$= .1536$$

b
$$P(\text{at least two}) = p(2) + p(3) + p(4)$$
$$= 1 - p(0) - p(1)$$
$$= 1 - C_0^4(.2)^0(.8)^4 - C_1^4(.2)(.8)^3$$
$$= 1 - .4096 - .4096$$
$$= .1808$$

c
$$p(4) = C_4^4(.2)^4(.8)^0$$
$$= \frac{4!}{4!0!}(.2)^4(1) = .0016 \quad \blacksquare$$

EXAMPLE 4.4 Large lots of incoming products at a manufacturing plant are inspected for defectives by means of a *sampling plan*. A random sample of n items is selected from each lot and inspected, and the number x of defectives in the sample is recorded. If x is less than or equal to some specified *acceptance number a*, the lot is accepted. If x is larger than a, the lot is rejected. Suppose that a manufacturer employs a sampling plan with $n = 10$ and $a = 1$. If a lot contains exactly 5% defectives, what is the probability that the lot will be accepted? rejected?

Solution Since the lot contains 5% defectives, the probability that an item drawn from the lot is defective is $p = .05$. Then the probability of observing x defectives in a sample of $n = 10$ items is

$$p(x) = C_x^{10}(.05)^x(.95)^{10-x}$$

The probability of accepting the lot is the probability that x is less than or equal to the acceptance number $a = 1$. Therefore,

$$P(\text{accept}) = p(0) + p(1) = C_0^{10}(.05)^0(.95)^{10} + C_1^{10}(.05)^1(.95)^9$$
$$= .914$$

$$P(\text{reject}) = 1 - P(\text{accept})$$
$$= 1 - .914$$
$$= .086$$

Although in a practical situation we would not know the exact value of p, we would want to know the probability of accepting bad lots (lots for which p is large) and good lots (for which p is small). This example shows how we could calculate this probability of acceptance for various values of p. ∎

A graph of the probability of lot acceptance versus the lot fraction defective p is called the **operating characteristic curve** for a lot acceptance plan. The operating characteristic curve for a sampling plan with $n = 5$ and $a = 0$ is given in Figure 4.2.

FIGURE **4.2**
Operating characteristic curve for
$n = 5, a = 0$

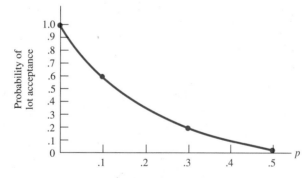

Note that acceptance sampling is an example of statistical inference' because the procedure implies a decision concerning the lot fraction defective p. If you accept a lot, you infer that the true lot fraction defective p is some relatively small, acceptable value. If you reject a lot, it is clear that you think p is too large. Consequently, lot acceptance sampling is a procedure that implies inference making concerning the lot fraction defective. The operating characteristic curve for the sampling plan provides a measure of the goodness of this inferential procedure.

Calculating binomial probabilities is a tedious task when n is large. To simplify our calculations, the sum of the binomial probabilities from $x = 0$ to $x = a$ is presented in Table 1 of Appendix II for $n = 2, 3, \ldots, 12, 15, 20$, and 25.

To illustrate the use of Table 1, let's find the sum of the binomial probabilities from $x = 0$ to $x = 3$ for $n = 5$ trials and $p = .6$. That is, we wish to find

$$P(x \le 3) = \sum_{x=0}^{3} p(x) = p(0) + p(1) + p(2) + p(3)$$

where

$$p(x) = C_x^5 (.6)^x (.4)^{5-x}$$

Since the table values give

$$P(x \le a) = \sum_{x=0}^{a} p(x)$$

we seek the table value in the row corresponding to $a = 3$ and the column for $p = .6$. The table value, .663, is shown in Table 4.1 as it appears in Table 1, Appendix II. Therefore, the sum of the binomial probabilities from $x = 0$ to $x = a = 3$ (for $n = 5$, $p = .6$) is .663.

T A B L E 4.1
Portion of Table 1, Appendix II, for $n = 5$

a	.01	.05	.10	.20	.30	.40	.50	.60	.70	.80	.90	.95	.99	a
0	—	—	—	—	—	—	—	—	—	—	—	—	—	0
1	—	—	—	—	—	—	—	—	—	—	—	—	—	1
2	—	—	—	—	—	—	—	—	—	—	—	—	—	2
3	—	—	—	—	—	—	—	.663	—	—	—	—	—	3
4	—	—	—	—	—	—	—	—	—	—	—	—	—	4

Table 1 can also be used to find individual binomial probabilities. For example, suppose we wish to find $p(3)$ when $n = 5$ and $p = .6$. Since $P(x = 3) = P(x \le 3) - P(x \le 2)$, we write

$$p(3) = \sum_{x=0}^{3} p(x) - \sum_{x=0}^{2} p(x) = .663 - .317 = .346$$

The values $\sum_{x=0}^{3} p(x)$ and $\sum_{x=0}^{2} p(x)$ are found directly from Table 1, indexing $n = 5$ with $p = .6$. In general, an individual binomial probability can be found by subtracting successive entries in the table for a given value of p.

E X A M P L E 4.5 Suppose that you are the personnel director of a company and that you wish to evaluate scores on a multiple-choice aptitude test. A score of zero on an objective test (questions requiring complete recall of the material) indicates that the person was unable to recall the test material at the time the test was given. In contrast, a person with little or no recall knowledge of the test material can score higher on a multiple-choice test because the person only needs to recognize (in contrast to recall) the correct answer and because some questions will be answered correctly, just by chance, even if the person does not know the correct answers. Consequently, the no-knowledge score for a multiple-choice test may well be above zero. Suppose that a multiple-choice test contains 20 questions, each with five possible answers. What is the probability that a person with no knowledge of the test material answers eight or more questions correctly?

Solution Assuming that the person has no knowledge of the answer to a question, the probability p that he or she answers the question correctly is $p = .2$. The probability distribution for x, the number of correct answers on a 20-question test, is

$$p(x) = C_x^{20}(.2)^x(.8)^{20-x}$$

Since direct calculation when $n = 20$ is very tedious, we alternately choose to use Table 1, with $n = 20$ and $p = .2$. Then

$$P(x \geq 8) = 1 - P(x \leq 7) = 1 - .968 = .032$$

where $P(x \leq 7)$ is found in Table 1 by indexing $n = 20$, $p = .2$, and $a = 7$. ∎

EXAMPLE 4.6 Refer to Example 4.5. What is the expected score for a person who has no knowledge of the test material? Within what limits would you expect a no-knowledge score to be?

Solution Recall that x is the number of correct answers on the $n = 20$-question test, with $p = .2$. A no-knowledge student's expected score is then

$$E(x) = np = 20(.2) = 4 \text{ correct answers}$$

To evaluate the variation of no-knowledge scores, we need to know σ, where

$$\sigma = \sqrt{npq} = \sqrt{(20)(.2)(.8)} = 1.79$$

Then, from our knowledge of Tchebysheff's Theorem,[†] we would expect x to fall within the interval $(\mu \pm 2\sigma)$ with a probability of at least .75, and within the interval $(\mu \pm 3\sigma)$ with probability at least .89. These intervals are

$$\mu \pm 2\sigma = 4 \pm 3.58 \qquad \text{or} \qquad .42 \text{ to } 7.58$$
$$\mu \pm 3\sigma = 4 \pm 5.37 \qquad \text{or} \qquad -1.37 \text{ to } 9.37$$

This compares with a score of zero for a no-knowledge student taking an objective recall test. ∎

[†]A histogram of $p(x)$ for $n = 20$, $p = .2$ will be mound-shaped. Hence, we would expect the Empirical Rule to work very well also. The reason that it does will be explained in Chapter 6.

Both individual and cumulative binomial probabilities are available through the Minitab package. Individual binomial probabilities associated with each value of x for any combination of n and p can be found by using the probability density function command PDF followed by a semicolon (;) and then the subcommand BINOMIAL N P followed by a period. The variable N is the given sample size, and P is the probability of success. Cumulative binomial probabilities can be found by using the cumulative distribution function command CDF followed by a semicolon (;) and then the subcommand BINOMIAL N P followed by a period.

The Minitab output for both the PDF and CDF commands when $n = 20$ and $p = .2$ is given in Table 4.2. The PDF command gives rise to the individual probabilities $P(x = K)$; the CDF command gives rise to the cumulative probabilities $P(x \leq K)$. (The letter K plays the role of the letter a in Table 1 of Appendix II.) Notice that, in the Minitab output, $P(x \leq 7) = .9679$ so that $P(x \geq 8) = (1 - .9679) = .0321$, which, to three-decimal accuracy, agrees with our earlier results using Table 1.

T A B L E **4.2**   Minitab output of binomial probabilities when $n = 20$ and $p = .2$

```
MTB > PDF;                              MTB > CDF;
SUBC> BINOMIAL 20 .2.                   SUBC> BINOMIAL 20 .2.

  BINOMIAL WITH N =  20  P = 0.200000     BINOMIAL WITH N =  20  P = 0.200000
     K              P( X = K)              K    P( X LESS OR = K)
     0                0.0115               0          0.0115
     1                0.0576               1          0.0692
     2                0.1369               2          0.2061
     3                0.2054               3          0.4114
     4                0.2182               4          0.6296
     5                0.1746               5          0.8042
     6                0.1091               6          0.9133
     7                0.0545               7          0.9679
     8                0.0222               8          0.9900
     9                0.0074               9          0.9974
    10                0.0020              10          0.9994
    11                0.0005              11          0.9939
    12                0.0001              12          1.0000
    13                0.0000
```

The Minitab output may not list the probabilities for all values of $x = 0, 1, 2, \ldots, n$ for various combinations of n and p, since the Minitab package has an internal check that stops the calculations when $P(x = K) = 0$ [or, equivalently, when $P(x \leq K) = 1$] to within a preassigned level of accuracy.

Exercises

Basic Techniques

4.1   A jar contains five balls—three red and two white. Two balls are randomly selected without replacement from the jar and the number x of red balls is recorded. Explain why x is or is not a binomial random variable. [*Hint:* Compare the characteristics of this experiment with the characteristics of a binomial experiment given in Section 4.2.] If the experiment is binomial, give the values of n and p.

4.2   Answer Exercise 4.1 by assuming that the sampling was conducted with replacement. That is, the first ball was selected from the jar and observed. It was then replaced, and the balls were mixed before the second ball was selected.

4.3   Find the following.

a $3!$ b $5!$ c C_4^6

d C_3^8 e C_2^{10} f C_6^6

g C_0^6 h C_1^9 i C_0^9

4.4   Calculate the value of $p(x)$ and construct a probability histogram for the following.

a $n = 5,\ p = .2$ b $n = 5,\ p = .5$ c $n = 5,\ p = .8$

4.5   Calculate $p(x)$ for $x = 0,\ 1,\ 2,\ \ldots,\ 5,\ 6$ for $n = 6$ and $p = .1$. Graph $p(x)$. Repeat these instructions for binomial probability distributions for $p = .5$ and $p = .9$. Compare the graphs. How does the value of p affect the shape of $p(x)$?

4.6   Use Table 1 in Appendix II to find the sum of the binomial probabilities from $x = 0$ to $x = a$ for the following.

a $n = 8,\ p = .1,\ a = 3$ b $n = 12,\ p = .6,\ a = 7$ c $n = 25,\ p = .5,\ a = 14$

4.7   Use the formula for the binomial probability distribution to calculate the values of $p(x)$ for $n = 5,\ p = .5$ (this calculation was done in Exercise 4.4). Next find $\sum_{x=0}^{a} p(x)$ for $a = 0,\ 1,\ 2,\ 3,\ 4$, using the values of $p(x)$ that you computed. Then compare these sums with the values given in Table 1 in Appendix II.

4.8   Use the information given in Table 1 in Appendix II to find $p(3)$ for $n = 5,\ p = .5$. Then compare this answer with the value of $p(3)$ calculated in Exercise 4.7.

4.9   Use the information given in Table 1 in Appendix II to find $p(3) + p(4)$ for $n = 5,\ p = .5$. Verify this answer by using the values of $p(x)$ that you calculated in Exercise 4.7.

4.10   Use Table 1 in Appendix II to find the following.

a $P(x < 12)$ for $n = 20,\ p = .5$ b $P(x \le 6)$ for $n = 15,\ p = .4$

c $P(x > 4)$ for $n = 10,\ p = .4$ d $P(x \ge 6)$ for $n = 15,\ p = .6$

e $P(3 < x < 7)$ for $n = 10,\ p = .5$

4.11   Use Table 1 in Appendix II to find the following.

a $P(x \le 5)$ for $n = 10,\ p = .4$ b $P(x < 3)$ for $n = 5,\ p = .6$

c $P(x \le 17)$ for $n = 20,\ p = .7$ d $P(x > 17)$ for $n = 20,\ p = .7$

e $P(x < 6)$ for $n = 15,\ p = .4$

4.12   Find the mean and standard deviation for the following binomial distributions.

a $n = 1000,\ p = .3$ b $n = 400,\ p = .01$

c $n = 500,\ p = .5$ d $n = 1600,\ p = .8$

4.13   Find the mean and standard deviation for a binomial distribution with $n = 100$ and the following values of p.

a $p = .01$ b $p = .9$ c $p = .3$

d $p = .7$ e $p = .5$

4.14   In Exercise 4.13 we calculated the mean and standard deviation for a binomial random variable for a fixed sample size, $n = 100$, and for different values of p. Graph the values of the standard deviation for the five values of p given in Exercise 4.13. For what value of p does the standard deviation seem to be a maximum?

Applications

4.15   A Gallup Poll conducted in January 1993 following the State of the Union address surveyed those who watched all or some of the State of the Union address. ("What Should Top 1994

Agenda?," 1994). Those interviewed were asked which of four issues from among the following should receive the highest priority in 1994: crime, health care, the budget deficit, or welfare reform. Explain why this sampling is or is not a binomial experiment.

4.16 There is a fear that managed health care organizations (HMOs) will limit choices. However, a new study suggests that people who belong to HMOs are more satisfied with their care than are members of traditional health insurance plans (Braus, 1994). The study compared 5000 households that were enrolled in either an HMO, an indemnity plan, or a preferred provider organization. This survey reported that 85% of HMO members were satisfied with their insurance policies. Explain why the survey of HMO members is or is not a binomial experiment.

4.17 In his article "High Court OKs Congress' Right to Regulate Cable Television," David Savage (1994) reported that 60% of American households have cable TV. Suppose that $n = 4$ households were polled as to whether or not they had cable TV. Assume that the 60% figure is correct in answering the following questions.

a What is the probability that x is exactly equal to 4?

b What is the probability that x is 1 or larger?

c What is the probability that x is exactly equal to 1?

4.18 In a study conducted by *Business Marketing, Advertising Age,* and *USA Chicago, Inc.* ("Survey Said . . . ," 1994) CEOs, senior marketing executives, and chief information officers across the country were asked how the information superhighway would affect their businesses, marketing practices, and individual responsibilities. When CEOs were asked whether they were aware of the national information superhighway, approximately 50% said they were. Suppose that this figure is representative of all CEOs nationwide and that 25 CEOs are randomly selected and asked if they are aware of the information superhighway.

a What is the probability that all 25 will indicate that they are aware of the information superhighway?

b What is the probability that at least 10 of the 25 CEOs will indicate that they are aware of the information superhighway?

c What is the probability that exactly 10 of the CEOs will indicate that they are aware of the information superhighway?

4.19 Many employers are finding that some of the people they hire are not who and what they claim to be. Detecting job applicants who falsify their application information has spawned some new businesses: credential-checking services. Suppose that you hired five new employees last week and that the probability that a single employee would falsify the information on his or her application form is .35. What is the probability that at least one of the five application forms has been falsified? two or more?

4.20 In a comprehensive look at Japanese and American attitudes toward each other, the Japanese show great pride in the quality of their products. However, they feel that the United States will play a stronger role in both leadership and economic power than Japan in the years ahead. In particular, 71% of the Japanese felt that their products are superior to American products, and 42% felt that the United States would be the number one economic power in the next century ("How the Japanese See Themselves . . . and Us," 1990). Suppose 50 Japanese citizens were randomly selected.

a What is the probability distribution of x, the number of Japanese who feel that their products are superior to American products?

b What is the probability distribution of x, the number of Japanese who feel that the United States will be the number one economic power in the next century?

c Find the mean and standard deviation of the random variable described in part (b).

d If only 5 of the 50 Japanese citizens feel that the United States will be the number one economic power in the next century, is it likely that the 42% figure is accurate? Explain.

4.21 Many utility companies have begun to promote energy conservation by offering discount rates to customers who keep their energy use below certain established subsidy standards. It is known that 70% of the residents of a midwestern town have reduced their electricity use sufficiently to qualify for discounted rates. Suppose five residents are randomly selected from this town.

a What is the probability that all five qualify for the favorable rates?

b What is the probability that at least four qualify for the favorable rates?

4.22 A machine is designed to fill cans with 12 oz. of soda. The variability of the fills causes each can to be filled with more or less than 12 oz. of soda. If the dispensing machine is set so that 12 oz. is the median fill, then the probability of overfill is .5 and the probability of underfill is also .5. Suppose that n cans are selected from the production line and the number of underfilled cans reported.

a What is the probability that all of the cans are underfilled when $n = 3$? when $n = 5$? when $n = 10$?

b If you were a supervisor in charge of monitoring the fills of the cans of soda from this machine, what would your conclusions be if, in fact, you observed 3, and then 5, and then 10 of the sampled soda cans underfilled?

c Explain how a run (a series of like items) of underfilled cans might be used as a process control variable.

4.23 According to an article in the *Los Angeles Times* (Horovitz, 1994), hot dogs rank among the least popular of items sold at fast-food outlets in Southern California. A survey of $n = 600$ Southern California residents were asked to name the last fast-food item they purchased as a meal. Among those surveyed, 34% purchased hamburgers, 19% purchased pizza, and 11% purchased Mexican food. If only 3.3% of those surveyed ordered hot dogs, within what limits would we expect the number of fast-food customers ordering hot dogs to lie with probability approximately equal to .95 if the 3.3% figure is, in fact, correct?

4.3 The Poisson Probability Distribution

The **Poisson probability distribution** is a good model for the relative frequency distribution of the number of rare events that occur in a unit of time, distance, space, and so on. For this reason it is often employed by business management to model the relative frequency distribution of the number of industrial accidents per unit of time (such as the nuclear accident at Three Mile Island), or by personnel managers to model the relative frequency distribution for the number of employee accidents or the number of insurance claims per unit of time. The Poisson probability distribution can also, in some situations, provide a good model for the relative frequency distribution of the number of arrivals per unit of time at a servicing unit (say, the number of orders received at a manufacturing plant or the number of customers at a servicing facility, a supermarket counter, etc.). The formulas for the Poisson distribution are given in the following display.

The Poisson Probability Distribution

$$p(x) = \frac{\mu^x e^{-\mu}}{x!}, \qquad x = 0, 1, 2, \ldots, \infty$$

> where x = Number of rare events per unit of time, distance, space, and
> so forth and e = 2.7182818 . . .
>
> Mean: symbol μ that appears in $p(x)$
> Variance: $\sigma^2 = \mu$
> Standard deviation: $\sigma = \sqrt{\mu}$

Values of $e^{-\mu}$ can be found using most scientific calculators.

Note that, in practice, x is usually small; theoretically, it could be large beyond all bound. Thus the Poisson random variable is an example of a discrete random variable that can assume an infinitely large (but countable) number of values. The Poisson distribution is unique among discrete distributions in that $\mu = \sigma^2$—that is, the mean is equal to the variance.

The Poisson probability distribution can also be used to approximate the binomial distribution when n is large and p is small[†] and when the binomial mean is less than 7. In this case, the value of $1 - p$ will be close to 1, the binomial mean np will be approximately equal to the binomial variance $np(1 - p)$, and the Poisson probabilities with $\mu = np$ will closely approximate binomial probabilities for given n and p.

We will illustrate these two types of applications in the following examples. Others will be suggested by the exercises.

EXAMPLE **4.7**   Serious worker injuries at a steel-fabricating company average 2.7 per year. Given that safety conditions at the plant remain the same next year, what is the probability that the number of serious injuries will be less than 2?

Solution   The event that fewer than two serious injuries will occur is the event that $x = 0$ or 1. Therefore,

$$P(x < 2) = p(0) + p(1) \quad \text{where } p(x) = \frac{(2.7)^x e^{-2.7}}{x!}$$

Substituting into the formula for $p(x)$ with $e^{-2.7} = .067206$, we obtain

$$P(x < 2) = p(0) + p(1) = \frac{(2.7)^0(.067206)}{0!} + \frac{(2.7)^1(.067206)}{1!}$$
$$= (.067206) + (2.7)(.067206)$$
$$= .249$$

[†]If p is near 1, interchange your definitions of success and failure so that p will be near zero.

(Recall that $0! = 1$.) Therefore, the probability that fewer than two serious worker injuries will occur next year in the steel-fabricating plant is .249. ∎

For your convenience, we provide in Table 2 of Appendix II the partial sums, $\sum_{x=0}^{a} p(x)$, for the Poisson probability distribution for values of μ from .25 to 5.0 in steps of .25. This table is constructed in the same manner as the table of partial sums for the binomial probability distribution, Table 1 of Appendix II. The following example illustrates the use of Table 2 and also demonstrates the use of the Poisson probability distribution to approximate the binomial probability distribution.

E X A M P L E **4.8** Suppose that you have a binomial experiment with $n = 25$ and $p = .1$. Find the exact value of $P(x \leq 3)$, using the table of partial sums for the binomial probability distribution, Table 1 of Appendix II. Then find the corresponding partial sum by using the Poisson approximation in Table 2 of Appendix II. Compare the exact and approximate values for $P(x \leq 3)$.

Solution From Table 1 in Appendix II, the exact value of $P(x \leq 3)$ is $\sum_{x=0}^{3} p(x) = .764$. The corresponding Poisson partial sum, for $\mu = np = (25)(.1) = 2.5$, given in Table 2 of Appendix II, is $P(x \leq 3) = \sum_{x=0}^{3} p(x) = .758$. Comparing the exact and approximate values of $P(x \leq 3)$, we see that the approximation is quite good. It differs from the exact value by only .006. ∎

The individual and cumulative probabilities for a Poisson distribution with mean μ can be found by using the PDF and the CDF Minitab commands, followed by the subcommand Poisson μ. The cumulative binomial probabilities for $n = 25$ and $p = .1$, together with the cumulative Poisson probabilities for $\mu = 2.5$, are given in Table 4.3. Comparing the actual binomial probabilities with the corresponding probabilities found by using the Poisson approximation, we see that they are quite accurate in this case. Furthermore, we see that the POISSON command also terminates when an individual probability equals zero within a preassigned level of accuracy.

T A B L E **4.3** Minitab printout of binomial and Poisson probabilities

```
MTB > CDF;                                          MTB > CDF;
SUBC> BINOMIAL 25 .1.                               SUBC> POISSON 2.5.

  BINOMIAL WITH N =  25  P = 0.100000                  POISSON WITH MEAN =    2.500
     K  P( X LESS OR = K)                                 K  P( X LESS OR = K)
     0           0.0718                                    0           0.0821
     1           0.2712                                    1           0.2873
     2           0.5371                                    2           0.5438
     3           0.7636                                    3           0.7576
     4           0.9020                                    4           0.8912
     5           0.9666                                    5           0.9580
     6           0.9905                                    6           0.9858
     7           0.9977                                    7           0.9958
     8           0.9995                                    8           0.9989
     9           0.9999                                    9           0.9997
    10           1.0000                                   10           0.9999
                                                          11           1.0000
```

Exercises

Basic Techniques

4.24 Suppose x is a Poisson random variable with $\mu = 1.2$. Find the following.

a $p(0)$

b $p(1)$

c $P(x \leq 2)$

d $P(x > 1)$

4.25 Suppose x is a Poisson random variable with $\mu = 2$. Find the following.

a $p(0)$

b $P(x > 1)$

c $P(x < 2)$

4.26 Use Table 2 in Appendix II to find $p(x)$ for a Poisson probability distribution with $\mu = 1$ and $x = 0, 1, 2, 3, 4, \ldots$. Then graph $p(x)$.

4.27 Repeat the instructions of Exercise 4.26 for $\mu = 3$. Notice how the distribution tends to become more mound-shaped as μ increases.

4.28 Use Table 2 in Appendix II to find the following.

a $P(x \leq 2)$ when $\mu = 3$.

b $P(x \geq 1)$ when $\mu = 1$.

c $P(x = 2)$ when $\mu = 2$. $\left[Hint: p(2) = \sum_{x=0}^{2} p(x) - \sum_{x=0}^{1} p(x). \right]$

4.29 A binomial experiment has $n = 20$ and $p = .2$.

a   Use Table 1 to find the exact value of $p(2)$.

b   Use Table 2 to find the Poisson approximation to $p(2)$.

Applications

4.30 Suppose that a random system of police patrol is devised so that a patrolman may visit a given location on his beat $x = 0, 1, 2, 3, \ldots$ times per half-hour period and that the system is arranged so that he visits each location on an average of once per time period. Assume that x

possesses, approximately, a Poisson probability distribution. Calculate the probability that the patrolman will miss a given location during a half-hour period. What is the probability that he will visit it once? twice? at least once?

4.31   The accidents in a particular industrial plant average 3.5 per week.

a   What is the probability that no accidents will occur in a given week?

b   Is it likely that the number of accidents per week would exceed 7? Explain.

c   If the number of accidents in a particular week were equal to 9, would you still believe that $\mu = 3.5$? Explain.

4.32   The number x, per week, of sales of a piece of large earth-moving equipment for a construction equipment company possesses a Poisson probability distribution with mean equal to 4.

a   What is the probability that the number of earth movers sold per week is equal to 1? less than or equal to 1?

b   Is it likely that x will exceed 9? Explain.

4.33   The sole proprietor of a residential real estate office notes that, on the average, telephone inquiries arrive randomly and independently at the rate of four per 8-hour workday. Since the real estate proprietor is often out with clients, she cannot offer immediate responses to some telephone inquiries to her office.

a   What is the probability that no telephone inquiries arrive during a 2-hour absence from her office during a typical 8-hour workday?

b   What is the probability that there will be at least five telephone inquiries during a typical 8-hour workday?

4.34   The number x of people entering the intensive care unit at a particular hospital on any one day possesses a Poisson probability distribution with mean equal to five persons per day.

a   What is the probability that the number of people entering the intensive care unit on a particular day is equal to 2? less than or equal to 2?

b   Is it likely that x will exceed 10? Explain.

4.4 Other Discrete Probability Distributions (Optional)

In Example 3.1 and Exercise 3.51, the experiment consisted of tossing a fair die and observing x, the number on the upper face. This is one example of a more general experiment in which the random variable can assume the values $x = 1, 2, \ldots, N$ with equal probabilities given by $p(x) = 1/N$. The resulting distribution for x is called the **discrete uniform probability distribution** because the resulting probability histogram has uniform height.

A second discrete probability distribution, which resembles the binomial distribution, occurs if you select a random sample of n consumers from a population containing N consumers. The number x of consumers favoring a specific product will possess a binomial probability distribution when the sample size n is small relative to the number N of consumers in the population (see Example 4.1). When n is large relative to N (as in Example 4.2), the number x favoring the product possesses a **hypergeometric probability distribution.** Its formulas are given in the following display.

Hypergeometric Probability Distribution

$$p(x) = \frac{C_x^r C_{n-x}^{N-r}}{C_n^N}$$

where N = Number of elements in the population

r = Number of elements possessing some specific characteristic, say, the number of persons favoring a particular product

n = Number of elements in the sample

Mean: $\mu = n\left(\dfrac{r}{N}\right)$

Variance: $\sigma^2 = n\left(\dfrac{r}{N}\right)\left(\dfrac{N-r}{N}\right)\left(\dfrac{N-n}{N-1}\right)$

Standard deviation: $\sigma = \sqrt{\sigma^2}$

When the sample size n is small relative to the population size N and n/N is less than .05, the hypergeometric probabilities can be well approximated by a binomial distribution with $p = r/N$.

EXAMPLE   **4.9**   A jury panel consists of 20 people, two of whom are Native Americans. If three people are randomly chosen from the jury panel, what is the probability that two will be Native Americans?

Solution   For this example,

$$N = 20, \qquad n = 3$$
$$r = 2 \text{ (Native Americans)}$$
$$x = \text{Number of Native Americans in the selection}$$

Then

$$p(x) = \frac{C_x^r C_{n-x}^{N-r}}{C_n^N}$$

and

$$p(2) = \frac{C_2^2 C_{3-2}^{20-2}}{C_3^{20}}$$

where

$$C_2^2 = \frac{2!}{2!0!} = 1, \qquad C_{3-2}^{20-2} = C_1^{18} = \frac{18!}{1!17!} = 18$$

and

$$C_3^{20} = \frac{20!}{3!17!} = \frac{(20)(19)(18)}{6} = 1140$$

The probability of having two Native Americans in the sample of size $n = 3$ is

$$p(2) = \frac{(1)(18)}{1140} = .016 \quad \blacksquare$$

Exercises

Basic Techniques

4.35 **a** Calculate $p(x)$, where x has a hypergeometric probability distribution with $N = 10$, $n = 2$, $r = 3$, and $x = 0, 1, 2$.

    b Graph $p(x)$.

4.36 **a** Calculate $p(x)$, where x has a hypergeometric probability distribution with $N = 20$, $n = 3$, $r = 3$, and $x = 0, 1, 2, 3$.

    b Graph $p(x)$.

4.37 Find the mean and standard deviation for the random variable x described in Exercise 4.36. What is the probability that x lies in the interval $(\mu \pm 2\sigma)$?

Applications

 4.38 A problem encountered by personnel directors and others faced with the selection of the best in a finite set of elements is indicated by the following situation: From a group of 20 Ph.D. engineers, ten are selected for employment. What is the probability that the ten selected include the five best engineers in the group of 20?

 4.39 A particular industrial product is shipped in lots of 20. Testing to determine whether an item is defective is costly; thus the manufacturer samples the production rather than using a 100% inspection plan. A sampling plan designed to minimize the number of defectives shipped to customers calls for sampling five items from each lot and rejecting the lot if more than one defective is observed. (If rejected, each item in the lot is tested.) If a lot contains four defectives, what is the probability that it will be rejected?

 4.40 Texaco became the latest big oil company to cut back on its work force and shed some of its U.S. oil fields (Craig, 1994). Exxon and Mobil recently announced similar restructuring to boost profits. Suppose that three of the ten leading U.S. petroleum refiners are, in fact, going to restructure their companies. If a *USA Today* reporter interviews the CEO of four randomly selected petroleum refining businesses, calculate the following probabilities.

    a Her selection includes all three of the CEOs whose companies recently restructured.

    b Her selection includes none of the CEOs whose companies recently restructured.

    c Her selection includes at least one of the CEOs whose companies recently restructured.

4.5 MORE ABOUT DRIVING RIGHTS

As we suggested in the case study, estimation of the proportion of American adults who favor a mandatory physical exam every three years for drivers over age 65 depends upon the probability distribution of x, the number in the survey who favor mandatory physical exams for seniors. Since the number of persons contacted in the survey constitutes a random sample from among a large number of people, x will possess, for all practical purposes, a binomial probability distribution.

Suppose that the 70% figure were actually the true value of p. Given a sample of size 1003, the number in the sample who indicate that they are in favor of mandatory physical exams for drivers over 65 would possess a binomial distribution with mean and standard deviation equal to

$$\mu = np = (1003)(.7) = 702.1$$

and

$$\sigma = \sqrt{npq} = \sqrt{(1003)(.7)(.3)} = 14.51$$

Thus if p were, in fact, .7, according to the Empirical Rule with approximate 95% probability, we would expect the number in the sample to lie in the interval

$$\mu \pm 2\sigma = 702.1 \pm 2(14.51) = (673.08, \ 731.12)$$

or from 674 to 731 respondents.

If p were, in fact, .7 and we observed $x = 600$ of the 1003 respondents who agreed with mandatory physical exams for drivers over 65, we would be inclined to think that the value of $p = .7$ is much too large and that the actual proportion of positive respondents is closer to $p = .6$. When we recognize that the actual value of p is not .7 but rather that the *sample estimate of p* is .7, we might want to find limits within which the true value of p lies. Since the number of positive responses lies between 674 and 731 approximately 95% of the time, we would estimate that p lies in the interval

$$\left(\frac{674}{1003}, \ \frac{731}{1003} \right) = (.672, \ .729)$$

or from .672 to .729.

4.6 Summary

Familiarity with discrete probability distributions and the properties of the experiments that generate them is extremely helpful. Rather than solve the same probability problem over and over again from first principles (as was done in Chapter 3), you need only recognize the type of random variable involved and then substitute into the formula for its probability distribution.

Several useful discrete probability distributions were presented in this chapter: the binomial, the Poisson, and the hypergeometric distributions. These probability distributions enable us to calculate the probabilities associated with events that are of interest in the sciences, in business, and in marketing.

The binomial probability distribution allows us to calculate the probability of x successes in a series of n identical independent trials, where the probability of a success in a single trial is equal to p. The binomial experiment is an excellent model for many sampling situations, particularly surveys that result in "yes" or "no" types of data.

The Poisson probability distribution is important because it can be used to approximate certain binomial probabilities when n is large and p is small. Consequently, it can greatly reduce the computations involved in calculating binomial probabilities. In addition, the Poisson probability distribution is important in its own right. It provides an excellent probabilistic model for the number of occurrences of rare events in time or space.

The hypergeometric probability distribution is also related to the binomial probability distribution. It gives the probability of drawing x elements of a particular type from a population where the number N of elements in the population is small in relation to the sample size n. The binomial probability distribution applies to the same situation except that it is appropriate only when N is large in relation to n.

Tips on Problem Solving

1 If the random variable is the number of occurrences of a specified event in a given unit of *time or space* for which the average number of occurrences per unit time or space is μ, then the random variable has a Poisson distribution.

2 Suppose that a random sample of n items is selected without replacement from a population of N items in which a proportion p of the items possess a specified property and $q = 1 - p$ do not. The flowchart in Figure 4.3 is provided to help you determine which distribution is appropriate for the situations presented.

F I G U R E **4.3**   Decision tree

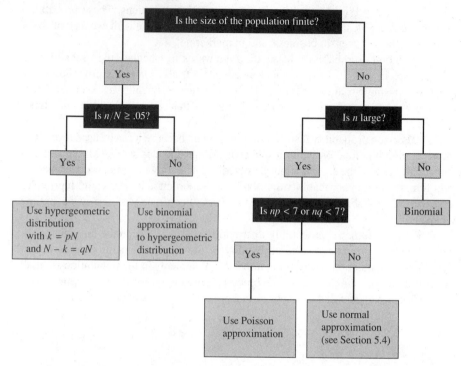

Supplementary Exercises

Starred (*) exercises are optional.

4.41 List the five identifying characteristics of the binomial experiment.

4.42 A balanced coin is tossed three times. Let x equal the number of heads observed.

   a Use the formula for the binomial probability distribution to calculate the probabilities associated with $x = 0, \ 1, \ 2$, and 3.

   b Construct a probability distribution similar to the ones shown in Figure 4.1.

   c Find the expected value and standard deviation of x, using the formulas

$$E(x) = np \qquad \sigma = \sqrt{npq}$$

   d Using the probability distribution you derived in (b), find the fraction of the population measurements lying within one standard deviation of the mean. Repeat for two standard deviations. How do your results agree with Tchebysheff's Theorem and the Empirical Rule?

4.43 Use Table 1 in Appendix II to find the partial sum $\sum_{x=0}^{a} p(x)$ for a binomial random variable x with the following characteristics.

   a $n = 10, \ p = .7, \ a = 8$

   b $n = 15, \ p = .05, \ a = 1$

   c $n = 20, \ p = .9, \ a = 14$

4.44 Use Table 1 in Appendix II to find $p(x)$ for a binomial random variable x with the following characteristics.

   a $n = 10, \ p = .6, \ x = 6$

b $n = 15$, $p = .5$, $x = 5$

c $n = 20$, $p = .2$, $x = 3$

4.45   Use Table 1 in Appendix II to find $\sum\limits_{x=a}^{b} p(x)$ for a binomial random variable x with the following characteristics.

a $n = 10$, $p = .1$, $a = 1$, $b = 10$

b $n = 10$, $p = .8$, $a = 7$, $b = 9$

c $n = 15$, $p = .4$, $a = 4$, $b = 15$

4.46   Use Table 1 in Appendix II to find the following probabilities for the binomial random variable x.

a $P(x \leq 1)$ for $n = 5$, $p = .2$

b $P(x > 1)$ for $n = 5$, $p = .2$

c $P(x < 1)$ for $n = 5$, $p = .2$

d $P(x = 1)$ for $n = 5$, $p = .2$

e $P(x = 1)$ for $n = 10$, $p = .2$

4.47   Use Table 1 in Appendix II to find the following probabilities for the binomial random variable x.

a $P(x \leq 4)$ for $n = 10$, $p = .5$

b $P(x < 4)$ for $n = 10$, $p = .5$

c $P(x > 4)$ for $n = 10$, $p = .5$

d $P(x = 4)$ for $n = 10$, $p = .5$

4.48   A recent study of commuter trains shows that they run more than 35 minutes late with probability equal to .5. If we are using a commuter train that exhibits these characteristics and we randomly select five days within the past year, what is the probability that the train is always late? What is the probability that the train is late more than three times out of five?

4.49   A warehouse contains ten printing machines, four of which are defective. A company selects five of the machines at random, thinking all are in working condition. What is the probability that all five of the machines are nondefective?

4.50   A union claims that 45 of the 80 employees of a company favor unionization. Suppose the union is correct, and the plant manager informally samples the opinion of 20 employees.

a   What is the expected value of the number x of employees in the sample that will favor unionization?

b   Find the standard deviation of x.

c   If the union is correct, is it likely that fewer than 9 employees in the sample will favor unionization? Explain.

4.51   A U.S. government study of telephone calls made by its employees suggests that one in every three calls is for nonbusiness purposes (*New York Times*, June 23, 1986). Suppose that you are a government employee and that three out of every ten of your telephone calls are for personal reasons. The government randomly sampled ten numbers that you dialed.

a   What is the probability that no more than one of the calls was for personal reasons?

b   What is the probability that more than five of the calls were for personal reasons?

c   What is the probability that exactly three of the calls were for personal reasons?

4.52   In Exercise 4.51, we noted that a government study suggests that one in every three telephone calls made by government employees is for nonbusiness purposes. Suppose that a government facility makes 10 million calls per year and that each call costs the taxpayers $1.50 for the use of the communications equipment and for the employees' time.

a   What is the expected number of nonbusiness calls?

b   What is the standard deviation?

c   Could the number x of nonbusiness calls be less than 3.3 million? Explain.

4.53 The Honda Corporation has increased its North American operations in the past few years, adding jobs, increasing production, and introducing new models at its North American facilities. Currently, about 60% of all Honda vehicles sold in North America are also made here (Prodis, 1994). Suppose 15 Hondas purchased in North America are randomly chosen and inspected.

a What is the probability that 8 or fewer Hondas were made in North America?

b What is the probability that between 8 and 12 Hondas (inclusive) were made in North America?

c What is the probability that exactly 10 Hondas were made in North America?

4.54 Refer to Exercise 4.53. A small company surveys its employees and finds that 15 of them drive Hondas. Of these 15 Hondas, 8 were made in North America. If five employees are randomly selected to give their opinion on the Hondas' performance, what is the probability that all five have cars that were made in North America?

4.55 You know that 90% of those who purchase a color television will have no claims covered by the guarantee during the duration of the guarantee. Suppose that 20 customers each buy a color television set from a certain appliance dealer. What is the probability that at least two of these 20 customers will have claims against the guarantee?

4.56 Suppose that you know that one out of ten undergraduate college textbooks is an outstanding financial success. A publisher has selected ten new textbooks for publication.

a What is the probability that exactly one will be an outstanding financial success?

b What is the probability that at least one will be an outstanding financial success?

c What is the probability that at least two will be outstanding financial successes?

4.57 The proportion of residential households in a small city that are heated by natural gas is approximately .2. A randomly selected city block within the city limits has 20 residential households. Assume that the properties of a binomial experiment are satisfied.

a Find the probability that none of the households are heated by natural gas.

b Find the probability that no more than four of the 20 are heated by natural gas.

c Why might the binomial experiment not provide a good model for this sampling situation?

4.58 Suppose, as noted in Exercise 4.19, that approximately 35% of all applicants for jobs falsify the information on their application forms. Suppose a company has 2300 employees.

a What is the expected value of the number x of application forms that have been falsified?

b Find the standard deviation of x.

c Calculate the interval $(\mu \pm 2\sigma)$.

d Suppose that the company had a credentials-checking firm verify the information on the 2300 application forms and that 249 application forms contained falsified information. Do you think that the company's application falsification rate is consistent with the contention that 35% of all job applicants falsify information on their applications? Explain.

4.59 A quality control engineer wishes to study the alternative sampling plans $n = 5$, $a = 1$ and $n = 25$, $a = 5$. On the same sheet of graph paper, construct the operating characteristic curves for both plans, making use of acceptance probabilities at $p = .05$, $p = .10$, $p = .20$, $p = .30$, and $p = .40$ in each case.

a If you were a seller producing lots with fraction defective ranging from $p = 0$ to $p = .10$, which of the two sampling plans would you prefer?

b If you were a buyer wishing to be protected against accepting lots with fraction defective exceeding $p = .30$, which of the two sampling plans would you prefer?

4.60 Consider a lot acceptance plan with $n = 20$ and $a = 1$. Calculate the probability of accepting lots having the following fraction defective values. Sketch the operating characteristic curve for the plan.

a $p = .01$

b $p = .05$

c $p = .10$

d $p = .20$

4.61 One consideration in selecting an occupation is the safety of workers in that industry. In 1991, the motor vehicles and equipment manufacturing industry had the highest rate of occupational injuries, with 18.6 injuries per 100,000 full-time employees, while the restaurant industry (including eating and drinking places) had the lowest incidence rate, with 7.4 accidents per 100,000 full-time employees (*World Almanac and Book of Facts,* 1994 ed.). Suppose that the number of accidents per 100,000 full-time employees in the restaurant industry follows a Poisson distribution with a mean equal to $\mu = 7.4$ per 100,000 employees, that a nationwide restaurant chain has approximately 100,000 full-time employees, and that the 7.4 figure is correct.

a   Use the Minitab printout to evaluate the probability that the chain has no accidents.

b   Use the Minitab printout to evaluate the probability that the chain has seven or fewer accidents.

c   Calculate the interval $\mu \pm 2\sigma$. Find the probability that the number of accidents per 100,000 will lie in this interval. Does this probability agree with the Empirical Rule?

```
POISSON WITH MEAN =    7.400
   K   P( X LESS OR = K)
   0          0.0006
   1          0.0051              11          0.9265
   2          0.0219              12          0.9609
   3          0.0632              13          0.9805
   4          0.1395              14          0.9908
   5          0.2526              15          0.9959
   6          0.3920              16          0.9983
   7          0.5393              17          0.9993
   8          0.6757              18          0.9997
   9          0.7877              19          0.9999
  10          0.8707              20          1.0000
```

4.62 In a random sample of 20 executive secretaries, 15 favor copy machine A over copy machine B. If the machines are equally desirable, the probability that a person will select machine A over B is .5. What is the probability that the number x, favoring machine A, in the sample of 20 is equal to 15 or larger if $p = .5$?

4.63 Refer to Exercise 4.62. Manufacturer A hired a marketing company to sample 500 executive secretaries to determine which of the two copy machines was preferred, A or B.

a   If $p = .5$, find the mean and standard deviation of the number x in the sample that prefer copy machine A.

b   Suppose that 280 executive secretaries in the sample of 500 prefer copy machine A. Is this sample result likely if, in fact, the two copy machines are equally desirable? Explain.

4.64 The U.S. airline industry has always been extremely concerned about the safety record of its air carriers. However, in 1989 the number of fatal accidents increased to 11, a large increase over the years 1981–1992 (excluding 1989), in which the average number of fatal accidents was only 3.5 per year (*Source:* National Transportation Safety Board, *The World Almanac and Book of Facts,* 1994 ed., p. 274). Assume that the number of fatal airplane accidents per year can be approximated by a Poisson random variable with mean equal to 3.5.

a   What is the probability of observing 11 or more fatal accidents in 1989 if, in fact, $\mu = 3.5$?

b   Given that 11 accidents actually did occur in 1989, is it likely that the average number of fatal accidents is still 3.5? Explain.

4.65 The proliferation of nuclear weapons is an important concern to all Americans. In fact, almost 70% of all Americans think that the top U.S. foreign policy goal should be preventing the spread of nuclear weapons (*U.S. News & World Report,* May 23, 1994). To check the accuracy of this statement, you take a random sample of 600 Americans.

a What is the expected number and standard deviation of x, the number in the sample who think that the top U.S. goal should be preventing the spread of nuclear weapons?

b Suppose that in your sample there are 365 respondents who feel that the top U.S. goal should be preventing the spread of nuclear weapons. Is this an unusual observation, assuming that the 70% figure is correct? [*Hint:* Calculate a z-score for this observation.]

c Based on the results of part (b), what conclusions might you draw?

4.66 A certified public accountant (CPA) has found that nine out of ten company audits contain substantial errors. The CPA begins with the first ten company accounts for audit.

a What is the probability that the CPA finds nine company accounts with substantial errors?

b What is the probability that the CPA finds at most nine company accounts with substantial errors?

c What is the probability that the CPA finds at least nine company accounts with substantial errors?

*4.67 A shipment of 200 portable television sets is received by a retailer. To protect himself against a "bad" shipment, he will inspect five sets and accept the entire lot if he observes zero or one defective. Suppose there are actually 20 defective sets in the shipment.

a What is the probability that he accepts the entire shipment?

b Given that the retailer accepts the entire lot, what is the probability that he observed exactly one defective set?

4.68 A study conducted by a bank found that the average number of transaction errors per cashier per day was 1.5. Is it likely that any one cashier will make more than four transaction errors per day? Explain.

*4.69 The binomial probability distribution receives its name from the fact that the values of $p(x)$ for n trials and probability of success p correspond to the terms of the expansion of $(q + p)^n$. For example, for $n = 2$ trials,

$$(q + p)^2 = q^2 + 2pq + p^2$$

where $p(0) = q^2$, $p(1) = 2pq$, and $p(2) = p^2$. Expand $(q + p)^3$. Then use the formula for $p(x)$ to show that $p(0) = q^3$, $p(1) = 3q^2p$, and so on.

*4.70 This exercise will give you an opportunity to see how the formula for a binomial probability distribution is derived.

a List the simple events for a binomial experiment with $n = 3$ and probability of success on a single trial equal to p.

b Calculate the probabilities associated with the simple events.

c Assign the appropriate simple events to the events $x = 0$, 1, 2, and 3.

d Calculate $p(x)$ for $x = 0$, 1, 2, and 3 by summing the probabilities of the appropriate simple events.

e Use the formula for $p(x)$, $n = 3$, to find $p(x)$ for $x = 0$, 1, 2, and 3. Compare the result with your answer to part (d).

4.71 In an attempt to minimize the number of complaints by purchasers of knockdown nightstands because of missing screws and fasteners, it was decided to use an acceptance plan with $n = 10$ units per shift checked for missing screws or fasteners, with an acceptance number of $a = 0$.

a Find the acceptance probabilities for this plan for $p = 0$, .05, .10, .20, and .30.

b Use the results of part (a) to construct an operating characteristic curve.

c If one or more units are found to have missing screws or fasteners, all units for a given shift will be checked and missing items replaced. If you were the manufacturer, how well would you expect this plan to work?

C H A P T E R

5

THE NORMAL AND OTHER CONTINUOUS PROBABILITY DISTRIBUTIONS

About This Chapter

Several discrete random variables and their probability distributions were presented in Chapter 4. The objective of this chapter is to introduce you to the normal random variable, one of the most important and most commonly encountered continuous random variables. We give its probability distribution, and we show how the probability distribution can be used.

Contents

How Much Is a Full Tank of Gas Worth?

Buying a new vehicle is always an exciting adventure, because each of us has different expectations of what our new vehicle—whether it be a car, a van, or a pickup truck—will be like and how it will perform. Once we have decided on the color, the type of vehicle, and the options that we want included with the vehicle that we purchase, we are faced with a number of other criteria that involve decisions we must make. Is our chosen vehicle fuel-efficient in city driving as well as freeway driving? How different is the braking distance when the roadway is wet compared to when it is dry? What is the driving range of the vehicle that we are going to choose?

In comparing the average miles per gallon (mpg) for city driving and freeway driving, the 20 vehicles that we selected from five issues of *Consumer Reports* (January–August 1994) ranged from 10 to 17 mpg for city driving and from 21 to 41 mpg for freeway driving. The driving range with a full tank of gas varied from 340 to 495 miles. In fact, the average driving range was 418.0 miles, the median and trimmed mean were 420.0 and 419.1, respectively, and the standard deviation was 45.8 miles. Since the median and the trimmed mean differ only slightly from the mean, we would expect the driving range shown in Figure 5.1 to be mound-shaped and, if many more vehicles were tested, most likely normally distributed. Variables such as those reported here tend to be normally distributed, as are other variables that reflect the many small but important factors that determine their values.

Apart from paint colors and other optional items that you might be able to include in your new purchase, do vehicles really differ in those characteristics that may ultimately save you money and, in the case of driving range, save you from the predicament of being stranded with an empty gas tank?

The normal curve, used as a model for the relative frequency distributions for many continuous random variables, is the topic of Chapter 5. We will examine its properties, learn how it can be used to calculate probabilities, and see how we might be able to decide which vehicle to select based upon rational decisions and the normal probability distribution.

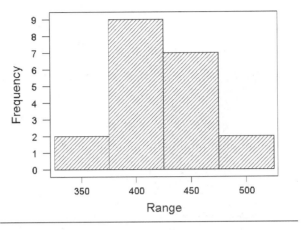

Source: Consumer Reports, January–August 1994.

5.1 Probability Distributions for Continuous Random Variables

When a random variable x is discrete, we can assign a positive probability to each value that x can take and obtain the probability distribution for x. The sum of all of the probabilities associated with the different values of x is 1. However, not all experiments result in random variables that are discrete. **Continuous random variables,** such as heights and weights, the length of life of a particular product, or the length of time between sales, can assume all the infinitely many values corresponding to points on a line interval. If we try to assign a positive probability to each of these uncountable values, the probabilities will no longer sum to 1, as with discrete random variables. Therefore, we must use a different approach to generate the probability distribution for a continuous random variable.

The probabilistic model for the frequency distribution of a continuous random variable involves the selection of a curve, usually smooth, called the **probability distribution** or **probability density function** of the random variable. If the equation of this continuous probability distribution is denoted as $f(x)$, then the probability that x falls in the interval $a < x < b$ is the area under the distribution curve for $f(x)$ between the two points a and b (see Figure 5.2). This agrees with the interpretation of a relative frequency histogram (Chapter 2), where areas over an interval under the histogram corresponded to the proportion of observations falling in that interval. Since the number of values that x may assume is infinitely large and uncountable, the probability that x equals some specific value, say a, is zero. Thus probability statements about continuous random variables always correspond to areas under the probability distribution over an interval, say a to b, and are expressed as $P(a < x < b)$. Note that the probability that $a < x < b$ is equal to the probability that $a \leq x \leq b$, because $P(x = a) = P(x = b) = 0$.

F I G U R E **5.2**
The probability distribution for a
continuous random variable

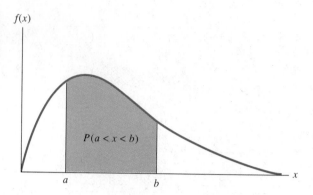

How do we choose the model—that is, the probability distribution $f(x)$—appropriate for a given experiment? Many types of continuous curves are available for modeling. Some are mound-shaped, like the one in Figure 5.2, but others are not. In general, we try to pick a model that

- fits the accumulated body of data
- allows us to make the best possible inferences using the data

Our model may not always fit the experimental situation perfectly, but we try to choose a model that *best fits* the population relative frequency histogram. The better our model approximates reality, the better our inferences will be. Fortunately, we will find that many continuous random variables have mound-shaped frequency distributions. A probability model that provides a good approximation to such a distribution is the **normal probability distribution,** the subject of Section 5.2.

5.2 The Normal Probability Distribution

In Section 5.1, we saw that the probabilistic model for the frequency distribution of a continuous random variable involves the selection of a curve, usually smooth, called the **probability distribution.** Although these distributions may assume a variety of shapes, a large number of random variables observed in nature possess a frequency distribution that is approximately bell-shaped or, as the statistician would say, is approximately a normal probability distribution. The formula that generates this distribution is shown below.

Normal Probability Distribution

$$f(x) = \frac{1}{\sigma\sqrt{2\pi}} e^{-(x-\mu)^2/(2\sigma^2)} \qquad -\infty < x < \infty$$

The symbols e and π are mathematical constants given approximately by 2.7183 and 3.1416, respectively; μ and σ ($\sigma > 0$) are parameters representing the population mean and standard deviation.

The graph of a normal probability distribution with mean μ and standard deviation σ is shown in Figure 5.3. The mean μ locates the *center* of the distribution, and the distribution is *symmetric* about its mean μ. Since the total area under the normal probability distribution is equal to 1, this implies that the area to the right of μ is .5 and the area to the left of μ is also .5. The *shape* of the distribution is determined by σ, the population standard deviation. Large values of σ reduce the height of the curve and increase the spread; small values of σ increase the height of the curve and reduce the spread. Figure 5.4 is an Execustat printout showing three normal probability distributions with different means and standard deviations. Notice the differences in shape and location.

F I G U R E **5.3**
Normal probability distribution

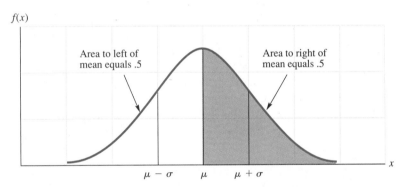

F I G U R E **5.4**
Execustat printout showing normal probability distributions with differing values of μ and σ

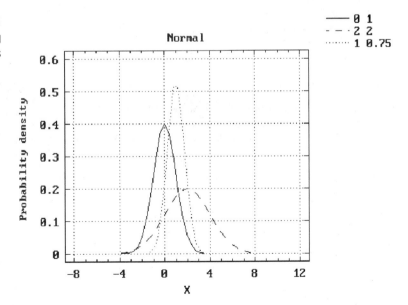

In practice, we seldom encounter variables that range from infinitely large negative values to infinitely large positive values. Nevertheless, many positive random variables (such as heights, weights, and times) generate a frequency histogram that is well approximated by a normal distribution. The approximation applies because almost all of the values of a normal random variable lie within three standard deviations of the mean, and in these cases ($\mu \pm 3\sigma$) almost always encompasses positive values.

5.3 Tabulated Areas of the Normal Probability Distribution

The probability that a continuous random variable assumes a value in the interval a to b is equal to the area under the probability density function between the points a and b (see Figure 5.2). However, since normal curves have different means and standard deviations (see Figure 5.4), we could generate an infinitely large number of normal distributions. A separate table of areas for each of these curves is obviously impractical. Instead, we would like to devise a standardization procedure that will allow us to use the same normal curve areas for all normal distributions.

Standardization is most easily accomplished by expressing the value of a normal random variable as the number of standard deviations to the left or right of the mean. In other words, the value of a normal random variable x with mean μ and standard deviation σ can be expressed as

$$z = \frac{x - \mu}{\sigma}$$

or, equivalently,

$$x = \mu + z\sigma$$

- When z is negative, x lies to the left of the mean μ.
- When $z = 0$, $x = \mu$.
- When z is positive, x lies to the right of the mean μ.

We will learn to calculate probabilities for x using $z = (x - \mu)/\sigma$, which is called the **standard normal random variable.** The probability distribution for z is called the **standardized normal distribution** because its mean is zero and its standard deviation is 1. It is shown in Figure 5.5. The area under the standard normal curve between the mean $z = 0$ and a specified value of z—say, z_0—is the probability $P(0 \leq z \leq z_0)$. This area is recorded in Table 3 of Appendix II and is shown as the shaded area in Figure 5.5. An abbreviated version of Table 3 in Appendix II is shown here in Table 5.1.

How can we find areas to the left of the mean? Since the standard normal curve is symmetric about $z = 0$ (see Figure 5.5), any area to the left of the mean can be found by using the equivalent area to the right of the mean.

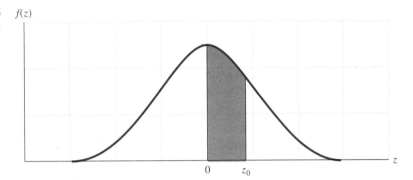

FIGURE **5.5**
Standardized normal distribution

TABLE **5.1**
Abbreviated version of Table 3 in
Appendix II

z_0	.00	.01	.02	.03	.04	.05	.06	.07	.08	.09
0.0	.0000	.0040	.0080	.0120	.0160	.0199	.0239	.0279	.0319	.0359
0.1	.0398	.0438	.0478	.0517	.0557	.0596	.0636	.0675	.0714	.0753
0.2	.0793	.0832	.0871	.0910	.0948	.0987	.1026	.1064	.1103	.1141
0.3	.1179	.1217	.1255	.1293	.1331	.1368	.1406	.1443	.1480	.1517
0.4	.1554	.1591	.1628	.1664	.1700	.1736	.1772	.1808	.1844	.1879
0.5	.1915	.1950	.1985	.2019	.2054	.2088	.2123	**.2157**	.2190	.2224
0.6	.2257									
		⋮	⋮	⋮	⋮	⋮	⋮	⋮	⋮	⋮
0.7	**.2580**									
⋮	⋮									
1.0	**.3413**									
⋮	⋮									
2.0	**.4772**									

Note that z, correct to the nearest tenth, is recorded in the left-hand column of the table. The second decimal place for z, corresponding to hundredths, is given across the top row. Thus the area between the mean and $z = .7$ standard deviation to the right, read in the second column of the table opposite $z = .7$, is found to be .2580. Similarly, the area between the mean and $z = 1.0$ is .3413. The area between $z = -1.0$ and the mean is also .3413. Thus the area lying within one standard deviation on either side of the mean would be two times .3413, or .6826. The area lying within two standard deviations of the mean, correct to four decimal places, is $2(.4772) = .9544$. These numbers agree with the approximate values, 68% and 95%, used in the Empirical Rule in Chapter 2.

To find the area between the mean and a point $z = .57$ standard deviation to the right of the mean, proceed down the left-hand column to the 0.5 row. Then move across the top row of the table to the .07 column. The intersection of this row–column combination gives the appropriate area, .2157.

Since the normal distribution is continuous, the area under the curve associated with a single point is equal to zero. Keep in mind that this result applies only to continuous random variables. Later in this chapter, we will use the normal probability distribution to approximate the binomial probability distribution. The binomial random variable x is a discrete random variable. Hence, as you know, the probability that x takes some specific value, say $x = 10$, will not necessarily equal zero.

E X A M P L E   **5.1**   Find $P(0 \leq z \leq 1.63)$. This probability corresponds to the area between the mean ($z = 0$) and a point $z = 1.63$ standard deviations to the right of the mean (see Figure 5.6).

F I G U R E **5.6**
Probability required for
Example 5.1

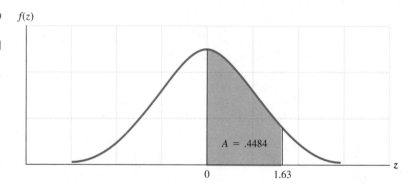

$f(z)$

$A = .4484$

0   1.63 z

Solution   The area is shaded and indicated by the symbol A in Figure 5.6. Since Table 3 in Appendix II gives areas under the normal curve to the right of the mean, we need only find the tabulated value corresponding to $z = 1.63$. Proceed down the left-hand column of the table to $z = 1.6$ and across the top of the table to the column marked .03. The intersection of this row and column combination gives the area $A = .4484$. Therefore, $P(0 \leq z \leq 1.63) = .4484$. ∎

E X A M P L E   **5.2**   Find $P(-.5 \leq z \leq 1.0)$. This probability corresponds to the area between $z = -.5$ and $z = 1.0$, as shown in Figure 5.7.

Solution   The area required is equal to the sum of A_1 and A_2 shown in Figure 5.7. From Table 3 in Appendix II we read $A_2 = .3413$. The area A_1 is equal to the corresponding area between $z = 0$ and $z = .5$, or $A_1 = .1915$. Thus the total area is

$$A = A_1 + A_2$$
$$= .1915 + .3413$$
$$= .5328$$ ∎

FIGURE **5.7**
Area under the normal curve in
Example 5.2

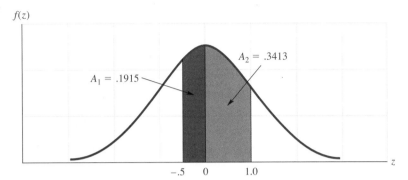

EXAMPLE **5.3** Find the value of z, say z_0, such that exactly (to four decimal places) .95 of the area is within $\pm z_0$ standard deviations of the mean.

Solution Half of the .95 area will lie to the left of the mean and half to the right, because the normal distribution is symmetrical. Therefore, we want to find the value z_0 corresponding to an area equal to .475. Referring to Table 3 in Appendix II, we see that the area .475 falls in the row corresponding to $z = 1.9$ and the .06 column. Therefore, $z_0 = 1.96$. Note that this result is very close to the approximate value, $z = 2$, used in the Empirical Rule. ■

EXAMPLE **5.4** Let x be a normally distributed random variable with mean equal to 10 and standard deviation equal to 2. Find the probability that x will lie between 11 and 13.6.

Solution As a first step, we must calculate the values of z corresponding to $x = 11$ and $x = 13.6$. Thus

$$z_1 = \frac{x_1 - \mu}{\sigma} = \frac{11 - 10}{2} = .5 \qquad z_2 = \frac{x_2 - \mu}{\sigma} = \frac{13.6 - 10}{2} = 1.8$$

The desired probability is therefore $P(.5 \le z \le 1.8)$ and is the area lying between z_1 and z_2, as shown in Figure 5.8. The area between $z = 0$ and z_1 is $A_1 = .1915$, and the area between $z = 0$ and z_2 is $A_2 = .4641$; these areas are obtained from Table 3. The desired probability is equal to the difference between A_2 and A_1; that is,

$$P(.5 \le z \le 1.8) = A_2 - A_1 = .4641 - .1915 = .2726 \quad ■$$

EXAMPLE **5.5** Studies show that gasoline use for compact cars sold in the United States is normally distributed, with a mean use of 30.5 miles per gallon (mpg) and a standard deviation of 4.5 mpg. What percentage of compacts obtain 35 or more mpg?

Solution The proportion of compacts obtaining 35 or more mpg is given by the shaded area in Figure 5.9.

FIGURE **5.8**
Area under the normal curve in Example 5.4

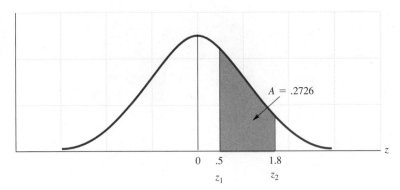

FIGURE **5.9**
Area under the normal curve in Example 5.5

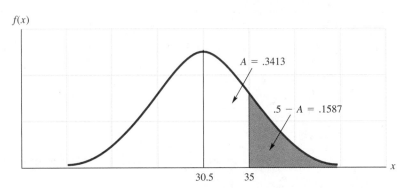

We must find the z value corresponding to $x = 35$. Substituting into the formula for z, we obtain

$$z = \frac{x - \mu}{\sigma} = \frac{35 - 30.5}{4.5} = 1.0$$

The area A to the right of the mean, corresponding to $z = 1.0$, is .3413 (from Table 3). Then the proportion of compacts having an mpg ratio equal to or greater than 35 is equal to the entire area to the right of the mean, .5, minus the area A:

$$P(x \geq 35) = .5 - P(0 \leq z \leq 1) = .5 - .3413 = .1587$$

The percentage exceeding 35 mpg is

$$100(.1587) = 15.87\% \quad \blacksquare$$

EXAMPLE **5.6** Refer to Example 5.5. In times of scarce energy resources, a competitive advantage is given to an automobile manufacturer who can produce a car obtaining substantially better fuel economy than the competitors' cars. If a manufacturer wishes to develop a compact car that outperforms 95% of the current compacts in fuel economy, what must the gasoline use rate for the new car be?

FIGURE **5.10**
The location of x_0 such that $P(x < x_0) = .95$

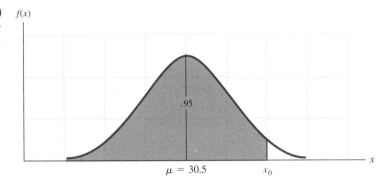

Solution Let x be a normally distributed random variable, with a mean of 30.5 and a standard deviation of 4.5. As shown in Figure 5.10, we want to find the value x_0 such that

$$P(x \le x_0) = .95$$

As a first step, we find

$$z_0 = \frac{x_0 - \mu}{\sigma} = \frac{x_0 - 30.5}{4.5}$$

and note that our required probability is the same as the area to the left of z_0 for the standardized normal distribution. Therefore,

$$P(z \le z_0) = .95$$

The area to the left of the mean is .5. The area to the right of the mean between z_0 and the mean is $.95 - .5 = .45$. Thus, from Table 3, we find that z_0 is between 1.64 and 1.65. Notice that the area .45 is exactly halfway between the areas for $z = 1.64$ and $z = 1.65$. Thus z_0 is exactly halfway between 1.64 and 1.65; that is, $z_0 = 1.645$. Substituting $z_0 = 1.645$ into the equation for z_0, we have

$$1.645 = \frac{x_0 - 30.5}{4.5}$$

Solving for x_0, we obtain

$$x_0 = (1.645)(4.5) + 30.5 = 37.9$$

The manufacturer's new compact car must therefore obtain a fuel economy of 37.9 mpg to outperform 95% of the compact cars currently available on the U.S. market. ∎

EXAMPLE **5.7** The salaries of MBA graduates who entered the field of marketing services averaged approximately $45,000, with a standard deviation of $2250. If these salaries were normally distributed, what proportion of MBA graduates who entered marketing services had salaries in excess of $47,500, which is the average salary for those graduates entering the field of brand/product management?

F I G U R E **5.11**
Area under the normal curve for
Example 5.7

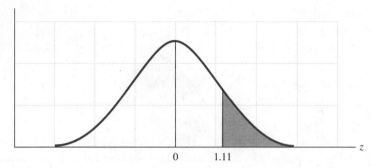

Solution   In order to find the proportion of MBA graduates whose salaries exceeded $47,500, we need the value of z corresponding to $47,500$. With $\mu = \$45,000$ and $\sigma = \$2250$,

$$z = \frac{x - \mu}{\sigma} = \frac{47,500 - 45,000}{2250} = 1.11$$

Next, we need to find the area under a normal curve to the right of $z = 1.11$, as shown in Figure 5.11. Hence, the required area is equal to .5, the total area to the right of zero, *minus* the area between 0 and 1.11.

$$P(x \geq 47,500) = .5 - .3665 = .1335$$

Therefore, 13.35% of the MBA graduates had salaries exceeding $47,500.   ▪

Exercises

Basic Techniques

5.1   Using Table 3 in Appendix II, calculate the area under the normal curve between these z values.
  a $z = 0$ and $z = 1.6$
  b $z = 0$ and $z = 1.83$

5.2   Repeat Exercise 5.1 for these z values.
  a $z = 0$ and $z = .90$
  b $z = 0$ and $z = -.90$

5.3   Repeat Exercise 5.1 for these z values.
  a $z = -1.3$ and $z = 1.8$
  b $z = .6$ and $z = 1.2$

5.4   Repeat Exercise 5.1 for these z values.
  a $z = -1.4$ and $z = 1.4$
  b $z = -2.0$ and $z = 2.0$
  c $z = -3.0$ and $z = 3.0$

5.5   Repeat Exercise 5.1 for these z values.
  a $z = -1.43$ and $z = .68$

b $z = .58$ and $z = 1.74$

c $z = -1.55$ and $z = -.44$

5.6 Find a z_0 such that $P(z > z_0) = .025$.

5.7 Find a z_0 such that $P(z < z_0) = .9251$.

5.8 Find a z_0 such that $P(z < z_0) = .2981$.

5.9 Find a z_0 such that $P(z > z_0) = .6985$.

5.10 Find a z_0 such that $P(-z_0 < z < z_0) = .4714$.

5.11 Find a z_0 such that $P(z < z_0) = .05$.

5.12 Find a z_0 such that $P(-z_0 < z < z_0) = .90$.

5.13 Find a z_0 such that $P(-z_0 < z < z_0) = .99$.

5.14 A variable x is normally distributed with mean $\mu = 10$ and standard deviation $\sigma = 2$. Find these probabilities.

a $P(x > 13.5)$

b $P(x < 8.2)$

c $P(9.4 < x < 10.6)$

5.15 A variable x is normally distributed with mean $\mu = 1.20$ and standard deviation $\sigma = .15$. Find the probability that x falls in the interval given.

a $1.00 < x < 1.10$

b $x > 1.38$

c $1.35 < x < 1.50$

5.16 A variable is normally distributed with unknown mean μ and standard deviation $\sigma = 2$. If the probability that x exceeds 7.5 is .8023, find μ.

5.17 A variable is normally distributed with unknown mean μ and standard deviation $\sigma = 1.8$. If the probability that x exceeds 14.4 is .3, find μ.

5.18 A variable is normally distributed with unknown mean and standard deviation. The probability that x exceeds 4 is .9772, and the probability that x exceeds 5 is .9332. Find μ and σ.

Applications

5.19 One method of arriving at economic forecasts is to use a consensus approach. A forecast is obtained from each of a large number of analysts; the average of these individual forecasts is the consensus forecast. Suppose that the individual 1995 January prime interest rate forecasts of all economic analysts are approximately normally distributed with mean equal to 7.75% and a standard deviation of 1.6%. A single analyst is randomly selected from among this group.

a What is the probability that the analyst's forecast of the prime interest rate will exceed 9%?

b What is the probability that the analyst's forecast of the prime interest rate will be less than 6%?

5.20 Suppose that you must establish regulations concerning the maximum number of people who can occupy an elevator. A study of elevator occupancies indicates that if eight people occupy the elevator, the probability distribution of the total weight of the eight people possesses a mean equal to 1200 pounds and a variance equal to 9800 (pounds)^2. What is the probability that the total weight of eight people exceeds 1300 pounds? 1500 pounds? (Assume that the probability distribution is approximately normal.)

5.21 The discharge of suspended solids from a phosphate mine is normally distributed, with a mean daily discharge of 27 milligrams per liter (mg/l) and a standard deviation of 14 mg/l. What proportion of days will the daily discharge exceed 50 mg/l?

5.22 Philatelists (stamp collectors) often buy stamps at or near retail prices, but when they sell the price is considerably lower. For example, it may be reasonable to assume that (depending on the collection, its condition, the demand, economic conditions, etc.) a collection may be expected to sell at x percent of retail price, where x is normally distributed with a mean equal to 45% and a standard deviation of 4.5%. A philatelist has a collection to sell that has a retail value of $30,000.

a What is the probability that the philatelist receives more than $15,000 for the collection?

b What is the probability that the philatelist receives less than $15,000 for the collection?

c What is the probability that the philatelist receives less than $12,000 for the collection?

5.23 How does the Internal Revenue Service (IRS) decide on the percentage of income tax returns to audit for each state? Suppose that it did so by randomly selecting 50 values from a normal distribution with a mean equal to 1.55% and a standard deviation equal to .45%. (Computer programs are available for this type of sampling.)

a What is the probability that a particular state will have more than 2.5% of its income tax returns audited?

b What is the probability that a state will have less than 1% of its income tax returns audited?

5.24 In an effort to boost the quality of production of its American workers, the Saturn Corporation is rewarding its workers with an average of $2800 in year-end bonuses for meeting quality production and profitability targets in 1993 ("Saturn Workers," 1994). Suppose that these bonuses are approximately normally distributed with a standard deviation of $500.

a What is the probability that a worker receives a year-end bonus of more than $3500?

b Ninety-five percent of all workers will receive year-end bonuses within what limits?

5.25 American consumers are becoming more and more aware of the costs of heating fuels. As these costs rise, consumers consider alternative fuels, home insulation improvements, and new heating systems altogether. Suppose that the cost of natural gas per metric cubic foot (MCF) is normally distributed with a mean of $6.00 and a standard deviation of $1.20.

a What is the probability that the cost of natural gas per MCF for a particular consumer is between $7.60 and $8.00?

b What is the median cost per MCF for natural gas?

c What are the upper and lower quartiles for the cost per MCF of natural gas?

5.4 The Normal Approximation to the Binomial Probability Distribution

Many probability distributions possess a useful characteristic. When certain conditions are satisfied, these distributions become approximately normal in shape. The binomial probability distribution is one of these. In particular, when the number n of trials in a binomial experiment is large and p is not too close to 0 or 1, the binomial probability distribution assumes a shape that is closely approximated by a normal curve with mean $\mu = np$ and standard deviation $\sigma = \sqrt{npq}$. This particular property of the binomial probability distribution is important when we have to calculate binomial probabilities $p(x)$ for large values of n. The great labor and tedium encountered in these calculations can be avoided by using the normal approximating curve.

Since the best way to show how and why the normal approximation works is to use graphs and a small value of n, we will illustrate the procedure for a binomial probability distribution with $n = 10$ and $p = 1/2$. The probability histogram for a

F I G U R E **5.12**
Comparison of a binomial
probability distribution and the
approximating normal
distribution, $n = 10$,
$p = 1/2(\mu = np = 5$;
$\sigma = \sqrt{npq} = 1.58)$

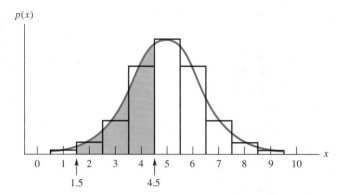

binomial probability distribution, $n = 10, \;\; p = 1/2$, is shown in Figure 5.12 along with an approximating normal curve with

$$\mu = np = 10\left(\frac{1}{2}\right) = 5$$

$$\sigma = \sqrt{npq} = \sqrt{(10)\left(\frac{1}{2}\right)\left(\frac{1}{2}\right)} = \sqrt{2.5} = 1.58$$

A visual inspection of the figure suggests that the approximation is reasonably good, even though a small sample, $n = 10$, was necessary for this graphical illustration.

Suppose that we wish to approximate the probability that x equals 2, 3, or 4. You can see in Figure 5.12 that this probability is exactly equal to the area of the three rectangles lying over $x = 2, \;\; 3$, and 4. We can approximate this probability with the area under the normal curve from $x = 1.5$ to $x = 4.5$, which is shaded in Figure 5.12. Note that the area under the normal curve between $x = 2$ and $x = 4$ *would not* be a good approximation to the probability that $x = 2, \;\; 3$, and 4 because it would exclude one-half of the probability rectangles corresponding to $x = 2$ and $x = 4$. To get a good approximation, you must remember to approximate the entire areas of the probability rectangles corresponding to $x = 2$ and $x = 4$ by including the area under the normal curve from $x = 1.5$ to $x = 4.5$.

Although the normal probability distribution provides a reasonably good approximation to the binomial probability distribution in Figure 5.12, this will not always be the case. When the mean np of a binomial probability distribution is near zero or n, the binomial probability distribution will be nonsymmetrical.[†] For example, when p is near zero, most values of x will be small, producing a distribution that is concentrated near $x = 0$ and that tails gradually toward n (see Figure 5.13). Certainly, when this is true, the normal distribution, symmetrical and bell-shaped, will provide a poor approximation to the binomial probability distribution. How, then, can we tell whether n and p are such that the binomial distribution will be symmetrical?

[†]A skewed binomial probability distribution can be approximated by a Poisson probability distribution. This approximation, which is discussed in Section 4.3, is satisfactory when n is large and np is small, say $np < 7$.

F I G U R E **5.13**
Comparison of a binomial
probability distribution (shaded)
and the approximating normal
distribution, $n = 10$,
$p = .1$ ($\mu = np = 1$;
$\sigma = \sqrt{npq} = .95$)

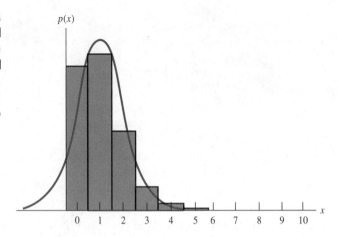

Recall the Empirical Rule from Chapter 2; approximately 95% of the measurements associated with a normal distribution will lie within two standard deviations of the mean and almost all will lie within three. We suspect that the binomial probability distribution would be nearly symmetrical if the distribution were able to spread out a distance equal to two standard deviations on either side of the mean, and this is, in fact, the case. **Hence, to determine when the normal approximation will be adequate, calculate $\mu = np$ and $\sigma = \sqrt{npq}$. If the interval $(\mu \pm 2\sigma)$ lies within the binomial bounds, 0 and n, the approximation will be adequate. The approximation will be good if the interval $(\mu \pm 3\sigma)$ lies in the interval 0 to n.** Note that this criterion is satisfied for the binomial probability distribution of Figure 5.12, but it is not satisfied for the distribution shown in Figure 5.13.

The formulas for the **normal approximation to the binomial probability distribution** are given in the following display.

The Normal Approximation to the Binomial Probability Distribution

Approximate the binomial probability distribution by using a normal curve with

$$\mu = np$$
$$\sigma = \sqrt{npq}$$

where $n = $ Number of trials
          $p = $ Probability of success on a single trial
          $q = 1 - p$

The approximation will be adequate when n is large and when the interval

$$\mu \pm 2\sigma$$

falls between 0 and n.

E X A M P L E   **5.8**   To see how well the normal curve can be used to approximate binomial probabilities, refer to the binomial experiment illustrated in Figure 5.12, with $n = 10$, $p = .5$. Calculate the probability that $x = 2$, 3, or 4, correct to three decimal places, using Table 1 in Appendix II. Then calculate the corresponding normal approximation to this probability.

Solution   The exact probability can be calculated with $n = 10$ by using Table 1 in Appendix II. Thus

$$\sum_{x=2}^{4} p(x) = \sum_{x=0}^{4} p(x) - \sum_{x=0}^{1} p(x)$$
$$= .377 - .011$$
$$= .366$$

In Figure 5.12, the binomial probability rectangles for $x = 2$, 3, and 4 correspond to the area between $x_1 = 1.5$ and $x_2 = 4.5$ under the approximating normal curve with $\mu = 5$ and $\sigma = 1.58$. The corresponding values of z are

$$z_1 = \frac{x_1 - \mu}{\sigma} = \frac{1.5 - 5}{1.58} = -2.22$$
$$z_2 = \frac{x_2 - \mu}{\sigma} = \frac{4.5 - 5}{1.58} = -.32$$

This probability is shown in Figure 5.14. The area between $z = 0$ and $z = 2.22$ is $A_1 = .4868$. Likewise, the area between $z = 0$ and $z = .32$ is $A_2 = .1255$. From Figure 5.14,

$$P(-2.22 \leq z \leq -.32) = .4868 - .1255 = .3613$$

Note that the normal approximation is quite close to the exact binomial probability, .366, obtained from Table 1.   ■

You must be careful not to exclude half of the two extreme probability rectangles when using the normal approximation to the binomial probability distribution. **The x values used to calculate z values always end in .5.**

FIGURE **5.14**
Area required for Example 5.8

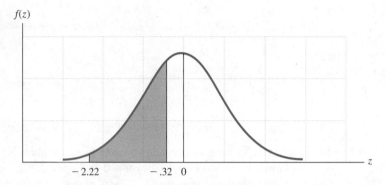

$f(z)$

-2.22 $-.32$ 0 z

Correction for Continuity

The procedure of adding or subtracting .5 in adjusting the values of x for the binomial distribution to those for the approximating normal distribution is called the **correction for continuity.**

To be certain that you have made the appropriate correction for continuity, always draw a rough sketch similar to Figure 5.12.

EXAMPLE **5.9**   The reliability of an electrical fuse is the probability that a fuse, chosen at random from production, will function under the conditions for which it has been designed. A random sample of 1000 fuses was tested and $x = 27$ defectives were observed. Calculate the probability of observing 27 or more defectives, assuming that the fuse reliability is .98.

Solution   The probability of observing a defective when a single fuse is tested is $p = .02$, given that the fuse reliability is .98. Then

$$\mu = np = 1000(.02) = 20$$
$$\sigma = \sqrt{npq} = \sqrt{1000(.02)(.98)} = 4.43$$

The probability of 27 or more defective fuses, given $n = 1000$, is

$$P = P(x \geq 27)$$
$$= p(27) + p(28) + p(29) + \cdots + p(999) + p(1000)$$

The normal approximation to P is the area under the normal curve to the right of $x = 26.5$. In making the correction for continuity, we must use $x = 26.5$ rather than $x = 27$ so as to include the entire probability rectangle associated with $x = 27$. The z value corresponding to $x = 26.5$ is

$$z = \frac{x - \mu}{\sigma} = \frac{26.5 - 20}{4.43} = \frac{6.5}{4.43} = 1.47$$

and the area between $z = 0$ and $z = 1.47$ is equal to .4292, as shown in Figure 5.15. Since the total area to the right of the mean is equal to .5,

$$P(x \geq 27) \approx .5 - .4292 = .0708 \quad \blacksquare$$

FIGURE 5.15
Normal approximation to the binomial for Example 5.9

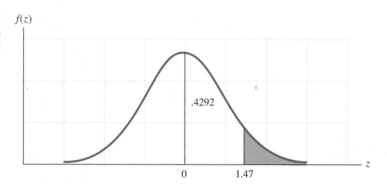

Exercises

Basic Techniques

5.26 Let x be a binomial random variable with $n = 25$, $p = .3$.

a Use Table 1 in Appendix II to find $P(8 \leq x \leq 10)$.

b Find μ and σ for the binomial distribution, and use the normal approximation to find $P(8 \leq x \leq 10)$. Compare the approximation with the exact value calculated in part (a).

5.27 Find the normal approximation to $P(x \geq 6)$ for a binomial probability distribution with $n = 10$, $p = .5$.

5.28 Find the normal approximation to $P(x > 6)$ for a binomial probability distribution with $n = 10$, $p = .5$.

5.29 Find the normal approximation to $P(x > 22)$ for a binomial probability distribution with $n = 100$, $p = .2$.

5.30 Find the normal approximation to $P(x \geq 22)$ for a binomial probability distribution with $n = 100$, $p = .2$.

5.31 Let x be a binomial random variable for $n = 25$, $p = .2$.

a Use Table 1 in Appendix II to calculate $P(4 \leq x \leq 6)$.

b Find μ and σ for the binomial probability distribution, and use the normal distribution to approximate the probability $P(4 \leq x \leq 6)$. Note that this value is a good approximation to the exact value of $P(4 \leq x \leq 6)$.

5.32 Consider a binomial experiment with $n = 20$, $p = .4$. Calculate $P(x \geq 10)$ by use of the following.

a Table 1 in Appendix II.

b The normal approximation to the binomial probability distribution.

5.33 Find the normal approximation to $P(355 \leq x \leq 360)$ for a binomial probability distribution with $n = 400$, $p = .9$.

Applications

5.34 Some brands of major appliances are more reliable than others. For example, approximately 10% of Maytag electric dryers purchased from 1986 to 1992 have ever needed repair ("Getting Things Fixed," 1994). Suppose that a consumer group surveys 56 owners of Maytag electric dryers.

a What is the probability that ten or more have ever needed repair?

b What is the probability that fewer than five have ever needed repair?

c What assumptions must you make in order for the probabilities found in parts (a) and (b) to be accurate?

d If the survey showed that 15 of the 56 dryers had needed repair, should you suspect that the 10% figure is incorrect? Explain.

5.35 Airlines and hotels often grant reservations in excess of capacity to minimize losses due to no-shows. Suppose that the records of a motel show that, on the average, 10% of their prospective guests will not claim their reservation. If the motel accepts 215 reservations and there are only 200 rooms in the motel, what is the probability that all guests who arrive to claim a room will receive one?

5.36 How old is an average board of directors? Sixty-seven percent of financial institutions have boards of directors whose average age is 57 years or more (adapted from *American Demographics,* November 1990, p. 22).

a In a random sample of $n = 400$ financial institutions, what is the probability that 300 or more boards of directors have an average age of 57 years or more?

b If the 67% figure is correct, the number of boards with an average age of 57 years or more should lie between what two values with probability .95? (Do not use the correction for continuity.)

5.37 Service and support have become a very important issue to PC users as the prices of personal computers have become more and more alike. Companies with quick expert technical support find that their customers are satisfied, even if they have a problem with their computer. For example, 82% of customers who have had problems with their Dell desktop computer would be willing to purchase another PC from Dell (Amirrezvani, 1994). Suppose a random sample of 200 Dell customers who have had problems with their desktop computer is interviewed; use the normal approximation to the binomial to answer the following questions.

a What is the probability of observing as few as 160 customers who would be willing to buy another Dell PC?

b Within what limits would you expect the number of customers willing to buy another Dell PC to lie?

5.38 The age distribution of householders is an important tool for marketers interested in providing age-appropriate advertising for the particular product they wish to market. A study conducted by the Joint Center for Housing Studies estimates that, in 1995, 31% of all householders will be between 45 and 64 (Darnay, 1994). Suppose a sample of 500 householders is taken in 1995.

a What is the probability that fewer than 135 householders will be between 45 and 64?

b What is the probability that between 135 and 180 householders inclusive will be between 45 and 64?

5.39 In the first quarter of 1994, the median income nationally was $39,900 ("Midwest, South," 1994). Suppose that 25 wage earners are randomly selected and their incomes are recorded.

a Use Table 1 in Appendix II to find the probability that at least 20 wage earners have incomes exceeding the nationwide median.

b Use the normal approximation to the binomial distribution to approximate the probability found in part (a). How does your approximation compare to the actual probability?

c   If the sample that you selected was restricted to wage earners living in one particular geographical area, what might the probability calculated in part (a) imply about the representativeness of your sample?

5.5   Other Useful Continuous Probability Distributions (Optional)

Not all relative frequency distributions are approximately normal, or even mound-shaped. Some distributions are relatively flat, others may be unimodal but skewed toward one tail, while others may be symmetric but U-shaped.

The Uniform Probability Distribution

The **uniform probability distribution** provides a simple model for a random variable that can randomly assume any value in an interval from a to b. The uniform probability density function has a rectangular shape over the interval from a to b with a height equal to $1/(b-a)$, as shown in Figure 5.16. The area under the uniform probability density function is equal to the length of the rectangle $(b-a)$ multiplied by its height $1/(b-a)$, or

$$(b-a)\left[\frac{1}{(b-a)}\right]=1$$

Probabilities associated with a uniform distribution are found simply as areas of rectangles with height $1/(b-a)$. Random variables that might be expected to have a uniform distribution are rounding errors, lengths of pieces of lumber remaining after lumber is cut into 8-foot lengths, and the predicted closing price of an OTC stock one month from today.

FIGURE **5.16**
The uniform probability
distribution

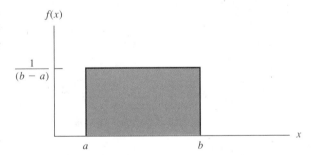

The probability density function, together with the mean and standard deviation of a uniform random variable x, is given in the display that follows. Notice that the mean is the midpoint of the interval a to b, falling exactly halfway between a and b, and that the standard deviation increases as the length of the interval increases.

The Uniform Probability Distribution

$$f(x) = \begin{cases} \dfrac{1}{(b-a)}, & a \le x \le b \\ 0, & \text{otherwise} \end{cases}$$

with

$$\mu = \frac{1}{2}(b+a) \quad \text{and} \quad \sigma = \frac{(b-a)}{\sqrt{12}}$$

EXAMPLE 5.10   With a sales tax of 6.75%, the amount of taxes paid on a purchase is rounded to the nearest cent. Assume that x, the rounding error, is uniformly distributed over the interval from $-.5$ to $.5$, where the rounding error is the difference between the rounded value and the actual value.

a   What is the probability that the amount will be rounded up to the nearest cent?

b   What is the probability that the rounding error is less than .2 in absolute value?

c   What is the probability that x lies in the interval $\mu \pm 2\sigma$? How does this compare with Tchebysheff's Theorem?

Solution   The probability distribution for x, the rounding error, is given by the uniform distribution over the interval $-.5$ to $.5$. Hence, $b = .5$ and $a = -.5$, so that

$$f(x) = \begin{cases} 1, & -.5 \le x \le .5 \\ 0, & \text{otherwise} \end{cases}$$

a   The amount of taxes will be rounded up if $0 \le x \le .5$. Therefore, $P(0 \le x \le .5)$ is equal to the area above the interval from 0 to .5, which is given by $1 \times (.5 - 0) = .5$.

b   If the rounding error is less than .2 in *absolute value,* we wish to find

$$P(-.2 < x < .2)$$

which is given by the area above the interval from $-.2$ to $.2$, or

$$1 \times [.2 - (-.2)] = .4$$

c   To find the endpoints of the interval $\mu \pm 2\sigma$, we first find

$$\mu = \frac{a+b}{2} = \frac{-.5 + .5}{2} = 0$$

and

$$\sigma = \frac{b-a}{\sqrt{12}} = \frac{.5 - (-.5)}{\sqrt{12}} = .29$$

Therefore, $\mu \pm 2\sigma = 0 \pm 2(.29)$ is the interval from $-.58$ to $.58$, and it contains all or 100% of the rounding errors. This is greater than the minimum 75% specified by Tchebysheff's Theorem.   ▪

The Exponential Probability Distribution

The **exponential probability distribution** is a model for random variables that represent waiting times. The time until a machine breaks down, the waiting time in a service line, and the length of life of a piece of industrial equipment are examples of random variables that could be expected to have exponential distributions. The probability density function and the mean and standard deviation of an exponential random variable are given in the display that follows.

The Exponential Probability Distribution

$$f(x) = \begin{cases} \lambda e^{-\lambda x}, & \lambda > 0, \; x \geq 0 \\ 0 & \text{otherwise} \end{cases}$$

with

$$\mu = \frac{1}{\lambda} \quad \text{and} \quad \sigma = \frac{1}{\lambda}$$

The exponential distribution depends upon only one parameter, λ. The mean and standard deviation for the exponential distribution are both equal to $1/\lambda$. Graphs of the probability density function of the exponential random variable x for several values of λ are shown in Figure 5.17.

When the mean number of events occurring in a unit time follows a Poisson distribution with mean λ, the waiting time between events follows an exponential distribution with mean $1/\lambda$. For example, if a stamping machine experiences an average of $\lambda = 2$ breakdowns per 80-hour workweek, then the average time between breakdowns would be $1/\lambda = 1/2$ of an 80-hour workweek, or 40 working hours.

Probabilities associated with an exponential random variable are most easily found in terms of right-tailed probabilities, as shown in the following display.

Right-Tailed Probabilities for the Exponential Distribution

$$P(x \geq a) = e^{-\lambda a}, \quad a \geq 0 \text{ and } \lambda > 0$$

These probabilities are easily found on a calculator with an exponential key.

FIGURE **5.17**
The exponential probability
density function

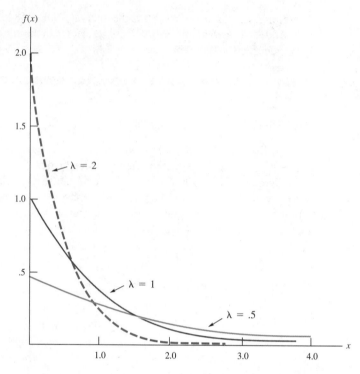

EXAMPLE 5.11 Suppose that the delivery time of an item after placing a factory order follows an exponential distribution with a mean of 10 days. What is the probability that it takes longer than 3 weeks (21 days) from the date of order to the day of delivery?

Solution If the mean of the delivery-time distribution is 10 days, then $\lambda = 1/10 = .1$. The probability of interest is

$$P(x > 21) = e^{-(.1)(21)} = e^{-2.1} = .122456$$

so the probability that delivery takes longer than 21 days is .122. ∎

In the next example, we explore the relationship between the Poisson and the exponential distributions.

EXAMPLE 5.12 Suppose that the number of breakdowns of a machine follows a Poisson distribution with a mean of two breakdowns per 40-hour week.

a What is the distribution of the time between breakdowns?

b What is the probability that the next breakdown will occur before 24 hours?

Solution   **a**   If the mean number of breakdowns has a Poisson distribution with a mean of 2, then the time between breakdowns follows an exponential distribution with mean $\frac{1}{2}$, or one-half of the 40-hour workweek. Therefore, the waiting time follows an exponential distribution with a mean $1/\lambda = 20$ hours, or $\lambda = .05$.

b   The required probability is $P(x < 24)$, which, when expressed in terms of its complement, becomes

$$P(x < 24) = 1 - P(x > 24)$$
$$= 1 - e^{-(.05)(24)} = 1 - e^{-1.2} = 1 - .301194 = .698806$$

so the probability that the next breakdown will occur within 24 hours is almost equal to .7.   ∎

The exponential distribution is said to have the *memoryless property,* which states that the probability distribution of the length of time between one event and the next is independent of what has occurred earlier. When this property is used to model the length of life of industrial equipment, the implication is that the equipment operates without fatigue or aging. This, of course, would be true if a repair following a breakdown restores the equipment to its condition when new.

Other Continuous Distributions

The normal, uniform, and exponential distributions are only three among a large number of probability distributions that are used and studied by statisticians, econometricians, and industrial engineers. In general, other distributions are not as simple to use in applications because of their dependence upon specialized tables that may or may not be widely available. For example, the **beta probability distribution,** which depends upon two parameters, is useful in modeling variables, such as percentages and reliabilities, that vary between 0 and 1. The distribution is very versatile; depending on the values of its parameters, this distribution can be mound-shaped, J-shaped, or U-shaped.

The **lognormal probability distribution** is another family of distributions that can be used as models for the distribution of incomes and other skewed distributions. If x has a lognormal distribution, then the logarithm of x, log x, has a normal distribution.

The beta, lognormal, and other families of distributions require a knowledge of integral and differential calculus and are usually covered as topics in advanced courses in statistics and econometrics.

5.6 HOW MUCH IS A FULL TANK OF GAS WORTH?

If the figures reported in the case study are typical of the averages across the different makes and models of cars for city-driving mpg, expressway mpg, average range per full tank of gas, and so on, what can we conclude about the car that we purchase? If we were to purchase a Pontiac Grand Prix, the average driving range is given as 340 miles. On the other hand, if we were to purchase a Toyota Camry LE, its average driving range is given as 495 miles, 155 miles more than the Grand Prix. Although the range in miles that we hope to achieve depends upon our speed, the number of red traffic lights we encounter in city driving, and the number of times we accelerate to pass that slow-moving vehicle in front of us on an expressway, the range that we can cover before refueling is obviously limited. If the overall mean driving range for all cars is 418.0 miles with a standard deviation of 45.8, what is the probability that our driving range will be less than 340 miles? If we designate x as the variable representing driving range, we wish to evaluate $P(x < 340)$. If the average driving range is 418.0 miles with a standard deviation of 45.8 miles, then

$$P(x < 340) = P\left(\frac{x - \mu}{\sigma} < \frac{340 - 418.0}{45.8}\right)$$
$$= P(z < -1.70) = .5 - .4554 = .0446$$

It seems apparent that the Pontiac Grand Prix has a driving range that is less than that of almost 96% of other cars in the study. What about the Toyota Camry LE? The reported driving range for this vehicle was 495 miles, and

$$P(x > 495) = P\left(\frac{x - \mu}{\sigma} > \frac{495 - 418.0}{45.8}\right) = P(z > 1.68) = .0465$$

In this case, only about 5% of the cars on the market have a driving range greater than that of the Toyota Camry LE.

Our calculations were based on the assumption that the mean and standard deviation in the driving ranges of the 20 vehicles in our study were, in fact, those for the population consisting of cars of all makes. Nonetheless, not only were the z-values that we calculated measures of relative standing of these vehicles, but the assumption that driving ranges were normally distributed allowed us to attach probabilities to these values. If times between refueling and the distance covered with a full tank of gas are important to you, then your selection should be a car with a driving range that exceeds almost all other cars with the options you wish to have.

5.7 Summary

Many continuous random variables observed in nature possess probability distributions that are bell-shaped and that can be approximated by the normal probability distribution of Section 5.2. As a case in point, the number x of successes associated with a binomial experiment can be approximated by a normal probability distribution when the number n of trials is large.

One explanation for this phenomenon is a mathematical result known as the Central Limit Theorem. In Chapter 6 we will learn why so many statistics possess probability distributions that are approximately normal and why the Central Limit Theorem and the normal distribution play such a prominent role in statistical inference.

Tips on Problem Solving

1   Always sketch a normal curve and locate the probability areas pertinent to the exercise. If you are approximating a binomial probability distribution, sketch in the probability rectangles as well as the normal curve.

2   Read each exercise carefully to see whether the data come from a binomial experiment or whether they possess a distribution that, by its very nature, is approximately normal. If you are approximating a binomial probability distribution, do not forget to make a half-unit correction so that you include the half rectangles at the ends of the interval. If the distribution is not binomial, *do not* make the half-unit corrections. If you make a sketch (as suggested in step 1), you will see why the half-unit correction is or is not needed.

Supplementary Exercises

Starred (*) exercises are optional.

5.40   Using Table 3 in Appendix II, calculate the area under the normal curve between these values.
   a $z = 0$ and $z = 1.2$
   b $z = 0$ and $z = -.9$

5.41   Repeat Exercise 5.40 for these values.
   a $z = 0$ and $z = 1.6$
   b $z = 0$ and $z = .75$

5.42   Repeat Exercise 5.40 for these values.
   a $z = 0$ and $z = 1.46$
   b $z = 0$ and $z = -.42$

5.43   Repeat Exercise 5.40 for these values.
   a $z = 0$ and $z = -1.44$
   b $z = 0$ and $z = 2.01$

5.44   Repeat Exercise 5.40 for these values.
   a $z = .3$ and $z = 1.56$
   b $z = .2$ and $z = -.2$

5.45   Repeat Exercise 5.40 for these values.

 a $z = .88$ and $z = 1.85$

 b $z = -.31$ and $z = 1.63$

5.46   Repeat Exercise 5.40 for these values.

 a $z = 1.21$ and $z = 1.75$

 b $z = -1.3$ and $z = 1.74$

5.47   Find the probability that z is greater than $-.75$.

5.48   Find the probability that z is less than 1.35.

5.49   Find a z_0 such that $P(z > z_0) = .9750$.

5.50   Find a z_0 such that $P(z > z_0) = .3594$.

5.51   Find a z_0 such that $P(-z_0 < z < z_0) = .8262$.

5.52   Find a z_0 such that $P(z < z_0) = .9505$.

5.53   Find a z_0 such that $P(-z_0 < z < z_0) = .7458$.

5.54   Find a z_0 such that $P(z < z_0) = .0968$.

5.55   Find a z_0 such that $P(z > z_0) = .5$.

5.56   Find a z_0 such that $P(z < z_0) = .8643$.

5.57   Find the probability that z lies between $z = .7$ and $z = 1.63$.

5.58   Let x be a normally distributed random variable with mean equal to 7 and standard deviation equal to 1.5. If a value of x is chosen at random from the population, find the probability that x falls between $x = 8$ and $x = 9$.

5.59   Find the probability that z lies between $z = -.2$ and $z = 1.83$.

5.60   Find the probability that z lies between $z = -1.48$ and $z = 1.48$.

5.61   Find a z_0 such that $P(-z_0 < z < z_0) = .5$.

5.62   The length of life of oil-drilling bits depends upon the types of rock and soil that the drill encounters, but it is estimated that the mean length of life is 75 hours. An oil exploration company purchases drill bits that have a length of life that is approximately normally distributed with mean equal to 75 hours and standard deviation equal to 12 hours.

 a What proportion of the company's drill bits will fail before 60 hours of use?

 b What proportion will last at least 60 hours?

 c What proportion will have to be replaced after more than 90 hours of use?

5.63   The influx of new ideas into a college or university, introduced primarily by hiring new young faculty, is becoming a matter of concern because of the increasing ages of faculty members. That is, the distribution of faculty ages is shifting upward, due most likely to a shortage of vacant positions and an oversupply of Ph.D.s. Thus faculty members are more reluctant to move and give up a secure position. If the retirement age at most universities is 65, would you expect the distribution of faculty ages to be normal?

5.64   A machine operation produces bearings whose diameters are normally distributed with mean and standard deviation equal to .498 and .002, respectively. If specifications require that the bearing diameter equal .500 inch plus or minus .004 inch, what fraction of the production will be unacceptable?

5.65   A used-car dealership has found that, for the cars it sells, the length of time before a major repair is required is normally distributed with a mean equal to ten months and a standard deviation of three months. If the dealer wants only 5% of the cars to fail before guarantee time, for how long (in months) should the cars be guaranteed?

5.66   Most users of automatic garage door openers activate their openers at distances that are normally distributed with a mean of 30 feet and a standard deviation of 11 feet. In order to minimize interference with other radio-controlled devices, the manufacturer is required to limit the operating distance to 50 feet. What percentage of the time will users attempt to operate the opener outside of its operating limit?

5.67 Consider a binomial experiment with $n = 25$, $p = .4$. Calculate $P(8 \leq x \leq 11)$ by use of the following.

a The binomial probabilities, Table 1 in Appendix II.

b The normal approximation to the binomial.

5.68 Consider a binomial experiment with $n = 25$, $p = .2$. Calculate $P(x \leq 4)$ by use of the following.

a Table 1 in Appendix II.

b The normal approximation to the binomial.

5.69 You are told that 30% of all calls coming into a telephone exchange are long-distance calls. If 200 calls come into the exchange, what is the probability that at least 50 will be long-distance calls?

5.70 An airline finds that 5% of the persons making reservations on a certain flight will not show up for the flight. If the airline sells 160 tickets for a flight with only 155 seats, what is the probability that a seat will be available for every person holding a reservation and planning to fly?

5.71 The admissions office of a small college is asked to accept deposits from a number of qualified prospective freshmen so that with probability about .95 the size of the freshman class will be less than or equal to 120. Consider that the applicants comprise a random sample from a population of applicants, 80% of whom would actually enter the freshman class if accepted.

a How many deposits should the admissions counselor accept?

b If applicants in the number determined in part (a) are accepted, what is the probability that the freshman class size will be less than 105?

5.72 Nearly half of the clothing bought for boys (49%) and girls (48%) in the 2–15 age category is purchased by people who are 35 to 44 years of age (Exter, 1990).

a If 100 randomly selected purchasers of clothing for girls in the 2–15 age category are interviewed, what is the probability that 40 or fewer are 35 to 44 years of age? 60 or more?

b If the sample revealed exactly 30 purchasers of girls' clothing to be in the 35–44 age category, what would you infer about the 48% figure?

5.73 Refer to Exercise 5.72. The article by Thomas Exter further stated that of the $5.9 billion spent on infants' (less than two years old) clothing 46% was purchased by people between 25 and 34 years old.

a If 200 purchasers of infants' clothing were randomly sampled, what is the mean and standard deviation of x, the number of purchasers between 25 and 34 years old?

b What is the probability of observing fewer than 80 of the 200 purchasers to be between 25 and 34 years old?

5.74 The retired, who are more willing to travel at all times of the year, have been instrumental in smoothing out seasonal differences in the travel industry. Fall foliage, winter snow, and a balmy southern climate have kept Americans on the go throughout the year. The average distance of a fall pleasure trip is 870 miles, 30 miles more than that for a summer pleasure trip (Waldrop, 1990). Assume that the distance traveled on a fall pleasure trip is normally distributed.

a If 60% of all fall pleasure trips are between 570 and 1170 miles, what is the standard deviation?

b What proportion of fall pleasure trips exceeds 1000 miles?

c What proportion is less than 500 miles?

*5.75 Assume that the yield per share of a group of stocks expressed as

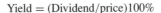

$$\text{Yield} = (\text{Dividend/price})100\%$$

follows an exponential distribution with a mean of 4%.

a   What is the probability that the yield will be greater than 10%?

b   Ninety-five percent of all yields will be less than what percentage?

*5.76   The repair time spent in fixing a flat tire is uniformly distributed between 5 and 15 minutes.

a   What is the probability that the repair time for a flat tire exceeds 10 minutes?

b   If two repairs are brought in together and the repair times are independent, what is the probability that both require less than 10 minutes repair time?

*5.77   The time until malfunction of a copying machine following service has an exponential distribution with a mean of 30 days.

a   What is the probability that the machine will malfunction before 30 days following repair?

b   At what time should the machine be serviced so that the probability of malfunction on or before the service time is equal to .2?

*5.78   The weekly amount of gasoline sold at a self-serve gasoline station and food mart has a uniform distribution on the interval 5000 to 15,000 gallons.

a   If the dealer begins the week with 12,000 gallons in his supply tanks, what is the probability that he will run out of gas before the end of the week?

b   What is the smallest amount that the dealer can have in his storage tanks to keep the probability of running out of gas as small as .001?

*5.79   Problems with the Internal Revenue Service's (IRS) new computer in 1985 created backlogs in the processing of income tax returns and in responding to inquiries. At one point, it was claimed that 53% of inquiries had not been answered 45 days after being received. Assume that the waiting time for the IRS to answer an inquiry has an exponential distribution. Then if $P(x > 45) = .53$,

$$e^{-\lambda(45)} = .53$$

a   Verify that the value of λ is .0141 and that the mean waiting time is 70.88 days.

b   Find the probability that the length of time to answer an inquiry is 60 days or more.

5.80   A newly designed portable radio was styled on the assumption that 50% of all purchasers are female. If a random sample of 400 purchasers is selected, what is the probability that the number of female purchasers in the sample will be greater than 175?

5.81   A salesperson has found that, on the average, the probability of a sale on a single contact is equal to .3. If the salesperson contacts 50 customers, what is the probability that at least 10 will buy? (Assume that x, the number of sales, follows a binomial probability distribution.)

5.82   A soft-drink machine can be regulated so that it discharges an average of μ ounces per cup. If the ounces of fill are normally distributed with standard deviation equal to .3 ounce, give the setting for μ so that 8-ounce cups will overflow only 1% of the time.

5.83   A manufacturing plant utilizes 3000 electric light bulbs that have a length of life that is normally distributed with mean and standard deviation equal to 500 and 50 hours, respectively. In an effort to minimize the number of bulbs that burn out during operating hours, all the bulbs are replaced after a given period of operation. How often should the bulbs be replaced if we want no more than 1% of the bulbs to burn out between replacement periods?

5.84   The daily sales (except Saturday) at a small restaurant have a probability distribution that is approximately normal, with mean μ equal to $530 per day and standard deviation σ equal to $120.

a   What is the probability that the sales will exceed $700 on a given day?

b   The restaurant must have at least $300 in sales per day in order to break even. What is the probability that on a given day the restaurant will not break even?

5.85 The length of life of a type of automatic washer is approximately normally distributed with mean and standard deviation equal to 3.1 and 1.2 years, respectively. If this type of washer is guaranteed for one year, what fraction of original sales will require replacement?

5.86 A grain loader can be set to discharge grain in amounts that are normally distributed with mean μ bushels and a standard deviation equal to 25.7 bushels. If a company wishes to use the loader to fill containers that hold 2000 bushels of grain and wants to overfill only one container in 100, at what value of μ should the company set the loader?

5.87 A publisher has discovered that the number of words contained in a new manuscript is normally distributed with a mean equal to 20,000 words in excess of that specified in the author's contract and a standard deviation of 10,000 words. If the publisher wants to be almost certain (say, with a probability of .95) that the manuscript will be less than 100,000 words, what number of words should the publisher specify in the contract?

Exercises Using the Data Disk

5.88 Refer to data set B: Broccoli Data, on the data disk. Examine the data summary in Appendix I and the four histograms corresponding to the variables fresh weight, market weight, dry weight, and head diameter.

   a Based on the histogram of head diameter measurements, what proportion is less than 7.5 cm? [*Hint:* Use the ratio of sum of the class frequencies with class midpoints less than or equal to 7 cm to the number of sample values, $n = 317$.]

   b Use the calculated mean (9.07) and standard deviation (1.89) to calculate $P(x < 7.5)$ if x represents the diameter of a head of broccoli.

   c How does the value found in part (a) compare to the value found in part (b)?

   d Repeat parts (a), (b), and (c) for different values of head diameters, say, 8.5 and 10.5 cm. How good are the approximations using a normal distribution?

5.89 Repeat Exercise 5.88 using one or more of the variables fresh weight, market weight, and dry weight from the broccoli data.

5.90 Refer to data set C on the data disk and the histograms in Appendix I corresponding to those variables.

   a Use the histogram for average days to maturity to evaluate $P(x < 25)$. [*Hint:* Use the number of observations falling in classes whose midpoints are less than 25 divided by the total number of observations.]

   b Using the calculated mean and standard deviation for average days to maturity, approximate $P(x < 25)$ using the table of normal curve areas. Comment on why this result either agrees with or differs from the value found in part (a).

C H A P T E R

6

SAMPLING DISTRIBUTIONS

About This Chapter

In the preceding chapters we discussed some useful random variables and their probability distributions. In practical sampling situations we do not usually sample a single value of x. Rather, we select a sample of n values and then use those values to calculate statistics such as the sample mean and standard deviation. Then we use these statistics to make inferences about the sampled population. The objective of this chapter is to study some useful statistics and their probability distributions. Then we will explain why, under rather general conditions, all of these statistics possess probability distributions that can be approximated by the normal curve. In subsequent chapters we will show how sample statistics and their probability distributions are used to make inferences about sampled populations.

Contents

SAMPLING THE ROULETTE AT MONTE CARLO

How would you like to try your hand at gambling without the risk of losing? You could do it by simulating the gambling process, making imaginary bets, and observing the results. If you were to repeat the simulation over and over again a large number of times, you would be able to see how your winnings might vary if you were to gamble "for real."

The technique of simulating a process that contains random elements and repeating the process over and over to see how it behaves is called a **Monte Carlo procedure.** It is widely used in business and other fields to investigate the properties of an operation that is subject to a number of random effects, such as weather or human behavior. For example, you can model the behavior of a manufacturing company's inventory by creating, on paper, daily arrivals and departures of manufactured product from the company's warehouse. Each day, a random number of items produced by the company will be received into inventory. Similarly, each day a random number of orders of varying random sizes will be shipped. Based on the input and output of items, you can calculate the inventory, the number of items on hand at the end of each day. The values of the random variables—the number of items produced, the number of orders, and the number of items per order needed for each day's simulation—can be obtained from theoretical distributions of observations that closely model the corresponding distributions of the variables that have been observed over time in the manufacturing operation. By repeating the simulation of the supply and shipping and the calculation of daily inventory for a large number of days (a sampling of what might really happen), you can observe the behavior of the plant's daily inventory. The Monte Carlo procedure is particularly valuable because it enables the manufacturer to see how the daily inventory will behave when certain changes are made in the supply pattern or in some other aspect of the operation that can be controlled.

In an article titled "The Road to Monte Carlo," Daniel Seligman notes that although the Monte Carlo technique is widely used in business schools to study capital budgeting, inventory planning, and management of cash flow, no one seems to have used the procedure to study how well we might do if we were to gamble at Monte Carlo.

To follow up on this thought, Seligman programmed his personal computer to simulate the game of roulette. Roulette consists of a wheel whose rim is divided into 38 pockets. Thirty-six of the pockets are numbered 1 to 36 and are alternately colored red and black. The two remaining pockets are colored green and are marked 0 and 00. To play the game, you bet a certain amount of money on one or more pockets. The wheel is spun and turns until it stops. A ball falls into a slot on the wheel to indicate the winning number. If you have money on that number, you win a specified amount. For example, if you were to play the number 20, the payoff is 35 to 1. If the wheel does not stop at your pocket, you lose your bet. Seligman decided to see how his nightly gains (or losses) would fare if he were to bet $5 on each turn of the wheel and to repeat the process 200 times each night. He repeated the process 365 times, thereby simulating the outcomes of 365 nights at the casino. Not surprisingly, the mean "gain" per $1000 evening for the 365 nights was a loss of $55, the average of the winnings retained by the gambling house. The surprise, according to Seligman, was the extreme variability of the nightly "winnings." Seven times out of the 365 evenings, the fictitious gambler lost the total $1000 stake, and he won a maximum of $1160 only once. One hundred forty-one of the losses exceeded $250.

So much for Monte Carlo and gambling. Our interest in the Monte Carlo procedure is its use in studying the behavior of **sample statistics.** Since we will use sample statistics to make inferences about population parameters, we will want to see how they behave in repeated sampling. This can be done by using the Monte Carlo procedure—sampling, observing the value of a statistic, and then repeating the process over and over again.

In this chapter we examine the properties of some useful statistics. In Section 6.6 we note that the value of a night's winnings in Seligman's simulation of Monte Carlo gambling is itself a statistic, the sum of the gains and losses incurred for 200 $5 bets. Then we use our knowledge of the behavior of a sample sum to decide whether Seligman observed an improbable number of large losses.

6.1 Random Sampling

In earlier chapters, we studied random variables and their probability distributions. We presented several discrete and continuous probability distributions that are possible models for practical situations. These probability distributions depended upon descriptive measures called *parameters,* such as a population mean or standard deviation.

How do we apply these probability models in the practice of statistics? Usually, we are able to decide which type of probability distribution might serve as a model in a given situation; however, the values of the parameters that specify the distribution exactly are not available. In situations such as these, we rely on the *sample* to provide information about these unknown population parameters.

The way a sample is selected is called the *sampling plan* or *experimental design* and determines the quantity of information in the sample. In addition, by knowing the sampling plan used in a particular situation, we can determine the probability of observing specific samples. These probabilities allow us to assess the reliability or goodness of the inferences that are based on the samples.

Simple random sampling is a commonly used sampling plan in which every sample of size n has the same chance of being selected. For example, suppose we want to select a sample of size $n = 2$ from a population containing $N = 4$ objects. If the four objects are identified by the symbols x_1, x_2, x_3, and x_4, there are six distinct pairs that could be selected:

Sample	Observations in Sample
1	x_1, x_2
2	x_1, x_3
3	x_1, x_4
4	x_2, x_3
5	x_2, x_4
6	x_3, x_4

If the sample of $n = 2$ observations is selected so that each of these six samples has the same chance of selection, given by 1/6, then the resulting sample would be called a **simple random sample,** or just a **random sample.**

It can be shown[†] that the number of ways of selecting a sample of size n elements from a population containing N elements is given by

$$C_n^N = \frac{N!}{n!(N-n)!}$$

where $n! = n(n-1)\cdots(3)(2)(1)$ and $0! = 1$. The symbol C_n^N stands for the number of distinct, unordered samples of size n selected *without replacement*. When $N = 4$ and $n = 2$, we have shown that there are

$$C_2^4 = \frac{4!}{2!2!} = \frac{4 \cdot 3 \cdot 2 \cdot 1}{(2 \cdot 1)(2 \cdot 1)} = 6$$

distinct samples. If we conduct an opinion poll of 5000 people based on a sample of size $n = 50$, there are C_{50}^{5000} different combinations of 50 people who could be selected in the sample. If each of these combinations has an equal chance of selection in the sampling plan, then the sample would be a *simple random sample.*

D E F I N I T I O N   ▪   If a sample of n elements is selected from a population of N elements using a sampling plan in which each of the C_n^N samples has the same chance of

---

[†]An understanding of this derivation is not essential to our discussion.

selection, the sampling is said to be **random** and the resulting sample is a **simple random sample.**   ▪

It is easy to understand what is meant by random sampling, but it is much more difficult to actually select a random sample in a practical situation. A knowledge of the concept of random sampling is necessary for some of the sampling situations in this chapter; however, the problem of actually selecting random samples is deferred to Section 14.2.

6.2 Sampling Distributions of Statistics

Numerical descriptive measures calculated from a sample are called **statistics.** Since the values of these sample statistics are unpredictable and vary from sample to sample, they are *random variables* and have a *probability distribution* that describes their behavior in repeated sampling. This probability distribution, called the **sampling distribution of the statistic,** allows us to determine the goodness of any inferences based on this statistic.

DEFINITION   ▪

The **sampling distribution** of a statistic is the probability distribution for all possible values of the statistic that results when random samples of size n are repeatedly drawn from the population.   ▪

For example, suppose that the $N = 4$ elements in the population described in Section 6.1 were given the numerical values $x_1 = 4$, $x_2 = 2$, $x_3 = 5$, and $x_4 = 1$. The sampling distribution for the sample mean, $\bar{x}$, when randomly sampling $n = 2$ elements *with replacement* from this population can be found by calculating $\bar{x}$ for each of the 16 samples, as shown in Table 6.1. Since each of the samples is equally likely, each of the 16 values of $\bar{x}$ has probability $p(\bar{x}) = 1/16$. The probability distribution or sampling distribution of $\bar{x}$ is shown in Table 6.2 and is graphed in Figure 6.1. The sampling distribution of $\bar{x}$ in Figure 6.1 is symmetric about the value $\bar{x} = 3$, which is, in fact, the mean or average value of this sampling distribution, since

$$E(\bar{x}) = \sum \bar{x}p(\bar{x})$$

$$= 1\left(\frac{1}{16}\right) + 1.5\left(\frac{2}{16}\right) + 2\left(\frac{1}{16}\right) + \cdots + 4.5\left(\frac{2}{16}\right) + 5\left(\frac{1}{16}\right) = 3$$

TABLE 6.1
Calculation of $\bar{x}$ for 16 possible samples of size $n = 2$

Sample	Observations in Sample	$\bar{x}$	Sample	Observations in Sample	$\bar{x}$
1	4,4	4	9	5,4	4.5
2	4,2	3	10	5,2	3.5
3	4,5	4.5	11	5,5	5
4	4,1	2.5	12	5,1	3
5	2,4	3	13	1,4	2.5
6	2,2	2	14	1,2	1.5
7	2,5	3.5	15	1,5	3
8	2,1	1.5	16	1,1	1

TABLE 6.2
Sampling distribution for $\bar{x}$

$\bar{x}$	$p(\bar{x})$
1	1/16
1.5	2/16
2	1/16
2.5	2/16
3	4/16
3.5	2/16
4	1/16
4.5	2/16
5	1/16

FIGURE 6.1
Sampling distribution for $\bar{x}$

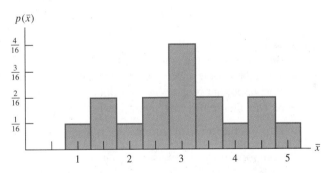

using the formula given in Section 3.6. Also notice that the average value of $\bar{x}$ is equal to μ, the population mean, which we can calculate as

$$\mu = \frac{x_1 + x_2 + x_3 + x_4}{4} = \frac{4 + 2 + 5 + 1}{4} = 3$$

The sample mean, standard deviation, median, and other descriptive measures computed from sample values can be used not only to describe the sample but also to make inferences in the form of estimates or tests abou : corresponding population parameters. However, we must know the *sampling distri ution of the statistic* in order to answer questions such as the following: Does the statistic consistently underestimate or overestimate the value of the parameter? Is this statistic less variable than other competitors, and hence more useful as an estimator?

The sampling distribution of a statistic may be derived mathematically or approximated empirically. Empirical approximations using the Monte Carlo technique described in the case study are found by drawing a large number of samples of size n from the specified population, calculating the value of the statistic for each sample, and tabulating the results in a relative frequency histogram. When the number of samples is large, the relative frequency histogram should closely approximate the theoretical sampling distribution. Alternatively, **for certain statistics that are sums or means of the sample values, an important theorem that we introduce in the next section will allow us to approximate their sampling distributions when the sample size is large.**

6.3 The Central Limit Theorem and the Sampling Distribution of the Sample Mean

The sampling distribution of the sample mean $\bar{x}$ possesses some unique properties. If a random sample of n observations is drawn from a population with mean μ and standard deviation σ, the sampling distribution of $\bar{x}$ will have a mean μ (the same as the mean of the sampled population) and a standard deviation equal to $\sigma/\sqrt{n}$. (The standard deviation of the sampling distribution of a statistic is sometimes called the **standard error** of the statistic. Thus the standard deviation of the sampling distribution of the sample mean is sometimes called the **standard error of the mean.**) But the most important property is a result known in statistics as the **Central Limit Theorem.** This theorem, which applies to both the sample mean $\bar{x}$ and the sample sum $\sum_{i=1}^{n} x_i$, states that when the sample size n is large, the sampling distribution of the sample mean (or sum) will possess approximately a normal distribution. The Central Limit Theorem is stated formally in the display that follows.

The Central Limit Theorem

If random samples of n observations are drawn from a nonnormal population with finite mean μ and standard deviation σ, then when n is large, the sampling distribution of the sample mean $\bar{x}$ is **approximately normally distributed with mean and standard deviation**[†]

$$\mu_{\bar{x}} = \mu \quad \text{and} \quad \sigma_{\bar{x}} = \frac{\sigma}{\sqrt{n}}$$

The approximation will become more and more accurate as n becomes larger and larger.

The Central Limit Theorem can be restated to apply to the **sum of sample measurements**,

$$\sum_{i=1}^{n} x_i$$

which, as n becomes large, also tends to possess a normal distribution, in repeated sampling, with mean $n\mu$ and standard deviation $\sigma\sqrt{n}$.

The mean and standard deviation of the sampling distribution of $\bar{x}$ can be derived and the Central Limit Theorem can be proved mathematically, but the actual proofs are beyond the scope of this text. We can, however, present some Monte Carlo experiments that lend support to our assertions.

Figure 6.2 gives the probability distribution for the number x observed in the toss of a single fair die. The mean of this distribution is $\mu = 3.5$, and its standard deviation is $\sigma = 1.71$ (found in Exercise 3.51). Thus Figure 6.2 is the theoretical distribution of a population of die tosses—that is, the distribution of observations obtained if a fair die were tossed over and over again an infinitely large number of times.

F I G U R E **6.2**
Probability distribution for x, the number appearing on a single toss of a die

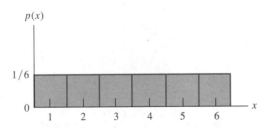

[†]When repeated samples of size n are randomly selected from a *finite* population with N elements whose mean is μ and whose variance is σ^2, the standard deviation of $\bar{x}$ is

$$\sigma_{\bar{x}} = \frac{\sigma}{\sqrt{n}} \sqrt{\frac{N-n}{N-1}}$$

where σ^2 is the population variance. When N is large relative to the sample size n, $\sqrt{(N-n)/(N-1)}$ is approximately equal to 1. Then

$$\sigma_{\bar{x}} = \frac{\sigma}{\sqrt{n}}$$

Now suppose that we want to approximate the sampling distribution for the mean $\bar{x}$ of a sample of $n = 5$ observations selected from the die-tossing population. We can obtain this approximation by conducting a Monte Carlo experiment. As a first step we draw a sample of $n = 5$ measurements from the population by tossing a die five times and observing the numbers $x = 3,\ 5,\ 1,\ 3$, and 2. We then repeat this sampling procedure, each time drawing $n = 5$ observations and recording them, for a total of 100 samples. These 100 sets of sample observations, along with the sample sums and means, are recorded in Table 6.3.

The relative frequency histogram for the 100 sample means, shown in Figure 6.3, is an approximation to the sampling distribution for the mean $\bar{x}$ of a random sample of $n = 5$ die tosses. The approximation would have been better (the shape of the histogram more regular) if we had repeated our Monte Carlo procedure a larger number of times, but the results of the 100 sample repetitions illustrate the properties of the sampling distribution of a sample mean. The relative frequency histogram of the 100 die-toss means in Figure 6.3 centers over the population mean, $\mu = 3.5$. You can also see in Figure 6.3 that the interval $(\mu \pm 2\sigma_{\bar{x}})$ where $\sigma_{\bar{x}} = \sigma/\sqrt{n} = 1.71/\sqrt{5} = .76$) includes most of the sample means. Most surprising is the shape of the sampling distribution. Even though we sampled only $n = 5$ observations from a population with a perfectly flat probability distribution (Figure 6.2), the distribution of sample means in Figure 6.3 is mound-shaped and gives the appearance of being approximately normal.

F I G U R E **6.3**

Histogram of sample means for the die-tossing experiments in Section 6.3

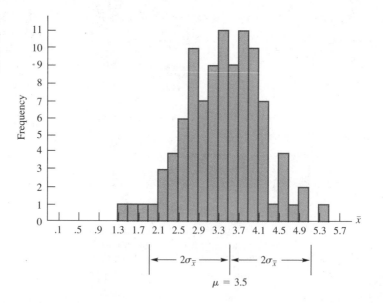

Figure 6.4 gives the results of some other Monte Carlo sampling experiments. We programmed a computer to select random samples of size $n = 2,\ 5,\ 10$, and 25 from each of three populations, the first possessing a normal probability distribution, the second a uniform probability distribution, and the third a negative exponential probability distribution. These population probability distributions are shown in the

TABLE **6.3**
Sampling from the population of die tosses

Sample Number	Sample Measurements	$\sum x_i$	$\overline{x}$	Sample Number	Sample Measurements	$\sum x_i$	$\overline{x}$
1	3,5,1,3,2	14	2.8	51	2,3,5,3,2	15	3.0
2	3,1,1,4,6	15	3.0	52	1,1,1,2,4	9	1.8
3	1,3,1,6,1	12	2.4	53	2,6,3,4,5	20	4.0
4	4,5,3,3,2	17	3.4	54	1,2,2,1,1	7	1.4
5	3,1,3,5,2	14	2.8	55	2,4,4,6,2	18	3.6
6	2,4,4,2,4	16	3.2	56	3,2,5,4,5	19	3.8
7	4,2,5,5,3	19	3.8	57	2,4,2,4,5	17	3.4
8	3,5,5,5,5	23	4.6	58	5,5,4,3,2	19	3.8
9	6,5,5,1,6	23	4.6	59	5,4,4,6,3	22	4.4
10	5,1,6,1,6	19	3.8	60	3,2,5,3,1	14	2.8
11	1,1,1,5,3	11	2.2	61	2,1,4,1,3	11	2.2
12	3,4,2,4,4	17	3.4	62	4,1,1,5,2	13	2.6
13	2,6,1,5,4	18	3.6	63	2,3,1,2,3	11	2.2
14	6,3,4,2,5	20	4.0	64	2,3,3,2,6	16	3.2
15	2,6,2,1,5	16	3.2	65	4,3,5,2,6	20	4.0
16	1,5,1,2,5	14	2.8	66	3,1,3,3,4	14	2.8
17	3,5,1,1,2	12	2.4	67	4,6,1,3,6	20	4.0
18	3,2,4,3,5	17	3.4	68	2,4,6,6,3	21	4.2
19	5,1,6,3,1	16	3.2	69	4,1,6,5,5	21	4.2
20	1,6,4,4,1	16	3.2	70	6,6,6,4,5	27	5.4
21	6,4,2,3,5	20	4.0	71	2,2,5,6,3	18	3.6
22	1,3,5,4,1	14	2.8	72	6,6,6,1,6	25	5.0
23	2,6,5,2,6	21	4.2	73	4,4,4,3,1	16	3.2
24	3,5,1,3,5	17	3.4	74	4,4,5,4,2	19	3.8
25	5,2,4,4,3	18	3.6	75	4,5,4,1,4	18	3.6
26	6,1,1,1,6	15	3.0	76	5,3,2,3,4	17	3.4
27	1,4,1,2,6	14	2.8	77	1,3,3,1,5	13	2.6
28	3,1,2,1,5	12	2.4	78	4,1,5,5,3	18	3.6
29	1,5,5,4,5	20	4.0	79	4,5,6,5,4	24	4.8
30	4,5,3,5,2	19	3.8	80	1,5,3,4,2	15	3.0
31	4,1,6,1,1	13	2.6	81	4,3,4,6,3	20	4.0
32	3,6,4,1,2	16	3.2	82	5,4,2,1,6	18	3.6
33	3,5,5,2,2	17	3.4	83	1,3,2,2,5	13	2.6
34	1,1,5,6,3	16	3.2	84	5,4,1,4,6	20	4.0
35	2,6,1,6,2	17	3.4	85	2,4,2,5,5	18	3.6
36	2,4,3,1,3	13	2.6	86	1,6,3,1,6	17	3.4
37	1,5,1,5,2	14	2.8	87	2,2,4,3,2	13	2.6
38	6,6,5,3,3	23	4.6	88	4,4,5,4,4	21	4.2
39	3,3,5,2,1	14	2.8	89	2,5,4,3,4	18	3.6
40	2,6,6,6,5	25	5.0	90	5,1,6,4,3	19	3.8
41	5,5,2,3,4	19	3.8	91	5,2,5,6,3	21	4.2
42	6,4,1,6,2	19	3.8	92	6,4,1,2,1	14	2.8
43	2,5,3,1,4	15	3.0	93	6,3,1,5,2	17	3.4
44	4,2,3,2,1	12	2.4	94	1,3,6,4,2	16	3.2
45	4,4,5,4,4	21	4.2	95	6,1,4,2,2	15	3.0
46	5,4,5,5,4	23	4.6	96	1,1,2,3,1	8	1.6
47	6,6,6,2,1	21	4.2	97	6,2,5,1,6	20	4.0
48	2,1,5,5,4	17	3.4	98	3,1,1,4,1	10	2.0
49	6,4,3,1,5	19	3.8	99	5,2,1,6,1	15	3.0
50	4,4,4,4,4	20	4.0	100	2,4,3,4,6	19	3.8

F I G U R E **6.4**   Probability distributions and approximations to the sampling distributions for three populations. [Note: Vertical scales are not constant.]

| Sample size, n | Normal distribution, $\mu = 0$, $\sigma = 1$ | Uniform distribution, $\mu = .5$, $\sigma = .29$ | Negative exponential, $\mu = 1$, $\sigma = 1$ |

top row of Figure 6.4. The computer printouts of the approximations to sampling distributions of the sample mean $\bar{x}$ for sample sizes $n = 2,\ 5,\ 10,$ and 25 are shown in rows 2, 3, 4, and 5 of Figure 6.4.

Figure 6.4 illustrates an important theorem of theoretical statistics. **The sampling distribution of the sample mean is exactly normally distributed (proof omitted), regardless of the sample size, when we are sampling from a population that possesses a normal distribution.** In contrast, the sampling distribution of $\bar{x}$ for samples selected from populations with uniform and negative exponential probability distributions tends to become more nearly normal as the sample size n increases from $n = 2$ to $n = 25$, rapidly for the uniform distribution and more slowly for the highly skewed exponential distribution. But note that the sampling distribution of $\bar{x}$ is normal or approximately normal for sampling from either the uniform or the exponential probability distribution when the sample size is as large as $n = 25$. This result suggests that for many populations the sampling distribution of $\bar{x}$ will be approximately normal for moderate sample sizes. There are exceptions to this rule. Consequently, we will give the appropriate sample size n for specific applications of the Central Limit Theorem as they are encountered in the text.

The properties of the sampling distribution of the sample mean are given in the following display.

The Sampling Distribution of the Sample Mean $\bar{x}$

1   If a random sample of n measurements is selected from a population with mean μ and standard deviation σ, the sampling distribution of the sample mean $\bar{x}$ will possess a mean

$$\mu_{\bar{x}} = \mu$$

and a standard deviation

$$\sigma_{\bar{x}} = \frac{\sigma}{\sqrt{n}}$$

2   If the population possesses a *normal* distribution, then the sampling distribution of $\bar{x}$ will be *exactly* normally distributed, *regardless of the sample size, n.*

3   If the population distribution is nonnormal, the sampling distribution of $\bar{x}$ will be, for large samples, approximately normally distributed (by the Central Limit Theorem). Figure 6.4 suggests that the sampling distributions of $\bar{x}$ will be approximately normal for sample sizes as small as $n = 25$ for most populations of measurements.

EXAMPLE   **6.1**   Suppose that you select a random sample of $n = 25$ observations from a population with mean $\mu = 8$ and $\sigma = .6$.

a   Find the approximate probability that the sample mean $\bar{x}$ will be less than 7.9.

b   Find the approximate probability that the sample mean $\bar{x}$ will exceed 7.9.

c   Find the approximate probability that the sample mean $\bar{x}$ will lie within .1 of the population mean $\mu = 8$.

Solution   **a**   Regardless of the shape of the population relative frequency distribution, the sampling distribution of $\bar{x}$ will possess a mean $\mu_{\bar{x}} = \mu = 8$ and a standard deviation

$$\sigma_{\bar{x}} = \frac{\sigma}{\sqrt{n}} = \frac{.6}{\sqrt{25}} = .12$$

For a sample as large as $n = 25$, it is likely (because of the Central Limit Theorem) that the sampling distribution of $\bar{x}$ is approximately normally distributed (we will assume that it is). Therefore, the probability that $\bar{x}$ will be less than 7.9 is approximated by the shaded area under the normal sampling distribution in Figure 6.5. To find this area, we need to calculate the value of z corresponding to $\bar{x} = 7.9$. This value of z is the distance between $\bar{x} = 7.9$ and $\mu_{\bar{x}} = \mu = 8.0$ expressed in standard deviations of the sampling distribution—that is, in units of

$$\sigma_{\bar{x}} = \frac{\sigma}{\sqrt{n}} = .12$$

Thus

$$z = \frac{\bar{x} - \mu}{\sigma_{\bar{x}}} = \frac{7.9 - 8.0}{.12} = -.83$$

From Table 3 in Appendix II, we find that the area corresponding to $z = .83$ is .2967. Therefore,

$$P(\bar{x} < 7.9) = .5 - .2967 = .2033$$

[Note that we must use $\sigma_{\bar{x}}$ (not σ) in the formula for z because we are finding an area under the sampling distribution for $\bar{x}$, not under the sampling distribution for x.]

F I G U R E 6.5
The probability that $\bar{x}$ is less than 7.9 for Example 6.1

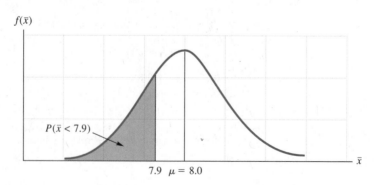

$f(\bar{x})$

$P(\bar{x} < 7.9)$

7.9 $\mu = 8.0$

$\bar{x}$

F I G U R E **6.6**
The probability that $\bar{x}$ lies within
.1 of $\mu = 8$ for Example 6.1

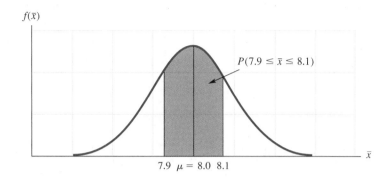

$f(\bar{x})$

$P(7.9 \leq \bar{x} \leq 8.1)$

$7.9 \quad \mu = 8.0 \quad 8.1$

$\bar{x}$

b   The event that $\bar{x}$ exceeds 7.9 is the complement of the event that $\bar{x}$ is less than 7.9. Thus the probability that $\bar{x}$ exceeds 7.9 is

$$P(\bar{x} > 7.9) = 1 - P(\bar{x} < 7.9) = 1 - .2033 = .7967$$

c   The probability that $\bar{x}$ lies within .1 of $\mu = 8$ is the shaded area in Figure 6.6. We found in part (a) that the area between $\bar{x} = 7.9$ and $\mu = 8.0$ is .2967. Since the area under the normal curve between $\bar{x} = 8.1$ and $\mu = 8.0$ is identical to the area between $\bar{x} = 7.9$ and $\mu = 8.0$, it follows that

$$P(7.9 < \bar{x} < 8.1) = 2(.2967) = .5934 \quad \blacksquare$$

E X A M P L E   **6.2**   To avoid difficulties with the Federal Trade Commission or state and local consumer protection agencies, a beverage bottler must make reasonably certain that 12-ounce bottles actually contain 12 ounces of beverage. To determine whether a bottling machine is working satisfactorily, one bottler randomly samples ten bottles per hour and measures the amount of beverage in each bottle. The mean $\bar{x}$ of the ten fill measurements is used to decide whether to readjust the amount of beverage delivered per bottle by the filling machine. If records show that the amount of fill per bottle is normally distributed with a standard deviation of .2 ounce, and if the bottle machine is set to produce a mean fill per bottle of 12.1 ounces, what is the approximate probability that the sample mean $\bar{x}$ of the ten test bottles is less than 12 ounces?

Solution   The mean of the sampling distribution of the sample mean $\bar{x}$ is identical to the mean of the population of bottle fills—namely, $\mu = 12.1$ ounces—and the standard deviation (or standard error) of $\bar{x}$ is

$$\sigma_{\bar{x}} = \frac{\sigma}{\sqrt{n}} = \frac{.2}{\sqrt{10}} = .063$$

[*Note:* σ is the standard deviation of the population of bottle fills, and n is the number of bottles in the sample.] Since the amount of fill is normally distributed, $\bar{x}$ is also normally distributed. Then the probability distribution of $\bar{x}$ will appear as shown in Figure 6.7.

F I G U R E **6.7**
Sampling distribution of $\bar{x}$, the
mean of the $n = 10$ bottle
fills, for Example 6.2

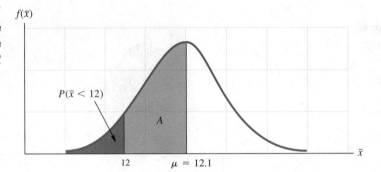

The probability that $\bar{x}$ will be less than 12 ounces will equal $(.5 - A)$, where A is the area between 12 and the mean $\mu = 12.1$. Expressing this distance in standard deviations, we have

$$z = \frac{\bar{x} - \mu}{\sigma_{\bar{x}}} = \frac{12 - 12.1}{.063} = -1.59$$

Then the area A over the interval $12 < \bar{x} < 12.1$, found in Table 3 of Appendix II, is .4441, and the probability that $\bar{x}$ will be less than 12 ounces is

$$P(\bar{x} < 12) = .5 - A = .5 - .4441 = .0559 \approx .056$$

Thus, if the machine is set to deliver an average fill of 12.1 ounces, the mean fill $\bar{x}$ of a sample of ten bottles will be less than 12 ounces with probability equal to .056. When this danger signal occurs ($\bar{x}$ is less than 12), the bottler takes a larger sample to recheck the setting of the filling machine. ∎

Tips on Problem Solving

Before attempting to calculate the probability that the statistic $\bar{x}$ falls in some interval, complete the following steps:

1   Calculate the mean and standard deviation of the sampling distribution of $\bar{x}$.

2   Sketch the sampling distribution. Show the location of the mean μ, and locate the intervals $\mu \pm 2\sigma_{\bar{x}}$ and $\mu \pm 3\sigma_{\bar{x}}$ on the horizontal axis.

3   Locate the interval on the sketch from part 2 and shade the area corresponding to the probability that you wish to calculate.

4   Find the z-score(s) associated with the value(s) of interest. Use Table 3 in Appendix II to find the probability.

> **5**   When you have obtained your answer, look at your sketch of the sampling distribution to see whether your calculated answer agrees with the shaded area. This provides a very rough check on your calculations.

Exercises

Basic Techniques

6.1   Random samples of size n were selected from populations with the following means and variances. Find the mean and standard deviation (standard error) of the sampling distribution of the sample mean.

    a $n = 25, \mu = 10, \sigma^2 = 9$ **b** $n = 100, \mu = 5, \sigma^2 = 4$

    c $n = 6, \mu = 120, \sigma^2 = 1$

6.2   Refer to Exercise 6.1.

    a If the sampled populations are normal, what is the sampling distribution of $\bar{x}$ for parts (a), (b), and (c)?

    b According to the Central Limit Theorem, if the sampled populations are *not* normal, what can be said about the sampling distribution of $\bar{x}$ for parts (a), (b), and (c)?

6.3   Refer to the sampling distribution described in Exercise 6.1(b).

    a Sketch the sampling distribution of $\bar{x}$. Locate the mean and the interval $(\mu \pm 2\sigma_{\bar{x}})$ along the $\bar{x}$ axis of the graph.

    b Shade the area under the curve that corresponds to the probability that $\bar{x}$ lies within .15 unit of the population mean μ.

    c Find the probability described in part (a).

6.4   Refer to the die-throwing experiment in Section 6.3 in which x is the number of dots observed when a single die is tossed. The probability distribution for x is given in Figure 6.2, and the relative frequency histogram for $\bar{x}$ is given in Figure 6.3 for 100 random samples of size $n = 5$.

    a Verify that the mean and standard deviation of x are $\mu = 3.5$ and $\sigma = 1.71$, respectively.

    b Look at the histogram in Figure 6.3. Guess the value of its mean and standard deviation. [*Hint:* The Empirical Rule states that approximately 95% of the measurements associated with a mound-shaped distribution will lie within two standard deviations of the mean.]

    c What are the theoretical mean and standard deviation of the sampling distribution of $\bar{x}$? How do these values compare to the guessed values of part (b)?

6.5   Refer to Exercise 6.4. Suppose the die-throwing experiment was repeated over and over again an infinitely large number of times. Find the mean and standard deviation (or standard error) for the sampling distribution of $\bar{x}$ if each sample consists of the following.

    a $n = 10$ measurements   **b** $n = 15$ measurements

    c $n = 25$ measurements

6.6   Refer to Exercises 6.4 and 6.5. What effect does increasing the sample size have on the sampling distribution of $\bar{x}$?

6.7   Exercises 6.5 and 6.6 demonstrate that the standard deviation of the sampling distribution decreases as the sample size increases. To see this relationship more clearly, suppose that a random sample of n observations is selected from a population with standard deviation $\sigma = 1$. Calculate $\sigma_{\bar{x}}$ for $n = 1, 2, 4, 9, 16, 25$, and 100. Then plot $\sigma_{\bar{x}}$ versus the sample size n, and connect the points with a smooth curve. Note the manner in which $\sigma_{\bar{x}}$ decreases as n increases.

6.8   Suppose that a random sample of $n = 5$ observations is selected from a population that is normally distributed with mean equal to 1 and standard deviation equal to .36.

    a   Give the mean and standard deviation of the sampling distribution of $\bar{x}$.

    b   Find the probability that $\bar{x}$ exceeds 1.3.

    c   Find the probability that the sample $\bar{x}$ will be less than .5.

    d   Find the probability that the sample mean will deviate from the population mean $\mu = 1$ by more than .4.

6.9   Suppose that a random sample of $n = 25$ observations is selected from a population that is normally distributed with mean equal to 106 and standard deviation equal to 12.

    a   Give the mean and the standard deviation of the sampling distribution of the sample mean $\bar{x}$.

    b   Find the probability that $\bar{x}$ exceeds 110.

    c   Find the probability that the sample mean will deviate from the population mean $\mu = 106$ by no more than 4.

Applications

6.10   Explain why the shipping weight of a truckload of oranges might be normally distributed.

6.11   Use the Central Limit Theorem to explain why a Poisson random variable, say, the number of employee accidents per year in a large manufacturing plant, possesses a distribution that is approximately normal when the mean μ is large. [*Hint:* One year is the sum of 365 days.]

6.12   A lobster fisherman's daily catch x is the total, in pounds, of lobster landed from a fixed number of lobster traps. What kind of probability distribution would you expect the daily catch to possess and why? If the mean catch per trap per day is 30 pounds with $\sigma = 5$ pounds, and the fisherman has 50 traps, give the mean and standard deviation of the probability distribution of the total daily catch x.

6.13   An important expectation of the recent federal income tax reduction is that consumers will save a substantial portion of the money that they receive. Suppose that estimates of the portion of total tax saved, based on a random sampling of 35 economists, possessed a mean of 26% and a standard deviation of 12%.

    a   What is the approximate probability that a sample mean, based on a random sample of $n = 35$ economists, will lie within 1% of the mean of the population of the estimates of all economists?

    b   Is it necessarily true that the mean of the population of estimates of all economists is equal to the percentage tax saving that will actually be achieved?

6.14   The 1993–94 Scholastic Aptitude Test (SAT) scores provide mixed results when compared to the same scores in 1989. The math test, taken by approximately one-third of the nation's high school seniors, showed an increase in the average score from 476 to 478, while the verbal test score decreased from 427 to 424. ("Using Your College Planning Report: 1993–94"). Why would these very small changes be viewed as significant by educators in measuring student achievement?

6.15   To obtain information on the volume of freight shipped by truck over a particular interstate highway, a state highway department monitored the highway for twenty-five 1-hour periods randomly selected throughout a one-month period. The number of truck trailers was counted for each 1-hour period, and $\bar{x}$ was calculated for the sample of 25 individual 1-hour periods. Suppose that the number of heavy-duty trailers per hour is approximately normally distributed, with $\mu = 50$ and $\sigma = 7$.

    a   What is the probability that the sample mean $\bar{x}$ for $n = 25$ individual 1-hour periods is larger than 55?

    b   Suppose you were to count the truck trailers for each of $n = 4$ randomly selected 1-hour periods. What is the probability that $\bar{x}$ would be larger than 55? [*Hint:* The distribution of the sample means will be normally distributed, regardless of the sample size, for the special case when the population possesses a normal distribution.]

   c   What is the probability that the total number of trucks for a 4-hour period would exceed 180?

6.16   A manufacturer of paper used for packaging requires a minimum strength of 20 pounds per square inch. As a check on the quality of the paper, a random sample of ten pieces of paper is selected each hour from the previous hour's production, and a strength measurement is recorded for each. The standard deviation σ of the strength measurements, computed by pooling the sum of squares of deviations of many samples, is known to equal 2 pounds per square inch. Assume that the strength measurements are normally distributed.

   a   What is the approximate probability distribution of the sample mean strength of $n = 10$ test pieces of paper?

   b   If the mean of the population of strength samples is 21 pounds per square inch, what is the probability that $\bar{x} < 20$ for a random sample of $n = 10$ test pieces of paper?

   c   What value would you want for the mean paper strength μ in order that $P(\bar{x} < 20)$ be equal to .001?

6.17   Execution time is a very important variable in the sales and advertising of personal computers (PCs). However, these execution times can be hard to quantify, even for a specific model, since they depend on the amount and type of software loaded on the PC, the amount of disk space available, and so on. Suppose that we wish to measure the amount of time (in seconds) required to load Ami Pro 2.0 on an IBM PS/2 Model 90 486DX/33 personal computer with a Standard Windows operating system ("Byte Windows," 1993).

   a   Explain why the time required to load Ami Pro 2.0 should be approximately normally distributed.

   b   If the time to load Ami Pro 2.0 has a mean of 1.33 seconds with a standard deviation of .2 second, what is the probability that it will take more than 1.4 seconds to load it on a randomly chosen PC?

   c   If five PCs are randomly chosen, what is the probability that the average time to load for these five machines exceeds 1.4 seconds?

6.4 The Sampling Distribution of a Sample Proportion

Many sampling problems involve consumer preference or opinion polls, which are concerned with estimating the proportion p of people in the population who possess some specified characteristic. These and similar situations provide practical examples of binomial experiments, if the sampling procedure has been conducted in the appropriate manner. If a random sample of n persons is selected from the population and if x of these possess the specified characteristic, then the sample proportion

$$\hat{p} = x/n$$

is used to estimate the population proportion p.[†]

    Since each distinct value of x results in a distinct value of $\hat{p} = x/n$, the probabilities associated with $\hat{p}$ are equal to the probabilities associated with the corresponding

---

[†]A "hat" placed over the symbol of a population parameter denotes a statistic used to estimate the population parameter. For example, the symbol $\hat{p}$ denotes the sample proportion.

values of x. Hence, the sampling distribution of $\hat{p}$ will be the same shape as the binomial probability distribution for x. Like the binomial probability distribution, it can be approximated by a normal distribution when the sample size n is large. The mean of the sampling distribution of $\hat{p}$ is

$$\mu_{\hat{p}} = p$$

and its standard deviation is

$$\sigma_{\hat{p}} = \sqrt{\frac{pq}{n}}.$$

where

$$q = 1 - p$$

Properties of the Sampling Distribution of the Sample Proportion $\hat{p}$

1 If a random sample of n observations is selected from a binomial population with parameter p, the sampling distribution of the sample proportion

$$\hat{p} = \frac{x}{n}$$

will have a mean

$$\mu_{\hat{p}} = p$$

and a standard deviation

$$\sigma_{\hat{p}} = \sqrt{\frac{pq}{n}}, \quad \text{where } q = 1 - p$$

2 When the sample size n is large, the sampling distribution of $\hat{p}$ will be approximately normal. The approximation will be adequate if $\mu_{\hat{p}} \pm 2\sigma_{\hat{p}}$ lies within the interval from 0 to 1, and the approximation will be good if $\mu_{\hat{p}} \pm 3\sigma_{\hat{p}}$ lies within the interval from 0 to 1.

EXAMPLE 6.3 A survey was taken of 313 children, ages 14 to 22 years, from among the children of the nation's top corporate executives. When asked to identify the best aspect of being one of this privileged group, 55% mentioned material and financial advantages. Describe the sampling distribution of the sample proportion $\hat{p}$ of children citing material advantage as being the best aspect of their privileged lives.

Solution We will assume that the 313 children represent a random sample of the children of all top corporate executives and that the true proportion in the population is equal to some unknown value that we will call p. Then the sampling distribution of $\hat{p}$ will be approximately normally distributed (because of the Central Limit Theorem) with a mean equal to p (see Figure 6.8) and a standard deviation

$$\sigma_{\hat{p}} = \sqrt{\frac{pq}{n}}$$

F I G U R E **6.8**
The sampling distribution of $\hat{p}$ based on a sample of $n = 313$ children for Example 6.3

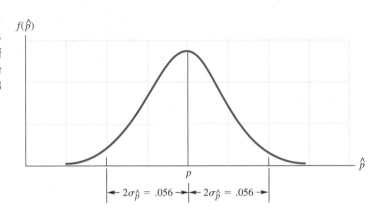

$f(\hat{p})$

p

$\hat{p}$

$\leftarrow 2\sigma_{\hat{p}} = .056 \rightarrow\!\leftarrow 2\sigma_{\hat{p}} = .056 \rightarrow$

Examining Figure 6.8, you can see that the sampling distribution of $\hat{p}$ centers over its mean p. Even though we do not know the exact value of p (the sample proportion $\hat{p} = .55$ may be larger or smaller than p), we can calculate an approximate value for the standard deviation of the sampling distribution by using the sample proportion $\hat{p} = .55$ to approximate the unknown value of p. Thus

$$\sigma_{\hat{p}} = \sqrt{\frac{pq}{n}} \approx \sqrt{\frac{\hat{p}\hat{q}}{n}} = \sqrt{\frac{(.55)(.45)}{313}} = .0283$$

Further, since the approximation to the interval $p \pm 3\sigma_{\hat{p}}$, given by

$$\hat{p} \pm 3\sigma_{\hat{p}}$$
$$.55 \pm 3(.028)$$
$$.55 \pm .084$$

or (.466, .634), lies within the interval from 0 to 1, the normal approximation to the distribution of $\hat{p}$ should be good. ∎

E X A M P L E **6.4** Refer to Example 6.3. Suppose that the proportion p of children in the population is actually equal to .5. What is the probability of observing a sample proportion as large as or larger than the observed value $\hat{p} = .55$?

Solution Figure 6.9 shows the sampling distribution of $\hat{p}$ when $p = .5$, with the observed value $\hat{p} = .55$ located on the horizontal axis. From Figure 6.9, you can see that the

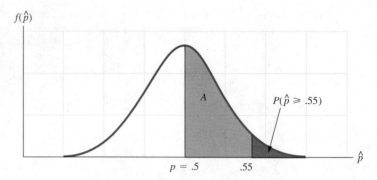

F I G U R E 6.9
The sampling distribution of $\hat{p}$
for $n = 313$ and $p = .5$ in
Example 6.4

probability of observing a sample proportion $\hat{p}$ equal to or larger than .55 is the shaded area in the upper tail of a normal distribution, with

$$\mu_{\hat{p}} = .5$$

and

$$\sigma_{\hat{p}} = \sqrt{\frac{pq}{n}} = \sqrt{\frac{(.5)(.5)}{313}} = .0283$$

To find this shaded area, we need to know how many standard deviations the observed value $\hat{p} = .55$ lies away from the mean of the sampling distribution $p = .5$. This distance is given by the z value,

$$z = \frac{\hat{p} - p}{\sigma_{\hat{p}}} = \frac{.55 - .5}{.0283} = 1.77$$

Table 3 in Appendix II gives the area A corresponding to $z = 1.77$ as

$$A = .4616$$

Therefore, the shaded area in the upper tail of the sampling distribution in Figure 6.9 is

$$P(\hat{p} \geq .55) = .5 - A = .5 - .4616 = .0384$$
$$\approx .04$$

This value tells us that if we were to select a random sample of $n = 313$ observations from a population with proportion p equal to .5, the probability that the sample proportion $\hat{p}$ would be as large as or larger than .55 is only .04. ∎

Exercises

Basic Techniques

6.18   Random sample of size n were selected from binomial populations with the following population parameters p. Find the mean and standard deviation of the sampling distribution of the sample proportion $\hat{p}$.

    a $n = 100, p = .3$ **b** $n = 400, p = .1$ **c** $n = 250, p = .6$

6.19   Sketch each of the sampling distributions listed in Exercise 6.18. For each, locate the mean p and the interval $(p \pm 2\sigma_{\hat{p}})$ along the $\hat{p}$ axis of the graph.

6.20   Refer to the sampling distribution given in Exercise 6.18(a).

    a   Sketch the sampling distribution for the sample proportion, and shade the area under the curve that corresponds to the probability that $\hat{p}$ lies within .08 of the population proportion p.

    b   Find the probability described in part (a).

6.21   If $n = 1000$ and $p = .1$, find the following probabilities.

    a $\hat{p} > .12$ **b** $\hat{p} < .10$ **c** $\hat{p}$ lies within .02 of p

6.22   Calculate $\sigma_{\hat{p}}$ for $n = 100$ and the following values of p.

    a $p = .01$ **b** $p = .1$ **c** $p = .3$ **d** $p = .5$

    e $p = .7$ **f** $p = .9$ **g** $p = .99$

    Plot $\sigma_{\hat{p}}$ versus p on graph paper, and sketch a smooth curve through the points. For what value of p is the standard deviation of the sampling distribution of $\hat{p}$ a maximum? What happens to $\sigma_{\hat{p}}$ when p is near 0 or near 1.0?

6.23   Assuming that p is some fixed value, what is the effect on $\sigma_{\hat{p}}$ of increasing the sample size? Does a change in the sample size n have the same effect on $\sigma_{\hat{p}}$ as on $\sigma_{\bar{x}}$? Explain.

6.24   If $p = .8$ and $n = 400$, find the following probabilities.

    a $\hat{p} > .83$

    b $.76 \leq \hat{p} \leq .84$

Applications

6.25   Male shoppers who live in high-income or PC-owning households have mixed opinions on the subject of shopping by computer as opposed to shopping in a store (Dholakia, 1994). In a recent survey of 1600 "upscale" male shoppers, 71% of the men find shopping by computer convenient, but only 22% find it satisfying. Assume that the 1600 men in the survey represent a random sample of "upscale" men in the United States.

    a   Describe the sampling distribution of $\hat{p}$, the proportion of men in the sample that find shopping by computer convenient. [*Hint:* Use $\hat{p}$ to approximate p when calculating $\sigma_{\hat{p}}$]

    b   Find the probability that $\hat{p}$ will lie within .03 of the proportion p of "upscale" men in the population who find shopping by computer convenient.

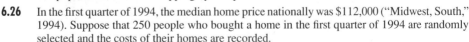

6.26   In the first quarter of 1994, the median home price nationally was $112,000 ("Midwest, South," 1994). Suppose that 250 people who bought a home in the first quarter of 1994 are randomly selected and the costs of their homes are recorded.

    a   Describe the sampling distribution of $\hat{p}$, the proportion of people whose home cost more than $112,000.

    b   What is the probability that the sample proportion $\hat{p}$ is 66% or greater?

    c   If you were to take a sample and you observed 165 people (66% of the sample) whose home cost more than $112,000, what conclusions might you draw? Explain.

6.27   Advertisers must be aware of the changing roles of men and women in society in order to properly target their advertisements toward the proper sector of the market. For example, women play an increasing role in automobile purchase decisions, and men are more involved than in the past in purchasing and preparing food. A recent study shows that 80% of all married people feel that they have an equal say in making major purchases in the home (Dortch, 1994). Suppose that a random sample of $n = 300$ married people is selected and asked if they feel that they have an equal say in making major purchases in the home.

a   What is the probability that more than 85% of the sample feel they have an equal say in major home purchases?

b   Within what limits would you expect 95% of the sample proportions to lie?

c   What is the probability that the sample proportion differs from the population proportion p by more than 5% in either direction?

6.5   A Sampling Application: Statistical Process Control

Statistical process control (SPC) methodology was developed to monitor, control, and improve products and services. Steel bearings must conform to size and hardness specifications, industrial chemicals must have a low prespecified level of impurities, and accounting firms must minimize and ultimately eliminate incorrect bookkeeping entries. It is often said that statistical process control consists of 10% statistics and 90% engineering and common sense. We can statistically monitor a process mean and tell when the mean falls outside preassigned limits, but we can't tell *why* it is out of control. Answering this last question requires knowledge of the process and problem-solving ability—the other 90%!

Product quality is usually monitored using statistical control charts. Measurements on a process variable to be monitored change over time. The cause of a change in the variable is said to be *assignable* if it can be found and corrected. Other variation—small haphazard changes due to alteration in the production environment—that is not controllable is regarded as *random variation*. If the variation in a process variable is solely random, the process is said to be *in control*. The first objective in statistical process control is to eliminate assignable causes of variation in the process variable and then get the process in control. The next step is to reduce variation and get the measurements on the process variable within *specification limits,* the limits within which the measurements on usable items or services must fall.

Once a process is in control and is producing a satisfactory product, the process variables are monitored by use of **control charts.** Samples of n items are drawn from the process at specified intervals of time, and a sample statistic is computed. These statistics are plotted on the control chart so that the process can be checked for shifts in the process variable that might indicate control problems.

A Control Chart for the Process Mean: The $\bar{x}$ Chart

Assume that n items are selected from the production process at equal intervals and that measurements are recorded on the process variable. If the process is in control, the sample means should vary about the population mean μ in a random manner. Moreover, according to the Empirical Rule and Tchebysheff's Theorem, we would expect that most of the values of $\bar{x}$ should fall in the interval $(\mu \pm 3\sigma_{\bar{x}}) = \mu \pm 3(\sigma/\sqrt{n})$. Although the exact values of μ and σ are unknown, we can obtain accurate estimates by using the sample measurements.

Every control chart has a *centerline* and *control limits*. The centerline is the estimate of μ, the grand average of all the sample statistics calculated from the measurements on the process variable. The upper and lower control limits are placed three standard deviations above and below the centerline. If we monitor the process mean based on k samples of size n taken at regular intervals, the centerline is $\bar{\bar{x}}$, the average of the sample means, and the control limits are at $\bar{\bar{x}} \pm 3(\sigma\sqrt{n})$, with σ estimated by s, the standard deviation of the nk measurements.

EXAMPLE **6.5**   A statistical process control monitoring system samples the inside diameters of $n = 4$ bearings each hour. Table 6.4 provides the data for $k = 25$ hourly samples. Construct an $\bar{x}$ chart for monitoring the process mean.

Solution   The sample mean was calculated for each of the $k = 25$ samples. For example, the mean for sample 1 is

$$\bar{x} = \frac{.992 + 1.007 + 1.016 + .991}{4} = 1.0015$$

The sample means are shown in column 6 of Table 6.4. The centerline is located at

$$\bar{\bar{x}} = \frac{99.87}{100} = .9987$$

The calculated value of s, the sample standard deviation of all $nk = 4(25) = 100$ observations, is $s = .011458$. The estimated standard error of the mean of $n = 4$ observations is then

$$\frac{s}{\sqrt{n}} = \frac{.011458}{\sqrt{4}} = .005729$$

The upper and lower control limits are found as

$$\text{UCL} = \bar{\bar{x}} + 3\frac{s}{\sqrt{n}} = .9987 + 3(.005729) = 1.015887$$

and

$$\text{LCL} = \bar{\bar{x}} - 3\frac{s}{\sqrt{n}} = .9987 - 3(.005729) = .981513$$

Figure 6.10 shows a Minitab printout of the $\bar{x}$ chart constructed from the data. Assuming that the samples used to construct the $\bar{x}$ chart were collected when the process was in control, the chart now can be used to detect changes in the process mean. Sample means are plotted periodically, and if a sample mean falls outside the control limits a warning should be conveyed. The process should be checked to locate the cause of the unusually large or small mean. ▪

T A B L E **6.4**
25 hourly samples of bearing diameters, $n = 4$ bearings per sample, for Example 6.5

Sample	Sample Measurements				Sample Mean, $\overline{x}$
1	.992	1.007	1.016	.991	1.00150
2	1.015	.984	.976	1.000	.99375
3	.988	.993	1.011	.981	.99325
4	.996	1.020	1.004	.999	1.00475
5	1.015	1.006	1.002	1.001	1.00600
6	1.000	.982	1.005	.989	.99400
7	.989	1.009	1.019	.994	1.00275
8	.994	1.010	1.009	.990	1.00075
9	1.018	1.016	.990	1.011	1.00875
10	.997	1.005	.989	1.001	.99800
11	1.020	.986	1.002	.989	.99925
12	1.007	.986	.981	.995	.99225
13	1.016	1.002	1.010	.999	1.00675
14	.982	.995	1.011	.987	.99375
15	1.001	1.000	.983	1.002	.99650
16	.992	1.008	1.001	.996	.99925
17	1.020	.988	1.015	.986	1.00225
18	.993	.987	1.006	1.001	.99675
19	.978	1.006	1.002	.982	.99200
20	.984	1.009	.983	.986	.99050
21	.990	1.012	1.010	1.007	1.00475
22	1.015	.983	1.003	.989	.99750
23	.983	.990	.997	1.002	.99300
24	1.011	1.012	.991	1.008	1.00550
25	.987	.987	1.007	.995	.99400

F I G U R E **6.10**
Minitab $\overline{x}$ chart for Example 6.5

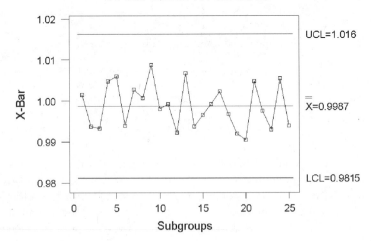

X-bar Chart for MEANS

Other commonly used control charts are the *p chart*, which is used to monitor *p*, the proportion of defectives in the population, the *R chart*, which is used to monitor variation in the process variable by using the sample range, and the *c chart*, which is used to monitor the number of defects per item. These quality control charts will be considered in more detail in Chapter 10.

Exercises

Basic Techniques

6.28   The sample means were calculated for 30 samples of size $n = 10$ for a process that was judged to be in control. The means of the 30 $\bar{x}$ values and the standard deviation of the combined 300 measurements were $\bar{\bar{x}} = 20.74$ and $s = .87$.

   a   Use the data to determine the upper and lower control limits for an $\bar{x}$ chart.

   b   What is the purpose of an $\bar{x}$ chart?

   c   Construct an $\bar{x}$ chart for the process and explain how it can be used.

6.29   The sample means were calculated for 40 samples of size $n = 5$ for a process that was judged to be in control. The means of the 40 values and the standard deviation of the combined 200 measurements were $\bar{\bar{x}} = 155.9$ and $s = 4.3$.

   a   Use the data to determine the upper and lower control limits for an $\bar{x}$ chart.

   b   Construct an $\bar{x}$ chart for the process and explain how it can be used.

Applications

 6.30   A gambling casino records and plots the mean of the daily gain or loss from five blackjack tables on an $\bar{x}$ chart. The overall mean of the sample means and the standard deviation of the combined data over 40 weeks were $\bar{\bar{x}} = \$10,752$ and $s = \$1605$.

   a   Construct an $\bar{x}$ chart for the mean daily gain per blackjack table.

   b   How might this $\bar{x}$ chart be of value to the manager of the casino?

 6.31   A coal-burning power plant tests and measures three specimens of coal each day to monitor the percentage of ash in the coal. The overall mean of 30 daily sample means and the combined standard deviation of all the data were $\bar{\bar{x}} = 7.24$ and $s = .07$. Construct an $\bar{x}$ chart for the process and explain how it can be of value to the manager of the power plant.

 6.32   The data given in the following table are measures of the radiation in air particulates at a nuclear power plant. Four measurements were recorded at weekly intervals over a 26-week period. Use the data to construct an $\bar{x}$ chart and plot the 26 values of $\bar{x}$. Explain how the chart will be used.

Week	Radiation				Week	Radiation			
1	.031	.032	.030	.031	14	.029	.028	.029	.029
2	.025	.026	.025	.025	15	.031	.029	.030	.031
3	.029	.029	.031	.030	16	.014	.016	.016	.017
4	.035	.037	.034	.035	17	.019	.019	.021	.020
5	.022	.024	.022	.023	18	.024	.024	.024	.025
6	.030	.029	.030	.030	19	.029	.027	.028	.028
7	.019	.019	.018	.019	20	.032	.030	.031	.030
8	.027	.028	.028	.028	21	.041	.042	.038	.039
9	.034	.032	.033	.033	22	.034	.036	.036	.035
10	.017	.016	.018	.018	23	.021	.022	.024	.022
11	.022	.020	.020	.021	24	.029	.029	.030	.029
12	.016	.018	.017	.017	25	.016	.017	.017	.016
13	.015	.017	.018	.017	26	.020	.021	.020	.022

CASE STUDY REVISITED

6.6 THE SAMPLING DISTRIBUTION OF WINNINGS AT ROULETTE

In the case study that introduces this chapter, we described a Monte Carlo experiment conducted by Daniel Seligman of *Fortune* magazine. Seligman simulated 365 evenings of gambling at Monte Carlo. On each of the 365 evenings Seligman placed 200 bets of $5 each with a payoff of 35 to 1 and with a probability of 1/38 of winning.

To evaluate the results of Seligman's Monte Carlo experiment, we note that each bet results in a gain of −$5 if he loses and $175 if he wins. Thus the probability distribution of the gain x on a single $5 bet is

x	$p(x)$
−5	37/38
175	1/38

Then from Chapter 3 the expected gain $E(x)$ and variance σ_x^2 are

$$E(x) = \mu_x = \sum x p(x) = (-5)\left(\frac{37}{38}\right) + (175)\left(\frac{1}{38}\right) = -.2632$$

$$\sigma_x^2 = \sum (x - \mu)^2 p(x) = \sum x^2 p(x) - \mu^2$$

$$= (-5)^2 \left(\frac{37}{38}\right) + (175)^2 \left(\frac{1}{38}\right) - (.2632)^2 = 830.1939$$

and

$$\sigma_x = \sqrt{830.1939} = \$28.81$$

Therefore, the mean gain for a $5 bet is a loss of approximately 26¢, and the standard deviation is $28.81. The 26¢ represents the mean amount that you lose to the "house."

The gain for an evening is the sum $S = \sum_{i=1}^{200} x_i$ of the gains or losses for two hundred \$5 bets. The properties of the sampling distribution for this sum are described in our statement of the Central Limit Theorem (see the display in Section 6.3). When the sample size n is large, the sampling distribution of the sum of the sample measurements will tend to normality. The mean and standard deviation of the sampling distribution are

$$\mu_S = n\mu \qquad \sigma_S = \sigma\sqrt{n}$$

where μ and σ are the mean and standard deviation of the gain x for a single \$5 bet. Therefore,

$$\mu_S = (200)(-.2632) = -\$52.64$$
$$\sigma_S = 28.81\sqrt{200} = 407.43$$

Therefore, the total winnings (or losses) for a single evening will vary from −\$1000 (if the gambler loses all 200 bets) to \$35,000 (if the gambler wins all 200 times), a range of \$36,000. The mean gain (actually, a loss) per evening is −\$52.64, and most of the nightly gains will fall (from the Empirical Rule) in the interval

$$\mu_S \pm 2\sigma_S, \quad \text{that is, } -52.64 \pm (2)(407.43)$$

or

$$-\$867.50 \quad \text{to} \quad \$762.22$$

Of course, the loss on any one night cannot exceed \$1000. Therefore, most of the large deviations from the mean will be observations in the upper tail of the distribution (improbably large gains).

Now that we know something about the sampling distribution of an evening's winnings at roulette, let us examine the results of Daniel Seligman's Monte Carlo experiment. We agree with Seligman that it is surprising that 7 of the 365 evenings resulted in losses of the total \$1000 stake. The probability of no wins in 200 (a single evening's betting) is less than .005, and the mean number of times this event would occur in a total of 365 evenings is less than 1.825. Based on a mean equal to 1.825, it can be shown that the observance of 7 evenings resulting in a loss of \$1000 is highly improbable.[†]

The largest evening's winnings, \$1160, lies 2.98 standard deviations away from the mean $\mu_S = -52.64$. It is improbable, but it is an event that might occur in one out of 365 evenings.

[†]The number x of evenings in a total of 365 that result in a \$1000 loss possesses a binomial probability distribution with $n = 365$ and $p = .005$. Using the Poisson approximation to the binomial probability distribution (Section 4.3), you can show that $x = 7$ lies almost four standard deviations away from the mean $\mu = np = 1.825$.

6.7 Summary

In a practical sampling situation, we draw a *single* random sample of n observations from a population, calculate a single value of a sample statistic, and use it to make an inference about a population parameter. But to interpret the statistic—to know how close to the population parameter the computed statistic might be expected to fall—we need to observe the behavior of the statistic in repeated sampling. Thus, if we were to repeat the sampling process over and over again an infinitely large number of times, the distribution of values of the statistic produced by this enormous Monte Carlo experiment would be the sampling (or probability) distribution of the statistic.

This chapter described the properties of the sampling distributions for two useful statistics that we will employ in the following chapters to make inferences about population parameters. First, sample means and sample proportions have sampling distributions that can be approximated by a normal distribution when the sample sizes are large. Second, these distributions are centered over their respective population parameters. Thus the mean of the sampling distribution of the sample mean $\bar{x}$ is the population mean μ, and the mean of the sampling distribution of the sample proportion $\hat{p}$ is the population proportion p. Third, the spread of the distributions, measured by their standard deviations, decreases as the sample size increases. As we will see in Chapter 7, this third characteristic is important when we wish to use a sample statistic to estimate its corresponding population parameter. By choosing a larger sample size, we can increase the probability that a sample statistic will fall close to the population parameter.

Supplementary Exercises

6.33 Review the die-tossing experiment of Section 6.3, where we simulated the selection of samples of $n = 5$ observations and obtained an approximation to the sampling distribution for the sample mean. Repeat this experiment, selecting 200 samples of size $n = 3$.

a Construct the sampling distribution for $\bar{x}$. Note that the sampling distribution of $\bar{x}$ for $n = 3$ does not achieve the bell shape that you observed for $n = 5$ (Figure 6.3).

b The mean and standard deviation of the probability distribution for x, the number of dots that appear when a single die is tossed, are $\mu = 3.5$ and $\sigma = 1.71$. What are the exact values of the mean and standard deviation of the sampling distributions of $\bar{x}$ based on samples of $n = 3$?

c Calculate the mean and standard deviation of the simulated sampling distribution of part (a). Are these values close to the corresponding values obtained for part (b)?

6.34 Refer to the sampling experiment of Exercise 6.33. Calculate the median for each of the 200 samples of size $n = 3$.

a Use the 200 medians to construct a relative frequency histogram that approximates the sampling distribution of the sample median.

b Calculate the mean and standard deviation of the sampling distribution of part (a).

c Compare the mean and standard deviation of this sampling distribution with the mean and standard deviation calculated for the sampling distribution of $\bar{x}$ of Exercise 6.33(b). Which statistic, the sample mean or the sample median, appears to fall closer to μ?

6.35   A finite population consists of the following four elements:

$$6, 1, 3, 2$$

a   How many different samples of size $n = 2$ can be selected from this population if we sample without replacement? [*Hint:* Sampling is said to be *without replacement* if an element cannot be selected twice for the same sample.]

b   List the possible samples of size $n = 2$.

c   Compute the sample mean for each of the samples given in part (b).

d   Find the sampling distribution of $\bar{x}$. Use a probability histogram to graph the sampling distribution of $\bar{x}$.

e   If all four population values are equally likely, calculate the value of the population mean μ. Do any of the samples listed in part (b) produce a value of $\bar{x}$ exactly equal to μ?

6.36   Refer to Exercise 6.35. Find the sampling distribution of $\bar{x}$ if random samples of size $n = 3$ are selected *without replacement*. Graph the sampling distribution of $\bar{x}$.

6.37   Refer to Exercise 6.35. Find the sampling distribution of the sample median if random samples of size $n = 3$ are selected *without replacement*. Graph the sampling distribution of the sample median.

6.38   The Central Limit Theorem implies that a sample mean $\bar{x}$ is approximately normally distributed for large values of n. Suppose that a sample of size $n = 100$ is drawn from a population with mean $\mu = 40$ and $\sigma = 4$.

a   What is $E(\bar{x})$?

b   What is the standard deviation of $\bar{x}$?

c   What is $P(\bar{x} > 41)$?

6.39   A survey of purchasing agents from 250 industrial companies found that 25% of the buyers reported higher levels of new orders in January than in earlier months. Assume that the 250 purchasing agents in the sample represent a random sample of company purchasing agents throughout the United States.

a   Describe the sampling distribution of $\hat{p}$, the proportion of buyers in the United States with higher levels of new orders in January. [*Hint:* Use $\hat{p}$ to approximate p when calculating $\sigma_{\hat{p}}$.]

b   What is the probability that $\hat{p}$ will differ from p by more than .01?

6.40   The length of time required for a local automobile dealer to run a 5000-mile check and service for a new automobile is approximately normally distributed with a mean of 1.4 hours and a standard deviation of .7 hour. Suppose that the service department plans to service 50 automobiles per 8-hour day and that in order to do so it must spend no more service time than an average of 1.6 hours per automobile. What proportion of all days will the service department have to work overtime?

6.41   Sony's model KV-27XBR26 27-inch TV set was rated first among 22 different brands and models based upon performance attributes such as picture quality, tone quality, and ease of use ("Ratings: 27-inch TV Sets," 1994). However, this was the most expensive model among the 22, with an average price of $1085 and a price range from $1005 to $1135.

a   If we assume that these are population values and that the range represents approximately six standard deviations, describe the sampling distribution of the average price of this Sony model in a random sample of $n = 100$ owners.

b   What is the probability that the sample mean exceeds $1090?

c   What is the probability that the sample mean is less than $1078? What would you conclude if, in fact, your sample mean was $1078?

6.42   According to a report by the U.S. Department of Commerce ("Is College Worth It," 1994), high school graduates over age 18 made an average of $17,072 in 1990, while those with four years of college made $31,256, almost twice as much as high school graduates. Assume that a random sample of $n = 25$ college graduates was polled in 1990 concerning their salary and that the standard deviation of salaries for college graduates was $1550.

a What are the mean and the standard deviation of $\bar{x}$, the sample mean?

b Would you expect the sampling distribution of $\bar{x}$ to be normally distributed or approximately so? Explain.

c Evaluate the probability that the sample mean salary for college graduates exceeds $32,000. Exceeds $33,000.

d Within what range would you expect the sample mean to lie with high probability, say, 95%?

 6.43 With the rising costs of a college education, most students depend on their parents or family for monetary support during their college years. The results of a freshman survey in 1993 ("Freshman Statistics," 1994) indicate that 86% of freshmen in the survey received financial aid from parents or family. Suppose that we were to survey the next freshman class by selecting a sample of $n = 1000$ freshmen and that 86% is the true percentage currently receiving financial aid from parents or family.

a Describe the approximate distribution of the sample proportion $\hat{p}$ of those receiving financial aid from parents or family.

b What is the probability that the sample proportion differs from .86 by less than .02?

c Would we be likely to see a sample proportion larger than 90%? Why or why not?

 6.44 How long will it be before you need to repair or replace your television set? According to a report in *Consumer Reports* ("Ratings: 27-inch TV Sets," 1994) dealing with the percentage of sets ever repaired for 15 brands of 25–27-inch television sets, the top three brands were General Electric, Panasonic, and JVC. For each of these brands, about 5% of the TV sets required repairs. Suppose a random sample of $n = 500$ owners of a 25–27-inch General Electric TV set is selected and the proportion of owners whose sets required repairs is recorded.

a What are the mean and standard deviation of the proportion of sets requiring repairs?

b Is the interval $p \pm 3\sqrt{\dfrac{pq}{n}}$ contained within the range of $\hat{p}$? Will the Central Limit Theorem apply to the distribution of $\hat{p}$?

c What is the probability that $\hat{p}$ will differ from p by more than .01?

 6.45 Refer to Exercise 6.44. In that same report, 10% of Zenith 24–27-inch TV sets required repairs. Suppose that a random sample of $n = 100$ owners of Zenith 24–27-inch TV sets is selected and the sample proportion of TV sets requiring repair is recorded.

a What is the approximate distribution of $\hat{p}$? Will this approximation be good or merely adequate? Explain.

b What is the probability that the sample proportion differs from the 10% figure by less than 5%?

Exercises Using the Data Disk

6.46 Random sampling is described in Section 6.1. Select 50 random samples of size $n = 10$ from the $N = 317$ observations on fresh weight from data set B: Broccoli Data on the data disk, and calculate $\bar{x}$ for each sample.

a Construct a histogram for the values of $\bar{x}$ in your 50 samples.

b Calculate the mean and standard deviation of the 50 sample means found in random sampling.

c Is the shape of the frequency histogram of the 50 values of $\bar{x}$ mound-shaped? How do the calculated mean and standard deviation found in part (b) compare to the approximate theoretical values of $\mu_{\bar{x}}$ and $\sigma_{\bar{x}}$ computed from the data summary?

6.47 Repeat Exercise 6.46 using 50 samples of size $n = 20$. The theoretical mean is still the same, but the theoretical standard deviation will be smaller.

6.48 A class Monte Carlo experiment: In the case study in Chapter 2, we presented data on money market mutual funds. As an experiment, regard the seven-day yields in Table 2.2 as a population. Have each member of the class select a random sample of $n = 4$ observations from this population and calculate the sample mean $\bar{x}$. (A procedure for selecting a random sample is described in Section 14.2.) Construct a relative frequency histogram for the sample means calculated by the class members. This relative frequency distribution of sample means provides an approximation to the sampling distribution of $\bar{x}$ for sample size $n = 4$.

   a Compare your sampling distribution of $\bar{x}$ with the population relative frequency histogram in Figure 2.30.

   b Calculate the theoretical mean and standard deviation of the sampling distribution of $\bar{x}$. The mean and standard deviation of this population of data are given in Appendix I as $\mu = 3.7324$ and $\sigma = .3337$. Locate the mean and the interval $(\mu \pm 2\sigma/\sqrt{n})$ along the horizontal axis. Is the mean near the center of the distribution of sample means? Does the interval $(\mu \pm 2\sigma/\sqrt{n})$ include most of the means? approximately 95%?

   c Calculate the mean and standard deviation of the sample means used to construct the relative frequency histogram. Are these values close to the values found for μ and $\sigma_{\bar{x}}$ in part (b)?

6.49 Refer to data set A on the data disk.

   a Randomly select 25 samples of size 5 from the list of salaries for male full professors and calculate the mean for each sample. Construct a histogram using the 25 sample means. Does the histogram appear to approximate a normal distribution?

   b Repeat part (a) for salaries of female full professors.

6.50 Refer to data set B on the data disk.

   a Using the histogram of maximum head diameters, find the proportion of maximum head diameters larger than 10.5 by summing the relative frequencies of all classes with a midpoint greater than 10 and dividing by 317, the number of observations. We will now assume that this proportion represents a population proportion.

   b Select 30 random samples of size 10 maximum head diameters. For each sample, determine the proportion having a maximum head diameter greater than 10.5. Construct a histogram of your sample proportions. Does it appear to be centered over the value of $p = .19$, the value calculated in part (a) for all observations?

   c Calculate the mean and standard deviation of the sample values of $\hat{p}$ found in part (b). How do they compare with the theoretical values found using $p = .19$?

ESTIMATION OF MEANS AND PROPORTIONS

About This Chapter

As we stated in Chapter 3, probability reasons from a population to a sample. Statistical inference, the reverse of this procedure, infers the nature of a population based on information contained in a sample. Chapters 2 through 6 present the basic concepts of probability, probability distributions, and probability (sampling) distributions of statistics. The objective of this chapter is to show you how sampling distributions can be used to make inferences about a population from the observed values of sample statistics. Thus we present one of two methods for making inferences about population parameters—statistical estimation.

Contents

SAMPLING: WHAT WILL THE IRS ALLOW?

The Internal Revenue Service (IRS) not only uses statistical sampling for examining large volumes of accounting data but also permits the use of statistical sampling and inference by corporations to estimate certain costs and other items when it is impractical to obtain exact data. Writing on this subject, W. L. Felix, Jr., and R. S. Roussey cite an example of a corporation that claimed $6 million in one year and $3.8 in the next for repair expense replacement and other costs (Felix, 1985).[†] These claims were based on samples of 350 items for the first year and 520 items for the second. The IRS did not dispute the use of sampling, the sampling procedure, or the sample sizes, but it did object to lack of information on the "sample error." Analyzing the corporation's sample data, the IRS concluded that the actual expense could have been as low as $3.5 million and $2.8 million for the first and second years, respectively, and therefore disallowed $3.4 million of the total $9.8 million in claims.

This example demonstrates one of the many ways that sampling and statistical inference can be of value in accounting. In this chapter we will study estimators for a number of useful population parameters, and we will use the sampling distribution of an estimator to determine how close to a population parameter an estimate is likely to be. Then in Section 7.10 we will examine the logic for the IRS's $3.4 million disallowance of the claim.

7.1 A Brief Summary

The preceding six chapters set the stage for the objective of this text: an understanding of statistical inference and how it can be applied to the solution of practical problems. In Chapter 1, we stated that statisticians are concerned with making inferences about populations of measurements based on information contained in samples. We showed you how to phrase an inference—that is, how to describe a set of measurements—in

[†]Copyright ©1985 by American Institute of Certified Public Accountants, Inc.

Chapter 2. In Chapter 3, we discussed probability and the mechanism for making inferences, and we followed that with a general discussion of probability distributions—discrete probability distributions in Chapter 4 and continuous probability distributions in Chapter 5.

In Chapter 6 we noted that statistics, quantities computed from sample measurements, are used to make inferences about population parameters, and we found an important use for the normal probability distribution of Chapter 5. In particular, you learned that some of the most important statistics—sample means and proportions—have sampling distributions that can be approximated by a normal distribution when the sample sizes are large, owing to the Central Limit Theorem. These statistics will now be used to make inferences about population parameters, and their sampling distributions will provide a means of assessing the reliability of these inferences.

7.2 Types of Estimators

Estimation procedures can be divided into two types, point estimation and interval estimation. Suppose that an Oldsmobile dealer wants to estimate the mean profit per sale of a new automobile. The estimate might be given as a single number, for instance $935, or we might estimate that the mean profit per sale will fall in an interval from $835 to $1035. The first type of estimate is called a **point estimate** because the single number representing the estimate may be associated with a point on a line. The second type, involving two points and defining an interval on a line, is called an **interval estimate.** We will consider each of these methods of estimation.

In order to construct either a point or an interval estimate, we use information from the sample in the form of an estimator. Estimators are functions of sample observations and hence, by definition, are also **statistics.**

DEFINITION ■

> An **estimator** is a rule that tells us how to calculate an estimate based on information in the sample and that is generally expressed as a formula. ■

For example, the sample mean

$$\overline{x} = \frac{\sum_{i=1}^{n} x_i}{n}$$

is an estimator of the population mean μ and explains exactly how the actual numerical value of the estimate can be obtained once the sample values $x_1, x_2, \ldots, x_n$ are known. The sample mean can be used to arrive at a single number to estimate μ or to construct an interval, two points that are intended to enclose the true value of μ.

DEFINITION ▪
> A **point estimator** of a population parameter is a rule that tells us how to calculate a single number based on sample data. The resulting number is called a **point estimate.** ▪

DEFINITION ▪
> An **interval estimator** of a population parameter is a rule that tells us how to calculate two numbers based on sample data, forming an interval within which the parameter is expected to lie. This pair of numbers is called an **interval estimate** or **confidence interval.** ▪

Both point and interval estimation procedures are developed using the sampling distribution of the best estimator of a specified population parameter. Moreover, many different statistics can be constructed to estimate the same parameter. For example, if we sample $n = 5$ measurements, 2, 7, 0, 1, and 4, from a symmetric population, we can estimate the population mean by using the sample mean,

$$\bar{x} = \frac{\sum_{i=1}^{n} x_i}{n} = \frac{14}{5} = 2.8$$

by using the sample median $m = 2$, or even by using the average of the smallest and largest measurements in the sample, $(0 + 7)/2 = 3.5$. How can we evaluate the properties of these estimators, compare one with another, and eventually decide which is "best"?

The goodness of an estimator is evaluated by observing its behavior in repeated sampling. Let us consider the following analogy. In many respects, point estimation is similar to the firing of a revolver at a target. The estimator, which generates estimates, is analogous to the revolver; a particular estimate is analogous to the bullet; and the parameter of interest is analogous to the bull's-eye. Drawing a sample from the population and estimating the value of the parameter is equivalent to firing a single shot at the target.

Suppose a man fires a single shot at the target and the shot pierces the bull's-eye. While this is an admirable feat, can we conclude that he is an excellent shot? The answer is no—not one of us would consent to hold the target while a second shot is fired. Not until his accuracy had been observed for **repeated firings,** with all shots coming close to the bull's-eye, would we declare him a good shot.

Suppose we consider an estimator of a population parameter such as μ, σ, or p. What are some desirable properties of an estimator? Essentially, there are two, and they can be seen by observing the sampling distributions given in Figures 7.1 and 7.2.

First, we would like the sampling distribution to be centered over the true value of the parameter. **Thus we would like the mean of the sampling distribution to equal the true value of the parameter.** Such an estimator is said to be **unbiased.**

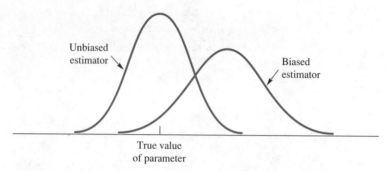

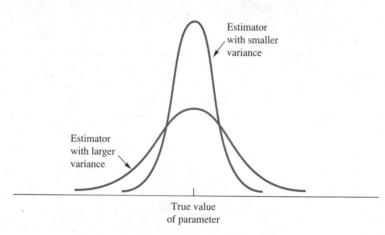

D E F I N I T I O N   ▪   An estimator of a parameter is said to be **unbiased** if the mean of its distribution is equal to the true value of the parameter. Otherwise, the estimator is said to be **biased.**   ▪

The sampling distributions for an unbiased estimator and a biased estimator are shown in Figure 7.1. The sampling distribution for the biased estimator in Figure 7.1 is shifted to the right of the true value of the parameter. This biased estimator is more likely than an unbiased one to overestimate the value of the parameter.

The second desirable property of an estimator is that **the spread (as measured by the variance) of the sampling distribution should be as small as possible.** This ensures that, with a high probability, an individual estimate will fall close to the true value of the parameter. The sampling distributions for two unbiased estimators, one with a small variance[†] and the other with a larger variance, are shown in Figure

---

[†]Statisticians usually use the term *variance of an estimator* when in fact they mean the variance of the sampling distribution of the estimator. This contractive expression is used almost universally.

7.2. Naturally, we would prefer the estimator with the smaller variance because the estimates tend to lie closer to the true value of the parameter than in the distribution with the larger variance.

In real-life sampling situations, you may know that the sampling distribution of a point estimator centers about the parameter that you are attempting to estimate, but all you have is the estimate computed from the n measurements contained in the sample. How far from the true value of the parameter will your estimate lie? The distance between the estimate and the true value of the parameter is called the **error of estimation** and provides a measure of the goodness of the point estimator.

DEFINITION ▪

> The distance between an estimate and the estimated parameter is called the **error of estimation.** ▪

The **goodness of an interval estimator** is analyzed in much the same way as that of a point estimator. Samples of the same size are repeatedly drawn from the population, and the interval estimate is calculated on each occasion. This process will generate a large number of intervals rather than points. **A good interval estimator would successfully enclose the true value of the parameter a large fraction of the time.** The "success rate" is referred to as the **confidence coefficient** and provides a measure of goodness for the interval estimator.

DEFINITION ▪

> The probability that a confidence interval will enclose the estimated parameter is called the **confidence coefficient.** ▪

The selection of a "best" estimator—the proper formula to use in calculating the estimates—involves the comparison of various methods of estimation. This is the task of the theoretical statistician and is beyond the scope of this text. Throughout the remainder of this chapter and succeeding chapters, populations and parameters of interest will be defined and the appropriate estimator indicated along with a measure of its goodness.

7.3 Large-Sample Estimation

Point Estimation

Suppose that we have an *unbiased* estimator whose sampling distribution is normal or can be approximated by a normal distribution. We know that 95% of the values of this estimator will fall within 1.96 standard deviations of its mean, the parameter of interest. Therefore, the **error of estimation,** defined as the difference between a particular point estimate and the parameter it estimates, should be less than 1.96 standard deviations of the estimator with probability approximately equal to .95. (Refer to Figure 7.3.) This quantity provides a bound on the error of estimation that

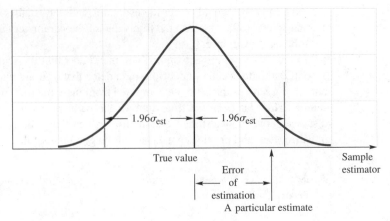

is usually called the **margin of error** in estimation. Although there is a 5% chance that the error of estimation will exceed this margin of error, it is very unlikely that such will be the case.

> **Point Estimator for a Population Parameter**
>
> Point estimator: a statistic calculated using sample measurements
> Margin of error: 1.96 × standard error of the estimator

Interval Estimation

In constructing an interval estimate for a parameter, we determine two points within which we expect the value of the unknown parameter to lie. Interval estimates are constructed so that in repeated sampling a large proportion (close to 1) of the intervals will enclose the parameter of interest. This proportion is called the **confidence coefficient,** and the resulting interval is called a **confidence interval.** For example, in estimating a population mean with a confidence interval, we speak of "the probability that the interval encloses μ," not "the probability that μ falls in the interval," because the value of μ is fixed but the interval consists of random endpoints.

A large-sample confidence interval based on a normally or approximately normally distributed unbiased estimator is obtained by measuring 1.96 × (standard error of the estimator) on either side of the point estimate. Since we know that 95% of the point estimates will lie within 1.96 standard errors of the population mean, it follows that 95% of the intervals constructed in this manner should enclose the population mean. An interval will fail to enclose the mean only if the point estimate lies further than 1.96 standard errors from the mean, and this will occur with probability .05. Figure 7.4 shows how this works when $\bar{x}$ is used as an estimator of μ.

In general, we can change the confidence coefficient by changing the value of $z_{.025} = 1.96$. If we wish to have a confidence coefficient equal to $1 - \alpha$, we select

FIGURE **7.4**
95% confidence limits for a
population mean

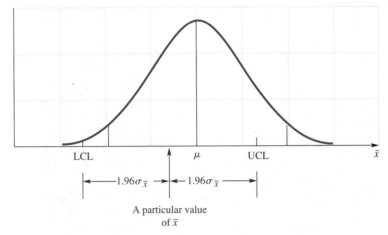

the value $z_{\alpha/2}$ that has $\alpha/2$ in the upper tail of the standard normal distribution. This value can be found in Table 3 of Appendix II.

When the sample size is large and an estimator is normally distributed or approximately normally distributed, a $(1 - \alpha)100\%$ confidence interval estimate for an unknown population parameter is given in the following display.

A $(1 - \alpha)100\%$ Large-Sample Confidence Interval

$$\text{(Point estimator)} \pm z_{\alpha/2} \times \text{(Standard error of the estimator)}$$

where $z_{\alpha/2}$ is the z value corresponding to an area $\alpha/2$ in the upper tail of the standard normal distribution. This formula generates two values, the **lower confidence limit (LCL)** and the **upper confidence limit (UCL)**.

Some common confidence intervals, confidence coefficients, and their z values are shown in Table 7.1.

TABLE **7.1**
Confidence limits for
large-sample interval estimation

Confidence Coefficient	α	$z_{\alpha/2}$
.90	.10	1.645
.95	.05	1.96
.99	.01	2.58

Throughout the remainder of this chapter, you will see how the general formulas for large-sample point and interval estimation apply to specific populations and parameters of interest. You will also learn how to modify the general formulas when the sample sizes are not large.

7.4 Large-Sample Estimation of a Population Mean

Practical problems very often lead to the estimation of a population mean μ. We may be concerned with the average achievement of MBA students in a particular university, with the average strength of a new type of steel, with the average number of deaths per capita in a given social class, or with the average demand for a new product. Many estimators are available for estimating the population mean μ, including the sample median, the average of the largest and smallest measurements in the sample, and the sample mean $\bar{x}$. Each estimator would have a sampling distribution and, depending on the population and practical problem involved, certain advantages and disadvantages. Although the sample median and the average of the sample extremes are easier to calculate, the sample mean $\bar{x}$ is usually superior in that, for some populations, its variance is a minimum and, regardless of the population, it is always unbiased.

The sampling distribution of the sample mean $\bar{x}$, discussed in Section 6.4, has four important properties:

- The sampling distribution of $\bar{x}$ will be *approximately normal* regardless of the probability distribution of the sampled population when n is large.

- If the sampled population is normal, the sampling distribution of $\bar{x}$ will be exactly *normal*.

- The *mean* of the sampling distribution of $\bar{x}$ will always equal μ. Thus $\bar{x}$ is an unbiased estimator of μ.

- The standard deviation of the sampling distribution of $\bar{x}$, also called the *standard error of the mean*, is $\sigma_{\bar{x}} = \sigma/\sqrt{n}$.

The estimator $\bar{x}$ satisfies all the conditions given in Section 7.3, so the general formulas can be applied for point and interval estimation.

Point Estimator of a Population Mean μ

   Point estimator: $\bar{x}$

   Margin of error: $1.96\sigma_{\bar{x}} = 1.96\sigma/\sqrt{n}$

A $(1 - \alpha)100\%$ Large-Sample Confidence Interval for a Population Mean μ

$$\bar{x} \pm z_{\alpha/2}\frac{\sigma}{\sqrt{n}}$$

where $z_{\alpha/2}$ is the z value corresponding to an area $\alpha/2$ in the upper tail of a standard normal z-distribution.

   $n = $ Sample size

   $\sigma = $ Standard deviation of the sampled population

If σ is unknown, it can be approximated by the sample standard deviation s when the sample size is large.[†]

Assumption: $n \geq 30$

E X A M P L E **7.1** A marketing research organization was hired to estimate the mean prime lending rate for banks located in the western region of the United States. A random sample of $n = 50$ banks was selected from within the region, and the prime rate was recorded for each. The mean and standard deviation of the 50 prime rates were

$$\bar{x} = 8.1\% \qquad \text{and} \qquad s = .24$$

Estimate the mean prime rate for the region, and find the margin of error associated with the estimate.

Solution The estimate of the mean prime rate is $\bar{x} = 8.1\%$. The margin of error is

$$1.96\sigma_{\bar{x}} = 1.96\frac{\sigma}{\sqrt{n}} = 1.96\frac{\sigma}{\sqrt{50}}$$

Although σ is unknown, the sample size is large, and we may approximate the value of σ by using s. Thus the margin of error is approximately

$$1.96\frac{s}{\sqrt{n}} = 1.96\frac{.24}{\sqrt{50}} = .0665 \approx .07$$

We can feel fairly confident that our estimate of 8.1% is within .07% of the true mean prime rate. ∎

E X A M P L E **7.2** Find a 90% confidence interval for the mean prime lending rate discussed in Example 7.1.

Solution The 90% confidence interval for the mean prime lending rate μ is

$$\bar{x} \pm z_{.05}\sigma_{\bar{x}}$$

or

$$\bar{x} \pm 1.645\frac{\sigma}{\sqrt{n}}$$

Substituting $\bar{x} = 8.1\%$ and $n = 50$ and using $s = .24\%$ to approximate σ, we obtain

$$8.1 \pm (1.645)\frac{.24}{\sqrt{50}}$$

[†]When we sample a normal distribution, the statistic $(\bar{x} - \mu)/(s/\sqrt{n})$ has a t distribution, which is discussed in Section 7.5. When the sample size is *large,* then this statistic is approximately normally distributed whether the sampled population is normal or nonnormal.

or

$$8.1 \pm .0558$$

Thus we estimate the mean prime lending rate to lie somewhere between 8.0442% and 8.1558%.

Can we say that this particular interval encloses μ? No, but we are fairly confident that it does, since intervals constructed in this manner enclose μ 90% of the time.   ∎

For a fixed sample size, the width of the confidence interval increases as the confidence coefficient increases, a result that is in agreement with our intuition. Certainly, if we wished to be more confident that the interval will enclose μ, we would increase the width of the interval. Since we prefer narrow confidence intervals and large confidence coefficients, we must reach a compromise in choosing the confidence coefficient.

The choice of the confidence coefficient to be used in a given situation is made by the experimenter and depends on the degree of confidence the experimenter wishes to place in the estimate. Most confidence intervals are constructed by using one of the three confidence coefficients shown in Table 7.1. The most popular seem to be 95% confidence intervals. Use of 99% confidence intervals is less common because of the wider interval width that results. Of course, you can always decrease the width by increasing the sample size n.

In addition to two-sided confidence intervals (which we will simply call confidence intervals), we can also construct **one-sided confidence intervals** for parameters. A **lower one-sided confidence interval** for a parameter gives a lower confidence limit (LCL) above which the parameter is expected to lie. An **upper one-sided confidence interval** will estimate the parameter to be less than some upper confidence limit (UCL). The z value to be used for a one-sided $(1 - \alpha)100\%$ confidence interval, z_α, locates α in a single tail of the normal distribution. Lower and upper one-sided $(1 - \alpha)100\%$ confidence limits for a population parameter *when the sample size is large* are

$$\text{LCL} = (\text{Point estimator}) - z_\alpha \times (\text{standard error of the estimator})$$

and

$$\text{UCL} = (\text{Point estimator}) + z_\alpha \times (\text{standard error of the estimator})$$

EXAMPLE   **7.3**   A corporation plans to issue some short-term notes and is hoping that the interest it will have to pay will not exceed 11.5%. To obtain some information on the mean interest rate that the corporation might expect to pay, the corporation marketed 40 notes, one through each of 40 brokerage firms. The mean and standard deviation for the 40 interest rates were $\overline{x} = 10.3\%$ and $s = .31\%$. Since the corporation is interested only in the upper limit on the interest rate that it must pay, find an upper one-sided 95% confidence interval for the mean interest rate that the corporation will have to pay for the notes.

Solution   Since the confidence coefficient is .95, $\alpha = .05$ and $z_{.05} = 1.645$. Therefore, the one-sided 95% confidence interval for μ is

$$\bar{x} + z_{.05}\sigma_{\bar{x}}$$

or

$$\bar{x} + 1.645\frac{\sigma}{\sqrt{n}}$$

Substituting $\bar{x} = 10.3$, $n = 40$, and $s = .31$ to approximate σ, we obtain the one-sided confidence interval

$$\text{UCL} = 10.3 + (1.645)\frac{.31}{\sqrt{40}}$$

or

$$\text{UCL} = 10.3 + .0806 = 10.3806$$

Thus we estimate that the mean interest rate that the corporation will have to pay on its notes is less than 10.3806%. How confident are we of this conclusion? We are fairly confident, because we know that intervals constructed in this manner enclose μ 95% of the time.   ▪

Exercises

Basic Techniques

7.1   Explain what is meant by "margin of error in estimation."

7.2   Give the margin of error in estimating a population mean μ for the following.
   a $n = 40$, $\sigma^2 = 4$ **b** $n = 100$, $\sigma^2 = .9$
   c $n = 50$, $\sigma^2 = 12$

7.3   Give the margin of error in estimating a population mean μ for the following.
   a $n = 50$, $\sigma = .1$ **b** $n = 100$, $\sigma = .9$
   c $n = 100$, $\sigma = .01$

7.4   Find a 95% confidence interval for a population mean μ for the following.
   a $n = 36$, $\bar{x} = 13.1$, $s^2 = 3.42$ **b** $n = 64$, $\bar{x} = 2.73$, $s^2 = .1047$
   c $n = 41$, $\bar{x} = 28.6$, $s^2 = 1.09$

7.5   Find a 90% confidence interval for a population mean μ for the following.
   a $n = 125$, $\bar{x} = .84$, $s^2 = .086$ **b** $n = 50$, $\bar{x} = 21.9$, $s^2 = 3.44$
   c $n = 46$, $\bar{x} = 907$, $s^2 = 128$

7.6   Find a $(1 - \alpha)100\%$ confidence interval for a population mean μ for the following.
   a $\alpha = .01$, $n = 38$, $\bar{x} = 34$, $s^2 = 12$ **b** $\alpha = .10$, $n = 65$, $\bar{x} = 1049$, $s^2 = 51$
   c $\alpha = .05$, $n = 89$, $\bar{x} = 66.3$, $s^2 = 2.48$

7.7 A random sample of n measurements is selected from a population with unknown mean μ and known standard deviation $\sigma = 10$. Calculate the width of a 95% confidence interval for μ for the following values of n.

a $n = 100$ **b** $n = 200$

c $n = 400$

7.8 Compare the confidence intervals in Exercise 7.7. What is the effect on the width of a confidence interval under the following conditions?

a You double the sample size.

b You quadruple the sample size.

7.9 Refer to Exercise 7.7.

a Calculate the width of a 90% confidence interval for μ when $n = 100$.

b Calculate the width of a 99% confidence interval for μ when $n = 100$.

c Compare the widths of the 90%, 95%, and 99% confidence intervals for μ. What effect does increasing the confidence coefficient have on the width of the confidence interval?

Applications

7.10 An increase in the rate of consumer savings is frequently tied to a lack of confidence in the economy and is said to be an indicator of a recessional tendency in the economy. A random sampling of $n = 200$ savings accounts in a local community showed a mean increase in savings account values of 7.2% over the past 12 months and a standard deviation of 5.6%. Estimate the mean percentage increase in savings account values over the past 12 months for depositors in the community. Calculate the margin of error in estimation.

7.11 Most of the claims on a small company's medical insurance are in the neighborhood of $800, but a few are very large. As a consequence, the distribution of claims is highly skewed to the right and possesses a standard deviation σ equal to $2000. The first 40 claims received this month possess a mean $\bar{x}$ equal to $930. Suppose that we were to regard this group of 40 claims as a random sample from the population of all potential claims, and use $\bar{x}$ to estimate the population mean μ.

a What is the margin of error in estimation?

b Can you make a precise statement about the probability that the error of estimation will be less than the margin in part (a)? Explain.

7.12 In his article on choice of a college major, Jake Batsell (1994) reports that most academic advisors encourage students to make their decision midway through their sophomore year. One factor that often enters into this decision is starting salaries for jobs associated with various majors. The hot areas are reported to be Computer Science and Chemical Engineering, for which average salaries are $41,800 and $39,400, respectively. The average salaries for other majors of interest are listed below.

Major	Average Starting Salary
Personnel Administration	$32,600
Accounting	28,600
Financial Administration	26,700
Marketing/Sales	24,100
Communications	22,909
Retailing	22,500
Advertising	21,400

We will assume that these averages are based upon samples of size 100.

a If the standard deviation for Personnel Administration is $1000, find a 95% confidence interval estimate for the true mean starting salary for majors in Personnel Administration.

b   Find a point estimate of the true mean starting salary for Marketing/Sales majors if the standard deviation is $800. What is the margin of error associated with this estimate?

c   Construct a 98% confidence interval for the true mean starting salary for Communications majors if the standard deviation is $\sigma = \$800$.

7.13   If you are renting an apartment and you think that your rent is too high, part of the rent that you are paying may be due to the high interest rate on borrowed money. What will be the prime rate of interest next September 1? A random sample of $n = 32$ economic forecasters produced a mean $\bar{x} = 11.7\%$ and standard deviation $s = 2.1\%$. If the forecasters' forecasts are "unbiased"—that is, if the mean of the population of forecasts of all economic forecasters will equal the actual interest rate next fall—find a 90% confidence interval for the September 1 prime interest rate.

7.14   A company personnel officer wants to estimate the mean time between occurrences of personnel accidents that might provide the potential for liability lawsuits. A random sample of $n = 30$ accidents from the company's records of the time x between an accident and the one preceding gave a sample mean of $\bar{x} = 42.1$ days and standard deviation of $s = 19.6$ days. Find a 90% confidence interval for the mean time between occurrences of personnel accidents possessing the potential for liability lawsuits.

7.15   A random sampling of a company's monthly operating expenses for a sample of $n = 36$ months produced a sample mean of $5474 and a standard deviation of $764. Find an upper one-sided 90% confidence interval for the company's mean monthly expenses.

7.5   Small-Sample Estimation of a Population Mean

The large-sample procedure for estimating a population mean that we discussed in Section 7.4 was based on two facts. First, when the sample size is large, the distribution of the sample mean is normally distributed with mean μ and standard deviation $\sigma/\sqrt{n}$, or approximately so due to the Central Limit Theorem. Second, when the value of the population standard deviation σ is not known and the sample size is large, the sample standard deviation s can be used as a reliable estimator of σ in $\sigma_{\bar{x}} = \sigma/\sqrt{n}$.

Cost limitations, time restrictions, and other factors, however, often limit the size of the sample that can be selected so that the large sample procedures do not apply. When the sample size is small, the sampling distribution of $\bar{x}$ will depend on the form of the sampled population, and $s/\sqrt{n}$ is much more variable as an estimator of $\sigma/\sqrt{n}$.

When the sampled population has a *normal distribution* and *σ is known,* the sample mean $\bar{x}$ has a normal distribution with mean μ and standard deviation $\sigma/\sqrt{n}$, and the statistic

$$\frac{\bar{x} - \mu}{\sigma/\sqrt{n}}$$

has a standard normal distribution for all sample sizes. What now can be said about the distribution of the statistic

$$\frac{\bar{x} - \mu}{s/\sqrt{n}}$$

when σ is unknown and n is small?

The distribution of the statistic

$$t = \frac{\bar{x} - \mu}{s/\sqrt{n}}$$

for samples drawn from a normally distributed population was discovered by W. S. Gosset and published (1908) under the pen name Student. He referred to the quantity under study as t, and it has since been known as **Student's t.** We omit the complicated mathematical expression for the density function for t but describe some of its characteristics.

The sampling distribution of the t test statistic, called a **t distribution,** is, like z, mound-shaped and perfectly symmetrical about $t = 0$. However, it is much more variable than z, tailing rapidly out to the right and left, a phenomenon that may readily be explained. The variability of z in repeated sampling is due solely to $\bar{x}$; the other quantities appearing in z (n, μ, and σ) are nonrandom. On the other hand, the variability of t is contributed by *two* random quantities, $\bar{x}$ and s, which can be shown to be independent of each other. Thus when $\bar{x}$ is very large, s may be very small, and vice versa. As a result, t will be more variable than z in repeated sampling (see Figure 7.5). Finally, as we might surmise, the variability of t decreases as n increases because s, the estimate of σ, will be based upon more and more information. When n is infinitely large, the t and z distributions will be identical. Thus Gosset discovered that the distribution of t depended upon the sample size n.

FIGURE 7.5
Standard normal z and a t distribution based on $n = 6$ measurements (5 d.f.)

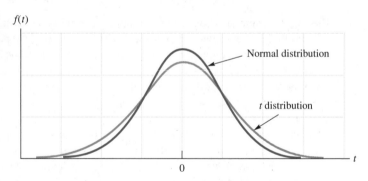

$f(t)$

Normal distribution

t distribution

0

t

The divisor of the sum of squares of deviations, $(n - 1)$, that appears in the formula for s^2 is called the number of **degrees of freedom** (d.f.) associated with s^2 and with the statistic t. The term *degrees of freedom* is linked to the statistical theory underlying the probability distribution of s^2 and deals with the number of independent squared deviations available for estimating σ^2.

The values of t having designated areas to their right are presented in Table 4 of Appendix II. Table 4 is partially reproduced in Table 7.2. The tabulated value t_a records the value of t such that an area a lies to its right, as shown in Figure 7.6. The degrees of freedom associated with t, d.f., are shown in the first and last columns of the table (see Table 7.2), and the t_a corresponding to various values of a appear in the top row. Thus, if we want to find the value of t such that 5% of the area lies to its right, we use the column labeled $t_{.05}$. For example, the value of $t_{.05}$ when $n = 6$,

FIGURE **7.6**
Tabulated values of Student's t

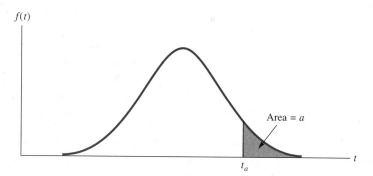

TABLE **7.2**
Format of the Student's t table,
Table 4 in Appendix II

d.f.	$t_{.100}$	$t_{.050}$	$t_{.025}$	$t_{.010}$	$t_{.005}$	d.f.
1	3.078	6.314	12.706	31.821	63.657	1
2	1.886	2.920	4.303	6.965	9.925	2
3	1.638	2.353	3.182	4.541	5.841	3
4	1.533	2.132	2.776	3.747	4.604	4
5	1.476	2.015	2.571	3.365	4.032	5
6	1.440	1.943	2.447	3.143	3.707	6
7	1.415	1.895	2.365	2.998	3.499	7
8	1.397	1.860	2.306	2.896	3.355	8
9	1.383	1.833	2.262	2.821	3.250	9
⋮	⋮	⋮	⋮	⋮	⋮	⋮
26	1.315	1.706	2.056	2.479	2.779	26
27	1.314	1.703	2.052	2.473	2.771	27
28	1.313	1.701	2.048	2.467	2.763	28
29	1.311	1.699	2.045	2.462	2.756	29
inf.	1.282	1.645	1.960	2.326	2.576	inf.

found in the $t_{.05}$ column opposite d.f. $= (n - 1) = (6 - 1) = 5$, is $t = 2.015$ (boxed in Table 7.2).

Notice that for a fixed-tail area, the right-tailed value of t will always be larger than the corresponding right-tailed value of z. For example, when $\alpha = .05$, the right-tailed value of t for $n = 2$ (d.f. $= n - 1 = 1$) is 6.314, which is very large compared with the corresponding value $z_{.05} = 1.645$. Proceeding down the $t_{.05}$ column, we note that the values of t decrease, reflecting the effect of a larger sample size (more degrees of freedom) on the estimation of σ. Finally, when n is infinitely large, the value of $t_{.05}$ equals $z_{.05} = 1.645$.

The reason for choosing $n = 30$ (an arbitrary choice) as the dividing line between large and small samples is now apparent. For $n = 30$ (d.f. $= 29$), the right-tailed value of $t_{.05} = 1.699$ is numerically quite close to $z_{.05} = 1.645$. With $a = .025$ in the right tail and $n = 30$, the right-tailed value of t is 2.045, which is very close to the value of $z_{.025} = 1.96$.

Remember that Student's t and corresponding tabulated values are based on the assumption that **the sampled population possesses a normal probability distribution.** This appears to be a very restrictive assumption, because in many sampling situations the properties of the population will be completely unknown and may well be nonnormal. If nonnormality of the population were to seriously affect the distribution of the t statistic, the application of the t test would be very limited. Fortunately, it can be shown that the distribution of the t statistic possesses nearly the same shape as the theoretical t distribution for populations that are nonnormal but possess a mound-shaped probability distribution. This property of the t statistic and the common occurrence of mound-shaped distributions of data in nature enhance the value of Student's t for use in statistical inference.

In constructing a confidence interval estimator of μ based on the t distribution, we simply replace the tabulated value of z with the equivalent tabulated value of t. The logic used in constructing a large-sample interval estimator for μ applies, except that the referenced distribution is t rather than z. In Figure 7.7(a), $(1 - \alpha)100\%$ of the values of t lie within the interval $(-t_{\alpha/2}, t_{\alpha/2})$; by analogy, in Figure 7.7(b), $(1 - \alpha)100\%$ of the values of the random variable $\bar{x}$ lie within $t_{\alpha/2}$ estimated standard deviations of the true value of μ. Therefore, there is a $(1 - \alpha)$ probability that the interval estimator

$$\bar{x} \pm t_{\alpha/2}\frac{s}{\sqrt{n}}$$

will enclose the true value of μ.

F I G U R E **7.7**
Sampling distributions of
(a) $t = \dfrac{\bar{x} - \mu}{s/\sqrt{n}}$ and (b) $\bar{x}$

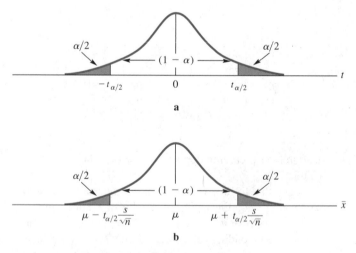

Small-Sample $(1 - \alpha)100\%$ Confidence Interval Estimator of the Mean μ

$$\bar{x} \pm t_{\alpha/2}\frac{s}{\sqrt{n}}$$

where s is the sample standard deviation, $s/\sqrt{n}$ is the *estimated* standard deviation of $\overline{x}$, and t has $(n-1)$ d.f.

Assumption: The sample has been *randomly selected* from a population that is *normally distributed* (or approximately so).

E X A M P L E **7.4**   An experiment has been conducted to evaluate a new process for producing synthetic diamonds. Six diamonds have been generated by the new process, with recorded weights of .46, .61, .52, .48, .57, and .54 carat. Find a 95% confidence interval estimate for μ, the true mean weight of synthetic diamonds produced using this process.

Solution   Using the methods of Chapter 2, you can verify that the sample mean and standard deviation for the six weights are

$$\overline{x} = .53 \quad \text{and} \quad s = .0559$$

The tabulated value of t with .025 in the right tail, based on $n - 1 = 6 - 1 = 5$ degrees of freedom, is found in Table 4 of Appendix II to be

$$t_{.025} = 2.571$$

Substituting these values into the formula for a confidence interval estimator yields

$$\overline{x} \pm t_{.025}\frac{s}{\sqrt{n}} = .53 \pm (2.571)\frac{.0559}{\sqrt{6}}$$

or

$$.53 \pm .059$$

The interval estimate for μ is therefore .471 to .589 with confidence coefficient equal to .95. If the experimenter wants to detect a small increase in mean diamond weight in excess of .5 carat, the width of the interval must be reduced by obtaining more diamond weight measurements. Increasing the sample size will decrease both $1/\sqrt{n}$ and $t_{\alpha/2}$ and thereby decrease the width of the interval.   ■

Exercises

Basic Techniques

7.16   Find the value of t.

a $t_{.05}$ for 5 degrees of freedom (d.f.)   b $t_{.025}$ for 8 d.f.

c $t_{.10}$ for 18 d.f.   d $t_{.025}$ for 30 d.f.

7.17 Find t_a given that $P(t > t_a) = a$.

    a $a = .10$, 12 d.f.   **b** $a = .01$, 25 d.f.

    c $a = .05$, 16 d.f.

7.18 A random sample of $n = 6$ observations from a normally distributed population produced the following data: 6.2, 5.8, 7.1, 6.3, 6.9, 5.7.

    a   Calculate $\bar{x}$ and s for the data.

    b   Find a 90% confidence interval for the population mean μ.

    c   Find a 95% confidence interval for the population mean μ. Compare the width of this interval to the interval calculated in part (b).

    d   Interpret the intervals found in parts (b) and (c).

7.19 Twelve measurements were randomly selected from a normal population, producing $\bar{x} = 125.12$ and $s = 12.3$.

    a   Find a 98% confidence interval for the population mean μ.

    b   Find a 99% confidence interval for the population mean μ.

    c   Interpret the intervals found in parts (a) and (b).

Applications

7.20 With home mortgage loan rates in a slow upward movement and a mean prevailing rate of 8.7%, a bank decided to survey the mortgage rate expectations of its current applicants. A random sample of the ten latest applicants found that the mean of the ten rate expectations that the applicants wanted to negotiate was 8.5%. The ten individual rate expectations varied from a low of 8.0% to a high of 8.9%, with a standard deviation equal to .23%.

    a   If the distribution of home mortgage rates is assumed to be approximately normally distributed, find a 90% confidence interval for the mean rate expectation of the bank's home mortgage loan applicants. Interpret this interval.

    b   Does the interval constructed in part (a) contain the mean 8.7% rate prevailing in the bank's market area? Would this lead you to believe that the mean mortgage rate expectation of the bank's mortgage loan applicants is less than the mean 8.7% prevailing rate? Explain.

7.21 Varying costs, primarily labor, make home building profits vary from one unit to the next. A builder of standard tract homes needs to make an average profit in excess of $8500 per home in order to achieve an annual profit goal. The profits per home for the builder's most recent five units are $8760, $6370, $9620, $8200, and $10,350.

    a   Find a 95% confidence interval for the builder's mean profit per unit. Interpret this interval.

    b   Does the interval constructed in part (a) contain $8500? Would you conclude that the builder is operating at the desired profit level?

7.22 The average prices in dollars for the remaining $n = 21$ 27-inch television sets mentioned in Exercise 6.41 ("Ratings: 27-inch TV Sets," 1994) are given in the following table.

380	590	585
670	699	530
560	430	610
565	705	475
565	580	405
385	465	425
390	450	390

    a   Find the mean and standard deviation of these prices.

    b   Construct a 99% confidence interval for the overall average price of the 21 brands/models of 27-inch TVs, not including the Sony KV-27XBR26.

c If the average price of the Sony, $1085, were included, how would your estimate in part (b) change?

d Could the average price of the Sony be considered an outlier?

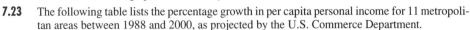 **7.23** The following table lists the percentage growth in per capita personal income for 11 metropolitan areas between 1988 and 2000, as projected by the U.S. Commerce Department.

Metro Area	Percentage Growth Per Capita Income
Anaheim	13.6
Atlanta	14.2
Cleveland	17.2
Detroit	15.3
Fort Lauderdale, FL	16.0
Jacksonville, FL	16.2
New Orleans	18.7
New York	15.3
Pittsburgh	15.8
San Diego	13.8
Seattle	15.5

Source: U.S. Commerce Department, *The Press Enterprise,* October 12, 1990, p. F–5.

a If these projections represent a random sample of the growth projections for all metropolitan areas in the United States, find a 98% confidence interval for the mean projected percent growth in per capita income for all metropolitan areas in the United States from 1988 to 2000. (The population of percent growth in per capita income is assumed to be approximately normally distributed.)

b Is it likely that the mean projected percent growth in part (a) will equal the actual mean percent growth in the year 2000? Explain.

7.6 Estimating the Difference Between Two Means

A problem of equal importance to the estimation of population means is the comparison of two population means. For instance, we might wish to estimate the difference between two states in the mean claim size for a type of automobile insurance. This estimate will be based on independent random samples of claims selected from among those filed in the two states. Or we might wish to compare the average yield in a chemical plant using raw materials furnished by two suppliers, A and B. Samples of daily yield, one for each of the two suppliers, will be recorded and used to make inferences concerning the difference in mean yield.

For each of these examples there are two populations, the first with mean and variance μ_1 and σ_1^2 and the second with mean and variance μ_2 and σ_2^2. A random sample of n_1 measurements is drawn from population 1 and n_2 from population 2, where the samples are assumed to have been drawn independently of each other. Finally, the estimates of the population parameters are calculated from the sample data using the estimators $\bar{x}_1$, s_1^2, $\bar{x}_2$, and s_2^2.

Intuitively, the difference between two sample means would provide the maximum information about the actual difference between two population means, and this is in fact the case. The best point estimator of the difference $(\mu_1 - \mu_2)$ between the population means is $(\bar{x}_1 - \bar{x}_2)$. The sampling distribution of this estimator is not difficult to derive, but we state it here without proof.

Properties of the Sampling Distribution of $(\bar{x}_1 - \bar{x}_2)$, the Difference Between Two Sample Means

When independent random samples of n_1 and n_2 observations have been selected from populations with means μ_1 and μ_2 and variances σ_1^2 and σ_2^2, respectively, the sampling distribution of the difference $(\bar{x}_1 - \bar{x}_2)$ will have the following properties:

1 The mean and the standard deviation[†] of $(\bar{x}_1 - \bar{x}_2)$ will be

$$\mu_{(\bar{x}_1 - \bar{x}_2)} = \mu_1 - \mu_2$$

and

$$\sigma_{(\bar{x}_1 - \bar{x}_2)} = \sqrt{\frac{\sigma_1^2}{n_1} + \frac{\sigma_2^2}{n_2}}$$

2 If the sampled populations are *normally distributed,* then the sampling distribution of $(\bar{x}_1 - \bar{x}_2)$ is *exactly normally distributed,* regardless of the sample size.

3 If the sampled populations are *not normally distributed,* then the sampling distribution of $(\bar{x}_1 - \bar{x}_2)$ is *approximately normally distributed* when n_1 and n_2 are large, due to the Central Limit Theorem.

Since $\mu_1 - \mu_2$ is the mean of sampling distribution, it follows that $(\bar{x}_1 - \bar{x}_2)$ is an unbiased estimator of $(\mu_1 - \mu_2)$. Hence, when the sample sizes are large, the general formulas of Section 7.3 can be used to construct point and interval estimates.

Point Estimation of $(\mu_1 - \mu_2)$

Estimator: $(\bar{x}_1 - \bar{x}_2)$

Margin of error: $1.96\sigma_{(\bar{x}_1 - \bar{x}_2)} = 1.96\sqrt{\frac{\sigma_1^2}{n_1} + \frac{\sigma_2^2}{n_2}}$

A $(1 - \alpha)100\%$ Confidence Interval for $(\mu_1 - \mu_2)$

$$(\bar{x}_1 - \bar{x}_2) \pm z_{\alpha/2}\sqrt{\frac{\sigma_1^2}{n_1} + \frac{\sigma_2^2}{n_2}}$$

[†]Finite population correction factors may be required if N_1 and N_2 are small and n_1/N_1 and n_2/N_2 are greater than .05.

If σ_1^2 and σ_2^2 are unknown, they can be approximated by the sample variances s_1^2 and s_2^2.

Assumption: n_1 and n_2 are both greater than or equal to 30.

EXAMPLE **7.5** Recognizing that court liability awards vary over time, an insurance company wants to compare the mean level of current personal liability awards with those of a year earlier. A random sample of $n = 30$ cases was selected from among cases tried during each of the two yearly periods. The sample means and variances of the liability awards (in millions of dollars) for each of the two years are shown in Table 7.3.

a Find a point estimate for the difference in the mean level of liability awards between the current and preceding years. Give the margin of error.

b Find a 90% confidence interval for the difference in the mean level of liability awards between the current and preceding years.

TABLE **7.3**
Sample means and variances for
Example 7.5

Year	Sample Size	Sample Mean ($ million)	Sample Variance ($ million)^2
Current	$n_1 = 30$	$\bar{x}_1 = 1.32$	$s_1^2 = .9734$
Previous	$n_2 = 30$	$\bar{x}_2 = 1.04$	$s_2^2 = .7291$

Solution a The point estimate for $\mu_1 - \mu_2$ is

$$\bar{x}_1 - \bar{x}_2 = 1.32 - 1.04 = .28$$

with margin of error given by

$$1.96\sqrt{\frac{\sigma_1^2}{n_1} + \frac{\sigma_2^2}{n_2}} \approx 1.96\sqrt{\frac{.9734}{30} + \frac{.7291}{30}} = 1.96(.2382) = .47$$

b Since we want to find a 90% confidence interval for $(\mu_1 - \mu_2)$, $(1 - \alpha) = .90$, $\alpha = .10$, $\alpha/2 = .05$, and $z_{.05} = 1.645$. The confidence interval is

$$(\bar{x}_1 - \bar{x}_2) \pm z_{\alpha/2}\sqrt{\frac{\sigma_1^2}{n_1} + \frac{\sigma_2^2}{n_2}}$$

$$(1.32 - 1.04) \pm 1.645\sqrt{\frac{.9734}{30} + \frac{.7291}{30}}$$

$$.28 \pm .392$$

Rounding to two places, we estimate the difference in mean liability awards to fall between −$110,000 and $670,000. You can see that this confidence interval is very wide, allowing the possibilities that the mean award this year could have been $670,000 larger than last year or $110,000 smaller. If the insurance company wants to estimate the difference in mean awards with a narrower confidence interval, it will have to obtain more information by increasing the samples sizes n_1 and n_2. ∎

Small-sample inference for the difference between population means is based on the assumption that *both populations are normally distributed* and, in addition, that they possess *equal variances*—that is, $\sigma_1^2 = \sigma_2^2 = \sigma^2$. For this special case, the large-sample confidence interval estimate is based on the statistic

$$z = \frac{(\bar{x}_1 - \bar{x}_2) - (\mu_1 - \mu_2)}{\sqrt{\dfrac{\sigma_1^2}{n_1} + \dfrac{\sigma_2^2}{n_2}}}$$

which with $\sigma_1^2 = \sigma_2^2 = \sigma^2$ reduces to

$$z = \frac{(\bar{x}_1 - \bar{x}_2) - (\mu_1 - \mu_2)}{\sqrt{\dfrac{\sigma^2}{n_1} + \dfrac{\sigma^2}{n_2}}} = \frac{(\bar{x}_1 - \bar{x}_2) - (\mu_1 - \mu_2)}{\sigma\sqrt{\dfrac{1}{n_1} + \dfrac{1}{n_2}}}$$

For small-sample inference, it seems reasonable to use the statistic

$$t = \frac{(\bar{x}_1 - \bar{x}_2) - (\mu_1 - \mu_2)}{s\sqrt{\dfrac{1}{n_1} + \dfrac{1}{n_2}}}$$

in which we substitute a sample standard deviation s for the population standard deviation σ. Surprisingly enough, this statistic has a Student's t distribution when the stated assumptions are satisfied.

The estimate s to be used in the t statistic could be either s_1 or s_2, the standard deviations for the two samples, although the use of either would be wasteful since both estimate σ. Since we want to obtain the best estimate available, it seems reasonable to use an estimator that pools the information from both samples. This **pooled estimator of σ^2**, using the sums of squares of the deviations about the mean for both samples, is given in the following display.

Pooled Estimator of σ^2

$$s^2 = \frac{\sum_{i=1}^{n_1}(x_{1i} - \overline{x}_1)^2 + \sum_{i=1}^{n_2}(x_{2i} - \overline{x}_2)^2}{n_1 + n_2 - 2}$$

or

$$s^2 = \frac{(n_1 - 1)s_1^2 + (n_2 - 1)s_2^2}{(n_1 - 1) + (n_2 - 1)}$$

with

$$s_1^2 = \frac{\sum_{i=1}^{n_1}(x_{1i} - \overline{x}_1)^2}{n_1 - 1} \quad \text{and} \quad s_2^2 = \frac{\sum_{i=1}^{n_2}(x_{2i} - \overline{x}_2)^2}{n_2 - 1}$$

The denominator in the formula for s^2, $(n_1 + n_2 - 2)$, is called the **number of degrees of freedom** associated with s^2. It can be proved either mathematically or experimentally that the expected value of the pooled estimator s^2 is equal to σ^2 and therefore that s^2 is an unbiased estimator of the common population variance. Finally, recall that the divisors of the sums of the squares of deviations in s_1^2 and s_2^2, $(n_1 - 1)$ and $(n_2 - 1)$, respectively, are the numbers of degrees of freedom associated with these two independent estimators of σ^2. Note that s^2 is a weighted average of s_1^2 and s_2^2, with the degrees of freedom as weights, and therefore is an estimator that uses the pooled information for both samples and possesses $(n_1 - 1) + (n_2 - 1)$, or $(n_1 + n_2 - 2)$, degrees of freedom.

The small-sample confidence interval estimator for $(\mu_1 - \mu_2)$ with confidence coefficient $(1 - \alpha)$ is given by the formula in the next display.

A $(1 - \alpha)100\%$ Small-Sample Confidence Interval for $(\mu_1 - \mu_2)$

$$(\overline{x}_1 - \overline{x}_2) \pm t_{\alpha/2}s\sqrt{\frac{1}{n_1} + \frac{1}{n_2}}$$

where s is obtained from the pooled estimate of σ^2, given above.

Assumptions: The samples were randomly and independently selected from normally distributed populations. The variances of the populations, σ_1^2 and σ_2^2, are equal.

Note the similarity in the procedures for constructing the confidence intervals for a single mean (Section 7.5) and the difference between two means. In both cases the interval is constructed by using the appropriate point estimator and then adding and subtracting an amount equal to $t_{\alpha/2}$ times the *estimated* standard deviation of the point estimator.

E X A M P L E   **7.6**   An assembly operation in a manufacturing plant requires approximately a one-month training period for a new employee to reach maximum efficiency. A new method of training was suggested and a test was conducted to compare the new method with the standard procedure. Two groups of nine new employees were trained for a period of three weeks, one group using the new method and the other following standard training procedure. The length of time in minutes required for each employee to assemble the device was recorded at the end of the three-week period. These measurements appear in Table 7.4. Construct a 95% confidence interval estimate for the difference between the mean time to assemble after a three-week training period for the standard procedure and the new procedure.

T A B L E **7.4**
Assembly time data for
Example 7.6

Standard Procedure	New Procedure
32	35
37	31
35	29
28	25
41	34
44	40
35	27
31	32
34	31

Solution   Let μ_1 and μ_2 equal the mean time to assemble for the standard and the new assembly procedures, respectively. Also, assume that the variability in mean time to assemble is essentially a function of individual differences and that the variability for the two populations of measurements will be approximately equal.

The sample means and standard deviations can be found using a calculator with a statistics function:

$$\bar{x}_1 = 35.22 \qquad s_1 = 4.9441$$
$$\bar{x}_2 = 31.56 \qquad s_2 = 4.4752$$

Then the pooled estimate of the common variance is

$$s^2 = \frac{(n_1-1)s_1^2 + (n_2-1)s_2^2}{n_1+n_2-2} = \frac{195.55 + 160.22}{9+9-2} = 22.24$$

and the standard deviation is $s = \sqrt{22.24} = 4.72$.

Referring to Table 4 in Appendix II, we find the value of t with area $\alpha/2 = .025$ to its right and $(n_1 + n_2 - 2) = 9 + 9 - 2 = 16$ degrees of freedom to be $t_{.025} = 2.120$. Substituting into the formula

$$(\bar{x}_1 - \bar{x}_2) \pm t_{\alpha/2} s \sqrt{\frac{1}{n_1} + \frac{1}{n_2}}$$

we find the interval estimate (or 95% confidence interval) to be

$$(35.22 - 31.56) \pm (2.120)(4.72)\sqrt{\frac{1}{9} + \frac{1}{9}}$$

or

$$3.66 \pm 4.72$$

Thus we estimate the difference in mean time to assemble, $(\mu_1 - \mu_2)$, to fall in the interval -1.06 to 8.38. Note that the interval width is considerable and that it would seem advisable to increase the size of the samples and reestimate. ▪

Before concluding our discussion, we comment on the two assumptions upon which our inferential procedures are based. Moderate departures from the assumption that the populations possess a normal probability distribution do not seriously affect the properties of the estimator or the confidence coefficient for the corresponding confidence interval. On the other hand, the population variances should be nearly equal in order that the aforementioned procedures be valid. A procedure will be presented in Section 8.9 for testing an hypothesis concerning the equality of two population variances. An alternative procedure for estimating $\mu_1 - \mu_2$ when the populations are normal but have unequal variances ($\sigma_1^2 \neq \sigma_2^2$) will be given in Chapter 8.

Exercises

Basic Techniques

7.24   Independent random samples were selected from two populations, 1 and 2. The sample sizes, means, and variances were as shown in the accompanying table. Give the margin of error for estimating the difference in population means $(\mu_1 - \mu_2)$.

	Population	
	1	2
Sample size	35	49
Sample mean	12.7	7.4
Sample variance	1.38	4.14

7.25   Independent random samples were selected from two populations, 1 and 2. The sample sizes, means, and variances were as shown in the following table. Find a 90% confidence interval for the difference in the population means, and interpret your result.

	Population	
	1	2
Sample size	64	64
Sample mean	2.9	5.1
Sample variance	.83	1.67

7.26   Give the number of degrees of freedom for s^2, the pooled estimator of σ^2, for the following sample sizes.

a $n_1 = 16$, $n_2 = 8$ b $n_1 = 10$, $n_2 = 12$
c $n_1 = 15$, $n_2 = 3$

7.27   Calculate s^2, the pooled estimator for σ^2, for the following sample data.

a $n_1 = 10$, $n_2 = 4$, $s_1^2 = 3.4$, $s_2^2 = 4.9$
b $n_1 = 12$, $n_2 = 21$, $s_1^2 = 18$, $s_2^2 = 23$

7.28   You are given the following two independent random samples drawn from each of two normal populations:

Sample 1	12	3	8	5	
Sample 2	14	7	7	9	6

Calculate s^2, the pooled estimator of σ^2.

7.29   Find a 95% confidence interval for $(\mu_1 - \mu_2)$ in Exercise 7.28.

7.30   Independent random samples of $n_1 = 16$ and $n_2 = 13$ observations were selected from two normal populations with equal variances. The sample means and variances are shown in the accompanying table. Find a 99% confidence interval for $(\mu_1 - \mu_2)$. Interpret this interval.

	Population	
	1	2
Sample size	16	13
Sample mean	34.6	32.2
Sample variance	4.8	5.9

Applications

7.31   A study was conducted to compare the mean number of police emergency calls per eight-hour shift in two districts of a large city. Samples of 100 eight-hour shifts were randomly selected from the police records for each of the two regions, and the number of emergency calls was recorded for each shift. The sample statistics are shown in the accompanying table. Find a 90% confidence interval for the difference in the mean number of police emergency calls per shift between the two districts of the city. Interpret the interval.

	Region	
	1	2
Sample size	100	100
Sample mean	2.4	3.1
Sample variance	1.44	2.64

7.32   One method suggested a number of years ago for solving the electric power shortage employed floating nuclear power plants located a few miles offshore in the ocean. Because there was concern about the possibility of a ship collision with the floating (but anchored) plant, an estimate of the density of ship traffic in the area was needed. The number of ships passing within 10 miles of the proposed power plant location per day, recorded for $n = 60$ days during July and August, possessed sample mean and variance equal to

$$\bar{x} = 7.2 \qquad s^2 = 8.8$$

a   Find a 95% confidence interval for the mean number of ships passing within 10 miles of the proposed power plant location during a one-day time period.

b   The density of ship traffic was expected to decrease during the winter months. A sample of $n = 90$ daily recordings of ship sightings for December, January, and February gave the following mean and variance:

$$\bar{x} = 4.7 \qquad s^2 = 4.9$$

Find a 90% confidence interval for the difference in mean density of ship traffic between the summer and winter months.

c   What is the population associated with your estimate in part (b)? What could be wrong with the sampling procedure in parts (a) and (b)?

7.33   An automobile manufacturer recently decided that the primary factor inhibiting sales was not the automobile or its service but the sales approach employed by its salespersons. In a test of this theory, 16 salespersons in a large dealership were randomly assigned to two groups of eight salespersons each. One group employed a hard-sell approach to customers for a one-month period; the other employed a slower-paced, soft-sell approach over the same period of time. The means and standard deviations of the dollar sales per salesperson per month for the two groups are shown in the accompanying table.

Sample Data	Hard Sell	Soft Sell
Sample size	8	8
Sample mean	106,200	111,900
Sample standard deviation	24,400	28,600

a   Find a 99% confidence interval for $(\mu_1 - \mu_2)$, the difference in the mean level of sales for the two approaches. Interpret this interval.

b   Does the confidence interval constructed in part (a) contain the value $(\mu_1 - \mu_2) = 0$? Would this confidence interval indicate that there is a difference in the mean level of sales for the two approaches? Explain.

7.34   A bank's loan department found that 57 home loans processed during April had a mean value of $78,100 and a standard deviation of $6300. An analysis of the loans in May, a total of 66, showed a mean value of $82,700 with a standard deviation of $7100. Suppose these home loans represent random samples of the values of home loan applications approved in the bank's service area. Find a 98% confidence interval for the difference in the mean level of approved home loan applications from April to May.

7.35 How do corporate executives and stock market analysts compare in their forecasts of increase in the gross national product (GNP) for the next year? Forecasts (in percentages) from five randomly selected corporate executives and five market analysts are shown below.

Corporate executives	3.4	2.8	3.9	3.7	3.4
Market analysts	3.3	3.9	3.4	3.8	4.0

Find a 90% confidence interval for the difference in mean GNP forecasts between corporate executives and stock market analysts. Interpret the interval.

7.7 Estimating a Binomial Proportion

Many surveys have as their objective the estimation of the proportion of people or objects in a large group that possess a particular attribute. Such a survey is a practical example of the binomial experiment discussed in Chapter 4. Estimating the proportion of sales that can be expected in a large number of customer contacts is a practical problem requiring the estimation of a binomial parameter p.

The best point estimator of the binomial parameter p is also the estimator that would be chosen intuitively. That is, the estimator of p, denoted by the symbol $\hat{p}$, is the total number x of successes divided by the total number n of trials:

$$\hat{p} = \frac{x}{n}$$

where x is the number of successes in n trials. [A "hat" ($\hat{\ }$) over a parameter is the symbol used to denote the estimator of the parameter.] By "best" we mean that $\hat{p}$ is unbiased and possesses a smaller variance than other possible estimators.

As noted in Chapter 6, the estimator $\hat{p}$ possesses a sampling distribution that can be approximated by a normal distribution because of the Central Limit Theorem. It is an unbiased estimator of the population proportion p, with mean and standard deviation given in the following display.

Mean and Standard Deviation of $\hat{p}$

$$E(\hat{p}) = p$$

$$\sigma_{\hat{p}} = \sqrt{\frac{pq}{n}}$$

Then, from Section 7.3, the point and interval estimation procedures for p are as given in the following display:

Point Estimator for p

Estimator: $\hat{p} = \dfrac{x}{n}$

Margin of error: $1.96\sigma_{\hat{p}} = 1.96\sqrt{\dfrac{pq}{n}}$

Estimated margin of error: $1.96\sqrt{\dfrac{\hat{p}\hat{q}}{n}}$

A $(1 - \alpha)100\%$ Confidence Interval for p

$$\hat{p} \pm z_{\alpha/2}\sqrt{\dfrac{\hat{p}\hat{q}}{n}}$$

Assumption: n must be sufficiently large so that the sampling distribution of $\hat{p}$ can be approximated by a normal distribution. The interval $p \pm 2\sigma_{\hat{p}}$ must be contained in the interval from 0 to 1.

The only difficulty encountered in our procedure will be in calculating $\sigma_{\hat{p}}$, which involves the unknown value of p (and $q = 1 - p$). You will note that we have substituted $\hat{p}$ for the parameter p in the standard deviation $\sqrt{pq/n}$. When n is large, little error will be introduced by this substitution. As a matter of fact, the standard deviation changes only slightly as p changes. This feature can be observed in Table 7.5, where $\sqrt{pq}$ is recorded for several values of p. Note that $\sqrt{pq}$ changes very little as p changes, especially when p is near .5.

TABLE **7.5**
Some calculated values of $\sqrt{pq}$

p	$\sqrt{pq}$
.5	.50
.4	.49
.3	.46
.2	.40
.1	.30

E X A M P L E   **7.7**   A random sample of $n = 100$ wholesalers who buy polyvinyl plastic pipe indicated that 59 plan to increase their purchases in the coming year. Estimate the proportion p of wholesalers in the population of all polyvinyl pipe wholesalers who plan to increase their purchases next year, and find the margin of error in estimation. Find a 95% confidence interval for p.

Solution   The point estimate is $\hat{p} = x/n = 59/100 = .59$, and the margin of error in estimation is

$$1.96\sigma_{\hat{p}} = 1.96\sqrt{\dfrac{pq}{n}} \approx 1.96\sqrt{\dfrac{(.59)(.41)}{100}} = .096$$

A 95% confidence interval for p is

$$\hat{p} \pm 1.96\sqrt{\frac{\hat{p}\hat{q}}{n}}$$

Substituting into this formula, we obtain

$$.59 \pm 1.96(.049)$$

or

$$.59 \pm .096$$

Thus we estimate that the proportion p of wholesalers who plan to increase their purchases lies in the interval .494 to .686, with confidence coefficient .95.   ▪

Exercises

Basic Techniques

7.36   A random sample of $n = 900$ observations from a binomial population produced $x = 655$ successes. Find a 99% confidence interval for p, and interpret the interval.

7.37   A random sample of $n = 300$ observations from a binomial population produced $x = 263$ successes. Find a 90% confidence interval for p, and interpret the interval.

7.38   A random sample of $n = 500$ observations from a binomial population produced $x = 140$ successes. Find a 95% confidence interval for p, and interpret the interval.

7.39   Suppose that the number of successes observed in $n = 500$ trials of a binomial experiment is 27. Find a 95% confidence interval for p. Why is the confidence interval narrower than the confidence interval in Exercise 7.38?

Applications

7.40   More and more women and men are making purchasing choices that were once left to the opposite sex, and in most married households husbands and wives have equal say in making major purchases. In her article in *American Demographics,* Shannon Dortch (1994) suggests that advertisers may have to change where they advertise certain items and what the pitch should be. Clearly, automobile advertising needs to reach women. Approximately 80% of husbands and wives both agree that they had equal say in purchasing a new home, while 58% of men and 56% of women agree that they had equal say in purchasing a new car. Assume that these percentages were based upon sample sizes of 3000.

    a   Find a 95% confidence interval estimate for the proportion of husbands who agree that both they and their wife had an equal say in purchasing their home. (The same numerical estimate would give the percentage of wives who agree that both they and their husband had an equal say in purchasing their home.)

    b   Find a 95% confidence interval for the percentage of husbands who agree that their spouse had an equal say in the purchase of their new car.

    c   Estimate the percentage of wives who agree that their spouse had an equal say in the purchase of their new car.

7.41   The Ford Motor Company commissioned a study to determine where professional leaders stand on environmental policy issues. Among the 7000 leaders in business, education, media, government, and environmental advocacy who took part, 80% felt that industry should be held liable for environmental damage and cleanup caused by their action (*Opinions '90,* June 1990, p. 64).

    a   Determine a point estimate of the true value of p with a 95% margin of error. Is the error less than three percentage points?

    b   Use the results in part (a) to construct a 95% confidence interval estimate for p.

7.42   In a recent survey of 1600 upscale households in five large metropolitan areas, 24% of men consider themselves the primary shoppers for their household's groceries and their own clothing (Dholakia, 1994). Although the male shoppers in the survey are highly computer-literate (87% have home computers), only 22% say that computerized shopping is satisfying.

    a   Find a 90% confidence interval estimate for the percentage of men who consider themselves the primary shoppers for their household's groceries and their own clothing.

    b   Find the margin of error in estimating the proportion of computer-literate male shoppers who find computerized shopping satisfying.

7.43   In a 1993 survey by the American Dietetic Association and Kraft General Food concerning nutrition and eating habits, 54% of the respondents saw nutrition as highly important, 52% indicated that eating habits were changing, and 44% said they do all they can in eating a healthy diet ("Crunching Numbers," 1994). If 1000 respondents were involved in this survey, how accurate are these estimates of the corresponding population percentages? Find the margin of error associated with each estimate.

7.44   The advertising industry tries to sell products by showcasing them in situations intended to appeal to various market segments. Unless specifically intended for senior citizens, most ads feature vibrant, young adults (Darnay, 1994). In a survey of 1002 Americans age 18 and older, the responses to the question "How fairly do you think the following are represented in advertising?" are summarized in the following table.

Population Segment	Underrepresented
Handicapped people	.57
Hispanics	.41
People 50 or older	.37
Blacks	.29
Teens/young adults	.09

    a   Find the margin of error for the estimate of the proportion of Americans who feel that handicapped people are underrepresented in advertising.

    b   What is the margin of error associated with the proportion of Americans who feel that teenagers and young adults are underrepresented in advertising? Why does the margin of error in this case differ from that in part (a)?

    c   Calculate a 95% confidence interval estimate for the proportion of Americans who feel that people 50 and older are underrepresented in advertising.

7.8   Estimating the Difference Between Two Binomial Proportions

The fourth and final estimation problem considered in this chapter is the estimation of the difference between the parameters of two binomial populations—that is, the difference between two binomial proportions. We may be interested in the difference between the production rates for each of two production lines, or in the difference in the proportion of voters favoring two leading mayoral candidates. Intuitively, the difference between two sample proportions would provide the maximum information about the corresponding population proportions.

Assume that the two binomial populations 1 and 2 have parameters p_1 and p_2, respectively. Independent random samples consisting of n_1 and n_2 trials are drawn from the respective populations, and the estimates $\hat{p}_1$ and $\hat{p}_2$ are calculated. The sampling distribution of the difference $\hat{p}_1 - \hat{p}_2$ is stated without proof in the following display.

Properties of the Sampling Distribution of the Difference $(\hat{p}_1 - \hat{p}_2)$ Between Two Sample Proportions

Assume that independent random samples of n_1 and n_2 observations have been selected from binomial populations with parameters p_1 and p_2, respectively. The sampling distribution of the difference between sample proportions

$$(\hat{p}_1 - \hat{p}_2) = \left(\frac{x_1}{n_1} - \frac{x_2}{n_2} \right)$$

will have the following properties:

1 The mean and the standard deviation of $(\hat{p}_1 - \hat{p}_2)$ will be

$$\mu_{(\hat{p}_1 - \hat{p}_2)} = p_1 - p_2$$

and

$$\sigma_{(\hat{p}_1 - \hat{p}_2)} = \sqrt{\frac{p_1 q_1}{n_1} + \frac{p_2 q_2}{n_2}}$$

2 The sampling distribution of $(\hat{p}_1 - \hat{p}_2)$ can be approximated by a normal distribution when n_1 and n_2 are large, due to the Central Limit Theorem.

When we use a normal distribution to approximate binomial probabilities, the interval $(\hat{p}_1 - \hat{p}_2) \pm 2\sigma_{(\hat{p}_1 - \hat{p}_2)}$ should be contained within the range of $(\hat{p}_1 - \hat{p}_2)$, which varies from -1 to 1 and not from 0 to 1, as in the case of a single proportion.

Since $(\hat{p}_1 - \hat{p}_2)$ is an unbiased estimator of $(p_1 - p_2)$, the point estimator of $p_1 - p_2$ is given in the following display.

Point Estimator of $(p_1 - p_2)$

Estimator: $(\hat{p}_1 - \hat{p}_2)$

Margin of error: $1.96\sigma_{(\hat{p}_1-\hat{p}_2)} = 1.96\sqrt{\dfrac{p_1q_1}{n_1} + \dfrac{p_2q_2}{n_2}}$

[*Note:* The estimates $\hat{p}_1$ and $\hat{p}_2$ must be substituted for p_1 and p_2 to estimate the margin of error.]

The $(1 - \alpha)100\%$ confidence interval, appropriate when n_1 and n_2 are large, is shown here.

A $(1 - \alpha)100\%$ Large-Sample Confidence Interval for $(p_1 - p_2)$

$$(\hat{p}_1 - \hat{p}_2) \pm z_{\alpha/2}\sqrt{\dfrac{\hat{p}_1\hat{q}_1}{n_1} + \dfrac{\hat{p}_2\hat{q}_2}{n_2}}$$

Assumption: n_1 and n_2 must be sufficiently large so that the sampling distribution of $(\hat{p}_1 - \hat{p}_2)$ can be approximated by a normal distribution. The interval $(p_1 - p_2) \pm 2\sigma_{(\hat{p}_1-\hat{p}_2)}$ must be contained in the interval -1 to 1.

EXAMPLE **7.8** A large clothing retailer conducted a study to compare the effectiveness of a news-paper advertisement in each of two large cities. A large advertisement was run in the major newspaper in each of the two large cities. Immediately thereafter, a marketing research organization conducted a telephone survey of 1000 randomly selected adults living in a middle- to upper-income suburban area in each of the cities to determine the proportion that had read the retailer's advertisement. The sample proportions were $\hat{p}_1 = .18$ and $\hat{p}_2 = .14$. Find a 95% confidence interval for the difference in the proportions of adults in the two populations who had read the advertisement.

Solution A 95% confidence interval for the difference $(p_1 - p_2)$ in the proportions of adults who had read the retailer's advertisement is

$$(\hat{p}_1 - \hat{p}_2) \pm z_{\alpha/2}\sqrt{\dfrac{\hat{p}_1\hat{q}_1}{n_1} + \dfrac{\hat{p}_2\hat{q}_2}{n_2}}$$

where $\alpha = .05$ and $z_{\alpha/2} = z_{.025} = 1.96$. Substituting this value, along with the esti-mates $\hat{p}_1 = .18$ and $\hat{p}_2 = .14$, into the formula for the confidence interval, we find the confidence interval to be

$$(.18 - .14) \pm 1.96\sqrt{\dfrac{(.18)(.82)}{1000} + \dfrac{(.14)(.86)}{1000}}$$

or

$$.04 \pm .0321$$

Thus we estimate the difference in the proportions of readers of the advertisement to lie between .0079 and .0721. If we wish to express this difference as a percentage, we multiply each proportion by 100. Then we can say that we estimate the difference in the percentages of adults reading the advertisement in the two marketing regions to be between .79% and 7.21%. ▪

Exercises

Basic Techniques

7.45 Samples of $n_1 = 500$ and $n_2 = 500$ observations were selected from binomial populations 1 and 2, and $x_1 = 120$ and $x_2 = 147$ were observed. Find the margin of error in estimating the difference in population proportions $(p_1 - p_2)$.

7.46 Samples of $n_1 = 800$ and $n_2 = 640$ observations were selected from binomial populations 1 and 2, and $x_1 = 337$ and $x_2 = 374$ were observed.

a Find a 90% confidence interval for the difference $(p_1 - p_2)$ in the two population parameters. Interpret the interval.

b What assumptions must you make in order for the confidence interval to be valid?

7.47 Samples of $n_1 = 1265$ and $n_2 = 1688$ observations were selected from binomial populations 1 and 2, and $x_1 = 849$ and $x_2 = 910$ were observed.

a Find a 99% confidence interval for the difference $(p_1 - p_2)$ in the two population parameters. Interpret the interval.

b What assumptions must you make in order for the confidence interval to be valid?

7.48 Samples of $n_1 = 314$ and $n_2 = 207$ observations were selected from binomial populations 1 and 2, and $x_1 = 108$ and $x_2 = 102$ were observed.

a Find a 95% confidence interval for the difference $(p_1 - p_2)$ in the two population parameters. Interpret the interval.

b What assumptions must you make in order for the confidence interval to be valid?

Applications

7.49 A bank survey of delinquent credit card payments found that the delinquency rate in a given month for 414 small-business owners was 5.8% versus only 3.6% for 1029 professionals. Assume that the data for these two types of cardholders can be regarded as independent random samples of monthly accounts over a relatively long period of time, say, one or two years. Find a 95% confidence interval for the difference in the proportions of delinquencies for these two types of credit card users.

7.50 During the recovery from the period of economic stagnation in the early 1990s, surveys by the Conference Board were used to gauge consumer behavior patterns and the economy's long-range health, with questions on subjects that ranged from household buying plans to local job conditions ("Consumer Confidence," 1994). One poll that appeared in July 1994 showed that 10% of 5000 households nationwide expected business conditions to get worse, an upward change from the 8% of 5000 households the month before.

a Construct a 95% confidence interval estimate of the difference in proportions of those respondents who expected business conditions to get worse.

b What assumptions underlie the validity of the confidence interval constructed in part (a)?

7.51   According to a study reported in the *American Enterprise* ("How the Japanese," 1990), Japanese satisfaction with economic accomplishment does not extend to their personal lives when compared with Americans, who express more satisfaction with their lives, their personal accomplishments, and their jobs. When asked "Would you describe your job as generally interesting?" 88% of the Americans in the survey responded positively compared to 58% of the Japanese in the survey. Suppose that we are told that 3000 Americans and 2500 Japanese were included in this survey. Use this information to construct a 99% confidence interval estimate for $p_1 - p_2$.

7.52   Age influences not only what you watch on television but also where you watch TV. A study (Darnay, 1994, p. 784) has shown that older Americans are less likely than younger ones to watch TV in bed and more likely to watch TV in the dining area. For the data given in the table, assume that the sample size for each age group was 1000.

Area	25 to 44	45 to 59	60 and older
Living room/family room/den	95%	95%	93%
Bedroom	58%	57%	45%
Kitchen	12%	20%	20%
Dining area	10%	10%	19%

a   Find a 95% confidence interval estimate of the difference in the proportions of Americans age 45 to 59 and those 60 and older who watch TV in the dining area.

b   Estimate the difference between the proportions of Americans in the 25 to 59 age group and those 60 years and older who watch television in the living room, family room, or den and find the margin of error for the estimate. [*Hint:* The proportion for the 25 to 59 age group will be the simple average of the individual proportions based upon a sample size of 2000.]

7.9   Choosing the Sample Size

The design of an experiment is essentially a plan for purchasing a quantity of information. This information, like any other commodity, may be acquired at varying prices depending on the manner in which the data are obtained. Some measurements contain a large amount of information concerning the parameter of interest; others may contain little or none. Since the sole product of research is information, we should try to purchase it at minimum cost.

The **sampling procedure,** or **experimental design** (as it is usually called), affects the quantity of information per measurement. This procedure, along with the sample size n, controls the total amount of relevant information in a sample. With a few exceptions, we will be concerned with the simplest sampling situation—random sampling from a relatively large population—and will focus our attention on the selection of the sample size n.

The researcher makes little progress in planning an experiment before encountering the problem of selecting the sample size. Indeed, perhaps one of the questions most frequently asked of the statistician is **How many measurements should be included in the sample?** Unfortunately, the statistician cannot answer this question without knowing how much information the experimenter wishes to buy. Certainly, the total amount of information in the sample will affect the measure of goodness of the method of inference and must be specified by the experimenter. Referring specifically to estimation, we would like to know how accurate the experimenter wishes the estimate to be. This accuracy may be stated by specifying a bound on the margin of error for the estimator.

For instance, suppose that we want to estimate the mean bank prime lending rate (Example 7.1), and we wish the error of estimation to be less than .06 with a probability of .95. Since approximately 95% of the sample means will lie within $1.96\sigma_{\bar{x}}$ of μ in repeated sampling, we are asking that $1.96\sigma_{\bar{x}}$ equal .06 (see Figure 7.8). Then

$$1.96\sigma_{\bar{x}} = .06$$

or

$$1.96\frac{\sigma}{\sqrt{n}} = .06$$

Solving for n, we obtain

$$n = \left(\frac{1.96}{.06}\right)^2 \sigma^2 \qquad \text{or} \qquad n = 1067.111\sigma^2$$

If we know σ, we can substitute its value into this formula and solve for n. If σ is unknown, we use the best approximation available, such as an estimate s obtained from a previous sample or knowledge of the range in which the measurements will fall. Since the range is approximately equal to 4σ (the Empirical Rule), one-fourth of the range will provide an approximate value for σ. For our example, we can use the sample standard deviation of Example 7.1, which provides a reasonably accurate estimate of σ equal to $s = .24$. Then

$$n = 1067.111\sigma^2$$
$$\approx (1067.111)(.24)^2 = 61.5$$

or

$$n = 62$$

Using a sample size $n = 62$, we will be reasonably certain (with probability approximately equal to .95) that our estimate will lie within $1.96\sigma_{\bar{x}} = .06$ of the true mean bank prime lending rate.

The solution $n = 62$ is only approximate because we had to use an approximate value for σ in calculating the value of n. Although this may bother you, it is the best method available for selecting the sample size, and it is certainly better than guessing.

Estimating μ with a $(1 - \alpha)100\%$ confidence interval whose width is to be less than $W = 2B$, where B is the equivalent bound on the margin of error of estimation, requires that

$$B = z_{\alpha/2}\frac{\sigma}{\sqrt{n}}$$

F I G U R E **7.8**
Approximate sampling
distribution of $\bar{x}$ for large
samples

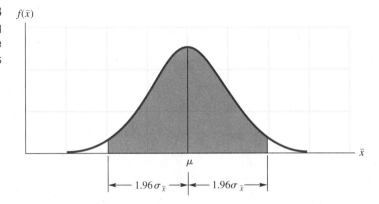

which in turn requires a sample size greater than or equal to

$$n = \frac{z_{\alpha/2}^2 \sigma^2}{B^2}$$

When the resulting sample size, n is less than 30 and an estimate s of σ is used in the calculations, the equation should be resolved using a tabulated value of $t_{\alpha/2}$ instead of $z_{\alpha/2}$. The degrees of freedom associated with $t_{\alpha/2}$ are taken to be $n_0 - 1$. This procedure is repeated iteratively using the previous value of n, until the sample size solution no longer changes. For example, if the bound is changed to .1, the required sample size would be greater than or equal to

$$n = \left(\frac{1.96}{.1}\right)^2 (.24)^2 = 22.13$$

or $n = 23$, which is less than 30. Using $t_{.025}$ with $23 - 1 = 22$ degrees of freedom, the appropriate value of t is 2.074 and

$$n = \left(\frac{2.074}{.1}\right)^2 (.24)^2 = 24.78$$

Now $n = 25$, d.f. $= 24$, $t_{.025} = 2.064$, and

$$n = \left(\frac{2.064}{.1}\right)^2 (.24)^2 = 24.54$$

so that $n = 25$. Since the sample size is the same as that found in the last iteration, we take n to be greater than or equal to 25.

The method of choosing the sample size for the estimation procedures discussed in preceding sections is identical to that described above.

Procedure for Choosing the Sample Size

Determine the parameter to be estimated and the standard error of its point estimator. Then proceed as follows:

1 Choose B, the bound on the margin of error, and a confidence coefficient $(1 - \alpha)$.

2 Solve the following equation for the sample size n:

$$z_{\alpha/2} \times (\text{standard error of the estimator}) = B$$

where $z_{\alpha/2}$ is the value of z having area $\alpha/2$ to its right.

3 If the resulting value of n is less than 30 and an estimate of σ is used, the value of $z_{\alpha/2}$ is replaced by $t_{\alpha/2}$ and the procedure is iterated until the sample size does not change.

[*Note*: For most estimators (all presented in this text), the standard error is a function of the sample size n.]

We will illustrate with examples.

EXAMPLE 7.9 In Example 7.7, a survey of 100 purchasing agents produced an estimate of the proportion of polyvinyl pipe wholesalers who plan to increase their purchases in the coming year. The margin of error, .096, was relatively large. Suppose that the marketing organization conducting the survey is asked to conduct a new survey and to obtain an estimate correct to within .04 with probability equal to .90. Approximately how many wholesalers would have to be included in the survey?

Solution For this particular example, the bound B on the margin of error is .04. Since the confidence coefficient is $(1 - \alpha) = .90$, α must equal .10 and $\alpha/2$ is .05. The z value corresponding to an area equal to .05 in the upper tail of the z distribution is $z_{.05} = 1.645$. We then require

$$1.645\sigma_{\hat{p}} = .04$$

or

$$1.645\sqrt{\frac{pq}{n}} = .04$$

In order to solve this equation for n, we must substitute an approximate value of p into the equation. We could use the estimate $\hat{p} = .59$ based on the sample of $n = 100$; or if we want to be certain that the sample is large enough, we could use $p = .5$

(substituting $p = .5$ will yield the largest possible solution for n, since the maximal value of pq occurs when $p = q = .5$). We will substitute $p = .5$. Then

$$1.645\sqrt{\frac{(.5)(.5)}{n}} = .04$$

or

$$\sqrt{n} = \frac{(1.645)(.5)}{.04} = 20.56$$

and

$$n = (20.56)^2 = 422.7$$

Therefore, the marketing organization must include approximately 423 wholesalers in its survey if it wants to estimate the proportion p correct to within .04.   ∎

E X A M P L E   **7.10**   A personnel director wishes to compare the effectiveness of two methods of training industrial employees to perform a certain assembly operation. A number of employees are to be divided into two equal groups, the first receiving training method 1 and the second training method 2. Each will perform the assembly operation, and the length of assembly time will be recorded. It is expected that the measurements for both groups will have a range of approximately 8 minutes. For the estimate of the difference in mean times to assemble to be correct to within 1 minute with probability equal to .95, how many workers must be included in each training group?

Solution   Equating $1.96\sigma_{(\bar{x}_1 - \bar{x}_2)}$ to $B = 1$ minute, we obtain

$$1.96\sqrt{\frac{\sigma_1^2}{n_1} + \frac{\sigma_2^2}{n_2}} = 1$$

Since we wish n_1 to equal n_2, we may let $n_1 = n_2 = n$ and obtain the equation

$$1.96\sqrt{\frac{\sigma_1^2}{n} + \frac{\sigma_2^2}{n}} = 1$$

As noted above, the variability (range) of each method of assembly is approximately the same, and hence $\sigma_1^2 = \sigma_2^2 = \sigma^2$. Since the range, equal to 8 minutes, is approximately equal to 4σ, then

$$4\sigma \approx 8$$

and

$$\sigma \approx 2$$

Substituting this value for σ_1 and σ_2 in the above equation, we obtain

$$1.96\sqrt{\frac{(2)^2}{n} + \frac{(2)^2}{n}} = 1$$

or

$$1.96\sqrt{\frac{8}{n}} = 1$$

and

$$\sqrt{n} = 1.96\sqrt{8}$$

Solving, we have $n = 31$. Thus each group should contain approximately $n = 31$ workers. ∎

Table 7.6 provides a summary of the formulas used to determine the sample sizes required for estimation with a given bound on the margin of error B or confidence interval width $W(W = 2B)$. Notice that in estimating p the sample size formula uses $\sigma^2 = pq$, while in estimating $p_1 - p_2$, the sample size formula uses $\sigma_1^2 + \sigma_2^2 = p_1 q_1 + p_2 q_2$.

TABLE 7.6
Sample size formulas

Parameter	Estimator	Sample Size	Assumptions
μ	$\bar{x}$	$n \geq \dfrac{z_{\alpha/2}^2 \sigma^2}{B^2}$	
$\mu_1 - \mu_2$	$\bar{x}_1 - \bar{x}_2$	$n \geq \dfrac{z_{\alpha/2}^2 (\sigma_1^2 + \sigma_2^2)}{B^2}$	$n_1 = n_2 = n$
p	$\hat{p}$	$n \geq \dfrac{z_{\alpha/2}^2 pq}{B^2}$ or $n \geq \dfrac{(.25)z_{\alpha/2}^2}{B^2}$	$p = .5$
$p_1 - p_2$	$\hat{p}_1 - \hat{p}_2$	$n \geq \dfrac{z_{\alpha/2}^2 (p_1 q_1 + p_2 q_2)}{B^2}$ or $n \geq \dfrac{2(.25)z_{\alpha/2}^2}{B^2}$	$n_1 = n_2 = n$ $n_1 = n_2 = n$ and $p_1 = p_2 = .5$

Exercises

Basic Techniques

7.53 Suppose that you wish to estimate a population mean from a random sample of n observations, and that prior experience suggests that $\sigma = 12.7$. If you wish to estimate μ correct to within 1.6 with probability equal to .95, how many observations should be included in your sample?

7.54 Suppose that you wish to estimate a binomial parameter p correct to within .04 with probability equal to .95. If you suspect that p is equal to some value between .1 and .3 and you want to be certain that your sample is large enough, how large should n be? [*Hint:* When calculating $\sigma_{\hat{p}}$, use the value of p in the interval $.1 < p < .3$ that will give the largest sample size.]

7.55 Independent random samples of $n_1 = n_2 = n$ observations are to be selected from each of two populations, 1 and 2. If you wish to estimate the difference between the two population means correct to within .17 with probability equal to .90, how large should n_1 and n_2 be? Assume that you know that $\sigma_1^2 \approx \sigma_2^2 \approx 27.8$.

7.56 Independent random samples of $n_1 = n_2 = n$ observations are to be selected from each of two binomial populations, 1 and 2. If you wish to estimate the difference in the two population binomial parameters correct to within .05 with probability equal to .98, how large should n be? (Assume that you have no prior information on the values of p_1 and p_2, but you want to make certain that you have an adequate number of observations in the samples.)

Applications

7.57 Reverse mortgages hold the promise of helping elderly homeowners use the equity in their homes to help with financing current consumption or to pay for emergencies (Mayer and Simons, 1994). A reverse mortgage would allow homeowners to borrow against the equity and receive monthly payments while still living in the home until they die or move. A study is proposed to estimate the amount of equity in homes owned by people 65 or over. The range of equities is expected to be about $100,000. If the investigators wish to estimate equity to within $5000, how large should their sample be?

7.58 In Exercise 7.34 we estimated the difference in the mean values of a bank's home loans processed in the month of April versus the month of May. Suppose that the bank desires a more accurate estimate of this difference, say, an estimate correct to within $1000 with probability equal to .95. How many home loans would have to be included in each sample? (Assume that the sample sizes for the two months are equal.)

7.59 An auditing firm wishes to estimate the mean error per account in accounts receivable for a plumbing supply company correct to within $20 with probability equal to .99. A small prior sample suggests that the error per account possesses a standard deviation approximately equal to $58. If the firm wishes to estimate the mean error per account correct to within $20, how many accounts would have to be sampled? What attribute(s) must the sample possess?

7.60 A food products company has hired a marketing research firm to sample two markets, I and II, to compare the proportions of consumers who prefer the company's frozen dinners over its competitors' products. No prior information is available on the magnitude of the proportions p_1 and p_2. If the food products company wishes to estimate the difference in proportions of consumers who prefer its products correct to within .04 with probability equal to .95, how many consumers must be sampled in each market? How must the samples be collected?

7.61 Refer to Exercise 7.41. If Ford Motor Company wished to estimate the percentage of professional leaders who felt that industry should be held liable for environmental damage and cleanup caused by their action to within .01 with probability .99, how large a sample should it take? Assume that there is no previous knowledge about the value of p.

7.62 Refer to Exercise 7.61.

  a Find the appropriate sample size if it is known that p is between .7 and .9.

  b Is the sample size found in part (a) larger or smaller than the sample size found in Exercise 7.61? Can you explain the difference?

7.63 Ethnic groups in America buy differing amounts of various food products because of their ethnic cuisines (Rickard, 1994). Asians buy fewer canned vegetables than do other groups, and Hispanics purchase more cooking oil. A researcher interested in market segmentation for these two groups would like to estimate the proportion of households selecting certain brands for various products. If the researcher wishes these estimates to be within .03 with probability .95, how many households should she include in the samples?

7.10 THE LOGIC BEHIND THE IRS'S $3.4 MILLION DISALLOWANCE

The case study in this chapter describes a corporation's claim for repair expense replacement and the disallowance of a large portion of the claim by the Internal Revenue Service (IRS). In the discussion that follows, we will explain the basis for the IRS's disallowance of a portion of the claim. Both the case study and the explanation in this section are based on information contained in an article by W. L. Felix, Jr., and R. S. Roussey ("Statistical Inference and the IRS," *Journal of Accountancy,* June 1985).

To review the details of the case: A corporation claimed $6 million in one year and $3.8 million the next for repair expense replacement. These figures were based on "random sampling techniques," specifically, samples of 350 items in the first year and 520 in the second. The IRS did not object to the sampling procedure, the sample sizes, or the use of statistical inference to estimate the total replacement claim. The IRS did, however, "refer to written standards for statistical sampling that would allow a mathematical expression of the margin of error associated with a sample (the calculation of a sample error)." Furthermore, the IRS took the position that the company "could and should have applied the sample error to calculate a lower limit of allowable repair allowance so that management could have said, for example, 'We are 95% certain that the allowable repair allowance is at least X dollars.' " The IRS did, in fact, calculate this lower limit and used it as the repair expense replacement that it would allow.

To understand the position of the IRS with regard to the disallowance, we need to translate these two quotations into the language of this chapter. In the first quotation the IRS is stating that the corporation should have provided information on the "sample error." By sample error, we assume that the agency means the error of estimation.

Interpreting the second quotation, we assume the IRS means that management should have located a **95% lower confidence limit on the total repair cost. The IRS performed this calculation and then chose this LCL as the maximum repair expense that it would allow.**

Although Felix and Roussey did not explain the details of the sampling and estimation procedures employed for this case study, we will explain how the IRS would have arrived at the lower confidence limit for the repair expense allowance based on random sampling.

We will assume that the total number of repair expenses for the year is N, and for purposes of illustration we assume that

$$N = 100,000$$

A random sample of $n = 500$ of these repair expenses is selected from among the 100,000 in the population, and the average cost of a repair expense for the $n = 500$ is calculated to be $\bar{x} = \$65$. Then if the estimate of a single repair expense is $65, the estimate of the total repair expense costs for the 100,000 repair expenses incurred during the year is $(100,000)(\$65) = \$6,500,000$. That is, letting $\hat{\tau}$ (Greek letter tau) represent the total repair expense cost, we have

$$\text{Estimate of total repair expense cost} = \hat{\tau} = N\bar{x}$$
$$= (100,000)(65)$$
$$= \$6,500,000$$

To find a one-sided lower confidence interval for the total repair expense cost, we need to determine the properties of the sampling distribution of $\hat{\tau} = N\bar{x}$. We know, from the Central Limit Theorem, that the sampling distribution of $\bar{x}$ will be approximately normally distributed. Then, since N is a constant, the sampling distribution of $\hat{\tau} = N\bar{x}$ is approximately normally distributed with mean $N\mu$ and standard deviation

$$\sigma_{\hat{\tau}} = N\sigma_{\bar{x}} = N\frac{\sigma}{\sqrt{n}}$$

Since $\hat{\tau}$ is an unbiased estimator of the total repair expense cost and it is approximately normally distributed, it satisfies all of the properties required for the one-sided lower 95% confidence interval given in Section 7.4. That is,

$$\text{LCL} = \hat{\tau} - z_{.05}\sigma_{\hat{\tau}}$$
$$= \hat{\tau} - 1.645\frac{N\sigma}{\sqrt{n}}$$

To illustrate the calculation of the LCL, suppose that the standard deviation s for the sample of 500 repair expenses is $342. Then a one-sided 95% lower confidence limit for the repair expense total for the year is

$$\text{LCL} = \hat{\tau} - z_{.05}\frac{N\sigma}{\sqrt{n}}$$
$$= 6,500,000 - (1.645)(100,000)\frac{(342)}{\sqrt{500}}$$
$$= 6,500,000 - 2,515,979$$
$$= \$3,984,021$$

Therefore, from this illustration the IRS would allow a repair expense replacement of $3,984,021.

The IRS is certainly not losing money by choosing the lower confidence limit on total repair expense as the maximum repair expense replacement that it will allow. As you can see from Figure 7.9, the unbiased point estimator $\hat{\tau}$ estimates the total repair expense costs to be $6,500,000. By allowing only the LCL (in this case, $3,984,021), the IRS will be allowing less than the actual total expense repair costs 95% of the time. Only 5% of the time will it shortchange the U.S. government. When it does, it will probably be by a small amount.

F I G U R E **7.9**
Location of the LCL with respect
to $\hat{\tau}$

LCL = \$3,984,021 $\hat{\tau}$ = \$6,500,000

Total repair expense costs

7.11 Summary

Chapter 7 presents the basic concepts of statistical estimation and demonstrates how these concepts can be applied to the solution of some practical problems.

Estimators are rules (usually formulas) that tell us how to calculate a parameter estimate based on sample data. Point estimators produce a single number (point) that estimates the value of a population parameter. The properties of a point estimator are contained in its sampling distribution. Thus we prefer a point estimator that is unbiased—that is, the mean of its sampling distribution is equal to the estimated parameter. The point estimator should possess a small, preferably a minimum, variance.

The reliability of a point estimator is usually measured by a 1.96-standard-deviation (of the sampling distribution of the estimator) margin of error in estimation. When we use sample data to calculate a particular estimate, the probability that the error in estimation will be less than the margin of error is approximately .95.

An interval estimator uses the sample data to calculate two points—a confidence interval—that we hope will enclose the estimated parameter. Since we will want to know the probability that the interval will enclose the parameter, we need to know the sampling distribution of the statistic used to calculate the interval.

Four estimators—a sample mean, a binomial proportion, and the differences between pairs of these statistics—were used to estimate their population equivalents and to demonstrate the concepts of estimation developed in this chapter. These estimators were chosen for a particular reason. Generally, they are "good" estimators for the respective population parameters in a wide variety of applications. Fortunately, they all possess, for large samples, sampling distributions that are approximately normal. This fact enabled us to use the same procedure to construct confidence intervals for the four population parameters μ, p, $(\mu_1 - \mu_2)$, and $(p_1 - p_2)$. Thus we were able to demonstrate an important role that the Central Limit Theorem (Chapter 6) plays in statistical inference.

When the sampled populations were normal and the sample sizes were not large, the estimators of μ and $\mu_1 - \mu_2$ were based on the Student's t distribution.

Tips on Problem Solving

In solving the exercises in this chapter, you are required to answer practical questions of interest to a businessperson, a professional person, a scientist, or a layperson. To find the answers to the questions, you need to make an inference about one or more population parameters. Consequently, the first step in solving a problem is to determine the objective of the exercise.

What parameters do you wish to make an inference about? Answering the following two questions will help you identify the parameter(s).

1 What *type of data* is involved? This question will help you decide which type of parameters you want to make inferences about: binomial proportions (ps) or population means (μs). Check to see if the data are of the yes/no (two-possibility) variety. If they are, the data are probably binomial, and you will be interested in proportions. If not, the data probably represent measurements on one or more quantitative random variables, and you will be interested in means. Look for key words such as "proportions," "fractions," and so on, which indicate binomial data. Binomial data often (but not exclusively) evolve from a "sample survey" or an opinion poll.

2 Do you wish to make an inference about a *single parameter p or μ*, or about the *difference between two parameters* $(p_1 - p_2)$ or $(\mu_1 - \mu_2)$? This is an easy question to answer. Check on the number of samples involved. One sample implies an inference about a single parameter; two samples imply a comparison of two parameters.

3 After identifying the parameters(s) involved in the exercise, you must identify the exercise objective. It will be either (a) choosing the sample size required to estimate a parameter with a specified bound on the margin of error or (b) estimating a parameter (or difference between two parameters). The objective will be very clear if it is (a), because the question will ask for or direct you to find the "sample size." Objective (b) also will be clear because the exercise will specifically direct you to estimate a parameter (or the difference between two parameters).

4 Check the conditions required for the sampling distribution of the parameter to be approximated by a normal distribution. For quantitative data, the sample size or sizes must be 30 or more. For binomial data, a large sample size will ensure that $p \pm 2\sigma_{\hat{p}}$ is contained in the interval 0 to 1 [or $(p_1 - p_2) \pm 2\sigma_{(\hat{p}_1 - \hat{p}_2)}$ contained in the interval -1 to 1 for the two-sample case].

To summarize these tips, your thought process should follow the decision tree shown in Figure 7.10.

FIGURE **7.10**
Decision tree

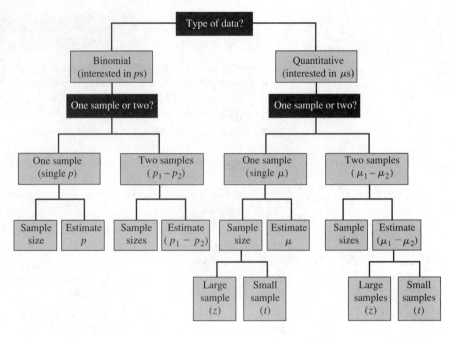

Supplementary Exercises

7.64   State the Central Limit Theorem. Of what value is the Central Limit Theorem in statistical inference?

7.65   What assumptions are made when using a Student's t test to construct a confidence interval for a single population mean?

7.66   What assumptions are made when using a Student's t test to construct a confidence interval for the difference between two population means?

7.67   A random sample of $n = 64$ observations possessed a mean $\bar{x} = 29.1$ and a standard deviation $s = 3.9$. Give the point estimate for the population mean μ, and give the margin of error in estimation.

7.68   Refer to Exercise 7.67, and find a 90% confidence interval for μ.

7.69   Refer to Exercise 7.67. How many observations would be required if you wanted to estimate μ with a margin of error in estimation equal to .5 with probability equal to .95?

7.70   Independent random samples of $n_1 = 50$ and $n_2 = 60$ observations were selected from populations 1 and 2, respectively. The sample sizes and computed sample statistics are shown in the accompanying table. Find a 90% confidence interval for the difference in population means, and interpret the interval.

	Population	
	1	**2**
Sample size	50	60
Sample mean	100.4	96.2
Sample standard deviation	.8	1.3

7.71   Refer to Exercise 7.70. Suppose that you want to estimate $(\mu_1 - \mu_2)$ correct to within .2 with probability equal to .95. If you plan to use equal sample sizes, how large should n_1 and n_2 be?

7.72   A random sample of $n = 500$ observations from a binomial population produced $x = 240$ successes.

a   Find a point estimate for p, and find the margin of error.

b   Find a 90% confidence interval for p.

7.73   Refer to Exercise 7.72. How large a sample would be required if you want to estimate p correct to within .025 with probability equal to .90?

7.74   Independent random samples of $n_1 = 40$ and $n_2 = 80$ observations were selected from binomial populations 1 and 2, respectively. The numbers of successes in the two samples were $x_1 = 17$ and $x_2 = 23$. Find a 90% confidence interval for the difference between the two binomial population proportions.

7.75   Refer to Exercise 7.74. Suppose that you want to estimate $(p_1 - p_2)$ correct to within .06 with probability equal to .90 and that you plan to use equal sample sizes, that is, $n_1 = n_2$. How large should n_1 and n_2 be?

7.76   A random sampling of bank customer-checking overdrafts at a local bank produced the following data (to the nearest dollar):

302	512	97	316
69	16	133	701
107	156	401	14
465	72	128	68

Suppose that the general level of the bank's overdrafts is expected to remain stable in the immediate future. Use the sample data to find a 90% confidence interval for the mean size of the bank's checking overdrafts.

7.77   A coin-operated soft-drink machine was designed to discharge, on the average, 7 ounces of beverage per cup. In a test of the machine, ten cupfuls of beverage were drawn from the machine and measured. The mean and standard deviation of the ten measurements were 7.1 ounces and .12 ounces, respectively. Find a 99% confidence interval for the mean discharge.

7.78   Two random samples, each containing 11 measurements, were drawn from normal populations possessing means μ_1 and μ_2, respectively, and a common variance σ^2. The sample means and variances are as follows:

Population I	Population II
$\bar{x}_1 = 60.4$	$\bar{x}_2 = 65.3$
$s_1^2 = 31.40$	$s_2^2 = 44.82$

Find a 90% confidence interval for the difference between the population means.

7.79   "Big investments in automation and technology won't bring manufacturing happiness. ... It's the small, simple improvements that have made Toyota the world leader in efficiency," according to the Toyota Supplier Support Center in Lexington, Kentucky (Chappell, 1994). This group travels across America discussing the tenets of the Toyota Production System with anyone who will listen. Of the thousands of American suppliers, only 22 from among the 391 contacted by Toyota have requested assistance. If these 391 suppliers can be deemed to represent a random sample of suppliers, find an estimate of the proportion of all American suppliers who might request assistance from this group. What is the margin of error?

7.80   Although most Americans say they prefer home-cooked meals, 56% say they ate dinner at a restaurant or a fast-food establishment sometime during the last week. A recent survey revealed that only 45% of those age 60 and older ate out during the previous week, compared with 66%

of adults under 30 ("Most Restaurant Meals," 1994). Suppose that these percentages are based on two samples each of size 1200.

a What are the margins of error associated with the estimates of the proportions of those age 60 and older and those under 30 who ate out during the previous week?

b Find a point estimate and the margin of error for the difference in proportions between those age 60 and older and those under 30 who ate out during the previous week.

7.81 A random sample of 400 nightlights was tested, and 40 lights were found to be defective. With confidence coefficient equal to .90, estimate the interval within which the true fraction defective lies.

7.82 Past experience shows that the standard deviation of the yearly income of textile workers in a certain state was $1000. How large a sample of textile workers would one need to take if one wished to estimate the population mean to within $100.00 with a probability of .95 of being correct? Given that the mean of the sample in this problem is $14,800, determine 95% confidence limits for the population mean.

7.83 An experimenter wants to use the sample mean $\bar{x}$ to estimate the mean of a normally distributed population with an error of less than .5 with probability .9. If the variance of the population is equal to 4, how large should the sample be to achieve this accuracy?

7.84 A time study is planned to estimate, correct to within 4 seconds with probability .90, the mean time for a worker to complete an assembly task. If past experience suggests that $\sigma = 16$ seconds measures the worker-to-worker variation in assembly time, how many workers, each performing a single assembly operation, will have to be included in the sample?

7.85 To estimate the proportion of unemployed workers in the state of Iowa, an economist selected at random 400 persons from the working class. Of these, 25 were unemployed.

a Estimate the true proportion of unemployed workers, and give the margin of error in estimation.

b How many persons must be sampled to reduce the margin of error to .02?

7.86 Sixty of 87 homemakers prefer detergent A. If the 87 homemakers represent a random sample from the population of all potential purchasers, estimate the fraction of total homemakers favoring detergent A. Use a 90% confidence interval.

7.87 A dean of students wishes to estimate the average cost of the freshman year at a particular college correct to within $500.00 with a probability of .95. If a random sample of freshmen is to be selected and requested to keep financial data, how many must be included in the sample? Assume that the dean knows that the range of expenditure will vary from approximately $4800 to $13,000.

7.88 The percentage of Ds and Fs awarded to students by two college history professors was duly noted by the dean. Professor I achieved a rate equal to 32% as opposed to 21% for professor II, based upon 200 and 180 students, respectively. Estimate the difference in the percentage of Ds and Fs awarded by the professors. What is the margin of error?

7.89 Suppose you wish to estimate the mean hourly yield for a process manufacturing an antibiotic. The process is observed for 100 hourly periods chosen at random, with the following results:

$$\bar{x} = 34 \text{ ounces per hour} \qquad s = 3$$

Estimate the mean hourly yield for the process, using a 95% confidence interval.

7.90 A quality control engineer wants to estimate the fraction of defectives in a large lot of light bulbs. From previous experience, he feels that the actual fraction of defectives should be somewhere around .2. How large a sample should he take if he wants to estimate the true fraction to within .01, using a 95% confidence interval?

7.91 Samples of 400 AA batteries were selected from each of two production lines, A and B. The numbers of defectives in the samples are given in the accompanying table. Estimate the difference in the actual fractions of defectives for the two lines with a confidence coefficient of .90.

Line	Number of Defectives
A	40
B	80

7.92 How much combustion efficiency should a homeowner expect from an oil furnace? The EPA states that 80% or above is excellent, 75% to 79% is good, 70% to 74% is fair, and below 70% is poor. A home-heating contractor who sells two makes of oil heaters (call them A and B) decided to compare their mean efficiencies. An analysis was made of the efficiencies for 8 heaters of type A and 6 of type B. The efficiency ratings, in percentages, for the 14 heaters are shown in the table.

Type A	72	78	73	69	75	74	69	75
Type B	78	76	81	74	82	75		

a Find a 90% confidence interval for $(\mu_A - \mu_B)$, and interpret the result.

b Based on the results of part (a), would you conclude that there is a difference in the mean efficiency ratings for the two types of heaters? Explain.

7.93 An audit was conducted to estimate the difference in mean percentage shrinkage (loss due to theft, damage, etc.) in inventory at two department stores. One hundred items were randomly selected within each store, and the percentage of each item actually on hand, in comparison to the total shown on inventory records, was recorded. The mean and standard deviation of the percentage shrinkage for the 100 items are shown in the accompanying table for each store. Estimate the difference in mean percentage shrinkage between the two department stores, using a 95% confidence interval. Based on this interval, would you conclude that there is a difference in the mean percentage shrinkage for the two stores?

	Department Store	
	1	2
Sample size	100	100
Sample mean	5.3	6.4
Sample standard deviation	2.7	2.9

7.94 Federal standards currently in effect require bumpers to keep damage away from car bodies in 2.5-mph front- and rear-into-flat-barrier impacts. However, no minimum federal standard covers the bumpers on vans, in spite of the fact that in 1992 more than one of every ten buyers chose to purchase a van ("Passenger Van Bumpers," 1994). In a consumer test of seven popular van brands/models, the following repair costs reflect the total cost of repairing damage resulting from a front and rear collision into a barrier, a front collision into an angle barrier, and a rear collision into a pole:

$$\$1862 \quad 2554 \quad 3963 \quad 4082 \quad 4505 \quad 4601 \quad 7643$$

a Find the mean and standard deviation of the $n = 7$ repair costs.

b Calculate a 95% confidence interval estimate of the population repair costs for vans represented in this sample.

c Use the Minitab printout shown below to confirm the results of parts (a) and (b).

```
MTB > TINT C3

              N      MEAN    STDEV  SE MEAN    95.0 PERCENT C.I.
COSTS         7      4173    1842     696    (    2469,    5877)
```

7.95 Now that energy conservation is so important, some scientists think we should give closer scrutiny to the cost (in energy) of producing various forms of food. One recent study compares the mean amount of oil required to produce 1 acre of different types of crops. For example, suppose that we want to compare the mean amount of oil required to produce 1 acre of corn versus 1 acre of cauliflower. The readings in barrels of oil per acre, based on 20-acre plots, seven for each crop, are shown in the table. Use these data to find a 90% confidence interval for the difference in the mean amounts of oil required to produce the two crops.

Corn	5.6	7.1	4.5	6.0	7.9	4.8	5.7
Cauliflower	15.9	13.4	17.6	16.8	15.8	16.3	17.1

7.96 The length of time between billing and receipt of payment was recorded for a random sample of 100 of a CPA firm's clients. The sample mean and standard deviation for the 100 accounts were 39.1 days and 17.3 days, respectively. Find a 90% confidence interval for the mean time between billing and receipt of payment for all of the CPA firm's accounts. Interpret the interval.

7.97 Refer to Exercise 7.96. A comparison of the mean time between billing and receipt of payment was made by selecting random samples of 30 clients each from the population of all individual clients and from the population of business clients. The sample sizes, means, and standard deviations for the two samples are shown in the accompanying table. Find a 95% confidence interval for the difference in mean times to pay between individual and business clients.

	Individual Clients	Business Clients
Sample size	30	30
$\bar{x}$	28.3	47.7
s	15.5	18.9

7.98 Refer to Exercise 7.97. How many clients of each type would have to be sampled (with equal sample sizes) if you wanted to estimate the difference in mean time to pay correct to within 5 days with probability approximately equal to .95?

7.99 Television advertisers may mistakenly believe that most viewers understand most of the advertising that they see and hear. A recent research study used 2300 viewers above age 13. Each viewer looked at 30-second television advertising excerpts. Of these, 1914 of the viewers misunderstood all or part of the excerpt. Find a 95% confidence interval for the proportion of all viewers (of which the sample is representative) that will misunderstand all or part of the television excerpts used in this study.

7.100 The new electric car *Tropica* has been clocked by the California Air Resources Board to have a range of 44.7 miles ("Thinking Cheaper," 1994). In Exercise 2.13, an independent test consisting of 30 trials produced the following driving ranges:

45.4	45.9	45.3	42.9	45.8
44.6	44.7	46.4	45.7	45.9
44.1	47.1	46.4	44.3	47.3
46.8	44.2	45.5	45.0	46.4
43.5	46.6	44.5	44.8	44.3
47.0	44.6	43.3	46.1	43.8

Based on the Minitab printout that follows, use an appropriate estimation procedure to verify or dispute the 44.7-mile driving range reported by the California Air Resources Board.

```
MTB > DESC C1

                   N      MEAN    MEDIAN    TRMEAN    STDEV    SEMEAN
RANGES            30    45.273    45.350    45.292    1.199     0.219

                 MIN       MAX        Q1        Q3
RANGES        42.900    47.300    44.300    46.400
```

7.101 The speed with which a personal computer can perform various tasks is a very important selling point for different brands and models of computers as well as for the platforms that support application programs. The data that follow reflect the time (in seconds) to load Ami Pro 2.0 on an IBM PS/2 Model 90 486DX/33 personal computer using the Standard Windows and Enhanced Windows programs. Use an appropriate estimation procedure to determine if there really is a difference in the average time to load Ami Pro 2.0 using Standard Windows and Enhanced Windows.

Standard		Enhanced	
1.56	1.20	1.59	.96
1.41	1.38	1.68	1.09
1.48	1.54	1.17	1.26
1.37	1.41	.94	1.23
1.39	1.16	1.56	1.30

Exercises Using the Data Disk

7.102 Refer to data set A on the data disk. For a *fixed* faculty rank, select two random samples of salaries of size $n = 50$, one from male faculty and one from female faculty.

  a Estimate the difference in average salaries for males versus females using a 95% confidence interval based on a large-sample (Central Limit Theorem) approach.

  b Select another faculty rank and repeat part (a) with sample sizes of $n = 30$ for both males and females using the large-sample approach.

7.103 Refer to data set B on the data disk. Select a random sample of $n_1 = 10$ observations on the fresh weights of variety 1 (coded 0) and $n_2 = 10$ observations on the fresh weights of variety 2 (coded 1).

  a Use the small-sample approach based on the t distribution to find a 99% confidence interval estimate of $\mu_1 - \mu_2$, the true difference in average fresh weights of broccoli heads.

  b Select another variable among market weight, dry weight, and head diameter and repeat part (a).

7.104 Refer to data set C on the data disk. The true mean and standard deviation of the average maturity in days are $\mu = 38.56$ and $\sigma = 14.224$.

a Consider the average maturities for the 604 money market funds to be the population of interest and draw a random sample of size 30 from this population (use the random number table, Table 13, in Appendix II).

b Use the sample from part (a) to construct a 99% confidence interval for μ. Does this confidence interval enclose the true value, $\mu = 38.56$?

c Draw another random sample of size 30 and construct another 99% confidence interval for μ. Does this interval enclose μ?

d In general, what proportion of intervals constructed in this way would you expect to enclose the value $\mu = 38.56$?

C H A P T E R **8**

TESTS OF HYPOTHESES FOR MEANS AND PROPORTIONS

About This Chapter

As we explained in Chapter 7, there are two methods for making inferences about population parameters based on sample data. The first method, statistical estimation, was the topic of Chapter 7. The objective of Chapter 8 is to present a second method for making inferences about population parameters—testing hypotheses about their values. As was the case in Chapter 7, we will demonstrate the procedure for situations where the sample sizes are large enough to produce approximate normality in the sampling distributions of the sample statistics used to make the inferences. We will also explain the procedure for situations where the sample sizes are small but the sampled populations are normal.

Contents

POST–COLD WAR DEFENSE CONTRACTORS: IS CHANGE POSSIBLE?

With the demise of the Communist bloc in Eastern Europe, downturns in U.S. defense spending have caused major problems in states whose revenues depended heavily on profitable defense industries. These problems were compounded by the closing of many military bases and the downsizing of others. Furthermore, the defense contracting industry is facing the challenges of excess production capacity, top-heavy corporate structures, and inefficient operations. In light of these circumstances, government contractors must become more cost-effective and efficient.

In a study to assess the cost accounting systems currently used by defense contractors, Rezaee and Elmore (1993) surveyed 112 companies. Usable questionnaires were completed by 50 respondents (supervisory accountants), who were then classified into two groups: 25 defense contractors and 25 nondefense contractors. The results of one part of their survey concerning responses (measured on a Likert Scale) to questions about the use of budgeting and planning procedures are shown in Table 8.1.

TABLE **8.1**

	Defense	Nondefense	T-Value
Strategic Planning			
1 Budgets examined for consistency with long-range goals	4.0425	4.2000	−.30
2 Formal statement of goals, strategies, etc., used in planning the direction of the firm	4.1625	4.8800	−1.45
Budget and Planning			
1 Budgets used in performance evaluation of individual members	3.1600	4.5200	−2.56*
2 Comparison of actual to budgeted costs	3.9891	5.1782	−2.64*
3 Individual department budgets	2.8800	4.6800	−3.23*
4 Participation of lower- and mid-management in budgeting	3.6879	5.0800	−3.53*
5 Flexible budgets	2.1861	3.6000	−2.64*

* Significant at .01.

Are there real differences in the responses to these parts of the questionnaire for defense versus nondefense industries? By what yardstick can these differences be measured? In Chapter 7, we were able to estimate differences in population means using either the standard normal or the t distribution as the reference distribution, depending upon the sample size, the method of sampling, and the nature of the underlying population that was sampled. In this chapter, you will learn formal methods for testing hypotheses about several population parameters. We will use those techniques in Section 8.12 to determine whether there are real differences between defense versus nondefense industries with regard to their accounting and budgeting processes.

8.1   Testing Hypotheses About Population Parameters

Some practical problems require that we estimate the value of a population parameter; others require that we make decisions concerning the value of the parameter. For example, if a pharmaceutical company were fermenting a vat of antibiotic, it would want to test the potency of samples of the antibiotic and use those samples to estimate the mean potency μ of the antibiotic in the vat. In contrast, suppose that there were no concern that the potency of the antibiotic would be too high; the company's only concern is that the mean potency exceed some government-specified minimum in order that the vat be declared acceptable for sale. In this case, the company would not wish to estimate the mean potency. Rather, it would want to show that the mean potency of the antibiotic in the vat exceeded the minimum specified by the government. Thus the company would want to decide whether or not the mean potency exceeded the minimum allowable potency. The pharmaceutical company's problem illustrates a **statistical test of hypothesis.**

The reasoning employed in testing an hypothesis bears a striking resemblance to the procedure used in a court trial. In trying a person for theft, the court assumes the accused innocent until proved guilty. The prosecution collects and presents all available evidence in an attempt to contradict the "not guilty" hypothesis and hence to obtain a conviction. However, if the prosecution fails to disprove the "not guilty" hypothesis, this does not prove that the accused is "innocent" but merely that there is not sufficient evidence to conclude that the accused is "guilty."

The statistical problem portrays the potency of the antibiotic as the accused. The hypothesis to be tested, called the **null hypothesis,** is that the potency does not exceed the minimum government standard. The evidence in this case is contained in the sample of specimens drawn from the vat. The pharmaceutical company, playing the role of the prosecutor, believes that an **alternative hypothesis** is true—namely, that the potency of the antibiotic does exceed the minimum standard. Hence, the company attempts to use the evidence in the sample to reject the null hypothesis (potency does not exceed minimum standard), thereby supporting the alternative hypothesis (potency exceeds minimum standard). You will recognize this procedure as an essential feature of the scientific method, in which all proposed theories must be compared with reality.

In this chapter, we will explain the basic concepts of a test of an hypothesis and demonstrate the concepts with some very useful statistical tests of the values of a population mean, a population proportion, the difference between a pair of population means, and the difference between two binomial proportions. We will employ the four point estimators discussed in Chapter 7—$\bar{x}$, $(\bar{x}_1 - \bar{x}_2)$, $\hat{p}$, and $(\hat{p}_1 - \hat{p}_2)$—as test statistics and, in doing so, will obtain a unity in these four statistical tests. All four test statistics will, for large samples, have sampling distributions that are normal or can be approximated by a normal distribution. For small samples, statistical tests concerning population means involve the Student's t distribution.

8.2 A Statistical Test of Hypothesis

A statistical test of hypothesis consists of four parts:

- a null hypothesis, denoted by the symbol H_0
- an alternative hypothesis, denoted by the symbol H_a
- a test statistic
- a rejection region

The specification of these four elements defines a particular test; changing one or more of the parts creates a new test.

The **alternative hypothesis** is the hypothesis that the researcher wishes to support. The **null hypothesis** is the negation of the alternative hypothesis; that is, if the null hypothesis is false, the alternative hypothesis must be true. For reasons you will subsequently see, it is easier to show support for the alternative hypothesis by presenting evidence (sample data) that indicates that the null hypothesis is false. Thus we are building a case in support of the alternative hypothesis by using a method that is analogous to proof by contradiction.

Even though we want to gain evidence in support of the alternative hypothesis, the null hypothesis is the hypothesis to be tested. Thus H_0 will specify hypothesized values for one or more population parameters.

EXAMPLE 8.1   We want to show that the average hourly wage of construction workers in the state of California is different from $14, which is the national average. This is the alternative hypothesis, written as

$$H_a : \mu \neq 14$$

The null hypothesis is written as

$$H_0 : \mu = 14$$

We would like to reject the null hypothesis, thus concluding that the California mean is not equal to $14. ▪

E X A M P L E   **8.2**   A milling process currently produces an average of 3% defectives. We are interested in showing that a simple adjustment on a machine will decrease p, the proportion of defectives produced in the milling process. Thus we write the alternative hypothesis as

$$H_a : p < .03$$

and the null hypothesis as

$$H_0 : p = .03$$

If we can reject H_0, we can conclude that the adjusted process produces fewer defectives. ▪

There is a difference in the forms of the alternative hypothesis given in Examples 8.1 and 8.2. In Example 8.1, there is no directional difference suggested for the value of μ; that is, μ could be either larger or smaller than $14 if H_a is true. This type of test is called a **two-tailed test of hypothesis.** In Example 8.2, however, we are specifically interested in detecting a directional difference in the value of p; that is, if H_a is true, the value of p will be smaller than .03. This type of test is called a **one-tailed test of hypothesis.**

The decision to reject or accept the null hypothesis is based on information contained in a sample drawn from the population of interest. The sample values are used to compute a single number, corresponding to a point on a line, that operates as a decision maker. This decision maker is called the **test statistic.** The entire set of values that the test statistic may assume is divided into two sets, or regions. One set, consisting of values that support the alternative hypothesis, is called the **rejection region.** The other, consisting of values that do not contradict the null hypothesis, is called the **acceptance region.**

The acceptance and rejection regions are separated by a **critical value** of the test statistic. If the test statistic computed from a particular sample assumes a value in the rejection region, the null hypothesis is rejected, and the alternative hypothesis H_a is accepted. If the test statistic falls in the acceptance region, either the null hypothesis is accepted or the test is judged to be inconclusive. In any case, the failure to reject H_0 implies that the data do not present sufficient evidence to support H_a. The circumstances leading to this decision will be explained subsequently.

E X A M P L E **8.3** For the test of hypothesis given in Example 8.1, the average wage $\bar{x}$ for a random sample of 100 California construction workers might provide a good test statistic for testing

$$H_0 : \mu = 14 \quad \text{vs.} \quad H_a : \mu \neq 14$$

Since the sample mean is the best estimator of the corresponding population mean, we would be inclined to reject H_0 in favor of H_a if the sample mean $\bar{x}$ is either much smaller than \$14 or much larger than \$14. Hence, the rejection region would consist of both large and small values of $\bar{x}$, as shown in Figure 8.1. ▪

F I G U R E **8.1**
Rejection and acceptance regions for Example 8.1

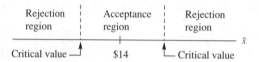

The decision procedure described above is subject to two types of errors that are prevalent in a two-choice decision problem.

D E F I N I T I O N ▪ A **type I error** for a statistical test is the error made by rejecting the null hypothesis when it is true. The probability of making a type I error is denoted by the symbol α.

A **type II error** for a statistical test is the error made by accepting (not rejecting) the null hypothesis when it is false and some alternative hypothesis is true. The probability of making a type II error is denoted by the symbol β. ▪

The two possibilities for the null hypothesis—that is, true or false— along with the two decisions the experimenter can make, are indicated in the two-way table, Table 8.2. The occurrences of the type I and type II errors are indicated in the appropriate cells.

T A B L E **8.2**
Decision table

Decision	Null Hypothesis	
	True	**False**
Reject H_0	Type I error	Correct decision
Accept H_0	Correct decision	Type II error

The goodness of a statistical test of hypothesis is measured by the probabilities of making a type I or a type II error, denoted by the symbols α and β, respectively.

The various parts of a statistical test of hypothesis are summarized in the following display.

Parts of a Statistical Test

- **Null hypothesis:** The hypothesis that is assumed true until proven false; the negation of the alternative hypothesis
- **Alternative hypothesis:** The hypothesis that the researcher wishes to support or prove true
- **One-tailed test of hypothesis:** A test that assumes a one-directional difference for the parameter of interest if the alternative hypothesis is true
- **Two-tailed test of hypothesis:** A test that assumes a two-directional difference (either larger or smaller) for the alternative hypothesis
- **Test statistic:** A statistic calculated from sample measurements that will be used as a decision maker
- **Rejection region:** Values of the test statistic for which H_0 will be rejected
- **Acceptance region:** Values of the test statistic for which H_0 will be accepted
- **Critical values of the test statistic:** The values of the test statistic that separate the rejection and acceptance regions
- **Conclusion:** The course of action to be followed, based upon the observed value of the test statistic
- **Type I error (with probability α):** Rejecting H_0 when H_0 is true
- **Type II error (with probability β):** Accepting H_0 when H_0 is false

Large-sample tests of hypothesis concerning population means and proportions are similar. The similarity lies in the fact that all of the point estimators discussed in Chapter 7 are unbiased and have sampling distributions that, for large samples, can be approximated by a normal distribution. Therefore, we can use the point estimators as test statistics to test hypotheses about the respective parameters. For small samples, the forms of the statistical tests of hypothesis are similar, but the sampling distributions of the test statistics follow a Student's t distribution. We will consider tests of hypothesis concerning the four population parameters μ, p, $\mu_1 - \mu_2$, and $p_1 - p_2$ separately in the following sections.

8.3   A Test of Hypothesis for a Population Mean

Consider a random sample of n measurements drawn from a population that has mean μ and standard deviation σ. We would like to test a hypothesis of the form[†]

$$H_0 : \mu = \mu_0$$

where μ_0 is some hypothesized value for μ, versus a one-tailed alternative hypothesis

$$H_a : \mu > \mu_0$$

We will use a subscript zero to indicate the value of the parameter specified by H_0. Notice that H_0 provides a value of the parameter to be tested, that is, μ equals μ_0, while H_a gives a range of possible values for μ. The sample mean $\bar{x}$ is the best estimate of the actual value of μ, which is presently in question. What values of $\bar{x}$ would lead us to believe that H_0 is false and μ is, in fact, greater than the hypothesized value? Those values of $\bar{x}$ that are extremely *large* would imply that μ is larger than hypothesized. Hence, we will reject H_0 if $\bar{x}$ is "too large."

The next problem is to define what we mean by "too large." Values of $\bar{x}$ that lie too many standard deviations to the right of the mean are not very likely to occur. Hence, we can define "too large" as being too many standard deviations away from μ_0. Remember that the standard deviation or standard error of $\bar{x}$ is calculated as

$$\sigma_{\bar{x}} = \frac{\sigma}{\sqrt{n}}$$

The Large-Sample Test

When the sample size n is large, the sampling distribution of $\bar{x}$ is approximately normal, and we can measure the number of standard deviations that $\bar{x}$ lies from μ_0 using the **standardized test statistic**

$$z = \frac{\bar{x} - \mu_0}{\sigma/\sqrt{n}}$$

which has a standard normal distribution when $H_0 : \mu = \mu_0$ is true. If the alternative hypothesis is $H_a : \mu > \mu_0$, the probability α of rejecting the null hypothesis, when it is true, is equal to the area under the normal curve lying above the rejection region. Thus, if we want $\alpha = .05$, we would reject H_0 when $\bar{x}$ is more than 1.645 standard

---

[†]Note that if the test rejects the null hypothesis $\mu = \mu_0$ in favor of the alternative hypothesis $\mu > \mu_0$, then it will certainly reject a null hypothesis of the form $\mu < \mu_0$, since this is even more contradictory to the alternative hypothesis. For this reason, in this text we state the null hypothesis for a one-tailed test as $\mu = \mu_0$ rather than $\mu \leq \mu_0$.

F I G U R E **8.2**

Distribution of $z = \dfrac{\overline{x} - \mu_0}{\sigma/n}$
when H_0 is true

F I G U R E **8.2**

Distribution of $z = \dfrac{\overline{x} - \mu_0}{\sigma/n}$
when H_0 is true

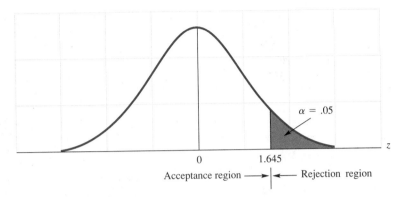

deviations to the right of μ_0. Equivalently, we would reject H_0 if the standardized test statistic z, defined above, is greater than 1.645 (Figure 8.2).

If we wish to detect departures either greater than or less than μ_0, the alternative hypothesis would be *two-tailed*, written as

$$H_a : \mu \neq \mu_0$$

which implies either $\mu > \mu_0$ or $\mu < \mu_0$. Values of $\overline{x}$ that are either "too large" or "too small" in terms of their distance from μ_0 will be placed in the rejection region. Since we still want $\alpha = .05$, the area in the rejection region is equally divided between the two tails of the normal distribution, as shown in Figure 8.3. Using the standardized test statistic z, we will reject H_0 if $z > 1.96$ or $z < -1.96$. For different values of α, the critical values of z that separate the rejection and acceptance regions will change.

F I G U R E **8.3**
The rejection region for a
two-tailed test with $\alpha = .05$

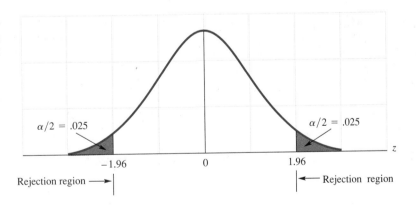

Large-Sample Statistical Test for μ

1 Null Hypothesis: $H_0 : \mu = \mu_0$

2 Alternative Hypothesis:

One-Tailed Test	Two-Tailed Test
$H_a : \mu > \mu_0$ (or $H_a : \mu < \mu_0$)	$H_a : \mu \neq \mu_0$

3 Test Statistic: $z = \dfrac{\bar{x} - \mu_0}{\sigma_{\bar{x}}} = \dfrac{\bar{x} - \mu_0}{\sigma/\sqrt{n}}$

If σ is unknown (which is usually the case), substitute the sample standard deviation s for σ.

4 Rejection Region:

One-Tailed Test	Two-Tailed Test
$z > z_\alpha$ (or $z < -z_\alpha$ when the alternative hypothesis is $H_a : \mu < \mu_0$)	$z > z_{\alpha/2}$ or $z < -z_{\alpha/2}$

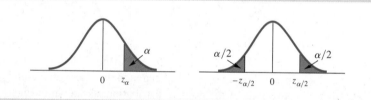

Assumptions: The n observations in the sample were randomly selected from the population, and n is large, say, $n \geq 30$.

E X A M P L E   **8.4**   The daily yield at a chemical plant, recorded for $n = 50$ days, possesses a sample mean and standard deviation of $\bar{x} = 871$ tons and $s = 21$ tons. Test the hypothesis that the average daily yield of the chemical is $\mu = 880$ tons per day against the alternative that μ is either greater than or less than 880 tons per day.

Solution   We want to test the null hypothesis

$$H_0 : \mu = 880 \text{ tons}$$

against the alternative hypothesis

$$H_a : \mu \neq 880 \text{ tons}$$

The point estimate for μ is $\bar{x}$. Therefore, the test statistic is

$$z = \frac{\bar{x} - \mu_0}{\sigma_{\bar{x}}} = \frac{\bar{x} - \mu_0}{\sigma/\sqrt{n}}$$

For $\alpha = .05$, the rejection region is $z > 1.96$ or $z < -1.96$ (shown in Figure 8.3). Using s to approximate σ, we obtain

$$z = \frac{871 - 880}{21/\sqrt{50}} = -3.03$$

Since the calculated value of z falls in the rejection region, we reject the hypothesis that $\mu = 880$ tons. (In fact, it appears that the mean yield is less than 880 tons per day.) The probability of rejecting H_0, assuming it to be true, is only $\alpha = .05$. Therefore, we are reasonably confident that our conclusion that $\mu \neq 880$ tons is correct. ∎

The statistical test based on a normally distributed test statistic, with given α, and the $(1 - \alpha)100\%$ confidence interval (Section 7.4) are clearly related. The interval $\bar{x} \pm 1.96\sigma/\sqrt{n}$, or approximately 871 ± 5.82 for Example 8.4, is constructed such that in repeated sampling $(1 - \alpha)100\%$ of the intervals will enclose μ. Noting that $\mu = 880$ does not fall in the interval, we would be inclined to reject $\mu = 880$ as a likely value and conclude that the mean daily yield was indeed different from 880.

There is another similarity between this test and the confidence interval of Section 7.4. The test is "approximate" because we substituted s, an approximate value, for σ. That is, the probability α of a type I error selected for the test is not *exactly* .05, but it is very close. This will be true for many statistical tests, since not all of the assumptions will be *exactly* satisfied.

Since α is the probability of rejecting H_0 when it is true, it is a measure of the chance of *incorrectly rejecting* H_0. Since β is the probability of accepting H_0 when it is false, its complement, $1 - \beta$, is the probability of rejecting H_0 when it is false and measures the chance of *correctly rejecting* H_0. This probability, $1 - \beta$, is called the **power of the test,** the chance that the test performs as it should.

DEFINITION ∎ The **power of a statistical test,** given as

$$1 - \beta = P(\text{reject } H_0 \text{ when } H_0 \text{ is false})$$

measures the ability of the test to perform as required. ∎

A graph of $1 - \beta$, the probability of rejecting H_0 when, in fact, H_0 is false, as a function of the true value of the parameter of interest is called the **power curve** for the statistical test. Ideally, we would like α to be small and the **power** $(1 - \beta)$ to be large. The experimenter should be able to specify values of α and β, measuring the risks of the respective errors he or she is willing to tolerate, as well as some deviation from the hypothesized value of the parameter he or she considers of practical importance and wishes to detect. The rejection region for the test will be located in accordance with the specified value of α; the sample size will be chosen large enough to achieve an acceptable value of β for the specified deviation the experimenter wishes to detect.

This choice could be made by consulting the power curves, corresponding to various sample sizes, for the chosen test.

E X A M P L E **8.5** Refer to Example 8.4. Calculate the probability β of accepting H_0 if μ is actually equal to 870 tons. Calculate the power of the test, $1 - \beta$.

Solution The acceptance region for the test in Example 8.4 is located in the interval $\mu_0 \pm 1.96\sigma_{\bar{x}}$. Substituting numerical values, we obtain

$$880 \pm 1.96 \left(\frac{21}{\sqrt{50}} \right)$$

or from

$$874.18 \quad \text{to} \quad 885.82$$

The probability of accepting H_0 if, in fact, $\mu = 870$ is equal to the area under the sampling distribution for the test statistic $\bar{x}$ above the interval from 874.18 to 885.82. Since $\bar{x}$ will be normally distributed with mean equal to 870 and $\sigma_{\bar{x}} \approx 21/\sqrt{50} = 2.97$, β is equal to the area under the normal curve between 874.18 and 885.82 (see Figure 8.4). Calculating the z values corresponding to 874.18 and 885.82, we obtain

$$z_1 = \frac{\bar{x}_1 - \mu}{\sigma/\sqrt{n}} \approx \frac{874.18 - 870}{21/\sqrt{50}} = 1.41$$

$$z_2 = \frac{\bar{x}_2 - \mu}{\sigma/\sqrt{n}} \approx \frac{885.82 - 870}{21/\sqrt{50}} = 5.33$$

Then

$$\begin{aligned} \beta &= P(\text{accept } H_0 \text{ when } \mu = 870) \\ &= P(874.18 < \bar{x} < 885.82 \text{ when } \mu = 870) \\ &= P(1.41 < z < 5.33) \end{aligned}$$

You can see from Figure 8.4 that the area under the normal curve above $\bar{x} = 885.82$ (or $z = 5.33$) is negligible. Therefore,

$$\beta = P(z > 1.41) = .5 - .4207 = .0793$$

and the power of the test is

$$1 - \beta = 1 - .0793 = .9207$$

The probability of correctly rejecting H_0, given that μ is really equal to 870, is .9207, or approximately 92 chances in 100.

Values of $(1 - \beta)$ can be calculated for various values of μ_a different from $\mu_0 = 880$. For example, if $\mu_a = 885$,

$$\begin{aligned} \beta &= P(874.18 < \bar{x} < 885.82 \text{ when } \mu = 885) \\ &= P(-3.64 < z < .28) = .5 + .1103 = .6103 \end{aligned}$$

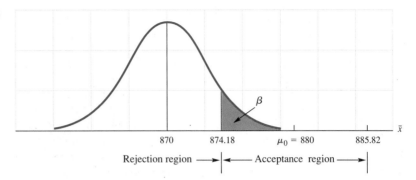

F I G U R E **8.4**
Calculating β in Example 8.5

870   874.18 $\mu_0 = 880$ 885.82

Rejection region ⟶|⟵ Acceptance region ⟶|

and the power is $(1 - \beta) = .3897$. Table 8.3 shows the power of the test for various values of μ_a, and a power curve is graphed in Figure 8.5. Note that the power of the test increases as the distance between μ_a and μ_0 increases. The result is a U-shaped curve for this two-tailed test.   ∎

T A B L E **8.3**
Values of $(1 - \beta)$ for various values of μ_a, Example 8.5

μ_a	$1 - \beta$	μ_a	$1 - \beta$
865	.9990	883	.1726
870	.9207	885	.3897
872	.7673	888	.7673
875	.3897	890	.9207
877	.1726	895	.9990
880	.0500		

F I G U R E **8.5**
Power curve for Example 8.5

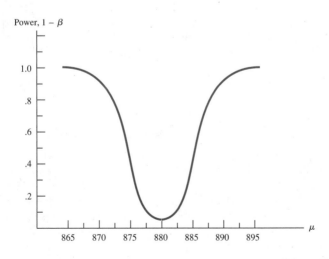

Power, $1 - \beta$

The Small-Sample Test

When the sample size is small and the population standard deviation σ is unknown, the test of hypothesis about a population mean μ is based on the test statistic

$$t = \frac{\bar{x} - \mu}{s/\sqrt{n}}$$

which has a Student's t distribution with $n - 1$ degrees of freedom when sampling from a normal population, as described in Section 7.5. The statistical test of an hypothesis concerning a population mean is given in the following display. Notice that the rejection regions are found using the critical values of t given in Table 4 of Appendix II.

Small-Sample Test of an Hypothesis Concerning a Population Mean

1 Null Hypothesis: $H_0 : \mu = \mu_0$

2 Alternative Hypothesis:

One-Tailed Test	Two-Tailed Test
$H_a : \mu > \mu_0$	$H_a : \mu \neq \mu_0$
(or $H_a : \mu < \mu_0$)	

3 Test Statistic: $t = \dfrac{\bar{x} - \mu_0}{s/\sqrt{n}}$

4 Rejection Region:

One-Tailed Test

$t > t_\alpha$
(or $t < -t_\alpha$ when the alternative hypothesis is $H_a : \mu < \mu_0$)

Two-Tailed Test

$t > t_{\alpha/2}$ or $t < -t_{\alpha/2}$

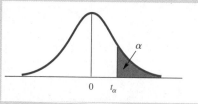 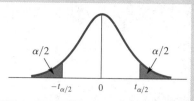

The critical values of t, t_α and $t_{\alpha/2}$, are based on $(n - 1)$ degrees of freedom. These tabulated critical values can be found in Table 4 of Appendix II.

Assumption: The sample has been randomly selected from a normally distributed population.

EXAMPLE **8.6**   In Example 7.4, we considered an experiment designed to evaluate a new process for producing synthetic diamonds. A study of the process costs indicates that the average weight of the diamonds must be greater than .5 carat in order that the process be operated at a profitable level. Do the weights of the six synthetic diamonds, .46, .61, .52, .48, .57, and .54 carat, provide sufficient evidence to indicate that the mean diamond weight produced by the process exceeds .5 carat? Test using $\alpha = .05$.

Solution   Since we want to detect values of $\mu > .5$, we will test the null hypothesis

$$H_0 : \mu = .5$$

against the alternative hypothesis

$$H_a : \mu > .5$$

The test statistic is

$$t = \frac{\bar{x} - \mu_0}{s/\sqrt{n}}$$

Since we wish to detect only large values of μ, we will conduct an upper one-tailed test. The rejection region for this test for $\alpha = .05$ and $(n - 1) = (6 - 1) = 5$ degrees of freedom is $t > 2.015$. This is the value of t, given in Table 4 of Appendix II, that places $\alpha = .05$ in the upper tail of the t distribution (see Figure 8.6). The sample mean and standard deviation for the six diamond weights are

$$\bar{x} = .53 \quad \text{and} \quad s = .0559$$

Substituting these quantities into the formula for the test statistic, we obtain

$$t = \frac{\bar{x} - \mu_0}{s/\sqrt{n}} = \frac{.53 - .5}{.0559/\sqrt{6}} = 1.31$$

Since the calculated value of the test statistic does not fall in the rejection region, we do not reject H_0. Nonrejection of H_0 implies that the data do not present sufficient evidence to indicate that the mean diamond weight exceeds .5 carat.   ∎

FIGURE **8.6**
Rejection region for the test in
Example 8.6

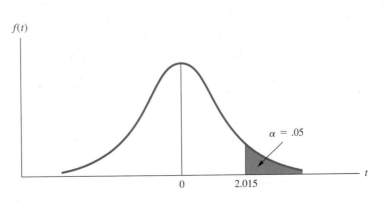

The calculation of the probability of a type II error, β, for the t test is very difficult and is beyond the scope of this text. We could obtain an interval estimate for μ (see Section 7.5), however, in order to determine a range of possible values for μ. If the experimenter is concerned about decreasing the value of β, he or she should increase the sample size.

The Minitab command TTEST is used for the small-sample test of a population mean; it also produces a small-sample confidence interval. The user must provide the value of the population mean to be tested and the column number in which the data are stored. The subcommand ALTERNATIVE followed by -1, 1, or 0 will implement a left-tailed, right-tailed, or two-tailed test procedure, respectively. (Without this subcommand, a two-tailed test is implemented.)

The results of using TTEST to conduct the test in Example 8.6 are given in Table 8.4. In addition to the observed value of the test statistic, $t = 1.32$, the output gives the sample mean, the sample standard deviation, and the standard error of the mean (SE MEAN $= s/\sqrt{n}$). Compared to our results in Example 8.6, the only difference is in the reported decimal accuracy of the results.

T A B L E **8.4**
Minitab output for the data of Example 8.6

```
MTB > TTEST .5 C1;
SUBC> ALTERNATIVE 1.

TEST OF MU = 0.5000 VS MU G.T. 0.5000

             N     MEAN    STDEV   SE MEAN        T    P VALUE
C1           6   0.5300  0.0559    0.0228     1.32       0.12

MTB > TINTERVAL 95 C1

             N     MEAN    STDEV   SE MEAN   95.0 PERCENT C.I.
C1           6   0.5300  0.0559    0.0228   ( 0.4714,  0.5886)

MTB >
```

The Minitab command TINTERVAL is available for constructing a confidence interval for a population mean. The command requires that the user supply the confidence coefficient and the column location of the data. The 95% confidence interval for μ using the data in Example 8.6 is also shown in Table 8.4, and the results agree with those given in Example 7.4.

E X A M P L E   **8.7**   The Federal Aviation Administration (FAA) provides a monthly report on aircraft utilization and propulsion reliability for U.S. air fleets. The data that follow give the average engine hours per aircraft for aircraft equipped with Pratt and Whitney model PW120 turboprop engines for each of $n = 7$ air carriers. (The number of aircraft per carrier varied from 5 to 15.)

389	364
359	308
408	295
393	

Do these data indicate that the average number of engine hours per aircraft using the PW120 model engine is less than 400 hours?

Solution   Testing the null hypothesis that $\mu = 400$ average engine hours against the alternative that μ is less than 400 will result in a one-tailed statistical test. Thus

$$H_0 : \mu = 400$$
$$H_a : \mu < 400$$

where μ is the average engine hours per aircraft. Using $\alpha = .05$ and placing .05 in the lower tail of the t distribution, we find the critical value of t for $n = 7$ measurements (or for $n - 1 = 6$ degrees of freedom) to be $t = 1.943$. Therefore, we will reject H_0 if $t < -1.943$ (see Figure 8.7).

You can verify that the sample mean and standard deviation for the $n = 7$ measurements in the table are

$$\bar{x} = 359.43 \qquad \text{and} \qquad s = 43.16$$

Substituting these values into the formula for the test statistic yields

$$t = \frac{\bar{x} - \mu_0}{s/\sqrt{n}} = \frac{359.43 - 400}{43.16/\sqrt{7}} = -2.487$$

Since the observed value of t falls in the rejection region, there is sufficient evidence to indicate that the average number of engine hours is less than 400. Furthermore, we will be reasonably confident that we have made the correct decision. In using our procedure, we should erroneously reject H_0 only $\alpha = .05$ of the time in repeated applications of the statistical test.   ■

FIGURE 8.7
Rejection region for Example 8.7

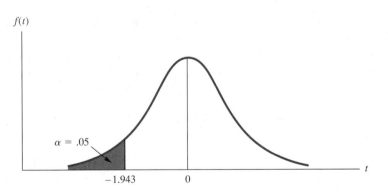

Comments Concerning Error Rates

Since α is the probability that the test statistic falls in the rejection region, assuming H_0 to be true, **an increase in the size of the rejection region increases** α and, at the same time, decreases β for a fixed sample size. Reducing the size of the rejection region decreases α and increases β. If the sample size n is increased, more information is available on which to base the decision, and, for fixed α, β will decrease.

The probability β of making a type II error varies depending on the true value of the population parameter. For instance, suppose that we wish to test the null hypothesis that the binomial parameter p is equal to $p_0 = .4$. Furthermore, suppose that H_0 is false and that p is really equal to an alternative value, say, p_a. Which will be more easily detected, a $p_a = .4001$ or a $p_a = 1.0$? Certainly, if p is really equal to 1.0, every single trial will result in success and the sample results will produce strong evidence to support a rejection of $H_0 : p = .4$. On the other hand, $p_a = .4001$ lies so close to $p_0 = .4$ that it would be extremely difficult to detect p_a without a very large sample. In other words, the probability β of accepting H_0 will vary depending on the difference between the true value of p and the hypothesized value p_0. Ideally, the farther p_a lies from p_0, the higher the probability of rejecting H_0. This probability is measured by $1 - \beta$, which is called the **power** of the test.

For constant values of n and α, **the power of a test should increase as the distance between the true and hypothesized values of the parameter increases.** An increase in the sample size n will increase the power, $1 - \beta$, for all alternative values of the parameter being tested. Thus we can create a power curve corresponding to each sample size.

In practice, β is often unknown, either because it was never computed before the test was conducted or because it may be extremely difficult to compute for the test. Then, rather than accept the null hypothesis when the test statistic falls in the acceptance region, you should withhold judgment. That is, you should not accept the null hypothesis unless you know the risk (measured by β) of making an incorrect decision. Notice that you will never be faced with this "no conclusion" situation when the test statistic falls in the rejection region. Then you can reject the null hypothesis (and accept the alternative hypothesis), because you always know the value of α, the probability of rejecting the null hypothesis when it is true. The fact that β is often unknown explains why we attempt to support the alternative hypothesis by rejecting the null hypothesis. When we reach this decision, the probability α that such a decision is incorrect is known.

In conclusion, keep in mind that **"accepting" a particular hypothesis means deciding in its favor.** Regardless of the outcome of a test, you are never *certain* that the hypothesis you "accept" is true. **There is always a risk of being wrong (measured by α and β).** Consequently, you never "accept" H_0 if β is unknown or its value is unacceptable to you. When this situation occurs, you should withhold judgment and collect more data.

Exercises

Basic Techniques

8.1 A random sample of $n = 35$ observations from a population produced a mean $\bar{x} = 2.4$ and a standard deviation equal to .29. Suppose that you wish to show that the population mean μ exceeds 2.3.

a Give the alternative hypothesis for the test.

b Give the null hypothesis for the test.

c   If you wish your probability of (erroneously) deciding that $\mu > 2.3$, when in fact $\mu = 2.3$, to equal .05, what is the value of α for the test?

d   Before you conduct the test, glance at the data and use your intuition to decide whether the sample mean $\bar{x} = 2.4$ implies that $\mu > 2.3$. Now test the null hypothesis. Do the data provide sufficient evidence to indicate that $\mu > 2.3$? Test using $\alpha = .05$.

8.2   Refer to Exercise 8.1. Suppose that you wish to show that the sample data support the hypothesis that the mean of the population is less than 2.9. Give the null and alternative hypotheses for this test. Would this test be a one- or a two-tailed test? Explain.

8.3   Refer to Exercises 8.1 and 8.2. Suppose that you wish to detect a value of μ that differs from 2.9, that is, a value of μ either greater than or less than 2.9. State the null and alternative hypotheses for the test. Would the alternative hypothesis imply a one- or a two-tailed test?

8.4   A random sample of $n = 40$ observations from a population produced a mean $\bar{x} = 83.8$ and a standard deviation equal to 2.9. Suppose that you wish to show that the population mean μ is less than 84.

a   Give the alternative hypothesis for the test.

b   Give the null hypothesis for the test.

c   If you wish your probability of (erroneously) deciding that $\mu < 84$, when in fact $\mu = 84$, to equal .05, what is the value of α for the test?

d   Before you conduct the test, glance at the data and use your intuition to decide whether the sample mean $\bar{x} = 83.8$ implies that $\mu < 84$. Now test the null hypothesis. Do the data provide sufficient evidence to indicate that $\mu < 84$? Test using $\alpha = .05$.

8.5   Refer to Exercise 8.4, where $H_0 : \mu = 84$ is tested against $H_a : \mu < 84$.

a   Find the critical value of $\bar{x}$ necessary for rejection of H_0.

b   Calculate $\beta = P[\text{accept } H_0 \text{ when } \mu = 82.8]$. Repeat the calculation for $\mu = 82.4$, 82.6, and 83.4.

c   Use the values of β calculated in part (b) to graph the power curve for the test.

8.6   A random sample of $n = 4$ observations from a normally distributed population produced the following data: 9.4, 12.2, 10.7, and 11.6. Do the data provide sufficient evidence to indicate that $\mu > 10$?

a   State H_a.

b   State H_0.

c   Give the rejection region for the test for $\alpha = .10$.

d   Conduct the test and state your conclusions.

8.7   A random sample of $n = 6$ observations from a normally distributed population produced the following data: 3.7, 6.4, 8.1, 8.8, 4.9, and 5.0. Do the data provide sufficient evidence to indicate that $\mu < 7$?

a   State H_a.

b   State H_0.

c   Give the rejection region for the test for $\alpha = .10$.

d   Conduct the test and state your conclusions.

8.8   Test the null hypothesis $H_0 : \mu = 3$ against $H_a : \mu > 3$ for $\alpha = .05$, $n = 12$, $\bar{x} = 3.18$, and $s^2 = .21$.

8.9   Test the null hypothesis $H_0 : \mu = 48$ against $H_a : \mu \neq 48$ for $\alpha = .10$, $n = 25$, $\bar{x} = 47.1$, and $s^2 = 4.7$.

8.10   The Minitab printout that follows resulted when the TTEST command was used on a set of data stored in C1. Use the printout to identify all four parts of the statistical test of hypothesis and draw the appropriate conclusions for $\alpha = .01$.

```
MTB > TTEST 5 C1;
SUBC> ALT 1.

TEST OF MU = 5.0000 VS MU G.T. 5.0000

            N      MEAN    STDEV    SE MEAN         T    P VALUE
C1         11     5.364    1.502     0.453       0.80      0.22
```

Applications

8.11   A money manager claims that her common stock selection for investment will, on average, beat the annual change in the Standard & Poor's stock average. A random selection of three of the manager's stock choices found annual increases of 22%, 12%, and 31% in comparison with an increase in the Standard & Poor's average of 19%. Does this sample of three stock picks provide sufficient evidence to indicate that the mean increase in all of the manager's stock selections exceeds 19%?

   a   State H_a.

   b   State H_0.

   c   Give the rejection region for the test for $\alpha = .05$.

   d   Conduct the test and state your conclusions.

8.12   A new-car dealer calculated that the company must average more than 4.8% profit on the sales of its allotted new cars. A random sampling of $n = 80$ cars gave a mean and standard deviation of the percentage profit per car of $\bar{x} = 4.87\%$ and $s = 3.9\%$. Do the data provide sufficient evidence to indicate that the sales manager's policy in approving sale prices is achieving a mean profit exceeding 4.8% per car?

   a   State the alternative hypothesis that the sales manager wants to show to be true.

   b   Examine the data. From your intuition only, do you think that the data support the alternative hypothesis of part (a)?

   c   State the null hypothesis to be tested.

   d   The company's owner wants to be reasonably certain that the decision is correct if, in fact, the data show that the company is operating at an acceptable profit level. To accomplish this, the owner wants to test the null hypothesis using $\alpha = .01$. Explain how this choice for α will accomplish the owner's objective.

   e   Give the rejection region for the test.

   f   Conduct the test and state your conclusions in a manner that will be understandable to the company's owner. Compare your answer with your intuitive guess in part (b).

8.13   A manufacturer of metal fasteners expects to ship an average of 1200 boxes of fasteners per day. An analysis of the shipments for the past 30 days possessed a mean of $\bar{x} = 1186$ boxes per day and a variance of $s^2 = 2480$ (boxes)^2 per day. Do the data provide sufficient evidence to indicate that the mean daily demand for the fasteners is slipping, that is, is below 1200 boxes per day?

   a   Give the alternative hypothesis that the manufacturer wishes to detect.

   b   Examine the data. From your intuition, do you think that the data support the alternative hypothesis of part (a)?

   c   Give H_0 for the test.

   d   Give the rejection region for the test for $\alpha = .10$.

   e   Conduct the test and state the practical conclusions to be derived from the test. Compare your own conclusions with your answer to part (b).

8.14   The tremendous growth of the Florida lobster (called spiny lobster) industry over the past 20 years has made it the state's second most valuable fishery industry. Several years ago, a declaration by the Bahamian government that prohibited U.S. lobstermen from fishing on the Bahamian portion of the continental shelf was expected to produce a dramatic reduction in the landings in pounds per lobster per trap. According to the records, the mean landing per trap is 30.31 pounds. A random sampling of 20 lobster traps since the Bahamian fishing restriction went into effect gave the following results (in pounds):

17.4	18.9	39.6	34.4	19.6
33.7	37.2	43.4	41.7	27.5
24.1	39.6	12.2	25.5	22.1
29.3	21.1	23.8	43.2	24.4

Do these landings provide sufficient evidence to support the contention that the mean landings per trap decreased after imposition of the Bahamian restrictions? Test using $\alpha = .05$.

8.15   In Exercise 2.13, we reported that the Tropica had an average driving range of 44.7 miles, according to the California Air Resources Board. An independent test consisting of 30 trials produced a sample of driving ranges for the Tropica. The following Minitab printout is obtained using those 30 observations.

```
TEST OF MU = 44.700 VS MU N.E. 44.700

               N      MEAN     STDEV    SE MEAN        T    P VALUE
RANGES        30    45.273     1.199      0.219     2.62      0.014
```

a   Based on the Minitab printout, explain the appropriate hypothesis-testing procedure to verify or dispute the 44.7-mile driving range. Use $\alpha = .05$.

b   Refer to Exercise 7.100. Do the results of part (a) agree with the conclusions you reached in that exercise? Explain.

8.4   Another Way to Report the Results of Statistical Tests: *p*-Values

The probability α of making a type I error is often called the significance level of a statistical test, since we declare a significant difference if the observed value of the test statistic falls in the rejection region determined by H_a and the value of α. Some experimenters prefer to use a variable level of significance. For example, if in a two-tailed test the observed value of z turned out to be $z = 2.03$, the null hypothesis could be rejected if $\alpha = .05$, since $z = 2.03$ is greater than $z_{.025} = 1.96$, but could not be rejected if $\alpha = .01$, since $z = 2.03$ is less than $z_{.005} = 2.58$. This would be reported by saying that test results were significant at the 5% level of significance but not at the 1% level. Other experimenters prefer to report their results by providing the smallest value of α for which the test results are significant. If we had used critical values of z equal to ± 2.03, we would have rejected H_0, and the value of α that we used would be

$$P(z \leq -2.03) + P(z \geq 2.03) = 2P(z \geq 2.03)$$
$$= 2(.5 - .4788)$$
$$= 2(.0212) = .0414$$

This value is called the **p-value**[†] or the **observed significance level** of the test.

DEFINITION ■

> The *p*-value or **observed significance level** is the smallest value of α for which test results are statistically significant. ■

Some statistical computer programs compute *p*-values for statistical tests correct to four or five decimal places. But if you are using statistical tables to determine a *p*-value, you will only be able to approximate its value, because most statistical tables give the critical values of test statistics only for large differential values of α (e.g., .01, .025, .05, .10, etc.). Consequently, the *p*-value reported by most experimenters is the smallest tabulated value of α for which the test remains **statistically significant.** For example, if a test result is statistically significant for $\alpha = .10$ but not for $\alpha = .05$, then the *p*-value for the test would be given as *p*-value = .10, or more precisely as

$$.05 < p\text{-value} < .10$$

Another way to use the *p*-value in decision making is to reject H_0 if the *p*-value is *less than* the value of α, since this will happen only if the observed value of the test statistic falls in the rejection region. For example, if in a right-tailed test with $\alpha = .05$ the observed value of a z statistic is 2.04, the *p*-value for the test would be $P(z > 2.04) = .0207$. Since the *p*-value of .0207 is less than .05, we can reject H_0, knowing that the value $z = 2.04$ lies in the rejection region when $\alpha = .05$.

Many scientific journals require researchers to report the *p*-values associated with statistical tests because these values provide a reader with *more information* than simply stating that a null hypothesis is or is not to be rejected for some value of α chosen by the experimenter. In a sense, it allows the reader of published research to evaluate the extent to which the data disagree with the null hypothesis. In particular, it enables each reader to choose his or her own personal value for α and then to decide whether or not the data lead to rejection of the null hypothesis.

The procedure for finding the *p*-value for a test is illustrated in the following examples.

EXAMPLE **8.8**   Find the *p*-value for the statistical test in Example 8.4. Interpret your results.

Solution   Example 8.4 presents a test of the null hypothesis $H_0 : \mu = 880$ against the alternative hypothesis $H_0 : \mu \neq 880$. The value of the test statistic, computed from the sample data, was $z = -3.03$. Therefore, the *p*-value for this two-tailed test is the probability that $z \leq -3.03$ or $z \geq 3.03$ (see Figure 8.8).

From Table 3 in Appendix II, you can see that the tabulated area under the normal curve between $z = 0$ and $z = 3.03$ is .4988, and the area to the right of $z = 3.03$ is $.5 - .4988 = .0012$. Then, because this was a two-tailed test, the value

---

[†] Users of statistics often call the observed significance level a "probability-value or *p*-value." The symbol *p* in the expression has no connection to the binomial parameter *p*.

F I G U R E **8.8**
Determining the *p*-value for the
test in Example 8.8

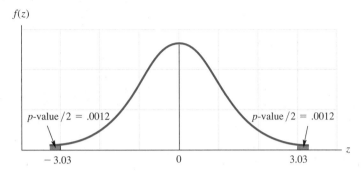

of α corresponding to a rejection region $z > 3.03$ or $z < -3.03$ is $2(.0012) = .0024$. Consequently, we report the *p*-value for the test as *p*-value $= .0024$. ∎

E X A M P L E   **8.9**   If you planned to report the results of the statistical test in Example 8.7, what *p*-value would you report?

Solution   The *p*-value for this test is the probability of observing a value of the t statistic at least as contradictory to the null hypothesis when H_0 is true as the one observed for this set of data, namely, a value of $t \leq -2.487$ (see Figure 8.9).

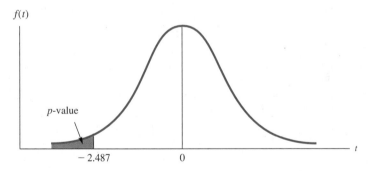

   Unlike the table of areas under the normal curve (Table 3 of Appendix II), Table 4 in Appendix II does not give the areas corresponding to various values of t. Rather, it gives the values of t corresponding to upper-tail areas equal to .10, .05, .025, .010, and .005. Since the t distribution is symmetric about its mean, we can use these upper-tail areas to approximate the probability that $t < -2.487$. The t statistic for this test is based on 6 degrees of freedom, so we consult the d.f. $= 6$ row of Table 4 and find that 2.487 falls between $t_{.025} = 2.447$ and $t_{.010} = 3.143$. Since the observed value of t, -2.487, is less than $-t_{.025} = -2.447$ but not less than $-t_{.01} = -3.143$, we reject H_0 for $\alpha = .025$ but not for $\alpha = .01$. Therefore, the *p*-value for the test is reported as $.01 \leq$ *p*-value $\leq .025$. ∎

To advocate that a researcher report the p-value for a test and leave its interpretation to the reader does not violate the traditional statistical test procedure described in the preceding sections. It simply leaves the decision of whether to reject the null hypothesis (with the potential for a type I or type II error) to the reader. Thus it shifts the responsibility for choosing the value of α, and possibly the problem of evaluating the probability β of making a type II error, to the person reading the report.

Exercises

Basic Techniques

8.16 Suppose that you tested the null hypothesis $H_0 : \mu = 94$ against the alternative hypothesis $H_a : \mu < 94$. For a random sample of $n = 52$ observations, $\bar{x} = 92.9$ and $s = 4.1$.

    a Give the observed significance level for the test.

    b If you wish to conduct your test using $\alpha = .05$, what would be your test conclusions?

8.17 Suppose that you tested the null hypothesis $H_0 : \mu = 94$ against the alternative hypothesis $H_a : \mu \neq 94$. For a random sample of $n = 52$ observations, $\bar{x} = 92.1$ and $s = 4.1$.

    a Give the observed significance level for the test.

    b If you wish to conduct your test using $\alpha = .05$, what would be your test conclusions?

8.18 Suppose that you tested the null hypothesis $H_0 : \mu = 15$ against the alternative hypothesis $H_a : \mu \neq 15$. For a random sample of $n = 18$ observations, $\bar{x} = 15.7$ and $s = 2.4$.

    a Give the approximate observed significance level for the test.

    b If you wish to conduct your test using $\alpha = .05$, what would be your test conclusions?

Applications

8.19 Find the p-value for the test of the mean demand for metal fasteners in Exercise 8.13, and interpret it.

8.20 If only the p-value for the test in Exercise 8.19 were reported to you, how could you use it to conduct the test for $\alpha = .05$?

8.21 Find the p-value for the test in Exercise 8.11, and interpret it.

8.22 Find the p-value for the test in Exercise 8.12, and interpret it.

8.23 Find the p-value for the test in Exercise 8.14, and interpret it.

8.24 In Exercise 7.20, we presented some results concerning mortgage loan rates in which the mean of ten rate expectations was 8.5% with a standard deviation of .23%.

    a Test the hypothesis $H_0 : \mu = 8.7$ versus $H_a : \mu < 8.7$ using $\alpha = .05$.

    b Find the p-value for this test and interpret its value.

8.25 In Exercise 7.11, the first 40 medical claims received for the month had a sample mean of $930.

    a If the population standard deviation is $\sigma = \$2000$, test the hypothesis $H_0 : \mu = \$800$ versus $H_a : \mu > \$800$ with $\alpha = .05$.

    b Find the p-value for this test and interpret its value.

8.5 Tests Concerning the Difference Between Two Population Means

In many situations, the statistical question to be answered involves a comparison of two population means. For example, the U.S. Postal Service is interested in reducing its massive 350 million gallons/year gasoline bill by replacing gasoline-powered trucks with electric-powered trucks. To determine whether significant savings in operating costs are achieved by changing to electric-powered trucks, a pilot study should be undertaken using, say, 100 conventional gasoline-powered mail trucks and 100 electric-powered mail trucks operated under similar conditions. The statistic that summarizes the sample information regarding the difference in population means $\mu_1 - \mu_2$ is the difference in sample means $\bar{x}_1 - \bar{x}_2$. Therefore, in testing whether the difference in sample means indicates that the true difference in population means differs from a specified value, $\mu_1 - \mu_2 = D_0$, we would use the number of standard deviations that $\bar{x}_1 - \bar{x}_2$ lies from the hypothesized difference D_0. The formal testing procedure **when the sample sizes are large** is given in the following display.

Large-Sample Statistical Test for $(\mu_1 - \mu_2)$

1 Null Hypothesis: $H_0 : (\mu_1 - \mu_2) = D_0$ where D_0 is some specified difference that you wish to test. For many tests, you will wish to hypothesize that there is no difference between μ_1 and μ_2—that is, $D_0 = 0$.

2 Alternative Hypothesis:

One-Tailed Test	**Two-Tailed Test**
$H_a : (\mu_1 - \mu_2) > D_0$	$H_a : (\mu_1 - \mu_2) \neq D_0$
[or $H_a : (\mu_1 - \mu_2) < D_0$]	

3 Test Statistic: $z = \dfrac{(\bar{x}_1 - \bar{x}_2) - D_0}{\sigma_{(\bar{x}_1 - \bar{x}_2)}} = \dfrac{(\bar{x}_1 - \bar{x}_2) - D_0}{\sqrt{\dfrac{\sigma_1^2}{n_1} + \dfrac{\sigma_2^2}{n_2}}}$

If σ_1^2 and σ_2^2 are unknown (which is usually the case), substitute the sample variances s_1^2 and s_2^2 for σ_1^2 and σ_2^2, respectively.

4 Rejection Region:

One-Tailed Test	**Two-Tailed Test**
$z > z_\alpha$	$z > z_{\alpha/2}$ or $z < -z_{\alpha/2}$
[or $z < -z_\alpha$ when the alternative hypothesis is $H_a : (\mu_1 - \mu_2) < D_0$]	

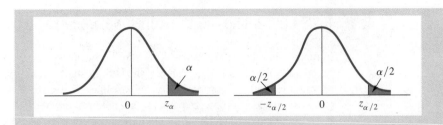

Assumptions: The samples were randomly and independently selected from the two populations, and $n_1 \geq 30$ and $n_2 \geq 30$.

E X A M P L E   **8.10**   A company employing a new sales-plus-commission compensation plan for its sales personnel wants to compare the annual salary expectations of its female and male sales personnel under the new plan. Random samples of $n_1 = 40$ female and $n_2 = 40$ male sales representatives were asked to forecast their annual incomes under the new plan. Sample means and standard deviations were

$$\bar{x}_1 = \$31{,}083 \qquad \bar{x}_2 = \$29{,}745$$
$$s_1 = \$2312 \qquad s_2 = \$2569$$

Do the data provide sufficient evidence to indicate a difference in mean expected annual income between female and male sales representatives? Test using $\alpha = .05$.

Solution   Since we wish to detect a difference in mean annual income between female and male sales representatives, either $\mu_1 > \mu_2$ or $\mu_1 < \mu_2$, we want to test the null hypothesis

$$H_0 : \mu_1 = \mu_2, \qquad \text{that is, } \mu_1 - \mu_2 = D_0 = 0$$

against the alternative hypothesis

$$H_a : \mu_1 \neq \mu_2, \qquad \text{that is, } \mu_1 - \mu_2 \neq 0$$

We use s_1^2 and s_2^2 to approximate σ_1^2 and σ_2^2, respectively. Substituting these values, along with $\bar{x}_1$ and $\bar{x}_2$, into the formula for the z test statistic, we obtain

$$z = \frac{(\bar{x}_1 - \bar{x}_2) - D_0}{\sqrt{\dfrac{\sigma_1^2}{n_1} + \dfrac{\sigma_2^2}{n_2}}} \approx \frac{(31{,}083 - 29{,}745) - 0}{\sqrt{\dfrac{(2312)^2}{40} + \dfrac{(2569)^2}{40}}} = 2.45$$

Using a two-tailed test with $\alpha = .05$, we place $\alpha/2 = .025$ in each tail of the z distribution and reject H_0 if $z > 1.96$ or $z < -1.96$ (see Figure 8.10). Since the observed value of $z = 2.45$ exceeds 1.96, the test statistic falls in the rejection region. We reject H_0 and conclude that there is a difference in the mean annual salary expectations between female and male sales representatives. We should feel very confident that we have made the correct decision. The probability that our test would lead us to reject H_0, when in fact it is true, is only $\alpha = .05$.

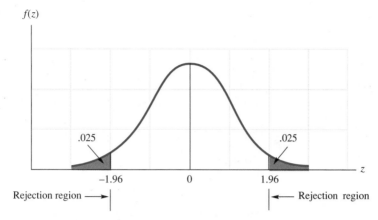

Alternatively, we could calculate the p-value for the test as

$$p\text{-value} = P(z \geq 2.45 \text{ or } z \leq -2.45) = 2P(z \geq 2.45)$$
$$= 2(.5 - .4929) = .0142$$

Since the p-value is *less than* the desired value of $\alpha = .05$, H_0 is rejected and the conclusions are the same. ■

The small-sample test for a difference between population means is based on the assumption that both populations are normally distributed and, in addition, that they possess equal variances—that is, $\sigma_1^2 = \sigma_2^2 = \sigma^2$. In this special case, the small-sample test of the hypothesis $H_0 : \mu_1 - \mu_2 = D_0$ utilizes the test statistic

$$t = \frac{(\bar{x}_1 - \bar{x}_2) - D_0}{s\sqrt{\dfrac{1}{n_1} + \dfrac{1}{n_2}}}$$

with s^2, the pooled estimate of σ^2, as given in Section 7.6. The test statistic has a Student's t distribution with $n_1 + n_2 - 2$ degrees of freedom, and the test procedure is as summarized in the display.

Small-Sample Statistical Test for $(\mu_1 - \mu_2)$

1 Null Hypothesis: $H_0 : (\mu_1 - \mu_2) = D_0$ is some specified difference that you want to test. For many tests, you will want to hypothesize that there is no difference between μ_1 and μ_2—that is, $D_0 = 0$.

2 Alternative Hypothesis:

One-Tailed Test

$H_a : (\mu_1 - \mu_2) > D_0$
[or $H_a : (\mu_1 - \mu_2) < D_0$]

Two-Tailed Test

$H_a : (\mu_1 - \mu_2) \neq D_0$

3 Test Statistic: $t = \dfrac{(\bar{x}_1 - \bar{x}_2) - D_0}{s\sqrt{\dfrac{1}{n_1} + \dfrac{1}{n_2}}}$

where s^2 is calculated as

$$s^2 = \frac{(n_1 - 1)s_1^2 + (n_2 - 1)s_2^2}{n_1 + n_2 - 2}$$

or

$$s^2 = \frac{\sum\limits_{i=1}^{n_1}(x_{1i} - \bar{x}_1)^2 + \sum\limits_{i=1}^{n_2}(x_{2i} - \bar{x}_2)^2}{n_1 + n_2 - 2}$$

4 Rejection Region:

One-Tailed Test

$t > t_\alpha$
[or $t < -t_\alpha$ when the alternative
hypothesis is $H_a : (\mu_1 - \mu_2) < D_0$]

Two-Tailed Test

$t > t_{\alpha/2}$
or $t < -t_{\alpha/2}$

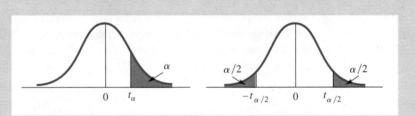

The critical values of t, t_α and $t_{\alpha/2}$, will be based on $(n_1 + n_2 - 2)$ degrees of freedom. The tabulated values can be found in Table 4 of Appendix II.

Assumptions: The samples were randomly and independently selected from normally distributed populations. The variances of the populations, σ_1^2 and σ_2^2, are equal.

E X A M P L E **8.11** Although union and nonunion wages tend to rise at the same rate in the long run, union wages usually advance faster during recessions and early in periods of recovery, and nonunion wages tend to advance more rapidly later in the business cycle when labor markets are tight. To examine this issue, an economist records the average hourly wages (including employee benefits) of employees with two years' experience for 11 randomly chosen consumer products manufacturing firms, 6 of which have nonunion shops and 5 of which have union shops. The data are as follows:

Nonunion shops	$8.26	$8.17	$8.45	$9.09	$8.85	$8.31
Union shops	$7.92	$8.39	$8.64	$8.04	$8.24	

Do these data suggest that union and nonunion wages differ for employees with two years' experience in the consumer products manufacturing industry?

Solution   Let μ_1 and μ_2 be the mean wages for nonunion and union shops, respectively. Also, assume that the variability in wages is essentially a function of individual differences and that the variability for the two populations of measurements is the same.

The sample means and standard deviations are

$$\bar{x}_1 = 8.522 \qquad s_1 = .3668$$
$$\bar{x}_2 = 8.246 \qquad s_2 = .2849$$

Then the pooled estimate of the common variance is

$$s^2 = \frac{(n_1 - 1)s_1^2 + (n_2 - 1)s_2^2}{n_1 + n_2 - 2} = \frac{.67271 + .32467}{6 + 5 - 2}$$
$$= .1108$$

The hypothesis to be tested is

$$H_0 : \mu_1 - \mu_2 = 0 \quad \text{versus} \quad H_a : \mu_1 - \mu_2 \neq 0$$

This alternative hypothesis implies that we should use a two-tailed statistical test and that the rejection region for the test will be located in both tails of the t distribution. Refer to Table 4 in Appendix II and note that the critical value of t for $\alpha/2 = .025$ and $n_1 + n_2 - 2 = 9$ degrees of freedom is 2.262. Therefore, we will reject H_0 when $t > 2.262$ or $t < -2.262$ (see Figure 8.11).

F I G U R E **8.11**
Rejection region for
Example 8.11

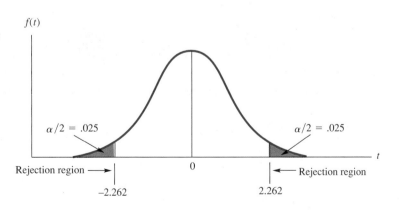

$f(t)$

$\alpha/2 = .025$

$\alpha/2 = .025$

Rejection region $\longrightarrow$

Rejection region

0

t

-2.262

2.262

The calculated value of the test statistic is

$$t = \frac{(\bar{x}_1 - \bar{x}_2) - D_0}{s\sqrt{\dfrac{1}{n_1} + \dfrac{1}{n_2}}} = \frac{8.522 - 8.246}{\sqrt{.1108\left(\dfrac{1}{6} + \dfrac{1}{5}\right)}} = 1.369$$

Comparing this value with the critical value, we see that the calculated value does not fall in the rejection region. Therefore, we cannot conclude that there is a difference in mean wages between union and nonunion shops. ▪

EXAMPLE **8.12** Find the p-value that would be reported for the statistical test in Example 8.11.

Solution The observed value of t for this two-tailed test was $t = 1.369$. Therefore, the p-value for the test would be two times the probability that $t > 1.369$ (see Figure 8.12). Since we cannot obtain this probability from Table 4 of Appendix II, we would report the p-value for the test as the smallest tabulated value for α that leads to the rejection of H_0. Consulting the row corresponding to 9 degrees of freedom in Table 4, we find that the observed value of $t = 1.369$ is less than $t_{.10} = 1.383$. Therefore, we would report that

$$p\text{-value} > 2(.10) = .20$$

Our conclusion to not reject H_0 is confirmed, since the p-value is greater than the value $\alpha = .05$. ▪

FIGURE **8.12**
p-value for the test in
Example 8.11

 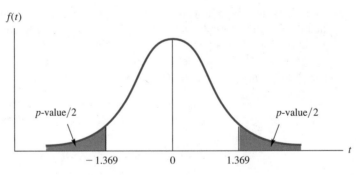

To implement a two-sample t procedure with a pooled estimate of variance using the Minitab package, enter the main command TWOSAMPLE, the columns identifying the two sets of sample values, and a semicolon. The subcommand POOLED, followed by a period, completes the instructions. The calculated value of t, its p-value, and its degrees of freedom are given as output together with a 95% confidence interval estimate of $(\mu_1 - \mu_2)$. Using the subcommand ALTERNATIVE, the user may also specify whether the test is to be left-tailed (-1), right-tailed (1), or two-tailed (0). The test is performed as two-tailed unless another alternative is specified.

The Minitab output for the TWOSAMPLE command using the data in Example 8.11 appears in Table 8.5. Notice that the entry SE MEAN (standard error of the mean), given for each column, is calculated as $s/\sqrt{n}$. The remaining entries are self-explanatory and can be compared with the results of Examples 8.11 and 8.12.

TABLE **8.5**
Minitab output for the data of Example 8.11

```
MTB > TWOSAMPLE C1 C2;
SUBC> POOLED.

TWOSAMPLE T FOR C1 VS C2
       N      MEAN     STDEV    SE MEAN
C1     6     8.522     0.367      0.15
C2     5     8.246     0.285      0.13

95 PCT CI FOR MU C1 - MU C2: ( -0.18,   0.73)

TTEST MU C1 = MU C2 (VS NE): T= 1.37   P=0.20   DF=   9

POOLED STDEV =        0.333
```

Before concluding our discussion, we should recall that **moderate departures from the assumption that the populations possess normal probability distributions do not seriously affect the distribution of the test statistic and the confidence coefficient for the corresponding confidence interval. On the other hand, the population variances should be nearly equal to ensure that the procedures given above are valid.**

If there is a reason to believe that the population variances are far from being equal, two changes must be made in the testing and estimation procedures. Since the pooled estimator s^2 is no longer appropriate, the sample variances s_1^2 and s_2^2 are used as estimators of σ_1^2 and σ_2^2. The resulting test statistic is

$$\frac{(\bar{x}_1 - \bar{x}_2) - D_0}{\sqrt{\dfrac{s_1^2}{n_1} + \dfrac{s_2^2}{n_2}}}$$

When the sample sizes are *small,* critical values for this statistic are found in Table 4 of Appendix II, using degrees of freedom approximated by the formula

$$\text{d.f.} \approx \frac{\left(\dfrac{s_1^2}{n_1} + \dfrac{s_2^2}{n_2}\right)^2}{\dfrac{\left(\dfrac{s_1^2}{n_1}\right)^2}{(n_1 - 1)} + \dfrac{\left(\dfrac{s_2^2}{n_2}\right)^2}{(n_2 - 1)}}$$

Obviously, this result must be rounded to the nearest integer. In the Minitab package, the TWOSAMPLE *t* command without the subcommand POOLED implements this procedure.

In Section 8.9, we present a procedure for testing an hypothesis concerning the equality of two population variances that can be used to determine whether or not the underlying population variances are equal.

If there is reason to believe that the normality assumptions have been violated, you can test for a shift in location of two population distributions using the nonparametric Mann-Whitney U test of Chapter 16. This test procedure, which requires fewer assumptions concerning the nature of the population probability distributions, is almost as sensitive in detecting a difference in population means when the conditions necessary for the t test are satisfied. It may be more sensitive when the assumptions are not satisfied.

Exercises

Basic Techniques

8.26 Independent random samples of $n_1 = 80$ and $n_2 = 80$ were selected from populations 1 and 2, respectively. The population parameters and the sample means and variances are shown in the accompanying table.

	Population	
Parameters and statistics	**1**	**2**
Population mean	μ_1	μ_2
Population variance	σ_1^2	σ_2^2
Sample size	80	80
Sample mean	11.6	9.7
Sample variance	27.9	38.4

a If your research objective is to show that μ_1 is larger than μ_2, state the alternative and the null hypotheses that you would choose for a statistical test.

b Is the test in part (a) a one- or a two-tailed test?

c Give the test statistic that you would use for the test in parts (a) and (b), and the rejection region for $\alpha = .10$.

d Look at the data. From your intuition, do you think that the data provide sufficient evidence to indicate that μ_1 is larger than μ_2?

e Conduct the test and draw your conclusions. Do the data present sufficient evidence to indicate that $\mu_1 > \mu_2$?

8.27 Refer to Exercise 8.26. Give the observed level of significance for the test.

8.28 Refer to Exercise 8.26.

a Explain the practical conditions that would motivate you to run a two-tailed z test.

b Give the alternative and null hypotheses.

c Use the data of Exercise 8.26 to conduct the test. Do the data provide sufficient evidence to reject H_0 and accept H_a? Test using $\alpha = .05$.

d What practical conclusions can be drawn from the test in part (c)?

8.29 Suppose that you wish to detect a difference between μ_1 and μ_2 (either $\mu_1 > \mu_2$ or $\mu_1 < \mu_2$) and that instead of running a two-tailed test, using $\alpha = .10$, you employ the following test procedure: You wait until you have collected the sample data and have calculated $\bar{x}_1$ and $\bar{x}_2$. If $\bar{x}_1$ is larger than $\bar{x}_2$, you choose the alternative hypothesis $H_a : \mu_1 > \mu_2$ and run a one-tailed test, placing $\alpha_1 = .10$ in the upper tail of the z distribution. If, on the other hand, $\bar{x}_2$ is larger than $\bar{x}_1$, you reverse the procedure and run a one-tailed test, placing $\alpha_2 = .10$ in the lower tail of the z distribution. If you use this procedure and if μ_1 actually equals μ_2, what is the probability α that you will conclude that μ_1 is not equal to μ_2 (i.e., what is the probability α that you will incorrectly reject H_0 when H_0 is true)? This exercise demonstrates why statistical tests should be formulated *prior* to observing the data.

8.30   Independent random samples of $n_1 = n_2 = 4$ observations were selected from two normal populations with equal variances. The data are shown below.

Sample 1	12	9	14	14
Sample 2	11	9	10	8

a   Suppose that you wish to determine whether μ_1 is larger than μ_2. Give the alternative hypothesis for the test.

b   State H_0.

c   Give the rejection region for the test for $\alpha = .10$.

d   Conduct the test and state your conclusions.

e   Give the approximate p-value for the test, and interpret it.

Applications

8.31   To compare the stock-picking abilities of two brokerage firms, we compared the annual gain (excluding brokerage fees) for a $1000 investment in each of 30 stocks listed on each of the two firms' "most recommended" lists of stocks. The means and standard deviations (in dollars) for each of the two samples are shown in the accompanying table. We want to determine whether the data provide sufficient evidence to indicate a difference between the two brokerage firms in the mean return per recommended stock.

	Firm	
Sample Statistic	1	2
Sample size	30	30
Mean	264	199
Standard deviation	157	111

a   State the alternative hypothesis that will best answer this question.

b   State H_0.

c   Give the rejection region for the test for $\alpha = .01$.

d   Conduct the test and state your conclusions.

8.32   In Exercise 7.34, we described a comparison of the mean level of a bank's approved home loan applications from April to May. The sample sizes, means, and standard deviations for the two months are reproduced in the accompanying table.

Sample Statistic	April	May
Sample size	57	66
Mean	$78,100	$82,700
Standard deviation	$6,300	$7,100

a   Do these data provide sufficient evidence to indicate a difference in the mean value of approved home loan applications from April to May? Test using $\alpha = .10$.

b   What difference does it make if you conduct the test in part (a) using $\alpha = .10$ versus $\alpha = .05$? Explain.

8.33   A supermarket chain sampled customer opinions on the service provided by the chain's supermarkets both before and after store personnel were exposed to three weekly ten-minute videotape training sessions that were aimed at the improvement of customer relations. Independent random samples of ten customers each were interviewed before and after the training sessions, and each person was asked to rate the store's service on a scale of 1 (poor) to 10 (excellent). The mean and standard deviation for each sample are shown in the accompanying table. We want to determine whether the data present sufficient evidence to indicate that the training course was effective in increasing customer service scores.

Before	After
$\bar{x}_1 = 6.82$	$\bar{x}_2 = 8.17$
$s_1 = .95$	$s_2 = .56$

a   State the alternative hypothesis that will best answer this question.

b   State H_0.

c   Give the rejection region for the test for $\alpha = .05$.

d   Conduct the test and state the practical conclusions to be derived from the test.

e   Describe the risk that you take in reaching an incorrect conclusion in part (d).

8.34   Refer to Exercise 7.101. The data reflecting the time (in seconds) to load Ami Pro 2.0 on an IBM PS/2 Model 90 486DX/33 personal computer using the Standard Windows and Enhanced Windows programs is reproduced below.

Standard		Enhanced	
1.56	1.20	1.59	.96
1.41	1.38	1.68	1.09
1.48	1.54	1.17	1.26
1.37	1.41	.94	1.23
1.39	1.16	1.56	1.30

If you *cannot* assume that the variances are equal, use an appropriate test of hypothesis to determine if there really is a difference in the average time to load Ami Pro 2.0 using Standard Windows and Enhanced Windows. Use $\alpha = .01$.

8.35   The Minitab printout shown below was run using the data in Exercise 8.34.

```
MTB > TWOSAMPLE C1 C2

TWOSAMPLE T FOR C1 VS C2
          N        MEAN       STDEV      SE MEAN
C1   10        1.390       0.129       0.041
C2   10        1.278       0.259       0.082

95 PCT CI FOR MU C1 - MU C2: ( -0.086,  0.310)

TTEST MU C1 = MU C2 (VS NE): T= 1.22  P=0.24  DF=  13
```

   a What are the null and alternative hypotheses for the test run in Minitab?

   b What is the p-value for the test in part (a)?

   c Based on the p-value, would you infer that there is a significant difference in the mean times to load Ami Pro 2.0 using the Standard Windows versus the Enhanced Windows operating system? Explain.

8.6   A Paired-Difference Test

A manufacturer wanted to compare the wearing qualities of two different types of automobile tires, A and B. In the comparison, a tire of type A and one of type B were randomly assigned and mounted on the rear wheels of each of five automobiles. The automobiles were then operated for a specified number of miles, and the amount of wear was recorded for each tire. These measurements appear in Table 8.6. Do the data present sufficient evidence to indicate a difference in the average wear for the two tire types?

T A B L E 8.6
Tire wear data

Automobile	Tire A	Tire B
1	10.6	10.2
2	9.8	9.4
3	12.3	11.8
4	9.7	9.1
5	8.8	8.3
	$\bar{x}_1 = 10.24$	$\bar{x}_2 = 9.76$

    Analyzing the data, we note that the difference between the two sample means is $(\bar{x}_1 - \bar{x}_2) = .48$, a rather small quantity considering the variability of the data and the small number of measurements involved. At first glance, it would seem that there is little evidence to indicate a difference between the population means, a conjecture that we can check by the method outlined in Section 8.5.

    The pooled estimate of the common variance σ^2 is

$$s^2 = \frac{\sum_{i=1}^{n_1}(x_{1i} - \bar{x}_1)^2 + \sum_{i=1}^{n_2}(x_{2i} - \bar{x}_2)^2}{n_1 + n_2 - 2}$$

$$= \frac{6.932 + 7.052}{5 + 5 - 2}$$

$$= 1.748$$

and

$$s = 1.32$$

The calculated value of t used to test the hypothesis that $\mu_1 = \mu_2$ is

$$t = \frac{\bar{x}_1 - \bar{x}_2}{s\sqrt{\dfrac{1}{n_1} + \dfrac{1}{n_2}}}$$

$$= \frac{10.24 - 9.76}{1.32\sqrt{\dfrac{1}{5} + \dfrac{1}{5}}}$$

$$= .57$$

a value that is not nearly large enough to reject the hypothesis that $\mu_1 = \mu_2$. The corresponding 95% confidence interval is

$$(\bar{x}_1 - \bar{x}_2) \pm t_{\alpha/2}s\sqrt{\frac{1}{n_1} + \frac{1}{n_2}} = (10.24 - 9.76) \pm (2.306)(1.32)\sqrt{\frac{1}{5} + \frac{1}{5}}$$

or -1.45 to 2.41. Note that the interval is quite wide considering the small difference between the sample means.

A second glance at the data reveals a marked inconsistency with this conclusion. We note that the wear measurement for the type A tire is larger than the corresponding value for the type B for *each* of the five automobiles. These differences, recorded as $d = A - B$, are as follows:

Automobile	$d = A - B$
1	.4
2	.4
3	.5
4	.6
5	.5
	$\bar{d} = .48$

If there is no difference in mean tire wear for the two tire types, then the probability that tire A shows more wear than tire B is equal to $p = .5$, and the five automobiles correspond to $n = 5$ independent binomial trials. Let x represent the number of times that the wear measurement for tire type A is larger than that for tire type B. A two-tailed test of the null hypothesis $p = .5$ would include a rejection region consisting of $x = 0$ and $x = 5$ and $\alpha = P(x = 0) + P(x = 5) = 2(1/2)^5 = 1/16 = .0625$. Since five of the differences are positive ($x = 5$), we have evidence to indicate that a difference exists in the mean wear of the two tire types.

You will note that we have used two different statistical tests to test the same hypothesis. Isn't it peculiar that the t test, which uses more information (the actual sample measurements) than the binomial test, fails to supply sufficient evidence for rejection of the hypothesis $\mu_1 = \mu_2$?

There is an explanation for this inconsistency. The t test described in Section 8.5 is not the proper statistical test to be used for our example. The statistical test procedure of Section 8.5 requires that the two samples be *independent and random*. Certainly, the independence requirement was violated by the manner in which the experiment was conducted. The (pair of) measurements, an A tire and a B tire, for a particular automobile are definitely related. A glance at the data shows that the readings have approximately the same magnitude for a particular automobile but vary markedly from one automobile to another. This, of course, is exactly what we might expect. Tire wear is largely determined by driver habits, the balance of the wheels, and the road surface. Since each automobile has a different driver, we would expect a large amount of variability in the data from one automobile to another. When samples are drawn in such a way that an observation in the second sample is related to an observation in the first sample, the samples are said to be *dependent*. For example, recording a person's blood pressure before and after taking medication for hypertension results in observations that are dependent, in the same way that two appraisers using the same criterion arrive at similar and therefore dependent appraised values for the same property.

The familiarity we have gained with interval estimation has shown us that the width of the large-sample confidence intervals depends on the magnitude of the standard deviation of the point estimator of the parameter. The smaller its value, the better is the estimate and the more likely it is that the test statistic will provide evidence to reject the null hypothesis if it is, in fact, false. Knowledge of this phenomena was utilized in *designing* the tire-wear experiment. The experimenter realized that the wear measurements would vary greatly from auto to auto and that this variability could not be separated from the data if the tires were assigned to the ten wheels in a random manner. (A random assignment of the tires would have implied that the data should be analyzed according to the procedure of Section 8.5.) Instead, a comparison of the wear between tire types A and B made for each automobile resulted in the five difference measurements. This design, called a **paired-difference test,** eliminates the effect of the car-to-car variability and yields more information on the mean difference in the wearing quality for the two tire types.

A proper analysis of the data would use the five different measurements to test the hypothesis that the average difference μ_d is equal to 0 or, equivalently, to test the null hypothesis $H_0 : \mu_d = \mu_1 - \mu_2 = 0$ against the alternative hypothesis $H_a : \mu_d = (\mu_1 - \mu_2) \neq 0$.

Paired-Difference Test for $(\mu_1 - \mu_2) = \mu_d$

1   Null Hypothesis: $H_0 : \mu_d = 0$

2   Alternative Hypothesis:

One-Tailed Test	Two-Tailed Test
$H_a : \mu_d > 0$	$H_a : \mu_d \neq 0$
(or $H_a : \mu_d < 0$)	

3   Test Statistic: $t = \dfrac{\bar{d} - 0}{s_d/\sqrt{n}} = \dfrac{\bar{d}}{s_d/\sqrt{n}}$

where n = Number of paired differences

$$s_d = \sqrt{\dfrac{\sum\limits_{i=1}^{n}(d_i - \bar{d})^2}{n-1}}$$

4 Rejection Region:

One-Tailed Test	**Two-Tailed Test**
$t > t_\alpha$	$t > t_{\alpha/2}$ or $t < -t_{\alpha/2}$
(or $t < -t_\alpha$ when the alternative hypothesis is $H_a : \mu_d < 0$)	

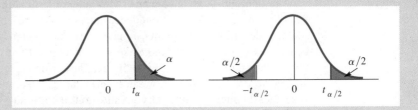

The critical values of t, t_α and $t_{\alpha/2}$, are based on $(n-1)$ degrees of freedom. These tabulated critical values are given in Table 4 of Appendix II.

Assumption: The n paired differences are randomly selected from a normally distributed population.

EXAMPLE **8.13** Do the data in Table 8.6 provide sufficient evidence to indicate a difference in mean wear for tire types A and B? Test using $\alpha = .05$.

Solution You can verify that the mean and standard deviation of the five difference measurements are

$$\bar{d} = .48 \quad \text{and} \quad s_d = .0837$$

Then

$$H_0 : \mu_d = 0 \quad \text{and} \quad H_a : \mu_d \neq 0$$

and

$$t = \frac{\bar{d} - 0}{s_d/\sqrt{n}} = \frac{.48}{.0837/\sqrt{5}} = 12.8$$

The critical value of t for a two-tailed statistical test, with $\alpha = .05$ and four degrees of freedom, is 2.776. Since the observed value of $t = 12.8$ falls far out in the rejection region, there is ample evidence of a difference in the mean amount of wear for tire types A and B.   ▪

You can also construct a $(1 - \alpha)100\%$ confidence interval for $(\mu_1 - \mu_2)$ based on data collected from a paired-difference experiment. The procedure is shown in the display.

$(1 - \alpha)100\%$ Small-Sample Confidence Interval for $(\mu_1 - \mu_2) = \mu_d$ (Based on a Paired-Difference Experiment)

$$\bar{d} \pm t_{\alpha/2}\frac{s_d}{\sqrt{n}}$$

where

$$n = \text{Number of paired differences}$$

and

$$s_d = \sqrt{\frac{\sum_{i=1}^{n}(d_i - \bar{d})^2}{n - 1}}$$

Assumption: The n paired differences are randomly selected from a normally distributed population.

EXAMPLE **8.14**   Find a 95% confidence interval for $(\mu_1 - \mu_2) = \mu_d$ using the data in Table 8.6.

Solution   A 95% confidence interval for the difference between the mean wear is

$$\bar{d} \pm t_{\alpha/2}\frac{s_d}{\sqrt{n}} = .48 \pm (2.776)\frac{.0837}{\sqrt{5}}$$

or $.48 \pm .10$.   ▪

When the units used to compare two or more procedures exhibit marked variability before any experimental procedures are implemented, the effect of this variability can be minimized by comparing the procedures *within* groups of relatively homogeneous units called **blocks.** In this way, the effects of the procedures are not masked by the initial variability among the units in the experiment. An experiment conducted in this manner is called a **randomized block design.** In an experiment involving daily sales, blocks may represent days of the week; in an experiment involving product marketing, blocks may represent geographic areas. (Randomized block designs are discussed in more detail in Section 9.6.)

The statistical design of the tire experiment is a simple example of a randomized block design, and the resulting statistical test is often called a paired-difference test. You will note that the pairing occurred when the experiment was planned and not after the data were collected. Comparisons of tire wear were made within relatively homogeneous blocks (automobiles), with the tire types randomly assigned to the two automobile wheels.

The amount of information gained by blocking the tire experiment can be measured by comparing the calculated confidence interval for the unpaired (and incorrect) analysis with the interval obtained for the paired-difference analysis. The confidence interval for $(\mu_1 - \mu_2)$ that might have been calculated had the tires been randomly assigned to the ten wheels (unpaired) is unknown but probably would have been of the same magnitude as the interval -1.45 to 2.41, which was calculated by analyzing the observed data in an unpaired manner. Pairing the tire types on the automobiles (blocking) and analyzing the resulting differences produced the interval estimate $.38$ to $.58$. Note the difference in the widths of the intervals, which indicates the very sizable increase in information obtained by blocking in this experiment.

Although blocking proved to be very beneficial in the tire experiment, it may not always be. We observe that the degrees of freedom available for estimating σ^2 are fewer for the paired than for the corresponding unpaired experiment. If there were actually no differences among the blocks, the reduction in the degrees of freedom would produce a moderate increase in the value of $t_{\alpha/2}$ employed in the confidence interval and hence would increase the width of the interval. This, of course, did not occur in the tire experiment because the large reduction in the standard error of $\bar{d}$ more than compensated for the loss in degrees of freedom.

Except for notation, the analysis of a paired-difference experiment is the same as that for a single sample presented in Section 8.3. This similarity enables us to use the Minitab commands TTEST and TINTERVAL to analyze the differences in a paired-difference experiment.

Before concluding, we want to reemphasize a point. **Once you have used a paired design for an experiment, you no longer have the option of using the unpaired analysis of Section 8.5. The assumptions on which that test is based have been violated. Your only alternative is to use the correct method of analysis, the paired-difference test (and associated confidence interval) of this section.**

Exercises

Basic Techniques

8.36 A paired-difference experiment was conducted using $n = 10$ pairs of observations. Test the null hypothesis $H_0 : \mu_1 - \mu_2 = 0$ against $H_a : \mu_1 - \mu_2 \neq 0$ for $\alpha = .05$, $\bar{d} = .3$, and $s_d^2 = .16$. Give the approximate p-value for the test.

8.37 Find a 95% confidence interval for $(\mu_1 - \mu_2)$ in Exercise 8.36.

8.38 How many pairs of observations would you need if you wanted to estimate $(\mu_1 - \mu_2)$ in Exercise 8.36 correct to within .1 with probability equal to .95?

8.39   For a paired-difference experiment consisting of $n = 18$ pairs, $\bar{d} = 5.7$ and $s_d^2 = 256$. We wish to detect $\mu_d > 0$.

    a   Give the null and alternative hypotheses for the test.

    b   Conduct the test and state your conclusions.

8.40   For a paired-difference experiment consisting of $n = 12$ pairs of observations, $\bar{d} = .13$ and $s_d^2 = .001$. Find a 90% confidence interval for $(\mu_1 - \mu_2)$.

8.41   A paired-difference experiment was conducted to compare the means of two populations. The data are shown in the accompanying table.

	Pairs				
Population	1	2	3	4	5
1	1.3	1.6	1.1	1.4	1.7
2	1.2	1.5	1.1	1.2	1.8

    a   Do the data provide sufficient evidence to indicate that μ_1 differs from μ_2? Test using $\alpha = .05$.

    b   Find the approximate observed significance level for the test, and interpret its value.

    c   Find a 95% confidence interval for $(\mu_1 - \mu_2)$. Compare your interpretation of the confidence interval with your test results in part (a).

    d   What assumptions must you make in order that your inferences be valid?

8.42   A paired-difference experiment was conducted to compare the means of two populations. The data are shown in the accompanying table.

	Pairs						
Population	1	2	3	4	5	6	7
1	8.9	8.1	9.3	7.7	10.4	8.3	7.4
2	8.8	7.4	9.0	7.8	9.9	8.1	6.9

    a   Do the data provide sufficient evidence to indicate that μ_1 differs from μ_2? Test using $\alpha = .01$.

    b   Find the approximate observed significance level for the test, and interpret its value.

    c   Find a 95% confidence interval for $(\mu_1 - \mu_2)$. Compare your interpretation of the confidence interval with your test results in part (a).

    d   What assumptions must you make in order that your inferences be valid?

Applications

8.43   In response to a complaint that a particular tax assessor (A) was biased, an experiment was conducted to compare the assessor named in the complaint with another tax assessor (B) from the same office. Eight properties were selected, and each was assessed by both assessors. The assessments (in thousands) are shown in the table.

Property	Assessor A	Assessor B
1	36.3	35.1
2	48.4	46.8
3	40.2	37.3
4	54.7	50.6
5	28.7	29.1
6	42.8	41.0
7	36.1	35.3
8	39.0	39.1

a Do the data provide sufficient evidence to indicate that assessor A tends to give higher assessments than assessor B? Test with $\alpha = .05$.

b Estimate the difference in mean assessments for the two assessors.

c What assumptions must you make to render the inferences in (a) and (b) valid?

d Suppose that assessor A had been compared with a more stable standard, say, the average $\bar{x}$ of the assessments given by four assessors selected from the tax office. Thus each property would be assessed by A and also by each of the four other assessors, and $x_A - \bar{x}$ would be calculated. If the test in part (a) is valid, could you use the paired-difference t test to test the hypothesis that the bias, the mean difference between A's assessments and the mean of the assessments of the four other assessors, is equal to zero? Explain.

8.44 A recent drop in the value of the dollar versus foreign currencies is expected to increase the value of U.S. exports. A comparison of the current year's versus last year's shipments (in thousands of cases) for each of six U.S. exporters is shown in the accompanying table. Assume that the exporters represent a random sample selected from among all U.S. exporters.

	Year	
Exporter	Current	Last
1	4.81	4.27
2	5.03	5.97
3	2.38	2.61
4	4.26	3.96
5	5.14	4.86
6	3.93	3.17

a Do the data provide sufficient evidence to indicate an increase in the mean number of cases exported from last year to the current year? Test using $\alpha = .05$.

b Find the approximate p-value for the test.

c Find a 95% confidence interval for the mean increase in the number of cases shipped.

8.45 Attempting to motivate customers to make early payment of bills, the manager of a consulting company offered customers a 2% discount for bills paid within 30 days of issuance of the bill. In order to assess the effect of the new policy on time to payment, the manager randomly sampled 15 accounts and recorded the number of days to payment for the last bill issued under the old system and the first bill issued under the incentive system. The data, in days, are shown in the accompanying table.

Billing	Company														
	1	2	3	4	5	6	7	8	9	10	11	12	13	14	15
Old system	92	88	65	85	95	64	65	62	90	89	65	75	84	90	80
Incentive	28	30	29	85	29	28	26	29	88	30	70	30	27	92	29

a Do the data provide sufficient evidence to indicate that the average payment time is reduced under the incentive system? What is the p-value for the test?

b Find a 95% confidence interval for the decrease in the mean time to payment per account after the incentive was introduced. Interpret the interval.

8.46 Does background music affect the behavior of supermarket shoppers? An experiment designed to answer this question was conducted in a supermarket during the relatively stable summer shopping months. Two days were selected in midweek. One day was randomly assigned to receive no background music. During the second day, slow-tempo background music was played. The daily sales (in dollars) for 12 weeks are shown in the accompanying table. Use the Minitab printout to answer the following questions.

Music	Week											
	1	2	3	4	5	6	7	8	9	10	11	12
None	$14,172	15,485	13,922	12,204	15,501	15,106	14,608	13,946	15,002	14,670	16,202	13,286
Slow tempo	15,917	16,110	13,818	14,709	13,982	16,416	14,727	14,823	14,825	15,949	15,488	14,955

```
TEST OF MU =    0 VS MU N.E.    0
                  N      MEAN    STDEV    SE MEAN      T     P VALUE
        DIFF     12      -635     1154        333   -1.90     0.083

MTB > TINT 90 C3
                  N      MEAN    STDEV    SE MEAN    90.0 PERCENT C.I.
        DIFF     12      -635     1154        333   ( -1233,     -36)
```

a Do the data provide sufficient evidence to indicate a difference in mean daily sales for days when no background music was played versus days when slow-tempo music was played? Test using $\alpha = .10$.

b Find a 90% confidence interval for the difference in mean daily sales for no-music days versus slow-tempo days.

8.7 A Large-Sample Test of Hypothesis for a Binomial Proportion

When a random sample of n identical trials is drawn from a binomial population, the sample proportion $\hat{p}$ has an approximately normal distribution when n is large, with mean p and standard deviation

$$\sigma_{\hat{p}} = \sqrt{\frac{pq}{n}}$$

To test an hypothesis of the form

$$H_0 : p = p_0$$

versus a one- or two-tailed alternative

$$H_a : p \neq p_0 \quad \text{or} \quad H_a : p > p_0 \quad \text{or} \quad H_a : p < p_0$$

the test statistic is constructed using $\hat{p}$, the best estimator of the true population proportion p. This large-sample test is summarized in the following display.

Large-Sample Test for a Population Proportion p

1 Null Hypothesis: $H_0 : p = p_0$
2 Alternative Hypothesis:

One-Tailed Test	Two-Tailed Test
$H_a : p > p_0$	$H_a : p \neq p_0$
(or $H_a : p < p_0$)	

3 Test Statistic: $z = \dfrac{\hat{p} - p_0}{\sigma_{\hat{p}}} = \dfrac{\hat{p} - p_0}{\sqrt{\dfrac{p_0 q_0}{n}}}$, with $\hat{p} = \dfrac{x}{n}$

where x is the number of successes in n binomial trials.[†]

4 Rejection Region:

One-Tailed Test	Two-Tailed Test
$z > z_\alpha$	$z > z_{\alpha/2}$ or $z < -z_{\alpha/2}$
(or $z < -z_\alpha$ when the alternative hypothesis is $H_a : p < p_0$)	

[†]An equivalent test statistic is found by multiplying the numerator and denominator of z by n to obtain

$$z = \frac{x - np_0}{\sqrt{np_0 q_0}}$$

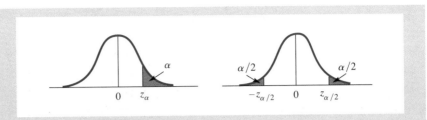

Assumptions: The sampling satisfies the assumptions of a binomial experiment (Section 4.2), and n is large enough so that the sampling distribution of $\hat{p}$ can be approximated by a normal distribution. The interval $p \pm 2\sigma_{\hat{p}}$ must be contained in the interval 0 to 1.

E X A M P L E   **8.15**   Approximately 1 in 10 consumers favor cola brand A. After a promotional campaign in a given sales region, 200 cola drinkers were randomly selected from consumers in the market area and were interviewed to determine the effectiveness of the campaign. The result of the survey showed that a total of 26 people expressed a preference for cola brand A. Do these data present sufficient evidence to indicate an increase in the acceptance of brand A in the region?

Solution   We assume that the sample satisfies the requirements of a binomial experiment. The question can be answered by testing the hypothesis

$$H_0 : p = .10$$

against the alternative

$$H_a : p > .10$$

A one-tailed test is used because we want to detect whether the value of p is greater than .1. (The promotional campaign has caused an *increase* in the value of p.)

The point estimator of p is $\hat{p} = x/n$, and the test statistic is

$$z = \frac{\hat{p} - p_0}{\sqrt{p_0 q_0/n}}$$

When H_0 is true, the value of p is $p_0 = .1$, and the sampling distribution of $\hat{p}$ has a mean equal to p_0 and a standard deviation of $\sqrt{p_0 q_0/n}$. **Hence, $\sqrt{\hat{p}\hat{q}/n}$ is not used to estimate the standard error of $\hat{p}$ in this case because the test statistic is calculated under the assumption that H_0 is true.** (When estimating the value of p using the estimator $\hat{p}$, the standard error of $\hat{p}$ is not known and is *estimated* by $\sqrt{\hat{p}\hat{q}/n}$.)

With $\alpha = .05$, we would reject H_0 when $z > 1.645$ (Figure 8.13). With $\hat{p} = 26/200 = .13$, the value of the test statistic is

$$z = \frac{\hat{p} - p_0}{\sqrt{\dfrac{p_0 q_0}{n}}} = \frac{.13 - .10}{\sqrt{\dfrac{(.10)(.90)}{200}}} = 1.41$$

FIGURE **8.13**
Location of the rejection region
in Example 8.15

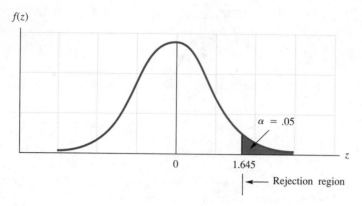

The p-value associated with this test is

$$p\text{-value} = P(z > 1.41) = .5000 - .4207 = .0793$$

The calculated value of the test statistic does not fall in the rejection region: the p-value is .0793, and hence *we do not reject* H_0.

Do we accept H_0? No, not until we have stated alternative values of p different from $p_0 = .1$ that are of *practical significance*. The probability of a type II error should be calculated using these alternative values. If β is sufficiently small, we would accept H_0 with full awareness of the risk of an erroneous decision. ∎

Tips on Problem Solving

When testing an hypothesis concerning p, use p_0 (not $\hat{p}$) to calculate $\sigma_{\hat{p}}$ in the denominator of the z statistic. The reason for this is that the rejection region is determined by the distribution of $\hat{p}$ when the null hypothesis is true, namely, when $p = p_0$.

Examples 8.4 and 8.15 illustrate an important point. **If the data present sufficient evidence to reject H_0, the probability of an erroneous conclusion α is known in advance because α is used in locating the rejection region. Since α is usually small, we are fairly certain that we have made a correct decision.** On the other hand, if the data present insufficient evidence to reject H_0, the conclusions are not so obvious. Ideally, following the statistical test procedure outlined in Section 8.2, we would have specified a practically significant alternative p_a in advance and chosen n such that β would be small. Unfortunately, many experiments are not conducted in this ideal manner. Someone chooses a sample size, and the experimenter or statistician is left to evaluate the evidence.

The calculation of β is not too difficult for the statistical test procedure outlined in this section but may be extremely difficult, if not beyond the capability of the beginner, in other test situations. **A much simpler procedure is to *not reject* H_0**

rather than to accept it, and then to estimate using a confidence interval. The interval will give you a range of plausible values for p.

Exercises

Basic Techniques

8.47   A random sample of $n = 1000$ observations from a binomial population produced $x = 279$.

    a   If your research hypothesis is that p is less than .3, what should you choose for your alternative hypothesis? your null hypothesis?

    b   Does your alternative hypothesis in part (a) imply a one- or two-tailed statistical test? Explain.

    c   Do the data provide sufficient evidence to indicate that p is less than .3? Test using $\alpha = .05$.

8.48   A random sample of $n = 2000$ observations from a binomial population produced $x = 1238$.

    a   If your research hypothesis is that p is greater than .6, what should you choose for your alternative hypothesis? your null hypothesis?

    b   Does your alternative hypothesis in part (a) imply a one- or two-tailed statistical test? Explain.

    c   Do the data provide sufficient evidence to indicate that p is greater than .6? Test using $\alpha = .05$.

8.49   A random sample of 120 observations was selected from a binomial population, and 72 successes were observed. Do the data provide sufficient evidence to indicate that p is larger than .5? Test using $\alpha = .05$.

8.50   Refer to Exercise 8.49. What is the p-value for the test? Does the p-value confirm your conclusion in Exercise 8.49? Explain.

Applications

8.51   In the survey of the 1993 freshman class reported in Exercise 6.43 ("Freshman Statistics," 1994), 86% of freshmen in the survey received financial aid from their parents or family. In a similar survey, a sample of $n = 1000$ college freshmen chosen at random from this year's freshman class revealed that 89% of the students in the survey received financial aid from their parents or family.

    a   Assume that the 1993 percentage, 86%, was in fact the population value and use a test of hypothesis with a .05 level of significance to determine whether this year's percentage indicates that the figure has changed over time and is no longer correct.

    b   What is the p-value associated with this test? Does your conclusion using the p-value agree with the results of the test in part (a)?

8.52   A check-cashing service has found that approximately 5% of all checks submitted to the service for cashing are bad. After instituting a check verification system to reduce its losses, the service found that only 45 checks were bad in a total of 1124 cashed.

    a   If you wish to conduct a statistical test to determine whether the check verification system reduces the probability that a bad check will be cashed, what should you choose for the alternative hypothesis? the null hypothesis?

    b   Does your alternative hypothesis in part (a) imply a one- or a two-tailed test? Explain.

    c   Noting the data, what does your intuition tell you? Do you think that the check verification system is effective in reducing the proportion of bad checks that were cashed?

    d   Conduct a statistical test of the null hypothesis in part (a), and state your conclusions. Test using $\alpha = .05$. Do the test conclusions agree with your intuition in part (c)?

8.53 From past experience, an appliance dealer has found that 10% of her customers who buy on installment pay off their bills before the last (the 24th) monthly installment is due. Suspecting an increase in this percentage, the dealer surveyed 200 installment buyers concerning their intentions. Of these, 33 stated that they planned to pay off their debt before the last installment. Do the data provide sufficient evidence to indicate that the percentage of installment buyers who will pay off their debt before the last installment exceeds 10%?

a State the alternative hypothesis for the test.

b State the null hypothesis.

c Give the rejection region for $\alpha = .05$.

d Conduct the test and state your conclusions.

8.54 A publisher of a news magazine had found through past experience that 60% of its subscribers renew their subscriptions. Because it was heading into a business recession, the company decided to randomly select a small sample of subscribers and, via telephone questioning, determine whether they planned to renew their subscriptions. One hundred eight of a sample of 200 indicated that they planned to renew their subscriptions.

a If you want to detect whether the data provide sufficient evidence of a reduction in the proportion p of all subscribers who will renew, what will you choose for your alternative hypothesis? your null hypothesis?

b Conduct the test using $\alpha = .05$. State the results.

c Find a 95% confidence interval for p.

d How many subscribers would have to be included in the publisher's sample in order to estimate p to within .01, with 95% confidence?

8.8 A Large-Sample Test of Hypothesis for the Difference Between Two Binomial Proportions

When the focus of an experiment or study is the difference in the proportion of individuals or items possessing a specified characteristic, the pivotal statistic for testing hypotheses about $p_1 - p_2$ is the difference in the sample proportions $\hat{p}_1 - \hat{p}_2$. The formal testing procedure concerning the difference in population proportions is given in the following display.

A Large-Sample Statistical Test for $(p_1 - p_2)$

1 Null Hypothesis: $H_0 : (p_1 - p_2) = D_0$ where D_0 is some specified difference that you wish to test. For many tests, you will wish to hypothesize that there is no difference between p_1 and p_2; that is,

$$D_0 = 0$$

2 Alternative Hypothesis:

One-Tailed Test	Two-Tailed Test
$H_a : (p_1 - p_2) > D_0$	$H_a : (p_1 - p_2) \neq D_0$
[or $H_a : (p_1 - p_2) < D_0$]	

3 In selecting the appropriate test statistic, we begin with

$$z = \frac{(\hat{p}_1 - \hat{p}_2) - D_0}{\sigma_{(\hat{p}_1 - \hat{p}_2)}} = \frac{(\hat{p}_1 - \hat{p}_2) - D_0}{\sqrt{\dfrac{p_1 q_1}{n_1} + \dfrac{p_2 q_2}{n_2}}}$$

where $\hat{p}_1 = x_1/n_1$ and $\hat{p}_2 = x_2/n_2$. Since p_1 and p_2 are unknown, we will need to approximate their values in order to calculate the standard deviation of $\hat{p}_1 - \hat{p}_2$ that appears in the denominator of the z statistic. Approximations are available for two cases.

Case I: If we hypothesize that p_1 equals p_2, that is,

$$H_0 : p_1 = p_2$$

or equivalently that

$$p_1 - p_2 = 0$$

then $p_1 = p_2 = p$ and the best estimate of p is obtained by pooling the data from both samples. Thus, if x_1 and x_2 are the numbers of successes obtained from the two samples,

$$\hat{p} = \frac{x_1 + x_2}{n_1 + n_2}$$

The test statistic would be

$$z = \frac{(\hat{p}_1 - \hat{p}_2) - 0}{\sqrt{\dfrac{\hat{p}\hat{q}}{n_1} + \dfrac{\hat{p}\hat{q}}{n_2}}} \qquad \text{or} \qquad z = \frac{\hat{p}_1 - \hat{p}_2}{\sqrt{\hat{p}\hat{q}\left(\dfrac{1}{n_1} + \dfrac{1}{n_2}\right)}}$$

Case II: On the other hand, if we hypothesize that D_0 is *not* equal to zero, that is,

$$H_0 : (p_1 - p_2) = D_0$$

where $D_0 \neq 0$, then the best estimates of p_1 and p_2 are $\hat{p}_1$ and $\hat{p}_2$, respectively. The test statistic would be

$$z = \frac{(\hat{p}_1 - \hat{p}_2) - D_0}{\sqrt{\dfrac{\hat{p}_1 \hat{q}_1}{n_1} + \dfrac{\hat{p}_2 \hat{q}_2}{n_2}}}$$

4 Rejection Region:

One-Tailed Test	**Two-Tailed Test**
$z > z_\alpha$	$z > z_{\alpha/2}$ or $z < -z_{\alpha/2}$
[or $z < -z_\alpha$ when the alternative hypothesis is $H_a : (p_1 - p_2) < D_0$]	

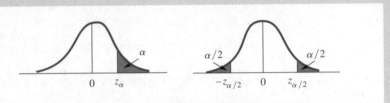

Assumptions: Samples were selected in a random and independent manner from two binomial populations, and n_1 and n_2 are large enough so that the sampling distribution of $\hat{p}_1 - \hat{p}_2$ can be approximated by a normal distribution. The interval $(p_1 - p_2) \pm 2\sigma_{(\hat{p}_1 - \hat{p}_2)}$ must be contained in the interval -1 to 1.

EXAMPLE **8.16**   A hospital administrator suspects that the delinquency rate in the payment of hospital bills has increased over the past year. Hospital records show that the bills of 48 of 1284 persons admitted in the month of April have been delinquent for more than 90 days. This number compares with 34 of 1002 persons admitted during the same month one year ago. Do these data provide sufficient evidence to indicate an increase in the rate of delinquency in payments exceeding 90 days? Test using $\alpha = .10$.

Solution   Let p_1 and p_2 represent the proportions of all potential hospital admissions in April of this year and last year, respectively, that would have allowed their accounts to be delinquent for a period exceeding 90 days, and let the $n_1 = 1284$ admissions this year and the $n_2 = 1002$ admissions last year represent independent random samples from these populations. Since we want to detect an increase in the delinquency rate, if it exists, we will test the null hypothesis

$$H_0 : p_1 = p_2, \qquad \text{that is, } p_1 - p_2 = D_0 = 0$$

against the one-sided alternative hypothesis

$$H_a : p_1 > p_2, \qquad \text{that is, } p_1 - p_2 > 0$$

To conduct this test, we will use the z test statistic and approximate the value of $\sigma_{(\hat{p}_1 - \hat{p}_2)}$ by using the pooled estimate of p described in Case I. Since H_a implies a one-tailed test, we will reject H_0 only for large values of z. Thus for $\alpha = .10$ we will reject H_0 if $z > 1.28$ (see Figure 8.14).

FIGURE 8.14
Location of the rejection region
in Example 8.16

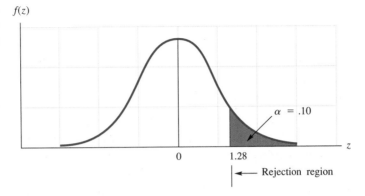

The estimates of p_1 and p_2 are

$$\hat{p}_1 = \frac{48}{1284} = .0374 \qquad \text{and} \qquad \hat{p}_2 = \frac{34}{1002} = .0339$$

The pooled estimate of p required for $\sigma_{(\hat{p}_1 - \hat{p}_2)}$ is

$$\hat{p} = \frac{x_1 + x_2}{n_1 + n_2} = \frac{48 + 34}{1284 + 1002} = .0359$$

The test statistic is

$$z = \frac{\hat{p}_1 - \hat{p}_2}{\sqrt{\hat{p}\hat{q}\left(\dfrac{1}{n_1} + \dfrac{1}{n_2}\right)}} = \frac{.0374 - .0339}{\sqrt{(.0359)(.9641)\left(\dfrac{1}{1284} + \dfrac{1}{1002}\right)}} = .45$$

The p-value for this test is

$$p\text{-value} = P(z > .45) = .5000 - .1736 = .3264$$

Since the computed value of z does not fall in the rejection region and the p-value is fairly large, we cannot reject the null hypothesis that $p_1 = p_2$. The data present insufficient evidence to indicate that the proportion of delinquent accounts in April of this year exceeds the corresponding proportion last year. ▪

Exercises

Basic Techniques

8.55   Independent random samples of $n_1 = 140$ and $n_2 = 140$ observations were randomly selected from binomial populations 1 and 2, respectively. The number of successes in the samples and the population parameters are shown in the accompanying table.

| | Population | |
Statistics and Parameters	1	2
Sample size	140	140
Number of successes	74	81
Binomial parameter	p_1	p_2

a   Suppose that you have no preconceived theory concerning which parameter, p_1 or p_2, is the larger and that you only wish to detect a difference between the two parameters, if it exists. What should you choose for the alternative hypothesis for a statistical test? the null hypothesis?

b   Does your alternative hypothesis in part (a) imply a one- or a two-tailed test?

c   Conduct the test and state your conclusions. Test using $\alpha = .05$.

8.56   Refer to Exercise 8.55. Suppose that for practical reasons you know that p_1 cannot be larger than p_2.

a   Given this knowledge, what should you choose as the alternative hypothesis for your statistical test? your null hypothesis?

b   Will your alternative hypothesis in part (a) imply a one- or a two-tailed test? Explain.

c   Conduct the test and state your conclusions. Test using $\alpha = .10$.

8.57   Independent random samples of $n_1 = 280$ and $n_2 = 350$ observations were randomly selected from binomial populations 1 and 2, respectively. The number of successes in the samples and the population parameters are shown in the accompanying table.

| | Population | |
Statistics and Parameters	1	2
Sample size	280	350
Number of successes	132	178
Binomial parameter	p_1	p_2

a   Suppose that you know that p_1 can never be larger than p_2, and you want to know if p_1 is less than p_2. What should you choose for your null and alternative hypotheses?

b   Does your alternative hypothesis in part (a) imply a one- or a two-tailed test?

c   Conduct the test and state your conclusions. Test using $\alpha = .05$.

Applications

8.58   A manufacturer modified a production line to reduce the mean fraction defective. To determine whether the modification was effective, the manufacturer randomly sampled 400 items before modification of the production line and 400 items after modification. The percentage defectives in the samples were

before: 5.25%
after: 3.5%

a   If the modification could not possibly increase the fraction defective, what should the manufacturer choose for an alternative hypothesis? the null hypothesis?

b   Conduct the test using $\alpha = .05$. Interpret the results.

8.59   In a poll (Gallup, 1994, p. 135) dealing with honesty and ethical standards, $n = 1000$ Americans were asked to rate the honesty and ethical standards of people in various fields as being very high, high, average, low, or very low. Although around 50% of those sampled rated people in most fields as average, over 50% rated druggists, pharmacists, doctors, and clergy as very high or high. In contrast, only 20% rated business executives as very high or high in 1993, up from 18% in 1992.

    a   If both polls involved 1000 Americans, determine whether this change reflects a basic increase in the rating for business executives. Use $\alpha = .05$.

    b   Does your conclusion using the p-value associated with your test agree with your findings in part (a)?

 8.60   In a poll regarding the amount of federal taxes Americans have to pay, 49% of males and 60% of females considered their taxes too high (Gallup, 1994, p. 69).

    a   If the respondents included 500 males and 500 females, can we conclude that there is a significant difference in the proportions of males and females who consider their taxes too high? Use $\alpha = .01$.

    b   Does the p-value support your conclusion in part (a)?

 8.61   Despite the fact that business travelers are the airlines' biggest customers, they pay the highest prices. Business travelers often plan trips on short notice and are likely to rearrange travel plans at the last minute. As a result, they generally cannot take advantage of discount fares. In their study of corporate air travel costs, Richard J. Fox and Frederick J. Stephenson (1990) reported that 45% of 56 big user firms (having annual travel costs of at least $10 million) and 76% of 146 small users (having annual travel costs less than $10 million) have established in-house travel agencies.

    a   Test for a significant difference in the proportion of big and small users that have established in-house travel agencies. Use $\alpha = .01$.

    b   Construct a 99% confidence interval estimate for the difference in the proportion of big and small users that have established in-house travel agencies. Does this estimate confirm the conclusion in part (a)?

8.62   According to a report by the American Cancer Society ("Profile of Smokers," 1990), more men than women smoke and twice as many smokers as nonsmokers die prematurely. In random samples of 200 males and 200 females, 62 of the males and 54 of the females surveyed were smokers.

    a   Is there sufficient evidence to conclude that the proportion of male smokers differs from the proportion of female smokers?

    b   What is the p-value for the test in part (a)?

8.9   Inferences Concerning Population Variances

We have seen in the preceding sections that an estimate of the population variance σ^2 is fundamental to procedures for making inferences about population means. Moreover, there are many practical situations where σ^2 is the primary objective of an experimental investigation; thus it may assume a position of far greater importance than that of the population mean.

    Scientific measuring instruments must provide unbiased readings with a very small error of measurement. An aircraft altimeter that measured the correct altitude on the *average* would be of little value if the standard deviation of the error of measurement were 5000 feet. Indeed, bias in a measuring instrument can often be corrected, but the precision of the instrument, measured by the standard deviation of the error of measurement, is usually a function of the design of the instrument itself and cannot be controlled.

Machined parts in a manufacturing process must be processed with minimum variability in order to reduce out-of-size and hence defective products. And, in general, it is desirable to maintain a minimum variance in the measurements of the quality characteristics of an industrial product in order to achieve process control and thereby minimize the percentage of poor-quality product.

The sample variance

$$s^2 = \frac{\sum_{i=1}^{n}(x_i - \bar{x})^2}{n - 1}$$

is an unbiased estimator of the population variance σ^2. Thus the distribution of sample variances generated by repeated sampling will have a probability distribution that commences at $s^2 = 0$ (since s^2 cannot be negative) with a mean equal to σ^2. Unlike the distribution of $\bar{x}$, the distribution of s^2 is nonsymmetrical, the exact form being dependent upon the probability distribution of the population.

For the methodology that follows, we will assume that the sample is drawn from a normal population and that s^2 is based on a random sample of n measurements and possesses $(n - 1)$ degrees of freedom. The next and obvious step is to consider the distribution of s^2 in repeated sampling from a specified normal distribution—one with a specific mean and variance—and to tabulate the critical value of s^2 for some of the commonly used tail areas. If this is done, we will find that the sampling distribution of s^2 is independent of the population mean μ but possesses a different distribution for each sample size and each value of σ^2. This task is quite laborious, but fortunately it can be simplified by *standardizing,* as was done by using z in the normal tables.

The quantity

$$\chi^2 = \frac{(n - 1)s^2}{\sigma^2}$$

called a **chi-square variable** by statisticians (χ is the Greek letter chi), admirably suits our purposes. Its distribution in repeated sampling is called, as we might suspect, a **chi-square probability distribution.** The equation of the density function for the chi-square distribution is well known to statisticians who have tabulated critical values corresponding to various tail areas of the distribution. These values are presented in Table 5 of Appendix II.

The shape of the chi-square distribution, like that of the t distribution, will vary with the sample size or, equivalently, with the degrees of freedom associated with s^2. Thus Table 5 in Appendix II is constructed in exactly the same manner as the t table, with the degree of freedom shown in the first and last columns. A partial reproduction of Table 5 in Appendix II is shown in Table 8.7. The symbol χ_a^2 indicates that the tabulated χ^2 value is such that an area a lies to its right. (See Figure 8.15.) Stated in probabilistic terms,

$$P(\chi^2 > \chi_a^2) = a$$

T A B L E **8.7**　Format of the chi-square table, Table 5 in Appendix II

d.f.	$\chi^2_{0.995}$	$\cdots$	$\chi^2_{0.950}$	$\chi^2_{0.900}$	$\chi^2_{0.100}$	$\chi^2_{0.050}$	$\cdots$	$\chi^2_{0.005}$	d.f.
1	0.0000393		0.0039321	0.0157908	2.70554	3.84146		7.87944	1
2	0.0100251		0.102587	0.210720	4.60517	5.99147		10.5966	2
3	0.0717212		0.351846	0.584375	6.25139	7.81473		12.8381	3
4	0.206990		0.710721	1.063623	7.77944	9.48773		14.8602	4
5	0.411740		1.145476	1.61031	9.23635	11.0705		16.7496	5
6	0.675727		1.63539	2.20413	10.6446	12.5916		18.5476	6
$\vdots$	$\vdots$		$\vdots$	$\vdots$	$\vdots$	$\vdots$		$\vdots$	$\vdots$
15	4.60094		7.26094	8.54675	22.3072	24.9958		21.8013	15
16	5.14224		7.96164	9.31223	23.5418	26.2962		34.2672	16
17	5.69724		8.67176	10.0852	24.7690	27.5871		35.7185	17
18	6.26481		9.39046	10.8649	25.9894	28.8693		37.1564	18
19	6.84398		10.1170	11.6509	27.2036	30.1435		38.5822	19
$\vdots$	$\vdots$		$\vdots$	$\vdots$	$\vdots$	$\vdots$		$\vdots$	$\vdots$

Thus 99% of the area under the χ^2 distribution lies to the right of $\chi^2_{.99}$. Note that the extreme values of χ^2 must be tabulated for both the lower and the upper tails of the distribution because it is nonsymmetrical.

F I G U R E **8.15**
A chi-square distribution

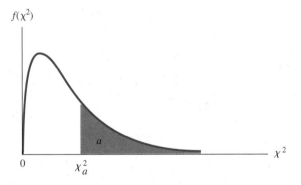

You can check your ability to use the table by verifying the following statements. The probability that χ^2, based upon $n = 16$ measurements (d.f. $= 15$), will exceed 24.9958 is .05. For a sample of $n = 6$ measurements (d.f. $= 5$), 95% of the area under the χ^2 distribution will lie to the right of $\chi^2 = 1.145476$. These values of χ^2 are boxed in Table 8.7.

The statistical test of a null hypothesis concerning a population variance,

$$H_0 : \sigma^2 = \sigma_0^2$$

will employ the test statistic

$$\chi^2 = \frac{(n-1)s^2}{\sigma_0^2}$$

If σ^2 is really greater than the hypothesized value σ_0^2, then the test statistic will tend to be large and will likely fall toward the upper tail of the distribution. If $\sigma^2 < \sigma_0^2$, the test statistic will tend to be small and will likely fall toward the lower tail of the χ^2 distribution. As in the other statistical tests, we can use either a one- or two-tailed statistical test, depending upon the alternative hypothesis that we choose. The procedure for the chi-square test is given in the display.

Test of an Hypothesis Concerning a Population Variance

1 Null Hypothesis: $H_0 : \sigma^2 = \sigma_0^2$

2 Alternative Hypothesis:

One-Tailed Test	**Two-Tailed Test**
$H_a : \sigma^2 > \sigma_0^2$	$H_a : \sigma^2 \neq \sigma_0^2$
(or $H_a : \sigma^2 < \sigma_0^2$)	

3 Test Statistic: $\chi^2 = \dfrac{(n-1)s^2}{\sigma_0^2}$

4 Rejection Region:

One-Tailed Test

$\chi^2 > \chi_\alpha^2$
(or $\chi^2 < \chi_{(1-\alpha)}^2$ when the alternative hypothesis is $H_a : \sigma^2 < \sigma_0^2$), where χ_α^2 and $\chi_{(1-\alpha)}^2$ are, respectively, the upper- and lower-tail values of χ^2 that place α in the tail areas

Two-Tailed Test

$\chi^2 > \chi_{\alpha/2}^2$ or $\chi^2 < \chi_{(1-\alpha/2)}^2$, where $\chi_{\alpha/2}^2$ and $\chi_{(1-\alpha/2)}^2$ are, respectively, the upper- and lower-tail values of χ^2 that place $\alpha/2$ in the tail areas

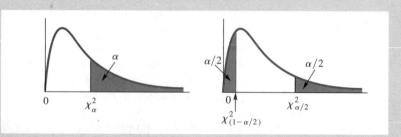

The critical values of χ^2 are based on $(n-1)$ degrees of freedom. These tabulated values are given in Table 5 of Appendix II.

> *Assumption:* The sample has been randomly selected from a normal population.

E X A M P L E 8.17   A cement manufacturer claimed that concrete prepared from his product would possess a relatively stable compressive strength and that the strength measured in kilograms per square centimeter would lie within a range of 40 kilograms per square centimeter. A sample of $n = 10$ measurements produced a mean and variance equal to, respectively,

$$\bar{x} = 312 \quad \text{and} \quad s^2 = 195$$

Do these data present sufficient evidence to reject the manufacturer's claim?

Solution   As stated, the manufacturer claimed that the range of the strength measurements would equal 40 kilograms per square centimeter. We will suppose that he meant that the measurements would lie within this range 95% of the time and, therefore, that the range would equal approximately 4σ and that $\sigma = 10$. We then want to test the null hypothesis

$$H_0 : \sigma^2 = (10)^2 = 100$$

against the alternative

$$H_a : \sigma^2 > 100$$

The alternative hypothesis requires a one-tailed statistical test, with the entire rejection region located in the upper tail of the χ^2 distribution. The critical value of χ^2 for $\alpha = .05$ and $(n - 1) = 9$ degrees of freedom is $\chi^2 = 16.9190$, which implies that we will reject H_0 if the test statistic exceeds this value.

Calculating, we obtain

$$\chi^2 = \frac{(n-1)s^2}{\sigma_0^2} = \frac{1755}{100} = 17.55$$

Since the value of the test statistic falls in the rejection region, we conclude that the null hypothesis is false and that the range of concrete strength measurements will exceed the manufacturer's claim. ∎

E X A M P L E 8.18   Find the approximate observed significance level for the test in Example 8.17.

Solution   Examining the row corresponding to 9 degrees of freedom in Table 5 in Appendix II, you will see that the observed value of chi-square, $\chi^2 = 17.55$, is larger than the tabulated value, $\chi^2_{.05} = 16.9190$, and less than $\chi^2_{.025} = 19.0228$. Therefore, the observed significance level (p-value) for the test lies between .025 and .05. We would report the observed significance level for the test as $.025 < p\text{-value} < .05$. This tells us that we would reject the null hypothesis for any value of α equal to .05 or larger. ∎

A confidence interval for σ^2 with a $(1 - \alpha)100\%$ confidence coefficient is given in the next display.

A $(1 - \alpha)100\%$ Confidence Interval for σ^2

$$\frac{(n-1)s^2}{\chi^2_{\alpha/2}} < \sigma^2 < \frac{(n-1)s^2}{\chi^2_{(1-\alpha/2)}}$$

where $\chi^2_{\alpha/2}$ and $\chi^2_{(1-\alpha/2)}$ are the upper and lower χ^2 values, respectively, that would locate one-half of α in each tail of the chi-square distribution.

Assumption: The sample has been randomly selected from a normal population.

E X A M P L E   **8.19**   Find a 90% confidence interval for σ^2 in Example 8.17.

Solution   The tabulated values of $\chi^2_{.95}$ and $\chi^2_{.05}$ corresponding to $(n-1) = 9$ degrees of freedom are

$$\chi^2_{(1-\alpha/2)} = \chi^2_{.95} = 3.32511 \qquad \text{and} \qquad \chi^2_{\alpha/2} = \chi^2_{.05} = 16.9190$$

Substituting these values and $s^2 = 195$ into the formula for the confidence interval,

$$\frac{(n-1)s^2}{\chi^2_{\alpha/2}} < \sigma^2 < \frac{(n-1)s^2}{\chi^2_{(1-\alpha/2)}}$$

yields the interval estimate for σ^2.

$$\frac{9(195)}{16.9190} < \sigma^2 < \frac{9(195)}{3.32511} \qquad \text{or} \qquad 103.73 < \sigma^2 < 527.80 \quad ∎$$

Comparing Two Population Variances

The need for statistical methods to compare two population variances is readily apparent from this discussion. We may frequently wish to compare the precision of one measuring device with that of another, the stability of one manufacturing process with that of another, or even the variability in the grading procedure of one college professor with that of another.

Intuitively, we might compare two population variances, σ_1^2 and σ_2^2, using the ratio of the sample variances s_1^2/s_2^2. If s_1^2/s_2^2 is nearly equal to 1, we would find little evidence to indicate that σ_1^2 and σ_2^2 are unequal. On the other hand, a very large or very small value for s_1^2/s_2^2 would provide evidence of a difference in the population variances.

How large or small must s_1^2/s_2^2 be in order that sufficient evidence exists to reject the null hypothesis $H_0 : \sigma_1^2 = \sigma_2^2$? The answer to this question can be acquired by studying the sampling distribution of s_1^2/s_2^2.

When independent random samples are drawn from two normal populations with equal variances—that is, $\sigma_1^2 = \sigma_2^2$—then s_1^2/s_2^2 possesses a sampling distribution that is known to statisticians as an **F distribution.** We need not concern ourselves with the equation of the probability distribution for F. It is well known, and the critical values have been tabulated. These values appear in Table 6 of Appendix II.

The shape of the F distribution is nonsymmetrical and will depend on the number of degrees of freedom associated with the numerator and denominator of $F = s_1^2/s_2^2$. We will represent these quantities as $v_1 = n_1 - 1$ and $v_2 = n_2 - 1$, respectively. This fact complicates the tabulation of critical values for the F distribution and necessitates the construction of a table to accommodate differing values of v_1, v_2, and a. (See Figure 8.16.)

FIGURE **8.16**
An F distribution with
$v_1 = 10$ and $v_2 = 10$

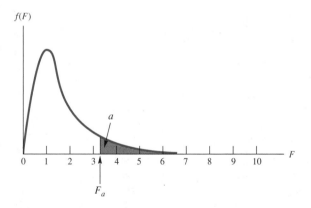

In Table 6 of Appendix II, critical values of F for right-tailed areas corresponding to $a = .10$, .05, .025, .010, and .005 are tabulated for various combinations of v_1 numerator degrees of freedom and v_2 denominator degrees of freedom. A portion of Table 6 is reproduced in Table 8.8. The numerator degrees of freedom v_1 are listed across the top margin, and the denominator degrees of freedom v_2 are listed along the left side margin. The values of a are listed in the second column from the left as well as the second column from the right. For a fixed combination of v_1 and v_2, the appropriate critical values of F are found in the line indexed by the value of a required.

Referring to Table 8.8, we note that $F_{.05}$ for sample sizes $n_1 = 7$ and $n_2 = 10$ (i.e., $v_1 = 6$, $v_2 = 9$) is 3.37. Similarly, the critical value $F_{.05}$ for sample sizes $n_1 = 9$ and $n_2 = 12(v_1 = 8$, $v_2 = 11)$ is 2.95. These values of F are boxed in Table 8.8.

T A B L E **8.8**   Format of the F table from Table 6 in Appendix II

v_2	a	1	2	3	4	5	6	7	8	9
						v_1				
1	.100	39.86	49.50	53.59	55.83	57.24	58.20	58.91	59.44	59.86
	.050	161.4	199.5	215.7	224.6	230.2	234.0	236.8	238.9	240.5
	.025	647.8	799.5	864.2	899.6	921.8	937.1	948.2	956.7	963.3
	.010	4052	4999.5	5403	5625	5764	5859	5928	5982	6022
	.005	16211	20000	21615	22500	23056	23437	23715	23925	24091
2	.100	8.53	9.00	9.16	9.24	9.29	9.33	9.35	9.37	9.38
	.050	18.51	19.00	19.16	19.25	19.30	19.33	19.35	19.37	19.38
	.025	38.51	39.00	39.17	39.25	39.30	39.33	39.36	39.37	39.39
	.010	98.50	99.00	99.17	99.25	99.30	99.33	99.36	99.37	99.39
	.005	198.5	199.0	199.2	199.2	199.3	199.3	199.4	199.4	199.4
3	.100	5.54	5.46	5.39	5.34	5.31	5.28	5.27	5.25	5.24
	.050	10.13	9.55	9.28	9.12	9.01	8.94	8.89	8.85	8.81
	.025	17.44	16.04	15.44	15.10	14.88	14.73	14.62	14.54	14.47
	.010	34.12	30.82	29.46	28.71	28.24	27.91	27.67	27.49	27.35
	.005	55.55	49.80	47.47	46.19	45.39	44.84	44.43	44.13	43.88
⋮	⋮	⋮	⋮		⋮		⋮			⋮
9	.100	3.36	3.01	2.81	2.69	2.61	2.55	2.51	2.47	2.44
	.050	5.12	4.26	3.86	3.63	3.48	3.37	3.29	3.23	3.18
	.025	7.21	5.71	5.08	4.72	4.48	4.32	4.20	4.10	4.03
	.010	10.56	8.02	6.99	6.42	6.06	5.80	5.61	5.47	5.35
	.005	13.61	10.11	8.72	7.96	7.47	7.13	6.88	6.69	6.54
10	.100	3.29	2.92	2.73	2.61	2.52	2.46	2.41	2.38	2.35
	.050	4.96	4.10	3.71	3.48	3.33	3.22	3.14	3.07	3.02
	.025	6.94	5.46	4.83	4.47	4.24	4.07	3.95	3.85	3.78
	.010	10.04	7.56	6.55	5.99	5.64	5.39	5.20	5.06	4.94
	.005	12.83	9.43	8.08	7.34	6.87	6.54	6.30	6.12	5.97
11	.100	3.23	2.86	2.66	2.54	2.45	2.39	2.34	2.30	2.27
	.050	4.84	3.98	3.59	3.36	3.20	3.09	3.01	2.95	2.90
	.025	6.72	5.26	4.63	4.28	4.04	3.88	3.76	3.66	3.59
	.010	9.65	7.21	6.22	5.67	5.32	5.07	4.89	4.74	4.63
	.005	12.23	8.91	7.60	6.88	6.42	6.10	5.86	5.68	5.54
12	.100	3.18	2.81	2.61	2.48	2.39	2.33	2.28	2.24	2.21
	.050	4.75	3.89	3.49	3.26	3.11	3.00	2.91	2.85	2.80
	.025	6.55	5.10	4.47	4.12	3.89	3.73	3.61	3.51	3.44
	.010	9.33	6.93	5.95	5.41	5.06	4.82	4.64	4.50	4.39
	.005	11.75	8.51	7.23	6.52	6.07	5.76	5.52	5.35	5.20

In a similar manner, the critical values for a tail area, $a = .01$, are presented in Table 6 in Appendix II. Thus, if $v_1 = 6$ and $v_2 = 9$,

$$P(F > F_{.01}) = P(F > 5.80) = .01$$

The statistical test of the null hypothesis

$$H_0 : \sigma_1^2 = \sigma_2^2$$

uses the test statistic

$$F = \frac{s_1^2}{s_2^2}$$

When the alternative hypothesis implies a one-tailed test, that is,

$$H_a : \sigma_1^2 > \sigma_2^2$$

we can use the tables directly. However, when the alternative hypothesis requires a two-tailed test,

$$H_a : \sigma_1^2 \neq \sigma_2^2$$

we note that the rejection region will be divided between the lower and upper tails of the F distribution and that tables of critical values for the lower tail are conspicuously missing. The reason for their absence is explained as follows: We are at liberty to identify either of the two populations as population I. If the population with the larger sample variance is designated as population II, then $s_2^2 > s_1^2$ and we will be concerned with rejection in the lower tail of the F distribution. Since the identification of the populations was arbitrary, we can avoid this difficulty by designating the population with the larger sample variance as population I. In other words, always place the larger sample variance in the numerator of

$$F = \frac{s_1^2}{s_2^2}$$

and designate that population as I. Then, since the area in the right-hand tail will represent only $\alpha/2$, we double this value to obtain the correct value for the probability of a type I error α.

Test of an Hypothesis Concerning the Equality of Two Population Variances

1   Null Hypothesis: $H_0 : \sigma_1^2 = \sigma_2^2$

2   Alternative Hypothesis:

One-Tailed Test	Two-Tailed Test
$H_a : \sigma_1^2 > \sigma_2^2$	$H_a : \sigma_1^2 \neq \sigma_2^2$
(or $H_a : \sigma_2^2 > \sigma_1^2$)	

3   Test Statistic:

One-Tailed Test	Two-Tailed Test

$$F = \frac{s_1^2}{s_2^2}$$ $$F = \frac{s_1^2}{s_2^2}$$

$$\left(\text{or } F = \frac{s_2^2}{s_1^2} \text{ for } H_a : \sigma_2^2 > \sigma_1^2 \right) \quad \text{where } s_1^2 \text{ is the larger sample variance.}$$

4   Rejection Region:

One-Tailed Test	Two-Tailed Test
$F > F_\alpha$	$F > F_{\alpha/2}$

When $F = s_1^2/s_2^2$, the critical values of F, F_α, and $F_{\alpha/2}$ are based on $v_1 = n_1 - 1$ and $v_2 = n_2 - 1$ d.f. These tabulated values, for $\alpha = .10, .05, .025, .01,$ and $.005$, can be found in Table 6 of Appendix II.

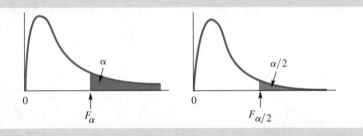

Assumptions: The samples were randomly and independently selected from normally distributed populations.

E X A M P L E   **8.20**   Two samples consisting of 10 and 8 measurements each were observed to possess sample variances equal to $s_1^2 = 7.14$ and $s_2^2 = 3.21$, respectively. Do the sample variances present sufficient evidence to indicate that the population variances are unequal?

Solution   Assume that the populations possess probability distributions that are reasonably mound-shaped and thus will satisfy, for all practical purposes, the assumption that the populations are normal.

We want to test the null hypothesis

$$H_0 : \sigma_1^2 = \sigma_2^2$$

against the alternative

$$H_a : \sigma_1^2 \neq \sigma_2^2$$

Using Table 6 in Appendix II and doubling the tail area, we will reject H_0 when $F > 3.68$ with $\alpha = .10$.

The calculated value of the test statistic is

$$F = \frac{s_1^2}{s_2^2} = \frac{7.14}{3.21} = 2.22$$

Noting that the test statistic does not fall in the rejection region, we do not reject $H_0 : \sigma_1^2 = \sigma_2^2$. Thus there is insufficient evidence to indicate a difference in the population variances. ∎

EXAMPLE **8.21**   The variability in the amount of impurities present in a batch of chemical used for a particular process depends upon the length of time the process is in operation. A manufacturer using two production lines, 1 and 2, has made a slight adjustment to process 2, hoping to reduce the variability as well as the average amount of impurities in the chemical. Samples of $n_1 = 25$ and $n_2 = 25$ measurements from the two batches yield means and variances as follows:

$$\begin{aligned} \bar{x}_1 &= 3.2 & s_1^2 &= 1.04 \\ \bar{x}_2 &= 3.0 & s_2^2 &= .51 \end{aligned}$$

Do the data present sufficient evidence to indicate that the process variability is less for process 2? Test the null hypothesis $H_0 : \sigma_1^2 = \sigma_2^2$.

Solution   The practical implications of this example are illustrated in Figure 8.17. We believe that the mean levels of impurities in the two production lines are nearly equal (in fact, that they may be equal) but that there is a possibility that the variation in the level of impurities is substantially less for line 2. Then distributions of impurity measurements for the two production lines would have nearly the same mean level, but they would differ in their variation. A large variance for the level of impurities increases the probability of producing shipments of chemical with an unacceptably high level of impurities. Consequently, we hope to show that the process change in line 2 has made σ_2^2 less than σ_1^2.

FIGURE **8.17**
Distributions of impurity
measurements for two
production lines in Example 8.21

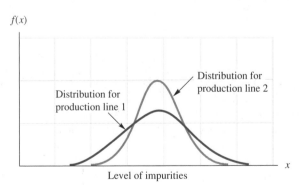

Testing the null hypothesis

$$H_0 : \sigma_1^2 = \sigma_2^2$$

against the alternative

$$H_a : \sigma_1^2 > \sigma_2^2$$

at an $\alpha = .05$ significance level, we will reject H_0 when F is greater than $F_{.05} = 1.98$; that is, we will employ a one-tailed statistical test.

The calculated value of the F test statistic,

$$F = \frac{s_1^2}{s_2^2} = \frac{1.04}{.51} = 2.04$$

falls in the rejection region. Therefore, we conclude that the variability of process 2 is less than that of process 1. ▪

Exercises

Basic Techniques

8.63 A random sample of $n = 25$ observations from a normal population produced a sample variance equal to 21.4. Do these data provide sufficient evidence to indicate that $\sigma^2 > 15$? Test using $\alpha = .05$.

8.64 A random sample of $n = 15$ observations was selected from a normal population. The sample mean and variance were $\bar{x} = 3.91$ and $s^2 = .3214$. Find a 90% confidence interval for the population variance σ^2.

8.65 A random sample of $n = 22$ observations was selected from a normal population. The sample mean and variance were $\bar{x} = 41.3$ and $s^2 = 14.14$. Do the data provide sufficient evidence to indicate that $\sigma^2 < 25$? Test using $\alpha = .05$.

8.66 Find a 90% confidence interval for the population variance in Exercise 8.65.

8.67 Independent random samples from two normal populations produced the following variances:

Sample Size	Sample Variance
16	55.7
20	31.4

   a Do the data provide sufficient evidence to indicate that σ_1^2 differs from σ_2^2? Test using $\alpha = .05$.

   b Find the approximate observed significance level for the test, and interpret its value.

8.68 Independent random samples from two normal populations produced the following variances:

Population	Sample Size	Sample Variance
1	13	18.3
2	13	7.9

   a Do the data provide sufficient evidence to indicate that $\sigma_1^2 > \sigma_2^2$? Test using $\alpha = .05$.

   b Find the approximate observed significance level for the test, and interpret its value.

Applications

8.69 The stability of measurements of the characteristics of a manufactured product is important in maintaining product quality. In fact, it is sometimes better to possess small variation in the measured value of some important characteristic of a product and have the process mean slightly off target than to suffer wide variation with a mean that perfectly fits requirements. The latter situation may produce a higher percentage of defective product than the former. A manufacturer of light bulbs suspected that one of her production lines was producing bulbs with a high variation in length of life. To test this theory, she compared the lengths of life of $n = 50$ bulbs randomly sampled from the suspect line and $n = 50$ from a line that seemed to be "in control." The sample means and variances for the two samples were as follows:

Suspect Line	Line in Control
$\bar{x}_1 = 1520$	$\bar{x}_2 = 1476$
$s_1^2 = 92{,}000$	$s_2^2 = 37{,}000$

a Do the data provide sufficient evidence to indicate that bulbs produced by the suspect line possess a larger variance in length of life than those produced by the line that is assumed to be in control? Use $\alpha = .05$.

b Find the approximate observed significance level for the test, and interpret its value.

8.70 The Environmental Protection Agency (EPA) has set a maximum noise level for heavy trucks at 83 decibels. This limit could be interpreted in several ways. One way to apply the limit would be to require all trucks to conform to the noise limit. A second but less satisfactory method would be to require the truck fleet's mean noise level to be less than the limit. If the latter were the rule, variation in the noise level from truck to truck would be important, because a large value of σ^2 would imply many trucks exceeding the limit, even if the mean fleet level was 83 decibels. The data for six trucks, in decibels, were

$$82.4, \ 83.8, \ 83.1, \ 82.3, \ 81.8, \ 83.0$$

Use these data to construct a 90% confidence interval for σ^2, the variance of the truck noise emission readings. Interpret your results.

8.71 A precision instrument is guaranteed to read accurately to within 2 units. A sample of four instrument readings on the same object yielded the measurements $353, 351, 351$, and 355. Test the null hypothesis that $\sigma = .7$ against the alternative hypothesis $\sigma > .7$. Conduct the test at the $\alpha = .05$ level of significance.

8.72 Find a 90% confidence interval for the population variance in Exercise 8.71.

8.73 A manufacturer of hard safety hats for construction workers is concerned about the mean and the variation of the force helmets transmit to wearers when subjected to a standard external force. The manufacturer desires the mean force transmitted by helmets to be 800 pounds (or less), well under the legal 1000-pound limit, and σ to be less than 40. A random sample of $n = 40$ helmets was tested, and the sample mean and variance were found to be equal to 825 pounds and 2350 (pounds)^2, respectively.

a If $\mu = 800$ and $\sigma = 40$, is it likely that any helmet, subjected to the standard external force, will transmit a force to a wearer in excess of 1000 pounds? Explain.

b Do the data provide sufficient evidence to indicate that when the helmets are subjected to the standard external force the mean force transmitted by the helmets exceeds 800 pounds?

c Do the data provide sufficient evidence to indicate that σ exceeds 40?

8.74 A personnel manager planning to use a Student's t test to compare the mean number of monthly absences for two categories of employees noticed a possible difficulty. The variation in the numbers of absences per month seemed to differ for the two groups. As a check, the personnel manager randomly selected five months and counted the number of absences for each group. The data are shown in the table.

Category A	20	14	19	22	25
Category B	37	29	51	40	26

a   About which assumption necessary for use of the t test was the personnel manager concerned?

b   Do the data provide sufficient evidence to indicate that the variances differ for the populations of absences for the two employee categories? Test with $\alpha = .10$, and interpret the results of the test.

8.75   A pharmaceutical manufacturer purchases a particular material from two different suppliers. The mean level of impurities in the raw material is approximately the same for both suppliers, but the manufacturer is concerned about the variability of the impurities from shipment to shipment. If the level of impurities tends to vary excessively for one source of supply, it could affect the quality of the pharmaceutical product. To compare the variation in percentage impurities for the two suppliers, the manufacturer selects ten shipments from each of the two suppliers and measures the percentage of impurities in the raw material for each shipment. The sample means and variances are shown in the table.

Supplier A	Supplier B
$\bar{x}_1 = 1.89$	$\bar{x}_2 = 1.85$
$s_1^2 = .273$	$s_2^2 = .094$
$n_1 = 10$	$n_2 = 10$

Do the data provide sufficient evidence to indicate a difference in the variability of the shipment impurity levels for the two suppliers? Test using $\alpha = .10$. Based on the results of your test, what recommendations would you make to the pharmaceutical manufacturer?

8.10 Some Comments on the Theory of Tests of Hypotheses

As outlined in Section 8.2, the theory of a statistical test of an hypothesis is indeed a very clear-cut procedure, enabling the experimenter to either reject or accept the null hypothesis with measured risks α and β. Unfortunately, as we noted, the theoretical framework does not suffice for all practical situations.

The crux of the theory requires that we be able to specify a meaningful alternative hypothesis that permits the calculation of the probability β of a type II error for all alternative values of the parameter(s). This calculation can be done for many statistical tests, including the large-sample test discussed in Section 8.3, although the calculation of β for various alternatives and sample sizes may be difficult in some cases. On the other hand, in some test situations it is difficult to clearly specify alternatives to H_0 that have practical significance. This may occur when we want to test an hypothesis concerning the values of a set of parameters, a situation that we will encounter in Chapter 15 in analyzing enumerative data.

The obstacle that we mention does not invalidate the use of statistical tests. Rather, it urges caution in drawing conclusions when insufficient evidence is available to reject the null hypothesis. The difficulty of specifying meaningful alternatives to the null hypothesis, together with the difficulty encountered in the calculation and tabulation of β for other than the simplest statistical tests, justifies skirting this issue in an introductory text. Thus we can adopt one of two procedures. We can present the p-value associated with a statistical test and leave the interpretation to the reader. Or we can agree to adopt the procedure described in Example 8.15 when tabulated values of β (the operating characteristic curve) are unavailable for the test. When the test statistic falls in the acceptance region, we will "not reject" rather than "accept" the null hypothesis. Further conclusions can be made by calculating an interval estimate for the parameter or by consulting one of several published statistical handbooks for tabulated values of β. We will not be too surprised to learn that these tabulations are inaccessible, if not completely unavailable, for some of the more complicated statistical tests.

The choice between a one- or a two-tailed test for a given situation is dictated by the practical aspects of the problem and will depend on the alternative value of the parameter the experimenter is trying to detect. If we were to sustain a large financial loss if μ were greater than μ_0 but not if it were less, we would concentrate our attention on the detection of values of μ greater than μ_0. Therefore, we would reject in the upper tail of the distribution. On the other hand, if we were equally interested in detecting values of μ that are either less than or greater than μ_0, we would employ a two-tailed test.

8.11   Assumptions

As noted earlier, the tests and confidence intervals based on the Student's t, the chi-square, and the F statistic require that the data satisfy specific assumptions in order that the error probabilities (for the tests) and the confidence coefficients (for the confidence intervals) be equal to the values that we have specified. For example, if the assumptions are violated by selecting a sample from a nonnormal population and the data are used to construct a 95% confidence interval for μ, the actual confidence coefficient might (unknown to us) be equal to .85 instead of .95. The assumptions are summarized in the following display for your convenience.

Assumptions for t, χ^2, and F statistics

1   For the small-sample tests and confidence intervals described in this chapter, we assume that samples are randomly selected from normally distributed populations.

2 When two samples are selected, we assume that they are selected in an independent manner, except in the case of the paired-difference experiment.

3 For tests or confidence intervals concerning the difference between two population means μ_1 and μ_2, based on independent random samples, we assume that $\sigma_1^2 = \sigma_2^2$.

In a practical sampling situation, you never know everything about the probability distribution of the sampled population. If you did, there would be no need for sampling or statistics. Furthermore, it is highly unlikely that a population would possess, exactly, the characteristics described above. Consequently, to be useful, the inferential methods described in this chapter must lead to good inferences when moderate departures from the assumptions are present. For example, if the population possesses a mound-shaped distribution that is nearly normal, we would like a 95% confidence interval constructed for μ to be one with a confidence coefficient close to .95. Similarly, if we conduct a t test of the null hypothesis $\mu_1 = \mu_2$, based on independent random samples from normal populations, where σ_1^2 and σ_2^2 are not exactly equal, we want the probability of incorrectly rejecting the null hypothesis, α, to be approximately equal to the value we used in locating the rejection region.

A statistical method that is insensitive to departures from the assumptions upon which the method is based is said to be *robust*. The t tests are quite robust to moderate departures from normality. In contrast, the chi-square and F tests are sensitive to departures from normality. The t test for comparing two means is moderately robust to departures from the assumption $\sigma_1^2 = \sigma_2^2$ when $n_1 = n_2$. However, the test becomes sensitive to departures from this assumption as n_1 becomes large relative to n_2 (or vice versa).

If you are concerned that your data do not satisfy the assumptions prescribed for one of the statistical methods described in this chapter, you may be able to use a nonparametric statistical method to make your inference. These methods, which require few or no assumptions about the nature of the population probability distributions, are particularly useful for testing hypotheses, and some nonparametric methods have been developed for estimating population parameters. Tests of hypotheses concerning the location of a population distribution or a test for the equivalence of two population distributions are presented in Chapter 16. If you can select relatively large samples, you can use the large-sample estimation or test procedures given in this chapter.

8.12   MORE ON POST–COLD WAR CONTRACTORS

In their study of the cost accounting systems currently being used by defense contractors as compared to nondefense contractors, Rezaee and Elmore summarized the responses of 25 defense contractors and 25 nondefense contractors. The information in Table 8.1 is taken from their Table 2. A Likert Scale is one in which responses are categorized according to the respondent's measure of belief in the statement. For example, the response to the statement "traditional financial management controls are important" could be as follows: 1, disagree strongly; 2, disagree; 3, no opinion; 4, agree; or 5, strongly agree. Since these responses are not normally distributed, we need to examine the sample sizes and the sampling distributions of the statistics that would be used in testing for significant differences between pairs of means. The data in Table 8.1 are reproduced in Table 8.9.

TABLE **8.9**

	Defense	Nondefense	T-Value
Strategic Planning			
1 Budgets examined for consistency with long-range goals	4.0425	4.2000	−.30
2 Formal statement of goals, strategies, etc., used in planning the direction of the firm	4.1625	4.8800	−1.45
Budget and Planning			
1 Budgets used in performance evaluation of individual members	3.160	4.520	−2.56*
2 Comparison of actual to budgeted costs	3.9891	5.1782	−2.64*
3 Individual department budgets	2.8800	4.6800	−3.23*
4 Participation of lower- and mid-management in budgeting	3.6879	5.0800	−3.53*
5 Flexible budgets	2.1861	3.6000	−2.64*

* Significant at .01.

For each of the categories listed in the table, the mean is based upon a sample of $n = 25$ responses. This sample size is large enough to assure approximate normality of the distribution of $\bar{x}$ for each of the categories. Can we assume equal variances of the two populations being compared? If we can, then the degrees of freedom with s^2, the pooled estimate of the common variance, will be $n_1 + n_2 - 2 = 25 + 25 - 2 = 48$, which indicates that the standard normal or Student's t tables of critical values could be used in determining critical values for the test.

If we wish to confirm the significance level of .01, then we will reference critical values with a right-tail area of $\alpha/2 = .005$ to find $t_{.005} = 2.576$ and $z_{.005} = 2.58$. On the other hand, if we cannot assume that the variances are equal, the sample sizes, although borderline, could be considered large enough to assure the approximate normality of the test statistic, and the critical value would again be taken to be $z_{.005} = 2.58$.

The test for each pair of means would proceed as follows. The hypotheses to be tested are

$$H_0 : \mu_1 - \mu_2 = 0 \qquad \text{versus} \qquad H_a : \mu_1 - \mu_2 \neq 0$$

The test statistic is

$$t = \frac{(\bar{x}_1 - \bar{x}_2)}{\sqrt{s^2 \left(\dfrac{1}{25} + \dfrac{1}{25} \right)}}$$

With d.f. = 48, and $\alpha = .005$, the critical value is $t_{.005} = 2.576$. Therefore, we will reject H_0 and conclude that the means are significantly different if the observed value of t is greater than 2.576 or less than -2.576.

If the calculations for the t statistics in the seven categories given in Table 8.9 are correct, then the first two differences are not significant, as indicated, and the next five differences are significant. However, the difference corresponding to category 1 under Budget and Planning is significant at the 5% level, not the 1% level as indicated; the remaining differences are significant at the 1% level.

What does all this mean? We have found that, although the mean differences were not significant, defense and nondefense industries both consider strategic planning to be very important. On the other hand, defense contractors give significantly less importance to areas dealing with budget and planning, indicating that they have a more highly centralized organizational control structure than do nondefense contractors.

8.13 Summary

In this chapter, we have presented the basic concepts of a statistical test of an hypothesis and have demonstrated the procedure for large and small samples. Some of the tests described in this chapter are based on the Central Limit Theorem and hence apply to large samples. When n is large, each of the respective test statistics possesses a sampling distribution that can be approximated by the normal distribution. This result, along with the properties of the normal distribution studied in Chapter 5, permits the calculation of α, β, and p-values for the statistical tests.

It is important to note that the t, χ^2, and F statistics employed in the small-sample statistical methods are based on the assumption that the sampled populations have a normal probability distribution. This requirement will be adequately satisfied for many types of experimental measurements.

You will observe the very close relationship connecting the Student's t and the z statistic and, therefore, the similarity of the methods for testing hypotheses and the construction of confidence intervals. The χ^2 and F statistics employed in making inferences concerning population variances do not, of course, follow this pattern, but the reasoning employed in the construction of the statistical tests and confidence intervals is identical for all the methods we have presented.

Tips on Problem Solving

When conducting a statistical test of an hypothesis, it is important to follow the same basic procedure for each problem.

1   Determine the type of data involved (quantitative or binomial) and the number of samples involved (one or two). Are you interested in a mean, a proportion, or a variance? This will allow you to identify the parameter of interest in the experiment.

2   Check the conditions required for the sampling distribution of the parameter to be approximated by a normal distribution. For quantitative data, the sample size (or sizes) must be 30 or more. For binomial data, a large sample size will ensure that $p \pm 2\sigma_{\hat{p}}$ is contained in the interval 0 to 1 [$(p_1 - p_2) \pm 2\sigma_{(\hat{p}_1 - \hat{p}_2)}$ contained in the interval -1 to 1 for the two-sample case].

3   State the null and alternative hypotheses (H_0 and H_a). The alternative hypothesis is the hypothesis that the researcher wishes to support; the null hypothesis is a contradiction of the alternative hypothesis.

4   State the test statistic to be used in the test of the hypothesis.

5   Locate the rejection region for the test. In this chapter, the rejection region will be found in the tail areas of the standard normal (z), the t, the χ^2, or the F distribution. The exact rejection region will be determined by the desired value of α and the form of the alternative hypothesis (one- or two-tailed).

6   Conduct the test, calculating the observed value of the test statistic based on the sample data.

7   Draw conclusions based on the observed value of the test statistic. If the test statistic falls in the rejection region, the null hypothesis is rejected in favor of the alternative hypothesis. The probability of an incorrect decision is α. However, if the test statistic does not fall in the rejection region, we cannot reject the null hypothesis. There is insufficient evidence to show that the alternative hypothesis is true. Judgment is withheld until more data can be collected.

Supplementary Exercises

Supplementary Exercises

8.76   Define α and β for a statistical test of an hypothesis.

8.77   What is the observed significance level of a test?

8.78   What conditions must be met in order that the z test may be used to test an hypothesis concerning a population mean μ?

8.79   What assumptions are made when using a Student's t test to test an hypothesis concerning a single population mean?

8.80   What assumptions are made when using a Student's t test to test an hypothesis concerning the difference between two population means?

8.81   The daily wages in a particular industry are normally distributed with a mean of $23.20 and a standard deviation of $4.50. A company in this industry employing 40 workers pays these workers an average of $21.20 daily. Based on this sample mean, could these workers be viewed as a random sample from among all workers in the industry?

   a   Find the observed significance level of the test.

   b   If you planned to conduct your test using $\alpha = .01$, what would be your test conclusion?

8.82   High airline occupancy rates on scheduled flights are essential to profitability. Suppose that a scheduled flight must average at least 60% occupancy in order to be profitable and that an examination of the occupancy rates for 120 10:00 A.M. flights from Atlanta to Dallas showed a mean occupancy rate per flight of 58% and a standard deviation of 11%.

   a   If μ is the mean occupancy per flight and if the company wishes to determine whether this scheduled flight is unprofitable, give the alternative and the null hypotheses for the test.

   b   Does the alternative hypothesis in part (a) imply a one- or a two-tailed test? Explain.

   c   Do the occupancy data for the 120 flights suggest that this scheduled flight is unprofitable? Test using $\alpha = .10$.

8.83   A manufacturer of automatic washers provides a particular model in one of three colors, A, B, or C. Of the first 1000 washers sold, 400 of the washers were of color A. Would you conclude that more than 1/3 of all customers have a preference for color A?

   a   Find the observed significance level of the test.

   b   If you planned to conduct your test using $\alpha = .05$, what would be your test conclusion?

8.84   A manufacturer claimed that at least 20% of the public preferred its product. A sample of 100 persons is taken to check this claim. With $\alpha = .05$, how small would the sample percentage need to be before the claim could be statistically refuted? (Note that this would require a one-tailed test of an hypothesis.)

8.85   Refer to Exercise 8.84. Sixteen people in the sample of 100 consumers expressed a preference for the manufacturer's product. Does this result present sufficient evidence to reject the manufacturer's claim? Test using $\alpha = .10$.

8.86   A manufacturer can tolerate a small amount (.05 milligram per liter) of impurities in a raw material needed for manufacturing its product. Because the laboratory test for the impurities is subject to experimental error, the manufacturer tests each batch ten times. Assume that the mean value of the experimental error is 0 and thus that the mean value of the ten test readings is an unbiased estimate of the true amount of the impurities in the batch. For a particular batch of the raw material, the mean of the ten test readings is .058 milligram per liter (mg/l), and the standard deviation is .012 mg/l. Do the data provide sufficient evidence to indicate that the amount of impurities in the batch exceeds .05 mg/l? Find the approximate p-value for the test, and interpret its value.

8.87   The temperature of operation of two paint-drying ovens associated with two manufacturing production lines was recorded for 20 days. (Pairing was ignored.) The means and variances of the two samples are

$$\bar{x}_1 = 164 \qquad \bar{x}_2 = 168$$
$$s_1^2 = 81 \qquad s_2^2 = 172$$

Do the data present sufficient evidence to indicate a difference in temperature variability for the two ovens? Test the hypothesis that $\sigma_1^2 = \sigma_2^2$ at the $\alpha = .10$ level of significance.

8.88   A production plant has two extremely complex fabricating systems, with one twice the age of the other. Both systems are checked, lubricated, and maintained once every two weeks. The number of finished products fabricated daily by each of the systems is recorded for 30 working days. The results are given in the table. Do these data present sufficient evidence to conclude that the variability in daily production warrants increased maintenance of the older fabricating system? Use a 5% level of significance.

New System	Old System
$\bar{x}_1 = 246$	$\bar{x}_2 = 240$
$s_1 = 15.6$	$s_2 = 28.2$

8.89 A manufacturer claimed that at least 95% of the equipment that it supplied to a factory conformed to specifications. An examination of a sample of 700 pieces of equipment revealed that 53 were faulty. Test the manufacturer's claim using $\alpha = .05$.

8.90 In deciding where to place its emphasis in advertising, the market research department for a major automobile manufacturer wished to compare the mean number of automobiles per family in two regions of the United States. Suppose that a preliminary study of the number of cars per family for $n = 200$ families from each of the two regions gave the means and variances for the two samples as shown in the accompanying table.

	Region 1	Region 2
Sample size	200	200
Sample mean	1.30	1.37
Sample variance	.53	.64

a Note that a small increase in the mean number of automobiles per family can represent a very large number of automobiles for a region. Do the data provide sufficient evidence to indicate a difference in the mean number of automobiles per family for the two regions?

b Picture the data associated with either of the two populations. What values will x assume? Imagine the probability distributions for these two populations. Will their nature violate the conditions necessary in order that the test in part (a) be valid? Explain.

c Find a 95% confidence interval for the mean number of automobiles per family for region 2. Interpret the interval.

d Find a 90% confidence interval for the difference in the mean number of automobiles for the two regions. Interpret the interval.

8.91 The mean lifetime of a sample of 100 fluorescent bulbs produced by a company is computed to be 1570 hours, and the standard deviation is 120 hours. If μ is the mean lifetime of all the bulbs produced by the company, test the hypothesis $\mu = 1600$ hours against the alternative hypothesis $\mu < 1600$.

a Find the observed significance level of the test.

b If you planned to conduct your test using $\alpha = .05$, what would be your test conclusion?

8.92 Presently 20% of potential customers buy a certain brand of soap, say, brand A. To increase sales, the company plans an extensive advertising campaign. At the end of the campaign, a sample of 400 potential customers will be interviewed to determine whether the campaign was successful.

a State H_0 and H_a in terms of p, the probability that a customer prefers soap brand A.

b The company will conclude that the advertising campaign was a success if at least 92 of the 400 customers interviewed prefer brand A. Find α. (Use the normal approximation to the binomial distribution to evaluate the desired probability.)

8.93 An experiment was conducted to compare the mean lengths of time required for two bank employees, A and B, to complete the paperwork for new customer personal checking accounts. Ten customers were randomly assigned to each employee, and the length of servicing time was recorded in minutes for each customer. The means and variances for the two samples are given in the accompanying table.

Employee A	Employee B
$\bar{x}_1 = 22.2$	$\bar{x}_2 = 28.5$
$s_1^2 = 16.36$	$s_2^2 = 18.92$

a Do the data provide sufficient evidence to indicate a difference in mean times required to complete the paperwork necessary for a new customer checking account? Test using $\alpha = .10$.

b Find the approximate observed significance level for the test, and interpret its value.

8.94 Refer to Exercise 8.93. Find a 95% confidence interval for the difference in mean servicing times.

8.95 Refer to Exercise 8.93. Suppose that you wanted to estimate the difference in mean servicing times correct to within 1 minute with probability approximately equal to .95. Approximately how large a sample would be required for each bank employee (assume that the sample sizes will be equal)? [*Hint:* To solve, use the method of Section 7.9.]

8.96 Suppose that an experiment has been designed to estimate the difference between two population means $(\mu_1 - \mu_2)$. Independent random samples of size n_1 and n_2 have been selected from the two populations, and the statistic $(\bar{x}_1 - \bar{x}_2)$ is used as the estimator. Would the amount of information extracted from the data be increased by pairing successive observations and analyzing the differences? Is this an appropriate method of analysis?

8.97 When should one employ a paired-difference analysis in making inferences concerning the difference between two means?

8.98 A utility company collected data to compare the length of time required to process an electric utility bill using two different processing methods. Eight billing clerks were each given a single utility bill and asked to process the bill by using both procedures 1 and 2. The processing times (in seconds) are shown in the accompanying table. Do the data present sufficient evidence to indicate a difference in mean processing time for the two processing methods?

	Process	
Processor	1	2
1	3	4
2	1	2
3	1	3
4	2	1
5	1	2
6	2	3
7	3	3
8	1	3

a Find the approximate observed significance level for the test.

b Test using $\alpha = .05$.

8.99 Place a 95% confidence interval on the difference in mean processing times for the two processing methods in Exercise 8.98.

8.100 A condominium apartment's monthly water consumption has been averaging 48,000 gallons per month over the past five years. The mean and standard deviation of the monthly consumption for the current 12 months are $\bar{x} = 51,102$ gallons and $s = 5127$ gallons. Do the data provide sufficient evidence to indicate that some unusual factor is causing a larger-than-expected water consumption for the condominium—that is, a consumption exceeding a mean of 48,000 gallons per month? Use $\alpha = .05$.

8.101 An industrial psychologist wanted to compare two methods, A and B, for indoctrinating new employees in the company's personnel policies. Twenty new employees were given a general intelligence test and were then matched, according to test scores, in ten pairs. From each pair, one employee was randomly assigned to indoctrination method A and the second to method B. Each employee was tested at the end of a four-week period. The achievement scores shown in the table were recorded.

Pair	Method A	Method B
1	36	35
2	37	35
3	41	40
4	42	41
5	36	36
6	35	34
7	42	40
8	33	31
9	40	39
10	38	37

a Do the data provide sufficient evidence to indicate that the mean achievement scores differ for the two indoctrination methods? (Use $\alpha = .05$.)

b Estimate the mean difference in achievement scores using a 98% confidence interval.

8.102 A manufacturer of a machine to package soap powder claimed that the machine could load cartons at a given weight with a range of no more than two-fifths of an ounce. The mean and variance of a sample of eight 3-pound boxes were found to equal 3.1 and .018, respectively. Test the hypothesis that the variance of the population of weight measurements is $\sigma^2 = .01$ against the alternative hypothesis $\sigma^2 > .01$. Use an $\alpha = .05$ level of significance.

8.103 Find a 90% confidence interval for σ^2 in Exercise 8.102.

8.104 Under what assumptions can the F distribution be used to make inferences about the ratio of population variances?

8.105 The closing prices of two common stocks were recorded for a period of 15 days. The means and variances are

$$\bar{x}_1 = 40.33 \qquad \bar{x}_2 = 42.54$$
$$s_1^2 = 1.54 \qquad s_2^2 = 2.96$$

Do these data present sufficient evidence to indicate a difference in variability of closing prices of the two stocks? Give the approximate p-value for the test, and interpret its value.

8.106 A chemical manufacturer claims that the purity of his product never varies more than 2%. Five batches were tested and gave purity readings of 98.2%, 97.1%, 98.9%, 97.7%, and 97.9%. Do the data provide sufficient evidence to contradict the manufacturer's claim? [*Hint:* To be generous, let a range of 2% equal 4σ.]

8.107 Refer to Exercise 8.106. Find a 90% confidence interval for σ^2.

8.108 A cannery prints "weight 16 ounces" on its label. The quality control supervisor selects nine cans at random and weighs them. She finds $\bar{x} = 15.7$ and $s = .5$. Do the data provide sufficient evidence to indicate that the mean weight is less than that claimed on the label? (Use $\alpha = .05$.)

8.109 A car dealer decided to compare the mean monthly sales of two salespersons, A and B. Because the strength of sales varies with the season and with people's opinions about the economy, the car dealer decided to make the comparison on a monthly basis. The data shown in the table give the monthly sales (to the nearest thousand dollars) for the two salespersons. Use the Minitab printout to answer the following questions.

Month	Salesperson A	Salesperson B
January	130	105
February	141	109
March	163	147
April	176	159
May	147	150
June	160	134
July	145	123
August	129	130
September	104	91
October	139	124
November	163	141
December	151	147

```
TEST OF MU =  0.00 VS MU N.E.  0.00

            N     MEAN    STDEV   SE MEAN       T    P VALUE
DIFF       12    15.67    10.92      3.15    4.97     0.0004

MTB > TINT C3

            N     MEAN    STDEV   SE MEAN    95.0 PERCENT C.I.
DIFF       12    15.67    10.92      3.15  (    8.72,   22.61)
```

a What type of experimental design has been used?

b Do the data provide sufficient evidence to indicate a difference in mean sales for the two salespersons? Test with $\alpha = .05$.

c Find a 95% confidence interval for $(\mu_A - \mu_B)$, and interpret your results.

8.110 In the past, a chemical plant has produced an average of 1100 pounds of chemical per day. The records for the past year, based on 260 operating days, show a mean and standard deviation of $\bar{x} = 1060$ pounds per day and $s = 340$ pounds per day. The plant manager wishes to test whether the average daily production has dropped significantly over the past year.

a Give the appropriate null and alternative hypotheses.

b If z is used as a test statistic, determine the rejection region corresponding to $\alpha = .05$.

c Do the data provide sufficient evidence to indicate a drop in average daily production?

8.111 Both union and management conducted surveys of worker opinion prior to a vote for or against unionization of a large industrial plant. The union survey, consisting of a sample of 500 workers, was reported to show 54% of the workers in favor of unionization. A corresponding management survey of 400 workers found only 46% in favor of unionization. Is it likely that surveys involving these sample sizes would produce percentages in favor of unionization that differ as much as those presented here? Or is it possible that something was wrong with the survey method or data analysis for either the union, management, or both? Find the p-value associated with the appropriate test and use it to make your decision.

8.112 A hotel needs a 60% occupancy rate in order to show a profit. A random sampling of 50 days produced a mean occupancy rate of 62% and a standard deviation of 8%. Do these data provide sufficient evidence to indicate that the mean occupancy rate (for the population of days representative of those in the sample) exceeds 60%? Test using $\alpha = .10$.

8.113 Refer to Exercise 8.110. Use the procedure described in Example 8.5 to calculate β for several alternative values of μ. (For example, $\mu = 1040, 1030, 1020$.) Use the computed values of β to construct a power curve for the statistical test.

8.114 Japan's All-Nippon Airways has found that painting menacing eyes on its aircraft jet engine intakes frightens away birds and saves maintenance money (*Gainesville Sun,* November 16, 1986). A study of multiengine passenger aircraft over a one-year period found an average of one bird hit per painted engine compared with an average of nine bird hits for unpainted engines.

    a What type of data would you expect the number of bird hits per engine to be? Explain.

    b Explain why the numbers of hits per engine on the same aircraft might be dependent and thus violate the assumption that the sample was randomly selected.

8.115 Refer to Exercise 8.114. Suppose that all aircraft in the study contained the same number of jet engines and that x represents the number of engine bird hits per aircraft.

    a If $n = 40$ aircraft were randomly selected to have their engines painted, would the 40 values of x represent a random sample?

    b Suppose that the mean value of bird hits per aircraft with unpainted engines is $\mu = 9$. If the sample mean number of bird hits per engine per aircraft is $\bar{x} = 1$, do you have sufficient evidence to indicate that the painted menacing eyes on the engines produced a reduction in the mean number of bird hits per engine per aircraft? Test using $\alpha = .05$. [*Hint:* The number of engine bird hits per aircraft is likely to be a Poisson random variable. Since the standard deviation σ of a Poisson random variable is equal to the square root of its mean μ— that is, $\sigma = \sqrt{\mu}$—it can be estimated by substituting $\bar{x}$ for μ, that is, $\hat{\sigma} = \sqrt{\bar{x}}$.]

8.116 Refer to the All-Nippon Airways study described in Exercises 8.114 and 8.115. All-Nippon found that painting menacing eyes on jet aircraft engine intakes seemed to produce a reduction in the mean number of bird hits per jet engine. (They estimated a maintenance cost reduction of almost $200,000 for the small number of aircraft included in their study.) Suppose that the study had involved independent random samples of aircraft, 40 with painted air intakes and 40 without, and that the sample mean number of bird hits per aircraft was $\bar{x}_1 = 1$ for the aircraft with painted intakes and $\bar{x}_2 = 9$ for those with unpainted intakes. Do these data provide sufficient evidence to indicate that the mean number of bird hits per engine per aircraft is less for aircraft with the painted air intakes? Test by using $\alpha = .05$. [*Hint:* The number of bird hits per engine per aircraft is likely to be a Poisson random variable. Since the standard deviation σ of a Poisson random variable is equal to the square root of its mean μ—that is, $\sigma = \sqrt{\mu}$—it can be estimated by substituting $\bar{x}$ for μ, that is, $\hat{\sigma} = \sqrt{\bar{x}}$.]

8.117 One way to compare the relative prices of two common stocks is to compare their price-earnings ratios, the ratios of a stock's price per share to the amount of money earned by the company per share per year. The table on the next page lists the price-earnings ratios (P/Es) for ten randomly selected electric power companies versus eight nonutility blue-chip stocks in July 1994.

    a Do the data provide sufficient evidence to indicate a difference in the variability of the P/E ratio between electric utility common stocks and nonutility blue-chip stocks? Test using $\alpha = .10$.

    b Give the approximate p-value for the test, and interpret its value.

8.118 Refer to Exercise 8.117. Based on the results of part (a), do the data provide sufficient evidence to indicate differences in the mean P/E ratios between electric utility common stocks and nonutility blue-chip common stocks?

    a Test using $\alpha = .05$.

    b Find the approximate p-value for the test, and interpret it.

Table for Exercise 8.117

Electric Utility	P/E	Nonutility Blue-Chip	P/E
Carolina Power & Light	13	IBM	14
Minnesota Power & Light	14	Abbot Labs	16
TECO Energy	14	Minnesota Mining	18
Duke Power	13	Safeway Stores	16
Wisconsin Energy Corp.	17	Reynolds Metals	18
Pacific Gas & Electric	10	Monsanto	15
Montana Power	12	Hilton Hotels	28
Houston Industries	11	Textron	12
Illinois Power	13		
Pennsylvania Power & Light	10		

Source: Data from *Press-Enterprise*, Riverside, Calif., July 29, 1994.

8.119 Owing to the variability of trade-in allowances, the profit per new car sold by an automobile dealer varies from car to car. The profit per sale (in hundreds of dollars), tabulated for the past week, was as follows:

6.3	9.4
6.2	7.7
4.4	8.3

Do these data provide sufficient evidence to indicate that the average profit per sale is less than $780? Test at an $\alpha = .05$ level of significance.

8.120 A manufacturer of television sets claimed that his product possessed an average defect-free life of 3 years. Three households in a community have purchased the sets, and all three sets are observed to fail before 3 years, with failure times equal to 2.5, 1.9, and 2.9 years, respectively.

a Do these data present sufficient evidence to contradict the manufacturer's claim? Test at an $\alpha = .05$ level of significance.

b Calculate a 90% confidence interval for the mean life of the television sets.

8.121 Refer to Exercise 8.120. Approximately how many observations would be required to estimate the mean life of the television sets correct to within .2 year with probability equal to .90?

8.122 An experiment is conducted to compare two new automobile designs. Twenty people are randomly selected, and each person is asked to rate each design on a scale of 1 (poor) to 10 (excellent). The resulting ratings will be used to test the null hypothesis that the mean level of approval is the same for both designs against the alternative hypothesis that one of the automobile designs is preferred. Would these data satisfy the assumptions required for the Student's t test of Section 8.5? Explain.

8.123 The data in the accompanying table on lost-time accidents (mean production-hours lost per month over a period of one year) were collected both before and after an industrial safety program was put into effect. Data were recorded for six industrial plants. Do the data provide sufficient evidence to indicate whether the safety program was effective in reducing lost-time accidents? (Use $\alpha = .10$.)

	Plant Number					
Data Collected	1	2	3	4	5	6
Before program	38	64	42	70	58	30
After program	31	58	43	65	52	29

8.124 To compare the demand for two different entrees, the manager of a cafeteria recorded the number of purchases for each entree on seven consecutive days. The data are shown in the table. Do the data provide sufficient evidence to indicate a greater mean demand for one of the entrees?

	Entree	
Day	A	B
Mon.	420	391
Tues.	374	343
Wed.	434	469
Thurs.	395	412
Fri.	637	538
Sat.	594	521
Sun.	679	625

 8.125 The EPA limit on the allowable discharge of suspended solids into rivers and streams is 60 milligrams per liter (mg/l) per day. A study of water samples selected from the discharge at a phosphate mine showed that over a long period of time the mean daily discharge of suspended solids was 48 mg/l, but the day-to-day discharge readings were very variable. State inspectors measured the discharge rates of suspended solids for $n = 20$ days and found $s^2 = 39(\text{mg/l})^2$. Find a 90% confidence interval for σ^2. Interpret your results.

 8.126 A manufacturer of electric motors compared the productivity of assembly workers for two types of 40-hour weekly work schedules, four ten-hour days (schedule 1) and the standard five eight-hour days (schedule 2). Twenty workers were assigned to each work schedule, and the number of units assembled was recorded for a one-week period. The sample means (in hundreds of units) and variances for the two schedules are shown in the accompanying table.

	Schedule	
Statistic	1	2
Sample mean	43.1	44.6
Sample variance	4.28	3.89

a Do the data provide sufficient evidence to indicate a difference in the mean productivity for the two work schedules? Test using $\alpha = .05$.

b Give the approximate p-value for the test, and interpret it.

c Find a 95% confidence interval for the difference in mean productivity for the two work schedules, and interpret the interval.

8.127 Suppose that the manufacturer of Exercise 8.126 wants to estimate the difference in mean weekly productivity for the two work schedules correct to within one unit. How many workers would have to be included in each sample? [*Hint:* To solve, use the method of Section 7.9.]

 8.128 The mean percentage (or dollar) profit per project is not the only concern of a real estate developer (or any type of investor). The developer must be concerned with a large variation in gain, because a large negative gain (a loss) could put the developer out of business. A particular developer plans projects so as to achieve a mean profit per project of 12% with a range no larger than 25%. A sampling of the percentage profit per project for the last 25 of the developer's projects produced a sample mean and standard deviation equal to 11.1% and 5.2%, respectively.

a   Suppose that the developer wants to be fairly certain that the range of the percentage profit per project is no more than 25%. What value of σ will achieve this goal? [*Hint:* Almost all of the observations in a population will fall within 3σ of the population mean μ.]

b   Do the data provide sufficient evidence to indicate that the variation in percentage profit per job is greater than the value of σ specified in part (a)? Test using $\alpha = .05$.

8.129   Refer to Exercise 8.128. Find a 95% confidence interval for the variance of the percentage profit per job for the developer, and interpret the interval.

Exercises Using the Data Disk

8.130   Refer to data set A on the data disk. For a fixed faculty rank (one different from the one you selected in Exercise 7.102) select a sample of size $n_1 = 10$ from among the 246 male salaries, and, independently, select another sample of size $n_2 = 10$ from among the 246 female salaries.

a   Use your sample results (assuming a common underlying variance) to test $H_0 : \mu_M - \mu_F = 0$ versus $H_a : \mu_M - \mu_F \neq 0$ with $\alpha = .05$.

b   Construct a 95% confidence interval estimate of $\mu_M - \mu_F$. Is your estimate consistent with the results of part (a)?

8.131   Refer to data set A on the data disk. Again, select a faculty rank (the same rank chosen in 8.130 would be best) and now select a paired sample of size $n = 10$ whereby you randomly select $n = 10$ colleges and record both the male and female salaries for each of the $n = 10$ colleges.

a   Using your paired samples, test the null hypothesis $H_0 : \mu_M - \mu_F = \mu_d = 0$ versus $H_a : \mu_M - \mu_F \neq 0$ with $\alpha = .05$. Would your conclusion change if the alternative hypothesis were $H_a : \mu_M - \mu_F = \mu_d > 0$?

b   Construct a 95% confidence interval estimate of $\mu_M - \mu_F$.

c   Compare the value of $s_{\bar{d}}$ found in part (a) with the value of $s_{\bar{x}_M - \bar{x}_F}$ found in Exercise 8.130. Does the paired-difference design provide more precision in inferences concerning $\mu_M - \mu_F = \mu_d$?

8.132   Refer to the average maturity data (data set C on the data disk). Draw a random sample of size $n = 30$ (you may use one of the samples you used in Exercise 7.104).

a   Since the mean of the population from which you are sampling is, in fact, $\mu = 38.56$, a test of hypothesis $H_0 : \mu = 38.56$ should not be rejected. Conduct this test using a two-tailed alternative, with $\alpha = .01$. Do you arrive at the correct conclusion?

b   A test of the hypothesis $H_0 : \mu = 20.00$ should be rejected (since we know that $\mu = 38.56$ for this population). Conduct this test using your sample information and a two-tailed alternative, with $\alpha = .01$. Do you arrive at the correct conclusion?

c   Is it likely that you will be able to reject a test of the hypothesis $H_0 : \mu = 37.00$ even though we know that this is not the true population mean? Explain.

9

THE ANALYSIS OF VARIANCE

About This Chapter

Methods for comparing two population means, based on independent random samples and on a paired-difference experiment, were presented in Chapter 8. Chapter 9 extends these analyses to the comparison of any number of population means, using a technique called an analysis of variance. In this chapter, we explain the logic of an analysis of variance and give the analyses for three experimental designs.

Contents

"NOBODY COMPARES TO LUCKY"

Television commercials and newspaper ads that appear regularly in Southern California report that "Nobody compares to Lucky" (*The Press-Enterprise,* Riverside, California, June 23, 1994, p. F-7). In fact, of full-service supermarkets, Lucky refers to itself as "The low-price leader. Every day." Lucky is one of several large grocery chains that operate in California in competition with Albertsons, Alpha Beta, Ralphs, and Vons. However, in an ad run the previous month (*The Press-Enterprise,* Riverside, California, May 26, 1994, p. F-3), Lucky's survey technique of initially selecting 1500 items but picking only 501 items for inclusion in their survey was criticized by one of their competitors, Alpha Beta. The point of their ad was that "anyone can win a survey when it's their own."

All of this followed the results of independent surveys conducted by various grocery chains in Southern California. One survey was conducted by Albertsons and was based on approximately 95 items selected by a survey firm in various Southern California cities on four dates beginning in December 1992 and ending in January 1993. The results of this survey, shown in Table 9.1, reported the prices for a market basket consisting of approximately 95 items.

TABLE 9.1
A comparison of prices for a market basket of a fixed number of grocery items

Week of	Albertsons	Ralphs	Vons	Alpha Beta	Lucky
Dec. 30, 1992	$254.26	$256.03	$267.92	$260.71	$258.84
Jan. 7, 1993	240.62	255.65	251.55	251.80	242.14
Jan. 14, 1993	231.90	255.12	245.89	246.77	246.80
Jan. 19, 1993	234.13	261.18	254.12	249.45	248.99
Average	240.23	256.99	254.87	252.18	249.19

Source: The Press-Enterprise, Riverside, California, February 11, 1993, p. E-8.

A cursory examination of Table 9.1 shows that Albertsons appears to have the lowest price among the five grocery chains surveyed, Lucky stores have the second lowest, and the other three have higher prices than Lucky. Are these real differences, or is the price of a market basket for the five stores really the same and we are simply observing random variation in the prices?

You will notice that this sampling situation is similar to the paired-difference experiment of Section 8.6. The major difference is that we are comparing five population means, rather than two. The pairing, or matching in this case, occurs because the stores are each surveyed during the same weeks. Since prices may vary from week to week, comparing the prices for all supermarkets during the same weeks enables us to remove between-weeks variation from the comparison. The experimental design used here is called a randomized block design, with the weeks as blocks and the supermarkets as treatments.

In this chapter, we extend the two-sample procedures of Chapter 8 to methods for comparing two or more means using a procedure called an analysis of variance.

9.1   The Motivation for an Analysis of Variance

Suppose that you want to compare the mean size of health insurance claims submitted by five groups of policyholders. Ten claims were randomly selected from among the claims submitted for each group. The data are shown in Table 9.2. Do the data contained in the five samples provide sufficient evidence to indicate a difference in the mean levels of claims among the five health groups?

TABLE **9.2**
Insurance claims submitted by
five health groups

Group 1	Group 2	Group 3	Group 4	Group 5
$ 763	$1335	$ 596	$3742	$1632
4365	1262	1448	1833	5078
2144	217	1183	375	3010
1998	4100	3200	2010	671
5412	2948	630	743	2145
957	3210	942	867	4063
1286	867	1285	1233	1232
311	3744	128	1072	1456
863	1635	844	3105	2735
1499	643	1683	1767	767

$\bar{x}_1 = 1959.8$ $\bar{x}_2 = 1996.1$ $\bar{x}_3 = 1193.9$ $\bar{x}_4 = 1674.7$ $\bar{x}_5 = 2278.9$

To answer this question, we might think of comparing the means in pairs by using repeated applications of the Student's t test of Section 8.5. If we were to detect a difference between any pair of means, then we would conclude that there is evidence of at least one difference among the means and it would appear that we had answered our question. The problem with this procedure is that there are $C_2^5 = 10$ different pairs of means that have to be tested. Even if all of the means are identical, we have a probability α of rejecting the null hypothesis that a particular pair of means are equal. When this test procedure is repeated ten times, the probability of incorrectly concluding that at least one pair of means differ is quite high. Because the risk of an erroneous decision may be quite large, we look for a single test of the null hypothesis, that the five group means μ_1, μ_2, ..., μ_5 are equal, against the alternative hypothesis, that at least one pair of means differ.

The procedure for comparing more than two population means is known as an **analysis of variance.** The logic underlying an analysis of variance can be seen by examining the dot diagrams for two sets of *sample* data for two different cases, (a) and (b). In case (a), the first sample contains $n_1 = 3$ measurements, 1, 8, and 3. The second sample contains $n_2 = 2$ measurements, 2 and 10. The dot diagrams for these two samples appear in Figure 9.1(a). For case (b), the first sample contains $n_1 = 3$ measurements, 4.5, 3, and 4.5. The second sample contains two measurements, 5.5 and 6.5. The dot diagrams for case (b) are shown in Figure 9.1(b). Examining the dot diagrams, you can see that for both cases, $\bar{x}_1 = 4$ and $\bar{x}_2 = 6$. Which case, do you think, suggests a difference between population means μ_1 and μ_2?

F I G U R E 9.1
Dot diagrams for comparing sample means

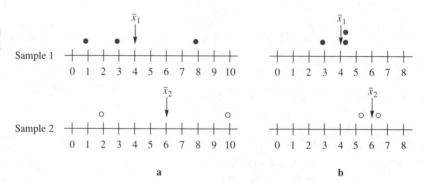

From a visual observation of the dot diagrams and your intuition, we think that you will agree that case (b) suggests a difference between the population means and case (a) does not. To arrive at this conclusion, we compared the *variation* (difference) *between the sample means* with the *variation within the samples*. Although $\bar{x}_1 - \bar{x}_2 = -2$ for both cases, this difference is small in case (a) in comparison to the large amount of variation within each of the two samples. In contrast, the difference in sample means in case (b) is very large in comparison to the variation within each of the two samples.

An analysis of variance to compare more than two population means formalizes the visual comparison of the variation between means with the variation within samples. This comparison of two sources of variation will lead us to the analysis of variance F test in Section 9.4.

9.2 The Assumptions for an Analysis of Variance

The assumptions upon which the test and estimation procedures for an analysis of variance are based are similar to those required for the Student's t statistic of

Chapter 8. Regardless of the sampling procedure employed to collect the data, we assume that the observations within each sampled population are normally distributed with a common variance σ^2.

In this chapter we will describe the analysis of variance for three different experimental designs. Two are based on independent random sampling from the respective populations. The third is an extension of the matched-pairs design of Chapter 8 and involves the random assignment of treatments within matched sets of observations. The sampling procedure for each of these designs will be restated in their respective sections.

The assumptions for an analysis of variance are given in the following display.

Assumptions Underlying Analysis of Variance Test and Estimation Procedures

1   The observations within each population are normally distributed with a common variance σ^2.

2   Assumptions regarding the sampling procedure are specified for each design in the sections that follow.

9.3 The Completely Randomized Design: A One-Way Classification

The analysis of experimental data depends on the design of the experiment, which refers to the way the data were collected. A very useful and relatively simple design, called the **completely randomized design,** is one in which random samples are independently selected from each of k populations. This design results in observations that are classified only according to the population from which they came, hence the designation as a one-way classification.

In some instances, this design is implemented by randomly selecting independent samples from the appropriate populations. For example, in assessing voter attitudes concerning various items on the ballot for the next state election, we may wish to select random samples of registered voters in each of k areas within the state. The k populations of registered voters exist in fact, and the design will specify how the registered voters in each of the k samples are to be selected. In this situation, we need a randomized device, such as a table of random numbers, to select those individuals who will be included in each of the samples from the k areas.

There are other situations in which we wish to determine the effect of various procedures, usually referred to as treatments, on a variable of interest. It may be that the populations exist only in concept, and our observations comprise samples from these conceptual populations. For example, we may be interested in evaluating several insecticides for use in agricultural pest control of the boll weevil in cotton. A completely randomized design in this situation would consist of the random assignment of the insecticides to experimental units consisting of a row or a fixed number of cotton plants. The sample observations from experimental units treated using insecticide number 1 constitute a sample from the conceptual population of all cotton plants treated with insecticide 1, and similarly for the other samples. If equal numbers of observations for each of the k treatments are to be taken, then a table of random numbers is used to first select the n experimental units to receive insecticide 1. The next n of the remaining units selected will receive insecticide 2, and so on. In both this example and the earlier example concerning voters, the design used was a completely randomized design resulting in a one-way classification of the data.

9.4 An Analysis of Variance for a Completely Randomized Design

The methodology, an analysis of variance, derives its name from the manner in which the quantities used to measure variation are acquired. Suppose that we want to compare k population means $\mu_1, \mu_2, \ldots, \mu_k$, based on independent random samples of $n_1, n_2, \ldots, n_k$ observations selected from the populations $1, 2, \ldots, k$, respectively. Thus for the health claims data in Table 9.2 we want to compare the means for $k = 5$ insurance groups based on random samples of $n_1 = n_2 = \cdots = n_5 = 10$ claims per group. Now let x_{ij} represent the jth measurement ($j = 1, \ldots, 10$) in the ith health group ($i = 1, \ldots, 5$). Then it can be shown (proof omitted) that the sum of squares of deviations of all $n = n_1 + n_2 + \cdots + n_5 = 50$ x-values about their overall mean $\bar{x}$, often called the **total sum of squares** (or Total SS),

$$\text{Total SS} = S_{xx} = \sum_{i=1}^{k} \sum_{j=1}^{n_i} (x_{ij} - \bar{x})^2$$

can be partitioned into two components. The first component, called the **sum of squares for treatments (SST),** is used to calculate a measure of the variation between sample means. For our example, the **treatments** are the five insurance groups. The second component, called the **sum of squares for error (SSE),** is used to measure the variation within samples. Thus

$$\text{Total SS} = \text{SST} + \text{SSE}$$

This partitioning of the Total SS into relevant components that measure sources of variation explains why the procedure is called an analysis of variance.

It is not difficult to calculate the sums of squares needed in an analysis of variance, but it is tedious. Since most statistical computer program packages contain programs to conduct analyses of variance for various experimental designs, we will discuss the procedure while explaining and interpreting the SAS printout for an analysis of variance of the health insurance claims data in Table 9.2. Corresponding Minitab and Excel printouts will also be presented so that you can compare the outputs for an analysis of variance for the same set of data. The computing formulas, which will enable you to perform the analysis of variance by using a hand calculator, are presented in optional Section 9.5.

E X A M P L E   **9.1**   Interpret the SAS analysis of variance printout and present corresponding Minitab and Excel printouts for the comparison of mean health insurance claims for the five health groups (see Table 9.2).

Solution   The SAS, Minitab, and Excel computer printouts for an analysis of variance of the health insurance claims data are shown in Tables 9.3, 9.4, and 9.5, respectively. Since the three printouts are similar, we will box and explain the meanings of the various quantities that appear in the SAS printout in Table 9.3, and we will box (where available) the corresponding quantities in the Minitab and Excel printouts in Tables 9.4 and 9.5.

T A B L E **9.3**   SAS computer printout for an analysis of variance of the health insurance data of Table 9.2

ANALYSIS OF VARIANCE PROCEDURE

DEPENDENT VARIABLE: COST

①

SOURCE	DF	SUM OF SQUARES	MEAN SQUARE	F VALUE	PR > F	R-SQUARE	C.V.
MODEL	4	6742554.48000000	1685638.62000000	0.98	0.4281	0.080122	72.0381
ERROR	45	77411264.40000000	1720250.32000000		ROOT MSE ④		COST MEAN
CORRECTED TOTAL	49	84153818.88000000			1311.58313499		1820.68000000

②

SOURCE	DF	ANOVA SS	F VALUE	PR > F ③
GROUP	4	6742554.48000000	0.98	0.4281

T A B L E **9.4**   Minitab computer printout for an analysis of variance of the health insurance data of Table 9.2

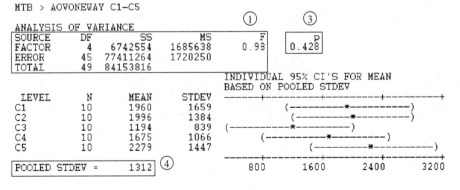

```
MTB > AOVONEWAY C1-C5

ANALYSIS OF VARIANCE              ①              ③
SOURCE    DF        SS       MS        F          P
FACTOR     4   6742554  1685638    0.98       0.428
ERROR     45  77411264  1720250
TOTAL     49  84153816
                                  INDIVIDUAL 95% CI'S FOR MEAN
                                  BASED ON POOLED STDEV
LEVEL      N      MEAN    STDEV   ------+---------+---------+---------+
C1        10      1960     1659            (----------*----------)
C2        10      1996     1384            (----------*----------)
C3        10      1194      839   (----------*----------)
C4        10      1675     1066      (----------*----------)
C5        10      2279     1447              (---------*----------)
                                  ------+---------+---------+---------+
POOLED STDEV =        1312  ④      800     1600     2400     3200
```

1   The table in area ① in the SAS printout is called an **analysis of variance** or **ANOVA table.** This table contains five columns. The three sources of variation are listed under SOURCE in column 1:

T A B L E **9.5**   Excel computer printout for an analysis of variance of the health insurance data of Table 9.2

	A	B	C	D	E	F	G	H
1			Analysis of Many Samples for CBS9-1.Claims by CBS9-1.Group					
2								
3	Class	Value	Sample Size		Mean		Standard Deviation	
4	1	1	10		1959.8		1658.8	
5	2	2	10		1996.1		1384.0	
6	3	3	10		1193.9		838.9	
7	4	4	10		1674.7		1066.3	
8	5	5	10		2278.9		1446.9	
9								
10			Oneway ANOVA for CBS9-1.Claims					③
11	Source of Variation		Sum of Squares		D.F.	Mean Square	F-Ratio	P Value
12	Group		6742550		4	1685640	0.9800	0.4281
13	Error		77411300		45	1720250		
14	Total (corr.)		84153850		49			①
15								
16			95% Confidence Intervals (pooled)					
17	Class	Value	Sample Size		Mean	+/-	Interval	
18	1	1	10		1959.8		835.4	
19	2	2	10		1996.1		835.4	
20	3	3	10		1193.9		835.4	
21	4	4	10		1674.7		835.4	
22	5	5	10		2278.9		835.4	

MODEL: This source, sometimes identified as TREATMENTS, represents variation among the sample means. Minitab calls it FACTOR, and Excel identifies it as Group.

ERROR: This source measures the variation within samples.

CORRECTED TOTAL: This source measures the variation of all x values about the overall mean of all $n = 50$ x values.

2   The sums of squares of deviations corresponding to the three sources of variation are shown in column 3 of area ① (column 2 in the Excel printout). The sum of squares corresponding to MODEL, the sum of squares for treatments (or, for this example, insurance groups), SST, is

$$\text{SST} = 6742554.48000000$$

The formula for computing SST is given in Section 9.5.

The sum of squares corresponding to ERROR is a measure of the variability of the x values within samples. Thus

$$\text{SSE} = 77411264.40000000$$

SSE is the pooled sum of squares of deviations of the observations about their respective sample means; that is,

$$\text{SSE} = \sum_{j=1}^{n_1}(x_{1j} - \bar{x}_1)^2 + \sum_{j=1}^{n_2}(x_{2j} - \bar{x}_2)^2 + \cdots + \sum_{j=1}^{n_5}(x_{5j} - \bar{x}_5)^2$$

This computing formula is given in Section 9.5. Finally, the sum of squares of deviations corresponding to the CORRECTED TOTAL is what we have called Total SS, that is,

$$\text{Total SS} = \sum_{i=1}^{k} \sum_{j=1}^{n_i} (x_{ij} - \bar{x})^2 = 84153818.88000000$$

You can verify that

$$\text{SST} + \text{SSE} = \text{Total SS}$$

that is,

$$6742554.48 + 77411264.40 = 84153818.88$$

3 Each sum of squares of deviations, divided by the appropriate number of degrees of freedom, will provide an estimate of σ^2 when the null hypothesis is true, in other words, when

$$\mu_1 = \mu_2 = \cdots = \mu_5$$

These degrees of freedom are shown in column 2 (column 3 in Excel).

a The degrees of freedom for the CORRECTED TOTAL (Total SS) will always be $(n - 1)$, or, for this example, $50 - 1 = 49$.

b The degrees of freedom for MODEL (or insurance groups) will always equal one less than the number k of populations, in this case, $(k - 1) = 5 - 1 = 4$.

c The number of degrees of freedom for ERROR will always equal $n_1 + n_2 + \cdots + n_k - k = n - k$, or, for this example, $50 - 5 = 45$. Note that the sum of the numbers of degrees of freedom for MODEL and ERROR will always equal the number of degrees of freedom for the CORRECTED TOTAL; that is, $4 + 45 = 49$.

4 Column 4 of the ANOVA table, headed MEAN SQUARE, gives the estimates of σ^2 based on the variation among the sample means (in the row corresponding to MODEL) and the variation within samples (in the row corresponding to ERROR) when the null hypothesis is true—that is, when $\mu_1 = \mu_2 = \cdots = \mu_5$. These estimates are calculated by dividing a sum of squares by its corresponding degrees of freedom. Thus the **mean square for treatments** (MODEL), denoted MST and shown in column 4, is

$$\text{MST} = \frac{\text{SST}}{k - 1} = \frac{6742554.48}{4} = 1685638.62000000$$

Similarly, the **mean square for error,** denoted MSE or s^2 and shown in column 4, is

$$\text{MSE} = s^2 = \frac{\text{SSE}}{n - k} = \frac{77411264.4}{45} = 1720250.32000000$$

This quantity s^2 is the pooled estimate of σ^2 based on the sum of squares of deviations of the x values about their respective sample means and is an extension (since it is based on $k = 5$ samples) of the pooled estimate of σ^2 given in Section 8.5.

5 The final step in testing $H_0 : \mu_1 = \mu_2 = \cdots = \mu_k$ is comparing the two estimates of σ^2: MST, which is based on the variation of the sample means about $\bar{x}$, and MSE $= s^2$, which is based on the variation of the x values about their respective sample means. We use the F statistic of Section 8.9. Thus, when H_0 is true, the sampling distribution of

$$F = \frac{\text{MST}}{\text{MSE}}$$

will be an F distribution with $v_1 = k - 1$ (for example, $k - 1 = 5 - 1 = 4$) numerator degrees of freedom and $v_2 = n - k$ (for our example, $n - k = 45$) denominator degrees of freedom. If H_0 is false—that is, if $\mu_1, \mu_2, \ldots, \mu_k$ are not all equal—the estimate of σ^2 based on MST will be overly large, and the calculated value of F will be larger than expected. Consequently, we reject H_0 for large values of F, that is, values of F larger than some critical value F_α (see Figure 9.2). The critical values of F corresponding to various values of v_1 and v_2 and for $\alpha = .10, .05, .025, .01,$ and $.005$ are shown in Table 6 of Appendix II. (The use of tables is explained in Section 8.9.)

FIGURE 9.2

Rejection region for the analysis of variance F test

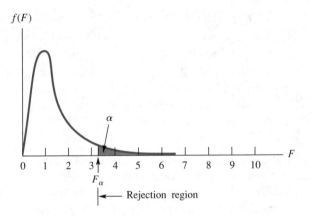

For example, if we want to test

$$H_0 : \mu_1 = \mu_2 = \mu_3 = \mu_4 = \mu_5$$

for $\alpha = .05$, we consult Table 6 in Appendix II and look for the F value corresponding to $v_1 = 4$ and $v_2 = 45$. Table 6 does not give this F value, but it does give the F value for $v_1 = 4$ and $v_2 = 40$ as 2.61 and the F value for $v_1 = 4$ and $v_2 = 60$ as 2.53. Consequently, we will reject H_0 if the computed value of F is larger than 2.61 (or actually a number slightly smaller). The computed value of F,

$$F = \frac{\text{MST}}{\text{MSE}} = .98$$

is shown in column 4 of area ②. You can see that this computed value of F, .98, does not exceed the critical value and therefore does not fall in the rejection region. Consequently, there is not sufficient evidence to indicate that the mean claim size differs among the five insurance groups.

6   The observed significance level, the probability of observing a value of the F statistic as large as or larger than .98, is shown in area ③ of the SAS printout. The p-value for the test is .4281. This large p-value is consistent with the results of the F test (step 5). For $\alpha = .05$, we will reject H_0 when the p-value is less than or equal to .05.

7   An analysis of variance can be conducted (as you will see in the following sections) to partition the Total SS into sums of squares corresponding to two or more sources of variation in addition to SSE. The SAS analysis of variance table in area ① always combines these sources into a single source designated as MODEL. The MODEL source is partitioned and shown in area ②. For our example, there is only one source in addition to SSE, namely, the sum of squares of deviations corresponding to treatments (insurance groups). Consequently, the sum of squares for MODEL is identical to the sum of squares for GROUP (insurance groups).

8   The standard deviation s, shown in area 4, is used to construct a confidence interval for a single mean or for the difference between a pair of means. Thus

$$s = \sqrt{\text{MSE}} = \sqrt{1720250.32} = 1311.58313499$$

For example, if we want to find a $(1 - \alpha)100\%$ confidence interval for a population mean—say, that for the mean size of a claim μ_4 for health group 4—we use the formula (given in Section 7.5)

$$\bar{x}_4 \pm t_{\alpha/2} \frac{s}{\sqrt{n_4}}$$

where $\bar{x}_4$ is the sample mean for insurance group 4 (given in Table 9.2), $s = 1311.58313499$, $n_4 = 10$, and $t_{\alpha/2}$ is based on $(n - k) = 50 - 5 = 45$ degrees of freedom (d.f.), the number of degrees of freedom associated with MSE $= s^2$. The t table, Table 4 in Appendix II, does not give the t values for 45 d.f., but you can see that the value $t_{.025}$ will be close to 2.0. Therefore, the 95% confidence interval for μ_4, the mean claim for health insurance group 4, is

$$\bar{x}_4 \pm t_{\alpha/2} \frac{s}{\sqrt{n_4}}$$

$$1674.7 \pm (2.0) \frac{1311.6}{\sqrt{10}} \qquad \text{or} \qquad \$845.20 \text{ to } \$2504.20$$

Means, standard deviations, and one-sample 95% confidence intervals are given on both the Minitab and Excel computer printouts.

If we want to estimate the difference in the size of the mean claims between health insurance groups 1 and 3 by using a $(1 - \alpha)100\%$ confidence interval, we use the formula (Section 7.6)

$$(\bar{x}_1 - \bar{x}_3) \pm t_{\alpha/2} s \sqrt{\frac{1}{n_1} + \frac{1}{n_3}}$$

For a 95% confidence interval, the value of $t_{.025}$ for d.f. = 45 will be approximately 2.0, $s = 1311.6$, and the values of $\bar{x}_1$ and $\bar{x}_3$ were shown in Table 9.2. Then the 95% confidence interval for $(\mu_1 - \mu_3)$ is

$$(\bar{x}_1 - \bar{x}_3) \pm t_{.025} s \sqrt{\frac{1}{n_1} + \frac{1}{n_3}}$$

$$(1959.8 - 1193.9) \pm (2.0)(1311.6)\sqrt{\frac{1}{10} + \frac{1}{10}}$$

$$765.9 \pm 1173.1$$

or from −$407.20 to $1939.00. Thus we estimate the difference in mean claims for groups 1 and 3 to be in the interval −$407.20 to $1939.00. Because this interval includes zero as a possible value, a t test of $H_0 : \mu_1 = \mu_3$ would not lead to rejection of H_0. There is not sufficient evidence to indicate that μ_1 and μ_3 differ.

9   The quantity R-SQUARE is related to a multiple regression analysis. The significance of R^2 will be explained in Section 12.3.

Now that we have explained the SAS output, compare the SAS output with the Minitab and the Excel outputs in Tables 9.4 and 9.5. You will be able to locate the relevant sources of variation, degrees of freedom, sums of squares of deviations, and mean squares on the Minitab and Excel printouts, and you will see the quantities necessary to conduct the F test for comparing the $k = 5$ population means.   ▪

The typical ANOVA table for an analysis of variance for k independent random samples is shown in the first display below. The other two displays summarize the analysis of variance F test and give the confidence intervals for treatment means.

ANOVA Table for k Independent Random Samples

Source	d.f.	SS	MS	F
Treatments	$k - 1$	SST	$\text{MST} = \text{SST}/(k - 1)$	MST/MSE
Error	$n - k$	SSE	$\text{MSE} = \text{SSE}/(n - k)$	
Total	$n - 1$	Total SS		

F Test for Comparing k Population Means

1   Null Hypothesis: $H_0 : \mu_1 = \mu_2 = \cdots = \mu_k$

2   Alternative Hypothesis: H_a : one or more pairs of population means differ.

3   Test Statistic: $F = \text{MST/MSE}$, where F is based on $v_1 = (k - 1)$ and $v_2 = (n - k)$ degrees of freedom.

4   Rejection Region: Reject if $F > F_\alpha$, where F_α lies in the upper tail of the F distribution (with $v_1 = k - 1$ and $v_2 = n - k$) and satisfies the expression $P(F > F_\alpha) = \alpha$.

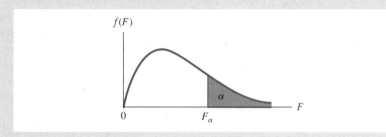

Assumptions:

1   The samples have been randomly and independently selected from their respective populations.

2   The populations are normally distributed with means μ_1, μ_2, ..., μ_k and equal variances, $\sigma_1^2 = \sigma_2^2 = \cdots = \sigma_k^2 = \sigma^2$.

Independent Random Samples: $(1 - \alpha)100\%$ Confidence Intervals for a Single Treatment Mean and the Difference Between Two Treatment Means

Single treatment mean:

$$\bar{x}_i \pm t_{\alpha/2} \frac{s}{\sqrt{n_i}}$$

Difference between two treatment means:

$$(\bar{x}_i - \bar{x}_j) \pm t_{\alpha/2} s \sqrt{\frac{1}{n_i} + \frac{1}{n_j}}$$

$$s = \sqrt{s^2} = \sqrt{\text{MSE}} = \sqrt{\frac{\text{SSE}}{n_1 + n_2 + \cdots + n_k - k}}$$

where

$$n = n_1 + n_2 + \cdots + n_k$$

and $t_{\alpha/2}$ is based on $(n - k)$ degrees of freedom.

9.5 Computing Formulas: An Analysis of Variance for a Completely Randomized Design (Optional)

Suppose that we want to compare k population means μ_1, μ_2, ..., μ_k, based on independent random samples, n_1 observations from population 1, n_2 observations from population 2, ..., and n_k observations from population k. We will employ the symbols given in the following display for the quantities used to perform an analysis of variance of the data.

Notation

	Population			
Statistic	1	2	$\cdots$	k
Sample size	n_1	n_2	$\cdots$	n_k
Total of all observations in the sample	T_1	T_2	$\cdots$	T_k
Sample mean	$\bar{x}_1$	$\bar{x}_2$	$\cdots$	$\bar{x}_k$

$$T = \text{Total of all } n = n_1 + n_2 + \cdots + n_k \text{ observations} = \sum_{ij} x_{ij}$$

$$= T_1 + T_2 + \cdots + T_k$$

A summary of the computing formulas is given in the following display.

Summary of Computing Formulas: Analysis of Variance for a Completely Randomized Design, k Treatments

$$\text{CM}^\dagger = \frac{(T)^2}{n}$$

---

†In the summary of formulas, the correction for the mean, abbreviated CM, is used to convert a sum of squares into a sum of squared deviations about the mean.

$$\text{Total SS} = \sum_{ij} x_{ij}^2 - \text{CM}$$

$$= \text{Sum of squares of all } x \text{ values} - \text{CM}$$

where

$$n = n_1 + n_2 + \cdots + n_k$$

and

$$T = \text{Total of all } n \text{ observations}$$

$$\text{SST} = \sum_{i=1}^{k} \frac{T_i^2}{n_i} - \text{CM} \qquad \text{MST} = \frac{\text{SST}}{k - 1}$$

$$\text{SSE} = \text{Total SS} - \text{SST} \qquad \text{MSE} = \frac{\text{SSE}}{n - k}$$

Tips on Problem Solving

The following suggestions apply to all the analyses of variance in this chapter.

1   When calculating sums of squares, be certain to carry at least six significant figures before performing subtractions.

2   Remember, sums of squares can never be negative. If you obtain a negative sum of squares, you have made a mistake in arithmetic.

3   Always check your analysis of variance table to make certain that the degrees of freedom sum to the total degrees of freedom $n - 1$ and that the sums of squares sum to the Total SS.

Exercises

Starred (*) exercises are optional.

Basic Techniques

9.1   Suppose that you wish to compare the means of six populations based on independent random samples, each of which contains ten observations. Insert in an ANOVA table the sources of variation and their respective degrees of freedom.

9.2   The values of Total SS and SSE for the experiment in Exercise 9.1 are Total SS = 21.4 and SSE = 16.2.

a   Complete the ANOVA table for Exercise 9.1.

b   How many degrees of freedom are associated with the F statistic for testing $H_0 : \mu_1 = \mu_2 = \cdots = \mu_6$?

c   Give the rejection region for the test in part (b) for $\alpha = .05$.

d   Do the data provide sufficient evidence to indicate differences among the population means?

9.3   The sample means corresponding to populations 1 and 2 in Exercise 9.1 are $\bar{x}_1 = 3.07$ and $\bar{x}_2 = 2.52$.

    a   Find a 95% confidence interval for μ_1.

    b   Find a 95% confidence interval for the difference $(\mu_1 - \mu_2)$.

9.4   Suppose that you wish to compare the means of four populations based on independent random samples, each of which contains six observations. Insert in an ANOVA table the sources of variation and their respective degrees of freedom.

9.5   The values of Total SS and SST for the experiment in Exercise 9.4 are Total SS = 473.2 and SST = 339.8.

    a   Complete the ANOVA table for Exercise 9.4.

    b   How many degrees of freedom are associated with the F statistic for testing $H_0 : \mu_1 = \mu_2 = \mu_3 = \mu_4$?

    c   Give the rejection region for the test in part (b) for $\alpha = .10$.

    d   Do the data provide sufficient evidence to indicate differences among the population means?

9.6   The sample means corresponding to populations 1 and 2 in Exercise 9.4 are $\bar{x}_1 = 88.0$ and $\bar{x}_2 = 83.9$.

    a   Find a 90% confidence interval for μ_1.

    b   Find a 90% confidence interval for the difference $(\mu_1 - \mu_2)$.

9.7   A portion of the ANOVA table for a completely randomized design is shown in the accompanying table.

Source	d.f.	SS	MS	F
Treatments	4	26.3		
Error		52.8		
Total	29			

    a   How many independent random samples were selected in this experiment?

    b   Does the ANOVA table provide the information necessary to determine the sample sizes?

    c   How many observations were involved in the complete design?

    d   Fill in the blanks in the ANOVA table.

    e   Give the rejection region for the analysis of variance F test.

    f   Do the data provide sufficient evidence to indicate a difference between at least two of the population means? Test using $\alpha = .05$.

*9.8   The following data are observations collected by using a completely randomized design.

Sample 1	Sample 2	Sample 3
3	4	2
2	3	0
4	5	2
3	2	1
2	5	

a Calculate CM and Total SS.

b Calculate SST and MST.

c Calculate SSE and MSE.

d Construct an ANOVA table for the data.

e State the null and alternative hypotheses for an analysis of variance F test.

f Give the rejection region for the test, using $\alpha = .05$.

g Conduct the test and state your conclusions.

*9.9 Refer to Exercise 9.8. Do the data provide sufficient evidence to indicate a difference between μ_2 and μ_3? Test using the t test of Section 8.5 with $\alpha = .05$.

*9.10 Refer to Exercise 9.8.

a Find a 90% confidence interval for μ_1.

b Find a 90% confidence interval for $(\mu_1 - \mu_3)$.

*9.11 The following data are observations collected by using a completely randomized design.

Sample 1	Sample 2	Sample 3	Sample 4
2	6	3	3
4	2	0	6
0	5	5	4
	4	2	
		3	

a Calculate CM and Total SS.

b Calculate SST and MST.

c Calculate SSE and MSE.

d Construct an ANOVA table for the data.

e State the null and alternative hypotheses for an analysis of variance F test.

f Give the rejection region for the test, using $\alpha = .05$.

g Conduct the test and state your conclusions.

*9.12 Refer to Exercise 9.11. Do the data provide sufficient evidence to indicate a difference between μ_1 and μ_2? Test using the t test of Section 8.5 with $\alpha = .05$.

*9.13 Refer to Exercise 9.11.

a Find a 90% confidence interval for μ_3.

b Find a 90% confidence interval for $(\mu_1 - \mu_3)$.

Applications

9.14 An experiment was conducted to compare the price of a loaf of bread (a particular brand) at four city locations. Four stores were randomly sampled in locations 1, 2, and 3, but only two were selected from location 4 (only two stores carried the brand). Note that a completely randomized design was employed. Conduct an analysis of variance for the data shown in the accompanying table. A Minitab computer printout of the analysis of variance for the data is shown below. Use the information in the printout to answer the following questions.

Location	Price (dollars)			
1	1.59	1.63	1.65	1.61
2	1.58	1.61	1.64	1.63
3	1.54	1.59	1.55	1.58
4	1.69	1.70		

```
MTB > AOVONEWAY C1-C4

ANALYSIS OF VARIANCE
SOURCE     DF        SS        MS         F         P
FACTOR      3  0.022871  0.007624     13.03     0.001
ERROR      10  0.005850  0.000585
TOTAL      13  0.028721
                                    INDIVIDUAL 95% CI'S FOR MEAN
                                    BASED ON POOLED STDEV
  LEVEL     N      MEAN     STDEV  ----+---------+---------+---------+--
C1          4   1.6200    0.0258              (---*----)
C2          4   1.6150    0.0265              (---*-----)
C3          4   1.5650    0.0238   (-----*----)
C4          2   1.6950    0.0071                    (-------*-----)
                                    ----+---------+---------+---------+--
POOLED STDEV =     0.0242           1.560     1.620     1.680     1.740
```

a Do the data provide sufficient evidence to indicate a difference in mean price of the bread in stores located in the four areas of the city?

b Suppose that prior to seeing the data we wanted to compare the mean prices between locations 1 and 4. Estimate the difference in means, using a 95% confidence interval.

9.15 An experiment was conducted to compare the effectiveness of three training programs, A, B, and C, in training assemblers of a piece of electronic equipment. Fifteen employees were randomly assigned, five each, to the three programs. After completion of the courses, each person was required to assemble four pieces of the equipment, and the average length of time required to complete the assembly was recorded. Owing to resignation from the company, only four employees completed program A and only three completed B. The data are shown in the accompanying table. An SAS computer printout of the analysis of variance for the data follows. Use the information in the printout to answer the following questions.

Training Program	Average Assembly Time (minutes)				
A	59	64	57	62	
B	52	58	54		
C	58	65	71	63	64

ANALYSIS OF VARIANCE PROCEDURE

DEPENDENT VARIABLE: TIME

SOURCE	DF	SUM OF SQUARES	MEAN SQUARE	F VALUE	PR > F	R-SQUARE	C.V.
MODEL	2	170.45000000	85.22500000	5.70	0.0251	0.559005	6.3802
ERROR	9	134.46666667	14.94074074		ROOT MSE		TIME MEAN
CORRECTED TOTAL	11	304.91666667			3.86532544		60.58333333

SOURCE	DF	ANOVA SS	F VALUE	PR > F
PROGRAM	2	170.45000000	5.70	0.0251

a Do the data provide sufficient evidence to indicate a difference in mean assembly times for people trained by the three programs?

b Find a 90% confidence interval for the difference in mean assembly times between persons trained by programs A and B.

c Find a 90% confidence interval for the mean assembly time for persons trained by program A.

d Do you think the data will satisfy (approximately) the assumption that they have been selected from normal populations? Why?

9.16 Larry R. Smeltzer and Kittie W. Watson conducted an experiment to investigate the effect of four different instructional approaches for improving learning. The treatments are listed next.

1 no instruction on listening

2 a 45-minute lecture on listening skills

3 a 30-minute video model on effective listening skills

4 both of the educational exposures described in treatments 2 and 3

Ninety-nine subjects were randomly assigned to receive the treatments: 19 to the control group (treatment 1), 31 to treatment 2, 27 to treatment 3, and 22 to treatment 4. The educational messages presented in treatments 2, 3, and 4 emphasized the importance of asking questions and taking notes during discussions. After the treatments were applied, the students were scored on the basis of the numbers of questions asked. The analysis of variance table for the data is shown below.

Source	d.f.	SS	MS	F
Between groups	3	63.21	21.07	8.11*
Within groups	95	350.55	3.69	
Total	98	413.76		

Source: Smeltzer and Watson, 1985.
* Significant at the .01 level.

a Why is "no instruction on listening" called the "control group"?

b Do the data present sufficient evidence to indicate differences in the mean number of questions asked among the four treatment groups? Test using $\alpha = .05$.

c The mean number of questions asked for each of the four treatments is shown on the next page. Compare the most intensive educational treatment, treatment 4, with the control, treatment 1. Do the data present sufficient evidence to indicate a difference in mean number of questions asked? Test using $\alpha = .05$.

Treatment	Sample Size	Mean
1 Control group	19	1.36
2 Lecture group	31	1.87
3 Video role model	27	2.97
4 Lecture plus video role model	22	3.18

*9.17 Perform the analysis of variance calculations for Exercise 9.14, and present them in an analysis of variance table. Compare your answers with those given in the computer printout in that exercise. Interpret the results of the analysis of variance.

*9.18 Perform the analysis of variance calculations for Exercise 9.15, and present them in an analysis of variance table. Compare your answers with those given in the computer printout in that exercise. Interpret the results of the analysis of variance.

9.6 The Randomized Block Design

The completely randomized design or one-way classification introduced in Sections 9.3 and 9.4 generalized the design involving two independent samples presented in Chapter 8. This design was deemed appropriate when the experimental material was homogeneous and no sources of variation other than those due to the treatments and experimental error were expected to influence the response. If the experimental material is not homogeneous, we may be able to find groups of homogeneous units, called **blocks,** within which the means associated with the treatments under investigation may be compared. This type of design, which is an extension of the paired-difference design of Chapter 8, is called a **randomized block design.** Its main purpose is to remove the block-to-block variability that might otherwise hide the effect of the treatments.

A randomized block design utilizes blocks of k-matched or homogeneous experimental units, with one unit within each block assigned to each treatment. The design is said to be randomized because the treatments are randomly assigned to the units within a block. If the randomized block design involves k treatments within each of b blocks, then the total number of observations in the experiment is $n = bk$.

Suppose that the chief executive of a large construction corporation employs three experienced construction engineers to perform the time-consuming cost analyses, estimates, and bids for the work on large construction projects. Do those "estimators" tend to estimate at the same mean level, or does one or another tend to always submit a high (or a low) bid on projects? To answer this question, the executive can select independent random samples of projects estimated by each of the three estimators and compare the three means. This procedure will be valid but will require very large sample sizes to detect differences in means because of the large amount of variation in the cost levels of the individual projects.

A much easier method for detecting differences in the mean level of estimates for the three estimators is to conduct an experiment. Each of the three estimators will be required to analyze, estimate, and provide a bid price for the same project for each of a set of b (say $b = 5$) projects. Then the bid prices of the three estimators can be compared for the same project, thereby eliminating the project-to-project variation in bid prices.

As a second example of the use of blocking to increase the information in an experiment, suppose that a production superintendent wants to compare the mean time for an assembly-line operator to perform a job using one of three methods, A, B, and C. Each of the $b = 5$ operators is to perform the job by using each of the methods, A, B, and C. The objective of the blocking is to eliminate the variation in time to assemble caused by operator-to-operator differences in manual dexterity, motivation, and so on. Since the sequence in which the operator performs the three assembly operations may be important (e.g., fatigue may be a factor), each assembly-line operator should be assigned a random sequencing of the three methods. For example, operator 1 might be assigned to perform method C first, followed by A and B. Operator 2 might be assigned to perform method A first, then C and B.

Matching (or blocking) can take place in many different ways. As we have illustrated, comparisons of treatments are often made within blocks of time, within blocks of people, or within similar external environments. The exercises and the case study revisited in Section 9.14 will provide other examples of the use of randomized block designs.

9.7 An Analysis of Variance for a Randomized Block Design

An analysis of variance for a randomized block design partitions the total sum of squares of deviations of all x values about the overall mean $\bar{x}$ into three parts, the first measuring the variation among treatment means, the second measuring the variation among block means, and the third measuring the variation of the differences among the treatment observations *within* blocks (which measures experimental error). Thus

$$\text{Total SS} = \text{SST} + \text{SSB} + \text{SSE}$$

where

$$\text{Total SS} = \sum_{ij}(x_{ij} - \bar{x})^2$$

$\quad$ SST = Sum of squares for treatments

$\qquad$ = b(sum of squares of deviations of the treatment means about $\bar{x}$)

$\quad$ SSB = Sum of squares for blocks

$\qquad$ = k(sum of squares of deviations of the block means about $\bar{x}$)

$\quad$ SSE = Sum of squares for error = Total SS $-$ SST $-$ SSB

$\qquad$ = Unexplained variation

To avoid distracting you with computational formulas and computations, we will explain how to perform an analysis of variance by explaining and interpreting an SAS analysis of variance printout for an example. We will present corresponding Minitab and Excel printouts in case you have access to these program packages. The computing formulas and the hand calculator computations for the example are given in optional Section 9.8.

EXAMPLE 9.2 Refer to the comparison of the mean project bid price levels for the three construction project estimators described in Section 9.6. Each of the three estimators was required to analyze and determine a bid price for each of $b = 5$ projects. The data are shown in Table 9.6, and the SAS, Minitab, and Excel analysis of variance printouts are shown in Tables 9.7, 9.8, and 9.9. Describe and interpret the SAS printout.

TABLE 9.6
Bid price data ($million) for three estimators for each of five projects for Example 9.2

Estimator	Project 1	2	3	4	5	Total
1	3.52	4.71	3.89	5.21	4.14	21.47
2	3.39	4.79	3.82	4.93	3.96	20.89
3	3.64	4.92	4.19	5.10	4.20	22.05
Total	10.55	14.42	11.90	15.24	12.30	64.41

TABLE 9.7 SAS ANOVA printout for Example 9.2

ANALYSIS OF VARIANCE PROCEDURE

DEPENDENT VARIABLE: PRICE ①

SOURCE	DF	SUM OF SQUARES	MEAN SQUARE	F VALUE	PR > F	R-SQUARE	C.V.
MODEL	6	5.02352000	0.83725333	99.32	0.0001	0.986753	2.1382
ERROR	8	0.06744000	0.00843000		ROOT MSE		PRICE TIME
CORRECTED TOTAL	14	5.09096000			0.09181503 ④		4.29400000

②

SOURCE	DF	ANOVA SS	F VALUE	PR > F
ESTIMATOR	2	0.13456000	7.98	0.0124 ③
PROJECT	4	4.88896000	144.99	0.0001

Solution 1 The SAS printout for an analysis of variance always presents the information for an ANOVA table in two stages. The first stage, area ① in Table 9.7, shows only two sources of variation, MODEL and ERROR. The source MODEL *includes all sources of variation other than ERROR.*

2 The source MODEL is broken down into its components, ESTIMATOR (treatments) and PROJECT (blocks), in the table area ②. This table gives the number of degrees of freedom for ESTIMATOR (treatments) and PROJECT (blocks). The number of degrees of freedom for treatments will always be one less than the number k of treatments, that is, $(k - 1)$. For our example there are $k = 3$ estimators (treatments). Therefore, $(k - 1) = 3 - 1 = 2$. This number appears

TABLE **9.8**

Minitab ANOVA Printout for Example 9.2

```
MTB > ANOVA C1 = C3 C2

Factor      Type Levels  Values
ESTIMATR   fixed     3      1     2     3
PROJECT    fixed     5      1     2     3     4     5


Analysis of Variance for C1                    ①

Source      DF        SS         MS        F       P      ③
ESTIMATR     2    0.13456    0.06728     7.98   0.012
PROJECT      4    4.88896    1.22224   144.99   0.000
Error        8    0.06744    0.00843
Total       14    5.09096
```

TABLE **9.9**   Excel ANOVA printout for Example 9.2

	A	B	C	D	E	F	G	H	I	J
1	Analysis of Variance for PRICE - Type III Sums of Squares									③
2	Source			Sum of Squares	Df		Mean Square	F-Ratio		P-Value
3	MAIN EFFECTS									
4	A:ESTIMATOR			0.13456	2		0.06728	7.98		0.0124
5	B:PROJECT			4.88896	4		1.22224	144.99		0.0000
6										
7	RESIDUAL			0.06744	8		0.00843			
8	TOTAL (Corrected)			5.09096	14				①	
9										
10										
11	Table of Least Squares Means for PRICE with 95.0% Confidence Intervals									
12	Level			Count	Mean		Stnd Error	Lower Limit		Upper Limit
13	GRAND MEAN			15	4.294					
14	ESTIMATOR									
15	1			5	4.294		0.041	4.19931		4.38869
16	2			5	4.178		0.041	4.08331		4.27269
17	3			5	4.410		0.041	4.31531		4.50469
18	PROJECT									
19	1			3	3.517		0.053	3.39443		3.63891
20	2			3	4.807		0.053	4.68443		4.92891
21	3			3	3.967		0.053	3.84443		4.08891
22	4			3	5.080		0.053	4.95776		5.20224
23	5			3	4.100		0.053	3.97776		4.22224
24										

in the DF column in area ② opposite ESTIMATOR. Similarly, if there are b blocks, the number of degrees of freedom for blocks is $(b - 1)$. For our example there are $b = 5$ blocks. Therefore, the number of degrees of freedom for blocks is $(b - 1) = 5 - 1 = 4$. This number appears in the DF column in area ② opposite PROJECT. The number of degrees of freedom for error will always equal $(n - b - k + 1)$. This number, $(n - b - k + 1) = 15 - 5 - 3 + 1 = 8$, is shown in the DF column in area ① opposite ERROR. The number of degrees of freedom corresponding to Total SS is always equal to $(n - 1) = 15 - 1 = 14$. This number appears under DF in the row corresponding to CORRECTED TOTAL. Note that the sum of the numbers of degrees of freedom corresponding to ESTIMATOR, PROJECT, and ERROR always equals the number of degrees of freedom corresponding to CORRECTED TOTAL; that is, $(2 + 4 + 8) = 14$.

The traditional ANOVA format combines areas (1) and (2) into a single table by replacing the source MODEL in area (1) with the sources in area (2). This traditional format is shown in the first display at the end of this section.

3 Column 3 of areas (1) and (2), labeled SUM OF SQUARES and ANOVA SS, respectively, shows the sums of squares for the sources of variation. Thus SSE, shown in area (1), is

$$\text{SSE} = .06744000$$

The Total SS is also shown in area (1) in the row corresponding to CORRECTED TOTAL. Thus

$$\text{Total SS} = 5.09096000$$

The sums of squares for ESTIMATOR (treatments) and PROJECT (blocks) are shown in area (2). Thus

$$\text{SST} = \text{SS(ESTIMATOR)} = .13456000$$

and

$$\text{SSB} = \text{SS(PROJECT)} = 4.88896000$$

4 Each mean square is obtained by dividing a sum of squares by its respective degrees of freedom. For example,

$$\text{MSE} = s^2 = \frac{\text{SSE}}{n - b - k + 1}$$

The mean squares for the sources of variation are shown in column 4 of area (1) for all three printouts. However, only Minitab and Excel provide the mean squares for treatments and blocks. Thus,

$$\text{MSE} = s^2 = .00843$$

is found on all three printouts, while

$$\text{MST} = \text{MS(ESTIMATOR)} = .06728$$

and

$$\text{MSB} = \text{MS(PROJECT)} = 1.22224$$

are found only on the Minitab and Excel printouts.

5 Under the null hypotheses that there are no differences among treatment means and that there are no differences among block means, the mean squares for treatments (MST), blocks (MSB), and error (MSE) provide independent estimates of the common population variance σ^2.

a To test H_0: no differences among the k treatment means, we use $F = \text{MST/MSE}$ as the test statistic, and we reject H_0 if $F > F_\alpha$, where F is based on the number of degrees of freedom associated with MST and MSE, namely, $v_1 = k - 1$ and $v_2 = n - b - k + 1$. The computed value of the F statistic,

$F = \text{MST/MSE}$, is shown in column 4 of area ② as $F = 7.98$.[†] For $\alpha = .05$, the critical value of $F_{.05}$ for $v_1 = k - 1 = 2$ and $v_2 = n - b - k + 1 = 8$ is 4.46. Since the computed value of F, 7.98, exceeds this value, there is sufficient evidence to indicate a difference between at least two of the treatment means. The observed significance level (p-value) for the test is shown in area ③ as .0124. Thus the probability of observing an F value as large as or larger than $F = 7.98$, assuming H_0 true, is only .0124.

b To test H_0: no difference among the b block means, we use $F = \text{MSB/MSE}$ as the test statistic, and we reject H_0 if $F > F_\alpha$, where F is based on the number of degrees of freedom associated with MSB and MSE, namely $v_1 = b - 1 = 4$ and $v_2 = n - b - k + 1 = 8$. The computed value of the F statistic, $F = \text{MSB/MSE}$, is shown in column 4 of area ② as $F = 144.99$.[†] For $\alpha = .05$, the critical value of F for $v_1 = 4$ and $v_2 = 8$ degrees of freedom is 3.84. Since the observed value of F, 144.99, greatly exceeds this critical value, there is ample evidence to indicate differences among the block means. Since the sizes of the construction projects (in dollars) are known to vary over a wide range, we should not be surprised to find differences in the mean values of the construction project bids.

6 The standard deviation, $s = \sqrt{\text{MSE}} = .09181503$, is shown in area ④. It can be used to construct a confidence interval for the difference between a pair of treatment means[‡] or between a pair of block means, and it can also be used to test for differences between pairs of means. The formulas and procedures are the same as those used for independent random samples. Since each treatment mean appears in each block, there are b observations per treatment. Therefore, a $(1 - \alpha)100\%$ confidence interval for the difference between a pair of treatment means, say i and j, is

$$(\bar{x}_i - \bar{x}_j) \pm t_{\alpha/2} s \sqrt{\frac{1}{b} + \frac{1}{b}}$$

or

$$(\bar{x}_i - \bar{x}_j) \pm t_{\alpha/2} s \sqrt{\frac{2}{b}}$$

where $t_{\alpha/2}$ is based on the number of degrees of freedom associated with s^2.

Similarly, since each block contains k treatments, each block mean will be based on k observations. And a $(1 - \alpha)100\%$ confidence interval for the difference between a pair of block means, say ℓ and m, is

$$(\bar{x}_\ell - \bar{x}_m) \pm t_{\alpha/2} s \sqrt{\frac{2}{k}}$$

[†]In the Minitab and Excel printouts, the calculated F values are in column 5 of area ①.

[‡]You cannot construct a confidence interval for a single mean unless the blocks have been randomly selected from among the population of all blocks. The procedure for constructing intervals for single means is beyond the scope of this text.

To illustrate, suppose that we want to construct a 95% confidence interval for the difference between the mean estimates for treatments (estimators) 3 and 1. The means for these treatments are $\bar{x}_1 = 4.294$ and $\bar{x}_3 = 4.410$ and, from the printout in Table 9.7, $s = .09181503 \approx .0918$. Since s^2 is based on 8 degrees of freedom, $t_{\alpha/2} = t_{.025} = 2.306$, and the 95% confidence interval for $(\mu_3 - \mu_1)$ is

$$(\bar{x}_3 - \bar{x}_1) \pm t_{\alpha/2} s \sqrt{\frac{2}{b}}$$

$$(4.410 - 4.294) \pm (2.306)(.0918)\sqrt{\frac{2}{5}}$$

or $.116 \pm .134$. Therefore, we estimate the difference in mean level of estimates between estimators 3 and 1 to be from $-\$.018$ to $\$.250$ million. [*Note:* Since this interval includes zero, there is not sufficient evidence to indicate a difference between μ_3 and μ_1.] ▪

The typical ANOVA table for k treatments in b blocks is shown in the first display that follows. The other two displays summarize the analysis of variance F tests for comparing treatment and block means and give the confidence intervals for the differences between pairs of treatment means and pairs of block means.

ANOVA Table for a Randomized Block Design, k Treatments and b Blocks

Source	d.f.	SS	MS	F
Treatments	$k-1$	SST	$\text{MST} = \text{SST}/(k-1)$	MST/MSE
Blocks	$b-1$	SSB	$\text{MSB} = \text{SSB}/(b-1)$	MSB/MSE
Error	$n-b-k+1$	SSE	$\text{MSE} = \text{SSE}/(n-b-k+1)$	
Total	$n-1$			

Tests for a Randomized Block Design

For Comparing Treatment Means

1 Null Hypothesis: H_0 : the treatment means are equal
2 Alternative Hypothesis: H_a : at least two of the treatment means differ
3 Test Statistic: $F = \text{MST}/\text{MSE}$, where F is based on $\nu_1 = k-1$ and $\nu_2 = n-b-k+1$ degrees of freedom.

4 Rejection Region: Reject if $F > F_\alpha$, where F_α lies in the upper tail of the F distribution (see the figure).

For Comparing Block Means

1 Null Hypothesis: H_0 : the block means are equal
2 Alternative Hypothesis: H_a : at least two of the block means differ
3 Test Statistic: $F = \text{MSB/MSE}$, where F is based on $\nu_1 = b - 1$ and $\nu_2 = n - b - k + 1$ degrees of freedom.
4 Rejection Region: Reject if $F > F_\alpha$, where F_α lies in the upper tail of the F distribution (see the figure).

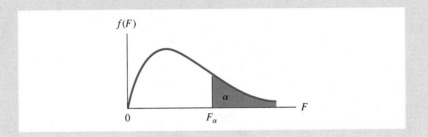

$(1 - \alpha)100\%$ Confidence Intervals for the Difference Between Pairs of Treatment and Block Means: A Randomized Block Design

Difference between treatment means i and j:

$$(\bar{x}_i - \bar{x}_j) \pm t_{\alpha/2} s \sqrt{\frac{2}{b}}$$

Difference between block means l and m

$$(\bar{x}_l - \bar{x}_m) \pm t_{\alpha/2} s \sqrt{\frac{2}{k}}$$

where

$$k = \text{Number of treatments}$$
$$b = \text{Number of blocks}$$
$$s = \sqrt{\text{MSE}}$$

and $t_{\alpha/2}$ is based on $(n - b - k + 1)$ degrees of freedom.

9.8 Computing Formulas: The Analysis of Variance for a Randomized Block Design (Optional)

We will use the notation given in the following display when conducting an analysis of variance for a randomized block design.

Notation

k = Number of treatments

b = Number of blocks

$n = bk$ = Total number of observations in the experiment

$T = \displaystyle\sum_{ij} x_{ij}$ = Total of all observations in the experiment

$\bar{x} = \dfrac{T}{n}$ = Mean of all observations in the experiment

T_i = Total of all observations receiving treatment i, $i = 1, 2, \ldots, k$

B_j = Total of all observations in block j, $j = 1, 2, \ldots, b$

A summary of the computing formulas is given in the next display.

Summary of Computing Formulas: Analysis of Variance for a Randomized Block Design, k Treatments in b Blocks

$$CM = \frac{(T)^2}{n}$$

where

$$n = bk \qquad \text{and} \qquad T = \text{Sum of all } n \text{ observations}$$

$$\text{Total SS} = \sum_{ij} x_{ij}^2 - CM$$

$$= \text{Sum of squares of all } x \text{ values} - CM$$

$$SST = \sum_{i=1}^{k} \frac{T_i^2}{b} - CM \qquad MST = \frac{SST}{k-1}$$

$$SSB = \sum_{j=1}^{b} \frac{B_j^2}{k} - CM \qquad MSB = \frac{SSB}{b-1}$$

Tips on Problem Solving

Be careful on this point: Unless the blocks have been randomly selected from a population of blocks, you cannot obtain a confidence interval for a single treatment mean. This limitation occurs because the sample treatment mean is biased by the positive and negative effects that the blocks have on the response.

Exercises

Starred (*) exercises are optional.

Basic Techniques

9.19 A randomized block design was conducted to compare the means of three treatments within six blocks. Construct an analysis of variance table showing the sources of variation and their respective degrees of freedom.

9.20 Suppose that the analysis of variance calculations for Exercise 9.19 gave SST = 11.4, SSB = 17.1, and Total SS = 42.7. Complete the analysis of variance table, showing all sums of squares, mean squares, and pertinent F values.

9.21 Do the data of Exercise 9.20 provide sufficient evidence to indicate differences among the treatment means? Test using $\alpha = .05$.

9.22 Refer to Exercise 9.20. Find a 95% confidence interval for the difference between a pair of treatment means A and B if $\bar{x}_A = 21.9$ and $\bar{x}_B = 24.2$.

9.23 Do the data of Exercise 9.20 provide sufficient evidence to indicate that blocking increased the amount of information in the experiment about the treatment means? Justify your answer.

9.24 A randomized block design was conducted to compare the means of six treatments within four blocks. Construct an analysis of variance table showing the sources of variation and their respective degrees of freedom.

9.25 Suppose that the analysis of variance calculations for Exercise 9.24 gave SST = 6.1, SSB = 2.2, and Total SS = 12.2. Complete the analysis of variance table, showing all sums of squares, mean squares, and pertinent F values.

9.26 Do the data of Exercise 9.25 provide sufficient evidence to indicate differences among the treatment means? Test using $\alpha = .10$.

9.27 Refer to Exercise 9.25. Find a 90% confidence interval for the difference between a pair of treatment means A and B if $\bar{x}_A = 291.2$ and $\bar{x}_B = 289.7$.

9.28 Do the data of Exercise 9.25 provide sufficient evidence to indicate that blocking increased the amount of information in the experiment about the treatment means? Justify your answer.

9.29 The partially completed ANOVA table for a randomized block design is shown on the next page.

a How many blocks were involved in the design?

b How many observations are in each treatment total?

c How many observations are in each block total?

d Fill in the blanks in the ANOVA table.

Source	d.f.	SS	MS	F
Treatments	4	14.2		
Blocks		18.9		
Error	24			
Total	34	41.9		

e Do the data provide sufficient evidence to indicate differences among the treatment means? Test using $\alpha = .10$.

f Do the data provide sufficient evidence to indicate differences among the block means? Test using $\alpha = .10$.

*9.30 The data shown in the accompanying table are observations collected from an experiment that compared four treatments, A, B, C, and D, within each of the three blocks by using a randomized block design.

Block	Treatment A	B	C	D	Total
1	6	10	8	9	33
2	4	9	5	7	25
3	12	15	14	14	55
Total	22	34	27	30	113

a Calculate CM and Total SS.

b Calculate SST and MST.

c Calculate SSB and MSB.

d Calculate SSE and MSE.

e Construct an ANOVA table for the data.

f Do the data provide sufficient evidence to indicate differences among the treatment means? Test using $\alpha = .05$.

g Do the data provide sufficient evidence to indicate differences among the block means? Test using $\alpha = .05$.

h Does it appear that the use of a randomized block design for this experiment was justified? Explain.

*9.31 Refer to Exercise 9.30. Find a 90% confidence interval for the difference $(\mu_A - \mu_B)$.

*9.32 The data shown in the accompanying table are observations collected from an experiment that compared three treatments, A, B, and C, within each of five blocks by using a randomized block design.

Treatment	Block 1	2	3	4	5	Total
A	2.1	2.6	1.9	3.2	2.7	12.5
B	3.4	3.8	3.6	4.1	3.9	18.8
C	3.0	3.6	3.2	3.9	3.9	17.6
Total	8.5	10.0	8.7	11.2	10.5	48.9

  a Calculate CM and Total SS.

  b Calculate SST and MST.

  c Calculate SSB and MSB.

  d Calculate SSE and MSE.

  e Construct an ANOVA table for the data.

  f Do the data provide sufficient evidence to indicate differences among the treatment means? Test using $\alpha = .05$.

  g Do the data provide sufficient evidence to indicate differences among the block means? Test using $\alpha = .05$.

  h Does it appear that the use of a randomized block design for this experiment was justified? Explain.

*9.33 Find a 99% confidence interval for the difference $(\mu_C - \mu_A)$.

Applications

9.34 In Exercise 8.46, we described a paired-difference experiment conducted to compare mean daily sales at a supermarket for days when no background music was present versus days when the supermarket provided slow-tempo background music. A follow-up experiment employed three treatments: no music, slow-tempo background music, and fast-tempo background music. Three midweek days, Tuesday, Wednesday, and Thursday, were selected for the experiment. The three treatments were randomly assigned, one to each of these three days for each of four weeks, and the daily sales were recorded as shown in the accompanying table. A Minitab analysis of variance printout for the data appears below.

	Week			
Treatment	**1**	**2**	**3**	**4**
No music	14,140	13,991	14,772	13,266
Slow tempo	15,029	14,546	15,029	14,783
Fast tempo	12,874	13,165	13,140	11,245

```
MTB > ANOVA C1 = C2 C3

Factor     Type Levels Values
TRTS       fixed    3      1     2    3
WEEKS      fixed    4      1     2    3    4


Analysis of Variance for C1
Source     DF         SS         MS       F       P
TRTS        2   10307993    5153997   22.03   0.002
WEEKS       3    2426157     808719    3.46   0.092
Error       6    1403851     233975
Total      11   14138001
```

  a Explain how the design of this experiment might increase the information available on the differences in mean daily sales for the three treatments.

  b Do the data provide sufficient evidence to indicate a difference in mean daily sales for the three treatments? Test using $\alpha = .05$.

  c Do the data indicate a difference in the mean daily sales for days when slow-tempo music was played versus days when the background music was fast-tempo? Test using $\alpha = .05$.

9.35 A fast-food chain wants to compare daily sales during three types of sales promotions. The sales promotions were employed in three different cities over a six-week period, with two weeks for each promotion. The promotions were randomly assigned to the two-week time periods within each city, and the amount of sales (in thousands of dollars) for one outlet in each city was measured for the last two days of the two-week period (to avoid carryover effects of the promotions). The data are shown in the accompanying table. An SAS analysis of variance printout follows.

		Promotion	
City	A	B	C
1	4.65	5.21	4.62
2	4.32	4.69	4.27
3	4.14	4.68	4.25

ANALYSIS OF VARIANCE PROCEDURE

DEPENDENT VARIABLE: SALES

SOURCE	DF	SUM OF SQUARES	MEAN SQUARE	F VALUE	PR > F	R-SQUARE	C.V.
MODEL	4	0.85640000	0.21410000	46.54	0.0013	0.978967	1.4950
ERROR	4	0.01840000	0.00460000		ROOT MSE		SALES MEAN
CORRECTED TOTAL	8	0.87480000			0.06782330		4.53666667

SOURCE	DF	ANOVA SS	F VALUE	PR > F
CITY	2	0.38580000	41.93	0.0021
PROMO	2	0.47060000	51.15	0.0014

a Identify the treatments and blocks in this experiment.

b How were the experimental units matched within blocks?

c Explain why you might expect the blocking for this experiment to increase the information available for comparing treatment means.

d Do the data provide sufficient evidence to indicate differences among mean sales for the three types of promotions? Test using $\alpha = .10$.

e Find a 90% confidence interval for the difference in mean sales for promotions A and C.

f Explain why you can or cannot construct confidence intervals for the mean sales for the individual promotions.

9.36 A service station manager wants to estimate the mean summer weekly demand for three types of gasoline: leaded, unleaded, and super unleaded. Data (in thousands of gallons) collected over an eight-week period are shown in the accompanying table. A Minitab analysis of variance printout for the data follows.

	Week							
Gasoline Type	1	2	3	4	5	6	7	8
Leaded	2.4	2.6	2.7	2.8	3.0	2.6	2.7	2.4
Unleaded	18.2	19.7	21.3	22.4	21.5	19.7	18.0	18.8
Super unleaded	4.3	4.6	4.7	5.5	5.0	4.8	4.6	5.0

```
Analysis of Variance for DEMAND

Source      DF          SS          MS         F       P
TYPE         2     1421.63      710.81   1003.25   0.000
WEEK         7        9.81        1.40      1.98   0.131
Error       14        9.92        0.71
Total       23     1441.36
```

a Explain why the sampling procedure employed in collecting these data is or is not a randomized block design.

b Would it be valid to analyze these data as though the data for the three gasolines represent independent random samples? Explain.

c Find a 95% confidence interval for the difference between unleaded and super unleaded mean weekly sales.

*9.37 Perform the analysis of variance calculations for Exercise 9.34, and present them in an analysis of variance table. Compare your answers with those given in the computer printout in Exercise 9.34. Interpret the results of the analysis of variance.

*9.38 Perform the analysis of variance calculations for Exercise 9.35, and present them in an analysis of variance table. Compare your answers with those given in the computer printout in Exercise 9.35. Interpret the results of the analysis of variance.

*9.39 Perform the analysis of variance calculations for Exercise 9.36, and present them in an analysis of variance table. Compare your answers with those given in the computer printout in Exercise 9.36. Interpret the results of the analysis of variance.

9.9 Factorial Experiments

Suppose the manager of a manufacturing plant suspects that the output (in number of units produced per 8-hour shift) of a production line depends upon two qualitative variables, the foreman supervising the line (of which there are two, A_1 and A_2) and the shift on which the production is measured. We will denote the three shifts, 8:00 A.M. to 4:00 P.M., 4:00 P.M. to midnight, and midnight to 8:00 A.M., as B_1, B_2, and B_3.

These two qualitative independent variables, "foreman" and "shift," are called **factors** in the language of statistics. The different settings for a given factor are called **levels.** Thus in this experiment the plant manager wants to investigate the effect of the first factor, "foreman," at two levels, A_1 and A_2, one level corresponding to each foreman. Similarly, the manager wants to investigate the effect of the second factor, "shifts," at three levels, B_1, B_2, and B_3, one level corresponding to each of the three shifts. What combination of the two factor levels should be included in the experiment?

One way to conduct the experiment is to hold one factor constant, say "foreman," and vary the shift. For example, we might use foreman A_1 on each of the three shifts B_1, B_2, and B_3. Thus we will collect output counts for each of three treatments, one corresponding to each of the factor level combinations $A_1 B_1$, $A_1 B_2$, and $A_1 B_3$. If the output counts for these three treatments are 600, 500, and 450, respectively (as represented by the plotted points in Figure 9.3), the data will lead us to believe

F I G U R E **9.3**
Output counts for three shifts
using foreman A_1

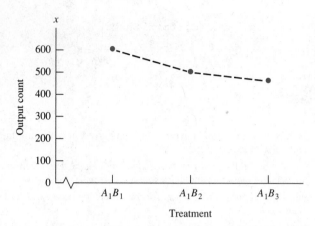

that the first (8:00 A.M. to 4:00 P.M.) shift seems to be the most productive and that productivity decreases for the second and third shifts.

Similarly, we might hold the second factor, "shift," constant and vary the foremen to see which foreman seems to be associated with the greater productivity. For example, we might use both foreman A_1 and foreman A_2 on day shift B_1, thus observing the output for the factor level combinations A_1B_1 and A_2B_1. Since we have already collected an observation for the factor level combination A_1B_1, we need to collect only one more observation, one corresponding to the factor level combination A_2B_1. Suppose that this combination gives an output count of 480 units. Comparing the outputs for A_1B_1 and A_2B_1 (see Figure 9.4), we may conclude that foreman A_1 achieves a greater productivity than foreman A_2.

F I G U R E **9.4**
Output counts for one shift using
foremen A_1 and A_2

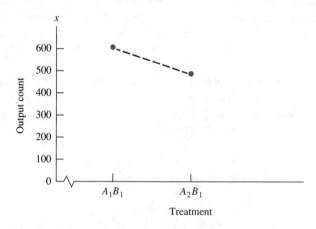

In the preceding discussion, we have investigated the effect of two factors on the output count x by using only four treatments, the factor level combinations A_1B_1,

$A_1 B_2$, $A_1 B_3$, and $A_2 B_1$. We concluded that the greatest output occurs when foreman A_1 manages the day shift B_1. Is there a flaw in our logic?

There are $2 \times 3 = 6$ possible combinations of levels of the two factors: $A_1 B_1$, $A_1 B_2$, $A_1 B_3$, and $A_2 B_1$, $A_2 B_2$, $A_2 B_3$. Suppose that we were to run the remaining two factor level combinations, $A_2 B_2$ and $A_2 B_3$, and found the outputs to be 380 and 330, respectively. A plot of all six outputs is shown in Figure 9.5.

F I G U R E **9.5**
Plots indicating that the factors affect output independently of each other

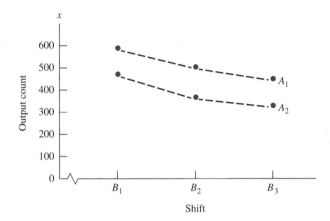

Figure 9.5 shows how the outputs would plot *if* the two factors "foreman" and "shift" affect output count *independently* of each other. They show, for example, that the output when A_1 is foreman is always 120 units higher than when A_2 is foreman, *regardless of the shift*. Similarly, they show that shift B_1 always produces 100 units more than shift 2 and 150 more than shift 3, regardless of who is the foreman. Of course, we have based this whole discussion on a single observation per treatment, and in practical situations we would select larger samples. Nevertheless, our example makes a point. If the two factors affect output x in an independent manner, we need to compare the means of only four treatments to reach this conclusion.

However, suppose that the output counts for the factor level combinations $A_2 B_2$ and $A_2 B_3$ were 600 and 650, respectively. A plot of the data (see Figure 9.6) leads us to a completely different conclusion. Although foreman A_1 achieves the larger output on the first shift, foreman A_2 (a night person) is superior on shifts B_2 and B_3. Thus we cannot generalize and say that one foreman is better than the other or that the output of one particular shift is best. The output *depends* upon the particular factor level combination that is employed. When this situation occurs, we say that the two factors **interact,** or that **interaction** exists between the two **factors.**

Multifactor experiments are conducted to determine which, if any, of the factors affect the mean response, and if they do whether they do so independently or interact. Therefore, the first step in an analysis of variance is to test the complete model— that is, to determine whether there is evidence of differences among the treatment means. This test appears at the top of the SAS analysis of variance printout. (For example, see the F test in the SAS analysis of variance printout in Table 9.3 in the line corresponding to MODEL.) If there is evidence of differences among the

FIGURE **9.6**
Output counts indicating factor interaction

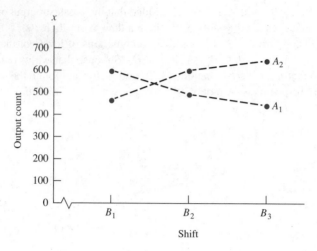

treatment means, the second step in the analysis is to test for factor interaction. If present, we ignore the factors themselves and focus our attention on the means for the individual factor level combinations. If factor interaction seems to be negligible, we can investigate the effect of the factors on the response x as if they affected x in an independent manner.

One safe way to detect factor interactions is to conduct a **factorial experiment,** one that collects sample data for every combination of factor levels. For example, a factorial experiment conducted to investigate the effect of two factors on a response— one at two levels A_1 and A_2 and the second at three levels B_1, B_2, and B_3—acquires data for each of the $2 \times 3 = 6$ treatments that correspond to factor level combinations $A_1 B_1$, $A_1 B_2$, $A_1 B_3$, $A_2 B_1$, $A_2 B_2$, $A_2 B_3$. Likewise, a two-factor experiment with one factor at three levels and one at four levels involves 12 treatments corresponding to the 3×4 factor level combinations. The first experiment is called a 2×3 factorial experiment and indicates that there are two factors, one at two levels and the other at three levels. A 3×4 factorial experiment is also a two-factor experiment, with one factor at three levels and the other factor at four levels.

Factorial experiments can also be employed to investigate the effects of three or more factors on a response. A $2 \times 2 \times 3$ factorial experiment involves three factors, two each at two levels and the third at three levels. The total number of treatments is $2 \times 2 \times 3 = 12$, one for each combination of the levels of the three factors. The factors in a three-factor factorial experiment may affect the response independently, one or more pairs of the factors may interact, or all three factors may interact. Thus the objective of experiments involving three or more factors is to test for factor interactions. If they exist, we know that we should examine treatment means individually. If factor interactions are negligible, we can study the effect of each factor on the response as if the factor affected the response independently of the levels of the other factors.

When comparing treatment means for a factorial experiment (or for any other experiment), we usually need more than one observation per treatment. For example, if we obtain two observations for each of the treatments of a complete factorial experiment, we say that we have two **replications** of the complete factorial experiment.

For all further discussions in this text, we will assume that the treatment samples are independent random samples of equal sample size and that the number r of replications is at least equal to 2. One can, under some circumstances, draw practical conclusions from a single replication of a factorial experiment when the number of factors is equal to 3 or larger, but that topic is beyond the scope of this text.

9.10   The Analysis of Variance for a Factorial Experiment

An analysis of variance for a replicated, two-factor factorial experiment partitions the total sum of squares of deviations into four parts, the first two representing the variation of the factor level means, the third representing factor interaction, and the fourth measuring the variation of the observations *within* a treatment about the treatment mean. Representing the factors as A and B, we have

$$\text{Total SS} = \text{SS}A + \text{SS}B + \text{SS}(AB) + \text{SSE}$$

where

$$\text{Total SS} = \text{S}_{xx} = \sum (x - \bar{x})^2$$
$$\text{SS}A = \text{Sum of squares for factor } A$$
$$\text{SS}B = \text{Sum of squares for factor } B$$

Sums of squares $\text{SS}A$ and $\text{SS}B$ are often called **main-effect** sums of squares for the respective factors to distinguish them from the interaction sum of squares $\text{SS}(AB)$. To continue,

$$\text{SS}(AB) = \text{Sum of squares measuring the interaction between factors } A \text{ and } B$$
$$\text{SSE} = \text{Total SS} - \text{SS}A - \text{SS}B - \text{SS}(AB)$$
$$= \text{Unexplained variation}$$

As the following example demonstrates, these sums of squares play the same role here as in the analyses of variance presented in earlier sections.

Once again, we will explain how to conduct an analysis of variance for a two-factor factorial experiment by describing and interpreting a computer output for an example. The computing formulas are presented in optional Section 9.11.

E X A M P L E   **9.3**   The manager of the manufacturing plant (see Section 9.9) conducted $r = 3$ replications of a 2×3 factorial experiment to investigate the effect of "foreman" (2 levels) and "shift" (3 levels) on the output of a production line. The observations are given in Table 9.10; the SAS, Minitab, and Excel printouts are shown in Tables 9.11, 9.12, and 9.13. Describe and interpret the SAS printout.

Solution   **1**   As noted in our discussion of the analysis of variance for a randomized block design, the SAS output for an analysis of variance presents the ANOVA table in two stages. The first stage, area ①ₙ in Table 9.11, separates the total variation of

TABLE **9.10**
Output per shift for the 2×3
factorial experiment of
Example 9.3

Foreman (factor A)	Shift (factor B)			Total
	B_1 (8 A.M.–4 P.M.)	B_2 (4 P.M.–midnight)	B_3 (midnight–8 A.M.)	
A_1	570 610 625	480 475 540	470 430 450	4650
A_2	480 515 465	625 600 580	630 680 660	5235
Total	3265	3300	3320	9885

TABLE **9.11** SAS ANOVA printout for Example 9.3

ANALYSIS OF VARIANCE PROCEDURE

DEPENDENT VARIABLE: OUTPUT

SOURCE	DF	SUM OF SQUARES	MEAN SQUARE	F VALUE	①	PR > F	R-SQUARE	C.V.
MODEL	5	100179.16666667	20035.83333333	27.85		0.0001	0.920659	4.8842
ERROR	12	8633.33333333	719.44444444			ROOT MSE		OUTPUT MEAN
CORRECTED TOTAL	17	108812.50000000				26.82246157 ④		549.16666667

②

SOURCE	DF	ANOVA SS	F VALUE	PR > F
FOREMAN	1	19012.50000000	26.43	0.0002 ③
SHIFT	2	258.33333333	0.18	0.8379
SHIFT*FOREMAN	2	80908.33333333	56.23	0.0001

TABLE **9.12**
Minitab ANOVA printout for
Example 9.3

```
MTB > ANOVA C1 = C2 C3 C2*C3

Factor      Type Levels Values
FOREMAN     fixed     2    1    2
SHIFT       fixed     3    1    2    3

Analysis of Variance for C1                    ①          ③

Source          DF       SS       MS       F      P
FOREMAN          1    19012    19012    26.43  0.000
SHIFT            2      258      129     0.18  0.838
FOREMAN*SHIFT    2    80908    40454    56.23  0.000
Error           12     8633      719
Total           17   108813
```

T A B L E **9.13**   Excel printout for Example 9.3

	A	B	C	D	E	F	G	H	I	J
1	Analysis of Variance for OUTPUT - Type III Sums of Squares									
2	Source			Sum of Squares	Df		Mean Square	F-Ratio		P-Value
3	MAIN EFFECTS									
4	A:FOREMAN			19012.5	1		19012.5	26.43		0.0002
5	B:SHIFT			258.3	2		129.2	0.18		0.8379
6										
7	INTERACTIONS									
8	AB			80908.3	2		40454.2	56.23		
9										
10	RESIDUAL			8633.3	12		719.4			
11	TOTAL (Corrected)			108813	17					
12										
13	Table of Least Squares Means for OUTPUT with 95.0% Confidence Intervals									
14	Level			Count	Mean		Stnd Error	Lower Limit		Upper Limit
15	GRAND MEAN			18	549.167					
16	FOREMAN									
17	1			9	516.667		8.941	497.186		536.147
18	2			9	581.667		8.941	562.186		601.147
19	SHIFT									
20	1			6	544.167		10.950	520.308		568.025
21	2			6	550.000		10.950	526.141		573.859
22	3			6	553.333		10.950	529.475		577.192
23	FOREMAN by SHIFT									
24	1	1		3	601.667		15.486	567.926		635.408
25	1	2		3	498.333		15.486	464.592		532.074
26	1	3		3	450.000		15.486	416.259		483.741
27	2	1		3	486.667		15.486	452.926		520.408
28	2	2		3	601.667		15.486	567.926		635.408
29	2	3		3	656.667		15.486	622.926		690.408
30										

the x values into two sources, MODEL and ERROR. The proportion for MODEL combines the three sources corresponding to main-effect factor A (foreman), main-effect factor B (shift), and the AB (foreman by shift) interaction. These three sources are shown in area (2) in the table.

2   The degrees of freedom for the respective sources are shown in column 2 of the tables (column 3 of Excel). If factor A is at a levels and factor B is at b levels, then the number of degrees of freedom for the main effects A and B will always equal $(a-1)$ and $(b-1)$, respectively, and the number of degrees of freedom for interaction will be $(a-1)(b-1)$. These numbers—$(a-1) = (2-1) = 1$, $(b-1) = (3-1) = 2$, and $(a-1)(b-1) = (1)(2) = 2$— are shown in the DF column in area (2). Note that the sum of these degrees of freedom, 5, is equal to the number of degrees of freedom for MODEL (shown in area (1)), the source that was partitioned to form sources A, B, and AB.

If the experiment is replicated r times—that is, there are r observations for each combination of factor levels (for our example, $r = 3$)—then the total number of observations in the experiment is $n = abr$, and the number of degrees of freedom associated with Total SS is $(n-1) = (abr-1)$. For our example, $(n-1) = (abr-1) = (2)(3)(3) - 1 = 17$. The number of degrees of freedom for ERROR will always equal $ab(r-1)$, that is, $(r-1)$ degrees of freedom for each of the ab factor level combinations. For our example, $ab(r-1) =$

$(2)(3)(3 - 1) = 12$. These numbers of degrees of freedom, for the CORRECTED TOTAL and for ERROR, are shown in column 2 in area ①.

3 Column 3 of areas ① and ②, labeled SUM OF SQUARES and ANOVA SS, respectively, shows the sums of squares for the sources of variation. Thus SSE, shown in area ①, is

$$SSE = 8633.33333$$

The Total SS is shown in area ① in the row corresponding to CORRECTED TOTAL:

$$\text{Total SS} = 108812.500000$$

The sums of squares for main effects A and B and for the AB interaction are shown in area ② as

$$SSA = 19012.5000$$
$$SSB = 258.3333$$
$$SS(AB) = 80908.3333$$

4 Mean squares are obtained by dividing a source sum of squares by its degrees of freedom. The mean squares for main effects A and B and the interaction AB are not given in the SAS printout. However, using the information in areas ① and ② of the SAS printout, we find that, to four-decimal accuracy,

$$MSE = s^2 = 719.4444$$
$$MSA = 19012.5000$$
$$MSB = 129.6667$$
$$MS(AB) = 40454.1667$$

5 The F values for testing the hypotheses "no interaction between factors A and B," "no main-effect factor A," and "no main-effect factor B" are shown in column 4 of area ②, and the observed significance levels (p-values) for the three tests are shown in area ③. The numerator number (v_1) of degrees of freedom for an F value will equal the number of degrees of freedom of the mean square appearing in the numerator of the F statistic. The denominator number (v_2) of degrees of freedom will always equal the number of degrees of freedom of the mean square appearing in the denominator of F, namely, MSE. For example, the F statistic used to test for AB interaction is

$$F = \frac{MS(AB)}{MSE}$$

This F value for our example possesses $v_1 = 2$ and $v_2 = 12$ degrees of freedom, the degrees of freedom associated with $MS(AB)$ and MSE, respectively.

The small p-value ($p = .0002$) indicates that there is sufficient evidence to indicate differences in the mean levels of factor A—that is, differences in the mean output per foreman. But this fact is overshadowed by the fact that there is strong evidence ($p = .0001$) of an AB interaction. This implies that the mean output for a given shift depends upon the supervising foreman. You can see this result by examining the means for the six factor level combinations shown in Table 9.14. The three largest mean outputs occur when foreman A_1 is on the day shift and when foreman A_2 is on one of the two night shifts. The largest mean output appears to occur when foreman A_2 is supervising the early-morning shift. The practical implications of these comparisons suggest that foreman A_1 should be scheduled for the day shift and foreman A_2 for the early-morning shift.

TABLE 9.14
Means of the factor level combinations

Foreman (factor A)	B_1 (8 A.M.–4 P.M.)	B_2 (4 P.M.–midnight)	B_3 (midnight–8 A.M.)
A_1	601.67	498.33	450.00
A_2	486.67	601.67	656.67

Shift (factor B)

6 The standard deviation, $s = \sqrt{MSE} = 26.822462$, is shown in area ④. It can be used to test or to construct confidence intervals for the individual treatment (factor level combination) means or for the difference between a pair of means. The formulas and procedures are the same as those used for the independent random sampling design. Since we obtained r observations for each treatment, $(1 - \alpha)100\%$ confidence intervals for a single treatment mean or the difference between two means are

$$\bar{x}_i \pm t_{\alpha/2}\frac{s}{\sqrt{r}}$$

and

$$(\bar{x}_i - \bar{x}_j) \pm t_{\alpha/2}s\sqrt{\frac{2}{r}}$$

For example, suppose that we want to construct a 95% confidence interval for the difference in mean output for foreman A_1 and A_2 on the second shift (B_2). We will denote these factor level combinations, A_1B_2 and A_2B_2, as μ_1 and μ_2, respectively. The sample means for these treatments are $\bar{x}_1 = 498.33$ and $\bar{x}_2 = 601.67$, with $s = 26.822462$. Since s^2 is based on 12 degrees of freedom, the tabulated value of t (Table 4 in Appendix II) is $t_{\alpha/2} = t_{.025} = 2.179$, and the confidence interval is

$$(\bar{x}_1 - \bar{x}_2) \pm t_{\alpha/2}s\sqrt{\frac{2}{r}}$$

$$(498.33 - 601.67) \pm (2.179)(26.822)\sqrt{\frac{2}{3}}$$

$$-103.34 \pm 47.72$$

or

$$-151.06 \quad \text{to} \quad -55.62$$

Therefore, we estimate the difference in mean output between foremen A_1 and A_2 on shift B_2 to be between -151.06 and -55.62 units. We estimate the mean output on the second shift to be 55.62 to 151.06 higher when foreman A_2 is supervising the shift. ∎

The typical ANOVA table for r replications of a two-factor factorial experiment, factor A at a levels and factor B at b levels, is shown in the first display that follows. The other two displays summarize the analysis of variance F tests and give the confidence intervals for an individual treatment mean and for the difference between a pair of treatment means.

ANOVA table for r Replications of a Two-Factor Factorial Experiment: Factor A at a Levels and Factor B at b Levels

Source	d.f.	SS	MS	F
A	$a-1$	SSA	$MSA = \dfrac{SSA}{a-1}$	$\dfrac{MSA}{MSE}$
B	$b-1$	SSB	$MSB = \dfrac{SSB}{b-1}$	$\dfrac{MSB}{MSE}$
AB	$(a-1)(b-1)$	SS(AB)	$MS(AB) = \dfrac{SS(AB)}{(a-1)(b-1)}$	$\dfrac{MS(AB)}{MSE}$
Error	$ab(r-1)$	SSE	$MSE = \dfrac{SSE}{ab(r-1)}$	
Total	$abr-1$	Total SS		

Tests for a Factorial Experiment

Testing for Interaction

1 Null Hypothesis: H_0 : factors A and B do not interact
2 Alternative Hypothesis: H_a : factors A and B interact
3 Test Statistic: $F = MS(AB)/MSE$, where F is based on $v_1 = (a-1)(b-1)$ and $v_2 = ab(r-1)$ degrees of freedom.
4 Rejection Region: Reject H_0 if $F > F_\alpha$, where F_α lies in the upper tail of the F distribution (see the figure).

Testing for Main Effects, Factor A

1 Null Hypothesis: H_0 : there are no differences among the factor A means
2 Alternative Hypothesis: H_a : at least two of the factor A means differ
3 Test Statistic: $F = MSA/MSE$, where F is based on $v_1 = a - 1$ and $v_2 = ab(r - 1)$ degrees of freedom.
4 Rejection Region: Reject H_0 if $F > F_\alpha$ (see the figure).

Testing for Main Effects, Factor B

1 Null Hypothesis: H_0 : there are no differences among the factor B means
2 Alternative Hypothesis: H_a : at least two of the factor B means differ
3 Test Statistic: $F = MSB/MSE$, where F is based on $v_1 = b - 1$ and $v_2 = ab(r - 1)$ degrees of freedom.
4 Rejection Region: Reject H_0 if $F > F_\alpha$ (see the figure).

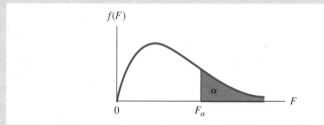

$(1 - \alpha)100\%$ Confidence Intervals for a Single Treatment Mean or the Difference Between a Pair of Treatment Means: A Two-Factor Factorial Experiment

Single treatment mean: $\bar{x}_i \pm t_{\alpha/2}\dfrac{s}{\sqrt{r}}$

Difference between two treatment means: $(\bar{x}_i - \bar{x}_j) \pm t_{\alpha/2}s\sqrt{\dfrac{2}{r}}$

where

$\quad\quad\quad r = $ Number of replications (observations per treatment)
$\quad\quad\quad s = \sqrt{MSE}$

and $t_{\alpha/2}$ is based on $ab(r - 1)$ degrees of freedom.

9.11 Computing Formulas: The Analysis of Variance for a Two-Factor Factorial Experiment (Optional)

We will use the symbols given in the next display when conducting an analysis of variance for a two-factor factorial experiment.

Notation

a = Number of levels of factor A

b = Number of levels of factor B

r = Number of replications, that is, the number of observations for each factor level combination (treatment)

$n = abr$ = Total number of observations

A_i = Total of all observations at the ith level of factor A, $i = 1, 2, \ldots, a$

B_j = Total of all observations at the jth level of factor B, $j = 1, 2, \ldots, b$

$(AB)_{ij}$ = Total of the r observations at the ith level of factor A and the jth level of factor B

A summary of the computing formulas is given in the next display.

Summary of Computing Formulas: Analysis of Variance for a Two-Factor Factorial Experiment

$$CM = \frac{(T)^2}{n} \qquad\qquad \text{Total SS} = \sum x^2 - CM$$

$$SSA = \sum_{i=1}^{a} \frac{A_i^2}{br} - CM \qquad\qquad SSB = \sum_{j=1}^{b} \frac{B_j^2}{ar} - CM$$

$$SS(AB) = \sum_{j=1}^{b} \sum_{i=1}^{a} \frac{(AB)_{ij}^2}{r} - SSA - SSB - CM$$

$$MSA = \frac{SSA}{a-1} \qquad\qquad MSB = \frac{SSB}{b-1}$$

$$MS(AB) = \frac{SS(AB)}{(a-1)(b-1)} \qquad\qquad MSE = s^2 = \frac{SSE}{ab(r-1)}$$

Exercises

Starred (*) exercises are optional.

Basic Techniques

9.40   Suppose that you were to conduct a two-factor factorial experiment, factor A at four levels and factor B at five levels, with three replications per treatment.

   a   How many treatments will be involved in the experiment?

   b   How many observations will be involved?

   c   List the sources of variation and their respective degrees of freedom.

9.41   Suppose that you were to conduct a two-factor factorial experiment, factor A at four levels and factor B at two levels, with r replications per treatment.

   a   How many treatments will be involved in the experiment?

   b   How many observations will be involved?

   c   List the sources of variation and their respective degrees of freedom.

9.42   The analysis of variance table for a 3×4 factorial experiment, factor A at three levels and factor B at four levels, with two observations per treatment, is shown in the accompanying table.

Source	d.f.	SS	MS	F
	2	5.3		
	3	9.1		
	6			
	12	24.5		
Total	23	43.7		

   a   Fill in the missing items in the table.

   b   Do the data provide sufficient evidence to indicate that factors A and B interact? Test using $\alpha = .05$. What are the practical implications of your answer?

   c   Do the data provide sufficient evidence to indicate that the factors A and B affect the response variable x? Explain.

9.43   Refer to Exercise 9.42. The means of the factor level combinations A_1B_1 and A_1C_1 are $\bar{x}_1 = 12.4$ and $\bar{x}_2 = 6.3$, respectively. Find a 95% confidence interval for the difference between the two corresponding population means.

9.44   The analysis of variance table for a 2×3 factorial experiment, factor A at two levels and factor B at three levels, with five observations per treatment, is shown in the accompanying table.

Source	d.f.	SS	MS	F
A		1.14		
B		2.58		
AB		.49		
Error				
Total		8.41		

a Do the data provide sufficient evidence to indicate interaction between factors A and B? Test using $\alpha = .05$. What are the practical implications of your answer?

b Give the approximate p-value for the test in part (a).

c Do the data provide sufficient evidence to indicate that factor A affects the response? Test using $\alpha = .05$.

d Do the data provide sufficient evidence to indicate that factor B affects the response? Test using $\alpha = .05$.

9.45 Refer to Exercise 9.44. The means of all observations at the factor A levels A_1 and A_2 were $\bar{x}_1 = 3.7$ and $\bar{x}_2 = 1.4$, respectively. Find a 95% confidence interval for the difference in mean response for factor levels A_1 and A_2.

*9.46 The following table gives data for a 3×3 factorial experiment, two replications per treatment.

	Levels of Factor A		
Levels of Factor B	1	2	3
1	5, 7	9, 7	4, 6
2	8, 7	12, 13	7, 10
3	14, 11	8, 9	12, 15

a Perform an analysis of variance for the data, and present the results in an analysis of variance table.

b What do we mean when we say that factors A and B interact?

c Do the data provide sufficient evidence to indicate interaction between factors A and B? Test using $\alpha = .05$.

d Find the approximate p-value for the test in part (c).

e Find a 95% confidence interval for the mean for the factor level combination $A_2 B_2$.

f Find a 95% confidence interval for the difference in means for factor level combinations $A_1 B_3$ and $A_3 B_1$.

*9.47 The following table gives data for a 2×2 factorial experiment, four replications per treatment.

	Levels of Factor A	
Levels of Factor B	1	2
1	2.1, 2.7, 2.4, 2.5	3.7, 3.2, 3.0, 3.5
2	3.1, 3.6, 3.4, 3.9	2.9, 2.7, 2.2, 2.5

a Perform an analysis of variance for the data, and present the results in an analysis of variance table.

b What do we mean when we say that factors A and B interact?

c Do the data provide sufficient evidence to indicate interaction between factors A and B? Test using $\alpha = .05$.

d Find the approximate p-value for the test in part (c).

e Find a 95% confidence interval for the mean for the factor level combination $A_2 B_2$.

f Find a 95% confidence interval for the difference in means for factor level combinations $A_1 B_2$ and $A_2 B_1$.

Applications

9.48   A builder of speculative houses uses one of three designs and assigns each house to the supervision of one of four foremen. Noticing variation in profit per house, the builder decided to investigate the effect of the two factors, "house design" and "foreman," on profit per house. The builder used each foreman as supervisor for each house design for three houses for each foreman-design combination. The data (in thousands of dollars profit per house) are shown in the accompanying table. The SAS analysis of variance printout for the data is also shown.

Design	Foreman			
	A_1	A_2	A_3	A_4
B_1	12.8	9.2	11.6	8.7
	9.4	7.8	12.9	7.4
	10.3	10.9	9.6	8.5
B_2	9.2	11.4	8.7	10.3
	7.4	9.6	7.5	10.9
	8.6	8.3	9.0	11.7
B_3	13.7	10.7	10.1	7.3
	12.0	10.2	8.7	8.6
	14.6	11.1	9.1	6.9

ANALYSIS OF VARIANCE PROCEDURE

DEPENDENT VARIABLE: PROFIT

SOURCE	DF	SUM OF SQUARES	MEAN SQUARE	F VALUE	PR > F	R-SQUARE	C.V.
MODEL	11	91.86972222	8.35179293	6.11	0.0001	0.736787	11.8687
ERROR	24	32.82000000	1.36750000		ROOT MSE		PROFIT MEAN
CORRECTED TOTAL	35	124.68972222			1.16940156		9.85277778

SOURCE	DF	ANOVA SS	F VALUE	PR > F
DESIGN	2	4.60055556	1.68	0.2072
FOREMAN	3	17.72750000	4.32	0.0143
DESIGN*FOREMAN	6	69.54166667	8.48	0.0001

a   What would be the practical implications if factor A and B interact?

b   Do the data provide sufficient evidence to indicate interaction between factors A and B? Test using $\alpha = .05$.

c   Give the approximate p-value for the test in part (b). Interpret the results.

d   Find a 95% confidence interval for the difference in mean profit between foremen A_1 and A_2 for houses built using plan B_1.

9.49   A chain of jewelry stores conducted an experiment to investigate the relationship between price and location and the demand for its diamonds. Six small-town stores were selected for the study, as well as six stores located in large suburban malls. Two stores in each of these location categories were assigned to each of three item percentage markups. The percentage gain (or loss) in sales for each store was recorded at the end of one month. The data are shown in the accompanying table. The Minitab analysis of variance printout for the data is also shown.

| | **Markup** | | |
Location	A_1	A_2	A_3
Small towns, B_1	10 4	−3 7	−10 −24
Suburban malls, B_2	14 18	8 3	−4 3

```
MTB > ANOVA C1 = C2 C3 C2*C3

Factor      Type Levels Values
LOCATION    fixed      2    1     2
MARKUP      fixed      3    1     2     3


Analysis of Variance for C1

Source               DF          SS         MS       F      P
LOCATION              1      280.33     280.33    7.97  0.030
MARKUP                2      835.17     417.58   11.87  0.008
LOCATION*MARKUP       2       85.17      42.58    1.21  0.362
Error                 6      211.00      35.17
Total                11     1411.67
```

a   Do the data provide sufficient evidence to indicate an interaction between markup and location? Test using $\alpha = .05$.

b   What are the practical implications of your test in part (a)?

c   Find a 95% confidence interval for the difference in mean change in sales for stores in small towns versus those in suburban malls if the stores are using price markup A_3.

9.50   Helena F. Barsam and Zita M. Simutis, at the U.S. Army Research Institute for the Behavioral and Social Sciences, conducted a study to determine the effect of two factors on terrain visualization training for soldiers. During the training programs, participants viewed contour maps of various terrains and then were permitted to view a computer reconstruction of the terrain as it would appear from a specified angle. The two factors investigated in the experiment were the participants' spatial abilities (abilities to visualize in three dimensions) and the viewing procedure (active or passive). Active participation permitted participants to view the computer-generated reconstructions of the terrain from any and all angles. Passive participation gave the participants a set of preselected reconstructions of the terrain. Participants were tested according to spatial ability, and from the test scores 20 were categorized as possessing high spatial ability, 20 medium, and 20 low. Then 10 participants within each of these groups were assigned to each of the two training modes, active or passive. The first table below shows the analysis of variance table computed by Barsam and Simutis, and the second table shows the six treatment means.

So.rce	d.f.	MS	Error d.f.	F	p
Main effects:					
Training condition	1	103.7009	54	3.66	.061
Ability	2	760.5889	54	26.87	.0005
Interaction:					
Training condition × ability	2	124.9905	54	4.42	.017
Within cells	54	28.3015			

	Training Condition	
Spatial Ability	**Active**	**Passive**
High	17.895	9.508
Medium	5.031	5.648
Low	1.728	1.610

Source: Barsam and Simutis, 1984.
Note: Maximum score = 36.

   a   Explain how the authors arrived at the degrees of freedom shown in the ANOVA table.

   b   Are the F values correct?

   c   Interpret the test results. What are their practical implications?

   d   Use Table 6 in Appendix II to approximate the p-values for the F statistics shown in the ANOVA table.

*9.51   Perform the analysis of variance calculations for the data in Exercise 9.48 and display the results in an ANOVA table. Compare your table with the SAS printout in Exercise 9.48.

*9.52   Perform the analysis of variance calculations for the data in Exercise 9.49 and display the results in an ANOVA table. Compare your table with the Minitab printout in Exercise 9.49.

9.12 Ranking Population Means

Many experiments are exploratory in nature, in that we have no preconceived notions about the results and have not decided (before conducting the experiment) to make specific treatment comparisons. Rather, we are searching for the treatment that possesses the largest treatment mean, possesses the smallest mean, or satisfies some other set of comparisons. When this situation occurs, we will want to rank the treatment means, determine which means differ, and identify sets of means for which no evidence of differences exists.

One way to achieve this goal is to order the sample means from the smallest to the largest and then to conduct t tests for adjacent means in the ordering. If two means differ by more than

$$t_{\alpha/2}s\sqrt{\frac{1}{n_1} + \frac{1}{n_2}}$$

you conclude that the pair of population means differ. The problem with this procedure is that the probability of making a type I error—that is, concluding that two means differ when, in fact, they are equal—is α for each test. If you compare a large number of pairs of means, the probability of detecting at least one difference in means when, in fact, none exists, is quite large.

A simple way to avoid the high risk of proclaiming differences in multiple comparisons when they do not exist is to use the **Studentized range,** the difference between the smallest and the largest in a set of k sample means, as the yardstick for determining whether there is a difference in a pair of population means. This method, often called **Tukey's method for paired comparisons,** makes the probability of declaring that a difference exists between at least one pair in a set of k treatment means when no difference exists equal to α.

Tukey's method for making paired comparisons is based on the usual analysis of variance assumptions. **In addition, it assumes that the sample means are independent and based upon samples of equal size.** The yardstick that determines whether a difference exists between a pair of treatment means is the quantity ω (Greek letter omega), which is presented in the following display.

Yardstick for Making Paired Comparisons

$$\omega = q_\alpha(k, v)\frac{s}{\sqrt{n_t}}$$

where

k = Number of treatments

s^2 = Estimator of the common variance σ^2 (calculated in an analysis of variance)

v = Number of degrees of freedom for s^2

n_t = Common sample size, that is, the number of observations in each of the k treatment means

$q_\alpha(k, v)$ = Tabulated value, from Tables 7 and 8 in Appendix II, for $\alpha = .05$ and .01, respectively, and for various combinations of k and v

Rule: Two population means are judged to differ if the corresponding sample means differ by ω or more.

As noted in the display, the values of $q_\alpha(k, v)$ are listed in Tables 7 and 8 in Appendix II for $\alpha = .05$ and .01, respectively. A portion of Table 7 in Appendix II is reproduced in Table 9.15. To illustrate the use of Tables 7 and 8, suppose that you want to make pairwise comparisons of $k = 5$ means with $\alpha = .05$ for an analysis of variance, where s^2 possesses $v = 9$ degrees of freedom. The tabulated value for $k = 5$, $v = 9$, and $\alpha = .05$, boxed in Table 9.15, is $q_{.05}(5, 9) = 4.76$.

The following example illustrates the use of Tables 7 and 8 in Appendix II in making paired comparisons.

T A B L E **9.15**   A partial reproduction of Table 7 in Appendix II: Upper 5% points

$$k$$

ν	2	3	4	5	6	7	8	9	10	11	12	...
1	17.97	26.98	32.82	37.08	40.41	43.12	45.40	47.36	49.07	50.59	51.96	...
2	6.08	8.33	9.80	10.88	11.74	12.44	13.03	13.54	13.99	14.39	14.75	...
3	4.50	5.91	6.82	7.50	8.04	8.48	8.85	9.18	9.46	9.72	9.95	...
4	3.93	5.04	5.76	6.29	6.71	7.05	7.35	7.60	7.83	8.03	8.21	...
5	3.64	4.60	5.22	5.67	6.03	6.33	6.58	6.80	6.99	7.17	7.32	...
6	3.46	4.34	4.90	5.30	5.63	5.90	6.12	6.32	6.49	6.65	6.79	...
7	3.34	4.16	4.68	5.06	5.36	5.61	5.82	6.00	6.16	6.30	6.43	...
8	3.26	4.04	4.53	4.89	5.17	5.40	5.60	5.77	5.92	6.05	6.18	...
9	3.20	3.95	4.41	4.76	5.02	5.24	5.43	5.59	5.74	5.87	5.98	...
10	3.15	3.88	4.33	4.65	4.91	5.12	5.30	5.46	5.60	5.72	5.83	...
11	3.11	3.82	4.26	4.57	4.82	5.03	5.20	5.35	5.49	5.61	5.71	...
12	3.08	3.77	4.20	4.51	4.75	4.95	5.12	5.27	5.39	5.51	5.61	...
⋮	⋮	⋮	⋮	⋮	⋮	⋮	⋮	⋮	⋮	⋮	⋮	

E X A M P L E **9.4**   In Example 9.3, we performed an analysis of variance for three replications of a 2×3 factorial experiment. The six cell means are shown in Table 9.16. Rank the $k = 6$ treatment means, make paired comparisons, and determine which treatment mean(s), if any, is (are) the largest.

T A B L E **9.16**
Treatment means for the factorial experiment in Example 9.3

	Shift (factor B)		
Foreman (factor A)	B_1 (8 A.M.–4 P.M.)	B_2 (4 P.M.–midnight)	B_3 (midnight–8 A.M.)
A_1	601.67	498.33	450.00
A_2	486.67	601.67	656.67

Solution   For this example, there are $k = 6$ treatment means, each based on a sample of $n_t = 3$ observations. The standard deviation s, obtained from the SAS analysis of variance printout in Table 9.11, is based on $\nu = 12$ degrees of freedom and is equal to 26.822462. Therefore, from Table 7 in Appendix II, $q_{.05}(k, \nu) = q_{.05}(6, 12) = 4.75$, and the yardstick for detecting a difference between a pair of treatment means is

$$\omega = q_{.05}(6, 12)\frac{s}{\sqrt{n_t}} = (4.75)\frac{(26.82)}{\sqrt{3}} = 73.55$$

The six treatment means are arranged in order from the smallest, 450.00, to the largest, 656.67, in Figure 9.7. The appropriate treatment (factor level combination) is shown above its mean. The next step is to check the difference between each pair of means. If the means differ by ω or more, there is sufficient evidence to indicate

		Treatment			
A_1B_3	A_2B_1	A_1B_2	A_1B_1	A_2B_2	A_2B_3
450.00	486.67	498.33	601.67	601.67	656.67

Means

a difference between the corresponding population means. If there is no evidence to indicate a difference between a pair of means, that fact is indicated by drawing a line under the means.

You can see that the three largest means in the ranking in Figure 9.7 differ by less than $\omega = 73.55$. Therefore, there is no evidence of differences among these three means, and this fact is indicated by the line drawn beneath them. Similarly, there is insufficient evidence to indicate differences among the three smallest means, which is also indicated by underlining. In contrast, each of the three largest means differs by more than $\omega = 73.55$ from the three smallest. Therefore, we conclude that there is sufficient evidence to indicate that the three largest means differ from the three smallest. The probability that we will make at least one error in making all of these multiple comparisons is only $\alpha = .05$.   ∎

Exercises

Starred (*) exercises are optional.

Basic Techniques

9.53   Suppose that you wish to use Tukey's method of paired comparisons to rank a set of population means. In addition to the analysis of variance assumptions, what other property must the treatment means satisfy?

9.54   Consult Tables 7 and 8 in Appendix II, and find the values of $q_\alpha(k, v)$ for the following.

  a $\alpha = .05$, $k = 5$, $v = 7$

  b $\alpha = .05$, $k = 3$, $v = 10$

  c $\alpha = .01$, $k = 4$, $v = 8$

  d $\alpha = .01$, $k = 7$, $v = 5$

9.55   If the sample size for each treatment is n_t and if s^2 is based on 12 degrees of freedom, find ω.

  a $\alpha = .05$, $k = 4$, $n_t = 5$

  b $\alpha = .01$, $k = 6$, $n_t = 8$

9.56   An independent random sampling design was employed to compare the means of six treatments based on samples of four observations per treatment. The pooled estimator σ^2 is 9.12, and the sample means follow:

$$\bar{x}_1 = 101.6 \qquad \bar{x}_2 = 98.4 \qquad \bar{x}_3 = 112.3$$
$$\bar{x}_4 = 92.9 \qquad \bar{x}_5 = 104.2 \qquad \bar{x}_6 = 113.8$$

   a Give the value of ω that you would use to make pairwise comparisons of the treatment means for $\alpha = .05$.

   b Rank the treatment means using pairwise comparisons.

*9.57 Use Tukey's pairwise comparison procedure to rank the means for the 3×3 factorial experiment in Exercise 9.46.

*9.58 Use Tukey's pairwise comparison procedure to rank the means for the 2×2 factorial experiment in Exercise 9.47.

Applications

9.59 Refer to Exercise 9.48. Rank the mean profits per house for the four foremen for house design B_1, and make pairwise comparisons for $\alpha = .05$. Which, if any, of the foremen appears to achieve the largest profit when building house design B_1? [*Note:* When calculating ω, use the value of s^2 calculated by using the complete data set.]

9.60 Refer to Exercise 9.48. Rank the mean profits per house for the four foremen for house design B_2, and make pairwise comparisons for $\alpha = .05$. Which, if any, of the foremen appears to achieve the largest profit when building house design B_2?

9.61 Refer to Exercise 9.48. Rank the mean profits per house for the four foremen for house design B_3, and make pairwise comparisons for $\alpha = .05$. Which, if any, of the foremen appears to achieve the largest profit when building house design B_3?

9.62 Rank the mean percentage gains in sales for the six markup price and location combinations and make paired comparisons for the data in Exercise 9.49 using $\alpha = .05$. What are the practical implications of your comparisons?

9.63 The means for the 2×3 factorial experiment conducted by Barsam and Simutis, discussed in Exercise 9.50, are shown in the accompanying table. Use Tukey's procedure to make paired comparisons of the treatment means, using $\alpha = .05$. What are the practical implications of your comparison?

	Training Condition	
Spatial Ability	Active	Passive
High	17.895	9.508
Medium	5.031	5.648
Low	1.728	1.610

Source: Barsam and Simitis, 1984.
Note: Maximum score = 36.

9.13 Satisfying the Assumptions for an Analysis of Variance: Variance-Stabilizing Transformations (Optional)

As with all statistical methods, the validity of the analysis of variance tests of hypotheses and confidence intervals is based on the assumptions specified in Section 9.2, namely, that the populations are normally distributed with common variance σ^2 and that samples have been selected according to certain specific designs (e.g., independent random samples, a randomized block design, etc.). In the real world we rarely know for certain if these assumptions of normality and a common variance have been satisfied. Therefore, we must know when the analysis of variance tests and confidence intervals will possess the theoretical properties that we expect.

Nonnormality does not seriously affect the methodology as long as the population distributions are not badly skewed. If the populations possess unequal variances, that is a more serious problem. If population variances do not differ greatly and the treatment sample sizes are equal, the properties of the statistical tests and confidence intervals will be approximately the same as if the assumptions were true. If the population variances differ substantially, we can **transform the data** before we conduct the analysis of variance.

For example, one way to transform the data is to take the square root of each value of x. Or we might take the logarithm of each value of x. The type of transformation to be employed depends upon the type of data involved in the experiment. Once the data are transformed, we perform the analysis of variance on the transformed data. Test results for differences in transformed treatment means and the like apply to the original untransformed treatment means.

The two most common types of data that violate the analysis of variance assumptions are those for which x represents the number of occurrences of some event or where x is a sample proportion (or percentage).

Count data often are generated by a random variable x that possesses a Poisson probability distribution. As explained in Section 4.3, the variance of a Poisson random variable is equal to its mean—that is, $\sigma^2 = \mu$. Therefore, the variances of populations of Poisson data are likely to vary from one treatment to another. The variance of sample proportions derived from binomial experiments will also vary with the mean. Thus, if x is a sample proportion, the mean value of the sample proportion is p, and its variance is

$$\sigma^2 = \frac{p(1-p)}{n}$$

A formula for transforming Poisson-type data so that the different treatment populations possess approximately the same variance is

$$y = \sqrt{x}$$

where x is the original Poisson observation and y is the new transformed observation. The following example illustrates the transformation procedure.

E X A M P L E   **9.5**   A manufacturer wishes to compare the mean number of accidents per month on $k = 5$ different production lines. Since the data for this experiment represent observations on five different Poisson random variables, the observations will be transformed by using $y = \sqrt{x}$, where the x values are the original Poisson counts of the number of accidents per month for a given production line and the y values are the new variance-stabilized observations. If three of the original x values are 2, 5, and 4, find the corresponding transformed observations that will be used in the analysis of variance.

Solution   Since $y = \sqrt{x}$, the transformed observations are

$$y_1 = \sqrt{2} = 1.414$$
$$y_2 = \sqrt{5} = 2.236$$
$$y_3 = \sqrt{4} = 2.000 \quad \blacksquare$$

If x is a sample proportion calculated from a binomial experiment, the variances of a set of different binomial populations can be stabilized by using the transformation

$$y = \sin^{-1}\sqrt{x}$$

where x is the original sample proportion and y, the new transformed observation, is equal to the angle (in radians) with a trigonometric sine equal to $\sqrt{x}$. We illustrate the transformation for binomial sample proportions in the following example.

E X A M P L E   **9.6**   A company plans to promote a new product by using one of three advertising plans. To investigate the extent of product recognition by consumers, 15 market areas were selected, and 5 were randomly assigned to each advertising plan. After the advertising plans were employed in the assigned market arenas, random samples of 400 adults were selected from each area, and the proportion x who were familiar with the new product was recorded for each sample. The data are shown in Table 9.17. Transform the data by using the transformation $y = \sin^{-1}\sqrt{x}$.

T A B L E **9.17**
Proportion of 400 adults per marketing area familiar with the new product; data for Example 9.6

Advertising Plan 1	Advertising Plan 2	Advertising Plan 3
.33	.28	.21
.29	.41	.30
.21	.34	.26
.32	.39	.33
.25	.27	.31

Solution   The first step in the transformation is to calculate $\sqrt{x}$ for each value of x. These values are shown in Table 9.18.

The final step in the transformation is to find the value of y that corresponds to each value of $\sqrt{x}$. These values can be found using the inverse sine function on your electronic calculator. For example, when $\sqrt{x} = .574$, $y = \sin^{-1}(.574) = .61$

Advertising Plan 1	Advertising Plan 2	Advertising Plan 3
.574	.529	.458
.539	.640	.548
.458	.583	.510
.566	.624	.574
.500	.520	.557

T A B L E **9.19**
Transformed x values for the data in Table 9.18:
$y = \sin^{-1} \sqrt{x}$

Advertising Plan 1	Advertising Plan 2	Advertising Plan 3
.61	.56	.48
.57	.69	.58
.48	.62	.54
.60	.67	.61
.52	.55	.59

(radians). Similarly, when $\sqrt{x} = .539$, $y = .57$. You can verify that the remaining transformed values for x are as shown in Table 9.19.

To determine whether differences exist among the means for the three advertising plans, we would perform an analysis of variance on the transformed data in Table 9.19. We will leave that project as an exercise for you (see Exercises 9.73 and 9.74).   ∎

In the preceding discussion, we explained how to transform Poisson counts and binomial proportions so that they satisfy the assumptions of an analysis of variance. Transformations can be developed for other types of data where the variance σ^2 of a population is some function of the population mean μ. The procedure for determining an appropriate transformation for a specific relationship between σ^2 and μ is explained in Mendenhall (1968).

In conclusion, the use of a transformation is not without its drawbacks. It is often difficult to assign a practical interpretation to the transformed variable and to the various treatment means. Consequently, many applied statisticians do not use transformations unless there is evidence to suggest sizable differences among the treatment population variances.

The following display gives some useful transformations for data.

Some Useful Data Transformations

Relation Between Variance and Mean of a Population	Application	Transformation
$\sigma^2 = \mu$	Poisson data	$y = \sqrt{x}$
$\sigma^2 = \mu(1 - \mu), \ (0 < \mu < 1)$	Binomial sample proportions: equal sample sizes	$y = \sin^{-1} \sqrt{x}$
$\sigma^2 = \mu^2$	Sample variances, for quantitative data: equal sample sizes	$y = \ln x$

CASE STUDY REVISITED

9.14   AN ANALYSIS OF THE DIFFERENCE IN GROCERY PRICES AT FOUR POINTS IN TIME

The data for this experiment were collected using a randomized block design, which was presented in Sections 9.6 and 9.7. To make valid comparisons of the price of a market basket of food among the five full-service supermarkets, it was necessary to control the effect of collecting the data at different points in time. This was accomplished by surveying all five stores during the same week for four different weeks between December 30, 1992, and January 19, 1993. What we would like to know is whether Lucky has lower prices than the other four supermarkets when averaged over a span of four weeks during the given time period.

A Minitab printout of an analysis of variance is shown in Table 9.20. The key portions of the printout are boxed and numbered.

1   The ANOVA table, marked as area ①, shows that the sum of squares for STORES is 684.64 with 4 degrees of freedom, the sum of squares for WEEKS is 571.71 with 3 degrees of freedom, and the sum of squares for ERROR is 276.38 with 12 degrees of freedom. The mean squares corresponding to these sources of variation, found in column 4, are used to construct the F tests in testing for significant differences in prices among the five stores and among the three-week span, found in column 5.

2   The value of the F statistic used in testing for significant differences among prices for the five stores is

$$F = \frac{\text{MS(Stores)}}{\text{MSE}} = \frac{171.16}{23.03} = 7.43$$

TABLE **9.20**
Minitab analysis of variance for
price comparison

```
MTB > ANOVA C1 = C2 C3;
SUBC> MEANS C2.

Factor      Type Levels Values
STORES      fixed     5      1      2      3      4      5
WEEKS       fixed     4      1      2      3      4

Analysis of Variance for C1                    ①              ③

Source      DF          SS          MS        F        P
STORES       4      684.64      171.16     7.43    0.003
WEEKS        3      571.71      190.57     8.27    0.003
Error       12      276.38       23.03
Total       19     1532.73

      MEANS

STORES    N          C1
   1      4      240.23
   2      4      256.99
   3      4      254.87
   4      4      252.18
   5      4      249.19
```

Similarly, the value of the F statistic in testing for significant differences among prices for the three-week span is found to be

$$F = \frac{\text{MS(Weeks)}}{\text{MSE}} = \frac{190.57}{23.03} = 8.27$$

Both of these tests have p-values equal to .003. This means that if our value of α was greater than .003, we would reject both hypotheses of no differences among the means; that is, we could conclude that the average price of a market basket differs for at least two supermarkets, and possibly for more than two. Further, with the same p-value $= .003$, we could conclude that the average price of a market basket was not constant from week to week.

3   If we really want to know if Lucky has the lowest price of a market basket, then we can continue our analysis by using Tukey's method of paired comparisons. To use Tukey's method, we need to enter the Studentized ranges found in Table 7 in Appendix II. For $k = 5$, $\alpha = .05$, and $\nu = 12$, the table value is 4.51, and MSE is 23.03. Therefore, the critical difference between two means is

$$\omega = 4.51\sqrt{\frac{23.03}{4}} = 10.82$$

When we arrange the market basket means (from Table 9.1) from small to large and display the differences in a systematic way, we generate Table 9.21.

In examining the pairwise differences, we can see that the smallest mean, which is the mean for Albertsons, differs significantly from the means for Alpha Beta, Vons, and Ralphs and that the means for Albertsons and Lucky did not

TABLE **9.21**
Differences in means

Smaller Mean	Larger Mean				
	240.23	249.19	252.18	254.87	256.99
240.23	—	8.96	11.95*	14.64*	16.76*
249.19	—	—	2.99	5.68	7.80
252.18	—	—	—	2.69	4.81
254.87	—	—	—	—	2.12
256.99	—	—	—	—	—

* Denotes a significant difference in means.

differ significantly. We can report the results of our findings by underlying those means that do not differ significantly from one another with the same line.

Albertsons   Lucky   Alpha Beta   Vons   Ralphs

The results indicate that the average prices charged by Albertsons and Lucky for a market basket of food during the time indicated in the survey time-frame did not differ by more than random variation. Further, the average price of a market basket of food did not vary by more than that attributed to random variation for Lucky, Alpha Beta, Vons, and Ralphs. What would you be willing to conclude on the basis of this analysis? It seems that during this time period, although Albertsons' average market basket price was lower than all the others, Albertsons and Lucky had market basket prices that were not significantly different. Further, there were no significant differences in average prices of a market basket of food among Lucky, Alpha Beta, Vons, and Ralphs. Given the results of our statistical analysis, would you be willing to agree that Lucky had the lowest average prices during the time that this survey was conducted? If not, why not?

9.15 Summary

This chapter introduces you to the methodology of an analysis of variance, the notion of partitioning the total sum of squares of deviations of the x values about their overall mean $\bar{x}$ into components pertinent to one or more qualitative variables and to a within-sample source of variation, SSE. The mean squares for these sources of variation can then be compared with the mean square for error, MSE, using the F test. If there is evidence to indicate that a particular source mean square is overly large compared with MSE, we conclude that there are differences in the population means associated with that source.

The analysis of variance not only enables us to test the equivalence of a set of population means but also enables us to place confidence intervals on the difference between pairs of means and, in some cases, on the individual means themselves. Tukey's method for paired comparisons enables us to compare any and all pairs of means, to rank them, and to group means for which no evidence of differences exists.

The validity of an analysis of variance is based on the assumption that the sampled populations are normally distributed (at least approximately) with a common variance σ^2. It is further assumed that the data have been collected according to a specific design. If data from a designed experiment have been lost or destroyed, the standard analysis of variance formulas will give incorrect answers for the sums of squares. In this situation, the sums of squares can be calculated and the analysis of variance F tests conducted by using a procedure known as a multiple regression analysis, presented in Chapter 12. Although this procedure requires much more computation than the standard analyses of variance, it can be performed on a computer by using Minitab, SAS, Excel, or other computer programs.

Supplementary Exercises

Starred (*) exercises are optional.

9.64 Four chemical plants, producing the same product and owned by the same company, discharge effluents into streams in the vicinity of their locations. To check on the extent of the pollution created by the effluents and to determine if it varies from plant to plant, the company collected random samples of liquid waste, five specimens for each of the four plants. The data are shown in the accompanying table. An SAS computer printout of the analysis of variance for the data is also shown. Use the information in the printout to answer the following questions.

Plant	Polluting Effluents (pounds per gallon of waste)				
A	1.65	1.72	1.50	1.37	1.60
B	1.70	1.85	1.46	2.05	1.80
C	1.40	1.75	1.38	1.65	1.55
D	2.10	1.95	1.65	1.88	2.00

ANALYSIS OF VARIANCE PROCEDURE

DEPENDENT VARIABLE: WASTE

SOURCE	DF	SUM OF SQUARES	MEAN SQUARE	F VALUE	PR > F	R-SQUARE	C.V.
MODEL	3	0.46489500	0.15496500	5.20	0.0107	0.493679	10.1515
ERROR	16	0.47680000	0.02980000		ROOT MSE		WASTE MEAN
CORRECTED TOTAL	19	0.94169500			0.17262677		1.70050000

SOURCE	DF	ANOVA SS	F VALUE	PR > F
PLANT	3	0.46489500	5.20	0.0107

a Do the data provide sufficient evidence to indicate a difference in the mean amount of effluents discharged by the four plants?

b If the maximum mean discharge of effluents is 1.5 pounds per gallon, do the data provide sufficient evidence to indicate that the limit is exceeded at plant A?

c Estimate the difference in the mean discharge of effluents between plants A and D, using a 95% confidence interval.

9.65   An investment advisor decided to compare the 1995 annual returns for three types of common stocks: (1) small-capitalization growth stocks, (2) high-capitalization, investment-grade, blue-chip stocks, and (3) electric utility stocks. The returns (in percentages) on random samples of eight stocks selected from among each of the stock-type populations are shown in the accompanying table. A Minitab analysis of variance printout is shown below.

Growth Stock	Blue-Chip Stock	Utilities
59	25	27
31	14	−6
120	40	36
−61	17	42
14	53	31
92	54	35
8	27	26
−51	35	32

```
ANALYSIS OF VARIANCE
SOURCE      DF        SS        MS        F        P
FACTOR       2       196        98      0.07     0.937
ERROR       21     31376      1494
TOTAL       23     31571
                                   INDIVIDUAL 95% CI'S FOR MEAN
                                   BASED ON POOLED STDEV
LEVEL       N      MEAN      STDEV   --+---------+---------+---------+-----
GROWTH       8     26.50     63.55   (----------------*---------------)
BLUECHIP     8     33.12     15.17        (--------------*--------------)
UTILITY      8     27.88     14.61   (--------------*--------------)
                                     --+---------+---------+---------+-----
POOLED STDEV =     38.65            0        20        40        60
```

a   Do the data provide sufficient evidence to indicate differences in the mean annual returns for the three types of stock? Test using $\alpha = .05$.

b   Find a 95% confidence interval for the difference in mean annual return between growth and utility stocks.

*9.66   Consider the accompanying one-way classification consisting of three treatments, A, B, and C, where the number of observations per treatment varies from treatment to treatment. The observations were randomly and independently selected from their respective treatment populations.

A	B	C
24.2	24.5	26.0
27.5	22.7	
25.9		
24.7		

a   Use the formulas of optional Section 9.5 to perform the analysis of variance for these data.

b   Do the data provide sufficient evidence to indicate a difference among the treatment means?

c   Find a 90% confidence interval for the mean for treatment B.

d   Find a 90% confidence interval for the difference between the means for treatments A and C.

9.67 A company wanted to study the differences among four sales-training programs on the sales abilities of their sales personnel. Thirty-two people were randomly divided into four groups of equal size, and the groups were then subjected to the different sales-training programs. Because there were some dropouts (illness, etc.) during the training programs, the number of trainees completing the programs varied from group to group. At the end of the training programs, each salesperson was randomly assigned a sales area from a group of sales areas that were judged to have equivalent sales potentials. The numbers of sales made by each of the four groups of salespeople during the first week after completing the training program are listed in the accompanying table. Use the Minitab printout below to answer the following questions.

	Training Program		
1	2	3	4
78	99	74	81
84	86	87	63
86	90	80	71
92	93	83	65
69	94	78	86
73	85		79
	97		73
	91		70
Total 482	735	402	588

```
MTB > ANOVA C1=C2;
SUBC> MEANS C2.

Factor      Type Levels Values
PROGRAM     fixed      4     1     2     3     4

Analysis of Variance for SALES

Source      DF        SS        MS       F      P
PROGRAM      3    1385.78    461.93   9.84  0.000
Error       23    1079.41     46.93
Total       26    2465.19


        MEANS

     PROGRAM   N     SALES
           1   6    80.333
           2   8    91.875
           3   5    80.400
           4   8    73.500
```

a Do the data provide sufficient evidence to indicate a difference in the mean achievement levels for the four training programs?

b Find a 90% confidence interval for the difference in the mean numbers of sales that would be expected for persons subjected to training programs 1 and 4. Interpret the interval.

c   Find a 90% confidence interval for the mean number of sales by persons subjected to training program 2.

9.68 An experiment was conducted to compare the effect of four different chemicals, A, B, C, and D, in producing water resistance in textiles. A strip of material, randomly selected from a bolt, was cut into four pieces, and the pieces were randomly assigned to receive one of the four chemicals A, B, C, or D. This process was replicated three times, thus producing a randomized block design. The design, with moisture resistance measurements, is shown in the table (low readings indicate low moisture penetration). An SAS computer printout of the analysis of variance for the data is presented below. Use the information in the printout to answer the following questions.

Blocks (bolt samples)

1	2	3
C	D	B
9.9	13.4	12.7
A	B	D
10.1	12.9	12.9
B	A	C
11.4	12.2	11.4
D	C	A
12.1	12.3	11.9

ANALYSIS OF VARIANCE PROCEDURE

DEPENDENT VARIABLE: MOISTURE

SOURCE	DF	SUM OF SQUARES	MEAN SQUARE	F VALUE	PR > F	R-SQUARE	C.V.
MODEL	5	12.37166667	2.47433333	27.75	0.0004	0.958549	2.5023
ERROR	6	0.53500000	0.08916667		ROOT MSE		MOISTURE MEAN
CORRECTED TOTAL	11	12.90666667			0.29860788		11.93333333

SOURCE	DF	ANOVA SS	F VALUE	PR > F
BLOCKS	2	7.17166667	40.21	0.0003
TRTMENTS	3	5.20000000	19.44	0.0017

a   Do the data provide sufficient evidence to indicate a difference in the mean moisture penetration for fabric treated with the four chemicals?

b   Do the data provide evidence to indicate that blocking increased the amount of information in the experiment?

c   Find a 95% confidence interval for the difference in mean moisture penetration for fabrics treated by chemicals A and D. Interpret the interval.

9.69 A building contractor employs three construction engineers, A, B, and C, to estimate and bid on jobs. To determine whether one tends to be a more conservative (or liberal) estimator than the others, the contractor selects four projected construction jobs and has each estimator independently estimate the cost (dollars per square foot) of each job. The data and the associated analysis of variance printout are shown in the accompanying tables.

Estimator	Construction Job (blocks)				
(treatments)	1	2	3	4	Total
A	35.10	34.50	29.25	31.60	130.45
B	37.45	34.60	33.10	34.40	139.55
C	36.30	35.10	32.45	32.90	136.75
Total	108.85	104.20	94.80	98.90	406.75

	A	B	C	D	E	F	G	H	I	J
1		Twoway Analysis of Variance for CBS9-69.COST								
2	Source			Sum of Squares	D.F.		Mean Square	F-Ratio		P-Value
3	TREATMENTS			10.8617	2		5.43083	7.196		0.0255
4	BLOCKS			37.6073	3		12.5358	16.610		0.0026
5	Error			4.5283	6		0.75472			
6	TOTAL (Corrected)			52.9973	11					
7										
8										
9	Table of Means									
10				Count	Mean		Stnd Error	Est. Effect		
11	TREATMENTS									
12	1			4	32.6125		0.4344	-1.283		
13	2			4	34.8875		0.4344	0.992		
14	3			4	34.1875		0.4344	0.292		
15	BLOCKS									
16	1			3	36.2833		0.5016	2.388		
17	2			3	34.7333		0.5016	0.838		
18	3			3	31.6000		0.5016	-2.296		
19	4			3	32.9667		0.5016	-0.929		
20	OVERALL			12	33.8958		0.2508			
21										

a Do the data provide sufficient evidence to indicate a difference in the mean building costs estimated by the three estimators? Test using $\alpha = .05$.

b Find a 90% confidence interval for the difference in the mean of the estimates produced by estimators A and B. Interpret the interval.

c Do the data support the contention that the mean estimate of the cost per square foot varies from job to job?

9.70 A production superintendent wants to compare the mean time to assemble a piece of electronic equipment for three different methods of assembly, A_1, A_2, and A_3. Six assemblers were chosen for the experiment. Each person was assigned to assemble one piece of equipment for each method of assembly, and the average assembly time (in minutes) was recorded for each. The methods of assembly were randomly assigned in sequence for each assembler. The data are shown in the accompanying table. Use the Excel printout to answer the following questions.

a What type of experimental design is this? Explain.

b Explain how the design may increase the amount of information in the experiment on treatment mean differences.

c Do the data provide sufficient evidence to indicate differences in the mean time to assemble for the three methods of assembly? Test using $\alpha = .05$.

Assembler

Sequence of Assembly	1	2	3	4	5	6
A_1	20.2	22.6	19.2	22.5	18.7	21.5
A_2	23.7	24.1	22.6	24.3	20.2	21.5
A_3	21.4	23.0	22.9	22.0	19.8	20.1

	A	B	C	D	E	F	G	H	I	J
1	\multicolumn{10}{c}{**Twoway Analysis of Variance for CBS9-70.TIMES**}
2	Source			Sum of Squares	D.F.		Mean Square	F-Ratio		P-Value
3	SEQUENCE			11.61	2		5.805	5.917		0.0201
4	ASSEMBLER			26.67	5		5.333	5.436		0.0113
5	Error			9.81	10		0.981			
6	TOTAL (Corrected)			48.085	17					
7										
8										
9	\multicolumn{3}{l}{**Table of Means**}							
10				Count	Mean		Stnd Error	Est. Effect		
11	SEQUENCE									
12	1			5	20.7833		0.4044	-0.900		
13	2			5	22.7333		0.4044	1.050		
14	3			5	21.5333		0.4044	-0.150		
15	ASSEMBLER									
16	1			3	21.7667		0.5718	0.083		
17	2			3	23.2333		0.5718	1.550		
18	3			3	21.5667		0.5718	-0.117		
19	4			3	22.9333		0.5718	1.250		
20	5			3	19.5667		0.5718	-2.117		
21	6			3	21.0333		0.5718	-0.650		
22	OVERALL			18	21.6833		0.2335			
23										

*9.71 A large food products company conducted an experiment to investigate the effects of two factors, package wrapper material and color of wrapper, on sales of one of the company's products. Two types of wrapping material were employed, a waxed paper and a plastic, in three colors. Eighteen supermarkets were selected for the experiment, and three were assigned to each of the six factor level combinations. After the product had been in the supermarkets for one week, the company recorded the percentage change in weekly sales over each supermarket's average weekly sales of the product for the past year. The data are shown in the accompanying table.

Wrapping Material	Package Color A_1	A_2	A_3
B_1	6	-3	7
	-2	7	3
	4	-2	10
B_2	5	3	12
	2	6	7
	5	4	10

a   Perform an analysis of variance for the data, and display the results in an ANOVA table.

b   Find the approximate p-values for the F tests.

c   What practical conclusions do you derive from the F test results?

9.72   A study was conducted to compare automobile gasoline mileage for three brands of gasoline, A, B, and C. Four automobiles, all of the same make and model, were employed in the experiment, and each gasoline brand was tested in each automobile. Using each brand within the same automobile has the effect of eliminating (blocking out) automobile-to-automobile variability. The accompanying table gives the data (in miles per gallon) for the 12 brand-automobile combinations, and the Minitab analysis of variance for the data is shown below.

Gasoline Brand	Automobile			
	1	2	3	4
A	25.7	27.0	27.3	26.1
B	27.2	28.1	27.9	27.7
C	26.1	27.5	26.8	27.8

```
MTB > ANOVA C1 = C2 C3

Factor      Type Levels Values
BRAND       fixed      3     1     2     3
AUTOTYPE    fixed      4     1     2     3     4

Analysis of Variance for MPG

Source      DF         SS         MS       F        P
BRAND        2     2.8950     1.4475    6.46    0.032
AUTOTYPE     3     2.5200     0.8400    3.75    0.079
Error        6     1.3450     0.2242
Total       11     6.7600
```

a   Do the data provide sufficient evidence to indicate a difference in mean mileage per gallon for the three gasolines?

b   Is there evidence of a difference in mean mileage for the four automobiles?

c   Suppose that *prior to looking at the data* we had decided to compare the mean mileage per gallon for gasoline brands A and B. Find a 90% confidence interval for this difference.

*9.73   In Example 9.6, we presented data representing consumer response to three different market advertising plans based on independent random samples, five observations per sample. Because the data represent sample proportions, we transformed each observation by using a $\sin^{-1} \sqrt{x}$ transformation. The transformed data of Table 9.20 are reproduced below.

Advertising Plan 1	Advertising Plan 2	Advertising Plan 3
.61	.56	.48
.57	.69	.58
.48	.62	.54
.60	.67	.61
.52	.55	.59

a Perform an analysis of variance on the transformed data, and present the results in an ANOVA table.

b Do the data provide sufficient evidence to indicate differences in mean responses to the three advertising plans? Test using $\alpha = .05$.

*9.74 The original untransformed sample proportions for Example 9.6 are shown in the table below.

Advertising Plan 1	Advertising Plan 2	Advertising Plan 3
.33	.28	.21
.29	.41	.30
.21	.34	.26
.32	.39	.33
.25	.27	.31

a Perform an analysis of variance for the data.

b Do the data provide sufficient evidence to indicate differences in mean responses to the three advertising plans? Test using $\alpha = .05$.

c Did the analysis of the transformed data lead to conclusions that differ from the analysis of the corresponding untransformed data?

Exercises Using the Data Disk

9.75 Consider the broccoli data, data set B on the data disk, with fresh weight and the 6-hour average ozone as the variables of interest.

a Let the variable AVG-6, the average 6-hour ozone level, represent the treatment applied and let fresh weight represent the dependent variable. Assume that the observations within a chamber are in random order. Now select the observations in rows 1–5 as having received treatment 1, an average of .029 ppm of ozone; next, select the observations in rows 40–44 as having received treatment 2, an average of .038 ppm of ozone, and select the observations in rows 83–87 as having received treatment 3, an average of .042 ppm of ozone. Notice that all these observations were made on the first variety, which was labeled zero. Perform an analysis of variance for these data and summarize your results.

b Following a similar procedure, or using random numbers to select your samples, select five observations on the second variety (labeled 1) from the market-weight column for each of the three 10-hour ozone averages: .030, .039, and .043 ppm. Perform an analysis of variance for these data and report your results.

c One way of analyzing the broccoli data for any of the four dependent variables is to examine the data using a two-way analysis of variance with interaction. The first factor is varieties, and the second is amount of ozone in ppm for *one* of the variables AVG-6, AVG-10, or AVG-13. Select random samples of size 5 in each of the 2×3 variety-ozone levels for, say, the AVG-13 variable with .014, .026, and .037 ppm. Perform an analysis of variance for your data and report your results.[†]

9.76 Refer to data set A on the data disk.

a Randomly and independently select a sample of five to ten colleges. For each college selected, record the salaries for male faculty in each of the four faculty ranks. Your data were collected using a randomized block design with colleges as blocks. Perform an analysis of variance for your data and report your results.

[†]*Note to the Instructor:* In using the selection procedure indicated in parts (a), (b), and (c), the chamber and ozone average are confounded effects. You may want to use an alternative sampling procedure.

b   Repeat part (a) using the salaries for female faculty. [You may use the same colleges that you selected for part (a).]

c   **Challenge Exercise:** If you have not recorded the salaries for the female faculty ranks for the colleges selected in part (a), do so now. You will have 2×4 or eight observations for each college chosen. The experiment now becomes one using a randomized block design (colleges are blocks) with eight treatments in which the treatment factors, gender and rank, constitute a factorial experiment. Use an analysis of variance to analyze your data. Have your instructor check your results. [*Hint:* In the analysis of variance, the line for treatments in the source column is replaced by three lines: one for factor A, gender; one for factor B, rank; and one for the interaction of A and B.]

QUALITY CONTROL

About This Chapter

The objective of this chapter is to present some very useful techniques used by industry to monitor and to improve the quality of manufactured products. Example 10.5 will show how some of these techniques can be used to improve the quality of business systems.

Contents

THE CASE OF THE
MISSING OIL STOCKS

Consider a management problem. You are the manager of a motor oil can filling operation. The 1-quart cans, the type that you purchase at your local filling station, are filled on a filling machine that contains 28 spindles (or nozzles). Each spindle releases oil into a single can, and the machine fills 28 cans, one can at each spindle.

The problem arises when your company accountants detect an unusual discrepancy. Although the oil stocks received for the filling operation amount to 10,000,000 quarts per month, the number of filled 1-quart cans is always considerably less. If the number of cans filled in a given month is 9,700,000 cans, what has happened to the missing 300,000 quarts of oil?

The problem that we have just described is similar to one encountered by V. Filimon and colleagues, who at the time were employed by the Standard Oil Company, Cleveland, Ohio (Filimon, Maggass, Frazier, and Klingel, 1955). Their operation consisted of three filling machines, a 1-quart machine containing 6 spindles, another 1-quart machine containing 28 spindles, and a 16-spindle, 1-gallon machine.

The search for the missing oil stocks quickly focused on the filling machines. Can you set a machine so that an individual spindle will discharge exactly 1 quart of oil into each can? The answer to this question is no, because the amount of oil discharged from a single spindle differs slightly from one discharge to another due to variation in the flow of oil through the spindle. Thus the amount x of oil discharged into a can, measured in either volume or weight, varies from can to can and possesses a relative frequency distribution similar to those described in Chapter 2. This variation in fill weight led Filimon and colleagues to suspect that the missing oil stocks left the plant in overfilled cans.

The can-filling operation, along with its associated problems, is typical of most ongoing business operations. Each operation yields a product that is judged to be acceptable or unacceptable depending upon one or more variables that measure the quality of the product. The product of an oil can filling operation is a can of oil, and one measure of its "quality" is the amount of oil in the can. The product of a machine producing electric light bulbs is a single bulb, and one measure of its quality is the amount of light that it emits. The product of a hospital is treatment and care for a single patient. The quality of patient care would undoubtedly be measured by a number of different quality variables.

In this chapter, we present some management and statistical techniques used to monitor, improve, and control the quality of the product of an ongoing business operation. Then in Section 10.12 we learn how Filimon and colleagues applied some of these techniques to reduce the loss of oil stocks in their can-filling operation.

10.1 Quality Control

As the title suggests, **quality control** methodology was developed to control and improve the product of a manufacturing process. Steel bars must possess a specified tensile strength, soap must be produced with a low level of impurities, a box of cereal must contain a specified weight, and the financial entries into a business computer must, with high probability, be accurate. Thus the objective of a quality control program is to ensure that the variables that measure a product's quality fall into ranges that are acceptable to prospective customers.

Quality control methods fall into one of three categories: (1) **monitoring techniques,** designed to track the level of quality variables and to detect undesirable shifts in product quality; (2) **troubleshooting techniques,** used to help locate the cause of undesirable changes in product quality; and (3) **screening techniques,** designed to remove defective or poor-quality products entering the process as raw materials and to perform the same job for finished products before shipment to a customer.

It is often said that quality control is 10% statistics and 90% engineering and common sense. As you will learn, most quality control methods are based on elementary statistical concepts presented in preceding chapters: the sampling distributions of Chapter 6 and the Empirical Rule of Chapter 2. The real problem arises when trouble is detected, in either the monitoring or screening process. Thus quality control methods can tell you *when* but not *why* trouble occurs. Finding the cause of poor product quality and correcting the situation require knowledge of the process and problem-solving ability.

This chapter will be concerned with two of the three types of quality control methods: methods for monitoring an ongoing production process and methods used to screen out unsatisfactory raw materials entering a process and/or defective product leaving a process. The third category of methodology, statistical methods for locating the cause of a downward shift in product quality, consists of all the methods described in the preceding chapters. Two of the most useful methods, regression analysis and contingency table analysis, may establish a correlation between one or more raw materials, or process or environmental variables, and product quality. Regression analysis and contingency table analysis are covered in Chapters 11, 12, and 15.

10.2 Monitoring Product Quality Using Control Charts

Measurements on a quality variable change from one point in time to another. For example, measurements on the inside diameter of a 1-inch-diameter bearing will vary slightly from one bearing to another. The bearing diameter may tend to become smaller over time due to wear in the cutting edge in the machining process. Variation of this type is said to be due to an *assignable cause.* Other variation—small haphazard changes due to the many unknown variables that affect the diameter, changes in raw materials, changes in environmental conditions, and so on—is regarded as *random variation.*

If the variation of a quality variable is solely random, the process is said to be *in control.* Being in control does not mean that a process is producing 100% good product. The values of the quality variable may or may not fall in a random pattern within the limits specified by the manufacturer's customers.

Figure 10.1(a) shows a plot of bearing diameters, one measured each hour, for a process that is in control and that is producing bearings that fall within customer specifications, say .980 to 1.020 inches. Figure 10.1(b) shows a similar plot for a process that is in control but is producing many bearings with diameters that fall outside customer specifications and would be judged defective. Figure 10.1(c) shows how a plot might look if an assignable cause of variation were present. Note that the plotted diameters no longer appear to vary in a random manner but trend downward over time.

The first objective of a manufacturer is to eliminate assignable causes of variation in a quality variable and to get the process in control. The next step is to reduce process variation and to get the distribution of quality measurements within specifications. The mean value of the distribution should fall at or near the center of the specifications interval, and the variance of the distributions should be as small as possible. A desirable distribution for the inside bearing diameters is one that falls entirely within specifications, as shown in Figure 10.2.

Once a process is in control and is producing satisfactory product, the process mean and variance are monitored by using **control charts.** Samples of n items are drawn from the process at specified intervals of time, say, every half hour or hour, and the sample mean $\bar{x}$ and range R are computed. These statistics are plotted on $\bar{x}$ and R charts similar to the charts shown in Figure 10.1. A control chart of the sample mean $\bar{x}$ is used to detect possible shifts in the distribution mean for a quality variable (see Section 6.5). Similarly, a control chart for the sample range R is used to detect changes in the distribution variance. Control charts for the mean $\bar{x}$ and the range R are discussed in Sections 10.3 and 10.4, respectively. Two other useful control charts are presented in Sections 10.6 and 10.7.

F I G U R E **10.1**

Plots of bearing diameters over time

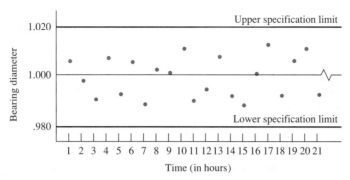

a   Process in control and within specifications

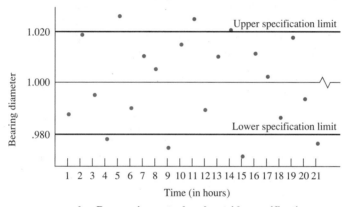

b   Process in control and outside specifications

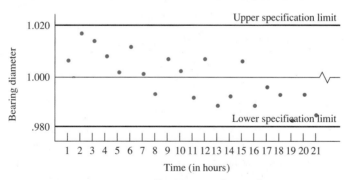

c   Process out of control

F I G U R E **10.2**
A desirable distribution

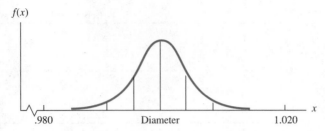

10.3 A Control Chart for the Process Mean: The $\bar{x}$ Chart

In order to monitor a process, n items are selected from the production at equal intervals of time or at a fixed number of items produced, and the measurements on a quantitative variable are recorded. For example, Figure 10.3 shows a plot of the sample mean $\bar{x}$ of the diameters of $n = 5$ bearings selected hourly from a machining process.

F I G U R E **10.3**
An $\bar{x}$ control chart

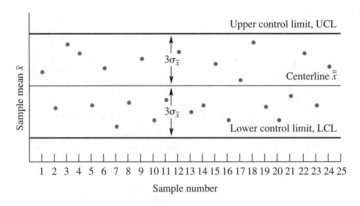

The logic behind an $\bar{x}$ **control chart** is that, if the process is in control, the sample means should vary about the population mean μ in a random manner and almost all values of $\bar{x}$ should fall in the interval $(\mu \pm 3\sigma_{\bar{x}})$. Although the exact value of the process mean μ is unknown, we can obtain an accurate estimate of it by averaging a large number k (at least 25)[†] of sample means. This estimate locates the centerline of the control chart. In traditional quality control notation, it is represented by the symbol $\bar{\bar{x}}$ (i.e., it is the mean of the sample means).

The upper and lower control limits are located

$$3\sigma_{\bar{x}} = \frac{3\sigma}{\sqrt{n}}$$

[†] The number $k \geq 25$ is recommended by Grant and Leavenworth (1979). This number is not critical; that is, k could be smaller, say, as small as 20. The larger the value of k, the better will be the chart.

above and below the centerline. The value of σ can be estimated by calculating the sample standard deviation s, using the combined set of data from the k samples as we did in Section 6.5. If the data collected each time period are entered into a computer, the estimate s can be obtained by a computer command. If a computer is unavailable (which was the case when these methods were developed), the calculation of s is time-consuming and, in some cases, beyond the arithmetic abilities of some production workers. For that reason, it has been traditional to calculate a range estimate of σ. To obtain this estimate, we calculate the range R for each sample, equal to the difference between the largest and smallest measurements in the sample; then we calculate the average $\bar{R}$ for the 25 (or more) sample R values. Then

$$\hat{\sigma} = \frac{\bar{R}}{d_2} = \frac{\sum\limits_{i=1}^{k} \dfrac{R_i}{k}}{d_2}$$

where d_2 is the constant that will make $\hat{\sigma}$ an unbiased estimate of σ when sampling from a normally distributed population.[†] Substituting this estimate of σ into the formula for $3\sigma_{\bar{x}}$, we obtain

$$3\hat{\sigma}_{\bar{x}} = 3\frac{\hat{\sigma}}{\sqrt{n}} = 3\frac{\bar{R}}{d_2\sqrt{n}} = A_2\bar{R}$$

where

$$A_2 = \frac{3}{d_2\sqrt{n}}$$

Values of A_2 and d_2 for $n = 2$ to 25 are given in Table 9 of Appendix II.

The details for an $\bar{x}$ chart are summarized in the following display and illustrated in Example 10.1.

A Control Chart for the Process Mean: The $\bar{x}$ Chart

Centerline: $\hat{\mu} = \bar{\bar{x}} = \dfrac{\sum\limits_{i=1}^{k} \bar{x}_i}{k}$

Upper control limit: $\text{UCL} = \bar{\bar{x}} + A_2\bar{R}$

Lower control limit: $\text{LCL} = \bar{\bar{x}} - A_2\bar{R}$

where $\bar{R} = \dfrac{\sum\limits_{i=1}^{k} R_i}{k}$

[†]This range estimate of σ can be biased if the population of quality measurements is not approximately normal. The large interval $6\sigma_{\bar{x}}$ (rather than $4\sigma_{\bar{x}}$) between the control limits probably compensates for error in estimating σ. Nevertheless, in this computer age it would be better to estimate σ^2 by using the sample variance s^2 based on the kn measurements from the k samples of n measurements each.

and the values of A_2 are given in Table 9 of Appendix II.

Assumption: $k \geq 25$

E X A M P L E   **10.1**   A quality control monitoring system samples the diameter of $n = 4$ bearings each hour. Table 10.1 provides data for 25 samples. Construct an $\bar{x}$ chart for the sample means.

T A B L E 10.1
25 hourly samples of bearing diameters, $n = 4$ bearings per sample, for Example 10.1

Sample	Sample Measurements				Sample Mean, $\bar{x}$	Sample Range, R
1	.992	1.007	1.016	.991	1.00150	.025
2	1.015	.984	.976	1.000	.99375	.039
3	.988	.993	1.011	.981	.99325	.030
4	.996	1.020	1.004	.999	1.00475	.024
5	1.015	1.006	1.002	1.001	1.00600	.014
6	1.000	.982	1.005	.989	.99400	.023
7	.989	1.009	1.019	.994	1.00275	.030
8	.994	1.010	1.009	.990	1.00075	.020
9	1.018	1.016	.990	1.011	1.00875	.028
10	.997	1.005	.989	1.001	.99800	.016
11	1.020	.986	1.002	.989	.99925	.034
12	1.007	.986	.981	.995	.99225	.026
13	1.016	1.002	1.010	.999	1.00675	.017
14	.982	.995	1.011	.987	.99375	.029
15	1.001	1.000	.983	1.002	.99650	.019
16	.992	1.008	1.001	.996	.99925	.016
17	1.020	.988	1.015	.986	1.00225	.034
18	.993	.987	1.006	1.001	.99675	.019
19	.978	1.006	1.002	.982	.99200	.028
20	.984	1.009	.983	.986	.99050	.026
21	.990	1.012	1.010	1.007	1.00475	.022
22	1.015	.983	1.003	.989	.99750	.032
23	.983	.990	.997	1.002	.99300	.019
24	1.011	1.012	.991	1.008	1.00550	.021
25	.987	.987	1.007	.995	.99400	.020

$$\bar{\bar{x}} = .9987 \qquad \bar{R} = .02444$$

Solution   The sample mean $\bar{x}$ and the sample range R were calculated for each of the $k = 25$ samples. For example, the sample mean $\bar{x}$ and range R for sample 1 are

$$\bar{x} = \frac{.992 + 1.007 + 1.016 + .991}{4} = 1.0015$$

and

$$R = 1.016 - .991 = .025$$

The sample means and ranges are shown in columns 6 and 7, respectively, of Table 10.1. The mean $\bar{\bar{x}}$ of the $k = 25$ values of $\bar{x}$ and the mean $\bar{R}$ of the $k = 25$ sample R values are shown at the bottom of their respective columns.

Figure 10.4 shows the $\bar{x}$ chart constructed from the data. The centerline is located at $\bar{\bar{x}} = .998700$, or $\bar{\bar{x}} \approx .9987$. The estimate of $3\sigma_{\bar{x}}$ is

$$3\hat{\sigma}_{\bar{x}} = A_2\bar{R}$$

where $\bar{R} = .02444$, or $\approx .0244$, and the value of A_2 for $n = 4$ (given in Table 9 of Appendix II) is .729. Therefore,

$$3\hat{\sigma}_{\bar{x}} = A_2\bar{R} = (.729)(.0244) = .0178$$
$$\text{UCL} = \bar{\bar{x}} + A_2\bar{R} = .9987 + .0178 = 1.0165$$

and

$$\text{LCL} = \bar{\bar{x}} - A_2\bar{R} = .9987 - .0178 = .9809$$

Therefore, lines locating the upper and lower control limits are located at 1.0165 and .9809, respectively. ▪

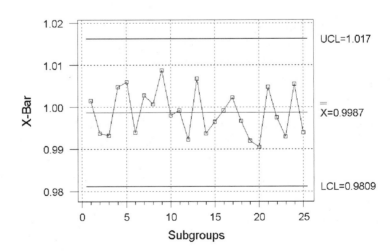

F I G U R E 10.4
An $\bar{x}$ chart for the bearing diameters of Table 10.1 (Example 10.1)

Assuming that the samples used to construct the $\bar{x}$ chart in Figure 10.4 were collected when the process was in control, the chart can now be used to detect changes in the process mean. Sample means can be plotted on the chart periodically. If a sample mean falls outside the control limits, it conveys a warning of a possible shift in the process mean. The process will be monitored closely, and efforts will be made to locate the cause of the errant mean. Do the sample means for Figure 10.4 vary about the centerline in a random manner, or do they suggest trends and the possibility of assignable causes in variation? A test for nonrandomness is presented in Section 10.8.

10.4 A Control Chart for Process Variation: The R Chart

Just as it is important to keep the mean value of a quality variable near the center of the specification interval, it is desirable to control the process variation. The smaller the variance of the quality measurements, the greater will be the probability that the measurements fall within the customer specification limits (assuming the process mean is within specifications).

The variation in a process quality variable is monitored by plotting the sample range R on an R **chart.** The R chart is constructed in essentially the same manner as the $\bar{x}$ chart. A centerline is located at the estimated value of μ_R, and control limits are located $3\sigma_R$ above and below μ_R.

The estimate of μ_R is $\overline{R}$, the mean of the ranges of the k samples used to construct the $\bar{x}$ chart. Calculation of the control limits,

$$\text{UCL} = \hat{\mu}_R + 3\hat{\sigma}_R \qquad \text{and} \qquad \text{LCL} = \hat{\mu}_R - 3\hat{\sigma}_R$$

has been reduced to a single calculation,

$$\text{UCL} = D_4 \overline{R} \qquad \text{and} \qquad \text{LCL} = D_3 \overline{R}$$

The values of D_4 and D_3, based on sampling from a normally distributed population of quality measurements, are given in Table 9 of Appendix II, for different values of n. The details for the R chart are summarized in the following display and illustrated in Example 10.2.

A Control Chart for Process Variation: The R Chart

Centerline: $\hat{\mu}_R = \overline{R} = \dfrac{\sum\limits_{i=1}^{k} R_i}{k}$

Upper control limit: $\text{UCL} = D_4 \overline{R}$

Lower control limit: $\text{LCL} = D_3 \overline{R}$

   The values of D_3 and D_4 for sample size n are given in Table 9 of Appendix II.

Assumption: $k \geq 25$

E X A M P L E   **10.2**   Construct an R chart based on the data in Table 10.1.

Solution   In Example 10.1, we found $\overline{R} = .0244$. For sample size $n = 4$, Table 9 in Appendix II gives $D_3 = 0$ and $D_4 = 2.282.^{\dagger}$ Therefore, the centerline for the control chart is located at $\overline{R} = .0244$, and the upper and lower control limits are

$$\text{UCL} = D_4\overline{R} = (2.282)(.0244) = .0557$$

and

$$\text{LCL} = D_3\overline{R} = (0)(\overline{R}) = 0$$

The R chart is shown in Figure 10.5.   ▪

F I G U R E **10.5**
An R chart for the bearing
diameters of Table 10.1
(Example 10.2)

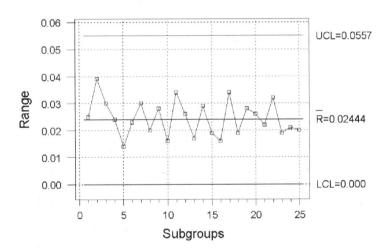

The R chart in Figure 10.5 is evaluated in the same manner as the $\bar{x}$ chart. The sample range R, calculated for samples taken periodically, should vary about the centerline $\overline{R}$ in a random manner and fall within the control limits when the process is in control. A sample range that falls outside the control limits will be taken as a warning of a possible change in process variation. The process will then be examined to determine whether the unusually large (or small) R value was caused by changes in the raw materials, the environment, or one of the many other variables that affect the process.

---

†The lower control limit, $\mu_R - 3\sigma_R$, is negative for sample size $n = 4$. Since a range R cannot be negative, the value for D_3 is given as 0.

10.5 Process Capability

After a process has been monitored with control charts and deemed to be in statistical control, the capability of the process can then be determined. **Process capability** refers to the ability of a process to stay within its specification limits. Histograms can be used to evaluate process capability by simply looking at how much of the histogram falls between the upper and lower specification limits. We can go a step further, though, by calculating a numerical measure of process capability in the following way. Assuming that the measurements generated by the process approximately follow a normal distribution, almost all of the readings should fall within a range of 6σ. This range is referred to as the **actual process spread;** the distance between the specification limits is called the **allowable process spread.** From these measures, the process capability index, denoted C_p, is derived.

DEFINITION ▪

The **process capability index** is defined by

$$C_p = \frac{\text{USL} - \text{LSL}}{6\hat{\sigma}}$$

where $\hat{\sigma}$ is an estimate of the standard deviation of the measurements from the process. ▪

The index C_p is interpreted as follows: If $C_p = 1.0$, then the process is said to be *capable* (but just barely) of meeting its specification limits. Values of C_p that exceed 1.0 are better, since then the probability is higher that the measurements will be able to stay within the specification limits. A value of C_p exceeding 1.33 (i.e., an 8σ range fits within the specification limits) is usually considered very good and is commonly used as a target in many applications. On the other hand, a value of C_p less than 1.0 implies that a process is not capable of meeting its specifications. Figure 10.6 illustrates typical values of the index C_p along with the associated distributions of measurements from the process.

The C_p is one of five measures, originally developed in Japan, that are now used in almost all quality control programs. These indexes are useful because they convey much information about the process being studied in a very simple fashion. Capability indexes are also unitless measures, which allows them to be used to compare two entirely different processes. For example, if copper-plating thicknesses (in inches) from a chemical-plating process have a C_p of .81 and resistance measurements (in ohms) on electronic components have a C_p of 2.30, then we can immediately conclude that the electronic component process is the better of the two, even though their measurement units, inches and ohms, are unrelated.

The index C_p does not take into account the location of the process, only its *potential* for meeting specifications. Figure 10.7 illustrates this characteristic by showing two processes with C_ps of 1.0, one centered between the specification limits and the other located closer to its upper specification limit. The latter process has the potential to be capable, since its C_p is 1.0, but something will have to be done to shift

FIGURE **10.6**
Interpretation of the process
capability index C_p

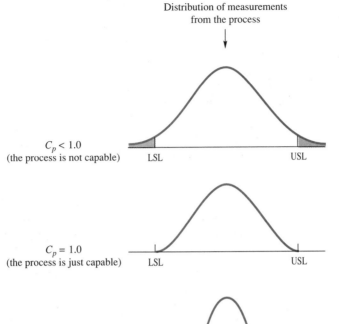

Distribution of measurements
from the process

$C_p < 1.0$
(the process is not capable)   LSL   USL

$C_p = 1.0$
(the process is just capable)   LSL   USL

$C_p > 1.0$
(the process is capable)   LSL   USL

the location of this process (estimated by $\bar{\bar{x}}$) closer to the center of the specification range.

A closely related index that *does* take the process mean into account is the index C_{pk}.

DEFINITION ▪ The **index C_{pk}** is defined by

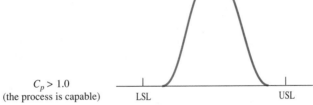

$$C_{pk} = \min\left[\frac{\text{USL} - \bar{\bar{x}}}{3\hat{\sigma}}, \frac{\bar{\bar{x}} - \text{LSL}}{3\hat{\sigma}}\right]$$ ▪

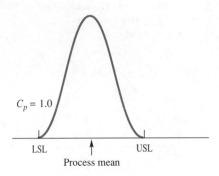

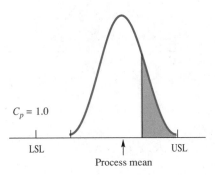

In the definition of C_{pk}, $\bar{\bar{x}}$ is the centerline of the process $\bar{x}$ chart, $\hat{\sigma} = \overline{R}/d_2$ is an estimate of the process variation, and min represents minimum. The k in the subscript of C_{pk} refers to the so-called k factor,

$$k = \frac{|(\text{USL} + \text{LSL})/2 - \bar{\bar{x}}|}{(\text{USL} - \text{LSL})/2}$$

which measures the extent to which the process location $(\bar{\bar{x}})$ differs from its desired value midway between the specification limits. When $\bar{\bar{x}}$ lies between USL and LSL, it can be shown that k always lies between 0 and 1 and that the indexes C_p and C_{pk} are related by the formula

$$C_{pk} = C_p(1 - k)$$

Since $0 \leq k \leq 1$, this formula shows that C_{pk} never exceeds C_p and that $C_{pk} = C_p$ precisely when the process is centered midway between its specification limits. Together, C_p and C_{pk} give a clear picture of how well a process is performing when compared with its specification limits. C_{pk} is generally preferred to C_p, since it gives a truer measure of capability when the process is not centered.

E X A M P L E   **10.3**   Suppose that acceptable bearing diameters in Example 10.1 were required to have diameters of $1.000 \pm .050$. The control charts in Examples 10.1 and 10.2 showed that the process was in statistical control, so its process capability can be measured. Calculate C_p and C_{pk} for the bearing diameter in Table 10.1.

Solution   The process specification limits of $1.000 \pm .050$ give rise to USL $= 1.050$ and LSL $= .950$. Since $\overline{R} = .0244$ for these data, the variation in the measurements can be estimated by

$$\hat{\sigma} = \frac{\overline{R}}{d_2} = \frac{.0244}{2.059} = .01185$$

so

$$C_p = \frac{\text{USL} - \text{LSL}}{6\hat{\sigma}} = \frac{1.050 - .950}{6(.01185)} = 1.406$$

In other words, this process is capable of staying within its specification limits.

Whether or not it actually *is* staying within these limits can be determined by calculating the index C_{pk}:

$$C_{pk} = \min\left[\frac{\text{USL} - \overline{\overline{x}}}{3\hat{\sigma}}, \frac{\overline{\overline{x}} - \text{LSL}}{3\hat{\sigma}}\right] = \min\left[\frac{1.050 - .9987}{3(.01185)}, \frac{.9987 - .950}{3(.01185)}\right]$$

$$= \min[1.443, \ 1.370] = 1.370$$

Alternatively, we could find the k factor

$$k = \frac{|(\text{USL} + \text{LSL})/2 - \overline{\overline{x}}|}{(\text{USL} - \text{LSL})/2} = \frac{|(1.050 + .950)/2 - .9987|}{(1.050 - .950)/2} = .026$$

and then calculate C_{pk} by

$$C_{pk} = C_p(1 - k) = 1.406(1 - .026) = 1.370$$

Since C_{pk} is greater than 1.0, the process appears to be capable of staying within these specification limits. The value of C_p is very close to the ideal value of 1.33 and therefore there should be no problem adhering to specification limits.   ∎

Exercises

Basic Techniques

10.1   The sample means and ranges were calculated for 30 samples of size $n = 10$ for a process that was judged to be in control. The means of the 30 $\overline{x}$ values and of the 30 R values were $\overline{\overline{x}} = 50.25$ and $\overline{R} = 16.80$.

a   Calculate a range estimate for the process standard deviation σ.

b   Explain why a sample standard deviation, calculated from the 300 observations in 30 samples, might provide a better estimate of σ than the range estimate.

c   Use the data to determine the upper and lower control limits for an $\bar{x}$ chart.

d   What is the purpose of an $\bar{x}$ chart?

e   Construct an $\bar{x}$ chart for the process, and explain how it can be used.

10.2   The sample means and ranges were calculated for 40 samples of size $n = 5$ for a process that was judged to be in control. The means of the 40 $\bar{x}$ values and of the 40 R values were $\bar{\bar{x}} = 70.38$ and $\bar{R} = 12.45$.

a   Calculate a range estimate for the process standard deviation σ.

b   Use the data to determine the upper and lower control limits for an $\bar{x}$ chart.

c   Construct an $\bar{x}$ chart for the process, and explain how it can be used.

10.3   Use the information in Exercise 10.1 to construct an R chart. What is the purpose of an R chart?

10.4   The specification limits for the process in Exercise 10.2 are 70 ± 10. Calculate C_p and C_{pk} for this process. Is the process capable of meeting specification limits?

Applications

10.5   A gambling casino records and plots the mean and range of the daily gain or loss from five blackjack tables on $\bar{x}$ and R charts. The means of the sample means and ranges for over 40 weeks were $\bar{\bar{x}} = \$10{,}752$ and $\bar{R} = \$6425$.

a   Construct an $\bar{x}$ chart for the mean daily gain per blackjack table.

b   How might this $\bar{x}$ chart be of value to the manager of the casino?

10.6   Refer to Exercise 10.5. Construct an R chart for the range of the daily gain per blackjack table. How might this chart be of value to the manager of the casino?

10.7   The manager of the gambling casino in Exercise 10.5 also plots the daily gain or loss for each blackjack table on a chart. This is, essentially, an $\bar{x}$ chart for the special case where $n = 1$. The mean gain for table 1, calculated over 40 days, was $\bar{x} = \$10{,}940$, and the sample standard deviation was \$5130.

a   If the gain x for the dealer at table 1 has a distribution with mean μ and standard deviation σ, within what limits would you expect x to fall on almost all days?

b   Since you have estimates of μ and σ, construct an $\bar{x}$ chart ($n = 1$) for the daily gains from table 1.

c   Of what value might the chart in part (b) be to the manager of the casino?

10.8   Refer to Exercise 6.32. The data given in the accompanying table measure the radiation in air particulates at a nuclear power plant. Four measurements were recorded at weekly intervals over a 26-week period. Use the data to construct an $\bar{x}$ chart, and plot the 26 values of $\bar{x}$. Explain how the chart will be used.

10.9   Construct an R chart for the data in Exercise 10.8, and explain how the chart will be used.

10.10   A coal-burning power plant tests and measures three specimens of coal each day to monitor the percentage of ash in the coal. The means of 30 daily sample means and ranges were $\bar{\bar{x}} = 7.24$ and $\bar{R} = .27$.

a   Construct an $\bar{x}$ chart for the process, and explain how it can be of value to the manager of the power plant.

b   Construct an R chart for the process. Explain how it can be of value to the manager.

Table for Exercise 10.8

Week	Radiation				Week	Radiation			
1	.031	.032	.030	.031	14	.029	.028	.029	.029
2	.025	.026	.025	.025	15	.031	.029	.030	.031
3	.029	.029	.031	.030	16	.014	.016	.016	.017
4	.035	.037	.034	.035	17	.019	.019	.021	.020
5	.022	.024	.022	.023	18	.024	.024	.024	.025
6	.030	.029	.030	.030	19	.029	.027	.028	.028
7	.019	.019	.018	.019	20	.032	.030	.031	.030
8	.027	.028	.028	.028	21	.041	.042	.038	.039
9	.034	.032	.033	.033	22	.034	.036	.036	.035
10	.017	.016	.018	.018	23	.021	.022	.024	.022
11	.022	.020	.020	.021	24	.029	.029	.030	.029
12	.016	.018	.017	.017	25	.016	.017	.017	.016
13	.015	.017	.018	.017	26	.020	.021	.020	.022

10.6 A Control Chart for the Proportion Defective: The p Chart

Sometimes, the observation made on manufactured items is simply whether an item meets the manufacturer's (or customer's) specifications. Thus each item is judged either defective or nondefective. If the proportion[†] of defectives produced by the process is p, then the number x of defectives in a random sample of n items possesses the binomial probability distribution of Section 4.2.

For the monitoring of items that are either defective or nondefective, samples of size n are selected at periodic intervals, and the sample proportion $\hat{p}$ is calculated. If the process is in control, the sample proportion $\hat{p}$ should fall in the interval $p \pm 3\sigma_{\hat{p}}$, where p is the process mean proportion of defectives and

$$\sigma_{\hat{p}} = \sqrt{\frac{p(1-p)}{n}}$$

The process mean proportion p of defectives is estimated by using the average of k sample proportions:

$$\bar{p} = \frac{\sum\limits_{i=1}^{k} \hat{p}_i}{k}$$

and $\sigma_{\hat{p}}$ is estimated by

$$\hat{\sigma}_{\hat{p}} = \sqrt{\frac{\bar{p}(1-\bar{p})}{n}}$$

[†]Quality control texts usually speak of "fraction defective" rather than "proportion defective."

The centerline for the **p chart** is located at $\overline{\hat{p}}$, and the upper and lower control limits are

$$\text{UCL} = \overline{\hat{p}} + 3\hat{\sigma}_{\hat{p}}$$

$$= \overline{\hat{p}} + 3\sqrt{\frac{\overline{\hat{p}}(1-\overline{\hat{p}})}{n}}$$

and

$$\text{LCL} = \overline{\hat{p}} - 3\hat{\sigma}_{\hat{p}}$$

$$= \overline{\hat{p}} - 3\sqrt{\frac{\overline{\hat{p}}(1-\overline{\hat{p}})}{n}}$$

The details for a p chart are summarized in the display and are illustrated in Example 10.4.

A Control Chart for the Proportion Defective: A p Chart

Centerline: $\overline{\hat{p}} = \dfrac{\displaystyle\sum_{i=1}^{k} \hat{p}_i}{k}$

Upper control limit: $\text{UCL} = \overline{\hat{p}} + 3\sqrt{\dfrac{\overline{\hat{p}}(1-\overline{\hat{p}})}{n}}$

Lower control limit: $\text{LCL} = \overline{\hat{p}} - \sqrt{\dfrac{\overline{\hat{p}}(1-\overline{\hat{p}})}{n}}$

Assumption: $k \geq 25$

EXAMPLE 10.4 A manufacturer of ballpoint pens randomly samples 400 pens per day and tests each to see whether the ink flow is acceptable. The proportions of pens judged defective each day over a 40-day period are shown in Table 10.2. Construct a control chart for the proportion $\hat{p}$ defective in samples of $n = 400$ pens selected from the process.

Solution The estimate of the process proportion defective is the average of the $k = 40$ sample proportions of Table 10.2. Therefore, the centerline of the control chart is located at

$$\overline{\hat{p}} = \frac{\displaystyle\sum_{i=1}^{k} \hat{p}_i}{k} = \frac{.0200 + .0125 + \cdots + .0225}{40} = \frac{.7600}{40} = .019$$

Day	**Proportion**	**Day**	**Proportion**	**Day**	**Proportion**	**Day**	**Proportion**
1	.0200	11	.0100	21	.0300	31	.0225
2	.0125	12	.0175	22	.0200	32	.0175
3	.0225	13	.0250	23	.0125	33	.0225
4	.0100	14	.0175	24	.0175	34	.0100
5	.0150	15	.0275	25	.0225	35	.0125
6	.0200	16	.0200	26	.0150	36	.0300
7	.0275	17	.0225	27	.0200	37	.0200
8	.0175	18	.0100	28	.0250	38	.0150
9	.0200	19	.0175	29	.0150	39	.0150
10	.0250	20	.0200	30	.0175	40	.0225

TABLE 10.2
Proportions of defectives in samples of $n = 400$ pens for Example 10.4

An estimate of $\sigma_{\hat{p}}$, the standard deviation of the sample proportions, is

$$\hat{\sigma}_{\hat{p}} = \sqrt{\frac{\bar{p}(1 - \bar{p})}{n}} = \sqrt{\frac{(.019)(.981)}{400}} = .00683$$

and $3\hat{\sigma}_{\hat{p}} = (3)(.00683) = .0205$. Therefore, the upper and lower control limits for the p chart are located at

$$\text{UCL} = \bar{p} + 3\hat{\sigma}_{\hat{p}} = .0190 + .0205 = .0395$$

and

$$\text{LCL} = \bar{p} - 3\hat{\sigma}_{\hat{p}} = .0190 - .0205 = -.0015$$

Or, since p cannot be negative, LCL $= 0$.

The p control chart is shown in Figure 10.8. Note that all 40 sample proportions fall within the control limits. If a sample proportion collected at some time in the future falls outside the control limits, the manufacturer will be warned of a possible increase in the value of the process proportion defective. Efforts will be initiated to seek possible causes for an increase in the process proportion defective. ∎

Although it is convenient to choose samples of the same size n at each sampling, there may be occasions when the number of items in a sample differs from the numbers collected at previous samplings. Since $\hat{\sigma}_{\hat{p}}$, UCL, and LCL depend upon the sample size n, you will not be able to calculate the limits for a single sample size. When the sample size varies for each sampling, the estimate of the mean process proportion p of defectives is equal to the sample proportion based on the number of defectives in the k samples. Then you compare each $\hat{p}_i$ with values for UCL and LCL calculated for that particular sample's sample size. This process will cause the control limits to change from sample to sample, and the graphs of the control limits will be a sequence of connected line segments.

F I G U R E **10.8**
A proportion defective chart for
ballpoint pens for Example 10.4

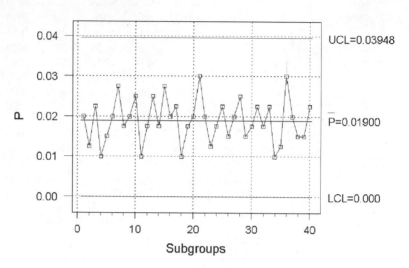

10.7 A Control Chart for the Number of Defects per Item: The c Chart

An important measure of quality for some products is the number of defects per manufactured item. A textile manufacturer often classifies irregularities in a woven product as defects. Since the eventual selling price of the material will depend upon its quality, the manufacturer wishes to reduce the number of defects per square yard of material to a minimum. Then the manufacturer will want to get the number of defects per square yard in control.

The number of defects per unit area, volume, or weight or on a single manufactured item is an important measure of quality for many products. Typical examples are the number of defects in the paint job on a new automobile, the number of defects on the varnish coating for a piece of furniture, the number of air holes in a cubic inch of cheese, and the number of incorrect billing entries per computer printout page for a plumbing company.

The number of defects per unit area, volume, weight, or single item, usually denoted by the symbol c, is monitored at equal intervals of time by using a **c chart.** For most applications, the probability distribution of c can be approximated by a Poisson probability distribution,[†] a distribution that possesses a very unique property. Its variance σ^2 is equal to its mean μ; that is,

$$\sigma_c^2 = \mu_c \quad \text{and} \quad \sigma_c = \sqrt{\mu_c}$$

Therefore, the number c of defects per item should fall in the interval

$$\mu_c \pm 3\sigma_c \quad \text{or} \quad \mu_c \pm 3\sqrt{\mu_c}$$

[†]The Poisson probability distribution is discussed in greater detail in Section 4.3.

To construct a c chart, we sample the process while it is in control and record the value of c for at least $k = 25$ points in time. The process mean μ_c is estimated by the sample mean:

$$\hat{\mu}_c = \bar{c} = \frac{\sum_{i=1}^{k} c_i}{k}$$

and the process standard deviation σ_c is estimated by

$$\hat{\sigma}_c = \sqrt{\bar{c}}$$

The centerline of the c chart is located at $\bar{c}$, and the upper and lower control limits are

$$\text{UCL} = \hat{\mu}_c + 3\hat{\sigma}_c = \bar{c} + 3\sqrt{\bar{c}}$$

and

$$\text{LCL} = \hat{\mu}_c - 3\hat{\sigma}_c = \bar{c} - 3\sqrt{\bar{c}}$$

The features of a c chart are summarized in the following display, and their use is illustrated in Example 10.5.

A Control Chart for the Number of Defects per Item: The c Chart

Centerline: $\hat{\mu}_c = \bar{c} = \dfrac{\sum_{i=1}^{k} c_i}{k}$

Upper control limit: $\text{UCL} = \bar{c} + 3\sqrt{\bar{c}}$

Lower control limit: $\text{LCL} = \bar{c} - 3\sqrt{\bar{c}}$

Assumption: $k \geq 25$

E X A M P L E   **10.5**   An auditor monitors a company's billing system each week. This monitoring is accomplished by comparing actual bills with computer entries for all entries that appear on ten computer printout pages. The number c of incorrect entries on ten pages of printout was recorded each week for 40 weeks. The data are shown in Table 10.3. Use the data to construct a c chart for the auditing process.

Solution   The value of $\bar{c}$, calculated for data collected over $k = 40$ weeks, is

$$\bar{c} = \frac{\sum_{i=1}^{k} c_i}{k} = \frac{49}{40} = 1.225$$

TABLE **10.3**
Number c of incorrect entries
per ten pages of computer
printout for Example 10.5

Week	1	2	3	4	5	6	7	8	9	10
c	1	3	2	0	0	1	4	2	1	1

Week	11	12	13	14	15	16	17	18	19	20
c	1	0	1	1	3	2	1	1	0	3

Week	21	22	23	24	25	26	27	28	29	30
c	0	2	1	0	1	1	2	2	1	0

Week	31	32	33	34	35	36	37	38	39	40
c	1	2	0	3	1	1	2	0	1	0

Then the centerline for the c chart is located at $\bar{c} = 1.225$, and the upper and lower control limits are located at

$$\text{UCL} = \bar{c} + 3\sqrt{\bar{c}} = 1.225 + 3\sqrt{1.225}$$
$$= 4.55$$

and

$$\text{LCL} = \bar{c} - 3\sqrt{\bar{c}} = 1.225 - 3\sqrt{1.225}$$
$$= -2.10$$

or, since c cannot be negative, LCL $= 0$.

The c chart for the data, along with the plotted values of c, is shown in the Minitab graphic, Figure 10.9. Note that all $k = 40$ of the c values fall within the upper and lower control limits, something that we would expect if the process is in control.

FIGURE **10.9**
A c chart for the number of
incorrect billings for
Example 10.5

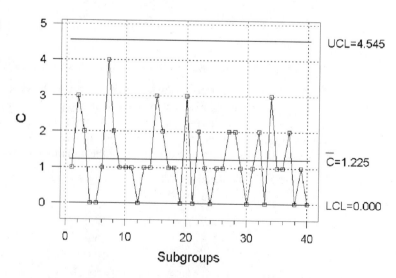

C Chart for Incorrect Entries

In the use of the *c* chart, values of *c* will be recorded each week in the future and plotted on the *c* chart. A value of *c* falling outside the control limits will be a warning of possible problems with the data entry operation.   ▪

Exercises

Basic Techniques

10.11   Explain the difference between a *p* chart and a *c* chart.

10.12   Samples of $n = 100$ items were selected hourly over a 100-hour period, and the sample proportion of defectives was calculated each hour. The mean of the 100 sample proportions was .035.

    a   Use the data to find the upper and lower control limits for a *p* chart.

    b   Construct a *p* chart for the process, and explain how it will be used.

10.13   Samples of $n = 200$ items were selected hourly over a 100-hour period, and the sample proportion of defectives was calculated each hour. The mean of the 100 sample proportions was .041.

    a   Use the data to find the upper and lower control limits for a *p* chart.

    b   Construct a *p* chart for the process, and explain how it will be used.

10.14   The number *c* of defects per item was recorded every hour over a period of 100 hours, and the mean number of defects per item was found to equal .7.

    a   Use the data to find the upper and lower control limits for a *c* chart.

    b   Construct a *c* chart, and explain how it will be used.

10.15   The number *c* of defects per item was recorded every hour over a period of 200 hours, and the mean number of defects per item was found to equal 1.3.

    a   Use the data to find the upper and lower control limits for a *c* chart.

    b   Construct a *c* chart, and explain how it will be used.

Applications

 10.16   A producer of brass rivets randomly samples 400 rivets each hour and calculates the proportion of defectives in the sample. The mean sample proportion, calculated from 200 samples, was equal to .021. Construct a control chart for the proportion of defectives in samples of 400 rivets. Explain how the control chart can be of value to a manager.

 10.17   A personnel manager plots the number of plant personnel accidents per month on a control chart. Over a period of 30 months the mean number of accidents per month was found to equal 3.7. Construct a control chart for the manager, and explain how it will be used.

 10.18   A company records and plots on a control chart the number of customer complaints received each week. The mean number of customer complaints, collected over a 52-week period, was 4.9 complaints per week. Construct a control chart for the number of customer complaints per week, and explain how the chart can be of value to a manager.

 10.19   The manager of a building supply company randomly samples incoming lumber to see whether it meets the manager's quality specifications. One hundred pieces of 2 × 4 lumber from each shipment are inspected and judged according to whether they are first (acceptable) or second (defective) grade. The proportions of second-grade 2 × 4s recorded for 30 shipments were

.14, .21, .19, .18, .23, .20, .25, .19, .22, .17, .21, .15, .23, .12, .19,
.22, .15, .26, .22, .21, .14, .20, .18, .22, .21, .13, .20, .23, .19, .26

Construct a control chart for the proportion of second-grade 2 × 4s in samples of 100 from shipments. Explain how the control chart can be of use to the manager of the building supply company.

10.8 A Test for Nonrandomness

We stated in earlier sections that a process is in control when all assignable causes of variation in a quality variable have been removed and all that remains is random variation. Therefore, if a process is in control, variables that measure a product's quality should vary in a random manner about their process means. Similarly, plotted sample means, ranges, proportions, and counts of defects on $\bar{x}$, R, p, and c control charts should vary in a random manner above and below their respective centerlines.

One way to detect nonrandomness in the behavior of a plotted quality variable is to record, for each point on a control chart, whether the point lies above (A) or below (B) the centerline. For example, the sequence of As and Bs for the $\bar{x}$ chart in Figure 10.10 is

B, A, A, B, A, B, A, A, A, B, A, B

FIGURE **10.10**

An $\bar{x}$ chart

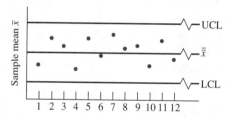

The sequence of As and Bs associated with a control chart often suggests nonrandomness and the possible presence of assignable causes of variation. For example, Figure 10.11(a) shows an $\bar{x}$ chart in which the mean appears to be trending upward. Figure 10.11(b) shows an $\bar{x}$ chart in which the mean appears to be cycling above and below the process mean.

The $\bar{x}$ charts shown in Figures 10.11(a) and 10.11(b) both suggest assignable causes of variation, both indicate trends in the values of $\bar{x}$, and both exhibit a sequence of deviations above and below the centerline that possesses a unique pattern. For the $\bar{x}$ chart in Figure 10.11(a), the sequence of As and Bs is

B, B, B, B, B, B, A, A, A, A, A, A, A

The sequence for the $\bar{x}$ chart in Figure 10.11(b) is

B, B, B, B, A, A, A, A, B, B, B, B

FIGURE **10.11**
$\bar{x}$ charts that suggest possible
assignable causes of variation

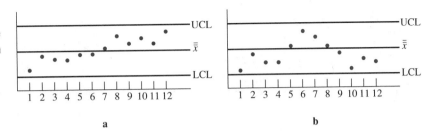

a   b

Notice the difference between these two sequences of As and Bs and the sequence associated with the apparent random behavior of $\bar{x}$ in Figure 10.10. Both consist of long runs of As and Bs. The sequence in Figure 10.11(a) contains two runs, a run of six Bs followed by a run of six As. The sequence in Figure 10.11(b) contains three runs: four Bs, then four As, then four Bs. In contrast, the sequence of As and Bs for the seemingly random behavior of $\bar{x}$ in Figure 10.10 contains nine runs. This discussion suggests that the number r of runs in a sequence of As and Bs can be used to detect nonrandom behavior in a process.

In a sequence of As and Bs, a **run** is defined as a maximal subsequence of like elements. For example, in the sequence A, A, A, B, A, B, B, A, the subsequence A, A, A is a run because it consists of a subsequence of like elements (all of them As), and it is maximal in the sense that it includes the maximal number of like elements before encountering a B. The first two elements in the sequence, A, A, form a subsequence of like elements, but this subsequence is not maximal because it is followed by another A. With this definition of a run, the sequence

$$\underbrace{A, A, A,}_{1} \quad \underbrace{B,}_{2} \quad \underbrace{A,}_{3} \quad \underbrace{B, B,}_{4} \quad \underbrace{A}_{5}$$

contains five runs.

DEFINITION ▪
> In a sequence of elements A and B, a **run** is a maximal subsequence of like elements. ▪

The number r of runs in a sequence of n_1 As and n_2 Bs can be used as a test statistic to test for nonrandomness. The null and alternative hypotheses for the test are

H_0 : the sequence of As and Bs has been generated by a random process

H_a : the sequence of As and Bs has been generated by a nonrandom process

Either a very large or a very small number of runs can be indicators of a nonrandom sequence of As and Bs, although trends in a process mean are usually indicated by a small number of runs. Therefore, the rejection region can be either one-tailed or two-tailed (see Figure 10.12).

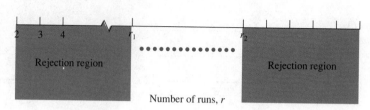

FIGURE 10.12
The rejection region for a runs test

Number of runs, r

The values of $P(r \leq r_0)$, given H_0 is true, are given in Table 10 in Appendix II for all combinations of n_1 and n_2, $n_1 \leq 10$ and $n_2 \leq 10$. A large-sample standard normal z test can be used when both n_1 and n_2 are larger than 10. The details of the test are summarized in the display, and Example 10.6 illustrates the use of Table 10 in Appendix II for small samples.

A Runs Test for Nonrandomness

1   Null Hypothesis: H_0 : the sequence of As and Bs has been generated by a random process

2   Alternative Hypothesis: H_a : the sequence of As and Bs has been generated by a nonrandom process

3   Test Statistic: $r =$ Number of runs in the sequence of As and Bs

4   Rejection Region:

One-Tailed Test	**Two-Tailed Test**
$r \leq r_1$	$r \leq r_1$ and $r \geq r_2$ (see Figure 10.12)

Values of $P(r \leq r_0)$ are given in Table 10 of Appendix II for $n_1 \leq 10$ and $n_2 \leq 10$.

E X A M P L E   **10.6**   Use the first 15 sample means in Table 10.1 to test for nonrandomness in the sequence of sample means used to construct the $\bar{x}$ chart in Figure 10.4. Use a two-tailed test with $\alpha \approx .10$.

Solution   The mean for the $k = 25$ samples used to locate the centerline of the $\bar{x}$ chart in Figure 10.4 was $\bar{\bar{x}} = .9987$. The $\bar{x}$ values for the first 15 samples are reproduced in Table 10.4. Each is identified by an A or B, depending upon whether the sample mean lies above or below the centerline $\bar{\bar{x}} = .9987$.

The sequence of the 15 As and Bs, extracted from Table 10.4,

$$A, B, B, A, A, B, A, A, A, B, A, B, A, B, B$$

contains $n_1 = 8$ As and $n_2 = 7$ Bs, and $r = 10$.

The two-tailed rejection region for the test will appear as shown in Figure 10.12, with r_1 and r_2 determined from Table 10 in Appendix II. Since we want α to be approximately .10, we will choose r_1 so that $\alpha/2 \approx .05$ falls in the lower tail of the distribution of r—that is, so that $P(r \leq r_1) \approx .05$.

T A B L E **10.4**   Sample means for samples 1–15 of Table 10.1 (Example 10.6)

Sample	$\bar{x}$	Deviation	Sample	$\bar{x}$	Deviation	Sample	$\bar{x}$	Deviation
1	1.00150	A	6	.99400	B	11	.99925	A
2	.99375	B	7	1.00275	A	12	.99225	B
3	.99325	B	8	1.00075	A	13	1.00675	A
4	1.00475	A	9	1.00875	A	14	.99375	B
5	1.00600	A	10	.99800	B	15	.99650	B

F I G U R E **10.13**   Rejection region for the runs test of Example 10.6

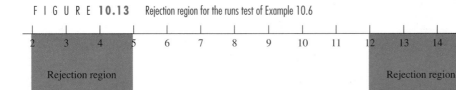

Number of runs, r

Table 10 in Appendix II gives values of $P(r \leq r_0)$ for combinations of n_1 and n_2 and for values of r_0. Values of r_0, from 2 to 20, are shown in the top row of the table. Combinations of (n_1, n_2) are given in the left column of the table. The value of $P(r \leq r_0)$ for $n_1 = 8$, $n_2 = 7$ is the same as for $n_1 = 7$, $n_2 = 8$. Therefore, we proceed down the (n_1, n_2) column to the row corresponding to (7,8). Moving across the (7,8) row in the table, we find the $P(r \leq r_0)$ for $r_0 = 4$ to be .015, for $r_0 = 5$ to be .051, and for $r_0 = 6$ to be .149. Since we want $\alpha/2$ to be approximately .05, we will reject H_0 if $r \leq 5$ (i.e., $r_1 = 5$). The upper-tail rejection value r_2 is a value such that $P(r < r_2) \approx .95$. Proceeding across the (7,8) row, we find $P(r \leq 10) = .867$ and $P(r \leq 11) = .949$. Therefore, we will reject H_0 when $r \geq 12$ (i.e., $r_2 = 12$). This places .051 in the lower tail and $1 - .949 = .051$ in the upper tail of the r distribution. Therefore, $\alpha = .051 + .051 = .102$. The rejection region for the test is shown in Figure 10.13. Since the observed number of runs in our sequence of $n_1 = 8$ As and $n_2 = 7$ Bs, $r = 10$, does not fall in the rejection region, there is insufficient evidence to indicate nonrandomness in the process.   ■

When n_1 and n_2 are both large, $n_1 > 10$ and $n_2 > 10$, we can conduct the runs test by using the familiar standard normal z test statistic of Chapter 8, where

$$z = \frac{r - \mu_r}{\sigma_r} \qquad \mu_r = \frac{2n_1 n_2}{n_1 + n_2} + 1$$

and

$$\sigma_r = \sqrt{\frac{2n_1 n_2 (2n_1 n_2 - n_1 - n_2)}{(n_1 + n_2)^2 (n_1 + n_2 - 1)}}$$

We summarize the test in the following display and illustrate it in Example 10.7.

Large-Sample Runs Test

1 Null Hypothesis: H_0 : the sequence of As and Bs has been generated by a random process

2 Alternative Hypothesis: H_a : the sequence of As and Bs has been generated by a nonrandom process

3 Test Statistic: $z = \dfrac{r - \mu_r}{\sigma_r}$

where

$$\mu_r = \frac{2n_1 n_2}{n_1 + n_2} + 1$$

and

$$\sigma_r = \sqrt{\frac{2n_1 n_2 (2n_1 n_2 - n_1 - n_2)}{(n_1 + n_2)^2 (n_1 + n_2 - 1)}}$$

4 Rejection Region:

Lower One-Tailed Test	Two-Tailed Test
$z < -z_\alpha$	$z < -z_{\alpha/2}$ or $z > z_{\alpha/2}$

Assumptions: $n_1 > 10$ and $n_2 > 10$.

E X A M P L E   10.7   Use the large-sample runs test to test for nonrandomness in the sequence of $k = 40$ observations on the number of incorrect billing entries per ten pages of computer printout in Table 10.3. Use a two-tailed test with $\alpha = .05$.

Solution   The mean number of incorrect billing entries per ten pages of computer printout for the $k = 40$ observations, calculated in Example 10.5, is $\bar{c} = 1.225$. The sequence of As (larger than $\bar{c} = 1.225$) and Bs (smaller than $\bar{c} = 1.225$) for the $k = 40$ weekly entries in Table 10.3 is

B, A, A, B, B, B, A, A, B, B, B, B, B, B, A, A, B, B, B, A,

B, A, B, B, B, B, A, A, B, B, B, A, B, A, B, B, A, B, B, B

The numbers of As and Bs in this sequence of $k = 40$ letters are, respectively, $n_1 = 13$ and $n_2 = 27$. The number of runs is $r = 19$. The mean and standard deviation for r, assuming H_0 true, are

$$\mu_r = \frac{2n_1 n_2}{n_1 + n_2} + 1 = \frac{2(13)(27)}{40} + 1 = 18.55$$

and

$$\sigma_r = \sqrt{\frac{2n_1 n_2 (2n_1 n_2 - n_1 - n_2)}{(n_1 + n_2)^2 (n_1 + n_2 - 1)}} = \sqrt{\frac{2(13)(27)[2(13)(27) - 13 - 27]}{(13 + 27)^2 (13 + 27 - 1)}}$$

$$= \sqrt{7.4475} = 2.73$$

Then

$$z = \frac{r - \mu_r}{\sigma_r} = \frac{19 - 18.55}{2.73} = .16$$

The rejection region for a two-tailed test with $\alpha = .05$ is $z > z_{.025} = 1.96$ or $z < -z_{.025} = -1.96$. Since the observed value of z does not fall in the rejection region, we conclude that there is insufficient evidence to reject the hypothesis of randomness. ∎

Exercises

Basic Techniques

10.20   For a random sequence of n_1 As and n_2 Bs, find the probability that the number r of runs is less than or equal to r_0 for the following.

    a $n_1 = 5$, $n_2 = 7$, $r_0 = 4$ **b** $n_1 = 6$, $n_2 = 7$, $r_0 = 4$

    c $n_1 = 8$, $n_2 = 10$, $r_0 = 7$ **d** $n_1 = 10$, $n_2 = 10$, $r_0 = 5$

10.21   For a random sequence of n_1 As and n_2 Bs, find the number of runs r_0 such that $P(r \leq r_0) \approx .05$ for the following.

    a $n_1 = 4$, $n_2 = 10$ **b** $n_1 = 6$, $n_2 = 10$

    c $n_1 = 7$, $n_2 = 8$ **d** $n_1 = 3$, $n_2 = 4$

10.22   For a random sequence of n_1 As and n_2 Bs, find the number of runs r_0 such that $P(r \leq r_0) \approx .10$ for the following.

    a $n_1 = 5$, $n_2 = 10$ **b** $n_1 = 4$, $n_2 = 5$

    c $n_1 = 8$, $n_2 = 8$ **d** $n_1 = 10$, $n_2 = 10$

10.23   Calculate μ_r and σ_r when $n_1 = 20$ and $n_2 = 15$.

10.24   Calculate μ_r and σ_r when $n_1 = 25$ and $n_2 = 30$.

10.25   Use the large-sample normal approximation to the distribution of the number of runs r in a random sequence of $n_1 = 10$ As and $n_2 = 10$ Bs to calculate $P(r \leq 8)$. Compare the approximation with the exact value given in Table 10 of Appendix II.

Applications

10.26   Use the runs test to test for nonrandomness in the deviations about the centerline for the 26 values of $\bar{x}$ in Exercise 10.8.

10.27   Use the runs test to test for nonrandomness in the deviations about the centerline for the 30 values of $\hat{p}$ in Exercise 10.19.

10.28   The home office of a corporation selects its executives from the staff personnel of its two subsidiary companies, Company A and Company B. During the past three years, nine executives have been selected by the parent company from the subsidiaries, the first selected from B, the second from A, and so on. The following sequence shows the order in which the executives have been selected and the subsidiary firms from which they came:

B, A, A, A, B, A, A, A, B

Does this selection sequence provide sufficient evidence to imply nonrandomness in the selection of executives by the parent company from its subsidiaries?

10.29 A quality control chart has been maintained for a certain measurable characteristic of items taken from a conveyor belt at a certain point in a production line. The measurements obtained today in order of time are

$$68.2, 71.6, 69.3, 71.6, 70.4, 65.0, 63.6, 64.7,$$
$$65.3, 64.2, 67.6, 68.6, 66.8, 68.9, 66.8, 70.1$$

a Classify the measurements in this time series as above or below the sample mean, and determine (using the runs test) whether consecutive observations suggest lack of stability in the production process.

b Divide the time period into two equal parts and compare the means, using Student's t test. Do the data provide evidence of a shift in the mean level of the quality characteristic?

10.9 Lot Acceptance Sampling for Defectives

Manufactured product emerging from a production process is usually divided into groups of items called **lots.** Thus every 1000 items leaving the production process might be identified as a lot and packaged in boxes, each marked with an identifying lot number. A lot might be a day's production, or it might be a fixed number of items, say one gross, packed into a shipping carton.

The reason for partitioning production into lots is to make it easy to locate faulty products in case a customer encounters a number of defective items. It also enables the manufacturer to identify the time period in which the lot was produced, information that helps in identifying the cause of defective products. The most noticeable use of lot numbers is by the Food and Drug Administration (FDA). If a can of food is found to contain botulism bacteria, the FDA releases the lot numbers to the news media and also issues orders to retailers to remove all items in those lot numbers from their shelves.

Suppose that each item in a lot must satisfy certain specifications and can, therefore, be classified as either **defective** or nondefective. Customers want to receive lots containing a very low lot fraction p of defectives. To reduce the risk of shipping lots containing a high lot fraction defective, a manufacturer employs a lot acceptance sampling plan for defectives.

A **lot acceptance sampling plan for defectives** is analogous to the screen on a door. The object of a screen door is to let air pass through the screen and keep the insects out. Similarly, a lot acceptance sampling plan for defectives is designed to allow lots containing a low fraction defective to pass through the screen and to reject lots containing a high fraction defective. Rejected lots can then be reinspected and the defectives removed prior to shipment. The result is an improvement in the quality of product emerging from the screening process.

Manufacturers use lot acceptance sampling plans to screen both incoming raw materials and outgoing product. A graphic depiction of a screen for incoming raw materials is shown in Figure 10.14.

In a lot acceptance sampling plan, a random sample of n items is selected from each lot. Each of the n items in the sample is inspected, and the number x of defectives

FIGURE **10.14**
Screening for defectives

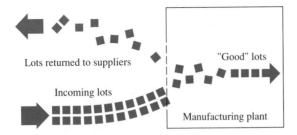

Lots returned to suppliers

Incoming lots

"Good" lots

Manufacturing plant

in the sample is recorded. If x is less than or equal to a predetermined **acceptance number a,** the lot is accepted. If x is larger than a, the lot is rejected. **Thus every sampling plan is specified by two numbers, the sample size n and the acceptance number a.**

E X A M P L E   **10.8**   A lot acceptance sampling plan employs a random sample of size $n = 25$ with acceptance number $a = 1$. Find the probability of accepting a lot that contains $p = .05$ fraction defective. Then calculate the probability of lot acceptance for $p = .1, .2, .5, .7, 0,$ and 1.0.

Solution   Assuming that the number N of items in the lot is large relative to the sample size n, the number x of defectives in a random sample of $n = 25$ items possesses the binomial probability distribution of Section 4.2:

$$p(x) = C_x^n p^x q^{n-x} \qquad \text{where } q = 1 - p$$

Since the acceptance number for the sampling plan is $a = 1$, a lot will be accepted if $x = 0$ or 1 and rejected if $x = 2, 3, \ldots, 25$. Therefore, the probability of accepting a lot containing a fraction defective p is

$$P(\text{accept lot}) = P(x = 0 \text{ or } 1) = p(0) + p(1)$$
$$= C_0^{25} p^0 q^{25} + C_1^{25} p q^{24}$$

We could calculate these probabilities by using a hand calculator, but we can avoid the tedium of these calculations by using Table 1 in Appendix II. Recall that Table 1 gives

$$\sum_{x=0}^{a} p(x) = p(0) + p(1) + \cdots + p(a)$$

for $n = 2, \ldots, 12, 15, 20,$ and 25. Using the table for $n = 25$, we find

$$P(\text{accept lot}) = p(0) + p(1)$$
$$P(\text{accept when } p = .05) = .642$$

Similarly, Table 1 gives the probabilities of lot acceptance when the fraction defective p is .1, .2, .5, and .7 as

$$P(\text{accept when } p = .1) = .271 \qquad P(\text{accept when } p = .5) = .000$$

$$P(\text{accept when } p = .2) = .027 \qquad P(\text{accept when } p = .7) = .000$$

When a lot contains no defectives—that is, $p = 0$—the probability of lot acceptance is always $P(\text{accept when } p = 0) = 1$. In contrast, if all of the items in a lot are defective, the probability of lot acceptance is always $P(\text{accept when } p = 1.0) = 0$. ∎

A graph of the probability of lot acceptance versus lot fraction defective p is called the **operating characteristic curve** for a sampling plan. The operating characteristic curve for the lot acceptance sampling plan in Example 10.8 is shown in Figure 10.15.

FIGURE **10.15**
Operating characteristic curve
($n = 25$, $a = 1$) for
Example 10.8

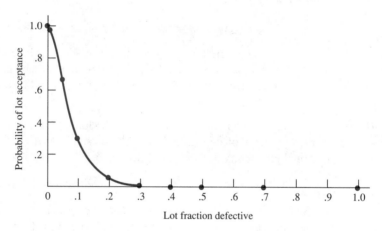

The operating characteristic curve for a sampling plan completely defines the characteristics of the screen, because it shows the probability of accepting lots with any specified probability p of fraction defective. For example, the operating characteristic curve in Figure 10.15 shows that the probability of accepting a lot with only 5% defectives is high, .642. As the fraction p of defectives in a lot increases, the probability of accepting lots containing 20% defectives is only .027, and the probability of accepting lots containing 50% or more defectives is .000.

If you are the producer, you will want the probability of accepting lots with a low fraction defective p to be high. Usually, the producer specifies that the lot fraction defective must be less than some value p_0. This fraction defective is called the **acceptable quality level (AQL)** for the plan.

DEFINITION ∎ The **acceptable quality level** (AQL) is an upper limit p_0 on the fraction defective that the producer (or consumer) is willing to accept. ∎

The probability of *rejecting good lots*—that is, lots for which $p = p_0$—is called the **producer's risk**:

$$\text{Producer's risk} = 1 - P(\text{accept when } p = p_0)$$

DEFINITION ▪

> The **producer's risk** is the probability of rejecting a lot when the lot fraction defective is equal to p_0, the AQL. ▪

In contrast, the purchaser of the product (the consumer) will want the probability of accepting bad lots (lots with high fraction defective) to be small. The consumer will have in mind some value p_1, usually larger than p_0, and will want to accept lots only if the fraction defective p is less than p_1. The probability of accepting lots, given that $p = p_1$, is called the **consumer's risk.** The producer s risk for AQL $= p_0 = .05$ and the consumer's risk for $p_1 = .10$ are shown on the operating characteristic curve in Figure 10.16.

DEFINITION ▪

> The **consumer's risk** is the probability of accepting lots when the fraction defective p is equal to p_1. ▪

FIGURE **10.16**
Operating characteristic curve showing the producer's risk for $AQL = p_o = .05$ and the consumer's risk for $p_1 = .10$

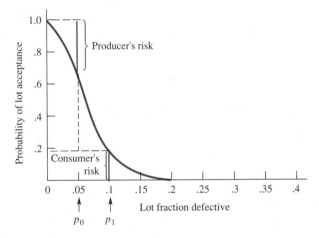

Both the producer's risk and the consumer's risk are quite large for sample sizes as small as $n = 25$. Both risks can be reduced by choosing a sampling plan based on a larger sample size n.

In theory, a manufacturer specifies the AQL, p_0, and a customer specifies p_1. Then the manufacturer selects a sampling plan (a combination of sample size n and acceptance number a) that yields producer and consumer risks satisfactory to both the manufacturer and the consumer. In practice, the manufacturer and the consumer agree on one of the many collections of sampling plans that have been developed over the years.

One of the most widely used sets of sampling plans, known as Military Standard 105D, or **MIL–STD–105D sampling plans,** was developed during World War II to control the quality of manufactured war materiel. These plans employ a sample size n that varies depending upon the *lot size*—that is, the number N of items in a lot. The sampling plans are also categorized according to levels of consumer risk: Level I is reduced (moderate consumer risk), Level II is normal, and Level III is tightened (low level of consumer risk). Two of the MIL–STD–105D tables are reproduced in Tables 11 and 12 in Appendix II. Another widely used set of sampling plans was developed by Dodge and Romig.

E X A M P L E   **10.9**   Find the MIL–STD–105D lot acceptance sampling plan for lot size $N = 1000$, a normal level of consumer risk, and an AQL $= .04$.

Solution   The first step is to locate the code letter corresponding to lot size $N = 1000$ and the normal level (II) of consumer risk. This code is given in Table 11 in Appendix II in the column labeled II, in the row corresponding to lot size 501 to 1200, as J.

The sample size n and acceptance number a for the sampling plan can be found in Table 12 in Appendix II. The lot size code letters are shown in the first column of Table 12, and the AQL is shown, in percentages, across the top of the table. Since we want the sampling plan corresponding to a lot size with code letter J and an AQL of $p_0 = .04$, or 4%, we move down the left-hand column to row J and find the required sample size to be $n = 80$. We then move across row J to the column corresponding to an AQL of 4.0% and read, under Ac (i.e., accept), the number 7. This is the acceptance number. Therefore, the MIL–STD–105D lot acceptance sampling plan for defectives, with AQL $= 4\%$ and a normal level of consumer risk, requires that you randomly select 80 items from each lot of $N = 1000$. If the number x of defectives is equal to 7 or less, you accept the lot. If $x = 8$ or more, you reject the lot.   ∎

Exercises

Basic Techniques

10.30   A buyer and a seller agree to use a sampling plan with sample size $n = 5$ and acceptance number $a = 0$. What is the probability that the buyer will accept a lot having the following fractions defective?

a $p = .1$ **b** $p = .3$ **c** $p = .5$
d $p = 0$ **e** $p = 1$

Construct the operating characteristic curve for this plan.

10.31   Repeat Exercise 10.30 for $n = 5$, $a = 1$.

10.32   Repeat Exercise 10.30 for $n = 10$, $a = 0$.

10.33   Repeat Exercise 10.30 for $n = 10$, $a = 1$.

10.34   Graph the operating characteristic curves for the four plans given in Exercises 10.30, 10.31, 10.32, and 10.33 on the same sheet of graph paper. What is the effect of increasing the acceptance number *a* when *n* is held constant? What is the effect of increasing the sample size *n* when *a* is held constant?

Applications

10.35 Find the producer's risk for a sampling plan with $n = 25$, $a = 0$, and AQL $= .05$.

10.36 Refer to the sampling plan in Exercise 10.35. Find the consumer's risk of accepting lots containing $p_1 = .20$ fraction defective.

10.37 Suppose that you want the MIL–STD–105D lot acceptance sampling plan for lot size $N = 2000$ and a normal inspection level. Find the appropriate sample size *n* and acceptance number *a* for AQL $= .015$.

10.38 Suppose that you want the MIL–STD–105D lot acceptance sampling plan for lot size $N = 400$ and the tightened inspection level, Level III. Find the appropriate sample size *n* and acceptance number *a* for AQL $= .025$.

10.39 Suppose that you want the MIL–STD–105D lot acceptance sampling plan for lot size $N = 5000$ and a normal inspection level. Find the appropriate sample size *n* and acceptance number *a* for AQL $= .01$.

10.10 Double and Sequential Lot Acceptance Sampling Plans

The lot acceptance sampling plan for defectives described in Section 10.9 is called a *single sampling plan*—that is, the decision to accept or reject a lot is based on the number of defectives in a *single* sample of *n* observations selected from the lot. Other sampling plans can sometimes provide the same protection as a single sampling plan at lower cost. One of them is a **double sampling plan.**

A single sampling plan reaches one of two decisions based on the number of defectives in a random sample of *n* items from each lot: either accept the lot or reject it. A double sampling plan reaches one of *three* decisions: accept the lot, reject the lot, or defer the decision and draw a second sample. Thus for a double sampling plan you select n_1 items from each lot. If the number x_1 in the sample of n_1 is less than or equal to an acceptance number a_1, you accept the lot. If it is equal to or larger than a rejection number r_1, you reject the lot. If x_1 falls between a_1 and r_1, you withhold judgment and draw a second sample of n_2 items from the lot and record the number x_2 of defectives in this second sample. If the total number of defectives $x = (x_1 + x_2)$ in the sample of $n = (n_1 + n_2)$ is less than or equal to an acceptance number a_2, the lot is accepted. Otherwise, it is rejected. The values of x_1 that imply accept the lot, reject the lot, or draw a second sample for the first stage of the double sampling plan are shown in Figure 10.17.

FIGURE **10.17**
The location of the acceptance number a_1 and rejection number r_1 for a double sampling plan

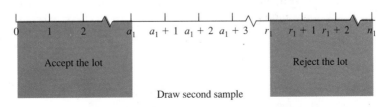

Multiple sampling plans are similar to double sampling plans except that they involve three or more stages of sampling, with each stage resulting in a decision to accept the lot, reject the lot, or continue sampling. The ultimate multiple sampling plan is called a **sequential sampling plan.** A sequential sampling plan is an extension of the concept embodied in a multiple sampling plan. A sequential sampling plan selects items from a lot, one by one, and the decision to accept the lot, reject the lot, or draw again is made after each draw. Thus the plan gives an acceptance number a_1 and a rejection number r_1 for the first draw. If the first item is less than or equal to a_1, the lot is accepted; if it is equal to or larger than r_1, the lot is rejected; otherwise, a second item is drawn from the lot. At this second stage the decision to accept the lot, reject the lot, or draw another item is based on the combined sample of $n = 2$ items and on a new pair of acceptance and rejection numbers, a_2 and r_2. If the second-stage decision is to draw again, the third-stage decision is based on the number x_3 of defectives in the combined sample of $n = 3$ items and a new pair of acceptance and rejection numbers, a_3 and r_3. This procedure is continued, stage by stage, until the number of defectives implies either acceptance or rejection of the lot.

Suppose that we were to compare a single sampling plan with a multiple sampling plan, say, a sequential sampling plan. In order to make a fair comparison, we will assume that the acceptance and rejection numbers for the sequential plan are such that the operating characteristic curves for the two plans are identical so that both plans possess the same screening properties. Which plan is better?

The number n of sampled items required to reach a final decision (to accept or reject the lot) is a random variable for a sequential sampling plan. Sometimes n will be smaller than the sample size required for the comparable single sampling plan, sometimes larger, but the mean sample size required for the sequential plan will always be (proof omitted) less than the sample size for the comparable single sampling plan. This is a very important advantage of sequential sampling plans. If the inspection of each item is costly or if inspection is destructive,[†] the sequential sampling plan will test, on the average, fewer items per lot and will be less expensive. Single sampling plans may be preferred if the cost of inspecting an item is relatively low, because it usually is easier to draw all n items from a lot at one time, and the single sampling plan is easier to use.

Double, multiple, and sequential sampling plans based on various values of AQL and with differing producer and consumer risks are available.

10.11 Acceptance Sampling by Variables

Most lot acceptance sampling plans involve sampling for defectives, but sometimes it is more appropriate to make a decision to reject or accept a lot based on the measured values of some quality variable. For example, a company that purchases large rolls of

---

[†] In the decision of whether some products are defective, the item must be destroyed. For example, in the decision of whether a photographic film is defective or nondefective, it must be exposed. Similarly, a test of an explosive charge will destroy the charge.

steel cable will want the cable to meet specifications on its tensile strength. Therefore, the decision to accept a roll will be based on the mean $\bar{x}$ of a sample of n tensile strength measurements made on the cable. Similarly, a purchaser of gasoline will specify certain octane ratings for the gasoline. Therefore, a shipment of gasoline will be accepted based on the mean octane rating of a sample of n octane readings made on the shipment. Both of these acceptance sampling plans are said to be **acceptance sampling by variables.**

The decision to accept or reject a lot based on a continuous variable x is essentially a test of an hypothesis about the process mean, using the sample mean $\bar{x}$ as a test statistic (see Section 8.3).

Suppose that good quality is associated with small values of μ_x. Then the producer will want a higher probability of accepting a lot when μ_x is less than some value μ_0. This value μ_0 is the **acceptable quality level (AQL)** for the sampling plan.

Assuming that n is large, the rejection region for a test of

$$H_0 : \mu = \mu_0$$

against the alternative hypothesis

$$H_a : \mu > \mu_0$$

is, for $\alpha = .05$,

$$\bar{x} > \mu_0 + z_{.05}\sigma_{\bar{x}}$$

or

$$\bar{x} > \mu_0 + 1.645\frac{\sigma}{\sqrt{n}}$$

If $\bar{x}$ is less than or equal to $\mu_0 + 1.645\sigma/\sqrt{n}$, the lot is accepted. Otherwise, it is rejected.

A graph of the probability β of accepting a lot versus μ_x is the operating characteristic curve for the sampling plan. It shows the probability of accepting a lot for given values of μ_x. These β probabilities are calculated by using the procedure employed in Example 8.5 in Section 8.3. A typical operating characteristic curve for a sampling plan is shown in Figure 10.18.

The consumer will want to accept lots with a small value of μ_x, say $\mu_x = \mu_1$. Therefore, the consumer's risk is defined to be the probability of accepting lots with $\mu_x = \mu_1$ (see Figure 10.18). The producer's risk is the probability of rejecting lots when $\mu = \mu_0$, that is, when μ_x equals the acceptable quality level (AQL) (see Figure 10.18).

You can see that lot acceptance by variables is very similar to lot acceptance sampling for defectives. The only difference is that lot acceptance sampling plans for defectives use the number x of defectives in a sample of n items to decide whether to accept or reject a lot. In contrast, lot acceptance sampling plans for variables use the sample mean $\bar{x}$ to make the decision. Acceptance sampling plans for variables have been developed for different levels of producer and consumer risks.

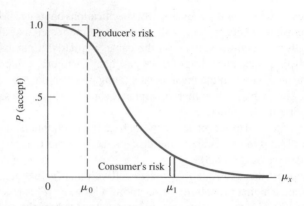

CASE STUDY REVISITED

10.12 A SOLUTION TO THE MISSING OIL STOCKS

The case study at the beginning of this chapter described a motor oil can filling operation and a problem brought to light by a monthly shortage of oil stocks. The number of filled cans was always less than the number of quarts of oil assigned to the filling operation.

Suspecting that the cause of the missing oil stocks was overfilling of the cans, the group assigned to solve the problem focused its attention on the filling machines. The first step taken to solve the problem was to convince both management and plant personnel that the amount of fill, measured in pounds, varied from can to can. To show this variance, they randomly selected 200 to 300 cans of oil from each filling machine, measured the weight of oil in each can, and constructed a relative frequency histogram for the oil weights. They found the resulting distributions to be approximately normal and to vary from one filling machine to another.

Figure 10.19(a) shows a graph (turned sideways) of the relative frequency distribution of the fill weights for the 1-gallon filling machine. You can see that there is substantial variation in the fill weights and that 16% of the cans are underfilled. One way to correct the problem of the underfilled cans is to reset the filling machine to shift the mean can fill weight upward [see Figure 10.19(b)] by .25 pound per gallon. This method will cause additional loss of oil stocks due to even greater overfill, a move estimated to cost an additional $25,000 per year. (This was no small sum in 1952.) The second and better option is to eliminate the underfill by reducing the variation in the distribution of fill weights, as shown in Figure 10.19(c). This option was achieved by constructing $\bar{x}$ and R control charts for the machine and, over the one-year period from June 1952 to May 1953, eliminating a number of assignable causes of variation.

The control charts were based on samples of five cans randomly selected at 5-minute intervals. Figures 10.20(a) and 10.20(b) show portions of the $\bar{x}$ and R charts, respectively, for the 1-gallon filling machine in June 1952. Figures 10.20(c) and 10.20(d) show portions of the $\bar{x}$ and R charts for May 1953. Note the reduction

F I G U R E **10.19**
Distribution of individual can
weights on the 1-gallon filling
machine

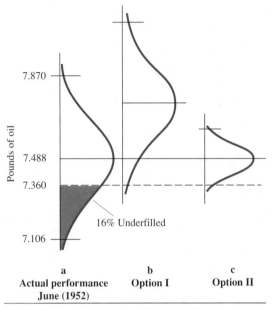

Source: Filimon, Maggass, Frazier, and Klingel, 1955.

in variation of the fill weights from June 1952 to May 1953, as indicated by the smaller distance between the control limits for the charts in 1953. This reduction in the variance of the distribution allowed the manufacturer to reduce the percentage of underfilled cans and, at the same time, reduce the amount of overfill.

The portions of the 1953 charts [Figures 10.20(c) and 10.20(d)] were chosen during a period of time when the process experienced a run of small values of $\bar{x}$. In fact, 12 of the 17 sample means are below the centerline. The authors concluded that the mean fill had drifted lower, and they adjusted the machine fill setting to increase the mean fill per can.

Figures 10.20(c) and 10.20(d) also show values of $\bar{x}$ and R that lie above the upper control limits of the charts. An investigation of the circumstances surrounding the filling operation when the errant $\bar{x}$ was observed revealed lower-than-desirable temperature in the oil stocks. A similar investigation of the R value that lies above the upper control limit found that it occurred when the oil tank level was low. Presumably, these indications of possible causes of variation in fill weight prompted management to maintain more careful control of the oil stock temperatures and storage tank levels.

In addition to using control charts to detect assignable causes of variation, Filimon and colleagues also conducted some experiments. They suspected that the mean fill weight varied from spindle to spindle and from day to day. To test this theory, they selected five cans on each of three days from each of the 28 spindles of the 1-quart filling machine. Although this design appears to represent a matched pairing of cans for the 28 spindles (matched by selection within the same day), the authors analyzed their data as if they had come from a two-factor factorial experiment with spindles as one factor and days as the other. Their analysis of variance table

FIGURE **10.20**
$\bar{x}$ and R charts for a 1-gallon
filling machine

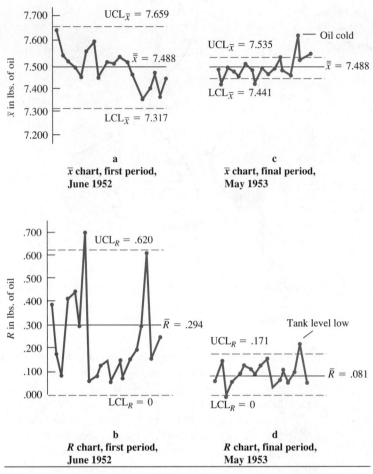

a
$\bar{x}$ chart, first period,
June 1952

c
$\bar{x}$ chart, final period,
May 1953

b
R chart, first period,
June 1952

d
R chart, final period,
May 1953

Source: Filimon, Maggass, Frazier, and Klingel, 1955.

is shown in Table 10.5. (For a review of the analysis of variance for a two-factor factorial experiment, see Section 9.10.)

Examine their ANOVA table. Spindles and days are the two factors. Since there were 28 spindles and 3 days, the degrees of freedom for these sources of variation are 27 and 2, respectively, and the number of degrees of freedom for spindle-by-day interaction is $27 \times 2 = 54$. The remaining degrees of freedom, 336, are the pooled variation of the sum of squares of deviations in weights of the five cans for each of the $(28)(3) = 84$ factor level combinations. Since there will be $(5 - 1) = 4$ degrees of freedom to estimate within-sample variation for each factor level combination, the total number of degrees of freedom for error is $(4)(84) = 336$.

Whether you view their design as a modified randomized block design or as a factorial experiment, it is clear from the F tests that the mean fill weight appears to differ from spindle to spindle and that the differences between pairs of spindles vary from one day to another. Thus efforts should be made to reduce the spindle-to-spindle variation in fill weights. The results also suggest that other unidentified variables that

TABLE **10.5**
Analysis of variance for the
28-spindle, 1-quart filling
machine

Source of Variation	Sum of Squares	Degrees of Freedom	Mean Square	F Ratio
Spindles	26,971	27	999	1.73*
Days	1,854	2	927	1.6[†]
Spindles × days	31,228	54	578	6.4[‡]
Error	30,285	336	90.1	
Total	90,338	419		

Source: Filimon, Maggass, Frazier, and Klingel, 1955.
* Significant at the .05 probability level.
[†] Not significant.
[‡] Significant at the .001 probability level.

vary from one day to another are having a decided effect on the variation in spindle fill weights.

Filimon and colleagues do not identify the specific assignable causes of variation that they discovered, but Figures 10.19 and 10.20 show that they achieved some success in reducing the variation in can fill weights over the period June 1952 to May 1953. This reduction in variation produced a decrease in the amount of overfill per can and resulted in smaller monthly losses of oil stocks in the filling operation.

10.13 Summary

Traditional quality control methods fall into two categories: control charts, which are designed to detect changes in product quality, and lot acceptance sampling plans, which screen out bad lots (lots containing product of poor quality) and allow good lots to pass through. A third and important quality control problem is locating the cause of poor product quality when it is detected. The statistical methodology presented in earlier chapters may sometimes help in this task.

We presented four types of control charts. An $\bar{x}$ chart is used to detect shifts in the mean for a process quality variable; the R chart is used to detect changes in process variation. The p chart is used to monitor the process percentage defective when the quality measurement is simply whether or not an item meets specifications. The c chart is used to monitor the number c of defects per item.

Each of these charts is constructed when the process is judged to be in control—that is, when assignable causes in the variation of the quality variable have been removed. After chart construction, sample values of the plotted variables should fall within the chart control limits. If an observation falls outside the control limits, it indicates a possible shift in the mean of the plotted statistic and is a warning of a possible change in the product quality.

Lot acceptance sampling procedures can be of two types: sampling for defectives or sampling by variables. The most commonly used plans are those that sample for defectives. They can be either single, double, multiple, or sequential sampling plans.

The single lot acceptance sampling plan for defectives involves the selection of a single sample of n items from each lot. If the number x of defectives in the lot is less than or equal to an acceptance number a, the lot is accepted. Otherwise, it is rejected. Double, multiple, and sequential sampling plans involve two or more stages of sampling, where the decision is made at every (all but the last) stage to accept the lot, reject the lot, or continue sampling.

The properties of every sampling plan can be acquired from its operating characteristic curve, a graph of the probability of lot acceptance versus the lot fraction defective p. From the operating characteristic curve the producer can determine the probability of rejecting good lots (i.e., lots containing a low fraction defective, say, $p \leq p_0$), and the consumer can determine the probability of accepting bad lots (say, $p \geq p_1$). These probabilities are called the producer's and consumer's risk, respectively, for the sampling plan.

Tables of sampling plans that give the sample sizes and acceptance number for values of p_0 [the acceptable quality level (AQL)] and for various levels of producer and consumer risks can be found in the literature.

Supplementary Exercises

10.40 Explain the concept behind a control chart.

10.41 List the elements of a control chart, and explain how a control chart is used.

10.42 What will be the intent of a study and what statistics will be measured by using each of the following types of quality control charts? Give an example of a business problem for which each chart will be useful in a quality control study.

    a an $\bar{x}$ chart   **b** an R chart   **c** a p chart   **d** a c chart

10.43 A bottle manufacturer has observed that, over a period of time when his manufacturing process was assumed to be in control, the average weight of the finished bottles was 5.2 ounces with a standard deviation of .3 ounce. The observed data were gathered in samples of six bottles selected from the production process at 50 different points in time. The average range of all samples was found to be .6 ounce, and the standard deviation of ranges was .2. During each of the next five days, samples of size $n = 6$ were selected from the manufacturing process, with the following results:

Day	$\bar{x}$	R
1	5.70	.43
2	5.32	.51
3	6.21	1.25
4	6.09	.98
5	5.63	.60

    a Construct the $\bar{x}$ chart and the R chart from the data obtained when the process was assumed to be in control, and plot data for the last five days.

    b Use the control charts constructed in part (a) to monitor the process for the sample data from the next five days.

    c Does the production of bottles appear to be out of control during any of these five days? Interpret your results.

10.44 Refer to Exercise 10.43. Suppose a soft-drink bottler specifies that the bottles she purchases from the manufacturer must weigh at least 4.8 ounces but not more than 5.5 ounces each. If the manufacturing process is in control, find the following.

  a The probability that the manufacturing process is capable of meeting the stated specifications

  b The number of bottles in a shipment of 10,000 bottles from the manufacturer that can be expected not to meet the bottler's stated specifications

10.45 The following table lists the number of defective 60-watt light bulbs found in samples of 100 light bulbs selected over 25 days from a manufacturing process. Assume that during these 25 days the manufacturing process was not producing an excessively large fraction of defectives.

Day	1	2	3	4	5	6	7	8	9	10	11	12	13
Defectives	4	2	5	8	3	4	4	5	6	1	2	4	3

Day	14	15	16	17	18	19	20	21	22	23	24	25
Defectives	4	0	2	3	1	4	0	2	2	3	5	3

  a Construct a p chart to monitor the manufacturing process, and plot the data.

  b How large must the fraction of defective items be in a sample selected from the manufacturing process before the process is assumed to be out of control?

  c During a given day, suppose a sample of 100 items is selected from the manufacturing process and that 15 defective bulbs are found. If a decision is made to shut down the manufacturing process in an attempt to locate the source of the implied controllable variation, explain how this decision might lead to erroneous conclusions.

10.46 A hardware store chain purchases large shipments of light bulbs from the manufacturer described in Exercise 10.45 and specifies that each shipment must contain no more than 4% defectives. When the manufacturing process is in control, what is the probability that the hardware store's specifications are met?

10.47 Refer to Exercise 10.45. During a given week, the number of defective bulbs in each of 5 samples of 100 was found to be 2, 4, 9, 7, and 11. Is there reason to believe that the production process has been producing an excess proportion of defectives at any time during the week?

10.48 A production process yields a long-run fraction defective of .03 when the process is in control. Once each hour, 100 items are selected from the production process, their measurements are recorded, and the fraction of defectives is noted.

  a Construct the control limits for the p chart to monitor the production process.

  b Suppose the true fraction of defectives within the production process suddenly shifts to .06. What is the probability that the shift will be detected in the first sample selected from the production process?

10.49 The following data represent the number of imperfections (scratches, chips, cracks, blisters) noted in 25 finished (4 × 8) walnut wall panels:

7, 5, 4, 10, 9, 5, 6, 3, 8, 8, 3, 5, 4, 9, 3, 3, 2, 4, 1, 5, 7, 3, 2, 6, 3

The total number of defects on 75 finished panels previously inspected was 375.

  a Assuming the manufacturing process was in statistical control during the period when the data were gathered, construct a c chart to monitor the process (use the total number of defects for the 100 panels to construct the chart). Plot the number of defects listed for the 25 panels.

  b If one wall panel is found to have more imperfections than the upper control limit allows, should the quality control engineer assume the manufacturing process is out of control, or should he wait until he finds repeated panels with an excessive number of imperfections before assuming the process is out of control? Explain.

10.50   A quality control engineer wishes to study the alternative sampling plans $n = 5$, $a = 1$ and $n = 25$, $a = 5$. On the same sheet of graph paper, construct the operating characteristic curve for both plans, making use of acceptance probabilities at $p = .05$, $p = .10$, $p = .20$, $p = .30$, and $p = .40$ in each case.

a   If you were a seller producing lots with fractions defective ranging from $p = 0$ to $p = .10$, which of the two sampling plans would you prefer?

b   If you were a buyer wishing to be protected against accepting lots with a fraction defective exceeding $p = .30$, which of the two sampling plans would you prefer?

10.51   A radio and television manufacturer who buys large lots of transistors from an electronics supplier wishes to accept all lots for which the fraction defective is less than 6%. The manufacturer's sampling inspector selects $n = 25$ transistors from each lot shipped by the supplier and notes the number of defectives.

a   On the same sheet of graph paper, construct the operating characteristic curves for the sampling plans $n = 25$, $a = 1$, 2, and 3.

b   Which sampling plan best protects the supplier from having acceptable lots rejected and returned by the manufacturer?

c   Which sampling plan best protects the manufacturer from accepting lots for which the fraction of defectives exceeds 6%?

d   How might the sampling inspector arrive at an acceptance level that compromises between the risk to the producer and the risk to the consumer?

10.52   Refer to Exercise 10.51 and assume that the manufacturer wishes the probability of his accepting lots containing 1% defective to be at least .90 and the probability of rejecting any lot with 10% or more defective to be about .90. If the manufacturer's sampling inspector samples $n = 25$ items from the supplier's incoming shipments, what is the acceptance level a that meets the requirements?

10.53   Find the MIL–STD–105D lot acceptance sampling plan for a lot size of $N = 800$, a normal inspection level, and AQL = .01.

10.54   Find the MIL–STD–105D lot acceptance sampling plan for a lot size of $N = 3000$, the tightened inspection level (level III), and AQL = .065.

C H A P T E R

11

LINEAR REGRESSION AND CORRELATION

About This Chapter

In Chapters 7 and 8, we presented methods
for making inferences about population
means based on large and small random
samples; in Chapter 9, we presented small-
sample methods for comparing more than
two population means. The object of this
chapter is to extend this methodology
to consider the case in which the mean
value of a variable y is related to another
variable, call it x. By making simultaneous
observations on y and the x variable,
we can use information contained in the
x measurements to estimate the mean
value of y and to predict particular values
of y for preassigned values of x. This
chapter is devoted to the case where y is
a linear function of one predictor variable,
x. The general case, where y is related
to one or more predictor variables, say
x_1, x_2, $\ldots$, x_k, will be discussed in
Chapter 12.

Contents

IS YOUR CAR "MADE IN THE U.S.A"?

T he phrase "made in the U.S.A." has become a battle cry in the last few years, as U.S. workers try to protect their jobs from overseas competition (*Automotive News*, 1994). For the last decade, a major trade imbalance in the United States has been caused by a flood of imported goods that enter the country and are sold at lower cost than comparable American-made goods. One prime source of competition is the automotive industry, in which the number of imported cars steadily increased during the 1970s and 1980s. The U.S. automobile industry has been besieged with complaints about product quality, worker layoffs, and high prices and has spent billions in advertising and research to produce an American-made car that will satisfy consumer demands. Have they been successful in stopping the flood of imported cars being purchased by American consumers? The data shown in Table 11.1 represent the number of imported cars y sold in the United States (in millions) for the years 1969–1993. To simplify the analysis, we have coded the year using the coded variable $x = \text{Year} - 1969$.

TABLE 11.1
Sales of imported cars, 1969–93

Year	(Year − 1969)x	Number of Imported Cars y	Year	(Year − 1969)x	Number of Imported Cars y
1969	0	1.1	1982	13	2.2
1970	1	1.3	1983	14	2.4
1971	2	1.6	1984	15	2.4
1972	3	1.6	1985	16	2.8
1973	4	1.8	1986	17	3.2
1974	5	1.4	1987	18	3.1
1975	6	1.6	1988	19	3.1
1976	7	1.5	1989	20	2.8
1977	8	2.1	1990	21	2.5
1978	9	2.0	1991	22	2.1
1979	10	2.3	1992	23	2.0
1980	11	2.4	1993	24	1.8
1981	12	2.3			

A scatterplot of the number of imported cars from 1969 to 1989 is given in Figure 11.1. Notice that an observation now consists of a pair of measurements, denoted by (x, y), for which x is time and y is the number of imported cars. Further, note that the plotted points exhibit an increasing trend; that is, the number of imported cars is increasing with time over the years 1969–89. In Chapter 2, we introduced the coefficient of correlation, a measure of the linear relationship between two random variables. However, in this problem, the value of x, unlike the value of y, is not random. Is it possible to use the values of x to estimate the values of y for the years between 1969 and 1989? Is it also possible to estimate the value of y for 1990? for 1992? for 1994? How accurate would these estimates be? Could we predict the value of y for 1998 with any degree of accuracy?

FIGURE **11.1**
Scatterplot of sales of import
cars, 1969–89

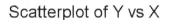
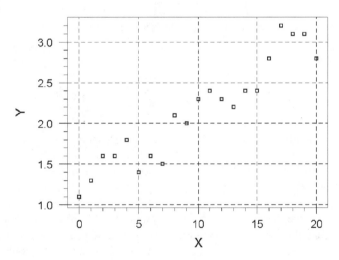

These and other interesting questions concerning pairs of observations are the topic of Chapter 11. We present methods to determine whether x and y are linearly related and whether the knowledge of x will allow us to make more accurate estimates of the mean of y or a prediction of some future value of y. The techniques developed in this chapter will be used to analyze these data when we revisit this case study in Section 11.9.

11.1 Introduction

In many practical situations, a random variable y may be related to one or more predictor variables, say $x_1, x_2, \ldots, x_k$. When this is true, we should be able to use the values of the predictor variables to more accurately estimate the mean value of y or to predict (forecast) some future value of y. For example, an automaker could estimate the weekly demand y for a new automobile based on an independent random sample of the weekly demands for 30 weeks. We could obtain a much more accurate prediction of demand by determining the relationship between the mean weekly demand $E(y)$ and the price x_1 of the automobile, the prevailing interest rate x_2, and so on. High automobile prices and high interest rates imply lower demand; the converse implies higher demand. Therefore, if we know what the price of the automobile will be next week and we know the prevailing interest rate, we should be able to obtain a more accurate prediction of next week's demand than we could obtain from the random sample of 30 weekly demands.

Practical examples of prediction problems are numerous in business, industry, and the sciences. The stockbroker wants to predict stock market behavior as a function of a number of "key indices," which are observable and serve as the predictor variables x_1, x_2, x_3, …. The manager of a manufacturing plant would like to relate yield of a chemical to a number of process variables. The manager would then use the prediction equation to find settings for the controllable process variables that would provide the maximum yield of the chemical. The personnel director of a corporation, like the admissions director of a university, wants to test and measure individual characteristics so that the corporation can hire the person best suited for a particular job. The biologist would like to relate body characteristics to the amounts of various glandular secretions. The political scientist may wish to relate success in a political campaign to the characteristics of a candidate, the nature of the opposition, and various campaign issues and promotional techniques. Certainly, all these prediction problems are, in many respects, one and the same.

In this chapter, we will be primarily concerned with the *reasoning* involved in acquiring a prediction equation based on one or more predictor variables. Thus we will restrict our attention to the simple problem of predicting y as a *linear* function of a *single* variable and observe that the solution for the multivariable problem— for example, predicting the weekly demand for a new automobile—will consist of a generalization of our technique. We will show you how to fit a simple linear model to a set of data, a process called a **regression analysis,** and we will show you how to use the model for estimation and prediction. The methodology for finding the multivariable predictor, called a multiple regression analysis, will be discussed in Chapter 12.

11.2 A Simple Linear Probabilistic Model

We introduce our topic by considering the problem of predicting the success rating that a management trainee will receive after five years of company employment; this prediction is based on the trainee's rating at the end of a training course. The success rating y is the average of the subjective ratings of five managers, scored from 0 (bad) to 100 (good). The trainee's rating x is the score (0 to 100) acquired on a combination achievement and personality test given at the end of a management training course. The company would like to use the management trainee's score x to predict management success rating y as a tool for eliminating poor management prospects from among their management trainees.

Ten managers currently in the company for a period of five years were randomly selected. Their success ratings and management trainee scores are shown in Table 11.2. Our initial approach to analyzing these data is to plot the data as points on a graph called a **scatterplot,** representing the manager success rating as y and the corresponding trainee test score as x. The graph generated by the Execustat software program is shown in Figure 11.2. You can see that y appears to increase as x increases. Do you think this arrangement of the points could have occurred due to chance even if x and y were unrelated?

TABLE 11.2
Management success ratings and trainee scores for ten company managers

Manager	Management Trainee Score, x	Management Success Rating, y
1	39	65
2	43	78
3	21	52
4	64	82
5	57	92
6	47	89
7	28	73
8	75	98
9	34	56
10	52	75

FIGURE 11.2
Execustat scatterplot of the data in Table 11.2

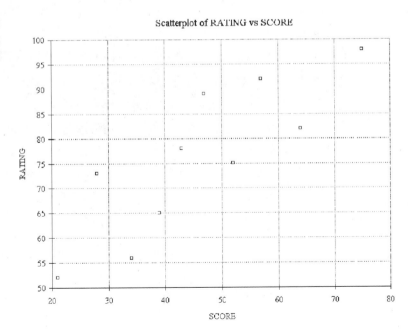

Scatterplot of RATING vs SCORE

One method of obtaining a prediction equation relating y to x is to place a ruler on the graph and move it about until it seems to pass through the points and provide what we might regard as the "best fit" to the data. Indeed, if we were to draw a line through the points, it would appear that our prediction problem was solved. We can now use the graph to predict a manager's success rating as a function of the management trainee score. In doing so, we have chosen a **mathematical model** that expresses the supposed functional relation between y and x.

Let us review several facts concerning the graphing of mathematical functions. First, the mathematical **equation of a straight line** is

$$y = \beta_0 + \beta_1 x$$

where β_0 is the **y intercept**, the value of y when $x = 0$, and β_1 is the **slope** of the line, the change in y for a one-unit change in x (Figure 11.3). Second, the line that we graph corresponding to any linear equation is unique. Each equation corresponds to only one line and vice versa. Thus, when we draw a line through the points, we automatically choose a mathematical equation

$$y = \beta_0 + \beta_1 x$$

where β_0 and β_1 have unique numerical values.

Equation of a Straight Line

$$y = \beta_0 + \beta_1 x$$

where

$$\beta_0 = y \text{ intercept}$$
$$= \text{Value of } y \text{ when } x = 0$$

and

$$\beta_1 = \text{Slope of the line}$$
$$= \text{Change in } y \text{ for a 1-unit increase in } x$$

F I G U R E **11.3**
The y intercept and slope for a line

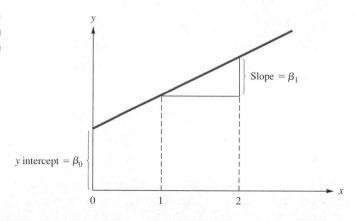

The linear model

$$y = \beta_0 + \beta_1 x$$

is said to be a **deterministic mathematical model,** because when a value of x is substituted into the equation, the value of y is determined and no allowance is made for error. Deterministic models are suitable for prediction only when the errors of prediction are small. When they are large, which is generally the case in business forecasting, we need to take the error of prediction into account and, particularly, to give some indication of its magnitude. We do so by constructing a **probabilistic mathematical model,** one that contains one or more random components that are added to the deterministic portion of the model to account for the random and unexplained error of prediction. Thus a probabilistic model relating success rating y to management trainee score x is given by the expression

$$y = \beta_0 + \beta_1 x + \epsilon$$

where ϵ is assumed to be a random error variable with expected value equal to zero and variance equal to σ^2. In addition, we will assume that any pair of random errors ϵ_i and ϵ_j corresponding to two observations y_i and y_j are independent. In other words, we assume that the *average* or expected value of y is linearly related to x and that observed values of y will deviate above and below this line by a random amount ϵ. Furthermore, we have assumed that the distribution of errors about the line will be identical, regardless of the value of x, and that any pair of errors will be independent of each other. The assumed line giving the expected value of y for a given value of x is indicated in Figure 11.4. The probability distribution of the random error ϵ is shown for several values of x.

FIGURE **11.4**
Linear probabilistic model

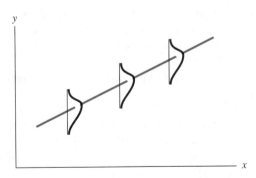

The straight-line or linear probabilistic model—the topic of this chapter—is only one of many types of probabilistic models that could be used to express the relationship between y and x. We will learn how to fit this model to a set of data in Section 11.3, and in subsequent sections we will see how it can be used for estimation and prediction. The procedure for choosing and fitting more complicated models is the topic of Chapter 12.

Exercises

Basic Techniques

11.1 Graph the line corresponding to the equation $y = 2x + 1$ by graphing the points corresponding to $x = 0$, 1, and 2. Give the y intercept and slope for the line.

11.2 Graph the line corresponding to the equation $y = -2x + 1$ by graphing the points corresponding to $x = 0$, 1, and 2. Give the y intercept and slope for the line. Give the similarities and differences between this line and the line of Exercise 11.1.

11.3 Graph the line corresponding to the equation $2y = -3x + 4$.

11.4 How is the line $2y = 3x + 4$ related to the line of Exercise 11.3?

11.5 Give the equation and graph for a line with a y intercept equal to 3 and a slope equal to -1.

11.6 Give the equation and graph for a line with a y intercept equal to -3 and a slope equal to 1.

11.7 What is the difference between deterministic and probabilistic mathematical models?

11.3 The Method of Least Squares

The statistical procedure for finding the "best-fitting" straight line for a set of points is, in many respects, a formalization of the procedure used to fit a line by eye. For instance, when we visually fit a line to a set of data, we move the ruler until we think that we have minimized the deviations of the points from the prospective line. If we denote the predicted value of y obtained from the fitted line as $\hat{y}$, the prediction equation is

$$\hat{y} = \hat{\beta}_0 + \hat{\beta}_1 x$$

where $\hat{\beta}_0$ and $\hat{\beta}_1$ represent estimates of the parameters β_0 and β_1.

Having decided that in some manner or other we will attempt to minimize the deviations of the points in choosing the best-fitting line, we must now define what we mean by "best." That is, we wish to define a criterion for "best fit" that will seem intuitively reasonable, that is objective, and that, under certain conditions, will give the best prediction of y for a given value of x.

We will employ a criterion of goodness that is known as the **principle of least squares,** which may be stated as follows: **Choose as the "best-fitting" line the line that minimizes the sum of squares of the deviations of the observed values of y from those predicted.** This sum of squares of deviations, commonly called the **sum of squares for error (SSE)** and defined as

$$\text{SSE} = \sum_{i=1}^{n}(y_i - \hat{y}_i)^2$$

is the sum of the squared distances represented by the vertical lines in Figure 11.5.

The method for finding the numerical values of β_0 and β_1 that minimize SSE uses differential calculus and hence is beyond the scope of this text. However, if you have a calculator with a statistics function or have access to a statistical software package, you can easily obtain the least-squares estimates, $\hat{\beta}_0$ and $\hat{\beta}_1$, once you

FIGURE **11.5**
Minitab graph of the
least-squares line,
$\hat{y} = \hat{\beta}_0 + \hat{\beta}_1 x$, and data
points from Table 11.2

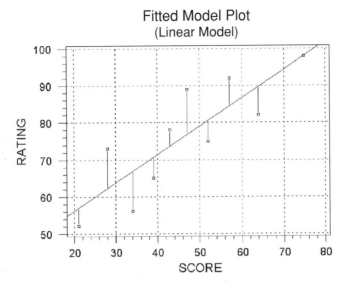

Fitted Model Plot
(Linear Model)

have entered the data using the appropriate commands. Each calculator and software program provides slightly different output, but all are similar in format. The line $\hat{y} = \hat{\beta}_0 + \hat{\beta}_1 x$ for the data of Table 11.2 is shown in Figure 11.5. The vertical lines drawn from the prediction line to each point represent the deviations of the points from the predicted value of y.

EXAMPLE   **11.1**   Obtain the least-squares prediction line for the data in Table 11.2 using an appropriate statistical software package. Predict the manager success rating if the trainee scored $x = 50$ on the test.

Solution   A portion of the Execustat output for a simple linear regression using the data in Table 11.2 is shown in Table 11.3. The least-squares line is given on the first line of the printout as Linear model, and the least-squares estimates $\hat{\beta}_0$ and $\hat{\beta}_1$ are given in the Table of Estimates in the column labeled Estimate and the rows labeled Intercept and Slope, respectively. Then, according to the principle of least squares, the best-fitting straight line relating the success rating to the test score is

$$\hat{y} = \hat{\beta}_0 + \hat{\beta}_1 x$$
$$= 40.784 + .766x$$

The graph of this equation is shown in Figure 11.5. Note that the y intercept, 40.784, is the value of y when $x = 0$. The slope of the line, .766, gives the estimated change in y for a one-unit change in x. Finally, if a trainee scores $x = 50$ on the test, his or her predicted rating would be

$$\hat{y} = \hat{\beta}_0 + \hat{\beta}_1 x$$
$$= 40.7842 + (.765562)(50) = 79.062 \quad \blacksquare$$

T A B L E 11.3   Execustat printout for Example 11.1

Simple Regression Analysis for CBS11-1

Linear model: RATING = 40.7842 + 0.765562*SCORE

Table of Estimates

	Estimate	Standard Error	t Value	P Value
Intercept	40.7842	8.50686	4.79	0.0014
Slope	0.765562	0.174985	4.38	0.0024

In general, the least-squares estimates will be labeled either "estimate" or "coefficient," and the intercept and slope may be labeled "constant" and "x," respectively. We will present computer printouts from both Minitab and Execustat software packages throughout this chapter so that you will become familiar with their formats.

Exercises

Basic Techniques

11.8   Given five points whose coordinates are

x	−2	−1	0	1	2
y	1	1	3	5	5

a computer software package produces the following information about the least-squares regression line:

SLOPE = 1.2000   INTERCEPT = 3.0000

a   Find the least-squares line for the data.

b   Plot the five points and graph the line in part (a). Does the line appear to provide a good fit to the data points?

11.9   Given the points whose coordinates are

x	1	2	3	4	5	6
y	5.6	4.6	4.5	3.7	3.2	2.7

a computer printout gives the least-squares estimates as

```
PREDICTOR     COEF
CONSTANT      6.0000
X            -.557143
```

a Find the least-squares line for the data.
b Plot the six points and graph the line. Does the line appear to provide a good fit to the data points?
c Use the least-squares line to predict the value of y when $x = 3.5$.

Applications

11.10 A manufacturer of soap powder conducted an experiment to investigate the effect of price per box on demand. Each of six different sales regions was assigned a wholesale unit price per box for sale to wholesalers or large supermarket chains in the region. After a one-month period, the percentage y of increase (or decrease) over the preceding month in unit sales per region was calculated. The unit prices assigned to the regions and the percentage increases in sales are shown in the following table.

Unit price, x ($)	6.40	6.45	6.50	6.55	6.60	6.65
Increase in sales, y (%)	9.8	7.6	6.3	4.5	4.2	1.7

Simple Regression Analysis for CBS11-10

Linear model: Y = 201.433 - 30*X

Table of Estimates

	Estimate	Standard Error	t Value	P Value
Intercept	201.433	16.3108	12.35	0.0002
Slope	-30	2.49952	-12.00	0.0003

a Use the Execustat printout to find the least-squares line relating increase in sales, y, to unit price, x.
b Plot the data points using a scatterplot, and graph the least-squares line on the same paper. Does it seem to provide a good fit to the data points?
c Use the least-squares line to estimate the mean change in unit sales for a unit price of $6.60 per unit.

11.11 The number of salespersons employed by an auto dealership has varied from a low of four salespersons to a high of eight. How does the number y of new cars sold depend on the number x of salespersons? To shed some light on this question, the sales manager examined sales records for the past four months and located eight weeks during which no special incentive programs were employed. The number y of cars sold per week and the number x of salespersons employed are shown in the accompanying table.

Variables	Week							
	1	2	3	4	5	6	7	8
Salesperson, x	5	6	5	4	7	6	5	8
Cars sold, y	10	20	18	10	21	15	13	22

```
The regression equation is
y = - 2.00 + 3.15 x

Predictor        Coef        Stdev      t-ratio         p
Constant       -2.000        4.930        -0.41     0.699
x               3.1522       0.8393        3.76     0.009

s = 2.846        R-sq = 70.2%      R-sq(adj) = 65.2%
```

a Plot the data points on graph paper.

b Find the least-squares line using the Minitab printout given.

c Graph the least-squares line to see how well it fits the data.

d Use the least-squares line to estimate the mean number of cars sold per week if the dealer employs six salespersons.

11.12 The federal government has spent billions of dollars in bailout costs for failed savings and loan institutions across the country. Following the passage of federal legislation enacted in 1989 requiring 1.5% of a savings and loan's assets to be tangible capital, the performance of savings and loan institutions with headquarters in Riverside or San Bernardino counties (California) was evaluated. At the conclusion of the three-month period from March 31–June 30, 1990, the following data were recorded.

Savings and Loan	Total Assets (x) (in millions)	Net Income (y) (in thousands)	Tangible Capital (% of assets)
Redlands Federal	$791.8	$2700	5.92
Hemet Federal	556.3	1261	5.69
Provident Federal	518.0	68	5.47
Palm Springs Federal	137.0	251	3.57
Inland Savings	111.1	748	5.14
Secure Savings	56.4	195	6.87
Mission Savings	35.6	68	5.53

Source: Hamilton, 1990.

a Plot the values of y (net income) against x (total assets).

b Do the data points appear to be linearly related? Are there any points that might be outliers?

c Fit the least-squares line for these data using the Execustat printout given.

d Use the fitted line to predict the net income of a savings and loan institution with total assets of $200 million.

Execustat output for Exercise 11.12

```
                 Simple Regression Analysis for CBS11-12

Linear model: Y = -17.8962 + 2.45502*X

                             Table of Estimates

                                 Standard          t           P
                   Estimate        Error         Value       Value

Intercept         -17.8962       383.864         -0.05       0.9646
Slope               2.45502        0.911933       2.69       0.0432
```

11.4 Inferences Concerning the Slope of the Line, β_1

The initial inference of importance in studying the relationship between y and x concerns the existence of the relationship. Does x contribute information for the prediction of y? That is, **do the data present sufficient evidence to indicate that y increases (or decreases) linearly as x increases over the region of observation?** Or is it probable that the points would fall on the graph in a manner similar to that observed in Figure 11.2 when y and x are completely unrelated?

Any inference concerning the least-squares line requires that we first estimate σ^2, the variability of the points about the line. It seems reasonable to use SSE, the sum of the squared deviations, about the predicted line for this purpose. Indeed, it can be shown that the formula given in the display provides an estimator for σ^2 that is unbiased, based on $(n - 2)$ degrees of freedom (d.f.).

An Estimator for σ^2

$$\hat{\sigma}^2 = s^2 = \frac{\text{SSE}}{n - 2}$$

The sum of squares of deviations, SSE, can be calculated directly by using the prediction equation to calculate $\hat{y}$ for each point, then calculating the deviations $(y_i - \hat{y}_i)$, and finally calculating

$$\text{SSE} = \sum_{i=1}^{n}(y_i - \hat{y}_i)^2$$

This procedure tends to be tedious and is rather poor from a computational point of view because the numerous subtractions tend to introduce computational rounding errors. An easier and computationally better procedure is to use the formula given in Section 11.6 or to obtain SSE and s^2 from a computer printout.

Table 11.4 shows the Minitab output for a linear regression analysis using the data in Table 11.2. Notice the clear specification of the regression equation in lines 1 and 2, and that the least-squares estimates $\hat{\beta}_0$ and $\hat{\beta}_1$ are given in the column labeled

Coef in lines 4 and 5. The estimator of σ is clearly labeled as s = 8.704 in the last line of the printout. In an Execustat printout, this estimator will be labeled Standard error of estimation.

T A B L E **11.4**
Minitab regression printout for
the data in Table 11.2

```
The regression equation is
y = 40.8 + 0.766 x

Predictor        Coef         Stdev      t-ratio         p
Constant       40.784        8.507         4.79      0.000
x               0.7656       0.1750        4.38      0.002

s = 8.704         R-sq = 70.5%        R-sq(adj) = 66.8%
```

The practical interpretation that can be given to s ultimately rests on the meaning of σ. Since σ measures the spread of the y values about the line of means $E(y|x) = \beta_0 + \beta_1 x$ (see Figure 11.4), we would expect (from the Empirical Rule) approximately 95% of the y values to fall within 2σ of that line. Since we do not know σ, $2s$ provides an approximate value for the half-width of this interval. Now return to Figure 11.5 and note the location of the data points about the least-squares line. Since we used the $n = 10$ data points to fit the least-squares line, we would not be too surprised to find that most of the points fall within $2s = 2(8.7) = 17.4$ of the line. If you check Figure 11.5, you will see that all ten points fall within $2s$ of the least-squares line. (You will find that, in general, most of the data points used to fit the least-squares line will fall within $2s$ of the line. This estimate provides you with a rough check for your calculated value of s.)

The initial question we pose concerns the value of β_1, which is the average change in y for a one-unit increase in x. Stating that y does not increase (or decrease) linearly as x increases is equivalent to saying that $\beta_1 = 0$. Thus we would wish to test an hypothesis that $\beta_1 = 0$ against the alternative that $\beta_1 \neq 0$. As we might suspect, the estimator $\hat{\beta}_1$ is extremely useful in constructing a test statistic to test this hypothesis. Therefore, we wish to examine the distribution of estimates $\hat{\beta}_1$ that would be obtained when samples, each containing n points, are repeatedly drawn from the population of interest. If we assume that the random error ϵ is normally distributed, in addition to the previously stated assumption, it can be shown that the test statistic

$$t = \frac{\hat{\beta}_1 - \beta_1}{s/\sqrt{S_{xx}}}, \qquad \text{where } S_{xx} = \sum_{i=1}^{n}(x_i - \bar{x})^2$$

follows a Student's t distribution in repeated sampling with $(n - 2)$ d.f. Note that the number of degrees of freedom associated with s^2 determines the number of degrees of freedom associated with t. Thus we observe that the test of an hypothesis that β_1 equals some particular numerical value, say, β_{10}, is the familiar t test encountered in Chapter 8.

Test of Hypothesis Concerning the Slope of a Line

1   Null Hypothesis: $H_0 : \beta_1 = \beta_{10}$

2   Alternative Hypothesis:

One-Tailed Test	Two-Tailed Test
$H_a : \beta_1 > \beta_{10}$ (or, $\beta_1 < \beta_{10}$)	$H_a : \beta_1 \neq \beta_{10}$

3   Test Statistic: $t = \dfrac{\hat{\beta}_1 - \beta_{10}}{s/\sqrt{S_{xx}}}$

When the random error is normally distributed and the assumptions given in Section 11.2 (and 11.8) are satisfied, the test statistic will have a Student's t distribution with $(n-2)$ d.f.

4   Rejection Region:

One-Tailed Test	Two-Tailed Test
$t > t_\alpha$ (or, $t < -t_\alpha$ when the alternative hypothesis is $H_a : \beta_1 < \beta_{10}$)	$t > t_{\alpha/2}$ or $t < -t_{\alpha/2}$

The values of t_α and $t_{\alpha/2}$ are given in Table 4 in Appendix II. Use the values of t corresponding to $(n-2)$ d.f.

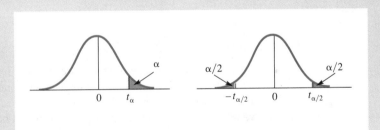

E X A M P L E   **11.2**   Use the data in Table 11.2 to determine whether the value of $\hat{\beta}_1$ provides sufficient evidence to indicate that β_1 differs from zero; that is, does a linear relationship exist between a trainee's test score x and his or her success rating y?

Solution   We wish to test the null hypothesis

$$H_0 : \beta_1 = 0 \quad \text{against} \quad H_a : \beta_1 \neq 0$$

for the success rating–test score data in Table 11.2.

Table 11.5 gives the Excel regression analysis printout for the data in Table 11.2 and contains the estimated line, the estimates of the intercept and the slope (together with their standard errors), and the value of the t statistic in testing $H_0 : \beta_1 = 0$

versus the two-tailed alternative $H_a : \beta_1 \neq 0$. The value of the t statistic is found in the column labeled t Value in the row corresponding to the Slope and is equal to Estimate/(Standard Error) $= 0.765562/0.17499 = 4.38$. The p-value associated with this test is the second entry in the column labeled p-value and is equal to .0024. Therefore, if we were using a value of α greater than .0024 we would reject $H_0 :$ $\beta_1 = 0$ and conclude that a significant linear relationship exists between a trainee's test score and the trainee's success rating. The value of $s = \sqrt{\text{MSE}}$ is given in the second part of the printout as the Standard error of estimation $= 8.70363$.   ∎

T A B L E 11.5   Excel printout for Example 11.2

	A	B	C	D	E	F	G	H
1			Simple Regression Analysis for CBS11-1					
2								
3	Linear Model: RATING = 40.7842 + 0.765562 * SCORE							
4								
5					Table of Estimates			
6			Estimate		Standard Error		t Value	P Value
7	Intercept		40.7842		8.50686		4.79	0.0014
8	Slope		0.765562		0.17499		4.38	0.0024
9								
10	R-Squared =		70.52 %					
11	Correlation Coef. =		0.84					
12	Std Error of Estimation =		8.70363					
13	Durbin-Watson Statistic =		1.17368					
14	Mean Absolute Error =		6.97801					
15	Sample Size (n) =		10					

Once we have decided that x and y are linearly related, we are interested in examining this relationship in detail. If x increases by one unit, what is the estimated change in y, and how much confidence can be placed in the estimate? In other words, we require an estimate of the slope β_1. You will not be surprised to observe a continuity in the procedures of Chapters 8 and 11; that is, the confidence interval for β_1, with confidence coefficient $(1 - \alpha)$, can be shown to be

$$\hat{\beta}_1 \pm t_{\alpha/2}(\text{estimated } \sigma_{\hat{\beta}_1})$$

or as expressed in the following display.

A $(1 - \alpha)100\%$ Confidence Interval for β_1

$$\hat{\beta}_1 \pm t_{\alpha/2}(\text{standard error of } \hat{\beta}_1)$$

where $t_{\alpha/2}$ is based on $(n - 2)$ d.f.

E X A M P L E   **11.3**   Find a 95% confidence interval for β_1 based on the data in Table 11.2.

Solution   The regression analysis printout in Table 11.5 contains the estimate of the slope and its standard error. Therefore, the 95% confidence interval for β_1 is $\hat{\beta}_1 \pm t_{.025}$ (standard error of $\hat{\beta}_1$), or

$$.766 \pm 2.306(.175) \qquad \text{or} \qquad .766 \pm .404 \quad \blacksquare$$

Several points concerning the interpretation of our results deserve particular attention. As we have noted, β_1 is the slope of the assumed line over the region of observation and indicates the linear change in $E(y|x)$ for a one-unit change in x. **Even if we do not reject the null hypothesis that the slope of the line β_1 equals zero, it does not necessarily mean that x and y are unrelated.** In the first place, we must be concerned with the probability of committing a type II error—that is, of accepting the null hypothesis that the slope equals zero when this hypothesis is false. Second, it is possible that x and y might be perfectly related in a nonlinear manner. For example, Figure 11.6 depicts a nonlinear relationship between y and x over the domain of $x : a \le x \le f$. We note that a straight line would provide a good predictor of y if fitted over a small interval in the x domain, say, $b \le x \le c$. The resulting line is line 1. On the other hand, if we attempt to fit a line over the region $c \le x \le d$, then β_1 equals zero and the best fit to the data is the horizontal line 2. This result would occur even if all the points fell perfectly on the curve and y and x possessed a functional relation as defined in Section 11.2. We must take care in drawing conclusions if we do not find evidence to indicate that β_1 differs from zero. Perhaps we have chosen the wrong type of probabilistic model for the physical situation.

F I G U R E **11.6**
Nonlinear relation

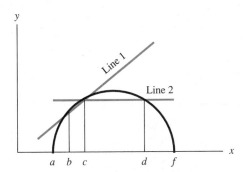

Note that the comments contain a second implication. **If the data provide values of x in an interval $b \le x \le c$, then the calculated prediction equation is appropriate only over this region. Extrapolation in predicting y for values of x outside the region $b \le x \le c$ for the situation indicated in Figure 11.6 would result in a serious prediction error.**

If the data present sufficient evidence to indicate that β_1 differs from zero, we do not conclude that the true relationship between y and x is linear. Undoubtedly, y is a function of a number of variables, which demonstrate their existence to a greater or lesser degree in terms of the random error ϵ that appears in the model. This random error, of course, is why we have been obliged to use a probabilistic model in the first place. Large errors of prediction imply curvatures in the true relation between y and x, the presence of other important variables that do not appear in the model, or both, as most often is the case. All we can say is that we have evidence to indicate that y changes as x changes and that we may obtain a better prediction of y by using x and the linear predictor than by simply using $\bar{y}$ and ignoring x. Note that this statement does not imply a causal relationship between x and y. Some third variable may have caused the change in both x and y, producing the relationship that we have observed.

Exercises

Basic Techniques

11.13   The Execustat printout below gives the least-squares regression analysis for the data in Exercise 11.8. Do the data provide sufficient evidence to indicate that y and x are linearly related? (Test the hypothesis that $\beta_1 = 0$; use $\alpha = .05$.)

```
           Simple Regression Analysis for CBS11-8

Linear model: Y = 3 + 1.2*X

                        Table of Estimates
```

	Estimate	Standard Error	t Value	P Value
Intercept	3	0.326599	9.19	0.0027
Slope	1.2	0.23094	5.20	0.0138

```
R-squared = 90.00%
Correlation coeff. = 0.949
Standard error of estimation = 0.730297
Durbin-Watson statistic = 2.6
Mean absolute error = 0.48
Sample size (n) = 5
```

11.14   Refer to Exercise 11.13. Find a 90% confidence interval for the slope of the line. Interpret this interval estimate.

11.15   Do the data in Exercise 11.9 provide sufficient evidence to indicate that y and x are linearly related? Use the Minitab printout provided below.

```
The regression equation is
y = 6.00 - 0.557 x

Predictor         Coef        Stdev      t-ratio          p
Constant        6.0000       0.1759        34.10      0.000
x              -0.55714      0.04518       -12.33      0.000

s = 0.1890        R-sq = 97.4%       R-sq(adj) = 96.8%
```

11.16 Refer to Exercise 11.15. Find a 90% confidence interval for the slope of the line. Interpret this interval estimate.

Applications

11.17 Refer to the data in Exercise 11.10 relating the percentage y of increase in sales to a manufacturer's wholesale price for a box of soap powder. Do the data provide sufficient information to indicate that the coded price per unit (over the range employed in this study) contributes information for the prediction of the percentage increase in sales? Test using $\alpha = .05$, and use the computer printout given in Exercise 11.10.

11.18 Refer to the data in Exercise 11.11 relating the number y of new cars sold per week by an automobile dealer to the number x of salespersons employed. Do the data provide sufficient information to indicate that the number of salespersons contributes information for the prediction of the number of cars sold per week? Test using $\alpha = .05$ and the computer printout given in Exercise 11.11.

11.19 Refer to Exercise 11.18. Find a 90% confidence interval for the increase (or decrease) in the mean number of cars sold per week for an increase of one salesperson. Interpret the interval.

11.20 Refer to the study of the relationship between net income y and total assets x for savings and loan institutions headquartered in Riverside and San Bernardino counties from Exercise 11.12.

a Does the amount of total assets x contribute information for the prediction of net income y for the savings and loans? Test using $\alpha = .05$ and the computer printout in Exercise 11.12.

b What assumptions must be made in order that the inference made in part (a) be valid? Do you think that these assumptions have been met?

 11.21 A marketing research experiment was conducted to study the relationship between the length of time necessary for a buyer to reach a decision and the number of alternative package designs of a product presented. Brand names were eliminated from the packages to reduce the effects of brand preferences. The buyers made their selections by using the manufacturers' product descriptions on the packages as the only buying guide. The length of time necessary to reach a decision is recorded for 15 participants in the marketing research study. Use the Minitab printout to answer the following questions.

Length of decision time, y (seconds)	5, 8, 8, 7, 9	7, 9, 8, 9, 10	10, 11, 10, 12, 9
Number of alternatives, x	2	3	4

```
The regression equation is
y = 4.30 + 1.50 x

Predictor        Coef        Stdev      t-ratio           p
Constant        4.300        1.216         3.53       0.004
x              1.5000       0.3913         3.83       0.002

s = 1.237         R-sq = 53.1%        R-sq(adj) = 49.5%
```

a   Find the least-squares line appropriate for these data.

b   Plot the points and graph the line as a check on your calculations.

c   Calculate s^2.

d   Do the data provide sufficient evidence to indicate that the length of decision time is linearly related to the number of alternative package designs? (Test at the $\alpha = .05$ level of significance.)

e   Find the approximate observed significance level for the test, and interpret its value.

  11.22   Housing markets in the United States are monitored using a "housing opportunity index" compiled by the National Association of Home Builders. This index "measures the ability of a typical family to purchase a home in its own market by comparing median family income with median home price." Could we predict this housing opportunity index if we knew the median family income or the median home price? The following table gives the values of these three variables measured in ten different metropolitan areas that have populations of less than 250,000. The Minitab printouts show two separate linear regressions, one relating the housing index to median family income and one relating the housing index to median home price.

Region	Housing Opportunity Index	Median Family Income ($ thousand)	Median Home Price ($ thousand)
Rochester, NY	88.4	45.4	88
Brockton, MA	84.1	45.6	115
Pittsfield, MA	77.5	41.6	98
Fall River, MA-RI	62.1	35.7	120
Kokomo, IN	92.1	41.9	73
Springfield, IL	90.3	42.6	83
Mansfield, OH	89.4	37.3	68
Benton Harbor, MI	80.3	36.7	83
Columbia, MO	74.5	39.1	93
Ocala, FL	87.0	29.2	69

Source: "Housing Opportunity," 1994.

```
The regression equation is
INDEX = 63.9 + 0.474 INCOME

Predictor          Coef        Stdev      t-ratio         p
Constant          63.86        25.12        2.54       0.035
INCOME           0.4736       0.6312        0.75       0.475

s = 9.472        R-sq = 6.6%        R-sq(adj) = 0.0%

The regression equation is
INDEX = 116 - 0.377 PRICE

Predictor          Coef        Stdev      t-ratio         p
Constant         116.15        11.20       10.37       0.000
PRICE           -0.3773       0.1236       -3.05       0.016

s = 6.659        R-sq = 53.8%       R-sq(adj) = 48.1%
```

a What is the least-squares line relating the housing index to median family income? Plot the points and graph the line. Does there appear to be a linear relationship between the variables?

b Repeat the instructions of part (a) using the housing index and median home price. Does there appear to be a linear relationship between the variables?

c Which of the two independent variables, family income or home price, is the better predictor of the housing index? Explain.

11.5 Estimation and Prediction Using the Fitted Line

In Chapter 7, we studied methods for estimating a population mean μ and encountered numerous practical applications of these methods in the examples and exercises. Now we consider a generalization of this problem.

Estimating the mean value of y for a given value of x—that is, estimating $E(y|x)$—can be a very important practical problem. If a corporation's profit y is linearly related to advertising expenditures x, the marketing director may wish to estimate the mean profit for a given expenditure x. Similarly, a research pharmacist might wish to estimate the mean response of a human to a specific drug dosage x, and a manager might wish to know the mean success rating for a trainee who acquired a test score of $x = 50$. The least-squares prediction equation can be used to obtain these estimates.

Assume that x and y are linearly related according to the probabilistic model defined in Section 11.2 and therefore that $E(y|x) = \beta_0 + \beta_1 x$ represents the expected value of y for a given value of x. **Since the fitted line**

$$\hat{y} = \hat{\beta}_0 + \hat{\beta}_1 x$$

attempts to estimate the line of means $E(y|x)$ (i.e., we estimate β_0 and β_1), then $\hat{y}$ can be used to estimate the expected value of y as well as to predict some value

of y that might be observed in the future. The errors of estimation and prediction differ for these two cases, as we will show.

Observe the two lines in Figure 11.7. One line represents the line of means

$$E(y|x) = \beta_0 + \beta_1 x$$

and the other is the fitted prediction equation

$$\hat{y} = \hat{\beta}_0 + \hat{\beta}_1 x$$

We observe from the figure that the error in estimating the expected value of y when $x = x_p$ is the deviation between the two lines above the point x_p. Also, this error increases as we move to the endpoints of the interval over which x has been measured. Although the expected value of y for a particular value of x is of interest for our example in Table 11.2, we are primarily interested in *using* the prediction equation $\hat{y} = \hat{\beta}_0 + \hat{\beta}_1 x$ based on our observed data to predict the success rating for some trainee selected from the population of interest. That is, we want to use the prediction equation obtained for the ten measurements in Table 11.2 to predict the success rating for a new trainee selected from the population. If the trainee's test score was x_p, we intuitively see that the error of prediction (the deviation between $\hat{y}$ and the actual rating y that the trainee will obtain) is composed of two elements. Since the trainee's rating will equal

$$y = \beta_0 + \beta_1 x_p + \epsilon$$

$(y - \hat{y})$ equals the deviation between $\hat{y}$ and the expected value of y, shown in Figure 11.7, *plus* the random amount ϵ that represents the deviation of the trainee's rating from the expected value (Figure 11.8). **Thus the variability in the error for predicting a single unit of y exceeds the variability for estimating the expected value of y.**

FIGURE **11.7**
Estimating $E(y|x)$ when
$x = x_p$

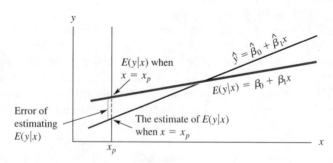

A $(1 - \alpha)100\%$ confidence interval for the expected value of y, given $x = x_p$, is given in the following display.

A $(1 - \alpha)100\%$ Confidence Interval for $E(y|x)$ When $x = x_p$

$$\hat{y} \pm t_{\alpha/2} s_{\hat{y}}$$

where $t_{\alpha/2}$ is based on $(n - 2)$ d.f. and $s_{\hat{y}}$, given in Section 11.6, is the standard error in estimating $E(y|x)$.

FIGURE **11.8**
Error in predicting a particular
value of y

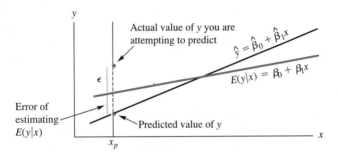

EXAMPLE **11.4**   Refer to Example 11.1. Find a confidence interval for the expected value of y, the success rating, given that the trainee test score is $x = 50$.

Solution   To estimate the success rating for a trainee whose test score was $x_p = 50$, we would use

$$\hat{y} = \hat{\beta}_0 + \hat{\beta}_1 x_p$$

to calculate $\hat{y}$, the estimate of $E(y|x = 50)$. Then, using values calculated previously,

$$\hat{y} = 40.7842 + (.765562)(50) = 79.0623$$

The Minitab printout in Table 11.6 shows that the REGRESS command followed by the subcommand PREDICT will produce an estimate of $E(y)$ for a given value of x, labeled Fit; the standard error of the prediction, $s_{\hat{y}}$, labeled Stdev.Fit; and the resulting 95% confidence interval estimate, labeled 95% C.I. Therefore, the estimate of $E(y|x = 50)$ is 79.06, with a standard error of 2.84, and the resulting 95% confidence interval estimate is (72.51, 85.61).   ∎

It can be shown that the variance of the error of predicting a particular value of y when $x = x_p$, that is, $(y - \hat{y})$, is

$$\sigma^2_{(y-\hat{y})} = \sigma^2_y + \sigma^2_{\hat{y}}$$

```
MTB > REGRESS C1 1 C2;
SUBC> PREDICT 50.

The regression equation is
Y = 40.8 + 0.766 X

Predictor        Coef        Stdev      t-ratio         p
Constant       40.784        8.507         4.79     0.000
X              0.7656       0.1750         4.38     0.002

s = 8.704         R-sq = 70.5%        R-sq(adj) = 66.8%
    .                   .                   .
    .                   .                   .
    .                   .                   .

  Fit   Stdev.Fit            95% C.I.                95% P.I.
79.06        2.84    (   72.51,    85.61)  (   57.94,    100.18)
```

When n is very large, the variance of the prediction error will approach σ^2. These results may be used to construct the following prediction interval for y, given $x = x_p$. The confidence coefficient for the prediction interval is $(1 - \alpha)$.

A $(1 - \alpha)100\%$ Prediction Interval for y When $x = x_p$

$$\hat{y} \pm t_{\alpha/2} s_{(y - \hat{y})}$$

where $t_{\alpha/2}$ is based on $(n - 2)$ d.f. and $s^2_{(y - \hat{y})} = s^2 + s^2_{\hat{y}}$

EXAMPLE   **11.5**   Refer to Example 11.1 and predict the success rating for some new trainee who scored $x = 50$ on the test.

Solution   The predicted value of y would be

$$\hat{y} = \hat{\beta}_0 + \hat{\beta}_1 x_p$$
$$= 40.7842 + (.765562)(50) = 79.0623$$

and the 95% prediction interval for the success rating would be

$$79.0623 \pm (2.306)\sqrt{(8.704)^2 + (2.84)^2}$$

or

$$79.06 \pm 21.11$$

T A B L E **11.7**   Execustat printout for Example 11.5

Table of Predicted Values

Row	SCORE	Predicted RATING	95.00% Prediction Limits		95.00% Confidence Limits	
			Lower	Upper	Lower	Upper
1	50	79.0622	57.9502	100.174	72.5133	85.6112

You can generate the Execustat printout shown in Table 11.7 by using a sub-command in the RELATE/SIMPLE REGRESSION menu of the program. The least-squares estimate and the prediction limits for $x = 50$ are given in the third, fourth, and fifth columns of the table and agree with the hand calculations.

Note that in a practical situation we would probably have the ratings and test scores for many more than the $n = 10$ students indicated in Table 11.2 and that this would reduce somewhat the width of the prediction interval. In fact, when n is large, the prediction interval approaches $\hat{y} \pm z_{\alpha/2}s$ or, for a 95% prediction interval, $\hat{y} \pm 1.96s$. ∎

Again, note the distinction between the confidence interval for $E(y|x)$ and the prediction interval for y presented in this section. $E(y|x)$ is a mean, a parameter of a population of y values, and y is a random variable that oscillates in a random manner about $E(y|x)$. The mean value of y when $x = 50$ is vastly different from some value of y chosen at random from the set of all y values for which $x = 50$. To make this distinction when making inferences, **we always *estimate* the value of a parameter and *predict* the value of a random variable.** As noted in our earlier discussion and as shown in Figures 11.7 and 11.8, the error of predicting y is different from the error of estimating $E(y|x)$. This is evident in the difference in widths of the two prediction and confidence intervals.

A graph of the confidence interval for $E(y|x)$ and the prediction interval for a particular value of y for the data in Table 11.2 is shown in Figure 11.9. The plot of the confidence interval is shown by dotted lines, and the prediction interval is identified by dashed lines. Note how the widths of the intervals increase as you move to the right or left of $\bar{x} = 46$. This occurs because the errors in estimation and prediction when $x = x_p$ both depend on $x_p - \bar{x}$, the distance from the center of the range over which x was measured. In particular, see the confidence interval and prediction interval for $x = 50$ calculated in Examples 11.4 and 11.5.

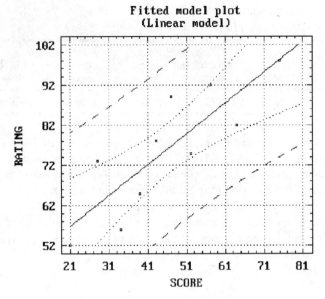

FIGURE **11.9**
Confidence intervals for
$E(y|x)$ and prediction
intervals for y based on data in
Table 11.2

Fitted model plot
(Linear model)

Exercises

Basic Techniques

11.23 Refer to Exercise 11.8. The Execustat printout shown below was generated using the ESTI-MATE command in the RELATE/SIMPLE REGRESSION menu.

Table of Predicted Values

Row	X	Predicted Y	90.00% Prediction Limits Lower	Upper	90.00% Confidence Limits Lower	Upper
1	-2	0.6	-1.57396	2.77396	-0.731273	1.93127
2	-1	1.8	-0.15958	3.75958	0.858648	2.74135
3	0	3	1.1173	4.8827	2.23139	3.76861
4	1	4.2	2.24042	6.15958	3.25865	5.14135
5	2	5.4	3.22604	7.57396	4.06873	6.73127

a Estimate the expected value of y when $x = 1$, using a 90% confidence interval.

b Find a 90% prediction interval for some value of y to be observed in the future when $x = 1$.

11.24 Refer to Exercise 11.9. If $s_{\hat{y}} = .102685$, find a 90% confidence interval for the mean value of y when $x = 2$.

11.25 Refer to Exercise 11.9 and use the computer printout in Exercise 11.15.

a Find the value of s_y in the computer printout.

b If $s_{\hat{y}} = .102685$, find $s_{y-\hat{y}}$.

c Find a 95% prediction interval for some value of y to be observed in the future when $x = 2$.

Applications

11.26   Refer to the simple linear regression analysis in Exercise 11.11, relating the number of cars sold per week y to the number of salespersons employed x. Part of a Minitab printout for the data is given here. Find a 95% confidence interval for the average number of cars sold per week if the automobile dealer employed seven salespersons.

```
MTB > Regress 'y' 1 'x';
SUBC>    Predict 7.

The regression equation is
y = - 2.00 + 3.15 x

Predictor       Coef       Stdev     t-ratio        p
Constant       -2.000      4.930       -0.41     0.699
x              3.1522      0.8393       3.76     0.009

s = 2.846      R-sq = 70.2%      R-sq(adj) = 65.2%
    .              .                 .
    .              .                 .
    .              .                 .
  Fit  Stdev.Fit          95% C.I.                95% P.I.
 20.07       1.45    (   16.51,    23.62)   (   12.24,    27.89)
```

11.27   Refer to Exercise 11.12, in which the net income y of a savings and loan institution was related to its total assets x. A portion of an Execustat printout for Exercise 11.27 is shown below.

Table of Predicted Values

Row	X	Predicted Y	90.00% Prediction Limits Lower	90.00% Prediction Limits Upper	90.00% Confidence Limits Lower	90.00% Confidence Limits Upper
1	200	473.109	-992.466	1938.68	-81.5812	1027.8
2	400	964.113	-494.454	2422.68	428.213	1500.01
3	500	1209.62	-279.839	2699.07	594.599	1824.63

a   Find a 90% confidence interval for the average net income of a savings and loan institution with total assets of $200 million.

b   Find a 90% prediction interval for the net income of a savings and loan institution with total assets of $500 million.

11.28   Following the gasoline shortage in the early 1970s, car manufacturers decreased the fuel usage of their vehicles by decreasing the size and weight of passenger cars and by decreasing the engine size while increasing its efficiency. For the data in the following table, the fuel consumption y and the mileage per gallon x are reported on a per-vehicle basis. Use the Excel printout to answer the following questions.

Year	Fuel Consumption (y)	Mileage per Gallon (x)
1976	723	13.5
1977	716	13.8
1978	701	14.0
1979	653	14.4
1980	591	15.5
1981	576	15.9
1982	566	16.7
1983	553	17.1
1984	536	17.8
1985	525	18.2
1986	526	18.3
1987	514	19.2
1988	509	19.9
1989	509	20.3
1990	502	20.0
1991	495	21.7

Source: Energy Facts, 1992.

	A	B	C	D	E	F	G	H
1			Simple Regression Analysis for CBS11-28					
2								
3	Linear Model: Y = 1071.42 - 28.765 * X							
4								
5					Table of Estimates			
6			Estimate		Standard Error		t Value	P Value
7	Intercept		1071.42		52.6283		20.36	0.0000
8	Slope		-28.765		3.0163		-9.54	0.0000
9								
10	R-Squared =		86.66 %					
11	Correlation Coef. =		-0.931					
12	Std Error of Estimation =		30.1105					
13	Durbin-Watson Statistic =		0.40218					
14	Mean Absolute Error =		24.8808					
15	Sample Size (n) =		16					
16								
17			Table of Predicted Values					
18			Predicted		99% Prediction		99% Confidence	
19	Row	X	Y		Lower Limit	Upper Limit	Lower Limit	Upper Limit
20	1	16	611.183		518.09	704.276	586.045	636.321

a Find the least-squares line appropriate for these data.

b Plot the points and graph the line as a check on your calculations.

c Do the data provide sufficient evidence to indicate that the mileage per gallon (x) contributes information for the prediction of fuel consumption (y) per vehicle?

d Find the p-value associated with the test in part (c) and interpret its value.

e Find a 99% confidence interval for the average fuel consumption per vehicle when the mileage per vehicle is 16 mpg.

f Find a 99% prediction interval for the average fuel consumption per vehicle when the mileage per vehicle is 16 mpg.

11.29 If you try to rent an apartment or buy a house, you will find that real estate representatives establish apartment rents and house prices on the basis of the square footage of the heated floor space. The data in the accompanying table give the square footages and sale prices of $n = 12$ houses randomly selected from those sold in a small city. Use the Minitab printout to answer parts (a) through (d).

Square Feet, x	Price, y	Square Feet, x	Price, y
1460	$78,700	1977	$95,400
2108	99,300	1610	87,000
1743	91,400	1530	82,400
1499	81,100	1759	88,200
1864	92,400	1821	94,300
2391	104,900	2216	101,700

```
MTB > Regress 'y' 1 'x';
SUBC>    Predict 2000.

The regression equation is
y = 41206 + 27.4 x

Predictor        Coef        Stdev      t-ratio         p
Constant        41206        3389        12.16       0.000
x              27.406        1.828       14.99       0.000

s = 1793        R-sq = 95.7%       R-sq(adj) = 95.3%
   .               .                  .
   .               .                  .
   .               .                  .

  Fit   Stdev.Fit         95% C.I.              95% P.I.
96018         602    (   94676,    97360)  (    91803,   100233)
```

a   Estimate the mean increase in the price for an increase of 1 square foot for houses sold in the city. Use a 90% confidence interval. Interpret your estimate.

b   Suppose that you are a real estate salesperson and you desire an estimate of the mean sale price of houses with a total of 2000 square feet of heated space. Use a 95% confidence interval and interpret your estimate.

c   Calculate the price per square foot for each house, and then calculate the sample mean. Why is this estimate of the mean cost per square foot not equal to the answer in part (a)? Should it be? Explain.

d   Suppose that a house containing 2000 square feet of heated floor space is offered for sale. Give a 95% prediction interval for the price at which the house will sell. Interpret this prediction.

 11.30 Does the number of years invested in schooling pay off in the job market? Apparently so—the better educated you are, the more money you will make. The information in the table that follows gives the median annual income of full-time workers age 25 years or older by the number of years of schooling completed. Use the Minitab printout to answer the questions that follow.

	Salary (dollars)	
Years of Schooling	Men	Women
8	18,000	11,500
10	20,500	13,000
12	25,000	16,100
14	28,100	18,300
16	34,500	22,100
19	39,700	27,600

Source: Data adapted from "Upward Mobility," 1989.

```
MTB > Regress 'y' 1 'x';
SUBC>   Predict 18.

The regression equation is
y = 648 + 2049 x

Predictor        Coef      Stdev     t-ratio          p
Constant          648       1532        0.42      0.694
x              2049.5       112.1       18.29      0.000

s = 1008        R-sq = 98.8%      R-sq(adj) = 98.5%
      .               .                 .
      .               .                 .
      .               .                 .

  Fit   Stdev.Fit          95% C.I.              95% P.I.
37539         680   (   35650,    39428)   (  34163,    40916)
```

a Find the least-squares prediction equation relating a man's salary y to the number of years of schooling x he has had.

b Do the data provide sufficient evidence to indicate that, with the straight-line linear model, x contributes information for the prediction of y? Test using $\alpha = .10$.

c Plot the data points. Does the distribution of plotted points appear to agree with the conclusions drawn in part (b)?

d Find a 95% confidence interval for the mean salary when a man has 18 years of schooling.

e Find a 95% prediction interval for the salary of a man who has 18 years of schooling.

11.6 Calculational Formulas

There are six basic quantities that are used in regression analysis calculations:

1 n

2 $\bar{y} = \dfrac{\sum_{i=1}^{n} y_i}{n}$

3 $\displaystyle \bar{x} = \frac{\sum\limits_{i=1}^{n} x_i}{n}$

4 $\displaystyle S_{yy} = \sum_{i=1}^{n}(y_i - \bar{y})^2 = \sum_{i=1}^{n} y_i^2 - \frac{\left(\sum\limits_{i=1}^{n} y_i\right)^2}{n}$

5 $\displaystyle S_{xy} = \sum_{i=1}^{n}(x_i - \bar{x})(y_i - \bar{y}) = \sum_{i=1}^{n} x_i y_i - \frac{\left(\sum\limits_{i=1}^{n} x_i\right)\left(\sum\limits_{i=1}^{n} y_i\right)}{n}$

6 $\displaystyle S_{xx} = \sum_{i=1}^{n}(x_i - \bar{x})^2 = \sum_{i=1}^{n} x_i^2 - \frac{\left(\sum\limits_{i=1}^{n} x_i\right)^2}{n}$

Relevant formulas for commonly used estimators and their standard errors follow:

1 Slope: $\hat{\beta}_1 = \dfrac{S_{xy}}{S_{xx}}$; intercept: $\hat{\beta}_0 = \bar{y} - \hat{\beta}_1 \bar{x}$; least-squares line: $\hat{y} = \hat{\beta}_0 + \hat{\beta}_1 x$

2 $\text{SSE} = S_{yy} - \dfrac{(S_{xy})^2}{S_{xx}}$; $s^2 = \dfrac{\text{SSE}}{n-2}$; $s = \sqrt{s^2}$

3 Standard error of $\hat{\beta}_1$: $s_{\hat{\beta}_1} = \sqrt{\dfrac{s^2}{S_{xx}}}$

4 Standard error of $\hat{\beta}_0$: $s_{\hat{\beta}_0} = \sqrt{s^2\left(\dfrac{1}{n} + \dfrac{\bar{x}^2}{S_{xx}}\right)}$

5 Standard error in estimating $E(y)$: $s_{\hat{y}} = \sqrt{s^2\left[\dfrac{1}{n} + \dfrac{(x_p - \bar{x})^2}{S_{xx}}\right]}$

6 Standard error of prediction: $s_{(y-\hat{y})} = \sqrt{s^2 + s_{\hat{y}}^2} = \sqrt{s^2\left[1 + \dfrac{1}{n} + \dfrac{(x_p - \bar{x})^2}{S_{xx}}\right]}$

Remember that, in general, a small-sample test statistic is constructed as

$$t = \frac{(\text{Estimator}) - (\text{parameter}|H_0)}{\text{SE(estimator)}}$$

based upon $(n-2)$ d.f. A $(1-\alpha)100\%$ interval estimator is found using (Estimator) $\pm t_{\alpha/2}\text{SE(estimator)}$.

Tips on Problem Solving

1 Be careful of rounding errors, which can greatly affect the answer you obtain in calculating S_{xx}, S_{xy}, and S_{yy}. If you must round a number, it is recommended that you carry six or more significant figures in the calculations. (Note also that, in working exercises, rounding errors might cause some slight discrepancies between your answers and the answers given in the back of the text.)

2 Always plot the data points and graph your least-squares line. If the line does not provide a reasonable fit to the data points, you may have made an error in your calculations.

11.7 A Coefficient of Correlation

Sometimes we wish to obtain an indicator of the strength of the linear relationship between two variables y and x that is independent of their respective scales of measurement. We call this a measure of the **linear correlation between y and x.**

The measure of the linear correlation commonly used in statistics is called the **Pearson product moment coefficient of correlation** between y and x. This quantity, denoted by the symbol r, was first presented in Section 2.11 and is computed as follows:

Pearson Product Moment Coefficient of Correlation

$$r = \frac{S_{xy}}{\sqrt{S_{xx}S_{yy}}}$$

where S_{xy}, S_{xx}, and S_{yy} are as defined in Section 11.6.

We will show you how to compute the Pearson product moment coefficient of correlation for the data in Table 11.2, and then we will explain how it measures the strength of the relationship between y and x.

EXAMPLE **11.6** Calculate the coefficient of correlation for the success rating–test score data in Table 11.2.

Solution The coefficient of correlation for the success rating–test score data in Table 11.2 can be obtained by using the formula for r and the quantities

$$S_{xy} = 1894$$

$$S_{xx} = 2474$$
$$S_{yy} = 2056$$

Then,

$$r = \frac{S_{xy}}{\sqrt{S_{xx}S_{yy}}} = \frac{1894}{\sqrt{(2474)(2056)}} = .84$$

Most regression programs will provide the value of r or r^2. The value of R-sq is found on the Minitab printout, and the correlation coeff. is found on the Execustat simple regression analysis printout. ■

A study of the coefficient of correlation r yields rather interesting results and explains the reason for its selection as a measure of linear correlation. We note that the denominators used in calculating r and $\hat{\beta}_1 = S_{xy}/S_{xx}$ will always be positive since they both involve sums of squares of numbers. Since the numerator used in calculating r is identical to the numerator of the formula for the slope $\hat{\beta}_1$, the coefficient of correlation r will assume exactly the same sign as $\hat{\beta}_1$ and will equal zero when $\hat{\beta}_1 = 0$. **Thus $r = 0$ implies no linear correlation between y and x. A positive value for r implies that the line slopes upward to the right; a negative value indicates that it slopes downward to the right.**

Figure 11.10 shows four typical scatterplots and their associated correlation coefficients. Note that $r = 0$ implies no linear correlation, not simply "no correlation." A pronounced pattern may exist, as in Figure 11.10(d), but its linear correlation coefficient may equal zero. In general, we can say that r measures the linear association of the two variables y and x. When $r = 1$ or -1, all the points fall on a straight line; when $r = 0$, they are scattered and give no evidence of a *linear* relationship. Any other value of r suggests the degree to which the points tend to be linearly related.

The interpretation of nonzero values of r can be obtained by comparing the errors of prediction for the prediction equation

$$\hat{y} = \hat{\beta}_0 + \hat{\beta}_1 x$$

with the predictor of y, $\bar{y}$, that would be employed if x were ignored. Figure 11.11(a) and (b) shows the lines $\hat{y} = \hat{\beta}_0 + \hat{\beta}_1 x$ and $\hat{y} = \bar{y}$ fit to the same set of data. Certainly, if x is of any value in predicting y, then SSE, the sum of squares of deviations of y about the linear model, should be less than the sum of squares of deviations about the predictor $\bar{y}$, which is

$$S_{yy} = \sum_{i=1}^{n}(y_i - \bar{y})^2$$

F I G U R E **11.10**
Some typical scatterplots with
approximate values of r

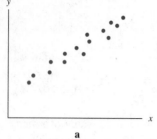

a
Strong positive linear correlation;
r is near 1

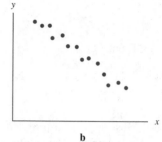

b
Strong negative linear correlation;
r is near -1

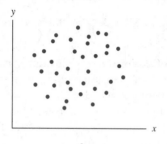

c
No apparent linear correlation;
r is near 0

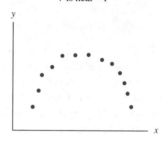

d
Curvilinear, but not linear, correlation;
r is near 0

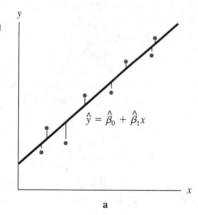

$$\hat{y} = \hat{\beta}_0 + \hat{\beta}_1 x$$

a

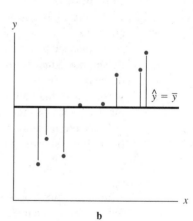

$$\hat{y} = \bar{y}$$

b

With the aid of a bit of algebraic manipulation, we can show that

$$r^2 = 1 - \frac{\text{SSE}}{\text{S}_{yy}} = \frac{\text{S}_{yy} - \text{SSE}}{\text{S}_{yy}}$$

In other words, r^2 lies in the interval

$$0 \le r^2 \le 1$$

and r will equal $+1$ or -1 only when all the points fall exactly on the fitted line, which occurs when SSE equals zero.

Actually, we see that r^2 **is equal to the ratio of the reduction in the sum of squares of deviations obtained by using the linear model to the total sum of squares of deviations about the sample mean $\bar{y}$, which would be the predictor of y if x were ignored.** Thus r^2, called the *coefficient of determination,* would seem to give a more meaningful interpretation of the strength of the relation between y and x than would the correlation coefficient r. Although this r^2 is simply the square of r given at the beginning of this section, in this form it is easier to assess the significance of the role of x in explaining y, and vice versa.

Coefficient of Determination

$$r^2 = \frac{S_{yy} - \text{SSE}}{S_{yy}}$$

You will observe that the sample correlation coefficient r is an estimator of a population correlation coefficient ρ (Greek letter rho), which would be obtained if the coefficient of correlation were calculated by using all the points in the population.

A test of the null hypothesis that no correlation exists between y and x, $H_0 : \rho = 0$, is exactly equivalent to a test of the hypothesis $H_0 : \beta_1 = 0$. The test statistic is given as

$$t = \frac{r\sqrt{n-2}}{\sqrt{1-r^2}} = \frac{\hat{\beta}_1}{s/\sqrt{S_{xx}}}, \quad \text{with } (n-2) \text{ d.f.}$$

For the data in Table 11.2, with $r = .84$, the value of the test statistic is

$$t = \frac{r\sqrt{n-2}}{\sqrt{1-r^2}} = \frac{.84\sqrt{8}}{\sqrt{1-(.84)^2}} = 4.38$$

which is identical to the value calculated for testing $H_0 : \beta_1 = 0$ in Example 11.2.

If in using either form of the t statistic the evidence in the sample suggests that y and x are related, it would seem that we would redirect our attention to the ultimate objective of our data analysis, using the prediction equation to obtain interval estimates for $E(y|x)$ and prediction intervals for y.

If the linear coefficients of correlation between y and each of two variables x_1 and x_2 were calculated to be .4 and .5, respectively, it does not follow that a predictor using both variables would account for $[(.4)^2 + (.5)^2] = .41$, or a 41% reduction in the sum of squares of deviations. Actually, x_1 and x_2 might be highly correlated and might therefore contribute virtually the same information for the prediction of y.

Finally, remember that r is a measure of **linear correlation** and that x and y could be perfectly related by some **nonlinear** function when the observed value of r is equal to zero.

Exercises

Basic Techniques

11.31   How does the coefficient of correlation measure the strength of the linear relationship between two variables y and x?

11.32   Describe the significance of the algebraic sign and the magnitude of r.

11.33   What value does r assume if all the data points fall on the same straight line and the following is true?

   a   The line has positive slope.

   b   The line has negative slope.

11.34   Consider the following data.

x	−2	−1	0	1	2
y	2	2	3	4	4

   a   Plot the data points. Based on your graph, what will be the sign of the sample correlation coefficient?

   b   If $S_{xx} = 10$, $S_{yy} = 4$, and $S_{xy} = 6$, calculate r and r^2 and interpret their values.

11.35   Consider the following data.

x	1	2	3	4	5	6
y	7	5	5	3	2	0

   a   Plot the six points on graph paper.

   b   If $S_{xx} = 17.5$, $S_{yy} = 31.3333$, and $S_{xy} = -23$, calculate the sample coefficient of correlation r and interpret.

   c   By what percentage was the sum of squares of deviations reduced by using the least-squares predictor $\hat{y} = \hat{\beta}_0 + \hat{\beta}_1 x$, rather than $\bar{y}$, as a predictor of y?

11.36   Reverse the slope of the line in Exercise 11.35 by reordering the y observations, as follows:

x	1	2	3	4	5	6
y	0	2	3	5	5	7

Repeat the steps of Exercise 11.35 with $S_{xy} = .23$. Notice the change in sign of r and the relation between the values of r^2 of Exercise 11.35 and this exercise.

Applications

11.37   Refer to the data of Exercise 11.10 relating the percentage y of increase in sales to a manufacturer's coded wholesale price x for a box of soap powder.

   a   If $S_{xx} = 17.5$, $S_{xy} = -26.25$, and $S_{yy} = 40.46833$, calculate the coefficient of correlation between y and x, and interpret its value.

   b   Calculate the coefficient of determination, and interpret its value.

11.38 Refer to the data of Exercise 11.11 relating the number y of new cars sold per week by an automobile dealer to the number x of salespersons employed.

a If $S_{xx} = 11.5$, $S_{xy} = 36.25$, and $S_{yy} = 162.875$, calculate the coefficient of correlation between y and x, and interpret its value.

b Calculate the coefficient of determination, and interpret its value.

c Compare the value of r^2 from part (b) to the value given on the computer printout in Exercise 11.11.

11.39 An experiment was conducted in a supermarket to observe the relation between the amount of display space allotted to a brand of coffee (brand A) and its weekly sales. The amount of space allotted to brand A was varied over 3-, 6-, and 9-square-feet displays in a random manner over 12 weeks; the space allotted to competing brands was maintained at a constant 3 square feet for each. The data in the accompanying table were observed. Use the Excel printout to answer the questions that follow.

Weekly Sales, y (dollars)	Space Allotted, x (square feet)	Weekly Sales, y (dollars)	Space Allotted, x (square feet)
526	6	434	6
421	3	443	3
581	6	590	9
630	9	570	6
412	3	346	3
560	9	672	9

	A	B	C	D	E	F	G	H
1			Simple Regression Analysis for CBS11-39					
2								
3	Linear Model: Y = 307.917 + 34.5833 * X							
4								
5					Table of Estimates			
6			Estimate		Standard Error		t Value	P Value
7	Intercept		307.917		39.4374		7.81	0.0000
8	Slope		34.5833		6.0853		5.68	0.0002
9								
10	R-Squared =		76.36 %					
11	Correlation Coef. =		0.874					
12	Std Error of Estimation =		51.6357					
13	Durbin-Watson Statistic =		2.3328					
14	Mean Absolute Error =		39.2361					
15	Sample Size (n) =		12					
16								
17			Table of Predicted Values					
18			Predicted		90% Prediction		90% Confidence	
19	Row	X	Y		Lower Limit	Upper Limit	Lower Limit	Upper Limit
20	1	6	515.417		418.007	612.826	488.400	542.433

a Find the least-squares line appropriate for the data.

b Calculate r and r^2. Interpret.

c Find a 90% confidence interval for the mean weekly sales given that 6 square feet are allotted for display.

d Use a 90% prediction interval to predict the weekly sales at some time in the future if 6 square feet are allotted for display.

e   By what percentage was the sum of squares of deviations reduced by using the least-squares predictor $\hat{y} = \hat{\beta}_0 + \hat{\beta}_1 x$ rather than $\bar{y}$ as a predictor of y for these data?

f   Would you expect the relation between y and x to be linear if x were varied over a wider range (say, $x = 1$ to $x = 30$)?

11.40   The data in Exercise 11.30 gave the median salary and number of years of schooling for both men and women. The data are reproduced below.

Years of Schooling	Salary (dollars)	
	Men	Women
8	18,000	11,500
10	20,500	13,000
12	25,000	16,100
14	28,100	18,300
16	34,500	22,100
19	39,700	27,600

Source: Data adapted from "Upward Mobility," 1989.

a   What type of correlation, if any, would you expect to see between the salaries of men and women? Plot the data. Does the correlation appear to be positive or negative?

b   The correlation between the salaries of men and women is $r = .995$. Is there a significant positive correlation? Test using $\alpha = .05$.

c   Interpret the value of r given in part (b). Comment on the value of this descriptive measure. Is there any other descriptive measure that might be of interest?

11.8   Assumptions

The assumptions for a regression analysis are given in the following display.

Assumptions for a Regression Analysis

1   The response y can be represented by the probabilistic model

$$y = \beta_0 + \beta_1 x + \epsilon$$

2 x is measured without error.

3 ϵ is a random variable such that, for a given value of x,

$$E(\epsilon) = 0 \quad \text{and} \quad \sigma_\epsilon^2 = \sigma^2$$

and all pairs ϵ_i, ϵ_j are independent in a probabilistic sense.

4   For a given value of x, ϵ possesses a normal probability distribution.

At first glance, you might fail to understand the significance of the first assumption. Models, deterministic or otherwise, are, as the name implies, only models for real relationships that occur in nature. Consequently, model misspecification is always a possibility. Even if you have obtained a good fit to the data, a large error in prediction is possible if you use the model to predict y for some value of x outside the range of values used to fit the least-squares equation. Of course, this problem will always occur if x is time and you attempt to forecast y at some point in the future. This problem occurs with any model for the future time predictions; consequently, you make the forecast but keep the model limitations in mind.

The assumption that the variance of the ϵs is constant and equal to σ^2 will not be true for all types of data. Furthermore, if x is time, it is possible that y values measured over adjacent time periods will tend to be dependent (an overly large value of y in 1995 might signal a large value of y in 1996). Substantial departure from either of these assumptions will affect the confidence coefficients of interval estimates and significance levels of tests described in this chapter.

Like unequal variances and correlation of the random errors, if the normality assumption (4) is not satisfied, the confidence coefficients and significance levels for interval estimates and tests will not be what we expect them to be. However, modest departures from normality will not seriously disturb these values.

CASE STUDY REVISITED

11.9   A COMPUTER ANALYSIS OF THE FOREIGN CAR IMPORTS

A plot of the number of imported cars from 1969 to 1989 is given in the case study. Notice that the last point in the plot, 2.8 million imported cars in 1989, represents what appears to be a deviation about a linear increasing trend in imports. A Minitab simple regression analysis printout using only the data from 1969 to 1989 is given in Table 11.8.

The least-squares line fitted to these data is

$$\hat{y} = 1.188 + .095x$$

The first portion of the printout provides the values of the t statistics when testing for a zero intercept and a zero slope, given as 13.45 and 12.63, respectively. Since both t values based on 19 degrees of freedom are highly significant (both p-values are reported as .000), there is a significant regression of y on x. Further, the value of R-squared = 89.4% indicates that approximately 89% of the variation in y is explained by the linear regression on time and hence that the fit of the line to the data is quite good.

The plot (Figure 11.12) of the regression line with the data points, the confidence bands, and the prediction bands superimposed indicates that all the observed points

T A B L E **11.8** Minitab output for sales of imported cars

```
The regression equation is
Y = 1.19 + 0.0955 X

Predictor         Coef        Stdev     t-ratio          p
Constant       1.18831      0.08835      13.45       0.000
X              0.095455     0.007557     12.63       0.000

s = 0.2097     R-sq = 89.4%     R-sq(adj) = 88.8%

Analysis of Variance

SOURCE         DF         SS          MS         F          p
Regression      1       7.0159      7.0159     159.54     0.000
Error          19       0.8355      0.0440
Total          20       7.8514


   Fit   Stdev.Fit          95% C.I.              95% P.I.
 3.1929    0.0949     ( 2.9942,  3.3915)    ( 2.7110,  3.6747)

 3.2883    0.1016     ( 3.0757,  3.5010)    ( 2.8005,  3.7761)

 3.3838    0.1084     ( 3.1569,  3.6107)    ( 2.8896,  3.8779)

 3.4792    0.1153     ( 3.2379,  3.7205)    ( 2.9782,  3.9802) X

X  denotes a row with X values away from the center
```

F I G U R E **11.12**

Plot of regression model for import-car data

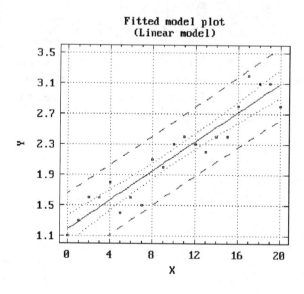

from 1969 through 1989 fall within the prediction limits (the outer bands) and that the regression fit is quite good.

Let us now check to see how well this regression model does in predicting the sales for 1990 through 1993. The predictions are also found in the bottom portion of the Minitab printout. For each value of x submitted to the program, the printout provides a point estimate, given in the column labeled Fit, 95% confidence limits, and 95% prediction limits. Since the four points in time—1990, 1991, 1992, and 1993—were not used in the regression analysis, we are really looking at predictions and should refer to the prediction intervals on the printout. We immediately see that each of the four prediction intervals fails to include the actual number of foreign-car imports for these four years. How can this happen when the fit of the regression line to the data from 1969 to 1989 was deemed to be very good, as evidenced by the significant regression and the fairly large value of r^2? If we were to plot all of the data from 1969 to 1993, as in Figure 11.13, we would see that the trend appears to change in 1989 and the last four points now show a decreasing linear trend.

FIGURE **11.13**
Scatterplot of foreign-car imports, 1969–93

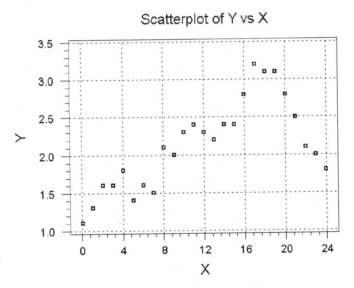

This decrease in sales explains why the predictions using only the first 21 data points failed to include the actual number of imports for 1990–93. (We will discuss alternate regression models that may include terms other than x alone in the next chapter.) However, the answer to the initial question is now at hand. It appears that the U.S. automobile industry has been successful in stopping the flood of imported cars being purchased by American consumers.

11.10 Summary

Although it was not stressed, you will observe that predicting the value of a random variable y was considered for the most elementary situation in Chapter 7. Thus, if we possessed no information concerning variables related to y, the sole information available for predicting y would be provided by its probability distribution. If we were to select one value as representative of the population, we would most likely choose μ or some other measure of central tendency. The estimation of the mean was considered in Chapter 7.

In this chapter, we were concerned with the problem of predicting y when auxiliary information is available on another variable that is related to y and hence assists in its prediction. Although here we have concentrated primarily on the problem of predicting y as a linear function of a single variable x, the more interesting case, where y is a linear function of a set of independent variables, is the subject of Chapter 12.

Supplementary Exercises

11.41   Graph the line corresponding to the equation $y = 3x + 2$ by locating points corresponding to $x = 0,\ 1$, and 2.

11.42   Do the following for the linear equation $2x + 3y + 6 = 0$.

   a   Give the y intercept and slope for the line.

   b   Graph the line corresponding to the equation.

11.43   Follow the instructions given in Exercise 11.42 for the linear equation $2x - 3y - 5 = 0$.

11.44   Follow the instructions given in Exercise 11.42 for the linear equation $x/y = 1/2$.

11.45   Given the following data for corresponding values of two variables, y and x, use an appropriate computer program to answer the following questions.

y	0	0	1	1	3
x	-2	-1	0	1	2

   a   Find the least-squares line for the data.

   b   As a check on the calculations in (a), plot the five points and graph the line.

   c   Do the data provide sufficient evidence to indicate that y and x are linearly related? (Test the hypothesis that $\beta_1 = 0$, using $\alpha = .05$.)

   d   Find a 90% confidence interval for the slope of the line.

   e   Obtain a 90% confidence interval for the expected value of y when $x = 1$.

   f   Calculate the coefficient of correlation r for the data. What is the significance of this particular value of r?

11.46   Given the following data for corresponding values of two variables, y and x, use an appropriate computer program to answer the following questions.

y	2	1.5	1	2.5	2.5	4	5
x	−3	−2	−1	0	1	2	3

a Find the least-squares line for the data.

b Plot the points and graph the line.

c Do the data provide sufficient evidence to indicate that y and x are linearly related? (Test the hypothesis that $\beta_1 = 0$, using $\alpha = .05$.)

d Find a 95% confidence interval for the slope of the line.

e Obtain a 95% confidence interval for the expected value of y when $x = -1$.

f Given that $x = 2$, find an interval estimate for a particular value of y. Use a confidence coefficient equal to .90.

g Calculate the coefficient of correlation for the data.

h By what percentage was the sum of squares of deviations reduced by using the least-squares predictor $\hat{y} = \hat{\beta}_0 + \hat{\beta}_1 x$ rather than $\bar{y}$ as a predictor of y for the data?

11.47 A psychological experiment was conducted to study the relationship between the length of time necessary for a human being to reach a decision and the number of alternatives presented. The questions presented to the participants required a classification of an object into two or more classes, similar to the situation that one might encounter in grading potatoes. Five individuals classified one item each for a two-class, two-decision situation. Five each were also allotted to three-class and four-class categories. The length of time necessary to reach a decision is recorded below for the 15 participants.

Length of reaction time, y (seconds)	1, 3, 3, 2, 4	2, 4, 3, 4, 5	5, 6, 5, 7, 4
Number of alternatives, x	2	3	4

```
The regression equation is
y = - 0.33 + 1.40 x

Predictor          Coef        Stdev      t-ratio          p
Constant         -0.333        1.095        -0.30       0.766
x                1.4000       0.3523         3.97       0.002

s = 1.114        R-sq = 54.9%      R-sq(adj) = 51.4%
```

a Find the least-squares line appropriate for these data.

b Plot the points and graph the line as a check on your calculations.

c Do the data provide sufficient evidence to indicate that the length of reaction time y and number of alternatives x are correlated? (Test at the $\alpha = .05$ level of significance.)

11.48 An experiment was conducted to investigate the effect of a training program on the length of time required for a production worker to assemble an electronic timing mechanism. Nine workers were placed in the program. The reduction y in time to complete the assembly was measured for three workers at the end of two weeks, for three at the end of four weeks, and for three at the end of six weeks of training. The data are shown in the table.

Reduction in time, y (seconds)	1.6, .8, 1.0	2.1, 1.6, 2.5	3.8, 2.7, 3.1
Length of training, x (weeks)	2	4	6

Use an appropriate computer software package to analyze these data. State any conclusions you can draw.

11.49 Wheat and corn production, both staples of the American agricultural industry, vary with the weather, rainfall, and the production of competing products. The data below show the annual U.S. production (in millions of bushels) for selected years between 1980 and 1992. Use an appropriate computer program to answer the following questions.

Year	Corn for Grain	All Wheat
1980	6639.4	2380.9
1982	8235.1	2765.0
1984	7672.1	2594.8
1985	8875.5	2424.1
1986	8225.8	2090.6
1987	7131.3	2107.7
1988	4928.7	1812.2
1989	7525.5	2036.6
1990	7934.0	2736.4
1991	7475.5	1981.1
1992	9478.9	2458.8

Source: World Almanac & Book of Facts, 1994.

a Let y represent the annual production of corn for grain. Find a least-squares line appropriate for the data.

b Plot the points and graph the line as a check on your calculations.

c Do the data provide sufficient evidence to indicate that the mean annual production of corn is linearly related to the year of production? Test using $\alpha = .05$.

d Construct a 90% confidence interval for the mean annual *change* in production of corn. Interpret this interval estimate.

e If you were a corn producer, how much of an increase would you expect in the mean production of corn in 1994 over 1993?

f Predict the annual corn production for 1993 using a 99% prediction interval. How confident can you be in this prediction? Are there any qualifications you might place on it? Explain.

11.50 Refer to Exercise 11.49, and perform the same type of data analysis for the annual wheat production.

11.51 Refer to Exercises 11.49 and 11.50. Find the correlation coefficient between corn and wheat production. Is there sufficient evidence to indicate a positive correlation between wheat and corn production over the years?

11.52 How closely do the values assigned to properties by appraisers agree with the actual market value of the properties? The accompanying table gives the tax-office appraised values and the sale prices of 12 residential properties sold in a mid-size city in 1995.

Appraised value, x	Sale price, y	Appraised value, x	Sale price, y
65.5	80.0	86.4	109.0
62.6	77.5	89.1	116.7
71.2	86.2	93.0	105.0
60.5	71.9	101.7	98.5
81.5	95.0	86.4	139.0
104.7	130.0	102.8	155.0

	A	B	C	D	E	F	G	H
1			Simple Regression Analysis for CBS11-52					
2								
3	Linear Model: Y = -4.58706 + 1.31176 * X							
4								
5					Table of Estimates			
6			Estimate		Standard Error		t Value	P Value
7	Intercept		-4.58706		26.7341		-0.17	0.8672
8	Slope		1.31176		0.3140		4.18	0.0019
9								
10	R-Squared =		63.57 %					
11	Correlation Coef. =		0.797					
12	Std Error of Estimation =		16.425					
13	Durbin-Watson Statistic =		1.6228					
14	Mean Absolute Error =		9.9415					
15	Sample Size (n) =		12					
16								
17			Table of Predicted Values					
18			Predicted		90% Prediction		90% Confidence	
19	Row	X	Y		Lower Limit	Upper Limit	Lower Limit	Upper Limit
20	1	80	100.354		69.294	131.414	91.494	109.213
21	2	90	113.471		82.285	144.658	104.178	122.765

a Plot the data points on graph paper.

b Fit a least-squares line to the data.

c Does the appraised value x contribute information for the prediction of y?

d Find r^2 and interpret its value.

e Find a 90% confidence interval for the mean sale price of a property with an appraised value of $80,000. Interpret the interval.

f Find a 90% prediction interval for the sale price of a property with an appraised value of $90,000. Interpret the interval.

11.53 A large retailer employs computer-operated record keeping from point of sale through customer billing. Theoretically, the only errors that can enter the system are caused by incorrect entries made for each sale by the salesperson making the sale. Does lack of sleep increase the incidence of incorrect entries by salespersons? A total of ten salespersons participated in a study designed to provide information on this question. Two each were assigned to each of five sleep deprivation periods. Following the assigned period of sleep deprivation, each salesperson worked for an 8-hour shift, and the number of incorrect computer entries was recorded. The data follow. Use an appropriate computer program to analyze these data. State any conclusions you can draw.

Number of errors, y	8, 6	6, 10	8, 14	14, 20	16, 12
Number of hours without sleep, x	8	12	16	20	24

11.54 A property insurer conducted a study to investigate the annual payout y for property damage claims (in millions of dollars) in Florida as a function of the number x of hurricanes hitting the Florida coast. A linear regression model, fit to data over a ten-year period, produced the following equation:

$$\hat{y} = 22.4 + 15.8x$$

with $\bar{x} = 1.5$, $S_{xx} = 4.95$, and $s^2 = 41.2$.

a Do the data provide sufficient evidence to indicate that the number x of hurricanes contributes information for the prediction of the insurer's annual property damage claims? Test using $\alpha = .05$. [*Hint:* Use the calculational formulas of Section 11.6.]

b Find a 90% confidence interval for the mean annual payout if one hurricane hits the Florida coast in a given year. Interpret the interval.

11.55 The property insurer of Exercise 11.54 is concerned about how large (or small) the payments might be for a given year. Find a 90% prediction interval for the annual payout if two hurricanes hit the Florida coast in a given year. Interpret the interval.

11.56 How well does the straight-line model in Exercise 11.54 fit the data relating the annual property damage payout y to the number of hurricanes that hit the Florida coast per year? Justify your answer.

11.57 The earnings y per share (EPS) for Wendy's International (the fast-food company) for the years 1987 through 1994 are shown in the accompanying table; the Excel regression analysis printout is also shown. To simplify calculations, we have coded the year variable by subtracting 1986 from each year; that is, $x =$ Year $- 1986$.

	Actual Year							
Variable	**1987**	**1988**	**1989**	**1990**	**1991**	**1992**	**1993**	**1994**
Coded year, x	1	2	3	4	5	6	7	8
EPS (dollars), y	.40	.30	.25	.40	.52	.63	.76	.90

Source: Value Line Investment Survey, 1994.

	A	B	C	D	E	F	G	H
1			Simple Regression Analysis for CBS11-57					
2								
3	Linear Model: Y = 0.141786 + 0.0840476 * X							
4								
5				Table of Estimates				
6			Estimate		Standard Error		t Value	P Value
7	Intercept		0.141786		0.082653		1.72	0.1371
8	Slope		0.084048		0.016368		5.13	0.0021
9								
10	R-Squared =		81.5 %					
11	Correlation Coef. =		0.903					
12	Std Error of Estimation =		0.1061					
13	Durbin-Watson Statistic =		0.9391					
14	Mean Absolute Error =		0.0725					
15	Sample Size (n) =		8					

a Locate $\hat{\beta}_0$ and $\hat{\beta}_1$ on the printout, and give the equation of the least-squares line.

b Plot the data points on graph paper and graph the least-squares line.

c Find r^2 on the printout and interpret its value.

d Does x contribute information for the prediction of y? Explain.

e Find a 95% confidence interval for the mean change in earnings per year.

 11.58 The following data from Exercise 2.49 represent the average seller's asking price, the average buyer's bid, and the average closing bid for each of ten types of used computer equipment ("Used Computer Prices," 1992). Use your knowledge of linear regression and correlation analysis to analyze these data. Use an appropriate statistical software package and present your results in the form of a report. Include any relevant graphs or computer output.

Machine	Average Seller's Asking Price	Average Buyer's Bid	Average Closing Bid
20MB PC XT	$ 400	$ 200	$ 300
20MB PC AT	700	400	575
IBM XT 089	450	200	325
IBM AT 339	700	350	600
20MB IBM PS/2 30	950	500	725
20MB IBM PS/2 50	1050	700	875
60MB IBM PS/2 70	2000	1600	1725
20MB Compaq SLT	1200	700	875
Toshiba 1600	1000	700	900
Toshiba 1200HB	1150	800	975

11.59 The table below, reproduced from Exercise 2.52, shows the number of curbside recycling programs and the number of landfills in each of seven regions of the United States (*EPA Journal,* July/August 1992). Use an appropriate statistical software package to analyze these data and draw any appropriate conclusions. Is the number of landfills in a given region useful for predicting the number of curbside recycling programs? Explain.

Region	Curbside Recycling Programs	Landfills
West	569	1374
Rocky Mountain	44	661
Midwest	108	1402
Great Lakes	1148	531
South	402	1007
Mid-Atlantic	1379	334
New England	305	503

Exercises Using the Data Disk

11.60 Refer to the broccoli data, set B on the data disk.

a Randomly select one observation on head diameter of the variety labeled 0 from each chamber. Let y denote head diameter and x denote the 6-hour average in ppm of the chamber. Plot the points and perform a regression analysis using the model

$$y_i = \beta_0 + \beta_1 x_i + \epsilon_i \quad i = 1, 2, \ldots, 17$$

b Repeat part (a) for those observations on the second variety, labeled 1.

c Repeat part (a) for variety 0 with y as the market-weight and x as the 13-hour average ozone level.

d *Correlation:* For each chamber number, record the values of the 6-hour and the 13-hour ozone averages, so that you have $n = 17$ pairs of observations. Find the correlation between these two variables. How would you assess its importance?

11.61 Refer to data set C, containing observations on average maturity in days, seven-day average yield, and assets. Select a sample of between 10 and 20 observations and record the values of the three variables for each observation. Use regression or correlation to investigate any linear relationship that may be present in the data. If you plan to use regression, let the seven-day average yield be denoted as the response variable y and each of the other two variables as a predictor variable.

12

MULTIPLE REGRESSION ANALYSIS

About This Chapter

In Chapter 11, we introduced the concepts of simple linear regression and correlation, the ultimate goal being to estimate the mean value of y or to predict a value of y by using information contained in a single independent (predictor) variable x. In this chapter, we expand this idea and relate the mean value of y to one or more independent variables x_1, x_2, ..., x_k in models that are more flexible than the straight-line model of Chapter 11. The process of finding the least-squares prediction equation, testing the adequacy of the model, and conducting tests about and estimating the values of the model parameters is called a multiple regression analysis.

Contents

"MADE IN THE U.S.A."— ANOTHER LOOK

The case study in Chapter 11 examined the effect of foreign competition in the automotive industry whereby the number of imported cars steadily increased during the 1970s and 1980s (*Automotive News,* 1994). The U.S. automobile industry has been besieged with complaints about product quality, worker layoffs, and high prices and has spent billions in advertising and research to produce an American-made car that will satisfy consumer demands. Has it been successful in stopping the flood of imported cars being purchased by American consumers? The data shown in Table 12.1 represent the number of imported cars (y) sold in the United States (in millions) for the years 1969–93. To simplify the analysis, we have coded the year using the coded variable $x = \text{Year} - 1969$.

TABLE **12.1**
Sales of imported cars,
1969–93

Year	(Year−1969) x	Number of Imported Cars y	Year	(Year−1969) x	Number of Imported Cars y
1969	0	1.1	1982	13	2.2
1970	1	1.3	1983	14	2.4
1971	2	1.6	1984	15	2.4
1972	3	1.6	1985	16	2.8
1973	4	1.8	1986	17	3.2
1974	5	1.4	1987	18	3.1
1975	6	1.6	1988	19	3.1
1976	7	1.5	1989	20	2.8
1977	8	2.1	1990	21	2.5
1978	9	2.0	1991	22	2.1
1979	10	2.3	1992	23	2.0
1980	11	2.4	1993	24	1.8
1981	12	2.3			

In Chapter 11, we fitted a simple linear regression model to the data corresponding to the years from 1969 to 1989 and found a good fit over those years. However, the model had poor predictability for the four years that followed. Further examination of the data over the total time span revealed that the increasing linear trend reversed and began a decline beginning in 1989. What models should we consider as alternatives to the simple linear regression model used in Chapter 11? How will we determine which of these models is better than others? Can we use r^2 as the primary criterion, or should we have one or more other criteria that we could use in conjunction with r^2?

These and other questions concerning the use of one or more predictor variables in a regression analysis are the thrust of this chapter, which deals with the use of several predictor variables or the use of power terms such as x, x^2, x^3, ..., x^p in the regression model. We will return to the analysis of the import-car data in Section 12.11.

12.1 Introduction

Multiple linear regression is an extension of the methodology of Chapter 11 to more than one independent variable. That is, instead of using only a single independent variable to explain the variation in y, multiple linear regression allows for the simultaneous use of several independent (or predictor) variables. By using more than one independent variable, we should do a better job of explaining the variation in y and hence be able to make more accurate predictions.

A common application of multiple linear regression is residential property value assessment by municipal assessors or private appraisers. In this instance, the objective is to estimate (or predict) the value of a residence y based on certain descriptive information. The assessor may begin by recording the size of the residence, x_1, as measured by the number of square feet of living space in the residence. But room size alone is an imprecise determinant of value. Two residences of equal size may have different characteristics with respect to total number of rooms (x_2), the number of bedrooms (x_3), the number of bathrooms (x_4), and the age since initial construction (x_5). All are important determinants of value and, when considered collectively with size in a multiple regression model, will almost certainly improve our ability to estimate market value accurately in comparison with a model that uses size alone.

As a further illustration, think of your special area of business (or your anticipated special area) and then think of some criterion variable y that measures success in the performance of that specialty. For example, if you are majoring in marketing, you might think of sales volume as a measure of success. A person operating a small business would probably use profit as a criterion of success, and the director of security for a large department store might measure performance by the value of merchandise lost by theft.

Now suppose that you possessed a multivariate prediction equation that gave an accurate prediction of values of y for given values of the xs. Think of the benefits to be derived from this tool. You would be able to predict values of the criterion variable for various values of the xs, and simply by noting when the variables enter into the equation you would likely develop a better understanding of how to control the criterion variable y and make it take values advantageous to you.

Finding a multivariate prediction equation is the subject of this chapter. The application of this method—the statistical tests and estimation procedures—to a set of data is often called **multiple regression analysis.** Our approach is to examine the information provided on computer regression printouts as they relate to the interpretation of results and to the use of the fitted regression model in estimation and prediction.

12.2 The Multiple Regression Model and Associated Assumptions

The **general linear model** for a multiple regression analysis will take the form shown in the following display.

The General Linear Model and Assumptions

$$y = \beta_0 + \beta_1 x_1 + \beta_2 x_2 + \cdots + \beta_k x_k + \epsilon$$

where the assumptions are as follows:

1 y is the response variable that you want to predict.

2 $\beta_0, \beta_1, \beta_2, \ldots, \beta_k$ are unknown constants.

3 $x_1, x_2, \ldots, x_k$ are independent **predictor variables** that are measured without error.

4 ϵ is a random error that for any given set of values for $x_1, x_2, \ldots, x_k$ is normally distributed with mean zero and variance equal to σ^2.

5 The random errors, say ϵ_i and ϵ_j, associated with any pair of y values are independent.

With these assumptions, it follows that the mean value of y for a given set of values for $x_1, x_2, \ldots, x_k$ is equal to

$$E(y) = \beta_0 + \beta_1 x_1 + \beta_2 x_2 + \cdots + \beta_k x_k$$

We will assume that the variables x_1, x_2, $\ldots$, x_k that appear in the general linear model need not represent *different* predictor variables. The development that follows requires that when we observe a value of y, the variables x_1, x_2, $\ldots$, x_k can be recorded without error. Furthermore, the random error ϵ associated with an observation y has a mean of zero and a constant variance σ^2, independent of the values of the predictor variables x_1, x_2, $\ldots$, x_k.

Suppose that we want to express y, the listed selling price of a home, as a function of several independent variables such as

$$x_1 = \text{Square footage of living space}$$
$$x_2 = \text{Number of bedrooms}$$

and

$$x_3 = \text{Number of bathrooms}$$

The multiple regression model relating y to x_1, x_2, and x_3 is

$$y = \beta_0 + \beta_1 x_1 + \beta_2 x_2 + \beta_3 x_3 + \epsilon$$

with

$$E(y) = \beta_0 + \beta_1 x_1 + \beta_2 x_2 + \beta_3 x_3$$

which describes a plane in four-dimensional space. The parameter β_0 is called the **intercept** and represents the average value of y when x_1, x_2, and x_3 are each zero. The parameters β_1, β_2, and β_3 are called the **partial slopes** or **partial regression coefficients** to distinguish their values from the **total slopes** obtained in the three simple linear regression equations relating y to x_1, y to x_2, and y to x_3. In the multiple linear regression equation relating y to x_1, x_2, and x_3, the partial slope β_1 represents the average increase in y for a one-unit increase in x_1 *when x_2 and x_3 are held constant*. The partial regression coefficients β_2 and β_3 have similar interpretations. For example, if x_2 and x_3 are held constant with $x_2 = 3$ and $x_3 = 2$, then

$$E(y) = \beta_0 + \beta_1 x_1 + \beta_2(3) + \beta_3(2)$$

or

$$E(y) = \beta_0^* + \beta_1 x_1$$

with $\beta_0^* = \beta_0 + 3\beta_2 + 2\beta_3$. In this form, we can see that β_1 is the increase in y for a one-unit increase in x_1 in the presence of x_2 and x_3 in the model. In general, the value of β_1 with x_2 and x_3 in the model **will not be the same** as the slope found when fitting x_1 alone. Notice that the effect of changing the value of x_2 or x_3 is to produce a line parallel to the first with a new intercept β_0^*.

The regression analysis methodology that we present in Section 12.3 is appropriate for the multiple linear regression model in which terms linear in x_1, x_2, ..., and x_k appear; however, it is also appropriate for models in which polynomial or cross-product terms such as x_1^3 or $x_1 x_2$ appear, since x_1^3 and $x_1 x_2$ are simply treated as new predictor variables. In this case, however, care must be taken in interpreting the fitted regression coefficients.

Exercises

Basic Techniques

12.1   Suppose that $E(y)$ is related to two predictor variables x_1 and x_2 by the equation

$$E(y) = 3 + x_1 - 2x_2$$

a   Graph the relationship between $E(y)$ and x_1 when $x_2 = 2$. Repeat for $x_2 = 1$ and for $x_2 = 0$.

b   What relationship do the lines in part (a) have to one another?

12.2   Refer to Exercise 12.1.

    a Graph the relationship between $E(y)$ and x_2 when $x_1 = 0$. Repeat for $x_1 = 1$ and for $x_1 = 2$.

    b What relationship do the lines in part (a) have to one another?

    c Suppose that in a practical situation you wanted to model the relationship between $E(y)$ and two predictor variables x_1 and x_2. What would be the implication of using the first-order model $E(y) = \beta_0 + \beta_1 x_1 + \beta_2 x_2$?

12.3 Suppose that $E(y)$ is related to two predictor variables x_1 and x_2 by the equation

$$E(y) = 3 + x_1 - 2x_2 + x_1 x_2$$

    a Graph the relationship between $E(y)$ and x_1 when $x_2 = 0$. Repeat for $x_2 = 2$ and for $x_2 = -2$.

    b Note that the equation for $E(y)$ is exactly the same as the equation in Exercise 12.1 except that we have added the term $x_1 x_2$. How does the addition of the $x_1 x_2$ term affect the graphs of the three lines?

    c What flexibility is added to the first-order model $E(y) = \beta_0 + \beta_1 x_1 + \beta_2 x_2$ by the addition of the term $\beta_3 x_1 x_2$, using the model $E(y) = \beta_0 + \beta_1 x_1 + \beta_2 x_2 + \beta_3 x_1 x_2$?

12.3 A Multiple Regression Analysis

A multiple regression analysis is performed somewhat like a simple linear regression analysis. A multiple regression model, say,

$$E(y) = \beta_0 + \beta_1 x_1 + \beta_2 x_2 + \cdots + \beta_k x_k$$

is fitted to a set of data by using the method of least squares, a procedure that finds the prediction equation

$$\hat{y} = \hat{\beta}_0 + \hat{\beta}_1 x_1 + \hat{\beta}_2 x_2 + \cdots + \hat{\beta}_k x_k$$

that minimizes SSE, the sum of squares of deviations of the observed values of y from their predicted values. This procedure is usually implemented using one of several regression programs available in the Minitab, SAS, Execustat, and other computer packages. In this section we present two sets of data and formulate models for each. We then present and discuss the printout of one or more regression programs. Although the printouts differ in format (selection and placement of relevant information), in general they contain the same essential information.

E X A M P L E   **12.1**   Moreno Valley, California, a bedroom community about 60 miles east of Los Angeles, has been described as one of the fastest-growing areas in the United States. Table 12.2 gives the listed selling price (y) in thousands of dollars, the living area (x_1) in thousands of square feet, and the number of floors (x_2), bedrooms (x_3), and baths (x_4) for $n = 29$ randomly selected residences on the market during the summer of 1990. Use a multiple linear regression model relating the listed selling price y and the independent variables x_1, x_2, x_3, and x_4.

	y	x_1	x_2	x_3	x_4
Observation	**LPRICE**	**SQFT**	**NUMFLRS**	**BDRMS**	**BATHS**
1	69.0	6	1	2	1.0
2	11.5	8	1	2	1.0
3	118.5	10	1	2	2.0
4	104.0	11	1	3	2.0
5	116.5	10	1	3	2.0
6	121.5	10	1	3	2.0
7	125.0	11	1	3	2.0
8	128.0	15	2	3	2.5
9	129.9	13	1	3	1.7
10	133.0	13	2	3	2.5
11	135.0	13	2	3	2.5
12	137.5	15	2	3	2.5
13	139.9	13	1	3	2.0
14	143.9	14	2	3	2.5
15	147.9	17	2	3	2.5
16	154.9	15	2	3	2.5
17	160.0	19	2	3	2.0
18	169.0	15	1	3	2.0
19	169.9	18	1	3	2.0
20	125.0	13	1	4	2.0
21	134.9	13	1	4	2.0
22	139.9	17	1	4	2.0
23	147.0	18	1	4	2.0
24	159.0	14	1	4	2.0
25	169.9	17	2	4	3.0
26	178.9	19	1	4	2.0
27	194.5	20	2	4	3.0
28	219.9	21	1	4	2.5
29	269.0	25	2	4	3.0

TABLE 12.2
Listed selling prices for 29 residential properties

Solution   The multiple linear regression model to be fitted is

$$E(y) = \beta_0 + \beta_1 x_1 + \beta_2 x_2 + \beta_3 x_3 + \beta_4 x_4$$

The computer printout in Table 12.3 resulted when the REGRESS command in the Minitab package was used to regress y, stored in column 8, on the *four* predictor variables x_1, x_2, x_3, and x_4, stored in C1, C2, C3, and C4.

Estimation of Parameters

The fitted regression equation, found in area ①, is given by

$$\hat{y} = -16.6 + 7.84 x_1 - 34.4 x_2 - 7.99 x_3 + 54.9 x_4$$

Information about the individual parameters of the model is found in the upper portion of the printout. The column headed Predictor identifies each parameter, with either the actual column number containing that variable or the column's assigned

T A B L E **12.3**
Minitab regression analysis for
the data in Example 12.1

```
MTB > REGRESS C8 4 C1-C4;
SUBC>   PREDICT 10 1 3 2;
SUBC>   PREDICT 14 2 3 2.5.      ①
```

```
The regression equation is
LPRICE = - 16.6 + 7.84 SQFT - 34.4 NUMFLRS - 7.99 BDRMS + 54.9 BATHS
```

Predictor	Coef ②	Stdev ③	t-ratio ④	p ⑤
Constant	-16.58	18.88	-0.88	0.389
SQFT	7.839	1.234	6.35	0.000
NUMFLRS	-34.39	11.15	-3.09	0.005
BDRMS	-7.990	8.249	-0.97	0.342
BATHS	54.93	13.52	4.06	0.000

s = 16.58 ⑥   R-sq = 88.2% ⑦   R-sq(adj) = 86.2% ⑧

Analysis of Variance   ⑨

SOURCE	DF	SS	MS	F	p
Regression	4	49359	12340	44.88	0.000
Error	24	6599	275		
Total	28	55958			

SOURCE	DF	SEQ SS ⑩
SQFT	1	44444
NUMFLRS	1	59
BDRMS	1	321
BATHS	1	4536

Unusual Observations

Obs.	SQFT	LPRICE	Fit	Stdev.Fit	Residual	St.Resid
1	6.0	69.00	35.02	9.45	33.98	2.49R
2	8.0	11.50	50.70	9.45	-39.20	-2.88R

R denotes an obs. with a large st. resid.

Fit	Stdev.Fit	95% C.I.	95% P.I.	⑪
113.32	5.80	(101.34, 125.30)	(77.05, 149.59)	
137.75	5.48	(126.44, 149.07)	(101.70, 173.81)	

name. The estimated coefficients are found in area ②, while their estimated standard deviations are found in area ③. For example, $\hat{\beta}_1 = 7.839$ with $s_{\hat{\beta}_1} = 1.234$, where $\hat{\beta}_1$ represents the estimated average increase in listed selling price for an increase of 1000 square feet of living area **when the other predictor variables are held constant.**

In area ⑥, the estimate of σ is given by $s = 16.58$, where

$$s^2 = \text{MSE} = \frac{\text{SSE}}{n - (k + 1)}$$

The number of degrees of freedom associated with s^2 is equal to $n - (k + 1)$ where $(k + 1)$ is the number of parameters (including β_0) in the model. In this case, s^2 has $29 - (4 + 1) = 24$ degrees of freedom. The values of SSE and MSE are found in the Error line of the analysis of variance table in area ⑨.

Assessing the Utility of the Model

In area ⑨, the analysis of variance table has two sources of variation: one due to *regression* with $k = 4$ degrees of freedom (corresponding to the four fitted parameters, $\hat{\beta}_1$, $\hat{\beta}_2$, $\hat{\beta}_3$, and $\hat{\beta}_4$) and the other due to *error* with $n - (k + 1) = 29 - (4 + 1) = 24$

degrees of freedom. The adequacy of the model using x_1, x_2, x_3, and x_4 is assessed by testing the hypothesis

$$H_0 : \beta_1 = \beta_2 = \beta_3 = \beta_4 = 0$$

against the alternative $H_a : \beta_i \neq 0$ for at least one value of $i = 1, 2, 3,$ or 4, using the statistic

$$F = \frac{\text{MSR}}{\text{MSE}} = \frac{12{,}340}{275} = 44.88$$

which, when compared to the critical value of F with 4 numerator and 24 denominator degrees of freedom given by $F_{.05} = 2.78$, is significant with $\alpha = .05$ and, in fact, has a p-value reported as 0.000.

In assessing the strength of the relationship between y and the predictor variables x_1, x_2, x_3, and x_4, we use the fact that the total sum of squared deviations of y about $\bar{y}$, S_{yy}, is equal to

$$S_{yy} = \text{SSR} + \text{SSE}$$

where SSR, the sum of squares due to regression, is the portion of S_{yy} explained by regression, while SSE, the sum of squares due to error, is the unexplained portion of S_{yy}. Hence, the larger the value of the ratio of SSR/S_{yy}, the stronger the relationship between y and the predictor variables. The **coefficient of determination** is defined as

$$R^2 = \left(\frac{\text{SSR}}{S_{yy}}\right)100\% \quad \text{or} \quad R^2 = \left(\frac{S_{yy} - \text{SSE}}{S_{yy}}\right)100\%$$

For this example,

$$R^2 = \left(\frac{49{,}359}{55{,}958}\right)100\% = 88.2\%$$

and 88.2% of the variation in y is explained by linear regression of y on x_1, x_2, x_3, and x_4. This value is found in area ⑦ of the printout. The value of the F statistic used for testing $H_0 : \beta_1 = \beta_2 = \beta_3 = \beta_4 = 0$ is related to R^2 in the following way:

$$F = \frac{R^2/k}{(1 - R^2)/[n - (k + 1)]}$$

so when R^2 is large, F is large, and vice versa.

The quantity R (the positive square root of R^2) is called the **multiple correlation coefficient** and measures the correlation between y and its predicted value $\hat{y}$. Notice that the value of R^2 can never decrease with the addition of one or more variables into the regression model. Hence, R^2 *can be artificially inflated by the inclusion of more and more predictor variables.*

Estimation and Testing of Individual Parameters

Confidence interval estimates and tests of hypotheses concerning individual partial slopes (β_i) are based on the estimated partial regression coefficients ($\hat{\beta}_i$) and their estimated standard errors. **A $(1 - \alpha)100\%$ confidence interval for β_i is given by**

$$\hat{\beta}_i \pm t_{\alpha/2} s_{\hat{\beta}_i}$$

where $t_{\alpha/2}$ is the value of t based on $n - (k + 1)$ degrees of freedom having an area of $\alpha/2$ to its right. For example, a 95% confidence interval for β_1 is

$$\hat{\beta}_1 \pm t_{.025} s_{\hat{\beta}_1}$$
$$7.839 \pm 2.064(1.234)$$
$$7.839 \pm 2.547$$

or from 5.292 to 10.386.

Tests of significance for individual partial regression coefficients are based on the Student's t statistic given by

$$t = \frac{\hat{\beta}_i - \beta_i}{s_{\hat{\beta}_i}}$$

The procedure is identical to the one used to test an hypothesis about the slope β_1 in a simple linear regression model.[†] In our example, testing $H_0 : \beta_1 = 0$ versus $H_a : \beta_1 \neq 0$ uses the statistic

$$t = \frac{\hat{\beta}_1 - 0}{s_{\hat{\beta}_1}} = \frac{7.839}{1.234} = 6.35$$

which is the t ratio given in area ④. This value exceeds the critical value of t with $\alpha = .05$ and 24 degrees of freedom given by $t_{.025} = 2.064$. The p-value of 0.000 for this test is found in the column labeled p in area ⑤. In rejecting $H_0 : \beta_1 = 0$, we conclude that x_1 contributes significant information above and beyond the information in x_2 and x_3 in explaining the listing price y of residences in Moreno Valley.

An alternative measure of the strength of the relationship between y and x_1, x_2, x_3, and x_4, adjusted for degrees of freedom by using mean squares rather than sums of squares, is defined as

$$R^2(\text{adj}) = \left(1 - \frac{\text{MSE}}{s_y^2} \right) 100\%$$

[†]Some packages use the t statistic just described, while others use the equivalent F statistic ($F = t^2$), since the square of a t statistic with ν degrees of freedom is equal to an F statistic with 1 degree of freedom in the numerator and ν degrees of freedom in the denominator.

where $s_y^2 = S_{yy}/(n-1)$. In this example,

$$R^2(\text{adj}) = \left(1 - \frac{275}{55{,}958/28}\right)100\% = (1 - .138)100\% = 86.2\%$$

which agrees with the value in area ⑧. The value of $R^2(\text{adj})$ is mainly used to compare two or more regression models using different numbers of independent predictor variables.

The decomposition of the sum of squares due to regression appears in area ⑩, in which the additional contribution of each predictor variable *given the variables already entered into the model* is given with the order specified in the REGRESS command. For example, the sum of squares due to regression on x_1 is 44,444; the additional sum of squares due to regression on x_2, *given that x_1 is already in the model*, is 59. The additional sum of squares due to regression on x_3, *given that x_1 and x_2 are already in the model*, is 321, and so on. In this example, it is interesting to note that the predictor variable x_1 alone accounts for $44{,}444/55{,}958 = .794$ or 79.4% of the total variation, compared to 88.2% using all four variables x_1, x_2, x_3, and x_4.

Estimating $E(y)$ and Predicting y

The subcommand PREDICT in the Minitab package, followed by fixed values of x_1, x_2, x_3, and x_4, implements the calculation of the estimated values of $\hat{y}$ (Fit), which is the point estimator for both $E(y)$ and y. This value is shown in area ⑪ together with $s_{\hat{y}}$, its estimated standard error (Stdev. Fit) and both a 95% confidence interval for $E(y)$ and a 95% prediction interval for y. (Recall from Section 11.5 that the prediction interval is always *wider* than the confidence interval.) For our example, using the values from the first PREDICT command, we can verify that when $x_1 = 10$, $x_2 = 1$, $x_3 = 3$, and $x_4 = 2$,

$$\hat{y} = -16.58 + 7.84(10) - 34.39(1) - 7.99(3) + 54.93(2)$$
$$= 113.32$$

and the 95% confidence interval for $E(y)$, based on 24 degrees of freedom, is

$$\hat{y} \pm t_{.025}s_{\hat{y}}$$
$$113.32 \pm 2.064(5.80)$$
$$113.32 \pm 11.97$$

or from 101.35 to 125.29, which to one decimal place agrees with the confidence interval given in area ⑪. ∎

E X A M P L E   **12.2**   One way of assessing productivity in the grocery trade is to use value added per man-hour. Value added is defined as the surplus money generated by the business available to pay for labor, furniture and fixtures, and equipment. For the data in Table 12.4, y is the value added per man-hour and x is the size of the store for each of ten grocery outlets. Choose a model to relate y to x.

T A B L E **12.4** Value added per man-hour versus size of store

Variable	Store									
	1	2	3	4	5	6	7	8	9	10
Value added per man-hour (dollars), y	4.08	3.40	3.51	3.09	2.92	1.94	4.11	3.16	3.75	3.60
Size of store (thousands of square feet), x	21.0	12.0	25.2	10.4	30.9	6.8	19.6	14.5	25.0	19.1

Solution The first step in choosing a model to describe the data is to plot the data points (see Figure 12.1). The relationship suggested by the data is one that depicts productivity rising as the size of a grocery outlet increases, until an optimal size is reached. Above that size, productivity tends to decrease. Since the relationship that we have described suggests curvature, we will fit a quadratic model,

$$E(y) = \beta_0 + \beta_1 x + \beta_2 x^2$$

to the data.

Keep in mind that in choosing a quadratic model to fit to our data, we are not saying that the true relationship between the mean value of y and the value x is defined by an equation of the type

$$E(y) = \beta_0 + \beta_1 x + \beta_2 x^2$$

Rather, we have chosen this equation to *model* the relationship. Presumably, it will provide a better description of the relationship than a simple linear model that graphs as a straight line, and it will enable us to estimate the mean productivity of a grocery retail outlet as a function of the outlet size. ■

F I G U R E **12.1**
Plot of the data points for
Example 12.2

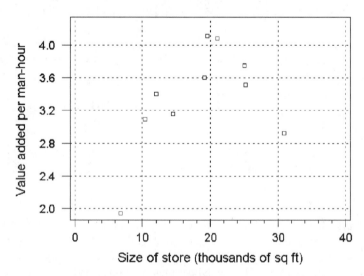

E X A M P L E   **12.3**   Refer to the data on grocery retail outlet productivity and outlet size given in Example 12.2.

      a   Fit a quadratic model to the data using the Minitab computer program package and discuss the adequacy of the fitted model.

      b   Graph the quadratic prediction curve, along with the plotted data points.

Solution   From the printout in Table 12.5, we see that the regression equation is

$$\hat{y} = -.159 + .392x - .00949x^2$$

The graph of this quadratic equation, together with the data points, is shown in Figure 12.2.

T A B L E **12.5**
Minitab regression analysis for
the data in Example 12.2

```
The regression equation is
Y = - 0.159 + 0.392 X - 0.00949 XSQ

Predictor        Coef         Stdev       t-ratio          p
Constant      -0.1594       0.5006         -0.32      0.760
X              0.39193      0.05801          6.76      0.000
XSQ           -0.009495     0.001535        -6.19      0.000

s = 0.2503        R-sq = 87.9%      R-sq(adj) = 84.5%

Analysis of Variance

SOURCE          DF            SS            MS            F          p
Regression       2        3.1989        1.5994        25.53      0.001
Error            7        0.4385        0.0626
Total            9        3.6374

SOURCE          DF        SEQ SS
X                1        0.8003
XSQ              1        2.3986
```

To assess the adequacy of the quadratic model, the test of

$$H_0 : \beta_1 = \beta_2 = 0$$

versus $H_a : \beta_i \neq 0$ for $i = 1$ or 2 is given in the analysis of variance section of the printout as

$$F = \frac{\text{MSR}}{\text{MSE}} = 25.53$$

with a p-value of 0.001. Hence, at least one of the two regression coefficients is highly significant. Quadratic regression accounts for $R^2 = 87.9\%$ of the variation in y [$R^2(\text{adj}) = 84.5\%$].

    In examining the individual analysis of variables section of the printout, we see that both $\hat{\beta}_1$ and $\hat{\beta}_2$ are highly significant with p-values equal to 0.000. It is interesting to note that, from the sequential sum of squares section, the sum of squares for linear regression is .8003 with an additional sum of squares of 2.3986 when the quadratic

FIGURE **12.2**

A graph of the least-squares prediction equation for the quadratic model of Example 12.3

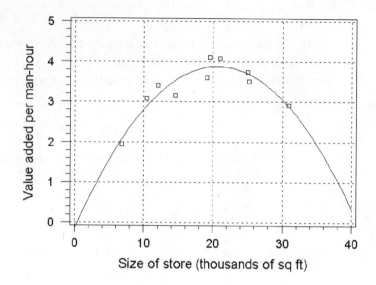

term is added. It is apparent from this analysis that a simple linear regression model would be quite inadequate in describing the data.   ∎

Exercises

Basic Techniques

12.4   Suppose that you were to fit the model

$$E(y) = \beta_0 + \beta_1 x_1 + \beta_2 x_2 + \beta_3 x_3$$

to 15 data points and found R^2 equal to .94.

a   Interpret the value of R^2.

b   Do the data provide sufficient evidence to indicate that the model contributes information for the prediction of y? Test using $\alpha = .05$.

12.5   The computer output for the multiple regression analysis for Exercise 12.4 provides the following information:

$$\hat{\beta}_0 = 1.04 \qquad \hat{\beta}_1 = 1.29 \qquad \hat{\beta}_2 = 2.72 \qquad \hat{\beta}_3 = .41$$
$$s_{\hat{\beta}_1} = .42 \qquad s_{\hat{\beta}_2} = .65 \qquad s_{\hat{\beta}_3} = .17$$

Which, if any, of the independent variables x_1, x_2, and x_3 contribute information for the prediction of y? Test using $\alpha = .05$.

12.6   Refer to Exercise 12.5.

a   Give the least-squares prediction equation.

b On the same sheet of graph paper, graph y versus x_1 when $x_2 = 1$ and $x_3 = 0$ and when $x_2 = 1$ and $x_3 = .5$. What relationship do the two lines have to each other?

c What is the practical interpretation of the parameter β_1?

12.7 Refer to Exercise 12.5. Find a 90% confidence interval for β_1 and interpret it.

12.8 Suppose you were to fit the model

$$E(y) = \beta_0 + \beta_1 x + \beta_2 x^2$$

to 20 data points and found $R^2 = .762$.

a What type of model have you chosen to fit the data?

b How well does the model fit the data? Explain.

c Do the data provide sufficient evidence to indicate that the model contributes information for the prediction of y? Test using $\alpha = .05$.

12.9 The computer output for the multiple regression analysis for Exercise 12.8 provides the following information:

$$\hat{\beta}_0 = 1.21 \qquad \hat{\beta}_1 = 7.60 \qquad \hat{\beta}_2 = -.94$$
$$s_{\hat{\beta}_0} = .62 \qquad s_{\hat{\beta}_1} = 1.97 \qquad s_{\hat{\beta}_2} = .33$$

a Give the prediction equation.

b Graph the prediction equation over the interval $0 \le x \le 6$.

12.10 Refer to Exercise 12.9.

a What is your estimate of the mean value of y when $x = 0$?

b Do the data provide sufficient evidence to indicate that the mean value of y differs from zero when $x = 0$? Test using $\alpha = .10$.

c Find a 90% confidence interval for $E(y)$ when $x = 0$.

12.11 Refer to Exercise 12.9.

a Suppose that the relationship between $E(y)$ and x is a straight line. What would you know about the value of β_2?

b Do the data provide sufficient evidence to indicate curvature in the relationship between y and x?

12.12 Refer to Exercise 12.9. Suppose that y is the profit for some business and x is the amount of capital invested and you know that the rate of increase in profit for a unit increase in capital invested can only decrease as x increases.

a The circumstances that we have described imply a one-tailed statistical test. Why?

b Do the data provide sufficient evidence to indicate a decreasing rate of increase in profit as the amount of capital invested increases? Conduct the test for $\alpha = .05$. State your conclusions.

Applications

12.13 A publisher of college textbooks conducted a study to relate profit per text to cost of sales over a six-year period when its sales force (and sales costs) were growing rapidly. The following inflation-adjusted data (in thousands of dollars) were collected:

Profit per text, y	16.5	22.4	24.9	28.8	31.5	35.8
Sales cost per text, x	5.0	5.6	6.1	6.8	7.4	8.6

Expecting profit per book to rise and then plateau, the publisher fitted the model $E(y) = \beta_0 + \beta_1 x + \beta_2 x^2$ to the data.

DEPENDENT VARIABLE: Y

SOURCE	DF	SUM OF SQUARES	MEAN SQUARE	F VALUE	PR > F	R-SQUARE	C.V.
MODEL	2	234.95514252	117.47757126	332.53	0.0003	0.995509	2.2303
ERROR	3	1.05985748	0.35328583		ROOT MSE		Y MEAN
CORRECTED							
TOTAL	5	236.01500000			0.59437852		26.65000000

SOURCE	DF	TYPE I SS	F VALUE	PR > F	DF	TYPE IV SS	F VALUE	PR > F
X	1	227.81864814	644.86	0.0001	1	15.20325554	43.03	0.0072
X*X	1	7.13649437	20.20	0.0206	1	7.13649437	20.20	0.0206

PARAMETER ESTIMATES

VARIABLE	ESTIMATE	T FOR HO: PARAMETER = 0	PR > |T|	STD ERROR
INTERCEPT	-44.19249551	-5.33	0.0129	8.28688218
X	16.33386317	6.56	0.0072	2.48991042
X*X	-0.81976920	-4.49	0.0206	0.18239471

OBSERVATION	OBSERVED VALUE	PREDICTED VALUE	RESIDUAL	LOWER 95% CL FOR MEAN	UPPER 95% CL FOR MEAN
1	16.50000000	16.98259044	-0.48259044	15.33066719	18.63451369
2	22.40000000	21.56917626	0.83082374	20.56714686	22.57120566
3	24.90000000	24.94045805	-0.04045805	23.93483233	25.94608377
4	28.80000000	28.97164643	-0.17164643	27.81528749	30.12800537
5	31.50000000	31.78753079	-0.28753079	30.65188830	32.92317327
6	35.80000000	35.64859804	0.15140196	33.81460946	37.48258661
7*		27.34236657		26.23203462	28.45269852

* OBSERVATION WAS NOT USED IN THIS ANALYSIS

SUM OF RESIDUALS	-0.00000000
SUM OF SQUARED RESIDUALS	1.05985748
SUM OF SQUARED RESIDUALS-ERROR SS	-0.00000000
PRESS STATISTIC	12.11621807
FIRST ORDER AUTOCORRELATION	-0.39797399
DURBIN-WATSON D	2.55457960

a What sign would you expect the actual value of β_2 to assume? The SAS computer printout is shown. Find the value of β_2 on the printout and check whether the sign agrees with your answer.

b Find SSE and s^2 on the printout.

c How many degrees of freedom do SSE and s^2 possess? Show that

$$s^2 = \frac{\text{SSE}}{\text{Degrees of freedom}}$$

d Do the data provide sufficient evidence to indicate that the model contributes information for the prediction of y? Test using $\alpha = .05$.

e Find the observed significance level for the test in part (d), and interpret its value.

f Do the data provide sufficient evidence to indicate curvature in the relationship between $E(y)$ and x (i.e., evidence to indicate that β_2 differs from zero)? Test using $\alpha = .05$.

g Find the observed significance level for the test part (f), and interpret its value.

h Find the prediction equation, and graph the relationship between $\hat{y}$ and x.

i Find R^2 on the printout and interpret its value.

j Use the prediction equation to estimate the mean profit per text when the sales cost per text is $6500. (Express the sales cost in thousands of dollars before substituting into the prediction equation.) We instructed the SAS program to print this confidence interval. The confidence interval when $x = 6.5$, 26.23203462 to 28.45269852, is shown at the bottom of the printout.

12.14 Refer to the printout given in Exercise 12.13.

a Verify that $S_{yy} = SSR + SSE$.

b Use the values of SSR and S_{yy} to calculate R^2. Compare this value with the value R-SQUARE given in the printout.

c Use the values given in the printout to calculate $R^2(adj)$. When would it be appropriate to use this value to measure the strength between y and x and x^2?

12.15 In Exercise 11.12, the performance of savings and loan institutions with headquarters in Riverside or San Bernardino counties (California) was evaluated. During the three-month period from March 31–June 30, 1990, the data in the following table were recorded.

Savings and Loan	Total Assets (x_1) (in millions)	Net Income (y) (in thousands)	Tangible Capital (x_2) (% of assets)
Redlands Federal	$791.8	$2700	5.92
Hemet Federal	556.3	1261	5.69
Provident Federal	518.0	68	5.47
Palm Springs Federal	137.0	251	3.57
Inland Savings	111.1	748	5.14
Secure Savings	56.4	195	6.87
Mission Savings	35.6	68	5.53

Source: Hamilton, 1990.

The Minitab command REGRESS was used to fit the model

$$y = \beta_0 + \beta_1 x_1 + \beta_2 x_2 + \epsilon$$

to these data. The printout is shown.

```
The regression equation is
Y = - 383 + 2.42 X1 + 69 X2

Predictor      Coef       Stdev      t-ratio        p
Constant       -383        1700       -0.23      0.833
X1            2.418       1.027        2.35      0.078
X2             69.1       311.4        0.22      0.835

s = 748.1      R-sq = 59.7%      R-sq(adj) = 39.5%

Analysis of Variance

SOURCE       DF          SS          MS         F        p
Regression    2     3312151     1656075      2.96    0.163
Error         4     2238508      559627
Total         6     5550659

SOURCE       DF      SEQ SS
X1            1     3284613
X2            1       27538
```

a Find the values of SSE and s^2.

b Find the prediction equation.

c Find R^2 and interpret its value.

d Find the value of R^2(adj). If the value of R^2(adj) for the linear regression performed in Exercise 11.12 was 51.0%, what conclusions might you draw about the value of x_2 in the model?

e Do the data provide sufficient evidence to indicate that the model contributes information for the prediction of y? Test using $\alpha = .10$.

f Note that the observed significance level for the test $H_0 : \beta_2 = 0$ is large. Does this mean that there is little evidence to indicate that x_2 contributes information for the prediction of y?

g What would you conclude regarding the adequacy of the model that we have fit in this exercise?

 12.16 In Exercise 11.58, we used methods of linear regression to predict the average closing bid for each of ten types of used computer equipment based on the average seller's asking price or the average buyer's bid ("Used Computer Prices," 1992). The data are shown in the following table. The Minitab printout shows a multiple regression analysis for the same data. Use the printout to answer the following questions.

Machine	Average Seller's Asking Price	Average Buyer's Bid	Average Closing Bid
20MB PC XT	$ 400	$ 200	$ 300
20MB PC AT	700	400	575
IBM XT 089	450	200	325
IBM AT 339	700	350	600
20MB IBM PS/2 30	950	500	725
20MB IBM PS/2 50	1050	700	875
60MB IBM PS/2 70	2000	1600	1725
20MB Compaq SLT	1200	700	875
Toshiba 1600	1000	700	900
Toshiba 1200HB	1150	800	975

a What multiple regression model has been fit to the data? What assumptions are necessary in order that our inferences be valid?

b Find the prediction equation.

c Find R^2 and interpret its value.

d Do the data provide sufficient evidence to indicate that the model contributes information for the prediction of y? Test using $\alpha = .05$.

e What is the observed significance level for the test in part (d)?

f What would you conclude regarding the adequacy of this model compared to the models used in Exercise 11.58?

Minitab output for Exercise 12.16

```
The regression equation is
y = 81.6 + 0.356 x1 + 0.591 x2

Predictor          Coef        Stdev     t-ratio          p
Constant          81.58        55.12        1.48      0.182
x1               0.3564       0.1767        2.02      0.084
x2               0.5915       0.1986        2.98      0.021

s = 41.46       R-sq = 99.2%    R-sq(adj) = 99.0%

Analysis of Variance

SOURCE         DF          SS          MS          F          p
Regression      2     1465780      732890     426.37      0.000
Error           7       12032        1719
Total           9     1477813

SOURCE         DF      SEQ SS
x1              1     1450531
x2              1       15249
```

 12.17   The following data (from Exercise 11.22) give a "housing opportunity index," the median family income, and the median home price for ten different metropolitan areas that have populations of less than 250,000. Could we predict this housing opportunity index if we knew the median family income or the median home price? Rather than use two separate linear regression analyses as in Exercise 11.22, we can use a multiple regression analysis. Use the Minitab printout to answer the following questions.

Region	Housing Opportunity Index	Median Family Income ($ thousand)	Median Home Price ($ thousand)
Rochester, NY	88.4	45.4	88
Brockton, MA	84.1	45.6	115
Pittsfield, MA	77.5	41.6	98
Fall River, MA-RI	62.1	35.7	120
Kokomo, IN	92.1	41.9	73
Springfield, IL	90.3	42.6	83
Mansfield, OH	89.4	37.3	68
Benton Harbor, MI	80.3	36.7	83
Columbia, MO	74.5	39.1	93
Ocala, FL	87.0	29.2	69

Source: "Housing Opportunity," 1994.

a   What multiple regression model has been fitted to the data? What assumptions are necessary in order that our inferences be valid?

b   Find the prediction equation.

c   Find R^2 and interpret its value.

d   Do the data provide sufficient evidence to indicate that the model contributes information for the prediction of y? Test using $\alpha = .05$.

e   What is the observed significance level for the test in part (d)?

f   What would you conclude regarding the adequacy of this model compared to the models used in Exercise 11.22?

Minitab output for Exercise 12.17

```
The regression equation is
INDEX = 83.9 - 0.471 PRICE + 1.03 INCOME

Predictor         Coef        Stdev      t-ratio          p
Constant         83.93        12.57         6.68      0.000
PRICE          -0.47067      0.08875       -5.30      0.000
INCOME          1.0258        0.3187        3.22      0.015

s = 4.520       R-sq = 81.4%      R-sq(adj) = 76.1%

Analysis of Variance

SOURCE         DF          SS          MS          F          p
Regression      2       625.15      312.57      15.30      0.003
Error           7       143.04       20.43
Total           9       768.18

SOURCE         DF       SEQ SS
PRICE           1       413.49
INCOME          1       211.65
```

12.4   The Use of Quantitative and Qualitative Variables in Linear Regression Models

The variables in a regression analysis can be one of two types: quantitative or qualitative. For example, the age of a person is a **quantitative variable** because its values express the quantity or amount of something—in this case, age. In contrast, the occupation of a person is a **qualitative variable** that varies from person to person; the values of the variable cannot be quantified but can only be classified. For the methods we present, the response variable y must always be (according to the assumptions of Section 12.2) a quantitative variable. In contrast, predictor variables may be either quantitative or qualitative. For example, suppose that we want to predict the annual income of a person. Both the age and the occupation of the person could be important predictor variables, and there are probably many others.

DEFINITION ▪

A **quantitative variable** is one whose values correspond to the quantity or the amount of something. ▪

DEFINITION ▪

A **qualitative variable** is one that assumes values that cannot be quantified but can only be categorized. ▪

If we want to relate the mean measure $E(y)$ of worker absenteeism to two quantitative predictor variables, say

$$x_1 = \text{Worker age}$$

$$x_2 = \text{Worker hourly wage rate}$$

we can construct any of a number of models. One possibility is

$$E(y) = \beta_0 = \beta_1 x_1 + \beta_2 x_2$$

This model graphs as a *plane,* a surface in the three-dimensional space defined by y, x_1, and x_2. Or, if we suspect that the response surface relating $E(y)$ to x_1 and x_2 possesses curvature, we might use the model

$$E(y) = \beta_0 + \beta_1 x_1 + \beta_2 x_2 + \beta_3 x_1 x_2$$

or

$$E(y) = \beta_0 + \beta_1 x_1 + \beta_2 x_2 + \beta_3 x_1 x_2 + \beta_4 x_1^2 + \beta_5 x_2^2$$

In addition to containing the *first-order terms,* those involving only x_1 or x_2, these models include *second-order terms,* such as x_1^2, x_2^2, and the two-variable cross product $x_1 x_2$.[†] The cross product $x_1 x_2$ can be thought of as a third variable, not just x_1 or x_2, and x_1^2 and x_2^2 can be thought of as the fourth and fifth variables in a model. Although only two predictor variables are used, the number of terms and interpretations of the models differ.

Choosing a good model relating y to a set of predictor variables is a difficult and important task, much more difficult than fitting the model to a set of data (this is usually done automatically by a computer software program). Even if your model includes all of the important predictor variables, it still may provide a poor fit to your data if the form of the model is not properly specified.

For example, suppose that $E(y)$ is *perfectly* related to a single predictor variable x_1 by the relation

$$E(y) = \beta_0 + \beta_1 x_1 + \beta_2 x_1^2$$

If you fit the first-order (straight-line) model

$$E(y) = \beta_0 + \beta_1 x_1$$

to the data, you will obtain the *best-fitting, least-squares line,* but it may still provide a poor fit to your data and may be of little value for estimation or prediction [see Figure 12.3(a)]. In contrast, if you fit the model

$$E(y) = \beta_0 + \beta_1 x_1 + \beta_2 x_1^2$$

you obtain a perfect fit to the data [see Figure 12.3(b)].

The lesson is quite clear. Including all of the important predictor variables in the model as first-order terms, that is, x_1, x_2, ..., x_k may not (and probably will not) produce a model that provides a good fit to your data. You may have to include second-order terms, such as x_1^2, x_2^2, x_3^2, $x_1 x_2$, $x_1 x_3$. Plots of y versus x_1, y versus

---

[†]The *order* of a term is determined by the sum of the exponents of variables making up that term. Terms involving x_1 or x_2 are first-order. Terms involving x_1^2, x_2^2, or $x_1 x_2$ are second-order.

FIGURE **12.3**
Two models, each using one
predictor variable

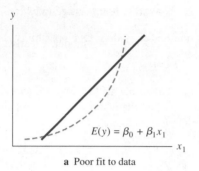

$$E(y) = \beta_0 + \beta_1 x_1$$

a Poor fit to data

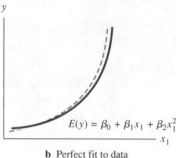

$$E(y) = \beta_0 + \beta_1 x_1 + \beta_2 x_1^2$$

b Perfect fit to data

x_2, and y versus x_3 are helpful tools in determining whether to include quadratic or other higher-order terms in the model.

In contrast to quantitative predictor variables, qualitative predictor variables are entered into a model by using **dummy (indicator) variables.** For example, suppose that you were attempting to relate the mean salary of a group of employees to a set of predictor variables and that one of the variables that you wanted to include was the employee's ethnic background. If each employee included in your study belongs to one of three ethnic groups, say A, B, or C, you will want to enter the qualitative predictor variable "ethnicity" into your model as follows:

$$E(y) = \beta_0 + \beta_1 x_1 + \beta_2 x_2$$

where

$$x_1 = \begin{cases} 1, & \text{if group B} \\ 0, & \text{if not} \end{cases} \qquad x_2 = \begin{cases} 1, & \text{if group C} \\ 0, & \text{if not} \end{cases}$$

If you want to find $E(y)$ for a group A, you examine the coding for the dummy variables x_1 and x_2 and note that $x_1 = 0$ and $x_2 = 0$. Therefore, for group A

$$E(y) = \beta_0 + \beta_1 x_1 + \beta_2 x_2 = \beta_0 + \beta_1(0) + \beta_2(0) = \beta_0$$

The value of $E(y)$ for group B is obtained by letting $x_1 = 1$ and $x_2 = 0$; that is, for group B

$$E(y) = \beta_0 + \beta_1 x_1 + \beta_2 x_2 = \beta_0 + \beta_1(1) + \beta_2(0) = \beta_0 + \beta_1$$

Similarly, the value of $E(y)$ for group C is obtained by letting $x_1 = 0$ and $x_2 = 1$. Therefore, for group C

$$E(y) = \beta_0 + \beta_1 x_1 + \beta_2 x_2 = \beta_0 + \beta_1(0) + \beta_2(1) = \beta_0 + \beta_2$$

The coefficient β_0 is the mean for group A, β_1 is the mean difference between groups B and A, and β_2 is the mean difference between groups C and A.

Models with qualitative predictor variables that may assume, say, k values can be constructed in a similar manner by using $(k - 1)$ terms involving dummy variables. These terms can be added to models containing other predictor variables, quantitative or qualitative, and you can also include terms involving the cross products (interaction terms) of the dummy variables with other variables that appear in the model.

This section introduced two important aspects of model formulation: the concept of predictor variable interaction (and how to cope with it) and the method for introducing qualitative predictor variables into a model. Clearly, this is not enough to make you proficient in model formulation, but it will help you to understand why and how some terms are included in regression models, and it may help you to avoid some pitfalls encountered in model construction.

EXAMPLE 12.4   A study was conducted to examine the relationship between university salary y, the number of years of experience of the faculty member, and the sex of the faculty member. If we expect a straight-line relationship between mean salary and years of experience for both males and females, write the model relating mean salary to the two predictor variables:

1   years of experience (quantitative)
2   sex of professor (qualitative)

Solution   Since we may suspect the mean salary lines for females and males to be different, we want to construct a model for mean salary $E(y)$ that may appear as shown in Figure 12.4.

FIGURE 12.4
Hypothetical relationship between mean salary $E(y)$, years of experience (x_1), and sex (x_2) for Example 12.4

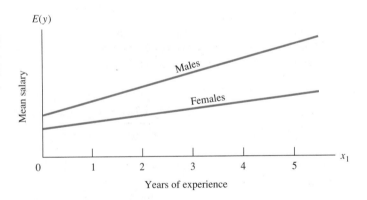

A straight-line relationship between $E(y)$ and years of experience x_1 implies the model

$$E(y) = \beta_0 + \beta_1 x_1 \qquad \text{(graphs as a straight line)}$$

The qualitative variable sex can assume only two "values," male and female. Therefore, we can enter the predictor variable sex into the model by using one **dummy** (or **indicator**) **variable,** x_2, as

$$E(y) = \beta_0 + \beta_1 x_1 + \beta_2 x_2 \qquad \text{(graphs as two parallel lines)}$$
$$x_2 = \begin{cases} 1, & \text{if male} \\ 0, & \text{if female} \end{cases}$$

The fact that we want to allow the slopes of the two lines to differ means that we think that the two predictor variables **interact**—that is, the change in $E(y)$ corresponding to a change in x_1 depends on whether the professor is male or female.

To allow for this interaction (difference in slopes), we introduce the interaction term $x_1 x_2$ into the model. The complete model that characterizes the graph in Figure 12.4 is

$$E(y) = \beta_0 + \beta_1 x_1 + \beta_2 x_2 + \beta_3 x_1 x_2$$

with labels: "dummy variable for sex" pointing to $\beta_2 x_2$; "years of experience" pointing to $\beta_1 x_1$; "interaction" pointing to $\beta_3 x_1 x_2$.

where

$$x_1 = \text{Years of experience}$$

$$x_2 = \begin{cases} 1, & \text{if male} \\ 0, & \text{if female} \end{cases}$$

The interpretation of the model parameters can be seen by assigning values to the dummy variable x_2. Thus, when you want to acquire the line for females, the dummy variable $x_2 = 0$ (according to our coding), and

$$E(y) = \beta_0 + \beta_1 x_1 + \beta_2(0) + \beta_3 x_1(0) = \beta_0 + \beta_1 x_1$$

Therefore, β_0 is the y intercept for the females' line, and β_1 is the slope of the line relating expected salary to years of experience for *females only.*

Similarly, the line for males is obtained by letting $x_2 = 1$. Then

$$\begin{aligned} E(y) &= \beta_0 + \beta_1 x_1 + \beta_2(1) + \beta_3 x_1(1) \\ &= \underbrace{(\beta_0 + \beta_2)}_{y \text{ intercept}} + \underbrace{(\beta_1 + \beta_3)}_{\text{slope}} x_1 \end{aligned}$$

The y intercept of the males' line is $(\beta_0 + \beta_2)$, and the slope, the coefficient of x_1, is equal to $(\beta_1 + \beta_3)$.

Because the slope of the males' line is $(\beta_1 + \beta_3)$ and the slope of the females' line is β_1, it follows that $(\beta_1 + \beta_3) - \beta_1 = \beta_3$ is the difference in the slopes of the two lines. Similarly, β_2 is equal to the difference in the y intercepts for the two lines. ∎

E X A M P L E   **12.5**   Random samples of six female and six male assistant professors were selected from among the assistant professors in a college of arts and sciences. The data on salary and years of experience are shown in Table 12.6. Note that both samples contained two professors with three years of experience, but no male professor had two years of experience.

a   Explain the output of the SAS GLM multiple regression computer printout given in Table 12.7.

b   Graph the predicted salary lines.

TABLE **12.6**	Years of experience, x_1	1	2	3	4	5
Salary versus sex and years of experience	Salary, y (males)	20,710		23,160 23,210	24,140	25,760 25,590
	Salary, y (females)	19,510	20,440	21,340 21,760	22,750	23,200

TABLE **12.7**   The SAS GLM multiple regression analysis computer printout for Example 12.5

DEPENDENT VARIABLE: Y

SOURCE	DF	SUM OF SQUARES	MEAN SQUARE	F VALUE	PR > F	R-SQUARE	C.V.
MODEL	3	42108777.02898556	14036259.00966185	346.24	0.0001	0.992357	0.8897
ERROR	8	324314.63768142	40539.32971018			ROOT MSE	Y MEAN
CORRECTED TOTAL	11	42433091.66666698				201.34380971	22630.83333333

② ① ⑦ ⑧

SOURCE	DF	TYPE I SS	F VALUE	PR > F	DF	TYPE IV SS	F VALUE	PR > F
X1	1	33294036.23595509	821.28	0.0001	1	9389610.00000008	231.62	0.0001
X2	1	8452796.51598297	208.51	0.0001	1	326808.74399183	8.06	0.0218
X1*X2	1	36944.27704750	8.93	0.0174	1	361944.27704750	8.93	0.0174

⑨

PARAMETER ESTIMATES

VARIABLE	ESTIMATE	T FOR HO: PARAMETER = 0	PR > |T|	STD ERROR
INTERCEPT	18593.00000000	89.41	0.0001	207.94699250
X1	969.00000000	15.22	0.0001	63.67050315
X2	866.71014493	2.84	0.0218	305.25678646
X1*X2	260.13043478	2.99	0.0174	87.05798112

③ ④ ⑤ ⑥

Solution   **a**   The SAS GLM multiple regression computer printout for the data in Table 12.6 is shown in Table 12.7. The explanation of this printout is similar to the explanation of the Minitab computer printout given in Example 12.1. The relevant portions of the printout are boxed and numbered.

The value of R^2, the **multiple coefficient of determination,** provides a measure of how well the model fits the data. As you can see in area ①, 99.2% of the sum of squares of deviations of the y values about $\bar{y}$ is explained by all of the terms in the model.

If the model contributes information for the prediction of y, at least one of the model parameters, β_1, β_2, or β_3, will differ from zero. Consequently, we want to test the null hypothesis $H_0 : \beta_1 = \beta_2 = \beta_3 = 0$ against the alternative hypothesis H_a : at least one of the parameters, β_1, β_2, or β_3, differs from zero. The test statistic for this test, an F statistic with $v_1 =$ (the number of parameters in H_0) = 3 and $v_2 = n -$ (number of parameters in the model) = 12 − 4 = 8, is shown in area ② to be 346.24. Since this value exceeds the tabulated value for

F—with $v_1 = 3$, $v_2 = 8$, and $\alpha = .05$, $F = 4.07$ (from Table 6 of Appendix II)—we reject H_0 and conclude that at least one of the parameters β_1, β_2, β_3 differs from zero. The observed significance level (p-value) for the test, PR > F, shown to the right of the value of the F statistic, is equal to .0001.

The estimates of the four model parameters are shown under the heading ESTIMATE in area ③. Rounding these estimates, we obtain the prediction equation

$$\hat{y} = 18593.0 + 969.0x_1 + 866.7x_2 + 260.1x_1x_2$$

The computed value of the t statistic for each of the model parameters is shown in area ④ under the column headed T FOR HO: PARAMETER = 0, and the observed significance levels (p-values) are shown in area ⑤ under the column headed PR > |T|. These values are calculated for two-tailed tests. The p-value for a one-tailed test is half of the value shown in the printout. Looking at the observed significance levels, we see that all of the regression coefficients are significantly different from zero, with p-values less than or equal to .03.

The estimated standard deviations of the estimators, used in constructing confidence intervals for the regression coefficients, are shown in area ⑥ under the heading STD ERROR. The values of SSE = 324214.638 and $s^2 =$ MSE = 40539.330 are shown in area ⑦, and the standard deviation $s = \sqrt{\text{MSE}} = 201.344$ is shown in area ⑧, labeled ROOT MSE. Area ⑨, labeled TYPE I SS, shows the sequential sum of squares given in the Minitab printout and discussed in Example 12.1.

b A graph of the two salary lines is shown in Figure 12.5. Note that the salary line corresponding to the male faculty members appears to be rising at a more rapid rate than the salary line for females. Is this difference real, or could it be due to chance? The following example will answer this question. ∎

E X A M P L E 12.6 Refer to Example 12.5. Do the data provide sufficient evidence to indicate that the annual rate of increase in male junior faculty salaries exceeds the annual rate of increase for women junior faculty salaries? That is, we want to know whether the data provide sufficient evidence to indicate that the slope of the men's faculty salary line exceeds the slope of the women's faculty salary line.

Solution Since β_3 measures the difference in slopes, the slopes of the two lines will be identical if $\beta_3 = 0$. Therefore, we want to test

$$H_0 : \beta_3 = 0$$

that is, the slopes of the two lines are identical, against the alternative hypothesis

$$H_a : \beta_3 > 0$$

that is, the slope of the male faculty salary line is greater than the slope of the female faculty salary line.

The calculated value of t corresponding to β_3 shown in area ④ of the computer printout (Table 12.7), is 2.99. Since we want to detect values $\beta_3 > 0$, we will conduct a

FIGURE **12.5**
A graph of the faculty salary
prediction lines for
Example 12.5

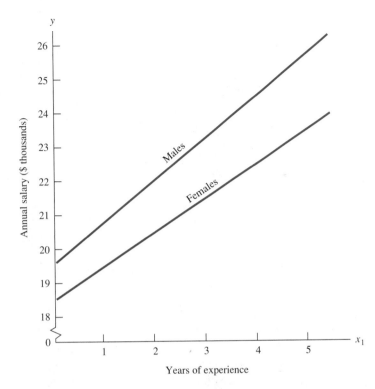

FIGURE **12.5** A graph of the faculty salary prediction lines for Example 12.5

one-tailed test and reject H_0 if $t > t_\alpha$. The tabulated t value from Table 4 of Appendix II, for $\alpha = .05$ and $\nu = 8$ degrees of freedom, is $t_{.05} = 1.860$. The calculated value of t exceeds this value, and thus there is evidence to indicate that the annual rate of increase in men's faculty salaries exceeds the corresponding annual rate of increase in faculty salaries for women.[†]   ∎

Most popular statistical program packages contain multiple regression programs. Some programs can be used only on specific computers; others are more versatile. If you plan to conduct a multiple regression analysis, you will want to determine the statistical program packages available at your computer center so that you can become familiar with their output.

The computer printouts for the Minitab, SAS, Execustat, and other multiple regression computer program packages are very similar. Once you can read one, it is likely that you will be able to read the output of the computer program available at your computer center. If you have difficulty understanding the output, consult the package instruction manual.

[†] If we want to determine whether the data provide sufficient evidence to indicate that the male faculty members start at higher salaries, we test $H_0 : \beta_2 = 0$ against the alternative hypothesis $H_a : \beta_2 > 0$.

Exercises

Basic Techniques

12.18   Suppose that you wish to predict production yield y as a function of several independent predictor variables. Indicate whether each of the following independent variables is qualitative or quantitative.

a   The prevailing interest rate in the area

b   The price per pound of one item used in the production process

c   The plant (A or B) at which the production yield is measured

d   The length of time that the production machine has been in operation

e   The shift (night or day) in which the yield is measured

12.19   A multiple linear regression model, involving one qualitative and one quantitative independent variable, produced the following prediction equation:

$$\hat{y} = 12.6 + .54x_1 - 1.2x_1x_2 + 3.9x_2^2$$

a   Which of the two variables is the quantitative variable? Explain.

b   If x_1 can take only the value 0 or 1, find the two possible prediction equations for this experiment.

c   Graph the two equations found in part (b). Compare the shapes of the two curves.

Applications

12.20   In Exercise 11.30, we presented data concerning the median annual income of full-time workers age 25 years or older by the number of years of schooling completed. The data are given in the following table.

	Salary (dollars)	
Years of schooling	Men	Women
8	18,000	11,500
10	20,500	13,000
12	25,000	16,100
14	28,100	18,300
16	34,500	22,100
19	39,700	27,600

Source: Data adapted from "Upward Mobility," 1989.

The Minitab command REGRESS was used to fit the model

$$y = \beta_0 + \beta_1 x_1 + \beta_2 x_2 + \epsilon$$

to these data, where

$$y = \text{Salary}$$
$$x_1 = \text{Years of schooling}$$
$$x_2 = \begin{cases} 0, & \text{if female} \\ 1, & \text{if male} \end{cases}$$

The Minitab printout is shown in the accompanying table.

```
MTB > REGRESS C1 2 C2 C3;
SUBC>    PREDICT 18 0;
SUBC>    PREDICT 18 1.

The regression equation is
Y = - 5125 + 1764 X1 + 9533 X2

Predictor        Coef       Stdev    t-ratio        p
Constant        -5125        1678      -3.05    0.014
X1             1763.9       118.5      14.88    0.000
X2             9533.3       870.1      10.96    0.000

s = 1507      R-sq = 97.4%    R-sq(adj) = 96.9%

Analysis of Variance

SOURCE        DF          SS           MS        F        p
Regression     2   775663808    387831904   170.74    0.000
Error          9    20442852      2271428
Total         11   796106688

SOURCE        DF      SEQ SS
X1             1   503010496
X2             1   272653344

Unusual Observations
Obs.        X1          Y     Fit   Stdev.Fit   Residual   St.Resid
 7         8.0      11500    8986         868       2514      2.04R

R denotes an obs. with a large st. resid.

   Fit   Stdev.Fit        95% C.I.            95% P.I.
 26626        841    ( 24723,  28528)    ( 22721,  30531)

 36159        841    ( 34257,  38061)    ( 32254,  40064)
```

a Find the prediction equation for these data.

b Test the adequacy of the complete model using the F statistic. What is the observed level of significance of the test?

c Which of the independent variables are significant? Test using $\alpha = .05$.

d Use the prediction equation in part (a) to predict the salary of a female who has 18 years of schooling. Compare with the prediction given in the printout.

e Use the prediction equation in part (a) to predict the salary of a male who has 18 years of schooling. Compare with the prediction given in the printout.

12.21 Americans are very vocal about their attempts to improve personal well-being by "eating right and exercising more." One desirable dietary change is to reduce our intake of red meat and to substitute poultry or fish in its place. Stang and Galifianakis (1994) produced a plot of beef and chicken consumption in annual pounds per person that shows that the consumption of beef is declining and the consumption of chicken is increasing, beginning with 1975 through 1995 and 2000. A summary of their data in annual pounds per person follows.

Year	Beef	Chicken
1970	85	37
1975	89	36
1980	76	47
1985	76	47
1990	68	62
1995[†]	67	74
2000[†]	60	79

[†]Projections.

Consider fitting the following model, which allows for simultaneously fitting two simple regression lines:

$$y = \beta_0 + \beta_1 x_1 + \beta_2 x_2 + \beta_3 x_1 x_2 + \epsilon$$

where

$$x_1 = \begin{cases} 1, & \text{if beef} \\ 0, & \text{if chicken} \end{cases}$$
$$x_2 = \text{Year} - 1969$$

A Minitab printout using this model appears below.

```
MTB > REGRESS 'Y' 3 'X1' 'X2' 'X1X2';
SUBC>    PREDICT 1 36 36.

The regression equation is
Y = 29.8 + 59.2 X1 + 1.55 X2 - 2.46 X1X2

Predictor       Coef       Stdev      t-ratio         p
Constant       29.771      2.972       10.02      0.000
X1             59.171      4.203       14.08      0.000
X2             1.5500      0.1575       9.84      0.000
X1X2          -2.4571      0.2227     -11.03      0.000

s = 4.167       R-sq = 95.4%      R-sq(adj) = 94.1%

Analysis of Variance

SOURCE         DF          SS           MS          F         p
Regression      3        3637.9       1212.6      69.83     0.000
Error          10         173.6         17.4
Total          13        3811.5

SOURCE         DF        SEQ SS
X1              1        1380.1
X2              1         144.6
X1X2            1        2113.1


     Fit   Stdev.Fit         95% C.I.               95% P.I.
   56.29        3.52    (   48.44,     64.13)  (   44.13,     68.45)
```

a   Describe the two predictor variables as qualitative or quantitative.

b   Interpret the value of R^2 and comment on its relevance to the regression analysis.

c   Do the data provide sufficient evidence to indicate that at least one variable in the model contributes information for predicting meat consumption?

d   Use the printout to find a 95% confidence interval estimate for the average beef consumption in 2005. What is the 95% prediction interval for the consumption of beef per person in 2005? (Notice that the values entered with the PREDICT subcommand are $x_1 = 1$, $x_2 = 8$, and $x_1 x_2 = 8$.)

12.22   One question facing many states today is whether their tax systems are discouraging businesses from locating and expanding within their borders. In a study of Massachusetts' tax system, Robert Tannenwald (1994) used multiple linear regression in summarizing business's share of state and local taxes (BS) as a function of capital/labor ratio (K/L), the value of farmland per capita (AGR), severance tax capacity per capita (SEV), and personal income per capita (PY). His analysis is summarized here:

---

1 The regression equation demonstrating the explanatory power of these variables is

$$BS = .1511 + .0036(K/L) - .0006AGR + .0004SEV + .0000022PY$$
$$\quad\quad (2.95^*) \quad\quad (3.67^{**}) \quad\quad (-2.13^*) \quad\quad (22.07^*) \quad\quad (1.66)$$

2 Degrees of freedom: 44 $R^2 = .68$

3 *Note:* Numbers in parentheses are t-statistics.
   ** Significant at the 1% level, two-tailed test.
   * Significant at the 5% level, two-tailed test.

Comment on the results of this regression analysis, focusing on the following points.

a Is there significant regression of the variable BS on the remaining variables?

b How would you characterize the strength of the explanatory value of the predictor variables?

c Can or should any variable be dropped from the model?

12.5 Testing Sets of Model Parameters

In the preceding sections, we found it useful to test a complete linear model to determine whether the model contributes information for the prediction of y. We also were able to test an hypothesis about an individual β parameter, using a Student's t test. In addition to these two important tests, we may want to test hypotheses about sets of parameters.

For example, suppose a company suspects that the demand y for some product could be related to as many as five independent variables x_1, x_2, x_3, x_4, and x_5. However, the cost of obtaining measurements on the variables x_3, x_4, and x_5 is very high. If, in a small pilot study, the company could show that these three variables contribute little or no information for the prediction of y, they could be eliminated from the study at great savings to the company.

If all five variables x_1, x_2, x_3, x_4, and x_5 are used to predict y, the multiple linear regression model would be written as

$$y = \beta_0 + \beta_1 x_1 + \beta_2 x_2 + \beta_3 x_3 + \beta_4 x_4 + \beta_5 x_5 + \epsilon$$

However, if x_3, x_4, and x_5 contribute no information for the prediction of y, then they would not appear in the model—that is, $\beta_3 = \beta_4 = \beta_5 = 0$—and the reduced model would be

$$y = \beta_0 + \beta_1 x_1 + \beta_2 x_2 + \epsilon$$

Hence, we want to test

$$H_0 : \beta_3 = \beta_4 = \beta_5 = 0$$

that is, the independent variables x_3, x_4, and x_5 contribute no information for the prediction of y, against the alternative hypothesis

$$H_a : \text{at least one of the parameters } \beta_3, \ \beta_4, \ \text{or } \beta_5 \text{ differs from zero}$$

that is, at least one of the variables x_3, x_4, or x_5 contributes information for the prediction of y. Thus, in deciding whether the complete model is preferable to the reduced model in predicting demand, we are led to a test of an hypothesis about a set of three parameters, β_3, β_4, and β_5.

To explain how to test an hypothesis concerning a set of model parameters, we will define two models:

Model 1 (reduced model)

$$E(y) = \beta_0 + \beta_1 x_1 + \beta_2 x_2 + \cdots + \beta_r x_r$$

Model 2 (complete model)

$$E(y) = \underbrace{\beta_0 + \beta_1 x_1 + \beta_2 x_2 + \cdots + \beta_r x_r}_{\text{terms in model 1}} + \underbrace{\beta_{r+1} x_{r+1} + \beta_{r+2} x_{r+2} + \cdots + \beta_k x_k}_{\text{additional terms in model 2}}$$

Suppose that we were to fit both models to the data set and calculate the sum of squares for error for both regression analyses. If model 2 contributes more information for the prediction of y than model 1, then the errors of prediction for model 2 should be smaller than the corresponding errors for model 1, and SSE_2 should be smaller than SSE_1. In fact, the greater the difference between SSE_1 and SSE_2, the greater is the evidence to indicate that model 2 contributes more information for the prediction of y than model 1.

The test of the null hypothesis

$$H_0 : \beta_{r+1} = \beta_{r+2} = \cdots = \beta_k = 0$$

against the alternative hypothesis

$$H_a : \text{at least one of the parameters } \beta_{r+1}, \beta_{r+2}, \cdots, \beta_k \text{ differs from zero}$$

uses the test statistic

$$F = \frac{(\text{SSE}_1 - \text{SSE}_2)/(k - r)}{\text{MSE}_2}$$

where F is based on $v_1 = k - r$ and $v_2 = n - (k + 1)$ degrees of freedom. Note that the $(k - r)$ parameters involved in H_0 are those added to model 1 to obtain model 2. The numerator number v_1 of degrees of freedom will always equal $(k - r)$, the number of parameters involved in H_0. The denominator number v_2 of degrees of freedom is the number of degrees of freedom associated with the sum of squares for error, SSE_2, for the complete model.

The rejection region for the test is identical to the rejection region for all of the analysis of variance F tests, namely,

$$F > F_\alpha$$

E X A M P L E   **12.7**   Refer to the real estate data of Example 12.1, in which the listed selling price
(y) is related to the square feet of living area (x_1), the number of floors (x_2), the
number of bedrooms (x_3), and the number of baths (x_4). The realtor suspects that the
square footage of living area is the most important predictor variable, and that the
other variables might be eliminated from the model without loss of much prediction
information. Test this claim with $\alpha = .05$.

Solution   The hypothesis to be tested is

$$H_0 : \beta_2 = \beta_3 = \beta_4 = 0$$

versus the alternative that at least one of β_2, β_3, or β_4 is different from zero. The
complete model (2), given as

$$y = \beta_0 + \beta_1 x_1 + \beta_2 x_2 + \beta_3 x_3 + \beta_4 x_4 + \epsilon$$

was fitted in Example 12.1, and the Minitab printout from Table 12.3 is reproduced
in Table 12.8 along with the Minitab printout for the simple linear regression analysis
of the **reduced model (1),** given as

$$y = \beta_0 + \beta_1 x_1 + \epsilon$$

Then $SSE_1 = 11514$ from Table 12.8(b), and $SSE_2 = 6599$ and $MSE_2 = 275$ from
Table 12.8(a). The test statistic is

$$F = \frac{(SSE_1 - SSE_2)/(k - r)}{MSE_2}$$

$$= \frac{(11514 - 6599)/(4 - 1)}{275} = 5.96$$

The critical value of F with $\alpha = .05$, $\nu_1 = 3$, and $\nu_2 = n - (k + 1) = 29 - (4 + 1) = 24$ degrees of freedom is $F_{.05} = 3.01$. Hence, H_0 is rejected. There is evidence
to indicate that at least one of the three variables—number of floors, bedrooms,
or baths—is contributing significant information for the prediction of listed selling
price.   ■

12.6   Residual Analysis

The deviations between the observed values of y and their predicted values are called
residuals. For example, the first three columns of Table 12.9 reproduce the data of
Table 11.2, and the ten predicted values of y, obtained from the prediction equation
(Example 11.1)

$$\hat{y} = 40.78 + .77x$$

a Complete model

```
The regression equation is
LPRICE = - 16.6 + 7.84 SQFT - 34.4 NUMFLRS - 7.99 BDRMS + 54.9 BATHS

Predictor        Coef        Stdev      t-ratio        p
Constant        -16.58       18.88       -0.88       0.389
SQFT              7.839       1.234        6.35       0.000
NUMFLRS         -34.39       11.15        -3.09       0.005
BDRMS            -7.990       8.249       -0.97       0.342
BATHS            54.93       13.52         4.06       0.000

s = 16.58        R-sq = 88.2%      R-sq(adj) = 86.2%

Analysis of Variance

SOURCE         DF          SS           MS          F         p
Regression      4         49359        12340       44.88     0.000
Error          24          6599          275
Total          28         55958

SOURCE         DF        SEQ SS
SQFT            1         44444
NUMFLRS         1            59
BDRMS           1           321
BATHS           1          4536
```

b Reduced model

```
The regression equation is
LPRICE = 3.0 + 9.61 SQFT

Predictor        Coef        Stdev      t-ratio        p
Constant         3.00       14.26        0.21       0.835
SQFT             9.6121      0.9416      10.21       0.000

s = 20.65        R-sq = 79.4%      R-sq(adj) = 78.7%

Analysis of Variance

SOURCE         DF          SS           MS          F         p
Regression      1         44444        44444      104.22     0.000
Error          27         11514          426
Total          28         55958
```

are shown in column 4. The predicted value of y for $x_1 = 39$ is $\hat{y}_1 = 40.78 + .77x = 40.78 + (.77)(39) = 70.81$, and the residual is $e_1 = y_1 - \hat{y}_1 = 65 - 70.81 = -5.81$. The ten residuals, one for each x value, appear in column 5 of Table 12.9 and are displayed graphically as the vertical line segments in Figure 12.6.

D E F I N I T I O N ▪

> The **residual** corresponding to the data point $(x_i, \ y_i)$ is
>
> $$e_i = y_i - \hat{y}_i \quad ▪$$

In **residual analysis,** plots of the residuals against $\hat{y}$ or against the individual independent variables often indicate departures from the assumptions required for an analysis of variance given in Section 12.2, and they also may suggest changes in the underlying model. Plots of the residuals against $\hat{y}$ are particularly useful for detecting nonuniformity in the variance of y for different values of $E(y)$ (assumption 4).

TABLE **12.9**
Management data of Table 11.2
with predicted values and
residuals

Manager	Management Trainee Score, x	Management Success Rating, y	$\hat{y} = 40.78 + .77x$	$e = y - \hat{y}$
1	39	65	70.81	−5.81
2	43	78	73.89	4.11
3	21	52	56.95	−4.95
4	64	82	90.06	−8.06
5	57	92	84.67	7.33
6	47	89	76.97	12.03
7	28	73	62.34	10.66
8	75	98	98.53	−.53
9	34	56	66.96	−10.96
10	52	75	80.82	−5.82

FIGURE **12.6**
Residuals for the least-squares
line $\hat{y} = 40.78 + .77x$

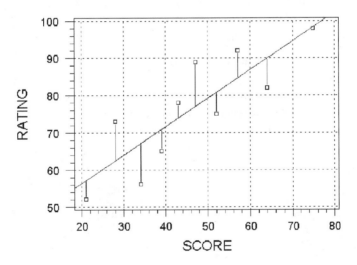

For example, the variance of some types of data—Poisson data, in particular—increases with the mean. A plot of the residuals for this type of data might appear as shown in Figure 12.7(a). Note that the range of the residuals increases as $\hat{y}$ increases, thus indicating that the variance of y is increasing as the mean value of y, $E(y)$, increases.

The variances for percentages and proportions calculated from binomial data increase for values of $p = 0$ to $p = .5$ and then decrease from $p = .5$ to $p = 1.0$. Plots of residuals versus $\hat{y}$ for this type of data would appear as shown in Figure 12.7(b). Therefore, if the data were small percentages, the plot of residuals would show the range of the residuals *increasing* as $\hat{y}$ increases. If the data were large percentages, the range of the residuals would appear to *decrease* as $\hat{y}$ increases.

If the range of the residuals increases as $\hat{y}$ increases and you know that the data are measurements on Poisson variables, you can stabilize the variance of the response

FIGURE **12.7** Plots of residuals against $\hat{y}$

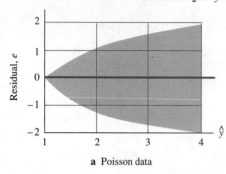

a Poisson data

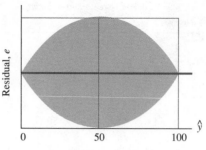

b Binomial percentages

by running the regression analysis on $y^* = \sqrt{x}$. Or, if percentages are calculated from binomial data, you can use the arc sin transformation, $y^* = \sin^{-1} \sqrt{x}$. Both of these transformations were discussed in Section 9.13.[†] If you have no prior reason to explain why the range of the residuals increases as $\hat{y}$ increases, you can still use a transformation on y that affects larger values of y more than smaller values, say $y^* = \sqrt{y}$ or $y^* = \ln y$. These transformations have a tendency both to stabilize the variance of y^* and to make the distribution of y^* more nearly normal when the distribution of y is highly skewed.

Plots of the residuals against the individual independent variables often indicate the problems with model selection. In theory, the residuals should vary in size and sign (positive or negative) in a random manner about $\hat{y}$ if the equation that you have selected for $E(y)$ is a good approximation to the true relationship between $E(y)$ and the independent variables. For example, if $E(y)$ and a single independent variable x are linearly related, that is,

$$E(y) = \beta_0 + \beta_1 x$$

and you fit a straight line to the data, then the observed y values should vary in a random manner about $\hat{y}$, and a plot of the residuals against x will appear as shown in Figure 12.8.

FIGURE **12.8**
Residual plot when the model provides a good approximation to reality

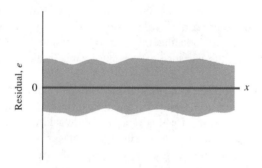

[†]In Chapter 10 and earlier chapters, we represented the response variable by the symbol x. In the chapters on regression analysis, Chapters 11 and 12, the response variable is represented by the symbol y.

In contrast, suppose that you fit the straight-line model

$$E(y) = \beta_0 + \beta_1 x$$

to the points plotted in Figure 12.9(a). You can see from the plotted points in Figure 12.9(a) that the relationship between y and x appears to be curvilinear. As a consequence, the residuals for small values of x tend to be negative; then they change to positive and back to negative, as shown in Figure 12.9(b). This particular type of nonrandom behavior in the residuals suggests what is apparent from Figures 12.9(a) and 12.9(b). A straight line does not provide a good approximation to the relationship between y and x.

FIGURE **12.9**
Data and residual plots for a
model that does not agree with
reality

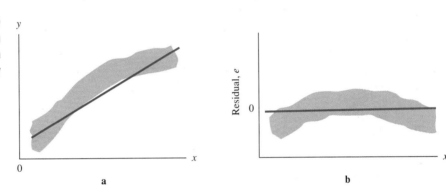

The residuals from the regression of assistant professors' salaries (y) on years of experience (x_1) and sex (x_2) described in Example 12.5 using the model

$$E(y) = \beta_0 + \beta_1 x_1 + \beta_2 x_2 + \beta_3 x_1 x_2$$

are plotted against the values of $\hat{y}$ in Figure 12.10. (Recall that this model plots as two straight lines, one for male salaries and one for female salaries.) Notice that there does not appear to be a pattern in the residuals from this model. On the other hand, when a common regression line is fitted for both male and female assistant professors, the model

$$E(y) = \beta_0 + \beta_1 x_1$$

underestimates salaries for males and overestimates salaries for females, and the plot of the residuals versus $\hat{y}$ given in Figure 12.11 appears to reveal one distinct set of positive residuals corresponding to the salaries of the males and a second distinct set of negative residuals corresponding to the salaries of the females. This nonrandom grouping indicates the need for the variable x_2, which was not included in the model.

In Example 12.3, we fitted a quadratic regression model relating the value added per man-hour (y) as a function of the size of the store (x). Suppose these data were fitted using the simple linear regression model

$$E(y) = \beta_0 + \beta_1 x$$

The residuals plotted against the values of $\hat{y}$ in Figure 12.12 display a distinct quadratic trend, indicating the omission of a quadratic term in the model. However,

FIGURE **12.10**
Residual plot based on two
regression lines, using the data
of Example 12.5

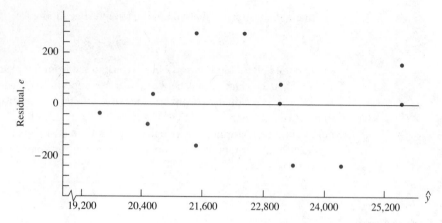

FIGURE **12.11**
Residual plot based on a
common regression line, using
the data of Example 12.5

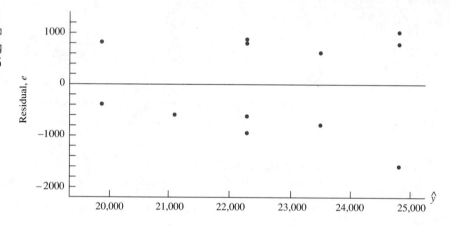

FIGURE **12.12**
Residual plot from a simple
linear regression model, using
the data of Example 12.3

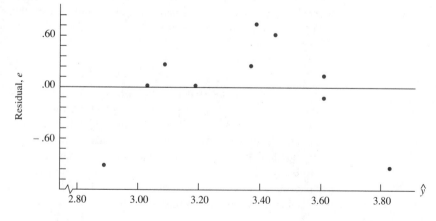

not all residual plots give as clear an indication of which terms or variables are missing as we observed in these two examples.

Most computer multiple regression analysis printouts give the predicted values of y and the residuals for each value of x when requested. In addition, you can request plots of the residuals versus $\hat{y}$ or versus the individual independent variables. **Examine the plots and look for nonrandomness in the behavior of the residuals. If nonrandomness is observed, seek an explanation for the behavior and attempt to correct it.**

12.7 Stepwise Regression Analysis

A company that hires college graduates as sales personnel wants to construct a model to predict a hired applicant's success (measured by the sales that the applicant will subsequently make during the third year of employment) from information acquired during the applicant's job interview. This information includes the applicant's high school and college grades, achievement test scores, ratings of personal interviews and letters of recommendation, outside activities, interests, achievements, and so on. **How does the company decide which among this long list of independent variables should be included in the model? One procedure for answering this question is to use a computer program package that performs a stepwise regression analysis.**

Suppose that we have data available on y and a number of possible independent variables x_1, x_2, $\ldots$, x_k. A **stepwise regression analysis** commences by fitting the model

$$E(y) = \beta_0 + \beta_1 x$$

for each of the k independent variables. The best-fitting of these k prediction equations is chosen (for most programs) as the one with the largest t values for β_1. We will denote this independent variable as x_1.

The second step in a stepwise regression analysis is to fit all possible two-variable models. Most computer programs retain x_1 and fit only models of the type

$$E(y) = \beta_0 + \beta_1 x_1 + \beta_2 x_2$$

where x_2 is one of the $(k-1)$ remaining variables. For these programs, the variable to be denoted as x_2 and retained in the model is again the one with the largest t value for β_2. Other programs try all possible two-variable models because **it is conceivable that the best two-variable model will not include x_1, the variable entered in step 1.** These programs usually choose as the best two-variable model the one with the largest value of R^2.

The third and succeeding steps in a stepwise regression analysis are identical to the second step. The third step fits the model

$$E(y) = \beta_0 + \beta_1 x_1 + \beta_2 x_2 + \beta_3 x_3$$

The variable x_3 chosen to be retained in the model is either the one for which the t value for β_3 is the largest or the value for which R^2 is the largest (depending upon the computer program). The procedure stops when the t value for the entering variable is less than some predetermined value that would imply statistical significance (or when R^2 exceeds some predetermined value).

A stepwise regression analysis is an easy way to locate some variables that contribute information for the prediction of y, but it is not a foolproof method for finding a good model. Stepwise regression procedures almost always fit first-order models composed of the variables entered into the computer—that is, models of the type

$$E(y) = \beta_0 + \beta_1 x_1 + \beta_2 x_2 + \cdots + \beta_k x_k$$

If you want to introduce second-order terms to account for curvature in the response surface, you will have to enter them into the computer as though they were new variables. In other words, if you had three independent variables, x_1, x_2, and x_3, you could enter these variables into the computer along with $x_4 = x_1^2$, $x_5 = x_2^2$, $x_6 = x_3^2$, $x_7 = x_1 x_2$, and so on.

If the number of independent variables is large, the total number of first-order and second-order terms may be unmanageable. One alternative, when this situation occurs, is to perform a stepwise regression analysis on the original set of independent variables, thereby acquiring a small set of information-contributing variables. If the model requires improvement, a second stepwise regression analysis could be performed, using these variables as well as their squares and cross products. This procedure is unlikely to yield the best second-order model to fit the data, but it is a reasonable alternative to the massive task involved in fitting all possible second-order models.

The computer printout for a stepwise regression analysis is a sequence of individual regression analyses, one for each step in the procedure. Most will be easy to read. They will differ, depending upon the computer program that you use, but the printout for each step will be similar to the printout for a standard multiple regression analysis.

12.8 Misinterpretations in a Regression Analysis

Several misinterpretations of the output of a regression analysis are common. We have already mentioned the importance of model selection. If a model does not fit a set of data, it does not mean that the variables included in the model contribute little or no information for the prediction of y. The variables may be very important contributors of information, but you may not have entered the variables into the model in an appropriate way. For example, a second-order model in the variables might provide a very good fit to the data when a first-order model appears to be completely useless in describing the response variable y.

Second, you must be careful not to deduce that a causal relationship exists between a response y and a variable x. Just because a variable x contributes information for the prediction of y, it does not imply that changes in x *cause* changes in y. It is possible to *design* an experiment to detect causal relationships. For example, if you randomly assign experimental units to each of two levels of a variable x, say $x = 5$

and $x = 10$, and the data show that the mean value of y is larger when $x = 10$, then you could say that the change in the level of x caused a change in the mean value of y. But in most regression analyses, where the experiments are not designed, there is no guarantee that an important predictor variable, say x_1, caused y to change. It is quite possible that some variable that is not even in the model caused *both* y and x_1 to change.

A third common misinterpretation concerns the magnitude of the regression co-efficients. Neither the size of a regression coefficient nor its t value indicates the importance of the associated variable as an information contributor. For example, suppose that you wish to predict a college student's calculus grade y as a function of the student's high school average mathematics grade x_1 and the student's score x_2 on a college mathematics placement test. A regression analysis using the first-order model $E(y) = \beta_0 + \beta_1 x_1 + \beta_2 x_2$ would likely show that both x_1 and x_2 contribute information for the prediction of y. However, it is conceivable that the t value associated with one of the regression coefficients would not be statistically significant, because much of the information contained in x_1 is the same information contained in x_2. This is called **collinearity** and is discussed under the general topic of multicollinearity, which is presented next. When collinearity occurs, the one-variable model

$$E(y) = \beta_0 + \beta_1 x_1 \qquad \text{or} \qquad E(y) = \beta_0 + \beta_2 x_2$$

may be almost as useful in predicting y as the model

$$E(y) = \beta_0 + \beta_1 x_1 + \beta_2 x_2$$

Multicollinearity

One or more of the predictor variables in a regression analysis may be highly cor-related with another. This situation and the problems that it causes in a regression analysis are referred to as **multicollinearity.** This situation arises when two predictor variables are linearly related and hence share the same predictive information or when successive powers of $x(x, x^2, x^3$, and so on) are used as predictor variables in polynomial regression.

When multicollinearity is present in a regression analysis, the estimated regression coefficients have standard errors that are quite large and thereby produce interval estimates of $E(y)$ or prediction intervals for y that are wide and hence imprecise. Furthermore, when multicollinearity is present, the addition or deletion of a predictor variable causes significant changes in the value of the other regression coefficients.

What are some obvious indications that multicollinearity may be present and therefore is distorting the results of a regression analysis? One easily recognized indication is a large value of R^2 and/or the presence of a highly significant F test when testing $H_0 : \beta_1 = \beta_2 = \cdots = \beta_k = 0$, together with nonsignificant t tests for individual regression coefficients. Another not so obvious indication is having signed values of the estimated regression coefficients that are contrary to the values dictated by theory or common sense. Alternatively, examining a matrix of correlations among the predictors $x_1, x_2, \ldots, x_k$ and the dependent variable y will allow you to determine which predictors are highly correlated with each other, as well as which predictors are most highly correlated with y.

The matrix of correlation coefficients for the independent variables square feet of living area (x_1), number of floors (x_2), number of bedrooms (x_3), number of baths (x_4), and listed selling price (y) for the data of Example 12.1 is given in Table 12.10. The elements in the diagonal positions are always equal to 1, and elements in symmetric locations with respect to the diagonal elements are always equal. The correlation between any two variables is found by cross-indexing the row corresponding to the first variable and the column corresponding to the second (or vice versa). For example, notice that the correlation between x_2 and x_4 is $r_{24} = .71765$; the next highest correlation among the predictor variables, between x_1 and x_3, is $r_{13} = .69002$, while the correlation between x_1 and x_4 is $r_{14} = .67142$, and so on. Therefore, x_1, x_2, x_3, and x_4 share a lot of information in common. Furthermore, we see that x_1 would be the best one-variable predictor of y since $r_{1y} = .89120$. Taken together, this information suggests the possible presence of multicollinearity in these data.

	x_1	x_2	x_3	x_4	y
x_1	1.00000	.39351	.69002	.67142	.89120
x_2	.39351	1.00000	.03924	.71765	.32092
x_3	.69002	.03924	1.00000	.51601	.67447
x_4	.67142	.71765	.51601	1.00000	.74095
y	.89120	.32092	.67447	.74095	1.00000

T A B L E 12.10 The matrix of correlation coefficients for the variables in Example 12.1

In examining the regression printout for Example 12.1 given in Table 12.3, we see that the signs of the coefficients of x_2 and x_3 are negative, while we would expect them to be positive. Also, t tests for individual βs are nonsignificant for both of these variables. In summary, we could conclude that some degree of multicollinearity may be distorting our interpretation of the regression results and that one or more of the predictor variables could be eliminated from the regression equation.

Since multicollinearity exists to some degree in most regression analyses, the individual terms in the model should be viewed as information contributors. The primary decision to be made is whether one or more of the terms contribute sufficient information for the prediction of y and whether they should be retained in the model.

12.9 Linear Models for Quantitative Variables (Optional)

We explained in Section 12.4 that quantitative predictor variables, say x_1, x_2, and so forth, are usually entered into a model by using first-order terms, those involving x_1, x_2, x_3, ..., and second-order terms, those involving x_1^2, x_2^2, x_3^2, $x_1 x_2$, $x_1 x_3$, $x_2 x_3$, ... In general, the second-order terms allow for curvature in the relationship between $E(y)$ and the dependent variables, but the cross-product, second-order terms possess a special significance. They are often used to model the

interaction between two predictor variables in their effect on the response variable. To understand the concept of interaction, consider the **first-order (planar) model.**

First-Order Model in Two Independent Variables

$$E(y) = \beta_0 + \beta_1 x_1 + \beta_2 x_2$$

If you were to graph $E(y)$ as a function of x_1 with x_2 held constant, you would obtain a straight line. Repeating this process for other values of x_2, you would obtain a set of parallel lines (with slopes β_1 but intercepts depending on the particular value of x_2) that might appear as shown in Figure 12.13. [Graphs of $E(y)$ versus x_2 for various values of x_1 would also produce a set of parallel lines, but with common slope β_2.]

F I G U R E **12.13**
No interaction between x_1 and x_2

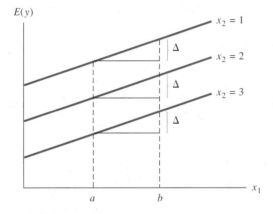

Figure 12.13 shows that, regardless of the value of x_2, a change in x_1, say from $x_1 = a$ to $x_1 = b$, will always produce the same change, Δ, in $E(y)$. When this situation occurs—that is, when the change in $E(y)$ for a change in one variable does not depend on the value of the second variable—we say that the variables **do not interact.**

In contrast, consider the **second-order (interaction) model.**

Second-Order Interaction Model in Two Independent Variables

$$E(y) = \beta_0 = \beta_1 x_1 + \beta_2 x_2 + \beta_3 x_1 x_2$$

If you were to again graph $E(y)$ versus x_1 for various values of x_2, you would obtain a set of lines, but they would not be parallel (see Figure 12.14). In Figure 12.14 you can see that $E(y)$ rises very slowly as x_1 increases when $x_2 = 1$, more rapidly when $x_2 = 2$, and even more rapidly when $x_2 = 3$. Therefore, the effect on $E(y)$ of a

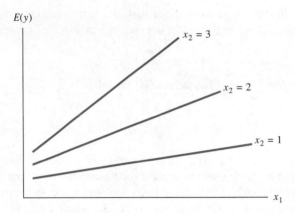

FIGURE 12.14
Interaction between x_1 and x_2

change in x_1 *depends* on the value of x_2. When this situation occurs, we say that the predictor variables *interact*. If you attempt to fit the noninteractive first-order model to data that graph as shown in Figure 12.14, you would obtain a very poor fit to your data. The warning is clear. You may need interaction terms in your model to obtain a good fit to a set of data.

D E F I N I T I O N   ▪

Two predictor variables are said to **interact** if the change in $E(y)$ corresponding to a change in one predictor variable depends upon the value of the other variable.   ▪

If the second-order terms $\beta_3 x_1 x_2$, $\beta_4 x_1^2$, and $\beta_5 x_2^2$ are added to a first-order model, we obtain a complete second-order model.

Complete Second-Order Model in Two Independent Variables

$$E(y) = \beta_0 + \beta_1 x_1 + \beta_2 x_2 + \overbrace{\beta_3 x_1 x_2 + \beta_4 x_1^2 + \beta_5 x_2^2}^{\text{second-order terms}}$$

First-order, second-order interaction, and complete second-order models for three or more independent variables are extensions of the corresponding two-variable models. These models, for three independent variables, are shown next.

First-Order Model in Three Independent Variables

$$E(y) = \beta_0 + \beta_1 x_1 + \beta_2 x_2 + \beta_3 x_3$$

Second-Order Interaction Model in Three Independent Variables

$$E(y) = \beta_0 + \beta_1 x_1 + \beta_2 x_2 + \beta_3 x_3 + \beta_4 x_1 x_2 + \beta_5 x_1 x_3 + \beta_6 x_2 x_3$$

Complete Second-Order Model in Three Independent Variables

$$E(y) = \beta_0 + \beta_1 x_1 + \beta_2 x_2 + \beta_3 x_3 + \beta_4 x_1 x_2 + \beta_5 x_1 x_3$$
$$+ \beta_6 x_2 x_3 + \beta_7 x_1^2 + \beta_8 x_2^2 + \beta_9 x_3^2$$

Graphs of linear models in two or more independent variables are called **response surfaces.** For example, a graph of a first-order model in two independent variables traces a plane in a three-dimensional space (see Figure 12.15). The second-order interaction model graphs as a twisted plane (see Figure 12.16), and the complete second-order model graphs as a paraboloid (see Figure 12.17). Models containing three (or more) independent variables would have to be graphed in a four- (or more) dimensional space. We cannot visualize response surfaces in four dimensions, but we can imagine that graphs of these models would possess the same types of shapes as their two-independent-variable counterparts.

F I G U R E **12.15**
Response surface for a first-order
model in two independent
variables

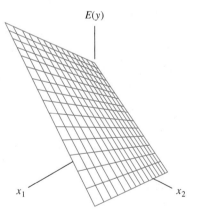

$E(y)$

x_1 x_2

$E(y) = 30 + x_1/3 - x_2$

$0 \leq x_1 \leq 25$
$0 \leq x_2 \leq 25$

F I G U R E **12.16**
Response surface for a
second-order interaction model
in two independent variables

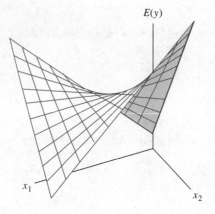

$$E(y) = 5 + x_1 + 2x_2 - x_1x_2/15$$

$$0 \leq x_1 \leq 50$$
$$0 \leq x_2 \leq 30$$

F I G U R E **12.17**
Response surface for a complete
second-order model in two
independent variables

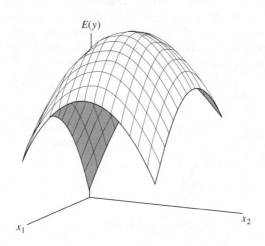

$$E(y) = 1 + 2x_1 + 2.5x_2 - .02x_1x_2 - .05x_1^2 - .06x_2^2$$

$$0 \leq x_1 \leq 32$$
$$0 \leq x_2 \leq 32$$

12.10 Steps to Follow When Building a Linear Model

The preceding sections describe most of the statistical tests and estimation procedures associated with a multiple regression analysis. Ultimately, the objective is to develop a model that will predict y as a function of a set of independent variables x_1, x_2, $\ldots$, x_k and to do so with a small error of prediction. The procedure for developing this model is summarized in the following steps:

1 Select the independent variables to be included in the model. Presumably, you have some understanding of the physical, biological, or social mechanisms that affect y. This knowledge should enable you to list some of the independent variables that may be important information contributors. Since some of these variables may contribute overlapping information, the list may be reduced by running a stepwise regression analysis (see Section 12.7) on the data. If the number of quantitative variables is not large, you may consider entering their squares and cross products as additional predictors to be considered in selecting model variables. You will want to keep the number of independent variables to a minimum so that the number of terms in your linear model does not become too large or the model unmanageable. Be aware that the number of observations in your data set must exceed the number of terms in your linear model—the greater the excess, the better.

2 Write a model relating y to the independent variables that you have selected in step 1. If the independent variables are qualitative, it is best to include terms representing factor interactions. Similarly, if the independent variables are quantitative, it is best to start with a complete second-order model to allow for curvature in the response curve (or surface). Unnecessary terms can be deleted later if their information contribution is negligible.

3 Fit the model to the data set. Check the F value to see whether the complete model contributes information for the prediction of y. Also note the value of R^2 to see how well the model fits the data set.

4 Test individual parameters and sets of parameters to detect factor interaction, curvature, and so on. If some terms appear to contribute little information for the prediction of y, you may wish to simplify the model by eliminating them. If you do eliminate terms, refit the model and recheck the model F value and the value of R^2.

5 Examine plots of the residuals as a final check to detect flaws in your model construction (see Section 12.6).

12.11   A MULTIPLE REGRESSION ANALYSIS FOR FOREIGN CAR IMPORTS

When confronted with a set of data collected over time with no other predictor variables available, we may get our best insight into what might be an appropriate model by examining a plot of the data against time. What should we look for in this plot? In general, we would try to determine the order of the polynomial that we should use. When the data appear to have only one decreasing or increasing trend, we fit a first-order or linear model. When the data increase and then decrease, a second-order model including x and x^2 is used. When there are three changes, a cubic model is used, and so on.

We have learned that the value of R^2 will never decrease when additional predictor variables or additional terms in x are included in the model. However, if we use residual plots in conjunction with R^2 for each of the models that we might fit, we have a second and very important criterion in model selection. Remember that a plot of residuals should have no trends or systematic patterns; the values of the residuals corresponding to a good model should appear as a random scatter in the residual plot.

The plot in Figure 12.18(a) shows the fitted quadratic model plotted on the same graph as the data points. The corresponding residual plot for the quadratic model, Figure 12.18(b), exhibits a very nonrandom distribution of residuals, indicating that the model is not adequate.

F I G U R E **12.18**   Fitted model plot and residual plot for a quadratic model

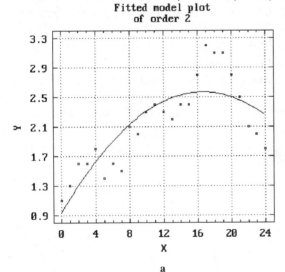

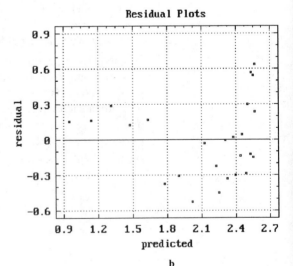

a

b

F I G U R E **12.19**   Fitted model plot and residual plot for a cubic model

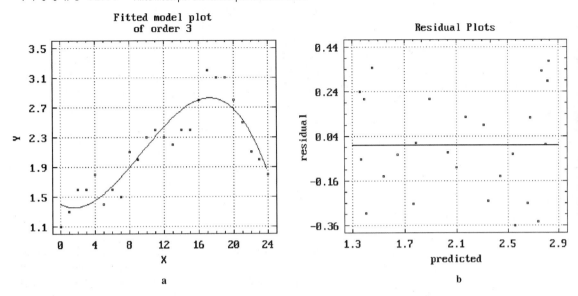

FIGURE **12.19**   Fitted model plot and residual plot for a cubic model

a

b

Figure 12.19(a) shows the plot of the fitted cubic model on the same graph as the data points. The corresponding plot of the residuals from a cubic polynomial model, Figure 12.19(b), exhibits no trends or discernible pattern of any kind. Hence, it appears that a cubic model will adequately describe the data.

An Execustat printout of the regression analysis for these data using the model

$$y = \beta_0 + \beta_1 x + \beta_2 x^2 + \beta_3 x^3 + \epsilon$$

is shown in Table 12.11.

T A B L E **12.11**
Execustat printout for the cubic model

Polynomial Regression Analysis for CASECH12

Dependent variable: Y

Table of Estimates

	Estimate	Standard Error	t Value	P Value
Intercept	1.40455	0.167695	8.38	0.0000
X	-0.0609133	0.0617702	-0.99	0.3353
X^2	0.0213263	0.00605279	3.52	0.0020
X^3	-0.000754873	0.000165622	-4.56	0.0002

R-squared = 84.88%
Adjusted R-squared = 82.72%
Standard error of estimation = 0.241789
Durbin-Watson statistic = 1.05778
Mean absolute error = 0.187032
Sample size (n) = 25
Note: 0 incomplete cases have been excluded.

Notice that all of the regression coefficients produce highly significant t values except for the coefficient of the linear term x. Although we might be tempted to remove this term, most regression analysts would retain the term in the model. Notice further that the value $R^2 = .8488$ is fairly large, indicating that we have accounted for about 85% of the variability in the number of foreign car imports. If we had any doubts about our earlier decision to conclude that the flow of foreign car imports had been stemmed, the fitted cubic model would now predict that the next few years would continue to show a decrease. It is important to notice that the only predictor variable used in this analysis is time (in years beyond 1969). We will introduce other methods for dealing with data measured over time in Chapter 13, which deals with time series.

12.12   Summary

Chapter 12 provides an introduction to an extremely useful area of statistical methodology, one that enables us to relate the mean value of a response variable y to a set of predictor variables. We start by postulating a model that expresses $E(y)$ as the sum of a number of terms, with each term involving the product of a single unknown parameter and a function of one or more of the predictor variables. Values of y are observed when the predictor variables assume specified values, and these data are used to estimate the unknown parameters in the model by using the method of least squares. The procedure for estimating the unknown parameters of the model, testing its utility, or testing hypotheses about sets of the model parameters is known as a multiple regression analysis.

Perhaps the most important result of a multiple regression analysis is the use of the fitted model—the prediction equation—to estimate the mean value of y or to predict a particular value of y for a given set of values of the predictor variables. Scientists, sociologists, engineers, and business managers all face the same problem: forecasting some quantitative response based on the values of a set of predictor variables that describe current or future conditions. When the appropriate assumptions are satisfied, a multiple regression analysis can be used to solve this important problem.

Supplementary Exercises

Starred (*) exercises are optional.

12.23   Utility companies, which must plan the operation and expansion of electricity generation, are vitally interested in predicting customer demand over both short and long periods of time. A short-term study was conducted to investigate the effect of mean daily temperature x_1 and cost per kilowatt x_2 on the mean daily consumption (kilowatt-hours, kWh) per household. One company expected the demand for electricity to rise in cold weather (due to heating), fall when the weather was moderate, and rise again when the temperature rose and there was a need for air-conditioning. It expected demand to decrease as the cost per kilowatt-hour increased, reflecting greater attention to conservation. Data were available for two years, a period in which the cost per kilowatt-hour, x_2, was increased owing to the increasing cost of fuel. The company fitted the model

$$E(y) = \beta_0 + \beta_1 x_1 + \beta_2 x_1^2 + \beta_3 x_2 + \beta_4 x_1 x_2 + \beta_5 x_1^2 x_2$$

to the data shown in the accompanying table. Before we analyze the data, let us examine the logic used in formulating the model.

Price per kWh, x_2

8¢	Mean daily temperature (°F), x_1	31 34 39 42 47 56 62 66 68 71 75 78	
	Mean daily consumption (kWh), y	55 49 46 47 40 43 41 46 44 51 62 73	
10¢	Mean daily temperature (°F), x_1	32 36 39 42 48 56 62 66 68 72 75 79	
	Mean daily consumption (kWh), y	50 44 42 42 38 40 39 44 40 44 50 55	

a Examine the relationship between $E(y)$ and temperature x_1 for a fixed price per kilowatt-hour (kWh), x_2. Substitute a value for x_2, say 10¢, into the equation for $E(y)$. What type of curve will the model relating $E(y)$ to x_1 represent?

b If the company's theory regarding the relationship between mean daily consumption $E(y)$ and temperature x_1 is correct, what should be the sign (positive or negative) of the coefficient of x_1^2?

c Examine the relationship between $E(y)$ and price per kilowatt-hour x_2 when temperature x_1 is held constant. Substitute a value for x_1 into the equation for $E(y)$, say $x_1 = 50°$F. What type of curve will the model relating $E(y)$ to x_2 represent?

d Refer to part (c). If the company's theory regarding the relationship between mean daily consumption $E(y)$ and price per kilowatt-hour x_2 is correct, what should be the sign of the coefficient of x_2?

e What effect do the terms $\beta_4 x_1 x_2$ and $\beta_5 x_1^2 x_2$ have on the curves relating $E(y)$ to x_1 for various values of price per kilowatt-hour?

12.24 The SAS multiple regression computer printout for the data of Exercise 12.23 is shown in the accompanying table.

DEPENDENT VARIABLE: Y

SOURCE	DF	SUM OF SQUARES	MEAN SQUARE	F VALUE	PR > F	R-SQUARE	C.V.
MODEL	5	1346.44751546	269.28950309	31.85	0.0001	0.898455	6.2029
ERROR	18	152.17748454	8.45430470		ROOT MSE		Y MEAN
CORRECTED							
TOTAL	23	1498.62500000			2.90762871		46.87500000

SOURCE	DF	TYPE I SS	F VALUE	PR > F	DF	TYPE IV SS	F VALUE	PR > F
X1	1	140.71101071	16.64	0.0007	1	104.43456772	12.35	0.0025
X1*X1	1	892.77914933	105.60	0.0001	1	125.58263370	14.85	0.0012
X2	1	192.44266992	22.76	0.0002	1	46.78392523	5.53	0.0302
X1*X2	1	57.83521130	6.84	0.0175	1	50.03623552	5.92	0.0256
X1*X1*X2	1	62.67947420	7.41	0.0140	1	62.67947420	7.41	0.0140

PARAMETER ESTIMATES

VARIABLE	ESTIMATE	T FOR H0: PARAMETER = 0	PR > |T|	STD ERROR
INTERCEPT	325.60644505	3.92	0.0010	83.06412820
X1	-11.38255606	-3.51	0.0025	3.23859476
X1*X1	0.11319735	3.85	0.0012	0.02944828
X2	-21.69920900	-2.35	0.0302	9.22432344
X1*X2	0.87302921	2.43	0.0256	0.35886030
X1*X1*X2	-0.00886945	-2.72	0.0140	0.00325742

a   Find SSE and s^2 on the printout.

b   How many degrees of freedom do SSE and s^2 possess? Show that $s^2 = $ SSE/(degrees of freedom).

c   Do the data provide sufficient evidence to indicate that the model contributes information for the prediction of mean daily kilowatt-hour consumption per household? Test using $\alpha = .10$.

d   Use Table 6 in Appendix II to find the approximate observed significance level for the test in part (c).

e   Find the Total SS, S_{yy}, on the printout. Find the value of R^2 and interpret its value. Show that $R^2 = 1- (SSE/S_{yy})$.

f   Use the prediction equation to predict mean daily consumption when the mean daily temperature is 60°F and the price per kilowatt-hour is 8¢.

g   If you have access to a computer, use it to perform a multiple regression analysis for the data in Exercise 12.23. Compare your computer output with the SAS output shown in the accompanying table.

12.25   Refer to Exercises 12.23 and 12.24.

a   Graph the curve depicting $\hat{y}$ as a function of temperature x_1 when the cost per kilowatt-hour is $x_2 = 8¢$. Construct a similar graph for the case when $x_2 = 10¢$ per kilowatt-hour. Does it appear that the consumption curves differ?

b   If cost per kilowatt-hour is unimportant in predicting use, then we do not need the terms involving x_2 in the model. Therefore, the null hypothesis H_0: "x_2 does not contribute information for the prediction of y" is equivalent to the hypothesis $H_0 : \beta_3 = \beta_4 = \beta_5 = 0$ (if $\beta_3 = \beta_4 = \beta_5 = 0$, the terms involving x_2 disappear from the model). This hypothesis is tested by using the procedure of Section 12.5. The SAS multiple regression computer printout, obtained by fitting the reduced model

$$E(y) = \beta_0 + \beta_1 x_1 + \beta_2 x_1^2$$

to the data, is shown in the accompanying table. The following steps enable you to conduct a test to determine whether the data provide sufficient evidence to indicate that price per kilowatt-hour, x_2, contributes information for the prediction of y. Test using $\alpha = .05$.

DEPENDENT VARIABLE: Y

SOURCE	DF	SUM OF SQUARES	MEAN SQUARE	F VALUE	PR > F	R-SQUARE	C.V.
MODEL	2	1033.49016004	516.74908002	23.33	0.0001	0.689626	10.0401
ERROR	21	465.13483996	22.14927809		ROOT MSE		Y MEAN
CORRECTED							
TOTAL	23	1498.62500000			4.70630196		46.87500000

SOURCE	DF	TYPE I SS	F VALUE	PR > F	DF	TYPE IV SS	F VALUE	PR > F
X1	1	140.71101071	6.35	0.0199	1	810.33211666	36.59	0.0001
X1*X1	1	892.77914933	40.31	0.0001	1	892.77914933	40.31	0.0001

PARAMETER ESTIMATES

VARIABLE	ESTIMATE	T FOR HO: PARAMETER = 0	PR > |T|	STD ERROR
INTERCEPT	130.00929147	8.74	0.0001	14.87580615
X1	-3.50171859	-6.05	0.0001	0.57893460
X1*X1	0.03337133	6.35	0.0001	0.00525631

i   Find SSE_1 and SSE_2.

ii   Give the values for v_1 and v_2.

iii   Calculate the value of F.

iv   Find the tabulated value of $F_{.05}$.

v   State your conclusions.

12.26 Refer to Example 12.5. The t value for testing the hypothesis $H_0 : \beta_1 = 0$ is

$$t = \frac{\hat{\beta}_1 - 0}{s_{\hat{\beta}_1}} = \frac{\hat{\beta}}{s_{\hat{\beta}_1}}$$

where $\hat{\beta}_1$ is the estimate of β_1 and $s_{\hat{\beta}_1}$ is the estimated standard deviation (or standard error) of $\hat{\beta}_1$. Both of these quantities are shown on the SAS printout in Table 12.7. Calculate $t = \hat{\beta}_1 / s_{\hat{\beta}_1}$. Verify that your calculated value of t is equal to the value shown on the printout, that is, $t = 15.22$.

12.27 The t test of Exercise 12.26 can also be conducted by using an F test (the test is described in Section 12.5). We can use an F test because the square of a t statistic with v degrees of freedom is equal to an F statistic with $v_1 = 1$ and $v_2 = v$ degrees of freedom; that is,

$$F_{1,v} = t_v^2$$

a To convince yourself that this relationship is valid, find the value $t_{.025}$ (Table 4 in Appendix II) that is used to locate the rejection region for the two-tailed $(\alpha = .05)$ t test of Exercise 12.26.

b Show that $t_{.025}^2$ [found in part (a)] is equal to the tabulated value $F_{.05}$ (Table 6 in Appendix II) for $v_1 = 1$ and $v_2 = 8$ degrees of freedom.

12.28 When will an F statistic equal the square of a t statistic?

12.29 A department store conducted an experiment to investigate the effects of advertising expenditures on the weekly sales for its men's wear, children's wear, and women's wear departments. Five weeks for observation were randomly selected from each department, and an advertising budget x_1 (hundreds of dollars) was assigned for each. The weekly sales (thousands of dollars) are shown in the accompanying table for each of the 15 one-week sales periods. If we expect weekly sales $E(y)$ to be linearly related to advertising expenditure x_1 and if we expect the slopes of the lines corresponding to the three departments to differ, then an appropriate model for $E(y)$ is

$$E(y) = \beta_0 + \underbrace{\beta_1 x_1}_{\substack{\text{quantitative} \\ \text{variable} \\ \text{"advertising} \\ \text{expenditure"}}} + \underbrace{\beta_2 x_2 + \beta_3 x_3}_{\substack{\text{dummy variables} \\ \text{used to introduce} \\ \text{the qualitative} \\ \text{variable "department"} \\ \text{into the model}}} + \underbrace{\beta_4 x_1 x_2 + \beta_5 x_1 x_3}_{\substack{\text{interaction terms} \\ \text{that introduce} \\ \text{differences in} \\ \text{slopes}}}$$

where

$$x_1 = \text{Advertising expenditure}$$
$$x_2 = \begin{cases} 1, & \text{if children's wear department B} \\ 0, & \text{if not} \end{cases}$$
$$x_3 = \begin{cases} 1, & \text{if women's wear department C} \\ 0, & \text{if not} \end{cases}$$

Advertising Expenditure (in $100)

Department	1	2	3	4	5
Men's wear, A	5.2	5.9	7.7	7.9	9.4
Children's wear, B	8.2	9.0	9.1	10.5	10.5
Women's wear, C	10.0	10.3	12.1	12.7	13.6

a Find the equation of the line relating $E(y)$ to advertising expenditure x_1 for the men's wear department A. [*Hint:* According to the coding used for the dummy variables, the model represents mean sales $E(y)$ for the men's wear department A when $x_2 = x_3 = 0$. Substitute $x_2 = x_3 = 0$ into the equation for $E(y)$ to find the equation of this line.]

b Find the equation of the line relating $E(y)$ to x_1 for the children's wear department B. [*Hint:* According to the coding, the model represents $E(y)$ for the children's wear department when $x_2 = 1$ and $x_3 = 0$.]

c Find the equation of the line relating $E(y)$ to x_1 for the women's wear department C.

d Find the difference between the intercepts of the $E(y)$ lines corresponding to the children's wear B and men's wear A departments.

e Find the difference in slopes between the $E(y)$ lines corresponding to the women's wear C and men's wear A departments.

f Refer to part (e). Suppose that you want to test the null hypothesis that the slopes of the lines corresponding to the three departments are equal. Express this as a test of an hypothesis about one or more of the model parameters.

*12.30 If you have access to a computer and a multiple regression computer program package, perform a multiple regression analysis for the data of Exercise 12.29.

a Verify that SSE $= 1.2190$, that $s^2 = .1354$, and that $s = .368$.

b How many degrees of freedom will SSE and s^2 possess?

c Verify that Total SS $= S_{yy} = 76.0493$.

d Calculate R^2 and interpret its value.

e Verify that the parameter estimates (values rounded) are approximately equal to

$$\hat{\beta}_0 \approx 4.10 \qquad \hat{\beta}_1 \approx 1.04 \qquad \hat{\beta}_2 \approx 3.53$$
$$\hat{\beta}_3 \approx 4.76 \qquad \hat{\beta}_4 \approx -.43 \qquad \hat{\beta} \approx -.08$$

f Find the prediction equation, and graph the three department sales lines.

g Examine the graphs in part (f). Do the slopes of the lines corresponding to the children's wear B and the men's wear A departments appear to differ? Test the null hypothesis that the slopes do not differ (i.e., $H_0 : \beta_4 = 0$) against the alternative hypothesis that the slopes do differ (i.e., $H_a : \beta_4 \neq 0$). Test using $\alpha = .05$. [*Note:* Verify on your computer printout that the estimated standard deviation of $\hat{\beta}_4$ is $s_{\hat{\beta}_4} \approx .165$.]

h Find a 95% confidence interval for the difference between the slopes of the lines for B and A.

i Do the data provide sufficient evidence to indicate a difference in the slopes between lines C and A? Test using $\alpha = .05$. [*Hint:* $s_{\hat{\beta}_5} \approx .165$.]

*12.31 Write a first-order linear model as a function of two quantitative variables x_1 and x_2. Describe the response surface.

*12.32 Suppose that y is a function of a single quantitative independent variable x and that $E(y)$ decreases, reaches a minimum, and then increases as x increases. Write a second-order model relating $E(y)$ to x.

*12.33 Write a first-order linear model as a function of three quantitative independent variables. Does this model imply that x_1, x_2, and x_3 affect $E(y)$ in an independent or a dependent manner? Explain.

*12.34 Suppose that $E(y)$ is related to a single qualitative variable at three levels. Write the model for $E(y)$.

*12.35 Write a complete second-order model for $E(y)$ as a function of two quantitative independent variables x_1 and x_2.

*12.36   Write a complete second-order model for $E(y)$ as a function of three quantitative independent variables x_1, x_2, and x_3.

*12.37   Suppose that $E(y)$ is related to two qualitative independent variables, one at three levels and the other at two levels. Write a linear model for $E(y)$. Assume that the two qualitative variables do not interact.

*12.38   Refer to Exercise 12.37 and assume that the two qualitative variables interact. Write a model for $E(y)$.

12.39   The earnings y per share (EPS) for Wendy's International (Exercise 11.57) are shown below for the years 1985–94, with the coded year $x =$ Year $- 1984$. Since the scatterplot indicates a quadratic relationship, an Execustat printout for fitting a second-order model to the data appears in the accompanying table.

	1985	1986	1987	1988	1989	1990	1991	1992	1993	1994
Coded year, x	1	2	3	4	5	6	7	8	9	10
EPS, y	.82	.49	.40	.30	.25	.40	.52	.63	.76	.90

Source: Value Line Investment Survey, 1994.

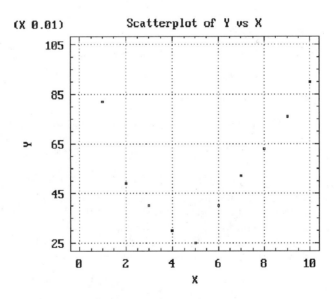

a   Plot the data points on graph paper.

b   Locate $\hat{\beta}_0$, $\hat{\beta}_1$, and $\hat{\beta}_2$ on the printout, and give the least-squares prediction equation.

c   Graph the least-squares prediction equation on your data plot for part (a). Does the least-squares curve seem to provide a good fit to the data?

d   Do the data provide sufficient evidence to indicate that the second-order term in the model contributes information for the prediction of y? [*Hint:* See the t value shown on the printout for testing $H_0 : \beta_2 = 0$.]

e   Find the observed significance level (p-value) for the test in part (e).

f   Find R^2 on the printout and interpret its value.

g   How does the number n of data points affect the significance that can be attached to a value of R^2?

Execustat output for Exercise 12.39

Polynomial Regression Analysis for CBSWENDY

Dependent variable: Y

Table of Estimates

	Estimate	Standard Error	t Value	P Value
Intercept	0.9555	0.0904451	10.56	0.0000
X	-0.25272	0.0377737	-6.69	0.0003
X^2	0.0254924	0.00334661	7.62	0.0001

R-squared = 90.76%
Adjusted R-squared = 88.12%
Standard error of estimation = 0.0768992
Durbin-Watson statistic = 1.31116
Mean absolute error = 0.0596121
Sample size (n) = 10
Note: 0 incomplete cases have been excluded.

Exercises Using the Data Disk

12.40 Refer to data set A on the data disk, describing faculty salaries by rank and gender at 185 colleges.

a Select a random sample of 15 colleges and record the salaries by rank for male faculty. Select another independent random sample of 15 colleges and record the salaries by rank for female faculty. Let the faculty salary be denoted as y. We wish to regress y on the variables rank and gender, which are both qualitative variables. Use x_1, x_2, and x_3 as dummy indicator variables representing rank, and x_4 as a dummy indicator variable representing gender. Analyze your data using the model

$$y = \beta_0 + \beta_1 x_1 + \beta_2 x_2 + \beta_3 x_3 + \beta_4 x_4 + \epsilon$$

b Repeat part (a), this time using a model that allows for a gender-by-rank interaction. Does interaction contribute any information in predicting faculty salary?

c Refer to part (a). Record the salaries for both male and female faculty for a random sample of 15 colleges. Now calculate the difference between male and female salaries and use these differences as y^*, the new response variable. Analyze your data using the model

$$y^* = \beta_0 + \beta_1 x_1 + \beta_2 x_2 + \beta_3 x_3 + \epsilon$$

in which x_1, x_2, and x_3 are the faculty rank dummy indicator variables. Comment on the results of the analysis compared to the results in part (a).

12.41 Refer to data set B on the data disk, concerning broccoli yield at various ozone levels.

a Let y represent the fresh weight of broccoli. Consider the variables variety, head diameter, and ten-hour ozone average as the predictor variables of interest. Select a random sample of $n = 25$ observations using these data. Analyze the fresh weights as a simple linear function of variety, head diameter, and the ten-hour ozone average. Is the variety of the broccoli a useful predictor variable?

b Using the data from part (a), fit a second-order model in the variables head diameter and ten-hour ozone average.

c   Now introduce the variable variety into the model in part (b), and allow the variable variety to interact with the second-order model variables. Discuss the results of the model fits in parts (a), (b), and (c) as they relate to one another and the selection of the model that best explains the data.

12.42   Refer to data set C on the data disk, describing the characteristics of 604 money market funds. Select a random sample of 15 to 20 observations on the three variables given as average maturity in days, seven-day average yields, and assets. Use a multiple regression program to regress seven-day average yield (y) on the average maturity in days (x_1) and assets (x_2). Comment on your results.

13

TIME SERIES AND INDEX NUMBERS

About This Chapter

The objectives of this chapter are to introduce index numbers and other variables that are used to monitor, over time, the health of a specific business and/or of the U.S. economy; to explain how sequences of data collected over time can be analyzed; and to identify the problems associated with the use of such data in business and economic forecasting.

Contents

MEASURING THE COST OF LIVING

A graph (Figure 13.1) of the Consumer Price Index for the period 1972 through 1991 shows that this measure of inflation has risen steadily over this period of time. So have the prices that you and I pay at the grocery store and the prices that we pay for housing, entertainment, and transportation. To what extent does the Consumer Price Index measure our cost of living? What is the Consumer Price Index? How is it computed, and what does it mean? How can we interpret the graph of Figure 13.1, and can we use the graph or a statistical model to forecast values of the Consumer Price Index at some point in the future?

As you will subsequently learn, the Consumer Price Index, like many other economic indexes, provides a measure of the change in consumer prices relative to the *level of prices* in some base year, in this case, 1982–84. Because it is calculated and reported monthly by the Bureau of Labor Statistics, it is called a time series variable; a sequence of monthly values of the index is called a *time series*. In this chapter, we introduce you to the difficult problem of business forecasting based on time series data. We tell you what an economic index is and explain how some of

FIGURE **13.1**
The Consumer Price Index,
1972 through 1991

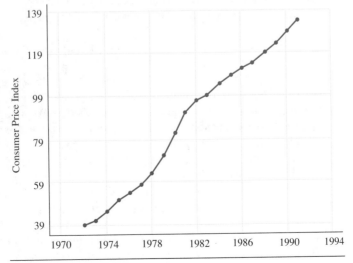

Source: U.S. Department of Commerce, Bureau of Economic Analysis, *Business Statistics, 1963–1991*, Washington, D.C.

the more common types are computed. Then in Section 13.9 we attempt to unravel some of the mystery surrounding the Consumer Price Index.

13.1 What Is a Time Series?

Many types of business and economic data are observations on a variable at equidistant points in time. A data set of this type is called a **time series,** and the variable is called a **time series variable.**

The Dow-Jones Industrial Average is a time series variable that provides a measure of the level of stock prices; the set of its daily readings is a time series. The Index of Leading Economic Indicators is a time series variable that purports to measure the health of the U.S. economy. A set of its monthly values is a time series. Other time series variables of interest to economists and business managers are the monthly sales of a particular corporation, the weekly sales of automobiles, the weekly starts in new housing construction, and the like.

DEFINITION ▪

> A **time series variable** is a variable that is observed at specific (usually equidistant) points in time. A **time series** is a set of sequential measurements on a time series variable. ▪

Most time series are graphed and analyzed so that economists and business managers can forecast the future state of the economy and of their respective businesses. Such forecasts, if moderately accurate, are useful in planning inventory size, production, and sales; they may also determine our thinking on investments. For example, Figure 13.2 is a graph of the value of the Dow-Jones Industrial Average (DJI) at the close of business at the end of each month, January 1992 to February 1994. Note that the graph is constructed so that the Dow-Jones Industrial Average y is plotted against time.

Figure 13.3 is a graph of another time series, the yields y of Treasury long bonds in the United States for the period 1950–94. Note that long bond yields peaked in the 1980s and then began to decline. Where will long bond yields be in 1998, 1999, or 2000? The analysis of a time series, such as those shown in Figures 13.2 and 13.3, is done ultimately for the purpose of prediction. We think that past and present observations of a time series variable can be used to predict its future value.

Can we fit a multiple regression model to a time series variable y and use time and other time series variables as predictor variables to forecast future values of y? The answer is that forecasts, by their very nature, violate the assumptions of a regression analysis. You may recall that the regression model is appropriate for estimating $E(y)$ and predicting y within the range of values of the predictor variables that were used in fitting the model. Because a forecast at some future time is outside the time interval used to fit the regression model, time series forecasts based on regression analysis will always be suspect. A second, but less obvious, difficulty is that observations

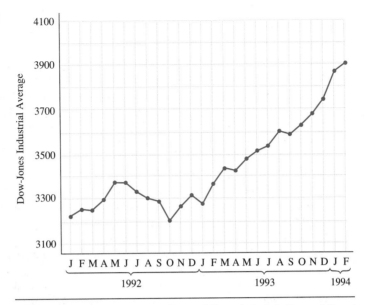

FIGURE **13.2**
The Dow-Jones Industrial
Average (end of month),
January 1992 to February 1994

Source: U.S. Department of Commerce, Bureau of Economic Analysis, *Survey of Current Business*, Washington D.C., March 1993 and March 1994.

FIGURE **13.3**   Treasury long bond yields, 1950–2000 (estimated)

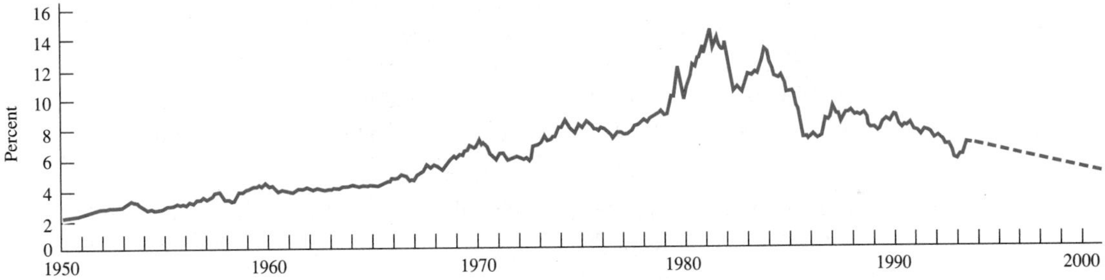

Source: Kerschner and Ryan, 1994.

over time often tend to be correlated and thus violate the assumption of independent errors.

The analysis of time series and the use of time series for forecasting are difficult and complex topics. In this chapter, we introduce you to an important type of time series, index numbers; we then describe some of the procedures used in a time series analysis. A more thorough discussion of this topic can be found in texts on the subject of time series analysis.

13.2 Index Numbers

An **index number** is a number that measures change in a time series variable in comparison to a **base year.** For example, the Consumer Price Index purports to measure the general level of consumer prices in comparison to the value of the index in the base year 1982–84. (The base-year value is always set to 100.) Thus a Consumer Price Index equal to 200 in a given year means that the general level of prices to the consumer, measured by the index, is double the level of prices in 1982–84.

Since an index number is reported at regular intervals of time, it is a time series variable. The Composite Index of Leading Economic Indicators is an index number that is constructed to provide a measure of the general level of the U.S. economy. The Dow-Jones Industrial Average is an index number that measures the general level of the prices of corporate common stocks on the New York Stock Exchange. The Consumer Price Index measures the general level of prices to the consumer. These are just three examples of the many index numbers used to measure the general level of economic and business conditions.

Index numbers used to measure the changes in some business or economic phenomena range from those that are very simple to those that are very sophisticated. For example, the ratio of the number of domestic automobiles sold annually to the number sold in 1970, in 1980, or in some other base year will provide a good measure of the relative levels of annual domestic automobile sales. In contrast, an index constructed to measure the general level of consumer prices cannot be constructed as the ratio of the prices of a single commodity, say new house prices; not everyone buys a new house every year. Consequently, this index will have to reflect the change in the prices of many different commodities, such as food, clothing, and transportation.

We will describe three types of indexes and give examples of their construction in this section. The first index we will examine is the **simple index number.**

DEFINITION ▪

A **simple index number** I_t is the ratio of the value of a single time series variable y_t at time t to its value y_0 at time t_0, multiplied by 100; that is,

$$I_t = \frac{y_t}{y_0}(100)$$

where

I_t = Value of the simple index number at time t

y_t = Value of the time series variable at time t

y_0 = Value of the time series variable at base time t_0 ▪

EXAMPLE   **13.1**   The factory sales of new U.S. automobiles for a 17-year period are shown in Table 13.1. Construct a simple index of domestic sales, using year 3 as the base year.

TABLE **13.1**
Annual new domestic
automobile sales

Year	Sales (millions)	Year	Sales (millions)	Year	Sales (millions)
1	8.8	7	9.2	13	7.6
2	9.7	8	8.4	14	8.0
3	7.3	9	6.4	15	7.5
4	6.7	10	6.2	16	7.1
5	8.5	11	5.0	17	7.1
6	9.2	12	6.7		

Solution   Since year 3 was chosen as the base year, $y_0 = 7.3$ and

$$I_1 = \frac{y_1}{y_0}(100) = \frac{8.8}{7.3}(100) = 121$$

$$I_2 = \frac{y_2}{y_0}(100) = \frac{9.7}{7.3}(100) = 133$$

$$I_3 = \frac{y_3}{y_0}(100) = \frac{7.3}{7.3}(100) = 100$$

You can verify that the values of I_t for years 4–17 are as follows:

$$I_4 = 92 \quad I_5 = 116 \quad I_6 = 126 \quad I_7 = 126 \quad I_8 = 115$$
$$I_9 = 88 \quad I_{10} = 85 \quad I_{11} = 68 \quad I_{12} = 92 \quad I_{13} = 104$$
$$I_{14} = 110 \quad I_{15} = 103 \quad I_{16} = 97 \quad I_{17} = 97$$

A graph of the new domestic car sales index for Example 13.1 is shown in Figure
13.4.   ■

FIGURE **13.4**
A graph of an index of domestic
car sales for Example 13.1

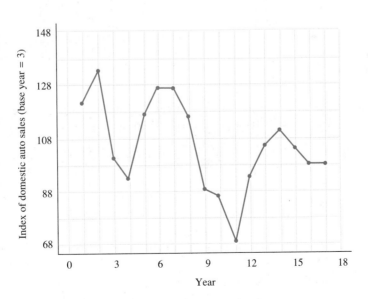

A second type of index, one that incorporates the value of more than one time series variable, is called a **simple composite index number.** To compute a simple composite index number, you sum the values of the time series variables entering into the index and then compute the yearly value of the index in the same way that you compute a simple index number.

DEFINITION ▪

A **simple composite index number** of k time series variables, call them P_{1t}, P_{2t}, ..., P_{kt}, is equal to

$$I_t = \frac{y_t}{y_0}(100)$$

where

$$y_t = P_{1t} + P_{2t} + \cdots + P_{kt}$$

and y_0 is the value of y_t for the base time period.   ▪

EXAMPLE   13.2   The year-end prices per pound of three commodities are shown in Table 13.2 for the years 1985 to 1994. Construct a simple composite index for the prices, using 1986 as the base year.

TABLE 13.2
Year-end prices for three commodities

Commodity	1985	1986	1987	1988	1989	1990	1991	1992	1993	1994
1	2.20	2.35	2.64	2.77	3.01	3.10	3.42	3.50	3.62	3.81
2	1.33	1.46	1.65	1.87	1.99	2.23	2.44	2.61	2.86	2.77
3	.91	.99	1.08	1.17	1.19	1.47	1.62	1.74	1.93	1.99
y_t	4.44	4.80	5.37	5.81	6.19	6.80	7.48	7.85	8.41	8.57
I_t	93	100	112	121	129	142	156	164	175	179

Solution   The sum y_t of the three commodity prices is shown in the fourth row of the table. Then $y_0 = 4.80$ for the base year 1986, and

$$I_{1985} = \frac{y_{1985}}{y_0}(100) = \frac{4.44}{4.80}(100) = 93$$
$$I_{1986} = 100$$
$$I_{1987} = \frac{y_{1987}}{y_0}(100) = \frac{5.37}{4.80}(100) = 112$$

and so on. You can verify that the remaining values of I_t for 1988 through 1994 are as shown in the bottom row of Table 13.2. A graph of this time series is shown in Figure 13.5.   ▪

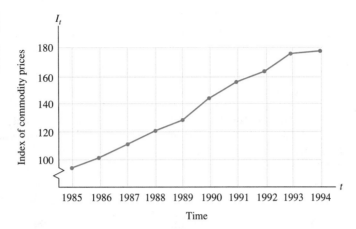

A third type of index number is a **weighted composite index.** Like the simple composite index, this index incorporates the values P_{1t}, P_{2t}, ..., P_{kt} of a number, say k, of time series variables. But rather than compute y_t as their sum, we multiply each of the values P_{1t}, P_{2t}, ..., P_{kt} by a weight that reflects their importance in the index, and then we sum these weighted products. The reason for this procedure is clear if you consider the problem of constructing an index to reflect the general level of consumer meat prices. If you were to include the prices of beef, pork, chicken, fish, and alligator tail in the index, you might want to assign different weights to these prices because of the substantial differences in the consumption of the products. Not many people eat alligator tail, so it would seem unreasonable to give the price of this product the same weight in the index as the price of beef. Weights for a weighted composite index can be chosen in different ways, but those used in constructing price indexes are usually the amounts of the commodities consumed. If the weights used to compute y_t are the respective commodity consumptions for the base year, the weighted composite index is said to be a **Laspeyres index.** If the weights are those for the time period t, the index is called a **Paasche index.**

D E F I N I T I O N ▪   A **weighted composite index number** of k time series variables, call them P_{1t}, P_{2t}, ..., P_{kt}, is equal to

$$I_t = \frac{y_t}{y_0}(100)$$

where

$$y_t = w_1 P_{1t} + w_2 P_{2t} + \cdots + w_k P_{kt}$$

$$w_1, w_2, \ldots, w_k = \text{Weights assigned to } P_{1t}, P_{2t}, \ldots, P_{kt}$$

$$y_0 = \text{Values of } y_t \text{ for the base time period}$$

[*Note:* A simple composite index is the special case of a weighted composite index where $w_1 = w_2 = \cdots = w_k = 1$.] ▪

DEFINITION ∎

> A **Laspeyres index number** is a weighted composite index in which the weights are assigned the values that they assumed for the base year. ∎

DEFINITION ∎

> A **Paasche index number** is a weighted composite index in which the weights used to compute the weighted sum for a given year are the values that the weights assumed for that year. ∎

EXAMPLE 13.3 Suppose that the consumption, in millions of tons, for the commodities of Example 13.2 are 2.1, 2.7, and 4.8 million tons, respectively, for the base year 1986. Find the values of the Laspeyres index for the commodity price data for 1985 and 1986.

Solution Since a Laspeyres index uses the values of the weights for the base period $w_1 = 2.1$, $w_2 = 2.7$, and $w_3 = 4.8$, then

$$y_{1985} = w_1 P_{1,1985} + w_2 P_{2,1985} + w_3 P_{3,1985}$$
$$= (2.1)(2.20) + (2.7)(1.33) + (4.8)(.91) = 12.579$$
$$y_{1986} = y_0 = w_1 P_{1,1986} + w_2 P_{2,1986} + w_3 P_{3,1986}$$
$$= (2.1)(2.35) + (2.7)(1.46) + (4.8)(.99) = 13.629$$

Then

$$I_{1985} = \frac{y_{1985}}{y_0}(100) = \frac{12.579}{13.629}(100) = 92$$

and

$$I_{1986} = \frac{y_{1986}}{y_0}(100) = \frac{y_0}{y_0}(100) = 100 \quad ∎$$

The extent to which index numbers actually measure the level of the economic or business phenomenon that they are intended to measure is often a point of debate. For example, recent complaints about the Consumer Price Index stress that the price of new housing is too heavily weighted in the index when you consider that most consumers do not buy a new house every month. Consequently, economic and business indexes should be taken for what they are—an attempt to achieve a simple measure for a very complicated phenomenon. Although economic indexes may not be perfect, most perform the function for which they were intended: to provide a rough but general measure of change over time.

Exercises

Applications

13.1   The total monthly production (to the nearest million short ton) of raw steel for each month in 1991 is shown in the table:

Month	Jan.	Feb.	Mar.	Apr.	May	June	July	Aug.	Sept.	Oct.	Nov.	Dec.
Production	7.6	6.7	7.3	7.1	7.1	7.0	7.3	7.4	7.5	7.7	7.5	7.3

Source: Business Statistics, 1963–91, 1992.

If January 1991 is taken as the base month, calculate the monthly simple index for this monthly time series.

13.2   The accompanying table gives the number of industrial and commercial business failures in the United States for the years 1972 through 1990. Calculate a simple index number I_t of business failures for the years 1980 through 1990, using 1975 as the base year.

Year	Number of Failures	Year	Number of Failures	Year	Number of Failures	Year	Number of Failures
1972	9566	1977	7919	1982	24,908	1987	61,384
1973	9345	1978	6619	1983	31,534	1988	57,099
1974	9915	1979	7564	1984	52,078	1989	50,361
1975	11,432	1980	11,742	1985	57,252	1990	60,432
1976	9628	1981	16,794	1986	61,601		

Source: Business Statistics, 1963–91, 1992.

13.3   The average annual price (in cents) per board foot and total annual production (in millions of board feet) for three grades of plywood are shown in the accompanying table for years 1, 2, 3, and 4. Consider year 1 as the base year.

Plywood Grade

	1		2		3	
Year	Price	Production	Price	Production	Price	Production
1	9.3	56.1	11.1	33.7	7.2	88.6
2	18.1	112.7	22.4	69.5	18.8	204.1
3	17.6	38.6	20.6	40.3	15.9	79.9
4	17.4	30.9	19.3	37.2	15.7	65.0

a   Find the simple index number for the price of plywood grade 1 for years 2, 3, and 4.

b   Compute the simple index numbers for plywood grade 2.

c   Compute the simple index numbers for plywood grade 3.

13.4   Refer to Exercise 13.3, and compute the simple composite index numbers to measure price changes in the three grades of plywood for the years 2, 3, and 4, relative to the base year 1.

13.5   Refer to Exercise 13.3, and compute the Laspeyres index number for the data for years 2, 3, and 4. Use year 1 as the base year and annual production as the weighting factor.

13.6   Refer to Exercise 13.3, and compute the Paasche index number for the data for years 2, 3, and 4. Use year 1 as the base year and annual production as the weighting factor.

13.7 The U.S. production (in millions of tons) of pig iron and raw steel is shown in the accompanying table for the years 1967 and 1983 through 1991. Calculate a simple composite index for iron and steel production for 1987 through 1991, using 1967 as the base year.

| | Year | | | | | | | | | |
Product	1967	1983	1984	1985	1986	1987	1988	1989	1990	1991
Pig iron	87.0	48.7	51.9	50.4	44.0	48.4	57.7	55.9	54.9	48.5
Steel	127.2	84.6	92.5	88.3	81.6	89.2	99.9	97.9	98.0	87.3

Source: Business Statistics, 1963–91, 1992.

13.8 Suppose that you wanted to construct an index to measure the price level of the common stock of major chemical companies. The accompanying table gives the approximate average price per share and the average number of shares outstanding (adjusted for stock splits) for the common stock of the Dow Chemical Company, Dupont, and Monsanto for the years 1985, 1990, and 1994.

| | Dow Chemical | | | Dupont | | | Monsanto | | |
Shares and Price	1985	1990	1994	1985	1990	1994	1985	1990	1994
Average number of shares (millions)	285	270	283	722	670	680	154	126	115
Average price per share (dollars)	23.0	56.4	61.5	19.5	36.9	54.1	24.0	49.5	77.2

Source: Data from Value Line Investment Survey, 1994.

a Construct a simple composite price level index for the three chemical company common stocks for the years 1990 and 1994. Use 1985 as the base year.

b Calculate a Laspeyres index for the prices of the chemical stocks for the years 1990 and 1994, using the average number of shares outstanding as weights. Explain why the Laspeyres index might provide a better measure of the general level of the stock prices of large chemical companies. Use 1985 as the base year.

c Calculate a Paasche index for the data for the years 1990 and 1994, using 1985 as the base year. Explain the difference between the Laspeyres and Paasche indexes.

13.3 The Components of a Time Series

Time series analysis is a complicated topic, and there is a diversity of opinion as to how analyses should be performed. One of the most accepted approaches is to view a time series as a composition of four components:

1 secular or long-term trend

2 cyclical fluctuation

3 seasonal variation

4 random, unpredictable variation

These components are then isolated and modeled by using one of several methods.

Secular or **long-term trend** is the tendency of a time series to gradually increase or decrease in a straight line or gradual curve over time (see Figure 13.6). For example,

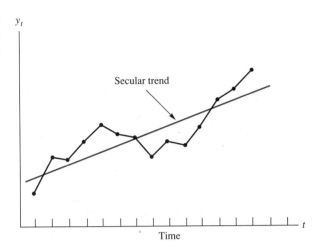

FIGURE **13.6**
The secular or long-term trend in a time series

Secular trend

Time

a time series of the price of housing might vary upward and downward over short intervals of time; but because of long-term inflation, it would trend gradually upward.

Cyclical components of a time series tend to rise and fall in a cyclical pattern about the secular curve (or line). For example, a time series of the price of housing would cycle about a long-term trend curve due to the cycling of supply, interest rates, the state of the economy, and so on. Builders tend to overbuild in a booming housing market, thus driving down the price of housing. As time continues, the market recovers, prices rise, more people buy, housing tends to become scarce, prices peak, and the cycle repeats itself. The cyclical component of a time series is illustrated in Figure 13.7.

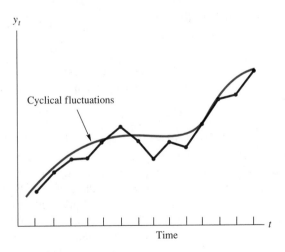

FIGURE **13.7**
The cyclical component of a time series

Cyclical fluctuations

Time

Seasonal components represent the tendency for the time series to jump upward or downward during specific seasons of the year. For example, the prices of fresh

F I G U R E **13.8**
The seasonal component of a
time series

Quarter

vegetables tend to fall during harvesting seasons and rise during the winter or the growing season. A seasonal component might appear as shown in Figure 13.8.

The fourth component of a time series is the unexplained **random variation** that exists in all economic and business time series. These random variations, which are added to the preceding three time series components, bob about in a haphazard manner and are sometimes correlated, owing to unexplained cyclical effects that have not been accounted for in the model.

One method of modeling a time series is to use one or more of the four components as either additive or multiplicative effects. For simplicity, let us denote the effects of the trend, the cyclical variation, the seasonal variation, and the random variation at time t as T_t, C_t, S_t, and ϵ_t, respectively. The value of a time series y_t at time t can be modeled using the **additive model,**

$$y_t = T_t + C_t + S_t + \epsilon_t$$

in which each component enters in an additive fashion. An alternative model is the **multiplicative model,** in which these components enter in a multiplicative fashion, whereby

$$y_t = T_t \cdot C_t \cdot S_t \cdot \epsilon_t$$

The choice of method for estimating the trend, cyclical, and seasonal effects depends on which of these two model formulations is used, together with any further available information concerning one or more of these effects. For example, sales for a new business usually rise rapidly and then level off, while established firms usually demonstrate steady growth reflecting possible inflation and a slowly growing market. In other words, the **trend** may be linear, quadratic, or possibly exponential. Hence, an estimate of the trend component will depend on how the trend is modeled when describing the response.

Economic data are often reported monthly, quarterly, or annually. Therefore, depending on the frequency with which data are reported, time series data are usually modeled using 12 or 4 **seasonal effects.** Rather than exhibiting seasonal effects, annual data may exhibit **cyclical behavior.** Cyclical effects are much less predictable in terms of the length or the strength of the underlying cycle. In light of this information, we are usually able to isolate trend and seasonal effects. With this information, we are able to *detrend* and *deseasonalize* data, thereby more clearly delineating the effects due to cyclical and random variation.

In the sections that follow, we discuss smoothing techniques, which are used to average out the effects of random variation and thereby better display trend, seasonal, and cyclical effects. In addition, we present methods for predicting future values of a time series based on estimates of the other components in a time series model. Some of these techniques, such as the *ratio-to-moving-average* method discussed in Section 13.6, use a multiplicative model, while others, which we discuss in Section 13.7, use an additive model.

13.4 Smoothing a Time Series: Moving Averages

Before you attempt to model a time series, it is useful to graph the time series to determine the nature of any secular, cyclical, and seasonal components, if they exist. These components can often be revealed by *smoothing* the time series—that is, averaging out the effects of the random variation. One method of smoothing is called the method of **moving averages.**

DEFINITION ▪

A *k*-point moving average M_t at time *t* is formed by averaging *k* sequential values of y_t. The time *t* is taken to be a point in the middle of these time intervals. ▪

Moving averages are easily calculated when *k* is odd. For example, a **centered moving average** for a time series y_1, y_2, ..., y_n of *n* observations with $k = 3$ is defined as the series of means given by

$$M_2 = \frac{y_1 + y_2 + y_3}{3}$$

$$M_3 = \frac{y_2 + y_3 + y_4}{3}$$

$$M_4 = \frac{y_3 + y_4 + y_5}{3}$$

$$\vdots$$

$$M_{n-1} = \frac{y_{n-2} + y_{n-1} + y_n}{3}$$

The first average (involving the first three observations) is centered at $t = 2$, the next at $t = 3$, and the last at $t = n - 1$. Notice that the moving averages are centered at time periods given in the original series and that we have lost one point at the beginning of the series and one at the end. In general, for k odd, we lose $k - 1$ points in the smoothing process.

E X A M P L E   **13.4**   A large manufacturing firm reports the production figures shown in Table 13.3 (in thousands of units) for a two-year period. Compute a three-point moving average for this time series, and graph the moving average together with the original time series.

T A B L E **13.3**
Production figures for Example 13.4

Year	Jan.	Feb.	Mar.	Apr.	May	June	July	Aug.	Sept.	Oct.	Nov.	Dec.
1	1947	1981	2044	2036	1989	2105	2104	2051	2080	2144	2099	2149
2	2235	2304	2283	2349	2440	2495	2554	2691	2693	2692	2642	2728

Solution   We compute

$$M_2 = \frac{y_1 + y_2 + y_3}{3} = \frac{1947 + 1981 + 2044}{3} = 1990.7$$

$$M_3 = \frac{y_2 + y_3 + y_4}{3} = \frac{1981 + 2044 + 2036}{3} = 2020.3$$

$$\vdots$$

$$M_{23} = \frac{y_{22} + y_{23} + y_{24}}{3} = \frac{2692 + 2642 + 2728}{3} = 2687.3$$

The moving averages for the data in Table 13.3 are shown in Table 13.4. Graphs of the moving average and of the original time series are shown in Figure 13.9. You can see in Figure 13.9 that much of the erratic variation in the original time series is removed by the three-point moving average.   ■

T A B L E **13.4**   Three-point moving averages for the data in Table 13.3

Year	Jan.	Feb.	Mar.	Apr.	May	June	July	Aug.	Sept.	Oct.	Nov.	Dec.
1	—	1990.7	2020.3	2023.0	2043.3	2066.0	2086.7	2078.3	2091.7	2107.7	2130.7	2161.0
2	2229.3	2274.0	2312.0	2357.3	2428.0	2496.3	2580.0	2646.0	2692.0	2675.7	2687.3	—

F I G U R E **13.9**
Graphs of the three-point moving
average and of the original time
series for the data of Table 13.3

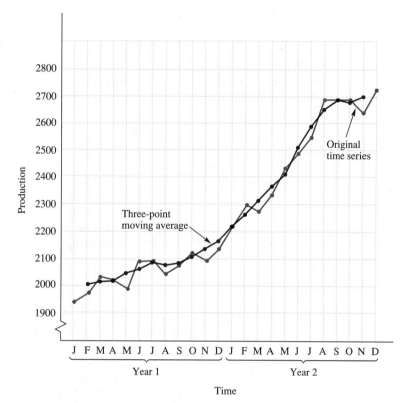

When k is even, as in the case of quarterly or monthly data, the series of moving averages are centered *between* the time periods given in the original series. For example, with $k = 4$, the first three moving averages,

$$M_{2.5} = \frac{y_1 + y_2 + y_3 + y_4}{4}$$

$$M_{3.5} = \frac{y_2 + y_3 + y_4 + y_5}{4}$$

$$M_{4.5} = \frac{y_3 + y_4 + y_5 + y_6}{4}$$

would be centered at $t = 2.5, \ 3.5,$ and 4.5. To avoid the problem of having the series of moving averages centered between the time periods used in the original series, the data are smoothed one more time, using a small even value of k, such as $k = 2$. With $k = 2$, a second smoothing would produce the centered values

$$M_3 = \frac{M_{2.5} + M_{3.5}}{2}$$

$$M_4 = \frac{M_{3.5} + M_{4.5}}{2}$$

$$M_5 = \frac{M_{4.5} + M_{5.5}}{2}$$

and so on. The doubly smoothed series has lost $k = 4$ points, two at the beginning and two at the end, and is now centered at time periods corresponding to the original series. When k is even and the series is doubly smoothed as indicated, the resulting smoothed series is referred to as a *centered k-point moving average*. Having the smoothed series centered at points corresponding to the original series facilitates the estimation of trend, seasonal, and cyclic components of the series.

E X A M P L E   **13.5**   The quarterly profits before taxes for a domestic nonfinancial corporation for the years 1990–94 are given in Table 13.5. To smooth out possible seasonal effects, compute a centered four-quarter moving average.

T A B L E   **13.5**
Profits before taxes for a
domestic nonfinancial
corporation, 1990–1994
(in $ millions)

Year	Quarter	Profits	Four-Quarter Moving Average	Centered Moving Average
1990	1	168.7		
	2	162.1		
			170.250	
	3	176.0		168.650
			167.050	
	4	174.2		167.688
			168.325	
1991	1	155.9		168.350
			168.375	
	2	167.2		170.475
			172.575	
	3	176.2		177.662
			182.750	
	4	191.0		187.838
			192.925	
1992	1	196.6		198.975
			205.025	
	2	207.9		207.600
			210.175	
	3	224.6		214.150
			218.125	
	4	211.6		222.200
			226.275	
1993	1	228.4		228.250
			230.225	
	2	240.5		234.600
			238.975	
	3	240.4		241.212
			243.450	
	4	246.6		
1994	1	246.3		

Solution   We begin by computing the four-quarter averages,

$$M_{2.5} = \frac{y_1 + y_2 + y_3 + y_4}{4}$$
$$= \frac{168.7 + 162.1 + 176.0 + 174.2}{4} = 170.25$$

$$M_{3.5} = \frac{y_2 + y_3 + y_4 + y_5}{4}$$
$$= \frac{162.1 + 176.0 + 174.2 + 155.9}{4} = 167.05$$

and so on. These quantities are given in Table 13.5 in the column labeled "four-quarter moving average." Notice that this series is centered at the values of $t = 2.5, 3.5$, etc. To center this series, we now compute a two-quarter moving average,

$$M_3 = \frac{M_{2.5} + M_{3.5}}{2} = \frac{170.25 + 167.05}{2} = 168.65$$

$$M_4 = \frac{M_{3.5} + M_{4.5}}{2} = \frac{167.05 + 168.325}{2} = 167.6875$$

and so on. The four-quarter *centered* moving averages are given in the column labeled "centered moving average" in Table 13.5. The original series y_t and the smoothed series are plotted in Figure 13.10. The smoothed series exhibits a dampening out of the quarterly variation that is present in the original series. The smoothing has brought out the strong linear trend in the data and also shows a possible long-term cyclical effect beginning with an upturn in profits in 1990 and a leveling out of profits in late 1993 and 1994.   ∎

FIGURE **13.10**
Graphs of the centered four-quarter moving average and the original series of Example 13.5

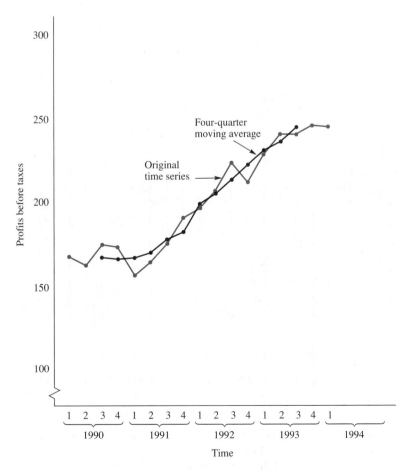

When the moving average technique is used to produce an estimate of the next value of the time series, y_{t+1}, the estimate is the average of the last k values of the series, given by

$$\hat{y}_{t+1} = \frac{y_t + y_{t-1} + \cdots + y_{t-k+1}}{k}$$

13.5 Exponential Smoothing

Another effective method for smoothing a time series is called **exponential smoothing.** Exponential smoothing is accomplished in the following way. Suppose that the values of the original time series are $y_1, y_2, \ldots, y_t$. Then the first, second, third, and tth values of the exponentially smoothed series, denoted by the symbols $E_1, E_2, E_3,$ and E_t, are

$$E_1 = y_1$$
$$E_2 = \omega y_2 + (1 - \omega)E_1$$
$$E_3 = \omega y_3 + (1 - \omega)E_2$$

$$\vdots$$

$$E_t = \omega y_t + (1 - \omega)E_{t-1}$$

The symbol ω that appears in the formulas for $E_1, E_2, \ldots, E_t$ is a weighting factor, a number between 0 and 1, that you must choose. Small values of ω give less weight to the current value of y_t and tend to produce a smoother series. Large values of ω give more weight to the current value of y_t and produce values of E_t that are very close to those for y_t.

DEFINITION ▪

An **exponentially smoothed value E_t** for a time series y_t is

$$E_t = \omega y_t + (1 - \omega)E_{t-1}$$

where

$$E_1 = y_1$$

and ω is a weight that can assume a value between 0 and 1. ▪

EXAMPLE 13.6 Refer to the data in Table 13.3. Calculate the exponentially smoothed time series for the original data, using a weight of $\omega = .4$. Then graph both the original time series and the exponentially smoothed time series on the same sheet of graph paper.

Solution The calculations for $E_1, E_2,$ and E_3 follow, and all values of E_t, $t = 1, 2, \ldots, 24,$ are shown in Table 13.6. The graphs of the exponentially smoothed time series and of the original time series are shown in Figure 13.11. Note the effect of the exponential

T A B L E **13.6**   Exponentially smoothed values for the time series in Table 13.3 with $\omega = .4$

Month

Year	Jan.	Feb.	Mar.	Apr.	May	June	July	Aug.	Sept.	Oct.	Nov.	Dec.
1	1947.0	1960.6	1994.0	2010.8	2002.1	2043.2	2067.5	2060.9	2068.6	2098.7	2098.8	2118.9
2	2165.3	2220.8	2245.7	2287.0	2348.2	2406.9	2465.8	2555.9	2610.7	2643.2	2642.7	2676.8

F I G U R E **13.11**
Graphs of the exponentially
smoothed ($\omega = .4$) and of
the original time series for the
data of Table 13.3

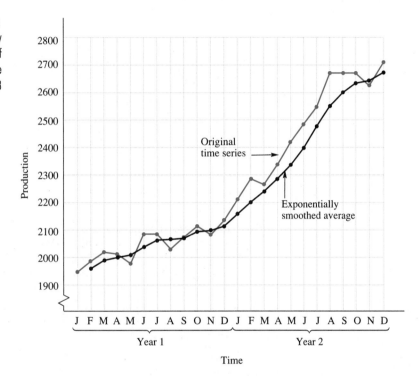

smoothing on the time series. Like the moving average (Figure 13.9), it clearly shows the rapidly rising upward trend during this two-year period.

$$E_1 = y_1 = 1947$$
$$E_2 = \omega y_2 + (1 - \omega)E_1 = (.4)(1981) + (.6)(1947) = 1960.6$$
$$E_3 = \omega y_3 + (1 - \omega)E_2 = (.4)(2044) + (.6)(1960.6) = 1994.0$$
$$\vdots$$
$$E_{24} = \omega y_{24} + (1 - \omega)E_{23} = (.4)(2728) + (.6)(2642.7) = 2676.8 \quad \blacksquare$$

When exponential smoothing is used as a predictive technique, the estimate of the value of the series at time $t + 1$, $\hat{y}_{t+1}$, is found as

$$\hat{y}_{t+1} = \omega y_t + (1 - \omega)\hat{y}_t$$

a weighted average of the last data value y_t and its estimate $\hat{y}_t$.

Exercises

Applications

13.9   Exercise 13.1 gave the total monthly production of raw steel (to the nearest million short ton) for each month in 1991. The data are reproduced below.

Month	Jan.	Feb.	Mar.	Apr.	May	June	July	Aug.	Sept.	Oct.	Nov.	Dec.
Production	7.6	6.7	7.3	7.1	7.1	7.0	7.3	7.4	7.5	7.7	7.5	7.3

Source: Business Statistics, 1963–91, 1992.

a   Plot the data points and connect them with line segments.

b   Calculate a two-point moving average for the data.

c   Calculate a three-point moving average for the data.

d   Graph the original time series along with the two- and three-point moving averages on the same sheet of graph paper. Compare the smoothing effects of the two- and the three-point moving averages.

13.10   The following time series data represent the total retail sales in the United States of both domestic and imported passenger cars (in thousands of units) for the years 1966 through 1991.

Year	Domestic	Imports	Year	Domestic	Imports
1966	8377	658	1979	8230	2329
1967	7568	779	1980	6581	2398
1968	8625	1030	1981	6209	2326
1969	8464	1117	1982	5758	2221
1970	7119	1283	1983	6793	2386
1971	8662	1566	1984	7952	2442
1972	9252	1621	1985	8205	2834
1973	9588	1762	1986	8215	3235
1974	7363	1412	1987	7081	3197
1975	6951	1587	1988	7539	3099
1976	8492	1502	1989	7078	2825
1977	8971	2075	1990	6898	2601
1978	9164	2000	1991	6137	2251

Source: Business Statistics, 1963–91, 1992.

a   Graph the time series for the annual domestic sales from 1966 to 1991.

b   Compute and graph a three-point moving average for the data from 1966 to 1991. Note the smoothing effect of the averaging.

c Compute and graph a centered four-point moving average for the data from 1966 to 1991 and note the additional smoothing obtained.

13.11 Refer to Exercise 13.10.

a Graph the time series for the annual imported sales from 1966 to 1991.

b Compute and graph a three-point moving average for the data from 1966 to 1991. Note the smoothing effect of the averaging.

c Compute and graph a centered four-point moving average for the data from 1966 to 1991 and note the additional smoothing obtained.

13.12 The mean wholesale prices of eggs, in cents per dozen, are shown in the accompanying table for each month from January 1985 through December 1993.

Year	Jan.	Feb.	Mar.	Apr.	May	June	July	Aug.	Sept.	Oct.	Nov.	Dec.	Average
1985	58.4	55.1	62.3	57.3	52.9	60.8	58.6	66.4	70.5	70.7	74.6	73.2	63.4
1986	70.6	65.7	76.9	62.6	62.0	57.3	69.4	70.0	69.4	66.3	74.1	72.8	68.1
1987	64.4	62.0	59.2	59.0	51.8	55.6	55.4	58.7	64.8	55.5	56.3	52.1	57.9
1988	51.2	48.9	53.6	47.9	47.1	52.8	69.8	65.4	71.4	63.1	62.2	66.1	58.3
1989	67.8	66.6	91.0	71.6	69.8	72.0	71.8	79.6	77.2	79.4	89.1	94.3	77.7
1990	88.6	75.1	86.1	78.6	60.3	66.9	64.2	73.9	75.4	80.0	80.0	83.1	76.0
1991	86.0	72.0	85.8	67.6	60.9	63.4	73.1	71.3	68.8	67.9	68.7	73.9	71.4
1992*	59.1	55.7	55.7	57.4	52.0	56.0	53.0	57.9	64.9	58.2	69.4	68.0	58.9
1993*	65.7	63.6	77.5	70.9	61.9	67.6	62.8	67.6	60.6	64.2	65.6	65.6	66.1

Source: 1994 CRB Commodity Year Book, 1994.
*Preliminary.

a Plot the data points for the 36 months from January 1991 through December 1993.

b Construct a three-point moving average for the data in part (a).

c Construct a graph showing the original time series and the three-point moving averages. Does the smoothing begin to show any secular, cyclical, or seasonal components in the time series for egg prices?

13.13 Perform exponential smoothing for the time series in Exercise 13.9.

a Use $\omega = .2$.

b Use $\omega = .5$.

c Use $\omega = .8$.

d Graph the original time series with the three exponentially smoothed series from parts (a), (b), and (c). Explain the effect of the value of ω on the smoothing.

13.14 Exponentially smooth the time series for the annual domestic car sales (Exercise 13.10), using a weighting factor of $\omega = .4$. Graph the original time series and the exponentially smoothed series to note the effect of the exponential smoothing on the original data. Is anything revealed by the smoothing?

13.15 Perform exponential smoothing for the egg price time series (Exercise 13.12), using a weight of $\omega = .3$. Construct a graph showing the original time series and the exponentially smoothed series. Does the smoothing begin to show any secular, cyclical, or seasonal components in the series?

13.16 The accompanying table gives the electric power production (in billions of kilowatt-hours) by electric utilities in the United States each month from January 1979 through November 1993.

Year	Jan.	Feb.	Mar.	Apr.	May	June	July	Aug.	Sept.	Oct.	Nov.	Dec.	Total
1979	209.7	186.3	182.8	170.0	178.1	186.7	202.3	204.9	180.8	179.7	177.5	188.7	2,247
1980	200.0	188.7	187.5	168.7	175.7	189.4	216.8	215.4	191.5	178.6	178.6	195.6	2,286
1981	206.5	179.6	185.6	172.5	177.8	202.7	220.4	210.4	186.8	181.4	175.6	195.6	2,295
1982	209.4	180.3	187.7	172.6	177.1	186.1	210.6	205.7	180.7	173.0	173.4	184.7	2,241
1983	195.6	172.5	182.5	170.4	174.4	191.0	220.2	230.0	195.6	182.9	182.9	212.3	2,310
1984	216.6	189.6	200.1	181.1	192.2	209.6	221.2	229.3	195.2	190.9	190.4	200.0	2,416
1985	227.9	198.2	195.0	184.9	196.8	205.4	226.7	226.1	202.5	194.8	192.4	219.3	2,470
1986	217.5	192.3	196.8	186.1	197.3	215.0	242.7	225.2	206.7	197.8	196.4	213.6	2,487
1987	222.7	194.0	201.8	189.5	206.1	225.6	247.9	247.6	213.0	203.0	200.3	220.5	2,572
1988	237.6	216.7	213.8	195.8	208.2	232.5	257.2	267.4	220.0	210.4	209.4	232.6	2,704
1989	231.3	219.1	226.4	207.7	219.8	235.4	256.7	258.3	226.9	219.1	219.0	258.6	2,784
1990	237.0	212.7	225.7	210.8	222.6	248.9	266.2	268.2	237.7	224.8	213.6	237.3	2,807
1991	248.0	210.5	221.1	208.9	233.9	248.2	271.5	267.7	233.9	223.2	221.2	233.6	2,823
1992*	244.0	217.8	224.7	210.8	220.4	236.8	266.1	255.2	234.8	221.3	221.3	244.1	2,797
1993*	245.8	224.7	234.6	211.3	222.4	249.6	282.3	279.1	236.5	223.6	225.8		

Source: CRB Commodity Year Book 1994, 1994.
*Preliminary.

a Plot the data for the months from January 1991 through November 1993.

b Smooth the data with a three-point moving average to reveal secular, cyclical, and/or seasonal trends.

c Graph the original time series along with the three-point moving average. Explain what, if anything, is revealed by smoothing.

d Smooth the data using a five-point moving average. Does this improve the smoothing?

13.17 Use exponential smoothing on the time series for Exercise 13.16 with $\omega = .3$. Was exponential smoothing with $\omega = .3$ successful in smoothing the time series and revealing secular, cyclical, and/or seasonal trends?

13.18 Refer to the electric power production data given in Exercise 13.16.

a Use the data from January 1989 to November 1993 to compute a 12-month moving average for this time series.

b Center the 12-point moving average found in part (a) using a second smoothing with $k = 2$.

13.19 The following table gives the quarterly earnings per share for the stocks of McDonald's Corporation for the years 1991 through 1995.

Year	Mar. 31	June 30	Sept. 30	Dec. 31
1991	.23	.32	.36	.27
1992	.25	.35	.40	.30
1993	.29	.39	.43	.35
1994	.33	.43*	.49*	.40*
1995	.38*	.50*	.57*	.45*

Source: Value Line Investment Survey, 1994.
*Estimated.

a Use these quarterly data to compute a four-month moving average for this time series.

b Center the four-point moving average found in part (a) using a second smoothing with $k = 2$.

13.6 Time Series Analysis: Multiplicative Models

When trend, cyclical, seasonal, and random variation are present in the data, the time series is modeled as

$$y_t = T_t \cdot C_t \cdot S_t \cdot \epsilon_t$$

Since seasonal variation refers to a pattern that recurs at the same time each year, seasonal components arise when data are collected *daily, monthly,* or *quarterly*. In contrast, data collected or reported yearly would exhibit a secular trend, a cyclical effect, and random variation. Therefore, *yearly* data would be modeled as

$$y_t = T_t \cdot C_t \cdot \epsilon_t$$

The secular or long-term trend in a time series is often modeled by using the first- or second-order polynomial regression models of Chapter 12. Straight-line, long-term trends in a time series, either up or down, are represented by the first-order model

$$T_t = \beta_0 + \beta_1 t$$

A second-order model

$$T_t = \beta_0 + \beta_1 t + \beta_2 t^2$$

is used when increases in T_t over time tend to decrease [see Figure 13.12(a)] or tend to increase [see Figure 13.12(b)]. A second-order model would also be appropriate if decreases in T_t over time tended to decrease, as shown in Figure 13.12(c).

F I G U R E 13.12
Second-order models to remove secular trend

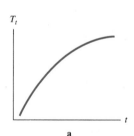 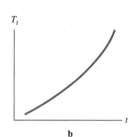 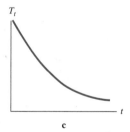

a   b   c

If we estimate the trend component using the regression methods of Chapters 11 and 12, this estimate, say $\hat{T}_t$, can be used to **detrend** the series whereby

$$\frac{y_t}{\hat{T}_t} \approx C_t \cdot \epsilon_t$$

The detrended series should more clearly exhibit any cyclical behavior present in the original series.

E X A M P L E   **13.7**   Fit a linear trend model to the new domestic automobile sales data given in Table 13.1. Use the fitted line to detrend the data using the multiplicative model

$$y_t = T_t \cdot C_t \cdot \epsilon_t \quad \text{where } t = \text{Year}$$

Solution   The Minitab regression analysis is given in Table 13.7. Notice that although the slope $\hat{\beta}_1 = -.112$ is significant at the .067 level of significance, the model accounts for only $R^2 = 20.6\%$ of the variation in y. The fitted trend line is given by

$$\hat{T}_t = 8.622 - .112t$$

The values of $\hat{T}_t$ are found in the column labeled Fit in the lower portion of the printout. The *detrended series* is found by dividing each value of y_t by $\hat{T}_t$. For example, when $t = 1$,

$$\frac{y_1}{\hat{T}_1} = \frac{8.8}{8.510} = 1.034$$

and when $t = 2$,

$$\frac{y_2}{\hat{T}_2} = \frac{9.7}{8.398} = 1.155$$

T A B L E   **13.7**
Minitab printout fitting a linear
trend line to the auto sales data
in Table 13.1*

MTB > REGRESS C1 1 C2

The regression equation is
Y = 8.62 - 0.112 T

Predictor	Coef	Stdev	t-ratio	p
Constant	8.6221	0.5831	14.79	0.000
T	-0.11225	0.05690	-1.97	0.067

s = 1.149   R-sq = 20.6%   R-sq(adj) = 15.3%

Analysis of Variance

SOURCE	DF	SS	MS	F	p
Regression	1	5.141	5.141	3.89	0.067
Error	15	19.816	1.321		
Total	16	24.958			

Obs.	T	Y	Fit	Stdev.Fit	Residual	St.Resid	$y_t/\hat{T}_t$
1	1.0	8.800	8.510	0.534	0.290	.29	1.034
2	2.0	9.700	8.398	0.486	1.302	1.25	1.155
3	3.0	7.300	8.285	0.441	-0.985	-0.93	.881
4	4.0	6.700	8.173	0.398	-1.473	-1.37	.820
5	5.0	8.500	8.061	0.360	0.439	0.40	1.054
6	6.0	9.200	7.949	0.327	1.251	1.14	1.157
7	7.0	9.200	7.836	0.301	1.364	1.23	1.174
8	8.0	8.400	7.724	0.285	0.676	0.61	1.088
9	9.0	6.400	7.612	0.279	-1.212	-1.09	.841
10	10.0	6.200	7.500	0.285	-1.300	-1.17	.827
11	11.0	5.000	7.387	0.301	-2.387	-2.15R	.677
12	12.0	6.700	7.275	0.327	-0.575	-0.52	.921
13	13.0	7.600	7.163	0.360	0.437	0.40	1.061
14	14.0	8.000	7.050	0.398	0.950	0.88	1.135
15	15.0	7.500	6.938	0.441	0.562	0.53	1.081
16	16.0	7.100	6.826	0.486	0.274	0.26	1.040
17	17.0	7.100	6.714	0.534	0.386	0.38	1.057

*$y_t/\hat{T}_t$ calculated separately.

The detrended values, which are plotted in Figure 13.13, exhibit a cyclical pattern that appears to have a seven-year cycle.   ▪

F I G U R E **13.13**
Detrended time series using the
data in Table 13.7

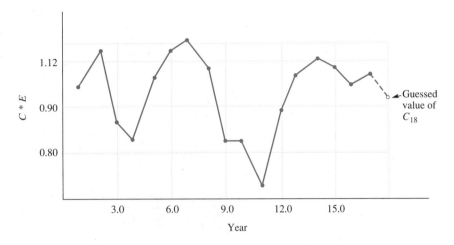

If the cyclical pattern in Figure 13.13 were to continue into the future unchanged, we would *approximate* the cyclical effect at time $t = 18$ to be $\hat{C}_{18} = .96$. The estimated trend component at $t = 18$ would be

$$\hat{T}_{18} = 8.622 - .112(18) = 6.606$$

Therefore, the estimated domestic auto sales for year 18 would be

$$\hat{y}_{18} = \hat{T}_{18} \cdot \hat{C}_{18} = 6.606(.96) = 6.34$$

How reliable is the estimate? Standard errors for estimators of this kind are not available in general, and the approximations to their standard errors are beyond the scope of this text. Furthermore, cyclical patterns may not remain stable in future time periods, and they are very difficult to predict. Hence, some forecasters prefer to use the trend estimate alone, with the standard error appropriate for the regression estimate.

Deseasonalizing a Time Series: Ratio-to-Moving-Average Method

When seasonal components are present in a time series, the **ratio-to-moving-average method** can be used to estimate the seasonal component S_t in the multiplicative model. Notice that when the estimated seasonal components, $\hat{S}_t$, are isolated the series can be *deseasonalized* as

$$\frac{y_t}{\hat{S}_t} = \frac{T_t \cdot C_t \cdot S_t \cdot \epsilon_t}{\hat{S}_t} \approx T_t \cdot C_t \cdot \epsilon_t$$

When this deseasonalized series is subsequently *detrended,* the resulting series should reveal the cyclical patterns and random variation that remain.

The ratio-to-moving-average method assumes that seasonal effects remain constant across the seasons. For example, if we consider quarterly data (one season consists of four quarters), then, over four seasons, it is assumed that

$$S_1 = S_5 = S_9$$
$$S_2 = S_6 = S_{10}$$
$$S_3 = S_7 = S_{11}$$
$$S_4 = S_8 = S_{12}$$

Furthermore, the seasonal effects S_1, S_2, S_3, and S_4 are normed to add to four. In this way, when four consecutive quarterly observations are summed and divided by four, the average is free of the seasonal effects. In general, if a season consists of M periods, then the average of M consecutive observations will be free of seasonal effects.

DEFINITION ▪

> A **specific seasonal index** s_t is found by dividing the original value of the series y_t by the corresponding *centered* moving average of order M at time t. ▪

Consider the data in Table 13.5. In the process of deseasonalizing this series, the first specific seasonal index to be calculated is for the third quarter of 1990:

$$s_3 = \frac{y_3}{M_3} = \frac{176.000}{168.650} = 1.043581$$

Each specific seasonal index for times $t = 3$ through $t = 15$ is calculated in the same way and is shown in Table 13.8. Since s_3, s_7, s_{11}, and s_{15} are each estimates of S_3, we use their average $\bar{s}_3$ as the initial estimator of S_3. Hence,

$$\bar{s}_3 = \frac{s_3 + s_7 + s_{11} + s_{15}}{4}$$
$$= \frac{1.043581 + .991768 + 1.048798 + .996632}{4} = 1.020195$$

Once $\bar{s}_1$, $\bar{s}_2$, $\bar{s}_3$, and $\bar{s}_4$ are calculated, the normed seasonalized indexes are found as

$$\hat{S}_i = \bar{s}_i \left(\frac{4}{S}\right) \quad \text{where } S = \bar{s}_1 + \bar{s}_2 + \bar{s}_3 + \bar{s}_4$$

DEFINITION ▪

> The **seasonal index** $\hat{S}_i$ for a particular month (quarter) is found by averaging all the specific indexes associated with that month (quarter) and then normalizing the resulting M indexes so that they sum to M. ▪

Year	Quarter	Profits	Four-Quarter Moving Average	Centered Moving Average	Specific Seasonal, s_t	$\dfrac{y_t}{\hat{S}_t}$
1990	1	168.7				173.5
	2	162.1	170.250			161.6
	3	176.0	167.050	168.650	1.043581	172.4
	4	174.2	168.325	167.688	1.038837	173.6
1991	1	155.9	168.375	168.350	.926047	160.3
	2	167.2	172.575	170.475	.980789	166.7
	3	176.2	182.750	177.662	.991768	172.6
	4	191.0	192.925	187.838	1.016836	190.3
1992	1	196.6	205.025	198.975	.988064	202.2
	2	207.9	210.175	207.600	1.001445	207.2
	3	224.6	218.125	214.150	1.048798	220.0
	4	211.6	226.275	222.200	.952295	210.9
1993	1	228.4	230.225	228.250	1.000657	234.9
	2	240.5	238.975	234.600	1.025149	239.7
	3	240.4	243.450	241.212	.996632	235.5
	4	246.6				245.8
1994	1	246.3				253.3

T A B L E 13.8
Profits before taxes for a domestic nonfinancial corporation, 1990–1994

Finally, the original series is deseasonalized by dividing each value y_t by its estimated seasonal index.

E X A M P L E 13.8   Use the information concerning profits before taxes for the domestic nonfinancial institution given in Table 13.8 to find the normed seasonal indexes. Use these indexes to deseasonalize this time series.

Solution   We have already calculated $\bar{s}_3 = 1.020195$. Since there are only three fourth-quarter specific seasonal indexes,

$$\bar{s}_4 = \frac{s_4 + s_8 + s_{12}}{3}$$

$$= \frac{1.038837 + 1.016836 + .952295}{3} = 1.002656$$

Similarly,

$$\bar{s}_1 = \frac{s_5 + s_9 + s_{13}}{3}$$

$$= \frac{.926047 + .988064 + 1.000657}{3} = .971589$$

and

$$\bar{s}_2 = \frac{s_6 + s_{10} + s_{14}}{3}$$

$$= \frac{.980789 + 1.001445 + 1.025149}{3} = 1.002461$$

To find the normed seasonal indexes, we first need the sum

$$S = \bar{s}_1 + \bar{s}_2 + \bar{s}_3 + \bar{s}_4 = 3.996901$$

Then

$$\hat{S}_1 = .971589 \left(\frac{4}{3.996901} \right) = .972342 \qquad \hat{S}_2 = 1.002461 \left(\frac{4}{3.996901} \right) = 1.003238$$

$$\hat{S}_3 = 1.020195 \left(\frac{4}{3.996901} \right) = 1.020986 \qquad \hat{S}_4 = 1.002656 \left(\frac{4}{3.996901} \right) = 1.003433$$

The first several values of the deseasonalized series are given by

$$\frac{y_1}{\hat{S}_1} = \frac{168.7}{.972342} = 173.5$$

$$\frac{y_2}{\hat{S}_2} = \frac{162.1}{1.003238} = 161.6$$

and so on. The deseasonalized series is shown in the last column of Table 13.8, labeled $y_t / \hat{S}_t$. ∎

The deseasonalized series should now be free of seasonal effects; the remaining variation is due to possible cyclical effects, trend, and random variation. The deseasonalized series, plotted in Figure 13.14, shows a marked linear trend. The results of a Minitab simple linear regression analysis using the deseasonalized series as the dependent variable and time ($t = 1, 2, \ldots, 17$) as the independent variable are given in Table 13.9. These results indicate a reasonably good fit ($R^2 = 90.2\%$). The three columns at the bottom of Table 13.9 give the value of the deseasonalized series, the estimated trend at time t, and the residuals after the series was detrended. For example, at time $t = 1$, the value of the residual is

$$\frac{173.50}{152.587} = 1.13706$$

FIGURE **13.14**
Deseasonalized time series,
Example 13.8

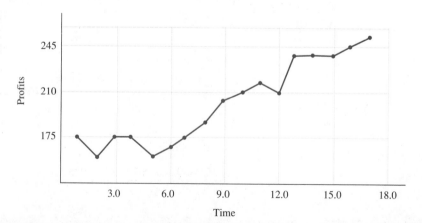

T A B L E **13.9**   Minitab printout of trend line for the deseasonalized data in Example 13.8

```
MTB > REGRESS C2 1 C1;
SUBC> PREDICT 18.

The regression equation is
PROFITS = 147 + 6.08 TIME

Predictor        Coef        Stdev      t-ratio         p
Constant      146.509        5.303        27.63     0.000
TIME           6.0775       0.5176        11.74     0.000

s = 10.45      R-sq = 90.2%      R-sq(adj) = 89.5%

Analysis of Variance

SOURCE         DF          SS          MS         F         p
Regression      1       15070       15070    137.89     0.000
Error          15        1639         109
Total          16       16709

Unusual Observations
Obs.    TIME     PROFITS        Fit   Stdev.Fit    Residual    St.Resid
  1      1.0      173.50     152.59        4.86       20.91       2.26R

R denotes an obs. with a large st. resid.

    Fit   Stdev.Fit         95% C.I.              95% P.I.
 255.90        5.30    ( 244.60,   267.21)   ( 230.91,   280.90)


    ROW    PROFITS      TREND      RESDLS

      1      173.5     152.59     1.13703
      2      161.6     158.66     1.01853
      3      172.4     164.74     1.04650
      4      173.6     170.82     1.01627
      5      160.3     176.90     0.90616
      6      166.7     182.97     0.91108
      7      172.6     189.05     0.91299
      8      190.3     195.13     0.97525
      9      202.2     201.21     1.00492
     10      207.2     207.28     0.99961
     11      220.0     213.36     1.03112
     12      210.9     219.44     0.96108
     13      234.9     225.52     1.04159
     14      239.7     231.59     1.03502
     15      235.5     237.67     0.99087
     16      245.8     243.75     1.00841
     17      253.3     249.83     1.01389
```

and when $t = 2$, the residual is

$$\frac{161.60}{158.664} = 1.01850$$

A plot of the residuals after deseasonalizing and detrending the series is given in Figure 13.15. It appears that, with the exception of the first residual, the remaining residuals do not exhibit any cyclical effects.

F I G U R E **13.15**
Minitab plot of residuals after
deseasonalizing and detrending
the profit data in Example 13.8

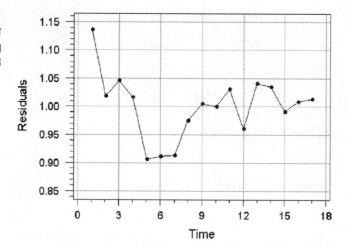

How would we produce a forecast of profit after taxes for the second quarter of 1994? Because we observed no strong cyclical effect, the multiplicative model that we have fitted is

$$y_t = S_t \cdot T_t \cdot \epsilon_t \quad \text{with } E(y_t) = S_t \cdot T_t$$

Now quarter two of 1994 corresponds to $t = 18$, and the estimated trend, $\hat{T}_{18}$, is given by

$$\hat{T}_{18} = 146.509 + 6.0775(18) = 255.90$$

(given by Fit in Table 13.9). The seasonal index for quarter two is estimated to be $\hat{S}_2 = 1.003238$. Therefore, the estimate of y_{18} is

$$\hat{y}_{18} = \hat{T}_{18} \cdot \hat{S}_2 = (255.90)(1.003238) = 256.73$$

Again, standard errors for estimators of this sort are not readily available. An alternative set of models using additive effects, presented in the next section, do have approximate standard errors that are easily calculated.

Before we conclude this section, we should point out that a *refined* ratio-to-moving-average method for deseasonalizing data is available through the Statistical Analysis Systems (SAS) package for analyzing Economic Time Series (SAS ETS). This program, known as X11 (PROC X11), is used by the U.S. Department of the Census to deseasonalize many sets of economic time series.

Exercises

Applications

13.20   The data in Exercise 13.10 represent the total U.S. retail sales of domestic and imported passenger cars for the years 1966 through 1991. The import sales data (in thousands of units) are reproduced in the accompanying table.

Table for Exercise 13.20

Year	Retail Sales	Year	Retail Sales	Year	Retail Sales	Year	Retail Sales
1966	658	1973	1762	1980	2398	1987	3197
1967	779	1974	1412	1981	2326	1988	3099
1968	1030	1975	1587	1982	2221	1989	2825
1969	1117	1976	1502	1983	2386	1990	2601
1970	1283	1977	2075	1984	2442	1991	2251
1971	1566	1978	2000	1985	2834		
1972	1621	1979	2329	1986	3235		

Source: Business Statistics, 1963–91, 1994.

Minitab output for Exercise 13.20

```
The regression equation is
Sales = 837 + 87.7 Year

Predictor        Coef        Stdev      t-ratio         p
Constant        836.6        121.1         6.91     0.000
Year           87.702        7.840        11.19     0.000

s = 299.8        R-sq = 83.9%     R-sq(adj) = 83.2%

Analysis of Variance

SOURCE        DF           SS           MS         F          p
Regression     1     11248994     11248994    125.14      0.000
Error         24      2157362        89890
Total         25     13406356


    ROW    Year    Sales      FITS1

     1       1       658     924.34
     2       2       779    1012.04
     3       3      1030    1099.75
     4       4      1117    1187.45
     5       5      1283    1275.15
     6       6      1566    1362.85
     7       7      1621    1450.55
     8       8      1762    1538.26
     9       9      1412    1625.96
    10      10      1587    1713.66
    11      11      1502    1801.36
    12      12      2075    1889.06
    13      13      2000    1976.77
    14      14      2329    2064.47
    15      15      2398    2152.17
    16      16      2326    2239.87
    17      17      2221    2327.57
    18      18      2386    2415.28
    19      19      2442    2502.98
    20      20      2834    2590.68
    21      21      3235    2678.38
    22      22      3197    2766.08
    23      23      3099    2853.79
    24      24      2825    2941.49
    25      25      2601    3029.19
    26      26      2251    3116.89
```

a   Use the accompanying Minitab printout to fit a linear trend model to the data. [*Note:* Time is coded as $t = $ Year $-$ 1965.] How well does the linear model fit the data?

b   Detrend the series using a multiplicative model and the results of part (a). Plot the detrended series. Are there are any cyclical effects?

c   Forecast the sales of imported passenger cars in 1992. How reliable is this estimate?

13.21   The numbers of business failures in the United States for the years 1972 through 1990 were given in Exercise 13.2 and are reproduced in the accompanying table.

Year	Failures	Year	Failures
1972	9566	1982	24,908
1973	9345	1983	31,534
1974	9915	1984	52,078
1975	11,432	1985	57,252
1976	9628	1986	61,601
1977	7919	1987	61,384
1978	6619	1988	57,099
1979	7564	1989	50,361
1980	11,742	1990	60,432
1981	16,749		

Source: Business Statistics, 1963–91, 1992.

a   Plot the data. Note the strong cubic effect that the data show.

b   Use the accompanying Minitab printout to fit a cubic trend model to the data. [*Note:* Time is coded as $t = $ Year $-$ 1971.]

c   Detrend the series using a multiplicative model and the results of part (b). Plot the detrended series. Are there any cyclical effects?

d   Forecast the number of business failures in the United States in 1991. How reliable is this estimate?

13.22   The electric power production data in Exercise 13.16 should show a pronounced seasonal effect. Use the ratio-to-moving average method to deseasonalize the time series. [*Hint:* The 12-month centered moving average was calculated in Exercise 13.18.]

13.23   Refer to Exercise 13.22.

a   Detrend the deseasonalized time series using a linear trend model. Do there appear to be any cyclical effects present?

b   Forecast the electric production for December 1993.

13.24   Refer to the McDonald's quarterly earnings in Exercise 13.19. Use the ratio-to-moving average method to deseasonalize the time series [*Hint:* The four-quarter centered moving average was calculated in Exercise 13.19.]

13.25   Refer to Exercise 13.24.

a   Detrend the deseasonalized time series using a linear trend model. Do there appear to be any cyclical effects present?

b   Forecast the quarterly earnings for the first quarter of 1996.

Minitab output for Exercise 13.21

```
The regression equation is
Y = 27449 - 11887 X + 1655 X-SQ - 49.7 X-CU

Predictor        Coef      Stdev    t-ratio        p
Constant        27449       7834       3.50    0.003
X              -11887       3305      -3.60    0.003
X-SQ           1655.2      378.9       4.37    0.001
X-CU           -49.70      12.48      -3.98    0.001

s = 6912       R-sq = 92.3%     R-sq(adj) = 90.8%

Analysis of Variance

SOURCE        DF          SS           MS         F        p
Regression     3  8645601280   2881867008     60.32    0.000
Error         15   716614144     47774276
Total         18  9362214912

SOURCE        DF      SEQ SS
X              1  7521741312
X-SQ           1   365591776
X-CU           1   758267776

    ROW        Y      FITS1

      1     9566    17167.6
      2     9345     9898.5
      3     9915     5343.2
      4    11432     3203.7
      5     9628     3181.7
      6     7919     4979.1
      7     6619     8297.5
      8     7564    12838.9
      9    11742    18305.0
     10    16749    24397.7
     11    24908    30818.6
     12    31534    37269.7
     13    52078    43452.7
     14    57252    49069.4
     15    61601    53821.7
     16    61384    57411.3
     17    57099    59540.0
     18    50361    59909.6
     19    60432    58221.9
```

13.7 Time Series Analysis: Additive Models

The multiplicative model we discussed in Section 13.6 is one of several different ways to develop a **time series forecasting model** and to analyze a time series. Another common method is to model the secular, cyclical, and seasonal components of a time series using an **additive regression model** given by

$$E(y_t) = T_t + C_t + S_t$$

This deterministic (nonrandom) portion of a time series model is fit to the time series data to obtain a prediction equation $\hat{y}_t$. The residuals—the deviations between the observed and the predicted values of y_t—are computed next. We then attempt to model these residuals by using a model that accounts for possible correlation between adjacent residual terms.

The long-term (or secular) trend is modeled using the methods of Chapter 12, as in Section 13.6. Either a linear, quadratic, or exponential model can be used.

Periodic movements of $E(y_t)$ about the secular trend line (curve) can be explained by the cyclical nature of the economy or (sometimes) by seasonal variations. Cyclical components can be added to the secular trend model by using trigonometric terms. For example, if we observe cyclical oscillation about a straight-line secular trend, we might use the model

$$E(y_t) = \overbrace{\beta_0 + \beta_1 t}^{\text{secular trend}} + \overbrace{\beta_2 \cos\left(\frac{2\pi}{b}t\right) + \beta_3 \sin\left(\frac{2\pi}{b}t\right)}^{\text{cyclical effect}}$$

The contribution to $E(y_t)$ provided by the trigonometric terms will produce a curve, measured from the secular trend line, as shown in Figure 13.16.

The time units, days, months, years, and so on, are measured along the horizontal axis of Figure 13.16. The terms $\beta_2 \cos(2\pi t/b) + \beta_3 \sin(2\pi t/b)$ trace a cyclical curve about the secular trend line that has an amplitude (maximum deviation from the secular trend line) equal to $a = \sqrt{\beta_2^2 + \beta_3^2}$. The standard position of the cyclical curve is for the maximum value of $E(y_t)$ (ignoring secular trend) to occur at $t = 0$. The curve will shift to the right of this position at a distance equal to $d = \tan^{-1}(\beta_3/\beta_2)$. The parameters β_0, β_1, β_2, and β_3 in the model

$$E(y_t) = \overbrace{\beta_0 + \beta_1 t}^{\text{secular trend}} + \beta_2 \cos\left(\frac{2\pi}{b}t\right) + \beta_3 \sin\left(\frac{2\pi}{b}t\right)$$

FIGURE 13.16
Contribution of cyclical terms,
$\beta_2 \cos(2\pi t/b) + \beta_3 \sin(2\pi t/b)$

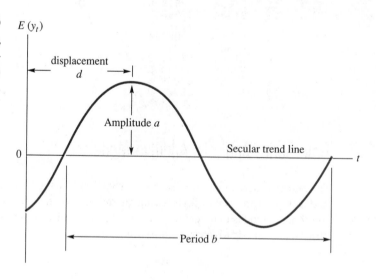

determine the amplitude a and the displacement d of the time series. These parameters can be estimated by the method of least squares. In contrast, the parameter b, which measures the period of the cycle, enters the model in a nonlinear manner and cannot be estimated by least squares. The value of b must be based on a visual observation of a graph of the time series and selected prior to estimating β_2 and β_3.

Seasonal effects that vary from the secular trend line (or curve) can be more easily inserted into a model. For example, new housing starts are usually low in the first and fourth quarters of the year (during the winter) but tend upward from April through September. If time is recorded in quarters, this type of seasonal effect is added to the long-term trend line by using the model

$$E(y_t) = \overbrace{\beta_0 + \beta_1 t}^{\text{secular trend}} + \beta_2 x_1 + \beta_3 x_2 + \beta_4 x_3$$

where

$$x_1 = \begin{cases} 1, & \text{if quarter 1} \\ 0, & \text{if not} \end{cases}$$

$$x_2 = \begin{cases} 1, & \text{if quarter 2} \\ 0, & \text{if not} \end{cases}$$

$$x_3 = \begin{cases} 1, & \text{if quarter 3} \\ 0, & \text{if not} \end{cases}$$

The independent variables x_1, x_2, and x_3 are dummy (indicator) variables that add β_2, β_3, and β_4 to the secular portion of $E(y_t)$ depending on whether the quarter is, respectively, the first, second, or third. When we want to model $E(y_t)$ for the fourth quarter, $x_1 = x_2 = x_3 = 0$ and $E(y_t) = \beta_0 + \beta_1 t$. Thus β_2, β_3, and β_4 are the positive or negative deviations of $E(y_t)$ from the secular trend line for quarters 1, 2, and 3, respectively. A graph depicting this situation is shown in Figure 13.17.

FIGURE 13.17
A model that adjusts for seasonal trends

$E(y_t)$

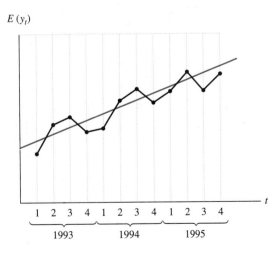

Almost all time series models will include terms to adjust for the secular trend, and they may include cyclical terms or terms to adjust for seasonal effects. We have described the nature of these terms and explained how they can be combined to form a model for the mean value of the time series variable y_t.

At this point, it is of interest to note that the multiplicative model of Section 13.6 given by

$$E(y_t) = T_t \cdot C_t \cdot S_t$$

is additive in the log-scale, since

$$\log(E(y_t)) = \log T_t + \log C_t + \log S_t$$

Therefore, as an alternative to using the techniques in Section 13.6, you may wish to fit an additive model to the transformed series $\log y_t$.

The final step in forming a time series model is to add a term to account for random variation of the values of y_t from $E(y_t)$, where $E(y_t)$ is the deterministic portion of the model. In contrast to the regression model of Chapter 11, we will want to allow for a correlation between pairs of random errors. We would expect the errors associated with adjacent values of y_t to be more highly correlated than the errors corresponding to two values of y_t that are far apart in time. For example, if the index of U.S. economic activity in March is higher than the value explained by $E(y_t)$—that is, the deviation is positive—we think there is a good chance that the deviation in the following month will also be positive.

One of the primary tools that is used in determining the correlation structure among the random errors in the model is the **autocorrelation function (acf)** of the time series. The autocorrelation function summarizes the various correlations that a time series has with *itself.* The components of the acf are calculated by the formula

$$r_k = \frac{\sum\limits_{t=k+1}^{n} (y_t - \bar{y})(y_{t-k} - \bar{y})}{\sum\limits_{t=1}^{n}(y_t - \bar{y})^2} \qquad \text{for } k = 1, 2, 3, \ldots$$

where r_k, the autocorrelation coefficient of lag k, measures the correlation of the y_t series that are k periods apart. The acf simply lists all such autocorrelations $r_1, r_2, r_3, \ldots$ together.

To interpret the acf, we need a method for testing which of the r_ks are statistically different from zero. Since the exact distribution of the r_ks is rarely known, we can perform *approximate tests.* One such test is given in the following display.

Tests Concerning Autocorrelation Coefficients

1 Null hypothesis: $H_0 : \rho_k = 0$

2 Alternative hypothesis: $H_a : \rho_k \neq 0$, where ρ_k is the *population* autocorrelation coefficient of lag k

> **3** Test statistic: r_k, the sample autocorrelation coefficient of lag k
>
> **4** Rejection region: At an approximate significance level of $\alpha \approx .05$ reject H_0 if $|r_k| > 2/\sqrt{n}$

The Minitab printout of the autocorrelation coefficients up to lag 10 for the data in Example 13.1 are given in Table 13.10.

T A B L E 13.10
Minitab printout of autocorrelation coefficients

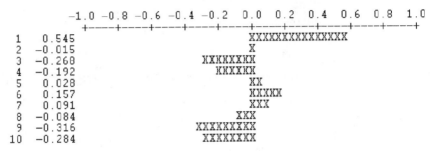

```
ACF of C1

               -1.0 -0.8 -0.6 -0.4 -0.2  0.0  0.2  0.4  0.6  0.8  1.0
               +----+----+----+----+----+----+----+----+----+----+
  1    0.545                                   XXXXXXXXXXXXXXX
  2   -0.015                                   X
  3   -0.268                      XXXXXXXX
  4   -0.192                        XXXXX
  5    0.028                                   XX
  6    0.157                                   XXXXX
  7    0.091                                   XXX
  8   -0.084                            XXX
  9   -0.316                     XXXXXXXXX
 10   -0.284                      XXXXXXXX
```

Applying the approximate test to the 17 observations from Example 13.1, only those autocorrelation coefficients whose values exceeded $2/\sqrt{17} = .485$ would be considered significantly different from zero at the $\alpha = .05$ level of significance. In examining the autocorrelation coefficients, only the correlation at lag 1 is significantly different from zero, since its value exceeds the critical value of .485.

The random error term at time t for our time series model will be represented by the symbol z_t. One way to model the type of error correlation described above is to use an **autoregressive model** for z_t. A *first-order* autoregressive model is given by the equation in the following display.

First-Order[†] Autoregressive Model

$$z_t = \phi z_{t-1} + \epsilon_t$$

The symbol ϕ is an unknown constant that must be estimated, and ϵ_t, known as *white noise,* is assumed to possess a normal distribution. A complete autoregressive

---

[†]The first-order autoregressive model is the simplest in the family of autoregressive models. For example, a pth-order autoregressive model is given by the expression

$$z_t = \phi_1 z_{t-1} + \phi_2 z_{t-2} + \cdots + \phi_p z_{t-p} + \epsilon_t$$

The parameters $\phi_1, \phi_2, \ldots, \phi_p$ are assumed to be unknown and must be estimated.

time series model, one that contains both deterministic and random components, is of the form shown in the next display.

Complete Autoregressive Time Series Model

$$y_t = \text{Deterministic portion of the model} + z_t \quad \text{where } z_t = \phi z_{t-1} + \epsilon_t$$

The correlation between pairs of values of z_t (and therefore y_t) separated by m units of time is given by the autocorrelation function. The population **autocorrelation function** for a first-order autoregressive model is given by the expression in the following display.

Autocorrelation Function for a First-Order Autoregressive Model

$$\rho_k = A(y_t, y_{t-k}) = \phi^k$$

For example, if $\phi = .6$, then the correlation between two adjacent ($k = 1$) values of y_t is

$$\rho_1 = A(y_t, y_{t-1}) = (.6)^1 = .6$$

and the autocorrelation between two values of y_t separated by $k = 2$ units is

$$\rho_2 = A(y_t, y_{t-2}) = (.6)^2 = .36$$

A graph of the autocorrelation function for the first-order autoregressive model, $\phi = .6$, is shown in Figure 13.18. Note that the autocorrelation decreases as the distance k between y_t and y_{t-k} increases.

F I G U R E **13.18**
A graph of the autocorrelation function for a first-order autoregressive model, $\phi = .6$

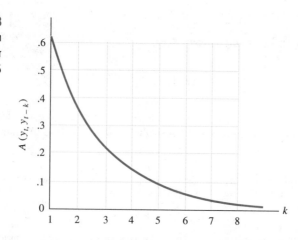

In order to forecast values of a time series to be observed in the future, we must fit the time series model to a set of data. We fit the model by using a modification of the least-squares method. This method fits the transformed variable y_t^* to x_t^*, where $y_t^* = y_t - \phi y_{t-1}$ and $x_t^* = x_t - \phi x_{t-1}$, for a given value of ϕ, using the method of least squares. The process is repeated for various values of ϕ until the sum of squares of the residuals is minimized. The resulting analysis will provide estimates of the parameters of $E(y_t)$—that is, β_0, β_1, ...—that differ slightly from those that would be obtained by using a standard regression analysis. It will also provide an estimate of the autocorrelation parameter ϕ as well as the values of the residuals, the difference between the observed and predicted values of y_t for each observation.

Because of the large amount of computation involved in fitting a time series model to a set of data, it is usually done on a computer by using a statistical program package. We illustrate the output of the popular SAS AUTOREG procedure in the SAS ETS package in the following example and then show how the resulting prediction equation can be used for forecasting.

E X A M P L E   **13.9**   Fit the autoregressive time series model

$$y_t = \beta_0 + \beta_1 t + z_t$$

to the domestic automobile sales data of Table 13.1.

Solution   The SAS printout for the SAS AUTOREG procedure package is shown in Table 13.11. The relevant parts of the printout are boxed and numbered.

1   The ordinary least-squares estimates are shown in area ①. Note that they agree with the values of the estimates shown in the Minitab simple linear regression printout in Table 13.7. These estimates are provided so that they can be compared with those obtained for the fitted autoregressive model.

2   The estimated value(s) of the autoregressive parameter(s) is (are) shown in area ②. Because we are fitting a first-order model, we will estimate only one ϕ parameter. The SAS AUTOREG package uses a slightly different model for the autoregressive term, one that produces estimates of the autocorrelation parameters that are the negatives of the estimates of our parameters. Therefore, since the printout gives $-.46752682$ as the estimate of the autocorrelation parameter, the estimate of ϕ for our model is $+.46752682$.

3   The AUTOREG estimates of the parameters of $E(y_t)$—in this case, β_0 and β_1—are shown in area ③ as

$$\hat{\beta}_0 = 8.64001004 \qquad \text{and} \qquad \hat{\beta}_1 = -.11073085$$

Note that these estimates vary very little from the standard regression estimates in area ①. If we round the estimates $\hat{\phi}$, $\hat{\beta}_0$, and $\hat{\beta}_1$, the fitted time series prediction equation is

$$\hat{y}_t = 8.640 - .1107t + \hat{z}_t$$

where

$$\hat{z}_t = .4675\hat{z}_{t-1}$$

T A B L E **13.11**
Domestic sales of U.S. cars: SAS
straight-line model with AR(1)
errors (Example 13.9)

SAS

AUTOREG PROCEDURE

DEPENDENT VARIABLE = Y

ORDINARY LEAST SQUARES ESTIMATES

SSE	19.81637	DFE	15
MSE	1.321092	ROOT MSE	1.149387
SBC	56.51635	AIC	54.84993
REG RSQ	0.2060	TOTAL RSQ	0.2060
DURBIN-WATSON	1.0532		

①

VARIABLE	DF	B VALUE	STD ERROR	T RATIO	APPROX PROB
INTERCPT	1	8.62205882	0.583083857	14.787	0.0001
T	1	-0.11225490	0.056903158	-1.973	0.0672

CORRELATION OF B-VALUES

	INTERCPT	T
INTERCPT	1.0000	-0.8783
T	-0.8783	1.0000

ESTIMATES OF AUTOCORRELATIONS

LAG	COVARIANCE	CORRELATION
0	1.16567	1.000000
1	0.544982	0.467527

PRELIMINARY MSE = 0.9108755

ESTIMATES OF THE AUTOREGRESSIVE PARAMETERS

②

LAG	COEFFICIENT	STD ERROR	T RATIO
1	-0.46752682	0.23625329	-1.978922

YULE-WALKER ESTIMATES

⑤

SSE	15.42817	DFE	14
MSE	1.102012	ROOT MSE	1.049768
SBC	55.34088	AIC	52.84124
REG RSQ	0.1097	TOTAL RSQ	0.3818

③

VARIABLE	DF	B VALUE	STD ERROR	T RATIO	APPROX PROB
INTERCPT	1	8.64001004	0.884887935	9.764	0.0001
T	1	-0.11073085	0.084312634	-1.313	0.2102

TABLE **13.11**
(continued)

CORRELATION OF B-VALUES

	INTERCPT	T
INTERCPT	1.0000	-0.8575
T	-0.8575	1.0000

DOMESTIC SALES OF U.S. CARS
STRAIGHT-LINE MODEL WITH AR(1) ERRORS
④

OBS	T	Y	YHAT	RESID	LP95	UPL95
1	1	8.8	8.52928	0.2707	5.44164	11.6169
2	2	9.7	8.54512	1.1549	5.78378	11.3065
3	3	7.3	8.90693	-1.6069	6.22367	11.5902
4	4	6.7	7.72590	-1.0259	5.11053	10.3413
5	5	8.5	7.38643	1.1136	4.82794	9.9449
6	6	9.2	8.16901	1.0310	5.65566	10.6824
7	7	9.2	8.43732	0.7627	5.95671	10.9179
8	8	8.4	8.37836	0.0216	5.91760	10.8391
9	9	6.4	7.94538	-1.5454	5.49127	10.3995
10	10	6.2	6.95136	-0.7514	4.49060	9.4121
11	11	5.0	6.79890	-1.7989	4.31828	9.2795
12	12	6.7	6.17890	0.5211	3.66555	8.6923
13	13	7.6	6.91474	0.6853	4.35625	9.4732
14	14	8.0	7.27655	0.7234	4.66118	9.8919
15	15	7.5	7.40460	0.0954	4.72134	10.0879
16	16	7.1	7.11188	-0.0119	4.35054	9.8732
17	17	7.1	6.86590	0.2341	4.01714	9.7147

4 The computer printout of the predicted values $\hat{y}_t$ and the residuals for the SAS AUTOREG fitted model is shown in area ④. For example, the predicted value of y_t for year 1, or $t = 1$, is 8.52928, the observed value is 8.8, and the residual, $y_t - \hat{y}_t$, is +.2707.

5 The SSE for the fitted autoregressive model, SSE = 15.42817, is shown in area ⑤. The value of the mean square error, MSE = 1.102012, is also shown. Note that this value is less than the value MSE = 1.321 obtained for the regression model shown in Table 13.7. ∎

A graph of the fitted first-order autoregressive prediction equation is shown in Figure 13.19. The autoregressive prediction equation is an improvement over the simple linear regression prediction equation. It adjusts for some of the cyclicality introduced by periodic rises and falls in economic activity, but it is not sufficient to avoid some large residuals.

FIGURE **13.19**
A graph of the fitted first-order
autoregressive model for car
sales, $\hat{y}_t = \hat{\beta}_0 + \hat{\beta}_1 t + \hat{z}_t$, where $\hat{z}_t = \phi\hat{z}_{t-1}$

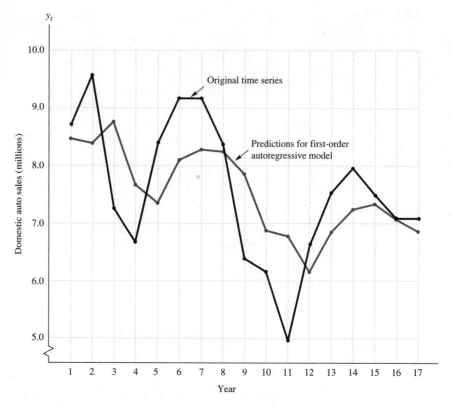

The introduction of trigonometric terms into $E(y_t)$ or the use of a second-order autoregressive model should provide a better fit to the data than is obtained by using the autoregressive model (call it model 1) of Example 13.9. Noting the distance between adjacent pairs of "high-point" years and between adjacent pairs of "low-point" years, we might choose a period $b = 7$ and fit the first-order autoregressive model

$$\text{model 2:} \quad y_t = \beta_0 + \beta_1 t + \beta_2 \cos\left(\frac{2\pi}{7}t\right) + \beta_3 \sin\left(\frac{2\pi}{7}t\right) + z_t$$

where

$$z_t = \phi z_{t-1} + \epsilon_t$$

to the data. Another possibility would be to use the second-order autoregressive model

$$\text{model 3:} \quad y_t = \beta_0 + \beta_1 t + z_t$$

where

$$z_t = \phi_1 z_{t-1} + \phi_2 z_{t-2}$$

The SAS AUTOREG printouts for fitting these models to the auto sales data are shown in Tables 13.12 and 13.13.

```
                                   SAS

                            AUTOREG PROCEDURE

DEPENDENT VARIABLE = Y
                     ORDINARY LEAST SQUARES ESTIMATES

           SSE            8.751625   DFE                  13
           MSE           0.6732019   ROOT MSE  0.8204888
           SBC            48.28921   AIC           44.95635
           REG RSQ          0.6493   TOTAL RSQ     0.6493
           DURBIN-WATSON    1.4720
```

VARIABLE	DF	B VALUE	STD ERROR	T RATIO	APPROX PROB
INTERCPT	1	8.68806876	0.418092252	20.780	0.0001
T	1	-0.11799401	0.040645135	-2.903	0.0123
COST	1	1.15068132	0.286085852	4.022	0.0015
SINT	1	0.15120332	0.281976484	0.536	0.6009

```
                     CORRELATION OF B-VALUES
```

	INTERCPT	T	COST	SINT
INTERCPT	1.0000	-0.8749	0.0494	-0.0799
T	-0.8749	1.0000	-0.0341	-0.0079
COST	0.0494	-0.0341	1.0000	0.0070
SINT	-0.0799	-0.0079	0.0070	1.0000

```
                    ESTIMATES OF AUTOCORRELATIONS
```

LAG	COVARIANCE	CORRELATION
0	0.514801	1.000000
1	0.068377	0.132822

```
                    PRELIMINARY MSE = 0.5057195
            ESTIMATES OF THE AUTOREGRESSIVE PARAMETERS
```

LAG	COEFFICIENT	STD ERROR	T RATIO
1	-0.13282198	0.28611745	-0.464222

```
                       YULE-WALKER ESTIMATES

           SSE            8.534803   DFE                  12
           MSE           0.7112335   ROOT MSE  0.8433466
           SBC            50.71374   AIC           46.54767
           REG RSQ          0.6077   TOTAL RSQ     0.6580
```

VARIABLE	DF	B VALUE	STD ERROR	T RATIO	APPROX PROB
INTERCPT	1	8.63751375	0.484357536	17.833	0.0001
T	1	-0.11150952	0.046875662	-2.379	0.0348
COST	1	1.12086322	0.314608626	3.563	0.0039
SINT	1	0.13803564	0.314941634	0.438	0.6690

T A B L E **13.12**
(continued)

CORRELATION OF B-VALUES

	INTERCPT	T	COST	SINT
INTERCPT	1.0000	-0.8710	0.0182	-0.0898
T	-0.8710	1.0000	0.0025	0.0006
COST	0.0182	0.0025	1.0000	-0.0005
SINT	-0.0898	0.0006	-0.0005	1.0000

DOMESTIC SALES OF U.S. CARS
SINE-COSINE MODEL WITH AR(1) ERRORS

OBS	T	Y	YHAT	RESID	LPL95	UPL95
1	1	8.8	9.33277	-0.5328	7.15134	11.5142
2	2	9.7	8.22889	1.4711	6.10886	10.3489
3	3	7.3	7.53901	-0.2390	5.43903	9.6390
4	4	6.7	7.11468	-0.4147	5.01537	9.2140
5	5	8.5	7.63996	0.8600	5.54164	9.7383
6	6	9.2	8.66617	0.5338	6.58559	10.7468
7	7	9.2	9.06290	0.1371	7.01703	11.1088
8	8	8.4	8.58172	-0.1817	6.57104	10.5924
9	9	6.4	7.49888	-1.0989	5.50304	9.4947
10	10	6.2	6.42381	-0.2238	4.41312	8.4345
11	11	5.0	6.29169	-1.2917	4.24582	8.3376
12	12	6.7	6.73726	-0.0373	4.65668	8.8178
13	13	7.6	7.75019	-0.1502	5.65187	9.8485
14	14	8.0	8.17349	-0.1735	6.07418	10.2728
15	15	7.5	7.74545	-0.2455	5.64547	9.8454
16	16	7.1	6.70245	0.3975	4.58242	8.8225
17	17	7.1	5.83990	1.2601	3.67241	8.0074

T A B L E **13.13**
Domestic sales of U.S. cars:
straight-line model with AR(2)
errors (model 3)

SAS

AUTOREG PROCEDURE

DEPENDENT VARIABLE = Y

ORDINARY LEAST SQUARES ESTIMATES

SSE	19.81637	DFE	15
MSE	1.321092	ROOT MSE	1.149387
SBC	56.51635	AIC	54.84993
REG RSQ	0.2060	TOTAL RSQ	0.2060
DURBIN-WATSON	1.0532		

VARIABLE	DF	B VALUE	STD ERROR	T RATIO	APPROX PROB
INTERCPT	1	8.62205882	0.583083857	14.787	0.0001
T	1	-0.11225490	0.056903158	-1.973	0.0672

T A B L E 13.13
(continued)

```
                    CORRELATION OF B-VALUES

                        INTERCPT        T
            INTERCPT    1.0000      -0.8783
            T          -0.8783       1.0000

           ESTIMATES OF AUTOCORRELATIONS

        LAG    COVARIANCE    CORRELATION
         0      1.16567       1.000000
         1      0.544982      0.467527
         2     -0.164336     -0.140980

        PRELIMINARY MSE = 0.7180175

     ESTIMATES OF THE AUTOREGRESSIVE PARAMETERS

        LAG    COEFFICIENT    STD ERROR     T RATIO
         1     -0.68265427   0.24624439   -2.772263
         2      0.46013926   0.24624439    1.868628

              YULE-WALKER ESTIMATES

        SSE      11.90736    DFE                 13
        MSE      0.9159508   ROOT MSE  0.9570532
        SBC      54.24627    AIC         50.91342
        REG RSQ   0.1497     TOTAL RSQ    0.5229

VARIABLE   DF     B VALUE     STD ERROR    T RATIO  APPROX PROB
INTERCPT    1   8.44244583  0.653852527   12.912      0.0001
T           1  -0.09731082  0.064316493   -1.513      0.1542


                    CORRELATION OF B-VALUES

                        INTERCPT        T
            INTERCPT    1.0000      -0.8853
            T          -0.8853       1.0000

            DOMESTIC SALES OF U.S. CARS
        STRAIGHT-LINE MODEL WITH AR(2) ERRORS

    OBS    T    Y     YHAT     RESID     LPL95     UPL95
     1     1   8.8  8.34514   0.4549   5.41132  11.2789
     2     2   9.7  8.46049   1.2395   5.85267  11.0683
     3     3   7.3  8.93255  -1.6325   6.60844  11.2567
     4     4   6.7  6.80439  -0.1044   4.52643   9.0824
     5     5   8.5  7.42348   1.0765   5.18398   9.6630
     6     6   9.2  8.85268   0.3473   6.64356  11.0618
     7     7   9.2  8.42663   0.7734   6.23947  10.6138
     8     8   8.4  8.02887   0.3711   5.85500  10.2028
     9     9   6.4  7.40709  -1.0071   5.23766   9.5765
    10    10   6.2  6.33424  -0.1342   4.16036   8.5081
    11    11   5.0  7.04233  -2.0423   4.85517   9.2295
    12    12   6.7  6.23951   0.4605   4.03040   8.4486
    13    13   7.6  7.87654  -0.2765   5.63704  10.1160
    14    14   8.0  7.63303   0.3670   5.35507   9.9110
    15    15   7.5  7.41631   0.0837   5.09220   9.7404
    16    16   7.1  6.81527   0.2847   4.43778   9.1928
    17    17   7.1  6.69662   0.4034   4.25899   9.1342
```

Extracting the parameter estimates from the printouts and reversing the signs of the autocorrelation estimates, we see that the prediction equations for the two models are

$$\text{model 2:} \quad \hat{y}_t = 8.638 - .1115t + 1.1209 \cos\left(\frac{2\pi}{7}t\right) + .1380 \sin\left(\frac{2\pi}{7}t\right) + \hat{z}_t$$

where

$$\hat{z}_t = .1328\hat{z}_{t-1}$$

and

$$\text{model 3:} \quad \hat{y}_t = 8.4424 - .0973t + \hat{z}_t$$

where

$$\hat{z}_t = .6827\hat{z}_{t-1} - .4601\hat{z}_{t-2}$$

Plots of the original data and graphs of the predicted values for the prediction equations of models 2 and 3 are shown in Figures 13.20 and 13.21, respectively. Note that both of these prediction equations fit the data better than the prediction equation for model 1. The values of MSE for both are lower than for model 1 and therefore confirm our visual comparisons:

$$\text{model 2:} \quad \text{MSE} = .71123 \qquad \text{model 3:} \quad \text{MSE} = .91595$$

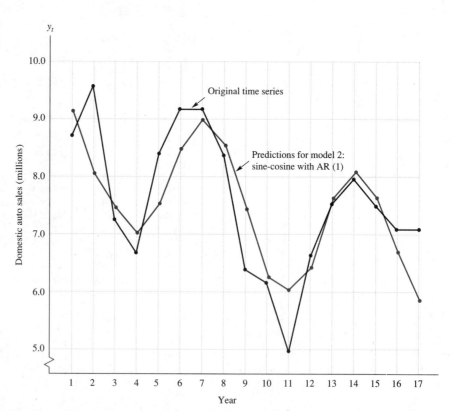

FIGURE 13.20
A graph of the fitted first-order autoregressive model for car sales, $\hat{y}_t = \hat{\beta}_0 + \hat{\beta}_1 t + \hat{\beta}_2 \cos[(2\pi/7)t] + \hat{\beta}_3 \sin[(2\pi/7)t] + \hat{z}_t$, where $\hat{z}_t = \hat{\phi}\hat{z}_{t-1}$

FIGURE **13.21**
A graph of the fitted second-order autoregressive model for car sales,
$$\hat{y}_t = \hat{\beta}_0 + \hat{\beta}_1 t + \hat{z}_t,$$
where
$$\hat{z}_t = \hat{\phi}_1 \hat{z}_{t-1} + \hat{\phi}_2 \hat{z}_{t-2}$$

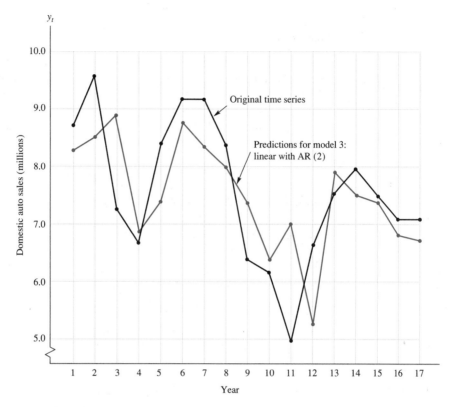

Once the time series prediction equation has been obtained, it can be used to forecast values of y_t to be observed in the future. The farther into the future you attempt to forecast, the greater will be the possibility for error. An approximate prediction interval for the next observation in the time series, the one immediately beyond the last observed data point, is given in the following display.

Approximate Prediction Interval for y_t

$$\hat{y}_t \pm 1.96\sqrt{\text{MSE}}$$

where MSE is the value of the mean square error shown in the SAS AUTOREG procedure printout.

We will illustrate the forecasting procedure with an example.

E X A M P L E   **13.10**   Forecast the year 18 new domestic automobile sales, using model 2 and the data in Table 13.1.

Solution   The prediction equation for model 2, obtained from the computer printout in Table 13.12, is

$$\hat{y}_t = 8.638 - .1115t + 1.1209 \cos\left(\frac{2\pi}{7}t\right) + .1380 \sin\left(\frac{2\pi}{7}t\right) + \hat{z}_t$$

where

$$\hat{z}_t = \hat{\phi}\hat{z}_{t-1} = .1328\hat{z}_{t-1}$$

Substituting $t = 18$ into this equation yields

$$\hat{y}_{18} = 8.638 - .1115(18) + 1.1209 \cos\left[\frac{2\pi}{7}(18)\right] + .1380 \sin\left[\frac{2\pi}{7}(18)\right] + \hat{z}_t$$

where

$$\hat{z}_{18} = .1328\hat{z}_{17}$$
$$\pi = 3.1415927$$
$$\left[\frac{2\pi}{7}(18)\right] = 16.1567 \text{ radians or } 205°$$

and

$$\cos(205°) = -.90631$$
$$\sin(205°) = -.42262$$

Substituting these values into the formula for $\hat{y}_t$ yields

$$\hat{y}_{18} = 8.638 - .1115(18) + 1.1209(-.906) + .1380(-.423) + \hat{z}_{18}$$
$$= 5.557 + \hat{z}_{18}$$

The next step is to calculate the estimated residual, $\hat{z}_{18} = .1328\hat{z}_{17}$, where $\hat{z}_{17} = 1.2601$ is the residual shown in the printout in Table 13.12 for the year 17. Therefore, the forecast value of the residual for year 18 is

$$\hat{z}_{18} = .1328\hat{z}_{17} = .1328(1.2601) = .1673$$

and the value of y_t forecast for year 18 is

$$\hat{y}_{18} = 5.557 + \hat{z}_{18} = 5.557 + .1673 = 5.724$$

or approximately 5.7 million automobiles.

The approximate prediction interval for $\hat{y}_{18}$ is

$$\hat{y}_{18} \pm 1.96\sqrt{\text{MSE}}$$
$$5.724 \pm 1.96\sqrt{.7112335}$$

or

$$5.724 \pm 1.653$$

Therefore, we forecast year 18 domestic new automobile sales to fall in the interval from 4.1 to 7.4 million automobiles.   ∎

13.8   Problems in Business Forecasting

We are well aware that business forecasting can be hazardous—that is, subject to substantial error. The great disparity in the current economic forecasts by leading economists in the United States is a clear indication of this fact. Nevertheless, statistical forecasting methods are of great value when combined with experience, intuition, good judgment, and an awareness that any forecasting method will be subject to error.

One of the major problems with forecasting, unlike a regression analysis, is that we are forecasting *beyond* the range of values of the independent time variable. This subjects us to the risk that the underlying economic or political conditions, upon which the model was constructed, may change. A set of data collected during the pre-1990 inflationary economic policy period is of limited value for forecasting economic activity, interest rates, and so on, during the post-1990 deflationary policy period. Consequently, the forecasts produced by a time series model depend on general economic and political stability.

As for the regression model, the quality of forecasts will depend on the model itself—that is, the selection of terms to be included in the model. For example, if a seasonal effect is present and you fail to account for it in the model, you will inflate the mean square error and increase your error of prediction.

Finally, note that Sections 13.6 and 13.7 present only an introduction to time series modeling. For example, we have presented only one of several models for the correlated errors z_t. A study of time series modeling and forecasting is an extensive subject. More information on this topic can be found in texts on time series analysis.

Exercises

Applications

13.26 The accompanying table gives data on quarterly U.S. investment (in billions of dollars) in new plants and equipment for three recent years.

Quarter	1	2	3
1	57.26	65.18	69.75
2	66.81	74.02	79.60
3	68.39	74.12	81.75
4	77.99	82.31	91.51

SAS AUTOREG printout for Exercise 13.26

```
DEPENDENT VARIABLE = DOLLAR

                    ORDINARY LEAST SQUARES ESTIMATES

          SSE            7.232867    DFE                7
          MSE            1.033267    ROOT MSE 1.016497
          SBC            40.4038     AIC         37.97927
          REG RSQ          0.9923    TOTAL RSQ  0.9923
          DURBIN-WATSON    0.7182

     VARIABLE   DF    B VALUE     STD ERROR    T RATIO   APPROX PROB
     INTERCPT   1   70.8966667   0.927930792   76.403      0.0001
     T          1    1.6300000   0.089846513   18.142      0.0001
     X1         1  -14.9833333   0.872637386  -17.170      0.0001
     X2         1   -7.2000000   0.849196107   -8.479      0.0001
     X3         1   -7.5533333   0.834815453   -9.048      0.0001

                     CORRELATION OF B-VALUES

                INTERCPT       T         X1        X2        X3
     INTERCPT    1.0000    -0.7746    -0.6646   -0.6010   -0.5280
     T          -0.7746     1.0000     0.3089    0.2116    0.1076
     COST       -0.6646     0.3089     1.0000    0.5301    0.5060
     SINT       -0.6010     0.2116     0.5301    1.0000    0.5086
                -0.5280     0.1076     0.5060    0.5086    1.0000

                   ESTIMATES OF AUTOCORRELATIONS

  LAG   COVARIANCE   CORRELATION   -1 9 8 7 6 5 4 3 2 1 0 1 2 3 4 5 6 7 8 9 1
   0     0.602739     1.000000     |                     |********************|
   1     0.336715     0.558641     |                     |***********          |

                     PRELIMINARY MSE = 0.4146361

             ESTIMATES OF THE AUTOREGRESSIVE PARAMETERS

           LAG  COEFFICIENT   STD ERROR    T RATIO
            1   -0.55864126   0.33860497   -1.649832

                      YULE-WALKER ESTIMATES

          SSE            4.180967    DFE                6
          MSE            0.6968278   ROOT MSE 0.8347621
          SBS            36.68568    AIC         33.77624
          REG RSQ          0.9948    TOTAL RSQ  0.9956

     VARIABLE   DF    B VALUE     STD ERROR    T RATIO   APPROX PROB
     INTERCPT   1   70.2999851   1.07522711    65.382      0.0001
     T          1    1.7044627   0.12446673    13.694      0.0001
     X1         1  -14.6078164   0.59763364   -24.443      0.0001
     X2         1   -6.9717083   0.62184782   -11.211      0.0001
     X3         1   -7.4449336   0.52711165   -14.124      0.0001
```

SAS AUTOREG printout for Exercise 13.26 (continued)

```
                      CORRELATION OF B-VALUES

                INTERCPT       T        X1        X2        X3
     INTERCPT    1.0000   -0.8310   -0.4906   -0.4214   -0.3106
     T          -0.8310    1.0000    0.2827    0.1868    0.1001
     X1         -0.4906    0.2827    1.0000    0.6270    0.3941
     X2         -0.4214    0.1868    0.6270    1.0000    0.5841
     X3         -0.3106    0.1001    0.3941    0.5841    1.0000

OBS    YHAT      RESID     LCL95     UCL95    T   QUARTER  DOLLARS  X1  X2  X3
 1    57.3966   -0.1333   54.1775   60.6158   1      1      57.26    1   0   0
 2    66.6609    0.1491   63.8167   69.5051   2      2      66.81    0   1   0
 3    68.0091    0.3809   65.2296   70.7886   3      3      68.39    0   0   1
 4    77.3533    0.6367   74.6605   80.0462   4      4      77.99    0   0   0
 5    64.7017    0.4783   62.1765   67.2269   5      1      65.18    1   0   0
 6    74.0944   -0.0744   71.5600   76.6289   6      2      74.02    0   1   0
 7    75.0460   -0.9260   72.5116   77.5805   7      3      74.12    0   0   1
 8    83.5635   -1.2535   81.0383   86.0887   8      4      82.31    0   0   0
 9    70.1242   -0.3742   67.4313   72.8170   9      1      69.75    1   0   0
10    79.6565   -0.0565   76.8771   82.4360  10      2      79.60    0   1   0
11    81.1724    0.5776   78.3282   84.0166  11      3      81.75    0   0   1
12    90.8350    0.6750   87.9247   93.7454  12      4      91.51    0   0   0
```

a Construct a graph of this time series. Note the upward secular and the distinct seasonal (quarterly) effect.

b Write a deterministic model for $E(y_t)$ that includes a straight-line secular component and three parameters, entered by dummy variables, to account for the four quarterly effects. [*Hint:* This model is discussed in Section 13.7.]

c Add the residual component to $E(y_t)$ in part (b) to form a complete first-order autoregressive model.

d On the accompanying SAS AUTOREG computer printout for the fitted first-order autoregressive model, find the prediction equation $\hat{y}_t$ for y_t.

e Graph the fitted time series on the graph of the original time series in part (a). Compare the fitted time series with the original.

13.27 Refer to Exercise 13.26.

a Use the prediction equation in Exercise 13.26 to forecast the quarterly U.S. investment for new plant and equipment for the first quarter of year 4.

b Find MSE on the computer printout accompanying Exercise 13.26.

c Find an approximate 95% prediction interval for y_t in part (a).

13.28 Use the information in the computer printout of Table 13.13 for the fitted second-order autoregressive model to forecast the domestic new automobile sales for year 18.

13.29 Refer to Exercise 13.28. Find an approximate 95% prediction interval for the domestic new automobile sales for year 18.

 13.30 The following table gives the prime interest rate charged by banks on short-term business loans each month for the period from January 1990 through October 1993.

Year	Jan.	Feb.	Mar.	Apr.	May	June	July	Aug.	Sept.	Oct.	Nov.	Dec.
1990	10.11	10.00	10.00	10.00	10.00	10.00	10.00	10.00	10.00	10.00	10.00	10.00
1991	9.52	9.05	9.00	9.00	8.50	8.50	8.50	8.50	8.20	8.00	7.58	7.21
1992	6.50	6.50	6.50	6.50	6.50	6.50	6.02	6.00	6.00	6.00	6.00	6.00
1993	6.00	6.00	6.00	6.00	6.00	6.00	6.00	6.00	6.00	6.00		

Source: Survey of Current Business, 1991–93.

SAS AUTOREG printout for Exercise 13.30

```
                        AUTOREG PROCEDURE

DEPENDENT VARIABLE = Y

                    ORDINARY LEAST SQUARES ESTIMATES

            SSE              9.709685    DFE                  43
            MSE              0.225807    ROOT MSE    0.475191
            SBC             70.47446     AIC           64.98854
            REG RSQ           0.9255     TOTAL RSQ     0.9255
            DURBIN-WATSON     0.1588

VARIABLE   DF       B VALUE    STD ERROR   T RATIO   APPROX PROB
INTERCPT    1    11.1404796      0.21967    50.714        0.0001
T           1    -0.1893490      0.02156    -8.783        0.0001
TSQ         1   0.0014596474     0.00044     3.282        0.0021

                    CORRELATION OF B-VALUES

                          INTERCPT         T          TSQ
            INTERCPT            1    -0.87631      0.76127
            T            -0.87631          1     -0.96958
            TSQ           0.76127   -0.96958            1

                 ESTIMATES OF AUTOCORRELATIONS

            LAG    COVARIANCE    CORRELATION
             0       0.21108      1.000000
             1       0.184093     0.872148

             PRELIMINARY MSE = 0.050524

        ESTIMATES OF THE AUTOREGRESSIVE PARAMETERS

        LAG    COEFFICIENT    STD ERROR      T RATIO
         1     -0.87214779    0.07549176   -11.552887

                    YULE-WALKER ESTIMATES

            SSE              1.526494    DFE                  42
            MSE              0.036345    ROOT MSE    0.190644
            SBC             -9.37402     AIC          -16.6886
            REG RSQ           0.6780     TOTAL RSQ     0.9883
            DURBIN-WATSON     1.2390

                        AUTOREG PROCEDURE

VARIABLE   DF       B VALUE    STD ERROR   T RATIO   APPROX PROB
INTERCPT    1    10.6131545      0.39897    26.601        0.0001
T           1    -0.1554008      0.03721    -4.176        0.0001
TSQ         1   0.0010718052     0.00075     1.421        0.1628

                    CORRELATION OF B-VALUES

                          INTERCPT         T         TSQ
            INTERCPT            1    -0.74346    0.568201
            T            -0.74346          1    -0.95278
            TSQ          0.568201   -0.95278           1
```

SAS AUTOREG printout for Exercise 13.30 (continued)

OBS	T	Y	YHAT	RESID	LPL95	UPL95
1	1	10.11	10.4588	-0.34883	9.37089	11.5468
2	2	10.00	10.0024	-0.00241	9.20060	10.8042
3	3	10.00	9.8892	0.11084	9.12485	10.6535
4	4	10.00	9.8721	0.12788	9.14011	10.6041
5	5	10.00	9.8554	0.14464	9.15052	10.5602
6	6	10.00	9.8389	0.16113	9.15624	10.5215
7	7	10.00	9.8226	0.17735	9.15753	10.4878
8	8	10.00	9.8067	0.19329	9.15473	10.4587
9	9	10.00	9.7910	0.20896	9.14826	10.4338
10	10	10.00	9.7756	0.22436	9.13862	10.4127
11	11	10.00	9.7605	0.23948	9.12631	10.3947
12	12	10.00	9.7457	0.25433	9.11187	10.3795
13	13	9.52	9.7311	-0.21110	9.09583	10.3664
14	14	9.05	9.2982	-0.24817	8.66004	9.9363
15	15	9.00	8.8742	0.12576	8.23230	9.5162
16	16	9.00	8.8169	0.18312	8.17061	9.4631
17	17	8.50	8.8034	-0.30340	8.15262	9.4542
18	18	8.50	8.3541	0.14588	7.69894	9.0093
19	19	8.50	8.3412	0.15880	7.68198	9.0004
20	20	8.50	8.3285	0.17146	7.66587	8.9912
21	21	8.20	8.3162	-0.11616	7.65076	8.9816
22	22	8.00	8.0424	-0.04241	7.37512	8.7097
23	23	7.58	7.8561	-0.27615	7.18790	8.5244
24	24	7.21	7.4783	-0.26829	6.81004	8.1465
25	25	6.50	7.1443	-0.64431	6.47702	7.8116
26	26	6.50	6.5141	-0.01407	5.84867	7.1795
27	27	6.50	6.5033	-0.00334	5.84066	7.1660
28	28	6.50	6.4929	0.00712	5.83366	7.1521
29	29	6.50	6.4827	0.01731	5.82750	7.1379
30	30	6.50	6.4728	0.02723	5.82199	7.1236
31	31	6.02	6.4631	-0.44313	5.81687	7.1094
32	32	6.00	6.0351	-0.03514	5.39321	6.6771
33	33	6.00	6.0086	-0.00860	5.37047	6.6467
34	34	6.00	5.9998	0.00021	5.36452	6.6351
35	35	6.00	5.9912	0.00876	5.35744	6.6250
36	36	6.00	5.9830	0.01703	5.34876	6.6172
37	37	6.00	5.9750	0.02502	5.33795	6.6120
38	38	6.00	5.9673	0.03274	5.32448	6.6100
39	39	6.00	5.9598	0.04019	5.30783	6.6118
40	40	6.00	5.9526	0.04736	5.28752	6.6178
41	41	6.00	5.9457	0.05426	5.26311	6.6284
42	42	6.00	5.9391	0.06089	5.23428	6.6439
43	43	6.00	5.9328	0.06724	5.20075	6.6648
44	44	6.00	5.9267	0.07332	5.16237	6.6910
45	45	6.00	5.9209	0.07912	5.11906	6.7227
46	46	6.00	5.9154	0.08465	5.07084	6.7599

a Plot the data and connect the points to form a time series.

b Calculate a three-point moving average for the series. Plot the moving average, and compare it with the original time series.

c From the SAS AUTOREG procedure printout for fitting a first-order autoregressive model to the data, write the equation of the fitted model and explain the significance of each component of the model.

d Obtain the predicted values of the prime interest rate from the computer printout and plot them on the graph in part (a). Compare the fitted time series with the observed values of the prime interest rate.

13.31 Refer to Exercise 13.30.

a Use the fitted first-order autoregressive model of Exercise 13.30 to forecast the prime rate for banks for November 1993.

b Find MSE on the computer printout accompanying Exercise 13.30.

c Find an approximate 95% prediction interval for y_t in part (a).

13.9 THE CONSUMER PRICE INDEX

The Consumer Price Index (CPI), introduced in the case study, is an extremely complicated index number, but its general interpretation is like that of any other index. It provides a general measure of the level of consumer prices relative to the level of the index in the base year 1982. In describing the Consumer Price Index and its applications, we now present a definition and brief description provided by the U.S. Department of Labor and follow it with some data to show how it has changed over the period 1967 through 1991.

According to the Bureau of Economic Analysis,[†]

The **Consumer Price Indexes** (CPI's) are statistical measures of the average changes in the cost of fixed, or constant, "market baskets" of consumer goods and services purchased by the index population, either all urban consumers or urban wage earners and clerical workers.

The most recent major revision of the indexes became effective with the release of CPI data for January 1987. The reference base for the indexes was updated to 1982–84 = 100 effective with the release of the CPI data for January 1988. All indexes that had been expressed on a base of 1967 = 100, or any other base through December, 1981, were rebased to 1982–84 = 100.

CPI's are based on prices of food, clothing, shelter, and fuels, transportation fares, charges for doctors' and dentists' services, drugs, and other goods and services that people buy for day-to-day living. Prices are collected in 85 urban areas across the country from about 18,000 tenants, 18,000 housing units for property taxes,

[†]*Business Statistics, 1963–91,* 1992.

and about 24,000 establishments—grocery and department stores, hospitals, filling stations, and other types of stores and service establishments.

The quantity and quality of items contained in the respective market baskets are held constant except at times of weight revisions. The CPI's reflect, therefore, only changes in prices and none of the other factors that affect family living expenses, such as change in family composition; they do not reflect changes in the kinds and amounts of goods and services families buy, or the total amount families spend for living, or the differences in living costs in different places.

The indexes are initially issued in a press release about four weeks following the month to which the data pertain. The *CPI Detailed Report* is issued about a month after the press release. Detailed data and quarterly articles analyzing price developments in the Nation's economy are presented in the *Monthly Labor Review.*

The Consumer Price Index for the period 1961 to 1991, as well as the Consumer Price Index for specific groups and subgroups of selected items, is reproduced in Table 13.14. Note that the annual average value of the Consumer Price Index for 1982–84 (CPI-W) is 100 and that it increased to 134.3 through 1991. Thus the general level of consumer prices in 1991, as measured by the CPI, is 1.34 times the general level in 1982–84.

A more extensive discussion of the statistical nature of the Consumer Price Index, provided by the Bureau of Labor Statistics (*1989 Supplement to Economic Indicators,* U.S. Government Printing Office), points out that the CPI prices for the market basket of goods and services are based on a probability sample of 24,000 retail stores and other outlets in 85 urban areas. Prices of medical, dental, and other health services are also acquired. Rental data on approximately 18,000 rental units and property tax data on 18,000 housing units are likewise included. The CPI is computed for each urban area and then combined to form a national index number by weighting the area CPI values by estimates of the 1970 population of urban consumers in each area.

The preceding descriptions of the Consumer Price Index should help you to understand how well the Consumer Price Index actually measures the increase in your cost of living. The answer is not very well, because the proportion of goods and services that we buy varies substantially from person to person and from one area of the country to another. However, that is not the intended purpose of the CPI. Rather, it is intended to provide an objective measure of the change in consumer prices that can be legally defined and used to make adjustments in salaries for labor contracts, Social Security benefits, and benefits from the many other federal entitlement programs. The CPI accomplishes this objective. It may not be perfect; it may overweight some expenditures (e.g., the cost of new housing); but it is clearly defined for legal purposes, and it does provide a satisfactory measure of the change in the cost of living.

TABLE 13.14 Consumer Price Index data: Commodity prices—consumer prices

(1982–1984 = 100)

CONSUMER PRICE INDEXES—Unadjusted for seasonal variation

YEAR AND MONTH	All items, all wage earners and clerical workers (CPI-W) ★	All items, all urban consumers (CPI-U) ★	All items less shelter	All items less food	All items less medical care	Special group indexes (CPI-U) Total ★	Commodities Total	Nondurables Total	Nondurables less food	Durables	Commodities less food	Services ★	Food Total ★	Food at home
1963	30.8	30.6	32.4	30.7	31.1	34.4	32.9		34.8	38.6	36.6	25.5	31.1	32.4
1964	31.2	31.0	32.8	31.1	31.5	34.8	33.2		35.1	39.0	36.9	26.0	31.5	32.7
1965	31.7	31.5	33.3	31.6	32.0	35.2	33.8		35.6	38.8	37.2	26.6	32.2	33.5
1966	32.6	32.4	34.3	32.3	33.0	36.1	35.1		36.4	38.9	37.7	27.6	33.8	35.2
1967	33.6	33.4	35.2	33.4	33.7	36.8	35.7		37.6	39.4	38.6	28.8	34.1	35.1
1968	35.0	34.8	36.7	34.9	35.1	38.1	37.1		39.1	40.7	40.0	30.3	35.3	36.3
1969	36.9	36.7	38.4	36.8	37.0	39.9	38.9		40.9	42.2	41.7	32.4	37.1	38.0
1970	39.0	38.8	40.3	39.0	39.2	41.7	40.8		42.5	44.1	43.4	35.0	39.2	39.9
1971	40.7	40.5	42.0	40.8	40.8	43.2	42.1		44.0	46.0	45.1	37.0	40.4	40.9
1972	42.1	41.8	43.3	42.0	42.1	44.5	43.5		45.0	46.9	46.1	38.4	42.1	42.7
1973	44.7	44.4	46.2	43.7	44.8	47.8	47.5		46.9	48.1	47.7	40.1	48.2	49.7
1974	49.6	49.3	51.4	48.0	49.8	53.5	54.0		52.9	51.5	52.8	43.8	55.1	57.1
1975	54.1	53.8	56.0	52.5	54.3	58.2	58.3		57.0	57.4	57.6	48.0	59.8	61.8
1976	57.2	56.9	59.3	56.0	57.2	60.7	60.5		59.5	60.9	60.5	52.0	61.6	63.1
1977	60.9	60.6	63.1	59.6	60.8	64.2	64.0		62.5	64.4	63.8	56.0	65.5	66.8
1978	65.6	65.2	67.4	63.9	65.4	68.8	68.6		65.5	68.6	67.5	60.8	72.0	73.8
1979	73.1	72.6	74.2	71.2	72.9	76.6	77.2		74.6	75.4	75.3	67.5	79.9	81.8
1980	82.9	82.4	82.9	81.5	82.8	86.0	87.6		88.4	83.0	85.7	77.9	86.8	88.4
1981	91.4	90.9	91.0	90.4	91.4	93.2	95.2		96.7	89.6	93.1	88.1	93.6	94.8
1982	96.9	96.5	96.2	96.3	96.8	97.0	97.8		98.3	95.1	96.9	96.0	97.4	98.1
1983	99.8	99.6	99.8	99.7	99.6	99.8	99.7		100.0	99.8	100.0	99.4	99.4	99.1
1984	103.3	103.9	103.9	104.0	103.2	103.2	102.5		101.7	105.1	103.1	104.6	103.2	102.8
1985	106.9	107.6	107.0	108.0	107.2	105.4	104.8		104.1	106.8	105.2	109.9	105.6	104.3
1986	108.6	109.6	108.0	109.8	108.8	104.4	103.5		98.5	106.6	101.7	115.4	109.0	107.3
1987	112.5	113.6	111.6	113.6	112.6	107.7	107.5		101.8	108.2	104.3	120.2	113.5	111.9
1988	117.0	118.3	115.9	118.3	117.0	111.5	111.8		105.8	110.4	107.7	125.7	118.2	116.6
1989	122.6	124.0	121.6	123.7	122.4	116.7	118.2		111.7	112.2	112.0	131.9	125.1	124.2
1990	129.0	130.7	128.2	130.3	128.8	122.8	126.0		119.9	113.4	117.4	139.2	132.4	132.3
1991	134.3	136.2	133.5	136.1	133.8	126.6	130.3		124.5	116.0	121.3	146.3	136.3	135.8

13.10 Summary

Time series, data sets collected sequentially over time, require different treatment because the data are likely to be correlated and, thereby, to violate the assumptions on which the preceding methodologies were based. Index numbers represent a particular type of time series variable, one that shows the general level of some economic or business phenomenon in relation to the level at some prechosen point in time.

A time series can be viewed as the sum or product of any or all of four components: a long-term or secular trend, a cyclical component, a seasonal component, and random unexplained error. Which (if any) of the first three components are present can sometimes be deduced by graphing the original time series or using moving averages or exponential smoothing to smooth the original time series and reveal long-term trends and cyclical or seasonal components.

The ultimate objective of most time series analyses is to construct a time series model, fit it to the data, and use the resulting prediction equation to forecast future values of y_t with a known bound on the error of prediction. One common method is to use a model with multiplicative effects; another method of modeling time series is to construct an additive regression model to represent secular, cyclical, and seasonal components of the time series. The residuals—the deviations between the observed and predicted values of y_t—are then modeled by using a model that accounts for autocorrelation between values of y_t separated in time. One of these, the autoregressive model, was described and its use illustrated in Section 13.7.

The problem of forecasting from time series is not one that is completely solved to the satisfaction of all economists, statisticians, and business managers. If you have any doubt on this point, we direct your attention to the wide variations in the economic and business forecasts provided by nationally recognized consulting firms.

Supplementary Exercises

13.32 The accompanying tables give data on the production, sales, and price for natural gas in the United States. Construct a simple composite consumer price index for gas based on two components: (1) price to residential consumers and (2) industrial sales price for the years 1980 to 1992. Use 1982 as the base year. Prices in the first table are given in dollars per thousand cubic feet.

Salient statistics of natural gas in the United States, Exercise 13.32

	Supply				Disposition						Average Price Delivered to Consumers						
Year	Marketed Production	Storage With-drawals	Import (con-sumed)	Total Supply	Consump-tion	Ex-ports	Stored	Extrac-tion Loss	Unac-counted for	Total Dispo-sition	Well-head Price	Im-ports	Ex-ports	Resi-dential	Comm-ercial	Industrial	Electric Utilities
	In Billions of Cubic Feet										$ Per Thous. Cu. Ft.						
1980	20,180	1972	985	22,515	19,877	49	1949	777	640	23,292	1.59	4.28	4.70	3.68	3.39	2.56	2.27
1981	19,956	1930	904	22,191	19,404	59	2228	775	501	22,967	1.98	4.88	5.90	4.29	4.00	3.14	2.89
1982	18,520	2164	933	21,000	18,001	52	2472	762	475	21,762	2.46	5.03	5.81	5.17	4.82	3.87	3.48
1983	16,822	2270	920	19,354	16,835	55	1822	790	642	20,144	2.59	4.78	5.10	6.06	5.59	4.18	3.58
1984	18,230	2098	843	20,443	17,953	55	2295	838	143	20,443	2.66	4.08	4.92	6.12	5.55	4.22	3.70
1985	17,198	2397	950	19,855	17,281	55	2163	816	356	19,855	2.51	3.19	4.77	6.12	5.50	3.95	3.55
1986	16,791	1837	750	18,692	16,221	61	1984	800	427	18,692	1.94	2.53	2.81	5.83	5.08	3.23	2.43
1987	17,349	1905	993	19,534	17,211	54	1911	812	359	19,534	1.67	2.17	3.07	5.54	4.77	2.94	2.32
1988	17,918	2270	1294	19,030	18,030	74	2211	816	368	20,668	1.69	1.84	2.74	5.47	4.63	2.95	2.33
1989	18,095	2854	1382	18,801	18,801	107	2528	785	380	21,782	1.69	1.82	2.51	5.64	4.74	2.96	2.43
1990	18,594	1986	1532	18,715	18,715	86	2499	784	40	21,656	1.71	1.94	3.10	5.80	4.83	2.93	2.38
1991	18,532	2752	1773	21,836	19,035	129	2672	835	272	22,199	1.64	1.82	2.59	5.82	4.81	2.69	2.18
1992[1]	18,712	2772	2138	22,360	19,544	216	2599	872	272	22,199	1.74	1.85	2.25	5.89	4.88	2.84	2.36
1993[2]	19,353	2840	2235	23,214	20,192	142	2879	902	294	22,846	1.98	1.98		6.16	5.14	3.04	2.60

Source: U.S. Department of Energy; in CRB Commodity Yearbook, 1994.
[1] Preliminary.
[2] Estimate.

Gas utility sales in the United States by types and class of service (in trillions of BTUs), Exercise 13.32

Year	Total Utility Sales	Num-ber of Cust. (mil.)	Class of Service					Revenue—$ Million from Sales to Customers					
			Resi-dential	Com-mercial	Indus-trial	Electric Genera-tion	Other	Total	Resi-dential	Com-mercial	Indus-trial	Electric Genera-tion	Other
1980	15,413	47.4	4826	2453	7957		177	48,303	17,432	8,183	22,215		473
1981	15,375	48.1	4610	2375	8239		150	56,110	19,180	9,286	27,124		520
1982	14,183	48.5	4770	2471	6794		147	63,200	23,700	11,666	27,200		634
1983	12,858	48.9	4450	2298	5970		140	65,837	26,173	12,659	26,315		690
1984	13,162	49.5	4628	2396	5991		146	67,496	27,485	13,205	26,093		713
1985	12,612	50.2	4513	2338	3686	1949	130	63,293	26,864	12,723	15,659	7428	620
1986	11,125	50.9	4381	2239	2890	1449	167	51,201	24,759	11,274	10,546	3949	673
1987	10,543	51.8	4385	2156	2541	1306	155	45,492	23,622	10,271	7500	3569	530
1988	10,691	52.7	4692	2304	2204	1331	160	46,109	24,812	10,670	6702	3387	539
1989	10,551	53.7	4798	2323	1962	1280	188	47,493	26,172	11,077	6211	3449	584
1990	9846	54.5	4471	2193	1890	1120	171	45,174	24,014	10,610	6034	2963	553
1991[1]	9605	55.4	4550	2198	1742	888	226	46,647	25,729	10,669	5326	2250	674
1992[2]	9757	56.2	4678	2215	1721	917	226	46,011	26,697	10,903	5540	2187	684

Source: American Gas Association; in *CRB Commodity Yearbook*, 1994.
[1] Preliminary.
[2] Estimate.

13.33   Refer to Exercise 13.32. Use the data to construct Laspeyres price index numbers for the years 1988 to 1992. Use 1982 as the base year, and use the sales (in trillions of BTUs) to (1) residential and (2) industrial customers as weights. Sales per year are given in the second table in columns 4 and 6, respectively. Plot the index numbers on graph paper, and connect the points to form a time series.

13.34   Refer to Exercises 13.32 and 13.33. Calculate Paasche index numbers for the gas price index. Plot the index numbers on graph paper, and connect the points to form a time series.

13.35   The following tables give the average price of mercury and the U.S. consumption of mercury for the time period 1983 through 1992.

Average price of mercury in New York [in dollars per flask of 76 pounds (34.5 kilograms)], Exercise 13.35

Year	Jan.	Feb.	Mar.	Apr.	May	June	July	Aug.	Sept.	Oct.	Nov.	Dec.	Average
1983	366.67	346.18	338.91	327.62	316.31	300.23	282.25	279.02	299.05	339.29	344.50	331.69	322.64
1984	318.67	295.68	300.45	324.26	328.27	325.48	314.64	302.09	317.18	329.07	327.13	320.08	316.92
1985	317.30	315.00	314.38	309.34	297.27	309.05	316.74	324.61	325.00	325.00	325.00	325.00	316.98
1986	285.68	261.97	252.44	272.05	277.50	276.07	260.00	218.10	181.91	193.80	215.19	219.43	242.85
1987	216.70	211.03	223.73	260.00	297.63	297.73	286.82	306.55	337.86	345.91	348.95	342.27	289.60
1988	354.00	357.00	350.00	337.14	326.90	352.28	367.13	360.43	349.29	322.74	299.34	293.45	339.14
1989	292.50	318.42	330.00	327.50	315.68	308.86	293.03	273.80	265.13	267.61	282.50	292.50	297.29
1990	292.50	292.50	287.84	285.00	285.00	285.00	281.45	262.50	259.34	235.54	198.75	182.50	262.33
1991	181.55	169.08	150.36	140.68	129.77	119.75	110.36	102.50	98.00	94.02	127.50	150.12	131.14
1992	162.86	177.24	180.00	180.00	190.63	202.50	202.50	203.45	207.50	207.50	207.50	207.50	211.39
1993	207.50	207.50	207.50	207.50	207.50	201.30	191.00	191.00	185.00	185.00	181.00	175.00	195.57

Source: American Metal Market; in *CRB Commodity Yearbook*, 1994.

Mercury consumed in the United States (in cubic metric tons), Exercise 13.35

Year	Batteries	Chlorine & Caustic Soda	Catalysts Misc.	Dental Equip.	Electrical Lighting	General Lab. Use	Measuring Control Instrum.	Paints	Wiring Devices & Switches	Other Uses	Grand Total
1982	24,880	6243	499	1019	826	281	3064	6794	2004	984	48,943
1983	23,350	8054	484	1597	1273	280	2465	6047	2316	1356	49,138
1984[3]	29,700	7347	359	1432	1487	269	2856	4651	2730	1404	54,669
1985	952	235	61	50	40	14	79	169	95	20	1718
1986	750	259	90	52	41	20	63	179	103	31	1588
1987	533	311	59	56	45	20	59	198	131	34	1446
1988	448	354	86	53	31	26	77	197	176	55	1503
1989	250	379	40	39	31	18	87	192	141	32	1212
1990	18	247	33	44	33	32	108	22	70	25	720
1991[1]	18	184	12	27	29	16	70	6	25	165	554
1992[2]	16	209	18	37	55	18	52	—	69	148	621

Source: U.S. Bureau of Mines; in *CRB Commodity Yearbook*, 1994.
[1] Preliminary.
[2] Estimate.
[3] Data prior to 1985 are in flasks.

a Graph the time series depicting the average price of mercury for the years 1983 through 1992. Compare it with a similar graph for the total U.S. consumption. Can you explain the relationship between price and consumption?

b Plot the quarterly price (March, June, September, December) for the years 1983 to 1992. Smooth the time series by using a four-point *centered* moving average.

c Smooth the series in part (b) by using exponential smoothing with $\omega = .3$.

13.36 Refer to the year-end prices of commodities in Table 13.2.

a Plot the time series for commodity 1 on graph paper. Note the apparent straight-line secular trend in the plotted points.

b Write a model for $E(y_t)$ to represent the straight-line secular trend in the commodity prices in part (a).

c Modify the model in part (b) to form a complete first-order autoregressive model for y_t.

d A SAS AUTOREG computer printout for the fitted first-order autoregressive model is shown in the accompanying table. Find the prediction equation for y_t.

SAS AUTOREG printout for Exercise 13.36

```
                    AUTOREG PROCEDURE

DEPENDENT VARIABLE = Y

              ORDINARY LEAST SQUARES ESTIMATES

          SSE           0.027759   DFE                 8
          MSE           0.00347    ROOT MSE    0.058905
          SBC          -25.8839    AIC         -26.4891
          REG RSQ        0.9897    TOTAL RSQ     0.9897
          DURBIN-WATSON  2.1796

VARIABLE   DF    B VALUE    STD ERROR   T RATIO   APPROX PROB
INTERCPT   1    2.05133333    0.04024   50.977       0.0001
T          1    0.18012121    0.00649   27.774       0.0001
```

SAS AUTOREG printout for Exercise 13.36 (continued)

```
                    CORRELATION OF B-VALUES

                         INTERCPT         T
            INTERCPT        1      -0.88641
            T            -0.88641       1

           ESTIMATES OF AUTOCORRELATIONS

           LAG    COVARIANCE    CORRELATION
            0      0.002776       1.000000
            1     -0.00039       -0.140217

             PRELIMINARY MSE = 0.002721

      ESTIMATES OF THE AUTOREGRESSIVE PARAMETERS

           LAG    COEFFICIENT    STD ERROR    T RATIO
            1     0.14021658    0.37423051   0.374680

                 YULE-WALKER ESTIMATES

         SSE           0.027146    DFE                   7
         MSE           0.003878    ROOT MSE       0.062274
         SBC          -23.7847     AIC           -24.6925
         REG RSQ        0.9917      TOTAL RSQ       0.9900
         DURBIN-WATSON 1.8852
```

VARIABLE	DF	B VALUE	STD ERROR	T RATIO	APPROX PROB
INTERCPT	1	2.05173936	0.03853	53.251	0.0001
T	1	0.18021702	0.00624	28.870	0.0001

```
                    CORRELATION OF B-VALUES

                         INTERCPT         T
            INTERCPT        1      -0.89107
            T            -0.89107       1
```

OBS	T	Y	YHAT	RESID	LPL95	UPL95
1	1	2.20	2.23196	-0.03196	2.06391	2.40001
2	2	2.35	2.41665	-0.06665	2.25522	2.57809
3	3	2.64	2.60111	0.03889	2.44377	2.75845
4	4	2.77	2.76593	0.00407	2.61139	2.92048
5	5	3.01	2.95319	0.05681	2.80006	3.10632
6	6	3.10	3.12502	-0.02502	2.97190	3.27815
7	7	3.42	3.31789	0.10211	3.16335	3.47244
8	8	3.50	3.47851	0.02149	3.32117	3.63585
9	9	3.62	3.67278	-0.05278	3.51134	3.83422
10	10	3.81	3.86144	-0.05144	3.69469	4.02819

13.37   Refer to Exercise 13.36.

   a   Forecast the year-end price for commodity 1 for 1995.

   b   Find an approximate 95% prediction interval for your forecast in part (a).

13.38   Exercise 13.30 gives data on the prime interest rate charged by banks on short-term business loans each month for the years 1990 through October 1993. Shown in the accompanying table is the SAS AUTOREG computer printout for fitting a second-order autoregressive model to the prime interest rate data.

SAS AUTOREG printout for Exercise 13.38

AUTOREG PROCEDURE

DEPENDENT VARIABLE = Y

ORDINARY LEAST SQUARES ESTIMATES

SSE	9.709685	DFE	43
MSE	0.225807	ROOT MSE	0.475191
SBC	70.47446	AIC	64.98854
REG RSQ	0.9255	TOTAL RSQ	0.9255
DURBIN-WATSON	0.1588		

VARIABLE	DF	B VALUE	STD ERROR	T RATIO	APPROX PROB
INTERCPT	1	11.1404796	0.21967	50.714	0.0001
T	1	-0.1893490	0.02156	-8.783	0.0001
TSQ	1	0.0014596474	0.00044	3.282	0.0021

CORRELATION OF B-VALUES

	INTERCPT	T	TSQ
INTERCPT	1	-0.87631	0.76127
T	-0.87631	1	-0.96958
TSQ	0.76127	-0.96958	1

ESTIMATES OF AUTOCORRELATIONS

LAG	COVARIANCE	CORRELATION
0	0.21108	1.000000
1	0.184093	0.872148
2	0.148853	0.705196

PRELIMINARY MSE = 0.047813

ESTIMATES OF THE AUTOREGRESSIVE PARAMETERS

LAG	COEFFICIENT	STD ERROR	T RATIO
1	-1.07417683	0.15192588	-7.070401
2	0.23164542	0.15192588	1.524727

YULE-WALKER ESTIMATES

SSE	1.399299	DFE	41
MSE	0.034129	ROOT MSE	0.184741
SBC	-9.43718	AIC	-18.5804
REG RSQ	0.7372	TOTAL RSQ	0.9893
DURBIN-WATSON	1.6059		

VARIABLE	DF	B VALUE	STD ERROR	T RATIO	APPROX PROB
INTERCPT	1	10.7592050	0.37968	28.337	0.0001
T	1	-0.1648549	0.03670	-4.492	0.0001
TSQ	1	0.0011758595	0.00075	1.569	0.1242

SAS AUTOREG printout for Exercise 13.38 (continued)

CORRELATION OF B-VALUES

	INTERCPT	T	TSQ
INTERCPT	1	-0.79714	0.643203
T	-0.79714	1	-0.95957
TSQ	0.643203	-0.95957	1

OBS	T	Y	YHAT	RESID	LPL95	UPL95
1	1	10.11	10.5955	-0.48553	9.53786	11.6532
2	2	10.00	10.0107	-0.01075	9.24910	10.7724
3	3	10.00	9.9213	0.07871	9.20532	10.6373
4	4	10.00	9.9235	0.07646	9.24285	10.6042
5	5	10.00	9.9007	0.09932	9.25005	10.5513
6	6	10.00	9.8782	0.12180	9.25249	10.5039
7	7	10.00	9.8561	0.14392	9.25035	10.4618
8	8	10.00	9.8343	0.16566	9.24391	10.4248
9	9	10.00	9.8130	0.18704	9.23357	10.3924
10	10	10.00	9.7920	0.20804	9.21980	10.3641
11	11	10.00	9.7713	0.22868	9.20314	10.3395
12	12	10.00	9.7511	0.24894	9.18413	10.3180
13	13	9.52	9.7312	-0.21117	9.16335	10.2990
14	14	9.05	9.1960	-0.14604	8.62573	9.7663
15	15	9.00	8.7832	0.21679	8.20930	9.3571
16	16	9.00	8.8196	0.18041	8.24142	9.3978
17	17	8.50	8.8128	-0.31276	8.23006	9.3955
18	18	8.50	8.2576	0.24237	7.67047	8.8448
19	19	8.50	8.3558	0.14422	7.76450	8.9471
20	20	8.50	8.3385	0.16152	7.74365	8.9333
21	21	8.20	8.3215	-0.12155	7.72391	8.9192
22	22	8.00	7.9827	0.01726	7.38315	8.5823
23	23	7.58	7.8212	-0.24120	7.22063	8.4218
24	24	7.21	7.4006	-0.19056	6.79998	8.0011
25	25	6.50	7.0850	-0.58495	6.48537	7.6845
26	26	6.50	6.3929	0.10708	5.79527	6.9906
27	27	6.50	6.5427	-0.04268	5.94784	7.1375
28	28	6.50	6.5283	-0.02834	5.93706	7.1196
29	29	6.50	6.5144	-0.01437	5.92721	7.1015
30	30	6.50	6.5008	-0.00077	5.91807	7.0835
31	31	6.02	6.4875	-0.46754	5.90937	7.0657
32	32	6.00	5.9591	0.04092	5.38517	6.5330
33	33	6.00	6.0363	-0.03630	5.46599	6.6066
34	34	6.00	6.0288	-0.02881	5.46100	6.5966
35	35	6.00	6.0171	-0.01707	5.45014	6.5840
36	36	6.00	6.0057	-0.00569	5.43750	6.5739
37	37	6.00	5.9947	0.00532	5.42253	6.5668
38	38	6.00	5.9840	0.01595	5.40466	6.5634
39	39	6.00	5.9738	0.02622	5.38336	6.5642
40	40	6.00	5.9639	0.03611	5.35815	6.5696
41	41	6.00	5.9544	0.04564	5.32865	6.5801
42	42	6.00	5.9452	0.05479	5.29458	6.5958
43	43	6.00	5.9364	0.06358	5.25573	6.6171
44	44	6.00	5.9280	0.07199	5.21205	6.6440
45	45	6.00	5.9200	0.08003	5.16352	6.6764
46	46	6.00	5.9123	0.08771	5.11022	6.7144

a Write the equation of the fitted second-order autoregressive model.

b Plot the predicted values of the prime interest rate (given on the printout), and connect the points with line segments. Repeat this process for the observed values of y_t (given in Exercise 13.30). Compare the fitted time series with the observed time series.

c Does the second-order autoregressive model provide much of an improvement over the first-order autoregressive model of Exercise 13.30? Explain.

13.39 Refer to Exercise 13.30. Use the ratio-to-moving-average method to deseasonalize the time series.

a Apply a 12-month centered moving average to the data.

b Calculate the specific seasonal indexes for the available data points.

c Use the results of part (b) to calculate the seasonal indexes.

d Normalize the seasonal indexes, and obtain the deseasonalized time series.

e Plot the deseasonalized series. Are any inherent components apparent?

13.40 Refer to Exercise 13.39.

a Use an appropriate Minitab or SAS computer program to fit a cubic trend line to the deseasonalized time series calculated in Exercise 13.39(d).

b Using the values of $\hat{T}_t$ from part (a), calculate the residuals for the multiplicative model. Are there any cyclical effects present?

c Forecast the prime interest rate for November 1993 using the multiplicative model. Can you give a measure of the goodness of your forecast?

Exercises Using the Data Disk

13.41 Refer to data set E: Total Manufacturer's Monthly Inventories for the years 1961–93 on the data disk.

a Select three consecutive years of data. Plot the data and identify the components that are apparent in the data.

b Use a moving average of order three to smooth the data. Does the plotted smoothed series reveal any components not identified in part (a)?

c Use exponential smoothing with a smoothing coefficient of $\omega = .4$. How does the plotted smoothed series compare with that in part (b)?

13.42 Refer to the last three years of data in data set E.

a Plot the data and identify the components in the series. Are they the same as those identified in Exercise 13.41(a)?

b Use a centered 12-month moving average to smooth these data.

c Use the results of part (b) to develop a multiplicative model to describe these data. Based on your model, determine the predicted value of the series for January 1994.

d Develop an additive model to describe these data. What is the predicted value of the series for January 1994 based on your fitted model?

e If you have the SAS package available, use the AUTOREG procedure to fit an autoregressive model to these data. Your model should include a deterministic portion that reflects the components identified in part (a).

CHAPTER 14

SAMPLING METHODS

About This Chapter

The estimation procedures and tests of hypotheses described in the preceding chapters were based on the assumption that the sample was randomly selected from the population. In this chapter, we note that selecting a simple random sample from a population may not be easy to accomplish. We emphasize the importance of the sampling procedure in evaluating the goodness of an inference, and we introduce you to some sampling procedures that will reduce the difficulties and the costs associated with sampling.

Contents

Sampling 58 Million Beer Cans

This case study is derived from a statistical sampling that formed a small part of the defense in a multimillion-dollar lawsuit. We will leave the litigants unnamed, and we will refrain from revealing some of the more sensitive details of the case.

The plaintiff in the case, a brewer, claimed that the defendant had, through negligence, damaged 58 million empty beer cans stored in the brewer's canning plant. Whether or not the damage occurred, who was responsible, and the extent of the damage are some of the uncertain aspects of the case upon which a just settlement must be based. Naturally, the defendants claimed that it had never happened, and, if it did, only a small portion of the cans were damaged and unusable. At that juncture they sought a statistician to instruct them how to select a random sample of cans from among the 58 million empty cans. The objective of the defense was to use the sample to estimate the proportion of the 58 million cans that had been damaged.

So how would you select a random sample of, say, 500[†] cans from among 58 million beer cans? Simple, you say! Remembering our definition of a random sample given in Section 6.1, we would select the sample of 500 cans in such a way that each distinctly different set of 500 cans would have an equal probability of being selected. But remember, a collection of 58 million beer cans is not a deck of cards or a bag full of beans. You cannot thoroughly mix them, select 500, and expect to acquire a random sample. Can you imagine the immense pile produced by 58 million beer cans?

This chapter presents some of the difficult problems encountered in selecting samples. First, we explain the relevance of the sampling procedure to statistical inference. Then we explain how (in principle) to select a random sample and explain, in Section 14.7, how we selected the 500 cans from among the 58 million cans in the warehouse. We survey some other sampling procedures and explain how they can reduce the cost of purchasing sample information.

[†]This sample size may seem small, but it was large enough to achieve the desired estimate. Incidentally, the cost of the testing was approximately $600 per can.

14.1 The Importance of the Sampling Procedure to Statistical Inference

We explained in Chapter 3 that probability is the vehicle that enables us to use sample data to make inferences about the population from which the sample was drawn. Because the sampling method is a key element of this process, we will review the logic involved.

Suppose that a soap manufacturer wants to estimate the proportion p of consumers in a given locale who favor soap brand A. A random sample of ten persons is selected from among the consumers, and it is observed that only one out of ten favors soap brand A. How does probability play a role in making an inference about p?

Since the number x in a sample of $n = 10$ from among the large number of consumers is a binomial random variable, we could calculate the probability of observing only $x = 1$ consumer in the sample for different values of p. These binomial probabilities, $p(1) = C_1^{10} pq^9$, for different values of p are shown in Table 14.1.

T A B L E **14.1**
The probability of observing $x = 1$ in a sample of $n = 10$

p	$p(1)$
1.0	0.000
.9	.000
.8	.000
.7	.000
.6	.002
.5	.010
.4	.040
.3	.121
.2	.268
.1	.387
.05	.315
0.0	0.000

You can see from Table 14.1 that if the population proportion p of consumers favoring brand A is .4 or larger, the probability of observing a value of x as small as 1 is very small. Since the probability of observing $x = 1$ is a maximum (from among the values of p included in Table 14.1) when the population proportion p is .1, we would be inclined to select .1 as our estimate of p. This principle—selecting, as the best estimate, the value of p that gives the highest probability for the observed sample—is the theoretical basis for many methods of statistical estimation. It is known as the **principle of maximum likelihood.**

How and why does the sampling procedure play a role in statistical inference? The answer is that we need to know how the sample is selected in order to be able to calculate the probability of observing specific sample outcomes. For example, we were able to calculate the probability of observing $x = 1$ consumer favoring soap brand A because the sample of $n = 10$ consumers was randomly selected from among a large population of consumers. If we had selected our sample of ten by choosing the first ten consumers to emerge from a supermarket, the probability of selecting consumers favoring brand A would not represent a random sample.

Simple random sampling—that is, choosing a sample in such a way that every sample of fixed size has an equal chance of being selected—is known as **probability sampling.** There are many other methods for choosing probability samples; most of them were devised to reduce the cost of sampling or to overcome some physical obstacle to acquiring the sample observations. Although most of the methods of inference based upon these sampling methods are beyond the scope of an introductory course, we will summarize some of the most common. You will then be aware of their availability in case you need them in the future.

14.2 How to Draw a Simple Random Sample

We defined a simple random sample in Section 6.1, but drawing one from a population is a lot easier said than done. Selecting 5 cards from a standard 52-card bridge deck in such a way that every set of 5 cards in the deck has an equal probability of inclusion in the sample is not too difficult. By thorough shuffling of the cards in the deck, we can ensure (to a reasonable degree of approximation) that the 5 cards selected from the deck represent a random sample. But selecting a random sample of n consumers from among all consumers in a specified region is a difficult task. We need to know who the consumers are, and because they cannot be shuffled, we must devise a system for drawing the sample so that every different sample of n consumers in the population will have equal probability of being selected.

The first step in selecting a sample from a population is to list the objects from which the sample is to be selected. These objects are called **sampling units,** and the list is called a **frame.** Most of the time, the sampling unit is a single **element,**[†] but as we note in Section 14.5, a sampling unit could be clusters of elements.

For example, suppose that we want to sample the choices of all adult household members in a community regarding the amount of a proposed sales tax. One way to do this is to treat each adult household member in the community as a sampling unit. Then the frame for the sample is a complete list of adult household members in the community. Each sampling unit is a single individual and therefore contains one element of the population.

A second and less expensive way to construct the frame and select the sample is to let each household represent a sampling unit. This will reduce the cost of sampling because you can obtain the opinions of all members in a household on a single visit. For this type of sampling, the frame is a list of all households in the community. The number of elements in a sampling unit will vary from one (for single-member households) to several, depending upon the number of adults in the household.

In order to be consistent with standard terminology, we will subsequently refer to the objects selected from the population as sampling units. However, you will understand that all of the methodology contained in the preceding chapters assumes

---

[†]In Chapter 1, we noted that the object upon which a measurement is made is called an **experimental unit** or, alternatively, an **element of the population.**

that a sampling unit and an element of the population are synonymous, that is, that each sampling unit contains only one element.

DEFINITION ▪ The objects to be selected from a population are called **sampling units.** ▪

DEFINITION ▪ A **frame** is a complete list of all sampling units contained in a population. ▪

The simplest and most reliable way to select a random sample of n sampling units from a large population is to employ a table of **random numbers** such as that shown in Table 13 of Appendix II. **Random number tables** are constructed so that integers occur randomly and with equal frequency. For example, suppose that the frame for a population contains $N = 1000$ sampling units, numbered in sequence from 0 to 999. Then turn to a table of random numbers such as the excerpt shown in Table 14.2.

TABLE **14.2**   Portion of a table of random numbers, Table 13 in Appendix II

Line	1	2	3	4	5	6	7	8	9	10	11	12	13	14
1	10480	15011	01536	02011	81647	91646	69179	14194	62590	36207	20969	99570	91291	90700
2	22368	46573	25595	85393	30995	89198	27982	53402	93965	34095	52666	19174	39615	99505
3	24130	48360	22527	97265	76393	64809	15179	24830	49340	32081	30680	19655	63348	58629
4	42167	93093	06243	61680	07856	16376	39440	53537	71341	57004	00849	74917	97758	16379
5	37570	39975	81837	16656	06121	91782	60468	81305	49684	60672	14110	06927	01263	54613
6	77921	06907	11008	42751	27756	53498	18602	70659	90655	15053	21916	81825	44394	42880
7	99562	72905	56420	69994	98872	31016	71194	18738	44013	48840	63213	21069	10634	12952
8	96301	91977	05463	07972	18876	20922	94595	56869	69014	60045	18425	84903	42508	32307
9	89579	14342	63661	10281	17453	18103	57740	84378	25331	12566	58678	44947	05585	56941
10	85475	36857	53342	53988	53060	59533	38867	62300	08158	17983	16439	11458	18593	64952
11	28918	69578	88231	33276	70997	79936	56865	05859	90106	31595	01547	85590	91610	78188
12	63553	40961	48235	03427	49626	69445	18663	72695	52180	20847	12234	90511	33703	90322
13	09429	93969	52636	92737	88974	33488	36320	17617	30015	08272	84115	27156	30613	74952
14	10365	61129	87529	85689	48237	52267	67689	93394	01511	26358	85104	20285	29975	89868
15	07119	97336	71048	08178	77233	13916	47564	81056	97735	85977	29372	74461	28551	90707
16	51085	12765	51821	51259	77452	16308	60756	92144	49442	53900	70960	63990	75601	40719
17	02368	21382	52404	60268	89368	19885	55322	44819	01188	65255	64835	44919	05944	55157

Select n of the random numbers in order. The numbers of the sampling units to be included in the random sample will be given by the first three digits of the random numbers. Thus if $n = 5$, we randomly select a starting point, say the random number in line 11, column 5 of Table 14.2. Using this random number and the four that follow in column 5, we will include sampling units numbered 709, 496, 889, 482, and 772. So as not to use the same sequence of random numbers over and over again, the experimenter should select different starting points in Table 13 to begin the selection of random numbers for different samples.

EXAMPLE   **14.1**   A county home builders association wanted to determine the proportion of home-owners in the county who planned major renovations, repairs, or additions during the coming year. The estimate of the proportion is to be based on a random sample of $n = 600$ homes to be selected from among the 712,524 homes in the county. Explain how the 600 homeowners should be selected in order to acquire a random sample.

Solution   The frame for the population is a complete numbered list of the 712,524 homes in the county. We are hoping this list will be available at the county tax office. To select the sample of 600, we will select six-digit numbers from the random number table. Six-digit numbers that exceed 712,524 will be discarded, as will any numbers that repeat themselves.

Consulting Table 14.2, we note that the random numbers in each line and column of the random number table contain five digits. We can obtain six-digit numbers by combining pairs of these five numbers and selecting the first six digits. For example, if we were to start our sample in line 5, columns 2 and 3, we would be using the ten-digit random numbers 3997581837, 0690711008, 7290556420, 9197705463, and so on. Since we plan to use the first six digits of these numbers and include only numbers equal to 712524 or less, we will discard two of the four numbers shown above (because they exceed 712524) and will retain only 399758 and 069071. Therefore, homes numbered 399758 and 69071 in the frame will be included in the sample. We will continue selecting ten-digit random numbers, retaining numbers equal to 712524 or less, until we have acquired the numbers of the remaining 598 homes needed for the sample.   ∎

Exercises

Basic Techniques

14.1   Explain how you can use a 52-card deck of bridge cards to draw a random sample of $n = 5$ from a population containing $N = 50$ sampling units.

14.2   Use the random number table, Table 13 in Appendix II, to draw a random sample of $n = 3$ sampling units from a population containing $N = 22$ sampling units. Explain precisely how you arrived at the sampling units to be included in your sample.

14.3   Use the random number table, Table 13 in Appendix II, to draw a random sample of $n = 12$ sampling units from a population containing $N = 12,500$ sampling units. Explain precisely how you arrived at the sampling units to be included in your sample.

14.4   Use the random number table, Table 13 in Appendix II, to draw a random sample of $n = 7$ sampling units from a population containing $N = 825$ sampling units. Explain precisely how you arrived at the sampling units to be included in your sample.

Applications

14.5   Suppose that a telephone company wishes to select a random sample of $n = 20$ (we select this small number to simplify this exercise) out of 7000 customers for a survey of customer attitudes concerning service. If the customers are numbered for identification purposes, indicate the customers you will include in your sample. Use the random number table, Table 13 in Appendix II, and explain how you selected your sample.

14.6   A small city contains 20,000 voters. Use the random number table, Table 13 in Appendix II, to identify the voters to be included in a random sample of $n = 15$.

14.7   If a survey of a newspaper's readership is obtained by requesting readers to respond to a questionnaire published in the newspaper, are the resulting responses likely to give a random sample of readership opinion? Explain.

14.8   A random sample of consumer preference for a new product was obtained by selecting and questioning every tenth person to pass by the busiest corner in a large city. Will this sample have the characteristics of a random sample selected from the consumers in the city? Explain.

14.9   Suppose that you decide to conduct a telephone public opinion survey, randomly sampling numbers from the telephone directory. The survey is conducted from 9 A.M. to 5 P.M. Will the resulting responses represent a random sample of adult public opinion in the community? Explain.

14.10   "Do workers in restaurants run by the House of Representatives in Washington want to join a union?" An article in the *New York Times* (July 6, 1987) describes a survey conducted by George M. White, the architect of the Capitol, which purports to answer this question. Questionnaires were sent to 235 eligible employees. Among several questions asked was the question "Do you believe your best interests would be served by becoming a member of a union?" Only 125 returned the questionnaire. "On the union question . . . there were 31 yeses, 68 noes, and 13 don't knows. Thirteen left the answer blank." William Raines, Mr. White's administrative assistant, noted that a majority had voted no on joining a union and stated that "the issue was 'closed'." Representative William L. Clay, Democrat of Missouri, said the poll was "a total farce." Do you think that the results of Mr. White's survey imply that the majority of the 235 eligible employees of the House of Representatives restaurant are opposed to joining a union? Explain.

14.3   Estimation Based on Simple Random Sampling

Most sample surveys have one of three objectives: to estimate a population mean μ, a population proportion p, or a population total τ. The total τ is defined to be the total of all x values in the population. If the population contains N values of x,—x_1, x_2, x_3, . . . , x_N—then

$$\tau = \sum_{i=1}^{N} x_i = N\mu$$

For example, a survey conducted to investigate the nature of household expenditures on food in a one-county area will be interested in the mean weekly expenditure μ on food by a single household in the county. It also may be interested in the total τ of the money spent on food by all households in the county during the given week.

In Chapter 7, we presented confidence intervals for a population mean μ and proportion p based on independent random sampling. These confidence intervals assume that the sample size n is large and that the number N of elements in the population is large relative to n.

This latter assumption is often violated in survey sampling. For example, a survey of the opinions of the chief executives of *Fortune 500* companies will involve sampling the opinions of 500 chief executives. If you sampled $n = 100$ of the $N = 500$, your sample would contain 20% of all elements in the population.

The estimator of the standard deviation of the mean is given by

$$s_{\bar{x}} = \frac{s}{\sqrt{n}}\sqrt{\frac{N - n}{N}}$$

The quantity $\dfrac{n}{N}$ is called the sampling fraction, and the quantity $\sqrt{\dfrac{N - n}{N}}$ is called the *finite population correction factor*. When the sample size n is small relative to N, say less than $.05N$, the quantity $\sqrt{(N - n)/N}$ will be close to 1 and can be ignored. For example, if $n/N = .05$,

$$s_{\bar{x}} = \frac{s}{\sqrt{n}}\sqrt{\frac{N - n}{N}} = \frac{s}{\sqrt{n}}\sqrt{1 - \frac{n}{N}} = \frac{s}{\sqrt{n}}\sqrt{1 - .05}$$
$$= \frac{s}{\sqrt{n}}\sqrt{.95} = \frac{s}{\sqrt{n}}(.97)$$

Since .97 is close to 1, the standard error of $\bar{x}$ reduces to

$$s_{\bar{x}} \approx \frac{s}{\sqrt{n}}$$

which is the standard error given in Chapter 7.

The displays that follow give approximate confidence intervals for a population mean μ, total τ, and proportion p for sampling from finite populations, that is, populations containing a finite number of elements.

Approximate 95% Confidence Interval for a Population Mean μ: Independent Random Sampling from a Finite Population

$$\bar{x} \pm 1.96\frac{s}{\sqrt{n}}\sqrt{\frac{N - n}{N}}$$

where

$$\bar{x} = \frac{\sum\limits_{i=1}^{n} x_i}{n} \qquad s = \sqrt{\frac{\sum\limits_{i=1}^{n}(x_i - \bar{x})^2}{n - 1}}$$

N = Number of sampling units in the population

n = Number of sampling units in the sample

Note: For simple random sampling, a sampling unit contains only one element.

Approximate 95% Confidence Interval for a Population Total τ: Independent Random Sampling from a Finite Population

$$\hat{\tau} \pm 1.96 \frac{Ns}{\sqrt{n}} \sqrt{\frac{N-n}{N}}$$

where

$$\hat{\tau} = N\bar{x} \qquad s = \sqrt{\frac{\sum_{i=1}^{n}(x_i - \bar{x})^2}{n-1}}$$

N = Number of sampling units in the population

n = Number of sampling units in the sample

Note: For simple random sampling, a sampling unit contains only one element.

Approximate 95% Confidence Interval for a Population Proportion p: Independent Random Sampling from a Finite Population

$$\hat{p} \pm 1.96 \sqrt{\frac{\hat{p}\hat{q}}{n-1}} \sqrt{\frac{N-n}{N}}$$

where

$$\hat{p} = \frac{x}{n} \qquad \hat{q} = 1 - \hat{p}$$

N = Number of sampling units in the population

n = Number of sampling units in the sample

Note: For simple random sampling, a sampling unit contains only one element.

EXAMPLE **14.2** The manager of an auto rental company wants to estimate the total number of miles put on its cars per month. A random sample of $n = 30$ cars was selected from the company's fleet of 280 cars, and the mileage was recorded for each car at the beginning and end of a particular month. The mean and standard deviation for the sample were $\bar{x} = 1342$ and $s = 227$. Find an approximate 95% confidence interval for the total mileage registered for the fleet during the month.

Solution For this example, $N = 280$, $n = 30$, $\bar{x} = 1342$, and $s = 227$. Then

$$\hat{\tau} = N\bar{x} = (280)(1342) = 375,760$$

and the approximate 95% confidence interval for τ is

$$\hat{\tau} \pm 1.96 \frac{Ns}{\sqrt{n}} \sqrt{\frac{N-n}{N}}$$

$$375,760 \pm (1.96) \frac{(280)(227)}{\sqrt{30}} \sqrt{\frac{280-30}{280}}$$

or

$$375,760 \pm 21,491.7$$

Therefore, we estimate the fleet mileage to be as low as 354,268.3 miles or as high as 397,251.7 miles per month. If the manager wants a smaller confidence interval for the estimate, a larger sample size must be used. ∎

Exercises

Basic Techniques

14.11 A random sample of $n = 50$ was selected from a population, and the sample mean and variance were $\bar{x} = 84.1$ and $s^2 = 122.44$.

a Calculate a 95% confidence interval for μ, assuming the number N of elements in the population to be very large.

b Calculate a 95% confidence interval for μ, assuming the number N of elements in the population to be equal to 100.

c Compare the two confidence intervals in parts (a) and (b), and note the effect of the finite population correction factor on the width of the interval in part (b).

14.12 Use the data in Exercise 14.11 to calculate an approximate 95% confidence interval for the population total τ if $N = 100$.

14.13 A random sample of $n = 100$ observations, each observation a success or failure, was selected from a population containing $N = 400$ elements. Find an approximate 95% confidence interval for the proportion p of successes in the population if the number of successes in the sample is 34.

14.14 Calculate the finite population correction factor for n/N equal to .001, .1, .3, .5, .7. Then graph the finite population correction factor as a function of n/N. Note how the correction factor decreases in value as n/N increases. What are the practical implications of this result?

Applications

14.15 A dealer in floor coverings has the opportunity to buy the inventory of another dealer who is going out of business. The dealer randomly sampled 50 of 421 different whole or partial rolls of carpet offered for sale and estimated the value of each. The mean and standard deviation of the 50 estimates were $\bar{x} = \$1248$ and $s = \$175$. Find an approximate 95% confidence interval for the mean estimated value per roll for the 421 rolls.

14.16 Refer to Exercise 14.15. Find an approximate 95% confidence interval for the total estimated value of the 421 rolls.

14.17 An auto dealership in a small town wanted to acquire information on the market for new cars during the coming year. The dealer obtained a list of the 8746 persons in the town who were

18 or older and randomly sampled 500. The number of persons in the sample who planned to buy a new car during the coming year was 29. Find an approximate 95% confidence interval for the proportion of all persons, 18 or older, in the town who plan to buy a new car during the coming year.

 14.18 A bank conducted a survey to investigate the potential market for home renovation loans among its customers. The bank randomly sampled 1000 from among its 9706 customers and asked each whether or not he or she planned home renovation (or addition) in the near future, whether the customer would seek a bank loan for the renovation, and what the approximate amount of money to be borrowed might be. Of the 1000 customers sampled, 46 indicated that they planned to borrow money for home renovation. The mean and standard deviation for the projected borrowings were $\bar{x} = \$6751$ and $s = \$1463$. Find an approximate 95% confidence interval for the mean value of a home loan for the bank's customers who wish to borrow money.

14.19 Refer to Exercise 14.18. Find an approximate 95% confidence interval for the proportion of the bank's 9706 customers who will seek a home renovation loan.

14.4 Stratified Random Sampling

Simple random sampling is not the only way to acquire a probability sample. Difficulties in constructing the frame and in contacting the elements of the population (particularly when the elements are people, households, etc.) suggest other methods of sampling that are easier to conduct and less costly. Methods for selecting a sample are called **sampling designs,** and the sample obtained is often called a **sample survey.**

One method for reducing the cost of public opinion or consumer polls is to split the geographic region in which the elements (people) of the population reside into segments called **strata.** Strata are selected so that the elements within a stratum are homogeneous in the variable of interest and elements in different strata are not homogeneous. Samples are selected from within each stratum, and then this information is combined to make an inference about the entire population.

For example, suppose that you want to sample the opinions of homeowners in a state. Rather than selecting a simple random sample from among all homeowners in the state, it would be easier to sample each individual county separately and then combine the information in the county samples to make inferences about the state. Thus we can split the state into strata, with one stratum corresponding to each county in the state. Frames for each county can easily be obtained from the county courthouses, and the costs of contacting people within the smaller county areas will be less than the costs encountered in nonstratified sampling.

Stratified random sampling has another advantage. Not only can you combine the information in the strata samples to make inferences about the complete population, but you can also use the sample information about the characteristics of each stratum. For example, you could compare homeowners' opinions in one county with those of another.

The formulas for approximate 95% confidence intervals for a population mean μ, a population total τ, and a population proportion p, based on stratified random sampling, are given in the following displays. We will illustrate the estimation procedures with examples.

Stratified Random Sampling

Notation

$$\mu = \text{Mean of the population}$$
$$N_i = \text{Number of elements in stratum } i, i = 1, 2, 3, \ldots, L$$
$$N = \text{Total number of elements in the population}$$
$$= N_1 + N_2 + N_3 + \cdots + N_L$$
$$n_i = \text{number of elements in the sample selected from}$$
$$\text{stratum } i, i = 1, 2, 3, \ldots, L$$
$$\overline{x}_i = \text{Mean of the sample selected from stratum } i,$$
$$i = 1, 2, 3, \ldots, L$$
$$s_i^2 = \text{Variance of the sample measurements from stratum } i,$$
$$i = 1, 2, 3, \ldots, L$$

An Approximate 95% Confidence Interval for the Population Mean μ: Stratified Random Sampling

$$\overline{x}_{\text{st}} \pm 1.96 \sqrt{\frac{1}{N^2} \sum_{i=1}^{L} N_i^2 \left(\frac{N_i - n_i}{N_i} \right) \frac{s_i^2}{n_i}}$$

where

$$\overline{x}_{\text{st}} = \frac{1}{N}(N_1 \overline{x}_1 + N_2 \overline{x}_2 + \cdots + N_L \overline{x}_L) = \frac{1}{N} \sum_{i=1}^{L} N_i \overline{x}_i$$

is the estimate of the population mean μ based on the stratified random sample. When the sample size n_i is small relative to the stratum size $N_i, i = 1, 2, \ldots, L$, this formula reduces to approximately

$$\overline{x}_{\text{st}} \pm 1.96 \sqrt{\frac{1}{N^2} \sum_{i=1}^{L} N_i^2 \frac{s_i^2}{n_i}}$$

An Approximate 95% Confidence Interval for the Population Total τ: Stratified Random Sampling

$$\hat{\tau} \pm 1.96 \sqrt{\sum_{i=1}^{L} N_i^2 \left(\frac{N_i - n_i}{N_i} \right) \frac{s_i^2}{n_i}}$$

where

$$\hat{\tau} = N\overline{x}_{\text{st}}$$

$$\overline{x}_{\text{st}} = \frac{1}{N}[N_1\overline{x}_1 + N_2\overline{x}_2 + \cdots + N_L\overline{x}_L] = \frac{1}{N}\sum_{i=1}^{L} N_i\overline{x}_i$$

When the sample size n_i is small relative to the stratum size $N_i, i = 1, 2, \ldots, L$, this formula reduces to

$$\hat{\tau} \pm 1.96 \sqrt{\sum_{i=1}^{L} N_i^2 \frac{s_i^2}{n_i}}$$

Approximate 95% Confidence Interval for a Population Proportion p: Stratified Random Sampling

$$\hat{p}_{\text{st}} \pm 1.96 \sqrt{\frac{1}{N^2} \sum_{i=1}^{L} N_i^2 \left(\frac{N_i - n_i}{N_i} \right) \frac{\hat{p}_i\hat{q}_i}{n_i - 1}}$$

where

$$\hat{p}_{\text{st}} = \frac{1}{N}[N_1\hat{p}_1 + N_2\hat{p}_2 + \cdots + N_L\hat{p}_L] = \frac{1}{N}\sum_{i=1}^{L} N_i\hat{p}_i$$

is the estimate of p based on the stratified random sample and $\hat{p}_i$ is the sample proportion from stratum $i, i = 1, 2, 3, \ldots, L$. When the sample size n_i is small relative to the stratum size $N_i, i = 1, 2, \ldots, L$, this formula reduces to approximately

$$\hat{p}_{\text{st}} \pm 1.96 \sqrt{\frac{1}{N^2} \sum_{i=1}^{L} N_i^2 \frac{\hat{p}_i\hat{q}_i}{n_i - 1}}$$

E X A M P L E   **14.3**   A television station serving a three-county area wished to estimate the mean number of hours per day of viewing time per household in its viewing area. The station decided to randomly select a 1% sample—that is, 1% of the total number of households in each county. A summary of the data is shown in Table 14.3. Find an approximate 95% confidence interval for the mean viewing time per household within the three-county viewing area.

T A B L E 14.3
TV viewing data for
Example 14.3

County	Number of Households in Stratum i, N_i	Stratum Sample Size, n_i	Stratum Sample Mean, $\bar{x}_i$	Stratum Sample Variance, s_i^2
1	12,473	125	2.92	1.96
2	35,241	352	2.14	1.21
3	23,178	232	3.63	3.24

$$N = 70{,}892$$

Solution   The stratified mean is

$$\bar{x}_{st} = \frac{1}{N}(N_1\bar{x}_1 + N_2\bar{x}_2 + N_3\bar{x}_3)$$

$$= \frac{1}{70{,}892}[(12{,}473)(2.92) + (35{,}241)(2.14) + (23{,}178)(3.63)]$$

$$= 2.76 \text{ hours per day}$$

Then the confidence interval for μ is

$$\bar{x}_{st} \pm 1.96\sqrt{\frac{1}{N^2}\sum_{i=1}^{L}N_i^2\left(\frac{N_i - n_i}{N_i}\right)\frac{s_i^2}{n_i}}$$

For our example, the quantity $(N_i - n_i)/N_i$ is approximately equal to 1 for each stratum because the sample size is so small (1%) in relation to the number of elements in the stratum. Therefore, the formula for our confidence interval reduces to

$$\bar{x}_{st} \pm 1.96\sqrt{\frac{1}{N^2}\sum_{i=1}^{L}N_i^2\frac{s_i^2}{n_i}}$$

$$2.76 \pm 1.96\sqrt{\frac{1}{(70{,}892)^2}\left[(12{,}473)^2\left(\frac{1.96}{125}\right) + (35{,}241)^2\left(\frac{1.21}{352}\right) + (23{,}178)^2\left(\frac{3.24}{232}\right)\right]}$$

$$2.76 \pm 1.96\sqrt{.0028}$$

or

$$2.76 \pm .10$$

Thus we estimate the mean number of viewing hours per household to be in the interval 2.66 to 2.86 hours. ▪

EXAMPLE 14.4 A large retailer serving two small towns conducted a survey to determine the total amount of money that households in the area planned to spend on "big-ticket" electric appliances during the coming year. A random sample of 200 households was selected from each town. The number of households in each town and the sample means and variances of the planned household expenditures are shown in Table 14.4. Find an approximate 95% confidence interval for the total planned expenditures for the two towns.

TABLE 14.4
Expenditures data for
Example 14.4

Town	Number of Households per Town, N_i	Sample Size, n_i	Sample Mean (in dollars), $\bar{x}_i$	Sample Variance, s_i^2
1	2149	200	134	40,122
2	1872	200	168	37,104

$N = 4021$

Solution The confidence interval for τ is

$$\hat{\tau} \pm 1.96\sqrt{\sum N_i^2 \left(\frac{N_i - n_i}{N_i}\right)\frac{s_i^2}{n_i}}$$

where

$$\hat{\tau} = N\bar{x}_{st} = N\left[\frac{1}{N}(N_1\bar{x}_1 + N_2\bar{x}_2)\right]$$
$$= N_1\bar{x}_1 + N_2\bar{x}_2 = (2149)(134) + (1872)(168) = \$602{,}462$$

Substituting into the formula for the confidence interval yields

$$(602{,}462) \pm 1.96\sqrt{(2149)^2\left(\frac{2149 - 200}{2149}\right)\left(\frac{40{,}122}{200}\right) + (1872)^2\left(\frac{1872 - 200}{1872}\right)\left(\frac{37{,}104}{200}\right)}$$

or

$$602{,}462 \pm 73{,}882$$

Therefore, we estimate the total expenditures on big-ticket electric appliances during the next year to be somewhere between \$528,580 and \$676,344. ▪

EXAMPLE 14.5 Refer to Example 14.3. Estimate the proportion of all households in the three-county area who prefer the television station's programs if the sample proportions preferring the station's programs are as shown in Table 14.5.

TABLE **14.5**	County	Number of Households in Stratum i, N_i	Stratum Sample Size, n_i	Stratum Sample Proportion, $\hat{p}_i$
TV viewing data for Example 14.5	1	12,473	125	.21
	2	35,241	352	.17
	3	23,178	232	.34

$$N = 70{,}892$$

Solution

$$\hat{p}_{st} = \frac{1}{70{,}892}[(12{,}473)(.21) + (35{,}241)(.17) + (23{,}178)(.34)]$$
$$= .23$$

Because the sample sizes are small relative to the strata sample sizes, $(N_i - n_i)/N_i \approx 1$ for $i = 1, 2, \ldots, L$, and the approximate 95% confidence interval for p is

$$\hat{p}_{st} \pm 1.96 \sqrt{\frac{1}{N^2} \sum_{i=1}^{L} N_i^2 \frac{\hat{p}_i \hat{q}_i}{n_i - 1}}$$

Substituting into this formula, we obtain

$$.23 \pm 1.96 \sqrt{\frac{1}{(70{,}892)^2}\left[(12{,}473)^2\frac{(.21)(.79)}{124} + (35{,}241)^2\frac{(.17)(.83)}{351} + (23{,}178)^2\frac{(.34)(.66)}{231}\right]}$$

$$.23 \pm 1.96\sqrt{.0002446}$$

or

$$.23 \pm .03$$

Therefore, we estimate the proportion of households who prefer the television station's programs to be in the interval .20 to .26. ∎

EXAMPLE **14.6** Refer to Example 14.4. Find a 95% confidence interval for the difference in the mean planned expenditures per household on big-ticket electric appliances between households in towns 1 and 2.

Solution The point estimator for the difference is $(\bar{x}_1 - \bar{x}_2)$, the difference in the sample means from strata 1 and 2. The variance of this estimator is

$$\sigma^2_{(\bar{x}_1 - \bar{x}_2)} = \sigma^2_{\bar{x}_1} + \sigma^2_{\bar{x}_2}$$

which, for finite populations, is equal to

$$\sigma^2_{(\bar{x}_1 - \bar{x}_2)} = \frac{\sigma_1^2}{n_1}\left(\frac{N_1 - n_1}{N_1}\right) + \frac{\sigma_2^2}{n_2}\left(\frac{N_2 - n_2}{N_2}\right)$$

Then the 95% confidence interval for the difference $(\mu_1 - \mu_2)$ in strata means is

$$(\bar{x}_1 - \bar{x}_2) \pm 1.96\sigma_{(\bar{x}_1 - \bar{x}_2)}$$

or approximately

$$(\bar{x}_1 - \bar{x}_2) \pm 1.96 \sqrt{\frac{s_1^2}{n_1}\left(\frac{N_1 - n_1}{N_1}\right) + \frac{s_2^2}{n_2}\left(\frac{N_2 - n_2}{N_2}\right)}$$

$$(134 - 168) \pm 1.96 \sqrt{\frac{40,122}{200}\left(\frac{2149 - 200}{2149}\right) + \frac{37,104}{200}\left(\frac{1872 - 200}{1872}\right)}$$

or

$$-\$34 \pm \$36.54$$

Therefore, we estimate the difference in mean planned expenditures per household on big-ticket electric appliances between towns 1 and 2 to be between −$70.54 and $2.54. The expenditures per household for town 2 could exceed those for town 1 by as much as $70.54 or could be less than those for town 1 by as much as $2.54.   ∎

Exercises

Basic Techniques

14.20 A stratified random sample was selected from among four strata. Use the data in the accompanying table to demonstrate the procedure for finding an approximate 95% confidence interval for the population mean μ. [*Note:* This exercise is presented for you to practice the mechanics of finding a confidence interval for μ. Because of the small sample sizes, it is unlikely that the confidence coefficient will be close to .95.]

	Stratum			
	1	2	3	4
Stratum size	40	30	30	50
Sample size	4	4	4	4
Sample observations	6, 8, 5, 6	7, 10, 8, 9	7, 8, 6, 6	5, 7, 6, 5

14.21 Use the data in Exercise 14.20 to find an approximate 95% confidence interval for the population total τ.

14.22 A stratified random sample was selected from among four strata. The pertinent data are shown in the accompanying table. Find an approximate 95% confidence interval for the population mean μ.

Stratum	Number of Sampling Units per Stratum, N_i	Stratum Sample Size, n_i	Stratum Sample Mean, $\bar{x}_i$	Stratum Sample Variance, s_i^2
1	1000	200	421	2410
2	3000	200	502	2938
3	2000	200	325	2047
4	1000	200	280	2214

14.23   Use the data in Exercise 14.22 to find an approximate 95% confidence interval for the population total τ.

14.24   A stratified random sample was selected from among four strata. Use the data shown in the accompanying table to find an approximate 95% confidence interval for the population proportion p.

	Stratum			
	1	2	3	4
Stratum size	1000	1200	800	1500
Sample size	100	100	100	100
Sample proportion	.3	.25	.29	.34

14.25   A stratified random sample was selected from among three strata. Use the data shown in the accompanying table to find an approximate 95% confidence interval for the population proportion p.

	Stratum		
	1	2	3
Stratum size	400	200	300
Sample size	100	100	100
Sample proportion	.62	.74	.55

Applications

14.26   A zoning commission is formed to estimate the mean appraised value of houses in a residential suburb of a city. It is convenient to use the two voting districts in the suburb as strata because separate lists of dwellings are available for each district. From the data given in the accompanying table, find an approximate 95% confidence interval for the mean appraised value for all houses in the suburb.

Stratum I	Stratum II
$N_1 = 110$	$N_2 = 168$
$n_1 = 20$	$n_2 = 30$
$\sum_{i=1}^{n_1} x_i = 240{,}000$	$\sum_{i=1}^{n_2} x_i = 420{,}000$
$\sum_{i=1}^{n_1} x_i^2 = 2{,}980{,}000{,}000$	$\sum_{i=1}^{n_2} x_i^2 = 6{,}010{,}000{,}000$

14.27   Refer to Exercise 14.26, and find an approximate 95% confidence interval for the total appraised value for all houses in the suburb.

14.28   A state energy office wished to estimate the proportion of public buildings that, based on energy analyses, were judged to be operating in an energy-efficient manner. The state was divided into three regions, two large urban areas and one rural area. The results of the survey are shown in the accompanying table. Find an approximate 95% confidence interval for the proportion p of all public buildings in the state that would be judged to be energy-efficient.

	Region		
	1	**2**	**3**
Number of state government buildings in region	249	432	316
Sample size	50	80	60
Number of energy-efficient buildings in sample	14	34	29

14.29 An antique dealer planned to purchase the complete inventories of three English dealers and export the entire stock to the United States. To obtain an approximate value of the stock, the dealer randomly selected 50 items from each of the three inventories and had each item appraised. The numbers of items in the three inventories, the sample means, and the sample variances are shown in the accompanying table. Find an approximate 95% confidence interval for the mean value per item of the dealer's purchases.

	Inventory		
	1	**2**	**3**
N_i	425	316	559
$\bar{x}_i$	$287	$389	$316
s_i^2	41,116	35,488	59,106

14.30 Refer to Exercise 14.29, and find an approximate 95% confidence interval for the total value of all three inventories.

14.31 Refer to Exercises 14.29 and 14.30. The estimate of the total value of inventory is

$$\hat{\tau}_i = N_i \bar{x}_i$$

and the variance of the estimate is

$$\sigma_{\hat{\tau}_i}^2 = N_i^2 \sigma_{\bar{x}_i}^2$$

where for a finite population

$$\sigma_{\bar{x}_i}^2 = \frac{\sigma_i^2}{n_i} \left(\frac{N_i - n_i}{N_i} \right)$$

Using this result and substituting s_i^2 for σ_i^2, we obtain an approximate 95% confidence interval for the difference in the total value of inventories i and j:

$$(\hat{\tau}_i - \hat{\tau}_j) \pm 1.96 \sqrt{ N_i^2 \frac{s_i^2}{n_i} \left(\frac{N_i - n_i}{N_i} \right) + N_j^2 \frac{s_j^2}{n_j} \left(\frac{N_j - n_j}{N_j} \right) }$$

Use this formula to find an approximate 95% confidence interval for the difference in the total value of inventories 1 and 2.

14.5   Cluster Sampling

A **cluster** is a collection of elements. Sometimes, it is less costly to construct a frame and to sample clusters rather than the individual elements of a population. For example, suppose that you wanted to estimate the mean dollar amount that adults in a large suburban residential area planned to spend per month on key items. One easy way to do this estimation is to randomly sample residential blocks and then survey the buying intentions of all adults within each of the sampled blocks. Thus each block will contain a cluster of elements, with the number of elements varying from one cluster to another. A frame can easily be constructed by using a map of the residential area and designating each block as a cluster. The travel costs involved in contacting the adults within a block will be minimal. Consequently, you will be able to contact many more people at a lower cost than you could by using a simple random sample of all adults in the area.

Cluster sampling can reduce the cost of sampling, but selecting large numbers of elements in clusters is not a substitute for selecting a reasonable number of clusters. The elements within a cluster may tend to give similar responses. For example, if the block (cluster) in the residential community happens to contain very expensive homes, many of the adults in the block may have plans to buy big-ticket items. If the block is in a lower-economic area, very few of the adults in the block may have buying intentions. Consequently, you need to sample enough clusters to obtain a measure of the variation in the responses from one cluster to another.

The formulas for approximate 95% confidence intervals for a population mean μ, a population total τ, and a population proportion p, based on cluster sampling, are given in the following displays. We will illustrate the estimation procedures with examples.

Cluster Sampling

Notation

$$N = \text{Number of clusters in the population}$$

$$n = \text{Number of clusters in the sample}$$

$$m_i = \text{Number of elements in cluster } i, i = 1, 2, \ldots, n$$

$$\overline{m} = \frac{\sum_{i=1}^{n} m_i}{n} = \text{Average cluster size in sample}$$

$$M = \sum_{i=1}^{N} m_i = \text{Number of elements in the population}$$

$$\overline{M} = \frac{M}{N} = \text{Average cluster size for the population}$$

$$x_i = \text{Total of all observations in the } i\text{th cluster}, i = 1, 2, \ldots, n$$

$$\overline{x} = \frac{\sum_{i=1}^{n} x_i}{\sum_{i=1}^{n} m_i}$$

An Approximate 95% Confidence Interval for μ: Cluster Sampling

$$\bar{x} \pm 1.96 \sqrt{\left(\frac{N-n}{Nn\overline{M}^2}\right) \frac{\sum_{i=1}^{n}(x_i - \bar{x}m_i)^2}{n-1}}$$

where

$$\sum_{i=1}^{n}(x_i - \bar{x}m_i)^2 = \sum_{i=1}^{n} x_i^2 - 2\bar{x}\sum_{i=1}^{n} x_i m_i + \bar{x}^2 \sum_{i=1}^{n} m_i^2$$

Note: $\overline{m}$ is used to approximate $\overline{M}$ when the number M of elements in the population is unknown.

An Approximate 95% Confidence Interval for τ: Cluster Sampling

$$\hat{\tau} \pm 1.96 \sqrt{N^2 \left(\frac{N-n}{Nn}\right) \frac{\sum_{i=1}^{n}(x_i - \bar{x}m_i)^2}{n-1}}$$

where

$$\hat{\tau} = M\bar{x}$$
$$\sum_{i=1}^{n}(x_i - \bar{x}m_i)^2 = \sum_{i=1}^{n} x_i^2 - 2\bar{x}\sum_{i=1}^{n} x_i m_i + \bar{x}^2 \sum_{i=1}^{n} m_i^2$$

Note: $\overline{m}$ is used to approximate $\overline{M}$ when the number M of elements in the population is unknown.

An Approximate 95% Confidence Interval for p: Cluster Sampling

$$\hat{p} \pm 1.96 \sqrt{\left(\frac{N-n}{Nn\overline{M}^2}\right) \frac{\sum_{i=1}^{n}(a_i - \hat{p}m_i)^2}{n-1}}$$

where

$$a_i = \text{Number of "successes" in cluster } i, i = 1, 2, \ldots, n$$

$$\hat{p} = \frac{\sum\limits_{i=1}^{n} a_i}{\sum\limits_{i=1}^{n} m_i}$$

$$\sum_{i=1}^{n}(a_i - \hat{p}m_i)^2 = \sum_{i=1}^{n} a_i^2 - 2\hat{p}\sum_{i=1}^{n} a_i m_i + \hat{p}^2 \sum_{i=1}^{n} m_i^2$$

Note: $\overline{m}$ is used to approximate $\overline{M}$ when the number M of elements in the population is unknown.

E X A M P L E   14.7   Twenty households were randomly selected from within a town in order to estimate the mean income per person. The town listed 12,205 households containing a total of 19,200 wage earners. The data are shown in Table 14.6. Find an approximate 95% confidence interval for the mean annual wage per person in the town.

T A B L E 14.6
Wage data for Example 14.7

Household i	Number of Wage Earners per Household, m_i	Annual Income ($)	Total Income per Household, x_i
1	2	12,100; 27,000	39,100
2	1	23,000	23,000
3	2	18,200; 12,800	31,000
4	2	20,900; 14,400	35,300
5	1	29,000	29,000
6	1	26,200	26,200
7	2	14,500; 18,300	32,800
8	2	16,900; 19,400	36,300
9	1	48,000	48,000
10	3	19,100; 12,000; 7,500	38,600
11	1	26,300	26,300
12	1	35,100	35,100
13	3	17,400; 18,900; 12,200	48,500
14	2	16,200; 19,900	36,100
15	1	13,200	13,200
16	1	18,400	18,400
17	2	13,100; 14,700	27,800
18	1	21,500	21,500
19	2	22,000; 8,000	30,000
20	2	14,000; 7,500	21,500

$$\sum_{i=1}^{20} m_i = 33 \qquad\qquad \sum_{i=1}^{20} x_i = 617,700$$

Solution   We are given the information that the number of households (clusters) is $N = 12,205$ and that the total number of wage earners in the town is $M = \sum\limits_{i=1}^{N} m_i = 19,200$.

From Table 14.6,

$$\sum_{i=1}^{n} m_i = 33 \qquad \text{and} \qquad \sum_{i=1}^{n} x_i = 617,700$$

Therefore,

$$\overline{m} = \frac{\sum_{i=1}^{n} m_i}{n} = \frac{33}{20} = 1.65 \qquad \text{and} \qquad \overline{M} = \frac{M}{N} = \frac{19,200}{12,205} = 1.5731258$$

The first step in finding the confidence interval for μ is to calculate

$$\overline{x} = \frac{\sum_{i=1}^{n} x_i}{\sum_{i=1}^{n} m_i} = \frac{617,700}{33} = 18,718.18$$

$$\sum_{i=1}^{n} x_i^2 = 20,669,130,000 \qquad \sum_{i=1}^{n} x_i m_i = 1,081,800 \qquad \sum_{i=1}^{n} m_i^2 = 63$$

$$\sum_{i=1}^{n}(x_i - \overline{x}m_i)^2 = \sum_{i=1}^{n} x_i^2 - 2\overline{x}\sum_{i=1}^{n} x_i m_i + \overline{x}^2 \sum_{i=1}^{n} m_i^2 = 2,243,802,645$$

Then substituting into the formula for the confidence interval, we have

$$\overline{x} \pm 1.96 \sqrt{\left(\frac{N-n}{Nn\overline{M}^2}\right) \frac{\sum_{i=1}^{n}(x_i - \overline{x}m_i)^2}{n-1}}$$

$$18,718.18 \pm 1.96 \sqrt{\frac{12,205 - 20}{(12,205)(20)(1.5731258)^2} \left(\frac{2,243,802,645}{19}\right)}$$

$$18,718.18 \pm 3025.08$$

Therefore, we estimate the mean salary per wage earner in the town to be in the interval \$15,693.10 to \$21,743.26.   ∎

E X A M P L E   **14.8**   Refer to Example 14.7, and find an approximate 95% confidence interval for the total annual wages earned by the town's 19,200 wage earners.

Solution   The formula for the confidence interval is

$$\hat{\tau} \pm 1.96 \sqrt{N^2 \left(\frac{N-n}{Nn}\right) \frac{\sum_{i=1}^{n}(x_i - \overline{x}m_i)^2}{n-1}}$$

where

$$\hat{\tau} = M\overline{x} = (19,200)(18,718.18) = \$359,389,056$$

Substituting into the formula for the confidence interval yields

$$\$359,389,056 \pm 1.96\sqrt{(12,205)^2\left[\frac{12,205-20}{(12,205)(20)}\right]\left(\frac{2,243,802,645}{19}\right)}$$

$$\$359,389,056 \pm \$58,081,558$$

Therefore, we estimate the total annual wages for all wage earners in the town to be in the interval \$301,307,498 to \$417,470,614. ∎

E X A M P L E   **14.9**   A large chain of retail stores purchases men's shirts in lots; each lot contains one dozen shirts. After receiving a shipment of 1000 lots, the retailer randomly selected 21 lots and counted the number of defective shirts per lot. Find an approximate 95% confidence interval for the proportion p of defective shirts in the 1000-lot shipment based on the data shown in Table 14.7.

T A B L E **14.7**
Shipment data for Example 14.9

Lot i	Number Defective, a_i	Lot i	Number Defective, a_i	Lot i	Number Defective, a_i
1	0	8	1	15	1
2	1	9	6	16	1
3	0	10	0	17	0
4	1	11	0	18	2
5	2	12	2	19	8
6	0	13	1	20	0
7	0	14	0	21	0

Solution   We are given the information that $N = 1000$ and $n = 21$, and since a cluster is a single lot and each lot contains 12 shirts, it follows that $m_i = 12$, $i = 1, 2, \ldots, 21$, and $\overline{M} = 12$. Then

$$\hat{p} = \frac{\sum\limits_{i=1}^{n} a_i}{\sum\limits_{i=1}^{n} m_i} = \frac{26}{(21)(12)}$$

$$= .1031746$$

$$\sum_{i=1}^{n}(a_i - \hat{p}m_i)^2 = \sum_{i=1}^{n} a_i^2 - 2\hat{p}\sum_{i=1}^{n} a_i m_i + \hat{p}^2 \sum_{i=1}^{n} m_i^2$$

$$= 118 - 2(.1031746)(312) + (.1031746)^2(3024)$$

$$= 85.809524$$

Then the approximate 95% confidence interval for p is

$$\hat{p} \pm 1.96 \sqrt{\left(\frac{N-n}{Nn\overline{M}^2}\right) \frac{\sum\limits_{i=1}^{n}(a_i - \hat{p}m_i)^2}{n-1}}$$

$$.103 \pm 1.96 \sqrt{\frac{1000-21}{(1000)(21)(12)^2}\left(\frac{85.809524}{21-1}\right)}$$

or

$$.103 \pm .073$$

Therefore, we estimate the proportion p of defective shirts in the 1000-lot shipment to lie in the interval .030 to .176. ∎

Exercises

Basic Techniques

14.32 Find an approximate 95% confidence interval for the mean of a population based on cluster sampling and upon the following information:

$$N = 20{,}000 \qquad n = 200$$

$$m_1 = m_2 = \cdots = m_{200} = 10 \qquad M = 200{,}000$$

$$\sum_{i=1}^{n} x_i = 4944 \qquad \sum_{i=1}^{n}(x_i - \bar{x}m_i)^2 = 483$$

14.33 Refer to Exercise 14.32, and find an approximate 95% confidence interval for the population total τ.

14.34 Find an approximate 95% confidence interval for a population mean μ based on the cluster sample data in the accompanying table. Assume that the total number of clusters in the population is $N = 2000$ and that the average cluster size in the population is $\overline{M} = 3.2$.

Cluster i	m_i	x_i	Cluster i	m_i	x_i
1	3	15	6	4	20
2	5	34	7	3	18
3	3	18	8	6	41
4	2	11	9	1	5
5	2	15	10	2	13

14.35 Refer to Exercise 14.34, and find an approximate 95% confidence interval for the population total τ.

14.36 Find an approximate 95% confidence interval for a population proportion p based on the cluster sample data in the accompanying table. Assume that the number of clusters in the population is 2000 and that the average cluster size in the population is $\overline{M} = 10.7$.

Cluster i	m_i	a_i	Cluster i	m_i	a_i
1	15	7	6	8	3
2	9	4	7	16	9
3	10	3	8	12	7
4	6	4	9	10	5
5	12	7	10	7	4

Applications

14.37 A survey was conducted to estimate the mean income of employed adult females in a small city. Because no list of adult females was available, the pollster divided the city into 415 blocks (clusters) and randomly selected $n = 25$ blocks from among the total of 415. The incomes of the employed adult females in the sample blocks are shown in the accompanying table. Find an approximate 95% confidence interval for the mean income of employed adult females in the city. [*Hint:* Since $\overline{M}$ is unknown, use $\overline{m}$ to approximate $\overline{M}$.]

Cluster i	Number of Adult Females, m_i	Total Income per Cluster, x_i	Cluster i	Number of Adult Females, m_i	Total Income per Cluster, x_i
1	8	$192,000	14	10	$ 98,000
2	12	242,000	15	9	106,000
3	4	84,000	16	3	100,000
4	5	130,000	17	6	64,000
5	6	104,000	18	5	44,000
6	6	80,000	19	5	90,000
7	7	150,000	20	4	74,000
8	5	130,000	21	6	102,000
9	8	90,000	22	8	60,000
10	3	100,000	23	7	78,000
11	2	170,000	24	3	94,000
12	6	86,000	25	8	82,000
13	5	108,000			

$$\sum_{i=1}^{25} m_i = 151 \qquad \sum_{i=1}^{25} x_i = \$2,658,000$$

14.38 A subsequent census of the city (Exercise 14.37) showed that 2562 adult females were employed. Find an approximate 95% confidence interval for the total income of all employed adult females in the city.

14.39 An industry is considering revision of its retirement policy and wants to estimate the proportion of employees who favor the new policy. The industry consists of 87 separate plants located throughout the United States. Since results must be obtained quickly and with little cost, the industry decides to use cluster sampling, with each plant as a cluster. A simple random sample of 15 plants is selected, and the opinions of the employees in these plants are obtained by questionnaire. The results are as shown in the accompanying table. Estimate the proportion of employees in the industry who favor the new retirement policy using a 95% confidence interval.

Plant	Number of Employees	Number Favoring New Policy	Plant	Number of Employees	Number Favoring New Policy
1	51	42	9	73	54
2	62	53	10	61	45
3	49	40	11	58	51
4	73	45	2	52	29
5	101	63	13	65	46
6	48	31	14	49	37
7	65	38	15	55	42
8	49	30			

14.6 Sampling: Some Problems

After constructing a frame and selecting the elements to be included in the sample, you will be faced with the problem of reaching the elements and making the observations. If an element is a human, as in the case of a consumer preference poll, you can send an interviewer to obtain a response or you can attempt to obtain the response by telephone or mail. Personal contact by an interviewer is the most reliable method, but it is often very costly. Telephone surveys are much less costly than personal interviews, but you must be certain that all of the elements in the population can be reached by telephone. Otherwise, you might fall into the trap that led to the demise of the *Literary Digest.*

In 1936 the *Literary Digest* conducted a telephone poll to determine the outcome of the Landon–Roosevelt presidential election. Unfortunately, most of the telephones in 1936 were owned by Republicans. Therefore, rather than surveying the voting intentions of all eligible voters, the population of interest to the readers of the *Literary Digest,* the researchers selected the random sample from among voting telephone owners. Needless to say, the sample provided ample evidence upon which to forecast a Landon victory (you know who won!).

An equally poor example of a telephone poll was one conducted by ABC Television immediately prior to the 1980 Carter–Reagan election. ABC invited its viewers to call in (long distance) their presidential preferences. Rather than achieving a random sample of voter sentiment, ABC obtained a sample that consisted of the preferences of those voters who were sufficiently interested in swaying the outcome of the poll that they were willing to invest in long-distance telephone calls. Clearly, ABC did not randomly sample the population of prospective voters. More Democrats made the long-distance calls, and ABC forecast a Carter victory.

Mail surveys are even more susceptible to error. Because mailing a questionnaire is relatively inexpensive, some pollsters send questionnaires to all persons in the population. Most of these questionnaires are relegated to the circular file and are never returned. The respondents to these surveys are often those who have an ax to grind—that is, they wish to influence the outcome of the poll. Thoughtful persons who may be in the majority may not have the time or the inclination to respond.

Even if the sample elements in a mail survey have been randomly selected, nonresponse is a problem. One way to cope with this problem is to contact the missing elements of the sample by other means, telephone or direct interview, in order to obtain a response. It is also possible to resample the nonrespondents. Then, based on the subsample, we can modify an estimate to adjust for the nonrespondents.

A problem more serious than nonresponse occurs when the questions in a survey induce the respondent to lie. For example, suppose that a merchandiser wanted to estimate the proportion of its clientele who were shoplifters. If asked whether they had ever shoplifted in the store, guilty or not, most people would say no. Questions about personal hygiene, thoughts, and habits often tend to be sensitive. Rather than risk being viewed unfavorably by the interviewer, the respondent may give a false response. Thus, in addition to the problem of constructing the frame and actually reaching the elements in the sample, we have the added problem of obtaining a correct response when an element is observed.

The method for coping with this problem is based on what is known as a **randomized response model.** Each person in the sample is presented two questions by the interviewer, one the sensitive question of interest to the pollster and the other a trivial, nonsensitive question. For example, the second question might be "Did you have cereal for breakfast?" The respondent is then given a mechanism for randomly selecting which of the two questions to answer. For example, the respondent might be asked to select a card from a deck of 52 cards. Without revealing the outcome to the interviewer, the respondent would be instructed to answer the sensitive question if the draw was a face card. Otherwise, the respondent answers the nonsensitive question. Thus the interviewer knows the *probability* that the respondent will select the sensitive question but does not know the outcome for any specific respondent. This allows the respondent to freely answer the question without risk of embarrassment. It also enables the statistician to use the sample data to estimate the value of the desired population parameter.

The study of survey sampling—the various procedures (designs) for collecting sample data and for estimating the values of population parameters—is a subject in itself. Almost all of the statistical methods for making inferences that are discussed in this text are based on simple random samples. To learn more about other sampling designs and the methods of inference associated with them, refer to a text on survey sampling.

CASE STUDY REVISITED

14.7 SAMPLING 58 MILLION BEER CANS: HOW TO DO IT

Using a random number table to select a random sample from a frame is an easy task. Constructing the frame and obtaining the actual sample observations may be difficult or, in the case of the beer cans (see the case study), almost impossible.

Fifty-eight million is a large number of beer cans. Located in one spot, stacked side by side, they occupy the best part of a city block and tower almost to the ceiling of a two-story building. The cans were stacked in a warehouse in layers on wooden pallets, 21 cans wide, 20 cans deep, and 19 cans high (see Figure 14.1). The pallets were then stacked by forklift trucks, one pallet on top of another, in stacks containing either two or three pallets. The stacks were arranged in long rows, from one end of the warehouse to the other. Space was allowed every four or five rows for a passageway large enough to accommodate a lift truck and permit removal of pallets. Passages perpendicular to the first set of passageways also occurred at irregular intervals.

FIGURE **14.1**
Locating a can in a pallet

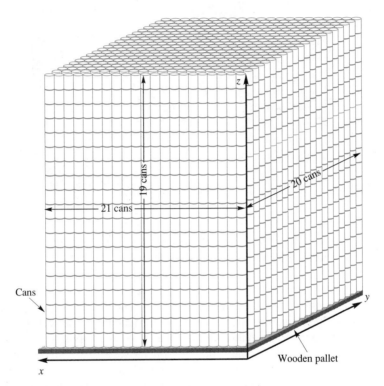

Two problems arise in the sampling procedure. For the construction of a frame, the cans (the elements of the population) need to be numbered so that the cans selected for the sample can be identified and located. Since the beer cans were not numbered, the alternative was to remove the 58 million cans from the warehouse and number them or to number the cans on a master drawing according to their location in the warehouse. The second problem involves the removal of the 500 sample cans from the midst of the 58 million. Fortunately, this difficulty was eliminated because the brewer planned to sell the cans for scrap aluminum. Therefore, cans in specific locations could be obtained as their respective pallets were trucked out of the warehouse.

Numbering each of the 58 million cans by location on a drawing seemed an impossible task. Instead, we decided to number *stacks* according to their location and then to record for each stack whether it contained two or three pallets. A plan of the warehouse showing the numbered stacks would be similar to the drawing in

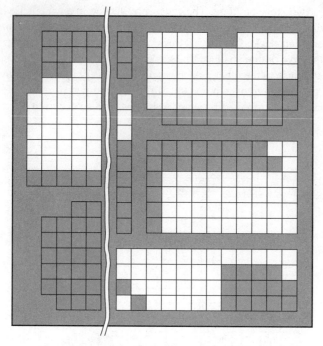

Figure 14.2. Each small square represents a stack. White squares represent two-pallet stacks; colored squares represent three-pallet stacks.

The final step in drawing a random sample is to devise a procedure for selecting a can so that every group of 500 cans has an equal probability of selection. To illustrate our procedure for doing this, let us assume that the warehouse contains 3200 stacks, 2300 containing two pallets and 900 containing three pallets (a total of 7300 pallets). Intuitively, it would seem reasonable to randomly select one of the 3200 stacks, randomly select one of the pallets in the stack, and then randomly select a can within the pallet. This procedure will not produce a simple random sample because cans in the two-pallet stacks will possess a higher probability of selection than cans in the three-pallet stacks.

One solution to this difficulty is to select a can according to the following multistep sampling procedure:

1   Decide whether to select a can from a two- or a three-pallet stack.

2   Choose the stack from among the stacks listed for the stack *type* chosen in step 1. Thus we must have a frame for each stack type. So all of the two-pallet stacks must be numbered and, separately, all of the three-pallet stacks must be numbered.

3   Once the stack has been chosen (steps 1 and 2), choose the pallet within the stack.

4   Once the pallet has been chosen (steps 1, 2, and 3), choose the can within the pallet. Since the cans are stacked in layers on the rectangular pallet (see Figure

14.1), the location of a can can be determined by its distance, x, y, z, from one of the corners of the pallet.

The probabilities of the various outcomes of each of the steps are as follows:

Step 1

   a   The probability that we select a type 2 stack (two-pallet stack) is p.

   b   The probability that we select a type 3 stack (three-pallet stack) is $1 - p$.

Step 2

   a   Given that a type 2 stack was selected in step 1, the probability of selecting a particular one of the $n_2 = 2300$ stacks of type 2 is 1/2300.

   b   Given that a type 3 stack was selected in step 1, the probability of selecting a particular one of the $n_3 = 900$ stacks of type 3 is 1/900.

Step 3

   a   Given the selection of a particular type 2 stack (steps 1 and 2), the probability of selecting a particular pallet from among the two in the stack is 1/2.

   b   Given the selection of a particular type 3 stack (steps 1 and 2), the probability of selecting a particular pallet from among the three in the stack is 1/3.

Step 4

   Given the selection of a particular pallet (steps 1, 2, and 3), the probability of selecting a particular can by randomly choosing values for x, y, and z (see Figure 14.1) will equal $1/N$, where N is the number of cans in a pallet. For example, each pallet contains $(21)(20)(19) = 7980$ cans (see Figure 14.1); therefore, the probability of selecting a particular can in the pallet is 1/7980.

Of the probabilities listed above, only one is unknown—the probability p of selecting a two-pallet stack. Since we want the probability of selecting a can in a two-pallet stack to equal the probability of selecting one in a three-pallet stack, we can equate these probabilities and solve for p.

The probability of selecting a particular can that is located in a two-pallet stack is equal to the probability of the intersection of four events corresponding to the four steps in the selection process:

$$P_2 = (p)\left(\frac{1}{2300}\right)\left(\frac{1}{2}\right)\left(\frac{1}{7980}\right)$$

$$\downarrow \qquad \downarrow \qquad \downarrow \qquad \downarrow$$

$$\text{step} \quad \text{step} \quad \text{step} \quad \text{step}$$
$$1 \qquad 2 \qquad 3 \qquad 4$$

Similarly, the probability of selecting a particular can in a three-pallet stack is

$$P_3 = (1 - p)\left(\frac{1}{900}\right)\left(\frac{1}{3}\right)\left(\frac{1}{7980}\right)$$

$$\downarrow \qquad \downarrow \qquad \downarrow \qquad \downarrow$$

$$\text{step} \quad \text{step} \quad \text{step} \quad \text{step}$$
$$1 \qquad 2 \qquad 3 \qquad 4$$

Since we want P_2 to equal P_3, we obtain

$$p\left(\frac{1}{2300}\right)\left(\frac{1}{2}\right)\left(\frac{1}{7980}\right) = (1-p)\left(\frac{1}{900}\right)\left(\frac{1}{3}\right)\left(\frac{1}{7980}\right)$$

Since $(2300)(2)$ represents the total number N_2 of pallets in two-pallet stacks and $(900)(3) = N_3$ equals the corresponding number of pallets in three-pallet stacks, it follows that

$$p\left(\frac{1}{N_2}\right) = (1-p)\left(\frac{1}{N_3}\right)$$

or

$$p = \frac{N_2}{N_2 + N_3} = \frac{\text{Total number of pallets in two-pallet stacks}}{\text{Total number of pallets in the warehouse}}$$

For our example,

$$p = \frac{(2300)(2)}{(2300)(2) + (900)(3)} = .63$$

Therefore, if we want to choose a random sample from among the 58 million cans, we will choose a can so that the probability of selecting a two-pallet stack is .63 and that of selecting a three-pallet stack is .37. Then we will choose the stack so that every stack in the stack type has an equal probability of being selected. Each pallet within a two-pallet stack will have a probability of 1/2 of being selected; each pallet within a three-pallet stack will have a probability equal to 1/3. Finally, since a pallet contains 7980 cans, we will select the appropriate can so that each of the 7980 cans will have an equal chance of selection. If each of the 500 cans is selected in this manner from among the 58 million cans in the warehouse, the result will be a simple random sample.

How can we use the random number table to perform one of the steps in the can selection? For example, how can we choose the stack type so that we give a probability of .63 to choosing a two-pallet stack and a probability of .37 to choosing a three-pallet stack? We will select three-digit random numbers from the table, retaining only those from 1 to and including 100. If the random number is less than or equal to 63, we decide to choose a two-pallet stack. If the number is equal to 64 or larger—up to 100—we choose a three-pallet stack. The random number table is used in a similar manner to perform the other steps in the selection process.

This case study stresses an important aspect of sampling. Constructing a frame for a population is often difficult and costly. The unusual aspect of this case study is that we circumvented this difficult and costly procedure by developing a multistep sampling procedure that satisfies the definition of simple random sampling.

14.8 Summary

This chapter stresses the role of the sampling procedure in making statistical inferences. Most statistical samples are probability samples, those selected in such a way that the probabilities of various sample outcomes are known in advance. By using these probabilities, we are able to infer the nature of the sampled population.

The most elementary type of probability sample, the one upon which almost all of the basic statistical methods are based, is the simple random sample. A simple random sample is one chosen so that every different sample of fixed size in the population has an equal probability of selection. The sampling units to be selected in a simple random sample can be identified by constructing a frame, that is, a list of all elements in the population. Then the elements to be included in the sample are chosen by using a random number table.

Constructing a frame and making observations on elements in the sample may be difficult and costly. It also may be difficult to obtain true readings of the responses. Sampling designs such as the stratified random sampling and cluster sampling designs will, in certain situations, reduce the labor and the cost of sampling. Sampling by using the randomized response model can often be employed to elicit correct responses to sensitive questions.

The sample survey designs discussed in this chapter provide a glimpse into a broad subject—survey sampling. Survey sampling encompasses the designs and associated estimation procedures required to cope with the many difficulties that arise in sampling. Further information on this topic can be found by consulting texts on survey sampling.

Supplementary Exercises

14.40 Suppose that a population contains 49 elements. Use a random number table to identify the $n = 5$ elements to be included in a random sample from among the population. Explain how you arrived at your selection.

 14.41 A city contains 474,159 homeowners listed on its tax rolls. Explain how you would use a random number table to draw a random sample of $n = 1000$ homeowners from among those listed on the tax rolls.

14.42 A stratified random sample was selected from among three strata. Find an approximate 95% confidence interval for the population mean μ, based on the data in the accompanying table.

	Stratum		
	1	2	3
Stratum size	2400	3000	900
Sample size	100	120	50
Sample mean	12.1	13.4	9.7
Sample variance	3.22	2.45	1.07

14.43 Use the data in Exercise 14.42 to find an approximate 95% confidence interval for the population total τ.

14.44 A stratified random sample was selected from among three strata. Use the data in the accompanying table to find an approximate 95% confidence interval for a population proportion p.

	Stratum		
	1	2	3
Stratum size	900	1200	600
Sample size	100	100	100
Sample population	.43	.52	.38

14.45   A survey was conducted to estimate the mean monthly expenditure for college room, board, books, supplies, and recreation for students at a university. Random samples of 100 male and 100 female students were randomly selected from among the 8435 males and 6453 females enrolled at the university. A summary of the information acquired in the survey is shown in the accompanying table.

Sex	Sample Size	Sample Mean ($)	Sample Variance
Female	100	717	22,000
Male	100	864	31,000

a   Find an approximate 95% confidence interval for the mean monthly expenditure per student.

b   Find an approximate 95% confidence interval for the mean monthly expenditure for female students. [*Hint:* Use the method of Section 7.5.]

14.46   How much money, in addition to tuition, does the university described in Exercise 14.45 bring into the community in which it is located? Find an approximate 95% confidence interval for the total amount of money expended per month on room, board, and so on by all students at the university.

14.47   Suppose that the persons in the survey of Exercise 14.45 were asked whether they would approve the budgeting of additional student funds for the school's athletic programs.

a   Given the data shown in the accompanying table, find an approximate 95% confidence interval for the proportion of students favoring the proposed budgeting change.

Sex	Sample Size	Number in Favor
Female	100	38
Male	100	61

b   Find an approximate 95% confidence interval for the proportion of male students who favor the proposed budgeting change. [*Hint:* Use the method of Section 7.7.]

14.48   Find an approximate 95% confidence interval for a population mean μ, based on the cluster sample data in the accompanying table. Assume that the total number of clusters in the population is 5000 and that the average cluster size in the population is $\overline{M} = 4.9$.

Cluster i	m_i	x_i	Cluster i	m_i	x_i
1	3	6	7	5	10
2	5	12	8	4	9
3	6	11	9	5	9
4	4	10	10	8	20
5	7	14	11	4	6
6	4	5	12	5	9

14.49 Use the data in Exercise 14.48 to find an approximate 95% confidence interval for the population total τ.

14.50 Find an approximate 95% confidence interval for a population proportion p, based on the cluster sample data in the accompanying table. Assume that the total number of clusters in the population is 5000 and that the average cluster size in the population is $\overline{M} = 15.2$.

Cluster i	m_i	a_i	Cluster i	m_i	a_i
1	16	4	7	19	5
2	21	6	8	18	3
3	10	4	9	13	5
4	15	3	10	15	6
5	18	8	11	11	3
6	14	4	12	18	6

14.51 An inspector wants to estimate the mean weight of fill for cereal boxes packaged in a certain factory. The cereal is available to him in cartons containing 12 boxes each. The inspector randomly selects five cartons and measures the weight of fill for every box in the sampled cartons, with the results (in ounces) as shown in the accompanying table. Find an approximate 95% confidence interval for the mean weight of fill per box packaged by the factory. Assume that the total number N of cartons packaged by the factory is large enough so that the quantity $(N - n)/N$ is close to 1 and can be ignored.

Carton	Ounces of Fill											
1	16.1	15.9	16.1	16.2	15.9	15.8	16.1	16.2	16.0	15.9	15.8	16.0
2	15.9	16.2	15.8	16.0	16.3	16.1	15.8	15.9	16.0	16.1	16.1	15.9
3	16.2	16.0	15.7	16.3	15.8	16.0	15.9	16.0	16.1	16.0	15.9	16.1
4	15.9	16.1	16.2	16.1	16.1	16.3	15.9	16.1	15.9	15.9	16.0	16.0
5	16.0	15.8	16.3	15.7	16.1	15.9	16.0	16.1	15.8	16.0	16.1	15.9

14.52 To emphasize safety, a taxicab company wants to estimate the proportion of unsafe tires on its 175 cabs. (Ignore spare tires.) It is impractical to select a simple random sample of tires, so cluster sampling is used, with each cab as a cluster. A random sample of 25 cabs gives the following number of unsafe tires per cab:

$$2, 4, 0, 1, 2, 0, 4, 1, 3, 1, 2, 0, 1,$$
$$1, 2, 2, 4, 1, 0, 0, 3, 1, 2, 2, 1$$

Estimate the proportion of unsafe tires being used on the company's cabs with an approximate 95% confidence interval.

Exercises Using the Data Disk

14.53 Refer to data set A on the data disk, the average salaries of instructional staff faculty.

  a Select a simple random sample of $n = 30$ colleges and record the salaries of male full professors. Use your data to estimate the average salary for male full professors.

  b For your sample in part (a), record the salaries of the female full professors at these same schools. Estimate the difference in salary for male versus female professors. What conclusions are possible, given your results?

14.54 Refer to data set A on the data disk.

  a Select a stratified sample of size 10 from each of the male faculty ranks. Use your sample to estimate the average salary for male faculty.

b Repeat part (a) for female faculty.

c Use the results of parts (a) and (b) to estimate the difference in average salaries for males versus females.

14.55 Refer to data set B on the data disk, concerning broccoli yields and the effects of ozone. Assume that the observations within a chamber are considered a cluster. Select five clusters at random and record the fresh weight of the variety coded 0. Use your sample of clusters to estimate the population fresh-weight average. Refer to the data summary for data set B in Appendix I. How does your estimate compare with the value reported there?

14.56 Refer to data set D on the data disk, containing financial information for the 100 best small companies in 1994. Select a one-in-five systematic sample and estimate the average current earnings (in millions of dollars), with a 95% confidence interval.

15

THE CHI-SQUARE GOODNESS-OF-FIT TEST

About This Chapter

Observations that can be classified, with each observation falling into one (and only one) of k categories, yield a set of counts, with each count representing the number of observations falling into a particular category. The objective of this chapter is to provide methods for testing hypotheses about the category probabilities and, as a particular application, to test the equivalence of more than two binomial proportions. Thus this chapter is an extension of the methodologies in Chapters 7 and 8 for comparing pairs of population means and proportions.

Contents

ARE WOMEN MORE SEXIST THAN MEN?

A s more and more women climb the executive ladder, they occupy leadership positions that were formerly held mainly by men. What effect has this had on the workplace? What are the attitudes of employees toward their bosses? If you were interviewing for a new job and could choose your boss, would you prefer to work for a male or a female? Moore and Gallup (1993) reported the responses to this last question for 1065 respondents selected at random. The results of their survey follow, with the respondents classified first according to sex, then according to age and sex, and finally by geographic region.

Sex	Preference				Total
	Male Boss	Female Boss	Either	Opinion	
Men	175	85	260	11	531
Women	235	155	128	16	534
Total	410	240	388	27	1065

Can we say that the differences between men and women in the preceding table are significant? How can we test to determine if these are significant differences or merely chance fluctuations? We might want to look at differences across age categories for the 531 men and the 534 women. The next two tables display these counts.

Ages—Men	Preference			Total
	Male Boss	Female Boss	Either	
18–29 yrs.	29	27	67	123
30–49 yrs.	69	49	125	243
50+ yrs.	77	9	79	165
Total	175	85	271	531

Ages—Women	Preference			Total
	Male Boss	**Female Boss**	**Either**	
18–29 yrs.	14	58	21	93
30–49 yrs.	98	63	71	232
50+ yrs.	123	34	52	209
Total	235	155	144	534

On the basis of what you observe here, would you be willing to say that men respond differently across these age categories than do women? We might then ask whether the preference for a male or female boss depends upon the region in which a respondent lives. The next table provides a cross-classification of respondents by geographical region and their responses to the question.

Region	Preference			Total
	Male Boss	**Female Boss**	**Either**	
East	102	58	102	262
Midwest	107	62	99	268
South	138	79	157	374
West	63	39	59	161
Total	410	238	417	1065

These and other questions can be answered by using a statistical technique that generalizes the test of the equality of two binomial proportions to a simultaneous test of several proportions. Further, we will be able to assess whether the classification of a respondent according to the row category depends upon his or her classification among the column categories. We return to the analysis of the information in these tables in Section 15.5.

15.1   A Multinomial Experiment

Many consumer and business surveys produce categorical (or count) data. For instance, the classification of people into five income brackets results in a count corresponding to each of the five income classes. Similarly, a traffic study might require a count and classification of the type of motor vehicles using a section of highway. An industrial process manufactures items that fall into one of three quality classes: acceptable, seconds, and rejects. The results of an advertising campaign yield count data indicating a classification of consumer reaction.

The illustrations in the preceding paragraph exhibit, to a reasonable degree of approximation, the following characteristics that define a **multinomial experiment.**

Characteristics of a Multinomial Experiment

1 The experiment consists of n identical trials.

2 The outcome of each trial falls into one of k categories or cells.

3 The probability that the outcome of a single trial will fall in a particular cell, say cell i, is $p_i (i = 1, 2, \ldots, k)$ and remains the same from trial to trial. Note that

$$p_1 + p_2 + p_3 + \cdots + p_k = 1$$

4 The trials are independent.

5 We are interested in $n_1, n_2, n_3, \ldots, n_k$, where $n_i (i = 1, 2, \ldots, k)$ is equal to the number of trials in which the outcome falls in cell i. Note that

$$n_1 + n_2 + n_3 + \cdots + n_k = n$$

A multinomial experiment is analogous to tossing n balls at k boxes, where each ball must fall in one of the boxes. The boxes are arranged such that the probability that a ball will fall in a box varies from box to box but remains the same for a particular box in repeated tosses. Finally, the balls are tossed in such a way that the trials are independent. At the conclusion of the experiment, we observe n_1 balls in the first box, n_2 in the second, $\ldots$, and n_k in the kth. The total number of balls is equal to

$$\sum_{i=1}^{k} n_i = n$$

Note the similarity between the binomial and multinomial experiments and note, in particular, that the binomial experiment represents the special case of the multinomial experiment when $k = 2$. The two cell probabilities, p and q, of the binomial experiment are replaced by the k cell probabilities, p_1, p_2, $\ldots$, p_k, of the multinomial experiment. The objective of this chapter is to make inferences about the cell probabilities p_1, p_2, $\ldots$, p_k. The inferences will be expressed in terms of a statistical test of an hypothesis concerning their specific numerical values or their relationship, one to another.

15.2 The Chi-Square Goodness-of-Fit Test

The simplest hypothesis about multinomial cell probabilities is one that specifies numerical values for each. For example, if the multinomial experiment involves $k = 4$ cells, we might want to test the hypothesis

$$H_0 : p_1 = .2, p_2 = .4, p_3 = .1 \text{ and } p_4 = .3$$

against the alternative hypothesis that at least two of the cell probabilities differ from the values specified in H_0.

The following example illustrates a common application of the **chi-square goodness-of-fit test** and shows how the test is conducted.

EXAMPLE   15.1   Suppose that customers can purchase one of three brands of milk at a supermarket. In a study to determine whether one brand is preferred over another, a record is made of a sample of $n = 300$ milk purchases. The data are shown in Table 15.1. Do the data provide sufficient evidence to indicate a preference for one or more brands?

TABLE 15.1
Milk purchases data for
Example 15.1

Brand 1	Brand 2	Brand 3
78	117	105

Solution   If all of the brands are *equally* preferred, then the probability that a purchaser will choose any one brand is the same as the probability of choosing any other—that is, $p_1 = p_2 = p_3 = 1/3$. Therefore, the null hypothesis of "no preference" is

$$H_0 : p_1 = p_2 = p_3 = 1/3$$

If p_1, p_2, and p_3 are not all equal, the brands are not equally preferred; in other words, the purchasers must have a preference for one (or possibly two) brands. The alternative hypothesis is

$$H_a : p_1, p_2, \text{ and } p_3 \text{ are not all equal}$$

Therefore, we seek a test statistic that will detect a **lack of fit** of the observed **cell counts,** or **frequencies,** to our hypothesized (null) expected cell counts based on the hypothesized cell probabilities.

If the observed cell counts fit the hypothesized cell probabilities, they should be close to their expected values. These expected values, for a multinomial experiment, are

$$E(n_1) = np_1, \qquad E(n_2) = np_2, \qquad \dots, \qquad E(n_k) = np_k$$

For $p_1 = p_2 = p_3 = 1/3$, the expected cell counts are

$$E(n_1) = np_1 = (300)\left(\frac{1}{3}\right) = 100$$

$$E(n_2) = np_2 = (300)\left(\frac{1}{3}\right) = 100$$

$$E(n_3) = np_3 = (300)\left(\frac{1}{3}\right) = 100$$

TABLE **15.2**
Observed and expected cell counts for milk purchases in Example 15.1

Brand 1	Brand 2	Brand 3
78	117	105
(100)	(100)	(100)

The observed cell counts are shown along with their respective expected counts (in parentheses) in Table 15.2.

The test statistic for comparing the observed and expected cell counts (and, consequently, testing $H_0 : p_1 = p_2 = p_3 = 1/3$) is the X^2 **statistic:**

$$X^2 = \sum_{i=1}^{k} \frac{[n_i - E(n_i)]^2}{E(n_i)} = \sum_{i=1}^{k} \frac{d_i^2}{E(n_i)}$$

where

$$n_i = \text{Observed count for cell } i$$
$$E(n_i) = np_i = \text{Expected count for cell } i$$
$$d_i = n_i - E(n_i)$$
$$= \text{Difference between the observed and expected counts for cell } i$$

The computation of X^2 for our example can be accomplished by using Table 15.3. The observed n_i and expected $E(n_i)$ cell counts are shown in columns 2 and 3, the difference $d_i = n_i - E(n_i)$ is shown in column 4, and d_i^2 is shown in column 5. The final step in calculating X^2 is performed in column 6, where we calculate $d_1^2/E(n_1)$, $d_2^2/E(n_2)$, and $d_3^2/E(n_3)$. The sum of these column 6 entries is X^2.

TABLE **15.3**
Table for calculating X^2 for Example 15.1

i	n_i	$E(n_i)$	d_i	d_i^2	$d_i^2/E(n_i)$
1	78	100	−22	484	4.84
2	117	100	17	289	2.89
3	105	100	5	25	.25
Total	300	300			$X^2 = 7.98$

If the null hypothesis is true, the statistic X^2 will possess, approximately, the **chi-square probability distribution** that we encountered in Section 8.9. The larger the sample size n, the better the approximation will be. As a rule of thumb, *we will require that all expected cell counts be greater than or equal to 5,* although Cochran[†] has noted that this value can be as low as 1 for some situations. Cells whose expected cell counts are less than 5 can be combined with other cells so that the combined cells have expected cell counts greater than 5. When the null hypothesis is false, the differences d_1, d_2, and d_3 between the observed and expected cell counts will be large and will make the X^2 statistic larger than expected. Consequently, we will reject H_0

[†]Cochran, W.G., "The χ^2 Test of Goodness of Fit." *Annals of Mathematical Statistics,* Vol. 23 (1952), pp. 315–45.

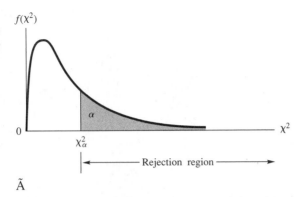

FIGURE **15.1**
Rejection region for the
chi-square test of goodness of fit

Ã

for large values of X^2, those that exceed χ_α^2, a value of χ^2 that places α in the upper tail of the chi-square distribution (see Figure 15.1).

Recall from Section 8.9 that the shape of the chi-square probability distribution depends on a single parameter, the **degrees of freedom** for the chi-square statistic. We will give you the number of degrees of freedom for the approximating chi-square distribution for all applications in this chapter. As a general rule, the degrees of freedom (d.f.) for the chi-square statistic will equal the number k of categories minus one degree of freedom for every linear restriction placed on the category counts. For example, you will always lose one degree of freedom because the sum of the category counts will equal the sample size n; that is,

$$n_1 + n_2 + \cdots + n_k = n$$

You will also lose one degree of freedom for each category probability that must be estimated to calculate the expected category frequencies. For the test described in this section, all of the cell probabilities are specified. Therefore, the number of degrees of freedom for this chi-square statistic will be $(k - 1)$.

The number k of cells for our example is 3. Therefore, the degrees of freedom for the approximating chi-square distribution will be

$$\text{d.f.} = k - 1 = 3 - 1 = 2$$

Consulting Table 5 in Appendix II for d.f. $= 2$, we find $\chi_{.05}^2 = 5.99147$. Consequently, we will reject the hypothesis of "no preference" in the milk brand study if

$$X^2 > 5.99147$$

Because the computed value of X^2, $X^2 = 7.98$ (see Table 15.3), exceeds the critical value, $\chi_{.05}^2 = 5.99147$, we reject $H_0 : p_1 = p_2 = p_3 = 1/3$ and conclude that the three brands of milk are not equally preferred.   ▪

A summary of the chi-square test of specific values of the cell probabilities is given in the following display.

Chi-Square Test of Specified Values for the Cell Probabilities

1   Null Hypothesis: $H_0 : p_1, p_2, \dots, p_k$ possess specified values

2   Alternative Hypothesis: H_a : at least two of the cell probabilities differ from the values specified in H_0

3   Test Statistic: $X^2 = \sum_{i=1}^{k} \dfrac{[n_i - E(n_i)]^2}{E(n_i)} = \sum_{i=1}^{k} \dfrac{d_i^2}{E(n_i)}$,

where n_i = Observed count for cell i
and $E(n_i) = np_i,\quad i = 1, 2, \dots, k$

4   Rejection Region: $X^2 > \chi_\alpha^2$, where χ_α^2 is based upon d.f. $= (k - 1)$ degrees of freedom (see Table 5 in Appendix II)

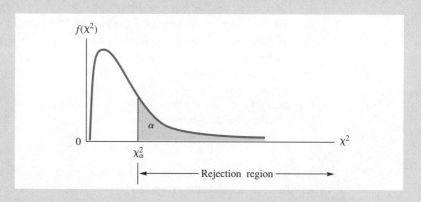

Assumption: All expected cell counts are greater than or equal to 5.

Exercises

Basic Techniques

15.1   List the characteristics of a multinomial experiment.

15.2   Give the value of χ_α^2 for each of the following.
    a $\alpha = .05$, d.f. $= 3$ **b** $\alpha = .01$, d.f. $= 8$
    c $\alpha = .10$, d.f. $= 15$ **d** $\alpha = .10$, d.f. $= 11$

15.3   Give the rejection region for a chi-square test of specified cell probabilities for the following if the experiment involves k cells.
    a $k = 7, \alpha = .10$ **b** $k = 10, \alpha = .01$
    c $k = 14, \alpha = .05$ **d** $k = 3, \alpha = .05$

15.4   Suppose that a response can fall into one of $k = 5$ categories with probabilities $p_1, p_2, \dots, p_5$, respectively, and that $n = 300$ responses produced the following category counts:

Category	1	2	3	4	5
Observed count	47	63	74	51	65

a   If you were to test $H_0 : p_1 = p_2 = \cdots = p_5$ using the chi-square test, how many degrees of freedom would the test statistic possess?

b   If $\alpha = .05$, find the rejection region for the test.

c   What is your alternative hypothesis?

d   Conduct the test in part (a), using $\alpha = .05$. State your conclusions.

e   Find the approximate observed significance level for the test, and interpret its value.

15.5   Suppose that a response can fall into one of $k = 3$ categories with probabilities p_1, p_2, and p_3, respectively, and that $n = 300$ responses produced the following category counts:

Category	1	2	3
Observed count	130	98	72

a   If you were to test $H_0 : p_1 = p_2 = p_3$ using the chi-square test, how many degrees of freedom would the test statistic possess?

b   If $\alpha = .05$, find the rejection region for the test.

c   What is your alternative hypothesis?

d   Conduct the test in part (a), using $\alpha = .05$. State your conclusions.

e   Find the approximate observed significance level for the test, and interpret its value.

Applications

15.6   A city expressway using four lanes in each direction was studied to see whether drivers preferred to drive on the inside lanes. A total of 1000 automobiles was observed during the heavy early-morning traffic, and their respective lanes were recorded. The results were as shown in the accompanying table. Do the data provide sufficient evidence to indicate that some lanes are preferred over others? Test using $\alpha = .05$.

Lane	1	2	3	4
Observed count	294	276	238	192

15.7   An occupant traffic study was conducted to aid in the remodeling of an office building that contains three entrances. The choice of entrance was recorded for a sample of 200 persons entering the building. Do the data shown in the table indicate that there is a difference in preference for the three entrances? Find a 90% confidence interval for the proportion of persons favoring entrance 1. [*Hint:* Use the method of Section 7.7.]

	Entrance		
	1	2	3
Number entering	83	61	56

15.8   Officials in a particular community are seeking a federal program that they hope will boost local income levels. As justification, the city claims that its local income distribution differs substantially from the national distribution and that incomes tend to be lower than expected. A random sample of 2000 family incomes was classified and compared with the corresponding national percentages. The results are shown in the accompanying table. Do the data provide sufficient evidence to indicate that the distribution of family incomes within the city differs from the national distribution? Test with $\alpha = .05$.

Income	National Percentages	City Salary Class Frequency
More than $50,000	2	27
$25,000 to $50,000	16	193
$20,000 to $25,000	13	234
$15,000 to $20,000	19	322
$10,000 to $15,000	20	568
$5,000 to $10,000	19	482
Below $5,000	11	174
Total	100	2000

15.9 The current video movie trends provide another avenue for marketers who seek new ways to advertise. What opinion does the consumer who rents the video have toward the included advertisements? A survey conducted by the Gallup organization (Miko, 1991) indicates that 36.3% of those surveyed find the advertisements very annoying, 31.2% find them somewhat annoying, and 32.5% do not find them annoying.

a The survey claims a margin of error of "plus or minus 3.1%." How large a sample would be necessary to estimate the true percentages to within ±.031?

b Although the survey size is not given, assume that $n = 1000$. Do the data provide sufficient evidence to indicate a predominance of people falling into one of the three opinion categories, or is each category equally likely? Use $\alpha = .01$.

15.10 A new technology called "interactive television" is currently in the developmental stage. Using this technology, you will soon be able to give your TV commands that order any movie at any time, do grocery shopping, and provide other services not currently available through today's television. However, not very many Americans are knowledgeable about interactive television. A recent survey produced the following percentages regarding knowledge about interactive TV:

Amount of Knowledge	Percentage
A great deal	12
A moderate amount	17
Only a little	41
Nothing at all	30

Source: "Gallup Short Subjects," 1994.

In an attempt to verify or disprove these percentages, a local cable TV company questioned 200 customers about their knowledge of interactive TV, with the following results: 30 knew a great deal, 38 knew a moderate amount, 75 knew a little, and 57 knew nothing. Does it appear that the survey percentages are inaccurate? Use $\alpha = .01$.

15.3 Contingency Tables

A problem frequently encountered in the analysis of count data concerns the independence of two methods of classification of observed events. For example, we might want to classify defects found on furniture produced in a manufacturing plant first according to the type of defect and second according to the production shift. Ostensibly, we want to investigate a contingency, that is, a dependence between the two classifications. Do the proportions of various types of defects vary from shift to shift?

A total of $n = 309$ furniture defects was recorded, and the defects were classified according to one of four types: A, B, C, or D. At the same time, each piece of furniture was identified according to the production shift in which it was manufactured. These counts are presented in Table 15.4, which is known as a **contingency table**. [*Note:* Numbers in parentheses are the expected cell frequencies.]

TABLE **15.4**
Contingency table

Shift	Type of Defect				Total
	A	**B**	**C**	**D**	
1	15 (22.51)	21 (20.99)	45 (38.94)	13 (11.56)	94
2	26 (22.99)	31 (21.44)	34 (39.77)	5 (11.81)	96
3	33 (28.50)	17 (26.57)	49 (49.29)	20 (14.63)	119
Total	74	69	128	38	309

Let p_A equal the unconditional probability that a defect will be of type A. Similarly, define p_B, p_C, and p_D as the probabilities of observing the three other types of defects. Then these probabilities, which we will call the **column probabilities** of Table 15.4, will satisfy the requirement

$$p_A + p_B + p_C + p_D = 1$$

In a like manner, let $p_i (i = 1, 2, \text{or } 3)$ equal the **row probability** that a defect will have occurred on shift i, where

$$p_1 + p_2 + p_3 = 1$$

Then, if the two classifications are independent of each other, a cell probability will equal the product of its respective row and column probabilities, in accordance with the multiplicative law of probability. For example, the probability that a particular defect will occur on shift 1 and be of type A is $(p_1)(p_A)$. Thus we observe that the numerical values of the cell probabilities are unspecified in the problem under consideration. The null hypothesis specifies only that each cell probability will equal the product of its respective row and column probabilities and therefore imply independence of the two classifications.

The contingency table in Table 15.4 contains $r = 3$ rows and $c = 4$ columns, and it is consequently called an $r \times c$ (in this case, a 3 × 4) contingency table. Similarly, a 2 × 5 contingency table contains 2 rows and 5 columns. We will denote the four column totals, proceeding from left to right, as C_1, C_2, C_3, and C_4 and the three row totals, proceeding from top to bottom, as R_1, R_2, and R_3. Thus $C_1 = 74$ and

$R_1 = 94$. Note that the sum of the column totals equals $n = 309$. Similarly, the sum of the row totals equals 309. These row and column totals will be used to estimate the unspecified cell probabilities and, consequently, the expected cell frequencies. In fact, it can be shown (proof omitted) that the appropriate estimate of the expected frequency for the cell in the ith row and jth column of a contingency table is as given in the following display.

Estimated Expected Frequency for the Cell in the ith Row and jth Column

$$\hat{E}(n_{ij}) = \frac{R_i C_j}{n}$$

where

$$R_i = \text{Total for row } i$$
$$C_j = \text{Total for column } j$$

For example, the estimated expected cell frequency for the cell corresponding to shift 1, defect A (row 1, column 1) is

$$\hat{E}(n_{11}) = \frac{R_1 C_1}{n} = \frac{(94)(74)}{309} = 22.51$$

Similarly,

$$\hat{E}(n_{23}) = \frac{R_2 C_3}{n} = \frac{(96)(128)}{309} = 39.77$$

The X^2 statistic is calculated in exactly the same manner as indicated in Section 15.2. Thus

$$X^2 = \sum_{j=1}^{4} \sum_{i=1}^{3} \frac{[n_{ij} - \hat{E}(n_{ij})]^2}{\hat{E}(n_{ij})} = \sum_{j=1}^{4} \sum_{i=1}^{3} \frac{d_{ij}^2}{\hat{E}(n_{ij})}$$

The approximating chi-square distribution for X^2 will always possess d.f. $= (r - 1)(c - 1)$ degrees of freedom.

Chi-Square Degrees of Freedom for a Contingency Table

$$\text{d.f.} = (r - 1)(c - 1)$$

where

$$r = \text{Number of rows}$$
$$c = \text{Number of columns}$$

The chi-square test is summarized in the next display.

Chi-Square Test for Independence of Row and Column Categories

1   Null Hypothesis: H_0 : The row and column categories are independent

2   Alternative Hypothesis: H_a : The row and column categories are dependent

3   Test Statistic: $X^2 = \sum\limits_{j=1}^{c} \sum\limits_{i=1}^{r} \dfrac{[(n_{ij} - \hat{E}(n_{ij})]^2}{\hat{E}(n_{ij})} = \sum\limits_{j=1}^{c} \sum\limits_{i=1}^{r} \dfrac{d_{ij}^2}{\hat{E}(n_{ij})}$,

where

$$n_{ij} = \text{Frequency for the cell in row } i \text{ and column } j$$
$$d_{ij} = n_{ij} - \hat{E}(n_{ij})$$
$$R_i = \text{Total for row } i$$
$$C_j = \text{Total for column } j$$
$$n = \text{Total for all observations}$$
$$\hat{E}(n_{ij}) = \frac{R_i C_j}{n}$$

4   Rejection Region: $X^2 > \chi_\alpha^2$, where χ_α^2 is based on d.f. $= (r-1)(c-1)$
degrees of freedom (see Table 5 in Appendix II).

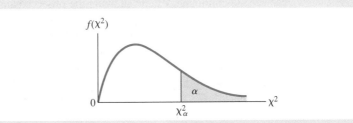

Assumption: All estimated expected cell frequencies are greater than or equal to 5.

E X A M P L E   **15.2**   Refer to the data in Table 15.4. Do the data provide sufficient evidence to indicate that the proportions of the four types of defects vary from shift to shift? Test using $\alpha = .05$.

Solution   The first step is to compute the estimated expected frequencies and insert them (as we have done in Table 15.4) into the contingency table. Then calculate

$$X^2 = \sum_{j=1}^{4} \sum_{i=1}^{3} \frac{[n_{ij} - \hat{E}(n_{ij})]^2}{\hat{E}(n_{ij})} = \sum_{j=1}^{4} \sum_{i=1}^{3} \frac{d_{ij}^2}{\hat{E}(n_{ij})}$$

$$= \frac{(15 - 22.51)^2}{22.51} + \frac{(26 - 22.99)^2}{22.99} + \cdots + \frac{(20 - 14.63)^2}{14.63}$$
$$= 19.18$$

The final step in our test is to locate the rejection region and complete the test. The number of degrees of freedom for this chi-square statistic is

$$\text{d.f.} = (r - 1)(c - 1) = (3 - 1)(4 - 1) = 6$$

If we choose $\alpha = .05$ for our test, the tabulated value of $\chi^2_{.05}$ corresponding to d.f. $= 6$ in Table 5 of Appendix II is $\chi^2_{.05} = 12.5916$. Therefore, we will reject the hypothesis of independence between the row and column classifications if

$$X^2 > 12.5916$$

Since the computed value, $X^2 = 19.18$, exceeds this value, we reject the null hypothesis and conclude that there is sufficient evidence to indicate that the proportions of the various types of defects vary from shift to shift. ▪

Most statistical packages include a program for analyzing data contained in an $r \times c$ contingency table. To illustrate, the SAS, Minitab, and Execustat computer printouts for a contingency table analysis of the data in Table 15.4 are shown in Tables 15.5, 15.6, and 15.7, respectively. Examine the SAS printout in Table 15.5, and you can see the contingency table showing the observed and the estimated expected cell frequencies. The computed value of X^2 is shown directly below the table in the row corresponding to CHI-SQUARE as 19.178 with DF $= 6$ degrees of freedom. The observed significance level is shown as .0039. The values printed below the value of CHI-SQUARE are not pertinent to our analysis.

The Minitab printout generated by the command CHISQUARE for data stored in columns C1, C2, ... is shown in Table 15.6; the observed and estimated expected frequencies are shown in each cell of the table. The value of the test statistic, $X^2 = 19.18$, is shown below the table along with the degrees of freedom for X^2, df $= 6$. The Minitab printout does not give the observed significance level for the test. Consequently, you must compare the observed value of X^2 with the tabulated value in Table 5 of Appendix II to determine whether the X^2 statistic is large enough to reject the null hypothesis of *independence* of the two qualitative variables, shift and type of defect.

The Minitab and Execustat printouts show the same 3×4 contingency table, except that the Execustat table displays the percentage of the total represented by each cell count rather than the expected cell count. The computed value, $X^2 = 19.178$, is shown below the table in both printouts along with its degrees of freedom.

The information shown on the printouts differs slightly from one printout to another, as do some of the computed numbers (due to rounding), but the basic information is the same. All three show the calculated value of X^2 and its degrees of freedom. These two quantities, along with the table of the critical values of chi-square, Table 5 in Appendix II, are all that we need to test in order to detect dependence between the two qualitative variables represented in the contingency table. The value of the observed significance level, given in the SAS and the EXECUSTAT printouts, eliminates the need for the chi-square table and gives us a measure of the weight of evidence favoring rejection of the null hypothesis.

TABLE 15.5
SAS contingency table analysis of Table 15.4

```
                          TABLE OF SHIFT BY TYPE

SHIFT                          TYPE

FREQUENCY
EXPECTED   A          B          C          D          TOTAL

   1           15         21         45         13         94
            22.5       21.0       38.9       11.6

   2           26         31         34          5         96
            23.0       21.4       39.8       11.8

   3           33         17         49         20        119
            28.5       26.6       49.3       14.6

TOTAL          74         69        128         38        309

                      STATISTICS FOR 2-WAY TABLES

CHI-SQUARE                      19.178    DF = 6    PROB = 0.0039
PHI                              0.249
CONTINGENCY COEFFICIENT          0.242
CRAMER'S V                       0.176
LIKELIHOOD RATIO CHISQUARE      20.336    DF = 6    PROB = 0.0024
```

TABLE 15.6
Minitab contingency table analysis of Table 15.4

```
MTB > CHISQUARE C1-C4

Expected counts are printed below observed counts

            C1         C2         C3         C4      Total
   1        15         21         45         13         94
         22.51      20.99      38.94      11.56

   2        26         31         34          5         96
         22.99      21.44      39.77      11.81

   3        33         17         49         20        119
         28.50      26.57      49.29      14.63

Total       74         69        128         38        309

ChiSq =  2.506 +   0.000 +   0.944 +   0.179 +
         0.394 +   4.266 +   0.836 +   3.923 +
         0.711 +   3.449 +   0.002 +   1.967 = 19.178
df = 6
```

T A B L E **15.7** Execustat contingency table analysis of Table 15.4

Crosstabulation

	A	B C	D	Row Total
1	15 4.9	21 6.8	45 14.6	13 4.2	94 30.42
2	26 8.4	31 10.0	34 11.0	5 1.6	96 31.07
3	33 10.7	17 5.5	49 15.9	20 6.5	119 38.51
Column Total	74 23.95	69 22.33	128 41.42	38 12.30	309 100.00

Summary Statistics for Crosstabulation

Chi-square	D.F.	P Value
19.18	6	0.0039

Exercises

Basic Techniques

15.11 Calculate the value and give the number of degrees of freedom for X^2 for the following contingency tables.

a

	Column			
Row	1	2	3	4
1	120	70	55	16
2	79	108	95	43
3	31	49	81	140

b

	Column		
Row	1	2	3
1	35	16	84
2	120	92	206

15.12 Suppose that a consumer survey summarizes the responses of $n = 307$ people in a contingency table that contains three rows and five columns. How many degrees of freedom will be associated with the chi-square test statistic?

15.13 A survey of 400 respondents produced the cell counts given in the accompanying 2×3 contingency table.

	Column			
Row	1	2	3	Total
1	37	34	93	164
2	66	57	113	236
Total	103	91	206	400

a If you wish to test the null hypothesis of "independence"—that the probability that a response falls in any one row is independent of the column it will fall in—and you plan to use a chi-square test, how many degrees of freedom will be associated with the χ^2 statistic?

b Find the value of the test statistic.

c Find the rejection region for $\alpha = .10$.

d Conduct the test and state your conclusions.

e Find the approximate observed significance level for the test, and interpret its value.

Applications

15.14 A carpet company was interested in comparing the fraction of new-home builders favoring carpet over other floor coverings for homes in three different areas of a city. The objective was to decide how to allocate sales effort to the areas. A survey was conducted, and the data are shown in the accompanying table.

	Areas		
Floor covering	**1**	**2**	**3**
Carpet	69	126	16
Other material	78	99	27

a Do the data indicate a difference in the percentage favoring carpet from one area of the city to another?

b Estimate the difference in the fractions of new-home builders who favor carpet between areas 1 and 2. Use a 95% confidence interval. [*Hint:* Use the procedure of Section 7.8.]

15.15 Is it likely that television viewers will believe an advertisement that includes a hidden camera interview? A random sample of 500 adults was taken; their opinions as well as their educational attainment are shown in the table below. Do the data provide sufficient evidence to indicate that advertising believability is dependent on the educational attainment of the viewer? Use $\alpha = .05$.

	Less Than High School	**High School Graduate**	**Some College**	**College Graduate**
Believable	11	42	61	24
Not believable	19	77	140	96
Not sure	2	13	11	4

Source: Data adapted from "American Voices," 1990.

15.16 In Exercise 3.19, we considered a survey of 100 cars, each of which was classified according to whether or not it had antilock brakes and whether or not it had been involved in an accident in the past year. The data are reproduced below in a different form.

	Antilock Brakes	**No Antilock Brakes**
Accident	3	12
No accident	40	45

Source: Data adapted from *Consumers' Research,* 1994.

a Do the data provide sufficient evidence to indicate that the proportion of cars that have had accidents depends on whether or not the car has antilock brakes? Test using a chi-square statistic with $\alpha = .05$.

b What is the approximate *p*-value for the test in part (a)? Does the *p*-value support the conclusions drawn in part (a)?

15.17 Refer to Exercise 15.16.

a Express the null hypothesis tested in part (a) of Exercise 15.16 as an hypothesis about the difference between two proportions (see Section 8.8).

b If it is believed that antilock brakes should reduce the proportion of cars that have accidents, state the appropriate alternative hypothesis.

c Based on the null and alternative hypotheses in parts (a) and (b), use an appropriate test to determine whether cars with antilock brakes have a lower proportion of accidents than those without. Base your conclusion on the observed significance level (p-value) of the test.

15.18 The Gallup Poll is used by economists to keep track of national trends in many areas of interest to them. In particular, one poll tracks Americans' approval or disapproval of the way the President handles the economy. One such poll, taken in April 1993, produced the results shown in the table, classified by response and by household income. Use the Minitab printout to determine whether there is a dependence between a person's opinion on the President's economic policy and his or her household income. Use $\alpha = .05$.

Income	Approve	Disapprove	No Opinion
$75,000 and over	72	120	8
$50,000–74,999	108	128	14
$30,000–49,999	117	182	26
$20,000–29,999	43	52	5
Under $20,000	66	51	8

Source: Adapted from Gallup, 1994.

```
MTB > ChiSquare 'Approve'-'No opin.'

Expected counts are printed below observed counts

          Approve   Do not No opin.    Total
      1        72        120        8      200
            81.20     106.60    12.20

      2       108        128       14      250
           101.50     133.25    15.25

      3       117        182       26      325
           131.95     173.23    19.83

      4        43         52        5      100
            40.60      53.30     6.10

      5        66         51        8      125
            50.75      66.62     7.62

Total        406        533       61     1000

ChiSq =   1.042 +   1.684 +   1.446 +
          0.416 +   0.207 +   0.102 +
          1.694 +   0.445 +   1.923 +
          0.142 +   0.032 +   0.198 +
          4.583 +   3.664 +   0.018 = 17.597
df = 8
```

15.19 Although women use office computers much more than men do, marketers target their advertisements toward men. Recent surveys show that more women than men think that computers are fun and that more women than men are concerned about learning to use a computer (Kaplan, 1994). However, more men than women read computer magazines or technical articles.

The accompanying table is based on a random sample of 150 men and 100 women who use office PCs. Do the data indicate that there is a difference in the proportion of men and women who use office PCs and read computer literature? Find the approximate *p*-value for the test you perform, and base your conclusions upon this observed significance level.

	Men	Women
Read computer literature	66	23
Do not read computer literature	84	77

15.20 Americans are keenly aware of the negative effects of poor diets on their health and life styles. This is reflected in a 1993 survey of 1000 adults commissioned by the American Dietetic Association and Kraft General Foods, in which respondents were queried about their eating habits in 1993. Specifically, they were asked if they were satisfied with their efforts to do all they could to have a healthy diet. A similar survey was done in 1991. The results of these surveys appear in the accompanying table. Has there been any significant change in the responses from 1991 to 1993?

	Response		
Year	Make Effort	Don't Make Effort	Total
1991	433	567	1000
1993	378	622	1000
Total	811	1189	2000

Source: Adapted from "Crunching Numbers," 1994.

15.4 *r* × *c* Tables with Fixed Row or Column Totals

In the previous section, we described the analysis of an *r* × *c* contingency table, using examples that, for all practical purposes, fit the multinomial experiment described in Section 15.1. Although the methods of collecting data in many surveys may adhere to the requirements of a multinomial experiment, other methods do not. For example, we might not want to randomly sample the population described in Example 15.1 because we might find that, owing to chance, one category is completely missing.

To illustrate, suppose that we want to determine whether differences exist among three different product markets in their attitudes to a particular form of product advertising. Say we were to randomly sample *n* = 600 persons and, just by chance, one of the groups, say Hispanics, does not appear in the sample. To ensure against this possibility, we might survey a random sample of *n* = 200 persons from each market group. A summary of the responses might appear as shown in Table 15.8.

Table 15.8 contains the results of three independent binomial experiments, one corresponding to each product market ($n = 200$, $k = 2$). The objective of the experiments is to test the equality of the binomial probabilities (proportions) p_1, p_2, and p_3. When these probabilities are equal, a person's inclination to favor the form of advertising is independent of the market from which the person is selected. When they are unequal, a person's opinion will depend upon the market. **Therefore, the**

TABLE 15.8
Consumer response to a form of product advertising for three product markets ($k = 2$ possible opinions)

| | Product Market | | | |
Opinion	1	2	3	Total
Favor	124	111	137	372
Do not favor or have no opinion	76	89	63	228
Total	200	200	200	600

equivalence of a set of binomial probabilities can be tested by using the contingency table analysis of Section 15.3. We refer to such a test as a test of *homogeneity of binomial distributions.*

Suppose that each opinion in the market surveys was classified into one of three categories, "favor" the form of product advertising, "do not favor," and "no opinion," and that the results of the surveys are as shown in Table 15.9. Table 15.9 contains the outcomes for three independent multinomial experiments, one corresponding to each market ($n = 200$, $k = 3$). If the distribution of opinions into the three categories is the same for all markets—that is, *opinion* is independent of *market*—the multinomial probabilities $p_1 = P(\text{favor})$, $p_2 = P(\text{do not favor})$, and $p_3 = P(\text{no opinion})$ will be the same for all three markets. Therefore, as in the comparison of binomial population proportions, we can compare multinomial populations by using a contingency table analysis. **The analysis of an $r \times c$ contingency table in which the column totals (or row totals) have been fixed is identical to the analysis of Section 15.3.**

TABLE 15.9
Consumer response to a form of product advertising for three product markets

| | Product Market | | | |
Opinion	1	2	3	Total
Favor	124	111	137	372
Do not favor	55	44	38	137
No opinion	21	45	25	91
Total	200	200	200	600

EXAMPLE 15.3 Do the data in Table 15.9 provide sufficient evidence to indicate that the proportions of responses in the "favor," "do not favor," and "no opinion" categories differ among the three product markets?

Solution The Minitab computer printout for a contingency table analysis of the data in Table 15.9 is shown in Table 15.10. You can verify that the number of degrees of freedom for the chi-square statistic is, as shown on the printout, d.f. $= (r - 1)(c - 1) = (3 - 1)(3 - 1) = 4$. You can also see that the computed value of the X^2 statistic is $X^2 = 16.882$ and that the observed level of significance is less than .005 from Table

TABLE **15.10**

Minitab computer printout for Table 15.9 and Example 15.3

```
MTB > CHISQUARE C1-C3

Expected counts are printed below observed counts

              C1        C2        C3      Total
    1         124       111       137      372
           124.00    124.00    124.00

    2          55        44        38      137
            45.67     45.67     45.67

    3          21        45        25       91
            30.33     30.33     30.33

Total         200       200       200      600

ChiSq =   0.000 +   1.363 +   1.363 +
          1.908 +   0.061 +   1.287 +
          2.872 +   7.092 +   0.938 = 16.882
df = 4
```

5 in Appendix II. Thus we would reject the hypothesis of homogeneity of the three product markets for all values of α larger than .005. In other words, there is ample evidence to indicate that the consumer response differs among the three markets. ∎

Exercises

Basic Techniques

15.21 Random samples of 200 observations were selected from each of three populations, and then each observation was classified according to whether it fell into one of three mutually exclusive categories. The cell counts are shown in the accompanying table. Do the data provide sufficient evidence to indicate that the proportions of observations in the three categories depend upon the population from which they were drawn?

Population	Category 1	2	3	Total
1	108	52	40	200
2	87	51	62	200
3	112	39	49	200

a Give the value of X^2 for the test.

b Give the rejection region for the test for $\alpha = .10$.

c State your conclusions.

d Find the approximate p-value for the test, and interpret its value.

15.22 Suppose that you want to test the null hypothesis that three binomial parameters p_A, p_B, and p_C are equal against the alternative that at least two of the parameters differ. Independent random samples of 100 observations were selected from each of the populations. The data are shown in the accompanying table.

Category	Population A	B	C	Total
Number of successes	24	19	33	76
Number of failures	76	81	67	224
Total	100	100	100	300

a Find the value of X^2, the test statistic.

b Give the rejection region for the test for $\alpha = .05$.

c State your test conclusions.

d Find the approximate observed significance level for the test, and interpret its value.

Applications

15.23 A study of the purchase decisions for three stock portfolio managers, A, B, and C, was conducted to compare the rates of stock purchases that resulted in profits over a time period that was less than or equal to one year. One hundred randomly selected purchases obtained for each of the managers gave the results shown in the accompanying table. Do the data provide evidence of differences among the rates of successful purchases for the three managers?

Purchase Result	Manager A	B	C
Profit	63	71	55
No profit	37	29	45
Total	100	100	100

15.24 A manufacturer of buttons wanted to determine whether the fraction of defective buttons produced by three machines varied from machine to machine. Samples of 400 buttons were selected from each of the three machines, and the number of defectives were counted for each sample. The results are as follows:

Machine number	1	2	3
Number of defectives	16	24	9

Do these data provide sufficient evidence to indicate that the fraction of defective buttons varies from machine to machine? Test with $\alpha = .05$.

15.25 Time is becoming a more and more important factor in marketing. Quite often, people now prefer to shop at stores where they can get in and out quickly, rather than shopping somewhere where they can get lower prices or greater selection. This is especially true of married shoppers who have dual incomes. In a survey, 500 shoppers classified as dual or single earners were asked whether they agreed with the statement, "I wish there were ways to reduce shopping time." The results are shown in the table that follows.

	Dual Earner	Single Earner
Agree	217	44
Disagree	158	81

Source: Data adapted from Eugene H. Fram and Joel Axelrod, "The Distressed Shopper," *American Demographics,* October 1990, p. 44

a   Do the data indicate a difference in the proportions who agree with the statement for the two groups of shoppers? Use the χ^2 test with $\alpha = .01$.

b   Use the large-sample *z* test of Section 8.8 to answer the question in part (a). Are the results the same?

15.26   The fierce economic competition between Japan and the United States for the lion's share of the world economic market has inevitably led to comparisons between American and Japanese workers. One such comparison survey, presented in the *American Enterprise,* dealt with the workers' satisfaction with their lives and their personal accomplishments. Random samples of 100 American and Japanese workers were sampled and asked about their personal accomplishments, and the results are shown in the table below. Use the Minitab printout to answer the following questions.

	American	Japanese
Have satisfied most ambitions in life	58	46
Have had to settle for less	41	53
No response	1	1

Source: Data adapted from *American Enterprise,* November/December 1990, Vol. 1, No. 6, p. 83.

```
Expected counts are printed below observed counts

        AMERICAN JAPANESE    Total
    1       58       46       104
         52.00    52.00

    2       41       53        94
         47.00    47.00

    3        1        1         2
          1.00     1.00

Total      100      100       200

ChiSq =  0.692 +  0.692 +
         0.766 +  0.766 +
         0.000 +  0.000 = 2.917
df = 2
2 cells with expected counts less than 5.0
```

a   Do the data indicate that there is a difference in personal satisfaction between American and Japanese workers?

b   Find the observed significance level associated with the test.

c   Have any assumptions been violated?

15.27 In Exercise 8.62, we referred to a report by the American Cancer Society ("Profile of Smokers," 1990) that indicates that more men than women are smokers. In random samples of 200 males and 200 females, 62 of the males and 54 of the females surveyed were smokers.

a Do the data provide sufficient evidence to indicate that there is a difference in the proportion of smokers between men and women? Test using the χ^2 test with $\alpha = .10$.

b Can the test in part (a) be used to conclude that more males than females are smokers? Explain.

CASE STUDY REVISITED

15.5 ATTITUDES IN THE WORKPLACE

Now that we have had an opportunity to learn how to analyze a contingency table, we can return to the case study that reported the results of a survey in which 1065 respondents were asked the following question: "If you were taking a new job and had your choice of a boss, would you prefer to work for a man or a woman?"

The first table in the case study was a cross-classification of the 1065 respondents by sex and by their selection of one of four responses: prefer a male boss, prefer a female boss, no difference/either, and no opinion. A Minitab printout of a chi-square analysis is given in Table 15.11. The table has two rows and four columns, and therefore the degrees of freedom are $(2 - 1)(4 - 1) = 3$. Since the calculated value of the test statistic, $X^2 = 75.022$, exceeds the .005 right-tailed critical value of $\chi^2_{.005} = 12.838$, its p-value is less than or equal to .005, and therefore we have found highly significant differences between the responses of men and women. Today, Americans prefer a male boss over a female boss by 38% to 23%, while, overall, women prefer a male boss by a margin of 44% to 29%.

TABLE 15.11
Minitab computer printout for workplace attitudes

```
MTB > ChiSquare 'Male' 'Female' 'Either' 'No-Opin'

Expected counts are printed below observed counts

            Male    Female   Either   No-Opin   Total
    1       175        85      260        11      531
          204.42    119.66   193.45     13.46

    2       235       155      128        16      534
          205.58    120.34   194.55     13.54

Total       410       240      388        27     1065

ChiSq =   4.235 + 10.040 + 22.891 +   0.450 +
          4.211 +  9.984 + 22.763 +   0.448 = 75.022
df = 3
```

However, the biggest differences are not between males and females but rather between respondents in different age groups. In the cross-classification of men by age and by their choice of boss, the resulting test statistic is $X^2 = 31.660$ with 4 d.f., significant at the .005 level. Similarly, in the cross-classification of women by age and choice of boss, the value of the test statistic is $X^2 = 78.097$, which with 4 d.f. is also significant at the .005 level of significance. If each of the cell counts in these two tables is converted to a percentage of the row totals, the largest differences are observed between young and old, especially between young and old women. Young women (18–29) prefer a boss of their own sex by a margin of 62% to 15%, while older women (50 and older) prefer a male boss by a margin of 59% to 16%.

Are there geographic differences in the response to our survey question? The Minitab printout of the chi-square test for independence of the row and column classifications has a value of the test statistic given as $X^2 = 2.363$, which with $(4 - 1)(3 - 1) = 6$ d.f. is not significant, indicating that the responses seem to be homogeneous across the country.

At this point, you are able to answer the question "Are women more sexist than men?" in the context of this case study. What will your answer be?

15.6   Summary

The preceding material has been concerned with a test of an hypothesis regarding the cell probabilities associated with a multinomial experiment. When the number of observations n is large, the test statistic X^2 can be shown to possess, approximately, a chi-square probability distribution in repeated sampling, with the number of degrees of freedom being dependent on the particular application. In general, we assume that n is large and that the minimum expected cell frequency is equal to or greater than 5.

Several words of caution concerning the use of the X^2 statistic as a method of analyzing enumerative-type data are appropriate. The determination of the correct number of degrees of freedom associated with the X^2 statistic is very important in locating the rejection region. If the number is incorrectly specified, erroneous conclusions may result. Also, note that nonrejection of the null hypothesis does not imply that it should be accepted. We would have difficulty stating a meaningful alternative hypothesis for many practical applications and, therefore, would lack knowledge of the probability of making a type II error. For example, we hypothesize that the two classifications of a contingency table are independent. A specific alternative would have to specify some measure of dependence, which may or may not possess practical significance to the experimenter. Finally, if parameters are missing and the expected cell frequencies must be estimated, the estimators of missing parameters should be of a particular type in order that the test be valid. In other words, the application of the chi-square test for other than the applications outlined in Sections 15.2, 15.3, 15.4, and 15.5 will require experience beyond the scope of this introductory presentation of the subject.

Supplementary Exercises

15.28   A die was rolled 600 times, with the following results:

Observed number	1	2	3	4	5	6
Frequency	89	113	98	104	117	79

Do these data provide sufficient evidence to indicate that the die is unbalanced? Test using $\alpha = .05$.

15.29   After inspecting the data in Exercise 15.28, we might want to test the null hypothesis that the probability of a "6" is 1/6 against the alternative that this probability is less than 1/6.

a   Carry out the test, using $\alpha = .05$.

b   What tenet of good statistical practice is violated in the test of part (a)?

15.30   Suppose that a government regulation on hiring requires that the proportions of workers of various races employed by a company be consistent with the corresponding racial proportions in the work force at large. In a particular region, the workers fall into one of four racial categories in the proportions $p_1 = .2$, $p_2 = .05$, $p_3 = .03$, and $p_4 = .72$. A company employs 300 workers who are racially distributed as follows: $n_1 = 45$, $n_2 = 21$, $n_3 = 8$, and $n_4 = 224$. Do these data disagree with the company's contention that the deviations of these numbers from the expected numbers (based on the distribution of the labor force) are solely due to random variation?

a   State H_0.

b   State H_a.

c   Calculate X^2.

d   Give the rejection region for the test for $\alpha = .10$.

e   Conduct the test and state your conclusions.

15.31   An analysis of industrial plant accident data was made to determine whether the distribution of numbers of fatal accidents differed among three departments of the plant. The data for 346 accidents are shown in the accompanying table. Do the data indicate that the frequency of fatal accidents is dependent on the plant department? Test using $\alpha = .05$ and the Minitab printout.

	Plant Department		
Accident Type	**1**	**2**	**3**
Fatal	67	26	16
Not fatal	128	63	46

```
Expected counts are printed below observed counts

              1            2            3      Total
    1        67           26           16        109
           61.43        28.04        19.53

    2       128           63           46        237
          133.57        60.96        42.47

 Total      195           89           62        346

ChiSq =   0.505 +   0.148 +   0.639 +
          0.232 +   0.068 +   0.294 = 1.886
 df = 2
```

15.32 A radio station conducted a survey to study the relationship between the number of radios per household and family income. The survey, based on $n = 1000$ interviews, produced the results shown in the accompanying table. Do the data provide sufficient evidence to indicate that the number of radios per household is dependent on family income? Test at the $\alpha = .10$ level of significance.

Number of Radios per Household	Family Income			
	Less than $9000	$9000–12,000	$12,000–15,000	More than $15,000
1	126	362	129	78
2	29	138	82	56

15.33 A trucking company claims that 30% of its trucks carry produce, 45% carry local freight, and 25% carry the bottled produce from neighboring vineyards. A group of investors who are interested in purchasing the trucking company decides to check this claim. The group randomly samples the bills of lading of 100 trucks and checks their cargos. The survey produced the following counts on the number of trucks carrying specific cargos:

produce: 21   local freight: 39   wine: 40

Do these data disagree with the trucking company's claim? Test using $\alpha = .10$.

15.34 One myth concerning small businesses is that most new firms fail early, and most people believe that small businesses fail in their first three years. In their study of the role of small businesses, Duncan and Handler (1994) found that surviving firms were most likely to be involved in agriculture, construction, or financial and nonfinancial services. The following information is taken from their Table 2, which tracked those small businesses started in 1985. Use an appropriate statistical technique to determine whether the survival rate among these three industries differs significantly.

Industry	Status			Total
	Survived	Discontinued	Failed	
Agriculture	3217	354	144	3715
Mining	1393	726	235	2354
Trans./pub. util.	5914	2225	775	8914
Total	10,524	3305	1154	14,983

15.35 The Toxic Release Inventory (TRI) is a voluntary public disclosure report by manufacturing facilities that use threshold amounts (or more) of 300 chemicals. These facilities disclose their estimated annual toxic emissions to land, air, and water or shipments of waste off-site. Of interest to Lynn and Kartez (1994) was the effectiveness of the promise of disclosure and the significance of public access to the TRI. In the table that follows, three TRI user groups— citizen groups (69), state agencies (51), and industry (19)—recorded the frequency with which business and industry groups contacted them. The respondents reported whether requests were made rarely/never, occasionally, or frequently. How would you test to determine whether the frequencies with which each agency group was contacted by business and industry users are statistically different? Are there problems because of small cell counts?

Frequency	Citizen	State	Industry
Rarely/never	55	35	12
Occasionally	12	19	6
Frequently	2	7	1

15.36 Aircraft accidents are most likely to occur during the beginning and ending of the flight, time periods that represent only 6% of the total flight time, according to an article in the *Press Enterprise* (December 5, 1990). In particular, the percentage of accidents that occur during different periods of the flight are shown in the accompanying table.

Period of Flight	Percentage of Accidents
Takeoff	18
Initial climb	11
Climb	7
Cruise	5
Descent	3
Initial approach	12
Final approach	16
Landing	25
Loading and taxiing	3

Source: Phelps, 1990.

Suppose that 200 accident reports are randomly selected from the files of a large airport, and the number of accidents occurring during each of these nine periods is found to be 40, 24, 15, 12, 3, 26, 35, 42, and 3, respectively. Are these figures significantly different from those reported in the table? Use $\alpha = .01$.

15.37 College graduates are still the most likely users of personal computers in our society, but most adults have at least some computer literacy. According to a survey in *American Demographics* (October 1990, p. 16), one-fourth of adults with only a high school diploma are also computer literate. In a similar survey, 150 adults were randomly selected and their use of the computer, as well as their educational background, was recorded. Do the data provide sufficient evidence to suggest that there is a difference in computer literacy for the four educational levels? Use $\alpha = .05$.

	Less Than High School	High School Graduate	Some College	College Graduate
Computer user	2	13	30	11
Not a computer user	13	41	35	5

Source: Data adapted from "American Voices," 1990.

15.38 A survey was conducted by an auto repairman to determine whether various auto ills were dependent on the make of the auto. His survey, restricted to this year's model, produced the results shown in the accompanying table. Do these data provide sufficient evidence to indicate a dependency between auto makes and type of repair for these new-model cars? Note that the repairman was not using all the information available when he conducted his survey. When one conducts a study of this type, what other factors should be recorded?

	Type of Repair		
Make	Electrical	Fuel Supply	Other
A	17	19	7
B	14	7	9
C	6	21	12
D	33	44	19
E	7	9	6

15.39 By tradition, U.S. labor unions have been content to leave the management of the company to the managers and corporate executives. But in Europe, worker participation in management decision making is an accepted idea and one that is continually spreading. In a study of the effect of worker satisfaction with worker participation in managerial decision making, 100 workers were interviewed in each of two separate German manufacturing plants. One plant had active worker participation in managerial decision making; the other did not. Each selected worker was asked whether he or she generally approved of the managerial decisions made within the firm. The results of the interviews are shown in the accompanying table.

Response	Participative Decision Making	No Participative Decision Making
Generally approve of the firm's decisions	73	51
Do not approve of the firm's decisions	27	49

a Do the data provide sufficient evidence to indicate that approval or disapproval of management's decisions depends on whether workers participate in decision making? Test using the X^2 test statistic. Use $\alpha = .05$.

b Do these data support the hypothesis that workers in a firm with participative decision making more generally approve of the firm's managerial decisions than those employed by firms without participative decision making? Test by using the z test presented in Section 8.8. This problem requires a one-tailed test. Why?

16

NONPARAMETRIC STATISTICS

About This Chapter

Chapters 7, 8, and 9 presented statistical techniques for comparing two or more populations by comparing their respective population parameters. The techniques are applicable to data measured on a continuum and to data possessing normal population relative frequency distributions. The purpose of this chapter is to present a set of statistical test procedures to compare populations for the many types of data that do not satisfy these assumptions.

Contents

ENVELOPE, PLEASE:
THE WINNER IS ...

More and more Americans are installing home filtration systems to improve their drinking water or are simply buying bottled water for drinking purposes. However, is it a fact that bottled water will automatically be better than the water that comes out of your tap? This was the question addressed when water samples from seven sources in Riverside County, California, were compared by four citizens who served as judges (Stokely, 1994, p. B-4). The water samples were scored in four categories—aroma, appearance, taste, and texture—on a scale of 1 to 5, with 5 being outstanding. In Table 16.1, the average score of the four judges is reported for each location on the four attributes considered in the informal test. The outcome of this informal survey was somewhat surprising to residents living in these cities or having water supplied by one of these sources, because the bottled water came in second when scores on the four attributes were averaged.

TABLE 16.1
Ratings of water samples from
seven California sources

Water	Aroma	Appearance	Taste	Texture	Average
Temecula	2.7	4.2	3.2	4.2	3.6
San Jacinto	3.7	4.2	2.7	3.7	3.6
Sparkletts[†]	4.7	3.7	4.2	4.2	4.2
Lake Elsinore	3.7	4.2	3.7	4.0	3.9
Norco	4.7	4.5	4.0	4.0	4.3
Metropolitan Water	2.5	4.7	2.2	4.2	3.4
Riverside	4.2	4.5	3.7	4.5	4.2

[†] Bottled water.

The scores on the appearance of the water samples do not appear to vary much and would probably not be a good predictor of the final rankings of the water samples. Aroma and taste exhibit more variation and, in fact, may be positively correlated; that is, we would expect a high score on aroma to be associated with a high score on taste. The problem here is that these data, which are scores, do not satisfy the normality assumptions required for using the correlation analysis presented in Chapter 11.

Studies have been conducted to determine the effect of moderate departures from assumptions on the measures of reliability associated with a method of inference. We would like to know whether moderate departures from the assumption of normality would seriously change the values of α and β, the probabilities of type I and type II errors, respectively. Estimation and test procedures whose properties remain fairly stable when the assumptions are relaxed are said to be **robust.**

In addition to studying the robustness of statistical procedures, statisticians have developed an alternate methodology of statistical inference that requires very few assumptions concerning the sampled population. Some of these methods, called **nonparametric** or **distribution-free statistical procedures,** are presented in this chapter. We reexamine the relationship between the attributes of aroma and taste in Section 16.8.

16.1 Introduction

Some experiments yield response measurements that defy quantification. That is, they generate response measurements that can be ordered (ranked), but the location of the responses on a scale of measurement is arbitrary. For example, the data may admit only pairwise directional comparisons (whether one observation is larger than another or vice versa). Or we may be able to rank all observations in a data set but may not know the exact values of the measurements. To illustrate, suppose that a judge is employed to evaluate and rank the sales abilities of four salespeople, the edibility and taste characteristics of five brands of cornflakes, or the relative appeal of five new automobile designs. Clearly, it is impossible to give an exact measure of sales competence, palatability of food, or design appeal, but it is usually possible to rank the salespeople, brands, or design appeal according to which one we think is best, second best, and so on. Thus the response measurements here differ markedly from those presented in preceding chapters. Experiments that produce this type of data occur in almost all fields of study, but they are particularly evident in social science research and in studies of consumer preference. The data that they produce can be analyzed by using *nonparametric statistical methods.*

Nonparametric statistical procedures are also useful in making inferences in situations where serious doubt exists about the assumptions that underlie standard methodology. For example, the t test for comparing a pair of means in Section 8.5 is based on the assumption that both populations are normally distributed with equal variances. The experimenter will never know whether these assumptions hold in a practical situation but will often be reasonably certain that departures from the assumptions are small enough so that the properties of the statistical procedure will be undisturbed. That is, α and β will be approximately what the experimenter thinks they are. On the other hand, it is not uncommon for the experimenter seriously to question the assumptions and wonder whether he or she is using a valid statistical procedure. This difficulty may be circumvented by using a nonparametric statistical test and thereby avoiding reliance on a very uncertain set of assumptions.

Research has shown that nonparametric statistical tests are almost as capable of detecting differences among populations as the parametric methods of preceding chapters when normality and other assumptions are satisfied. They may be, and often are, more powerful in detecting population differences when the assumptions are not satisfied. For this reason, many statisticians advocate the use of nonparametric statistical procedures in preference to their parametric counterparts.

In this chapter, we discuss only the most common situations in which nonparametric methods are used.

16.2 The Sign Test for Comparing Two Populations

Without emphasizing the point, we used a nonparametric statistical test as an alternative procedure for determining whether evidence existed to indicate a difference in the mean wear for the two types of tires in the **paired-difference experiment** in Section 8.6. Each pair of responses was compared, and x (the number of times A exceeded B) was used as the test statistic. This nonparametric test is known as the **sign test** because x is the number of positive (or negative) signs associated with the differences. The implied null hypothesis is that the two population distributions are identical, and the resulting technique is completely independent of the form of the distribution of differences. Thus, regardless of the distribution of differences, the probability that A exceeds B for a given pair will be $p = .5$ when the null hypothesis is true (that is, when the distributions for A and B are identical). Then x will possess a binomial probability distribution, and a rejection region for x can be obtained using the binomial probability distribution of Chapter 4.

The sign test is summarized in the following display and is illustrated by an example. The summary and example will help you to recall how the sign test was constructed and how it is used in a practical situation.

The Sign Test for Comparing Two Populations

1 Null Hypothesis: $H_0 : P(x_A > x_B) = 1/2$

2 Alternative Hypothesis:

One-Tailed

$H_a : P(x_A > x_B) > 1/2$; the A distribution is shifted to the right of the B distribution [or $H_a : P(x_A > x_B) < 1/2$; that is, that A distribution is shifted to the left of the B distribution]

Two-Tailed

$H_a : P(x_A > x_B) \neq 1/2$

3 Test Statistic: $x = $ Number of pairs of observations for which x_A exceeds x_B

4 Rejection Region:

One-Tailed	**Two-Tailed**
For $H_a : P(x_A > x_B) > 1/2$, reject H_0 for very large values of x.	Reject H_0 for very large or very small values of x.
For $H_a : P(x_A > x_B) < 1/2$, reject H_0 for very small values of x.	

Assumptions: The observations x_A and x_B are randomly and independently selected in matched pairs. Tied observations (i.e., when $x_A = x_B$) are eliminated from the data, and the number of pairs n is reduced accordingly.

EXAMPLE 16.1 Table 16.2 gives the numbers of employee health insurance claims per month for ten randomly selected months at each of two physically identical automobile assembly plants. Do the data provide sufficient evidence to indicate a difference in the numbers of health insurance claims per month at the two assembly plants? Test using α near .10.

TABLE 16.2
Numbers of health insurance claims per month at two assembly plants
(Example 16.1)

Month	Plant A	Plant B	Month	Plant A	Plant B
1	170	201	6	142	170
2	164	179	7	191	183
3	140	159	8	169	179
4	184	195	9	161	170
5	174	177	10	200	212

Solution Since we want to detect a shift in the A distribution either to the right or to the left of the B distribution—that is, $P(x_A > x_B) > 1/2$ or $P(x_A > x_B) < 1/2$—we will conduct a two-tailed test and reject $H_0 : P(x_A > x_B) = 1/2$ when x, the number of months for which the number of claims at plant A exceeded the number at plant B, is very small or very large. We will choose the values of x to be included in the rejection region so that half of the x values will fall in each tail of the binomial probability distribution and so that α is approximately equal to .10. Consult Table 1 in Appendix II for $n = 10$, and you will find that for $x = 0, 1, 9,$ and 10,

$$\alpha = P(x \text{ is in the rejection region if in fact } H_0 \text{ is true})$$
$$= p(0) + p(1) + p(9) + p(10) = .022$$

Since this value of α is too small, the rejection region is expanded by including the next pair of x values that are most contradictory to the null hypothesis, namely, $x = 2$ and $x = 8$. The value of α for this rejection region ($x = 0, 1, 2, 8, 9, 10$) is obtained from Table 1 in Appendix II:

$$\alpha = p(0) + p(1) + p(2) + p(8) + p(9) + p(10) = .110$$

Since this value of α is close to .10, we will employ $x = 0, 1, 2, 8, 9, 10$ as the rejection region for the test. (See Figure 16.1.)

FIGURE **16.1**
Rejection region for
Example 16.1

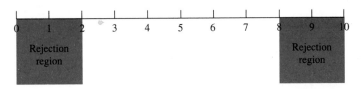

From the data, we observe that $x = 1$, and therefore we reject the null hypothesis. Thus we conclude that sufficient evidence exists to indicate that the population distributions differ in their relative locations. In fact, the data suggest that the number of health insurance claims in any one month at plant B tends to exceed the corresponding number at plant A. ∎

When the number of pairs in a paired-difference experiment is large, the probability distribution for x, the number of times x_A is larger than x_B, can be approximated by a normal distribution. Therefore, with $p = P(x_A > x_B)$ we can test

$$H_0 : p = .5$$

using the test for a binomial proportion described in Section 8.7. When H_0 is true, that is, when $p = .5$, then

$$\sigma_{\hat{p}} = \sqrt{\frac{pq}{n}} = \sqrt{\frac{(.5)(.5)}{n}} = \sqrt{\frac{.25}{n}}$$

and

$$z = \frac{\hat{p} - p}{\sqrt{\frac{pq}{n}}} = \frac{\frac{x}{n} - .5}{\sqrt{\frac{.25}{n}}} = \frac{x - .5n}{\sqrt{.25n}}$$

The test is summarized in the following display.

Sign Test for Large Samples, $n \geq 25$

1 Null Hypothesis: $H_0 : p = .5$ (i.e., neither treatment A nor treatment B is preferred over the other)

2   Alternative Hypothesis:

Two-Tailed Test   **One-Tailed Test**

$H_a : p \neq .5$ $H_a : p > 5$

(or $H_a : p < .5$)

3   Test Statistic: $z = \dfrac{x - .5n}{\sqrt{.25n}}$

4   Rejection Region:

Two-Tailed Test   **One-Tailed Test**

Reject H_0 if $z > z_{\alpha/2}$ or if   Reject H_0 if $z > z_\alpha$

$z < -z_{\alpha/2}$ (for $H_a : p < .5$, reject H_0 if $z < -z_\alpha$)

Assumptions: The observations x_A and x_B are randomly and independently selected in matched pairs. Tied observations (i.e., when $x_A = x_B$) are eliminated from the data and the number of pairs n is reduced accordingly.

Note: To find a z value, use the same procedure used in Chapter 7. The z values are given in Table 3 of Appendix II.

Exercises

Basic Techniques

16.1   Suppose that you wish to use the sign test to test $H_0 : p = .5$ against $H_a : p > .5$ for a paired-difference experiment with $n = 25$ pairs.

    a   State the practical situation that would dictate the alternative hypothesis given.

    b   Use Table 1 in Appendix II to find values of α ($\alpha < .15$) available for the test.

16.2   Repeat the instructions of Exercise 16.1 for $H_a : p \neq .5$.

16.3   Suppose that you wish to use the sign test to test $H_0 : p = .5$ against $H_a : p > .5$ for a paired-difference experiment with $n = 15$ pairs.

    a   State the practical situation that would dictate the alternative hypothesis given.

    b   Use Table 1 in Appendix II to find values of α ($\alpha < .15$) available for the test.

16.4   Repeat the instructions of Exercise 16.3 for $H_a : p \neq .5$.

16.5   Give the values of x in the rejection region for the sign test in Exercise 16.2 for $\alpha = .05$. Then find the values of x in the rejection region for the sign test for large samples based on the z statistic. Compare the two rejection regions. Do they agree?

16.6   Refer to Exercise 16.1. Suppose that you were to conduct the test for $\alpha = .05$ and observed $x = 17$.

    a   Do the data provide sufficient evidence to indicate that the distribution of x_A is shifted to the right of the distribution of x_B? Explain.

    b   Find the p-value for the test and interpret its value.

16.7   The following data were collected in a paired-difference experiment. Do the data provide sufficient evidence to indicate that $P(x_A > x_B)$ is larger than .5?

x_A	x_B	x_A	x_B
3.4	3.2	4.3	4.0
4.1	4.1	4.3	4.2
3.9	3.8	2.9	2.5
3.5	3.6	4.0	3.8
3.8	3.4	3.5	3.4

a State H_0.

b State H_a.

c Test H_0 for $\alpha = .05$, using a sign test.

d Find the p-value for the test.

16.8 Repeat the instructions of Exercise 16.7 if you wish to detect whether $P(x_A > x_B)$ is either less than .5 or greater than .5.

16.9 A paired-difference experiment was conducted, using $n = 30$ pairs. The number of times that x_A exceeded x_B was $x = 10$. Do the data provide sufficient evidence to indicate that $p = P(x_A > x_B)$ differs from .5? Test using $\alpha = .05$.

a State the null and alternative hypotheses.

b Give the test statistic and rejection region for the large-sample z test.

c Conduct the test and state the practical conclusions to be derived from the test results. Use $\alpha = .05$.

16.10 A paired-difference experiment was conducted, using $n = 35$ pairs. The number of times that x_A exceeded x_B was $x = 18$. Do the data provide sufficient evidence to indicate that the distribution for x_A is shifted to the right of the distribution for x_B?

a State the null and alternative hypotheses.

b Give the test statistic and rejection region for the test.

c Conduct the test and state the practical conclusions to be derived from the test results. Use $\alpha = .05$.

d Give the p-value for the test in part (c), and interpret its value.

Applications

16.11 In Exercise 8.43, we compared property evaluations of two tax assessors, A and B. Their assessments (in thousands) for eight properties are shown in the accompanying table.

	Assessor	
Property	A	B
1	36.3	35.1
2	48.4	46.8
3	40.2	37.3
4	54.7	50.6
5	28.7	29.1
6	42.8	41.0
7	36.1	35.3
8	39.0	39.1

a   Use the sign test to determine whether the data provide sufficient evidence to indicate that one of the assessors tends to be consistently more conservative than the other—that is, $P(x_A > x_B) \neq 1/2$. Test by using a value of α near .05. Find the p-value for the test and interpret its value.

b   Exercise 8.43 uses the t statistic to test the null hypothesis that there is no difference in the mean level of property assessments between assessors A and B. Check the answer for Exercise 8.43 and compare it with your answer to part (a). Do the test results agree? Explain why the answers are (or are not) consistent.

16.12   Two gourmets, A and B, rated 20 meals on a scale of 1 to 10. The data are shown in the accompanying table. Do the data provide sufficient evidence to indicate that one of the gourmets tends to give higher ratings than the other? Test using the sign test with a value of α near .05.

Meal	A	B	Meal	A	B
1	6	8	11	6	9
2	4	5	12	8	5
3	7	4	13	4	2
4	8	7	14	3	3
5	2	3	15	6	8
6	7	4	16	9	10
7	9	9	17	9	8
8	7	8	18	4	6
9	2	5	19	4	3
10	4	3	20	5	5

16.13   In preparing to add new inventory, the owner of a women's clothing store needs to know whether women have a preference for one or the other of two different dress styles. Twenty women customers were randomly selected and asked to choose the style they preferred. Six chose style A and 14 chose style B.

a   Do these data provide sufficient evidence to indicate a difference in preference for the two dress styles? Test using α near .05.

b   If the store owner wants to estimate the proportion of all customers who prefer dress style A correct to within .10 with probability equal to .95, approximately how many customers would the store owner have to sample?

16.14   The number of defective fuses proceeding from each of two production lines, A and B, was recorded daily for a period of ten days, with the results shown in the accompanying table. Assume that both production lines produced the same daily output. Compare the number of defectives produced by A and B each day, and let x equal the number of days when B exceeded A. Do the data provide sufficient evidence to indicate that production line B tends to produce more defectives than A? State the null and alternative hypotheses. Use x as a test statistic.

Day	Line A	Line B	Day	Line A	Line B
1	172	201	6	142	170
2	165	179	7	190	182
3	206	159	8	169	179
4	184	192	9	161	169
5	174	177	10	200	210

16.15   Establishing the value of an art object is subjective in nature and difficult at best. To determine whether two art appraisers tend to give different levels of appraisals, two appraisers, A and B, were asked to appraise each of seven art objects. If appraiser A tends to give smaller appraisals (or larger appraisals) than appraiser B, the probability p that the appraisal of A will exceed the appraisal for B will be less than 1/2 (or greater than 1/2). Under the assumption that there is no difference in the appraisal techniques of the two appraisers, $p = 1/2$. Let x, the number of art objects for which A's appraisal exceeds B's appraisal, be a test statistic.

a Find an appropriate rejection region to test the null hypothesis $p = 1/2$ for $\alpha \approx .10$.

b If B tends to give conservative evaluations, so that p actually is equal to .9, calculate β for the test.

c If A appraises five of the seven art objects to be worth more than B appraises them for, what do you conclude?

 16.16 A manufacturer of a liquid cleaner has developed a new spray intended to reduce the accumulation of dirt on windows. Two adjacent windows in each of 50 houses were employed in the experiment, and one window in each pair was randomly selected and treated with the dirt retardant. After the windows were exposed to the elements for two months, each homeowner was asked to identify the cleaner window panel. (Ties were permitted.) Two homeowners found no difference in the appearance of the panels. Thirty-eight of the remaining 48 selected the treated panels as the cleaner of the two. Do these data provide sufficient evidence to indicate that the treated glass panels were preferred over the untreated panels? Test using $\alpha = .01$.

16.3 The Mann-Whitney U Test: Independent Random Samples

A statistical test for comparing the relative locations of two populations A and B (a test for shift) based on independent random samples was proposed by Wilcoxon in 1945 and in slightly different form by Mann and Whitney in 1947. The test procedure involves **ranking** the n_1 and n_2 observations, randomly and independently selected from populations A and B, from the smallest (rank $= 1$) to the largest (rank $= n_1 + n_2$). Rankings for tied observations are averaged, and the average rank is assigned to each of the tied observations. Then the sum T_A of the ranks for sample A and the sum T_B of the ranks for sample B are calculated. The **rank sums** are used in constructing the test statistic.

The logic is that if distribution A is shifted to the right of distribution B, then the rank sum T_A should exceed T_B. The **Mann-Whitney U test statistic** will use one of the two quantities U_A or U_B shown in the following display.

Formulas for the Mann-Whitney U Statistic

$$U_A = n_1 n_2 + \frac{n_1(n_1 + 1)}{2} - T_A$$

$$U_B = n_1 n_2 + \frac{n_2(n_2 + 1)}{2} - T_B$$

where

$$n_1 = \text{Number of observations in sample } A$$
$$n_2 = \text{Number of observations in sample } B$$
$$U_A + U_B = n_1 n_2$$
$$T_A \text{ and } T_B = \text{Rank sums for samples } A \text{ and } B, \text{ respectively}$$

As you can see from the formulas for U_A and U_B, U_A will be small when T_A is large, a situation that likely will occur when the population distribution of the A measurements is shifted to the right of the population distribution for the B measurements. Consequently, to conduct a one-tailed test to detect a shift in the A distribution to the right of the B distribution, you will reject the null hypothesis of "no difference in the population distributions" if U_A is less than some specified value U_0. That is, you will reject H_0 for small values of U_A. Similarly, to conduct a one-tailed test to detect a shift of the B distribution to the right of the A distribution, you will reject H_0 if U_B is less than some specified value, say U_0. Consequently, the rejection region for the Mann-Whitney U test will appear as shown in Figure 16.2.

FIGURE **16.2**
Rejection region for a
Mann-Whitney U test

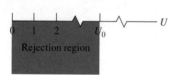

Table 14 in Appendix II gives the probability that an observed value of U will be less than some specified value, say U_0. This is the value of α for a one-tailed test. To conduct a two-tailed test—that is, to detect a shift in the population distributions for the A and B measurements in either direction—we will agree to always use U, the smaller of U_A or U_B, as the test statistic and reject H_0 or $U < U_0$ (see Figure 16.2). The value of α for the two-tailed test will be double the tabulated value given in Table 14 in Appendix II.

To see how to locate the rejection region for the Mann-Whitney U test, suppose that $n_1 = 4$ and $n_2 = 5$. Then we consult the third table in Table 14 in Appendix II, the one corresponding to $n_2 = 5$. The first few lines of Table 14, $n_2 = 5$, are shown here in Table 16.3. Note that the table is constructed on the assumption that $n_1 \leq n_2$.

TABLE **16.3**
An abbreviated version of Table
14 in Appendix II:
$P(U \leq U_0)$ for $n_2 = 5$

U_0	\multicolumn{5}{c}{n_1}
	1	2	3	4	5
0	.1667	.0476	.0179	.0079	.0040
1	.3333	.0952	.0357	.0159	.0079
2	.5000	.1905	.0714	.0317	.0159
3		.2857	.1250	.0556	.0278
4		.4286	.1964	.0952	.0476
5		.5714	.2857	.1429	.0754
⋮		⋮	⋮	⋮	⋮

Across the top of Table 16.3, we see values of n_1. Values of U_0 are shown down the left side of the table. The entries give the probability that U will assume a small value, namely, the probability that $U \leq U_0$. Since for our example $n_1 = 4$, we will move across the top of the table to $n_1 = 4$. Move to the third row of the table corresponding to $U_0 = 2$ for $n_1 = 4$. Then we see that the probability that U will be less than or equal to 2 is .0317. Similarly, moving across the row for $U_0 = 3$, we see that the probability that U is less than or equal to 3 is .0556. (This value is boxed in Table 16.3.) So, if we want to conduct a one-tailed Mann-Whitney U test with $n_1 = 4$ and $n_2 = 5$ and would like α to be near .05, we would reject the null hypothesis of equality of population relative frequency distributions when $U \leq 3$. The probability of a type I error for the test would be $\alpha = .0556$. If we use this same rejection region for a two-tailed test, that is, $U \leq 3$, α will be double the tabulated value, or $\alpha = 2(.0556) = .1112$.

Table 14 in Appendix II can also be used to find the observed significance level for a test. For example, if $n_1 = 5$, $n_2 = 5$, and $U = 4$, then from Table 16.3 the p-value for a one-tailed test is

$$P(U \leq 4) = .0476$$

If the test is two-tailed, the p-value is

$$2(.0476) \quad \text{or} \quad .0952$$

The details of the test are summarized in the display on pages 712–713.

EXAMPLE   16.2   An experiment was conducted to compare the strength of two types of kraft papers, one a standard kraft paper of a specified weight and the other the same standard kraft paper treated with a chemical substance. Ten pieces of each type of paper, randomly selected from production, produced the strength measurements shown in Table 16.4. Test the hypothesis of "no difference in the distributions of strengths for the two types of paper" against the alternative hypothesis that the treated paper tends to be of greater strength (i.e., its distribution of strength measurements is shifted to the right of the corresponding distribution for the untreated paper).

T A B L E **16.4**
Paper strength measurements
for Example 16.2

Standard, A	Treated, B
1.21 (2)	1.49 (15)
1.43 (12)	1.37 (7.5)
1.35 (6)	1.67 (20)
1.51 (17)	1.50 (16)
1.39 (9)	1.31 (5)
1.17 (1)	1.29 (3.5)
1.48 (14)	1.52 (18)
1.42 (11)	1.37 (7.5)
1.29 (3.5)	1.44 (13)
1.40 (10)	1.53 (19)

Rank sums	$T_A = 85.5$	$T_B = 124.5$

Solution   The ranks are shown in parentheses alongside the $n_1 + n_2 = 10 + 10 = 20$ strength measurements; the rank sums, T_A and T_B, are shown below the columns. Since we want to detect a shift in the distribution of the B measurements to the right of the distribution for the A measurements, we will reject the null hypothesis of "no difference in population strength distributions" when T_B is excessively large. Because this situation will occur when U_B is small, we will conduct a one-tailed statistical test and reject the null hypothesis when $U_B \leq U_0$.

Suppose that we choose a value of α near .05. Then we can find U_0 by consulting the portion of Table 14 in Appendix II corresponding to $n_2 = 10$. The probability $P(U \leq U_0)$ nearest .05 is .0526 and corresponds to $U_0 = 28$. Hence, we will reject H_0 if $U_B \leq 28$.

Calculating U_B, we have

$$U_B = n_1 n_2 + \frac{n_2(n_2 + 1)}{2} - T_B$$
$$= (10)(10) + \frac{(10)(11)}{2} - 124.5$$
$$= 30.5$$

As you can see, U_B is not less than $U_0 = 28$. Therefore, we cannot reject the null hypothesis. At the $\alpha = .05$ level of significance, there is not sufficient evidence to indicate that the treated kraft paper is stronger than the standard.   ∎

The Mann-Whitney U Test

1   Null Hypothesis: H_0 : the population relative frequency distributions for A and B are identical

2   Alternative Hypothesis: H_a : the two population relative frequency distributions are shifted with respect to their relative locations (a two-tailed test). Or H_a : the population relative frequency distribution for A is shifted to the right of the relative frequency distribution for population B (a one-tailed test).[†]

3   Test Statistic: For a two-tailed test, use U, the smaller of

$$U_A = n_1 n_2 + \frac{n_1(n_1 + 1)}{2} - T_A$$

[†]For the sake of convenience, we will describe the one-tailed test as one designed to detect a shift in the distribution of the A measurements to the right of the distribution of the B measurements. To detect a shift in the B distribution to the right of the A distribution, just interchange the letters A and B in the discussion.

and

$$U_B = n_1 n_2 + \frac{n_2(n_2 + 1)}{2} - T_B$$

where T_A and T_B are the rank sums for samples A and B, respectively. For a one-tailed test, use U_A.

4 Rejection Region:

Two-Tailed Test	**One-Tailed Test**
For a given value of α, reject H_0 if $U \le U_0$, where $P(U \le U_0) = \alpha/2$. [*Note:* Observe that U_0 is the value such that $P(U \le U_0)$ is equal to half of α.]	For a given value of α, reject H_0 if $U_A \le U_0$, where $P(U_A \le U_0) = \alpha$.

Assumptions: Samples have been randomly and independently selected from their respective populations. Ties in the observations can be handled by averaging the ranks that would have been assigned to the tied observations and assigning this average to each. Thus if three observations are tied and are due to receive ranks 3, 4, 5, we assign the rank of 4 to all three.

The Mann-Whitney U test can be implemented by using the MANN-WHITNEY command in the Minitab package, after storing the values from the first and second samples in two separate columns. The Minitab printout for the data of Example 16.2 is shown in Table 16.5. The value $W = 85.5$ in line 6 is the sum of the ranks for the data stored in column 1—and, in general, for the column number given first in the program command. The value of W can be used to calculate U_A and U_B if desired. The significance level of the test is p-value $= .0755$ from line 7, and the printout advises that we cannot reject H_0 at $\alpha = .05$.

TABLE 16.5
Minitab output for Example 16.2

```
MTB > MANN-WHITNEY -1 C1 C2

Mann-Whitney Confidence Interval and Test

C1           N =   10      Median =        1.3950
C2           N =   10      Median =        1.4650
Point estimate for ETA1-ETA2 is      -0.0800
95.5 Percent C.I. for ETA1-ETA2 is (-0.2000,0.0399)
W = 85.5
Test of ETA1 = ETA2   vs.   ETA1 < ETA2 is significant at 0.0755
The test is significant at 0.0753 (adjusted for ties)

Cannot reject at alpha = 0.05
```

A simplified large-sample test ($n_1 \ge 10$ and $n_2 \ge 10$) can be obtained by using the familiar z statistic of Chapter 8. When the population distributions are identical, it can be shown that the U statistic has expected value and variance

$$E(U) = \frac{n_1 n_2}{2} \quad \text{and} \quad V(U) = \frac{n_1 n_2 (n_1 + n_2 + 1)}{12}$$

and the distribution of

$$z = \frac{U - E(U)}{\sigma_U}$$

tends to normality with mean zero and variance equal to 1 as n_1 and n_2 become large. This approximation will be adequate when n_1 and n_2 are both greater than or equal to 10. Thus, for a two-tailed test with $\alpha = .05$, we will reject the null hypothesis if $|z| > 1.96$.

The next display gives the details of the Mann-Whitney U test for large samples.

The Mann-Whitney U Test for Large Samples, $n_1 \geq 10$ and $n_2 \geq 10$

1 Null Hypothesis: H_0: the population relative frequency distributions for A and B are identical

2 Alternative Hypothesis: H_a: the two population relative frequency distributions are not identical (a two-tailed test). Or H_a: the population relative frequency distribution for A is shifted to the right (or left) of the relative frequency distribution of population B (a one-tailed test).

3 Test Statistic: $z = \dfrac{U - (n_1 n_2/2)}{\sqrt{n_1 n_2 (n_1 + n_2 + 1)/12}}$. Suppose $U = U_A$.

4 Rejection Region: Reject H_0 if $z > z_{\alpha/2}$ or $z < -z_{\alpha/2}$ for a two-tailed test. For a one-tailed test, place all of α in one tail of the z distribution. To detect a shift in the distribution of the A observations to the right of the distribution of the B observations, let $U = U_A$ and reject H_0 when $z < -z_\alpha$. To detect a shift in the opposite direction, let $U = U_A$ and reject H_0 when $z > z_\alpha$. Tabulated values of z are given in Table 3 in Appendix II.

Observe that the z statistic will reach the same conclusion as the exact U test for Example 16.2. Thus, with $U = U_B$,

$$z = \frac{30.5 - [(10)(10)/2]}{\sqrt{[(10)(10)(10 + 10 + 1)]/12}} = \frac{30.5 - 50}{\sqrt{2100/12}} = -\frac{19.5}{\sqrt{175}}$$

$$= -\frac{19.5}{13.23} = -1.47$$

For a one-tailed test with $\alpha = .05$ located in the lower tail of the z distribution, we will reject the null hypothesis if $z < -1.645$. You can see that $z = -1.47$ does not fall in the rejection region and that this test reaches the same conclusion as the exact U test of Example 16.2.

Exercises

Basic Techniques

16.17   Suppose that you wish to detect a shift in distribution A to the right of distribution B based on sample sizes $n_1 = 6$ and $n_2 = 8$.

a   Should you use U_A or U_B for your test statistic?

b   Give the rejection region for the test if you wish α to be close to but less than .10.

c   Give the value of α for the test.

16.18   Suppose that the alternative hypothesis for Exercise 16.17 is that distribution A is shifted either to the left or to the right of distribution B.

a   Should you use U_A or U_B for your test statistic?

b   Give the rejection region for the test if you wish α to be close to but less than .10.

c   Give the value of α for the test.

16.19   Suppose that you wish to detect a shift in distribution A to the left of distribution B based on sample sizes $n_1 = 4$ and $n_2 = 5$.

a   Should you use U_A or U_B for your test statistic?

b   Give the rejection region for the test if you wish α to be close to but less than .10.

c   Give the value of α for the test.

16.20   Suppose that the alternative hypothesis for Exercise 16.19 is that distribution A is shifted either to the left or to the right of distribution B.

a   Should you use U_A or U_B for your test statistic?

b   Give the rejection region for the test if you wish α to be close to but less than .10.

c   Give the value of α for the test.

16.21   Suppose that you wish to detect a shift in distribution A to the right of distribution B based on sample sizes $n_1 = 12$ and $n_2 = 14$. If $T_A = 193$, what do you conclude? Use $\alpha = .05$.

16.22   Suppose that you wish to detect a difference in the location of two population distributions based on samples of $n_1 = 15$ and $n_2 = 15$ observations, respectively, selected from the populations. If $T_B = 251$, what do you conclude? Use $\alpha = .10$.

Applications

16.23   Is consumer reaction to a product marketing display different from one market to another? A company that manufactures kitchen appliances constructed the same product display in large department stores in each of two different markets, A and B. Ten persons viewing the display were randomly selected at each location and asked to rate the display on a scale of 1 to 20. The twenty ratings are shown below.

Market A	15	11	20	14	9	12	5	17	13	18
Market B	17	6	15	10	6	8	10	16	8	7

Do the data provide sufficient evidence to indicate that the levels of ratings differ between the two markets?

16.24   The life, in months of service, before a failure of the color television circuit board in eight television sets manufactured by firm A and ten sets manufactured by firm B is as follows:

Firm A	32	25	40	31	35	29	37	39		
Firm B	41	39	36	47	45	34	48	44	43	33

Use the U test to analyze the data, and test to see whether the life, in months of service, before failure of the circuit board is the same for the boards manufactured by each firm. Use $\alpha = .10$.

16.25   In Exercise 7.101, the times (in seconds) to load Ami Pro 2.0 on an IBM PS/2 Model 90 486DX/33 personal computer using the Standard Windows and Enhanced Windows programs were given. The data are reproduced in the accompanying table.

Standard	Enhanced
1.56	1.59
1.41	1.68
1.48	1.17
1.37	0.94
1.39	1.56
1.20	0.96
1.38	1.09
1.54	1.26
1.41	1.23
1.16	1.30

a   Use the Mann-Whitney U test to determine whether there really is a difference in the average time to load Ami Pro 2.0 using Standard Windows and Enhanced Windows.

b   Compare the results of the Mann-Whitney U test to the two-sample t test in Exercise 8.34.

16.26   A manufacturer of lawn mowers buys a particular bolt in boxes of 10,000 bolts each from two different suppliers. In order to compare the percentages of defectives shipped by the suppliers, the manufacturer sampled 100 bolts from each of ten cartons for each supplier. The number x of defective bolts in each sample of 100 is shown in the table. Do the data provide sufficient evidence to indicate that the proportions of defectives shipped by the two suppliers differ?

Supplier A	7	12	29	8	15	11	17	15	22	20
Supplier B	19	24	14	17	25	21	13	23	18	18

a   Explain why the data violate the assumptions required for a Student's t test.

b   Test for a difference, using the Mann-Whitney U test with $\alpha = .05$.

16.27   A real estate brokerage firm compared customer satisfaction with recent sales for two different brokers. The brokerage contacted the real estate purchasers for each of the broker's last month's sales, eight for broker A and seven for broker B, and asked them to rate their satisfaction in dealing with the broker on a scale of 1 (low) to 20 (high). The ratings are shown in the table. Do the data provide sufficient evidence to indicate that the level-of-satisfaction ratings differ from one broker to another? Test using $\alpha = .10$.

Broker A	13	18	17	20	16	20	14	151
Broker B	16	10	12	15	19	17	11	

16.4 The Wilcoxon Signed Rank Test for a Paired Experiment

Wilcoxon proposed a nonparametric test, similar to the test of Section 16.3, to analyze the paired-difference experiment of Section 8.6. The test uses the paired differences of the two treatments A and B to test the null hypothesis that $P(x_A > x_B) = 1/2$ versus either a one- or a two-sided alternative hypothesis.

To carry out the **Wilcoxon signed rank test,** calculate the differences $(x_A - x_B)$ for each of the n pairs. Differences equal to zero are eliminated, and the number of pairs n is reduced accordingly. Rank the *absolute values* of the differences, assigning a 1 to the smallest, a 2 to the second smallest, and so on. Then calculate the rank sum for the negative differences, and also calculate the rank sum for the positive differences. For a two-tailed test we use the smaller of these two quantities, T, as a test statistic to test the null hypothesis that the two population relative frequency histograms are identical. The smaller the value of T, the greater will be the weight of evidence favoring rejection of the null hypothesis. Therefore, we will reject the null hypothesis if T is less than or equal to some value, say T_0.

To detect the one-sided alternative—that the distribution of the A observations is shifted to the right of the B observations—use the rank sum T^- of the negative differences and reject the null hypothesis for small values of T^-, say $T^- \le T_0$. If you want to detect a shift of the distribution of B observations to the right of the distribution of A observations, use the rank sum T^+ of the positive differences as a test statistic and reject for small values of T^+, say $T^+ \le T_0$.

The probability that T is less than or equal to some value T_0 has been calculated for a combination of sample sizes and values of T_0. These probabilities, given in Table 15 of Appendix II, can be used to find the rejection region for the T test.

An abbreviated version of Table 15 of Appendix II is shown here in Table 16.6. Across the top of the table, you see the number of differences (the number of pairs), n. Values of α for a one-tailed test appear in the first column of the table. The second column gives values of α for a two-tailed test. Table entries are the critical values of T. You will recall that the critical value of a test statistic is the value that locates the boundary of the rejection region.

TABLE 16.6
An abbreviated version of Table 15 in Appendix II: Critical values of T

One-Sided	Two-Sided	$n=5$	$n=6$	$n=7$	$n=8$	$n=9$	$n=10$
$\alpha=.05$	$\alpha=.10$	1	2	4	6	8	11
$\alpha=.025$	$\alpha=.05$		1	2	4	6	8
$\alpha=.01$	$\alpha=.02$			0	2	3	5
$\alpha=.005$	$\alpha=.01$				0	2	3

One-Sided	Two-Sided	$n=11$	$n=12$	$n=13$	$n=14$	$n=15$	$n=16$
$\alpha=.05$	$\alpha=.10$	14	17	21	26	30	36
$\alpha=.025$	$\alpha=.05$	11	14	17	21	25	30
$\alpha=.01$	$\alpha=.02$	7	10	13	16	20	24
$\alpha=.005$	$\alpha=.01$	5	7	10	13	16	19

For example, suppose that you have $n = 7$ pairs and you are conducting a two-tailed test of the null hypothesis that the two population relative frequency distributions are identical. Checking the $n = 7$ column of Table 15 and using the second row (corresponding to $\alpha = .05$ for a two-tailed test), you see the entry 2 (boxed in Table 16.6). This is T_0, the critical value of T. As noted earlier, the smaller the value of T, the greater will be the evidence to reject the null hypothesis. Therefore, you will reject the null hypothesis for all values of T less than or equal to 2. The rejection region for the Wilcoxon signed rank test for a paired experiment is always of this form: Reject H_0 if $T \leq T_0$, where T_0 is the critical value of T. The rejection region is shown symbolically in Figure 16.3.

F I G U R E **16.3**
Rejection region for the
Wilcoxon signed rank test for a
paired experiment (reject H_0 if
$T \leq T_0$)

The details of the Wilcoxon test are summarized in the following display.

Wilcoxon Signed Rank Test for a Paired Experiment

1 Null Hypothesis: H_0: the two population relative frequency distributions are identical

2 Alternative Hypothesis: H_a: the two population relative frequency distributions differ in location (a two-tailed test). Or H_a: the population relative frequency distribution for A is shifted to the right[†] of the relative frequency distribution for population B (a one-tailed test).

3 Test Statistic:

Two-Tailed Test	**One-Tailed Test**
Use T, the smaller of the rank sum for positive differences and the rank sum for negative differences.	Use the rank sum T^- of the negative differences to detect the alternative hypothesis described above.

4 Rejection region:

Two-Tailed Test	**One-Tailed Test**
Reject H_0 if $T \leq T_0$, where T_0 is the critical value given in Table 15 of Appendix II.	Use the rank sum T^- of the negative differences. Reject H_0 if $T^- \leq T_0$ to detect the alternative hypothesis described above.

[†]To detect a shift of the distribution of B observations to the right of the distribution of A observations, use the rank sum T^+ of the positive differences as the test statistic and reject H_0 if $T^+ \leq T_0$.

EXAMPLE 16.3 The manager of a money fund compared the percentage monthly returns for two different types, A and B, of short-term loans. The data, collected over a six-month period, are shown in Table 16.7. Do the data provide sufficient evidence to indicate that the distributions of monthly returns for the two types of loans differ in location?

TABLE 16.7
Percentage monthly returns for
Example 16.3

		Month					
Statistic		1	2	3	4	5	6
x_A		10.5	7.2	6.8	11.1	10.1	11.4
x_B		9.9	9.0	8.2	12.2	10.5	13.3
Difference $(x_A - x_B)$		.6	−1.8	−1.4	−1.1	−.4	−1.9
Rank		2	5	4	3	1	6

Solution The percentage monthly returns, x_A and x_B, their differences, and the ranks of their absolute differences for the two types of short-term loans are shown in Table 16.7.

As with our other nonparametric tests, the null hypothesis to be tested is that the two population frequency distributions of percentage monthly returns are identical. The alternative hypothesis, which implies a two-tailed test, is that one of the distributions is shifted to the right of the other.

Because the amount of data is small, we will conduct our test using $\alpha = .10$. From Table 15 in Appendix II, the critical value of T for a two-tailed test, $\alpha = .10$, is $T_0 = 2$. Thus we will reject H_0 if $T \leq 2$ (see Figure 16.4).

FIGURE 16.4
Rejection region for Example
16.3, $\alpha = .10$

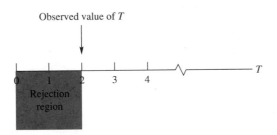

Now let us find the observed value of the test statistic T. Checking the ranks of the absolute differences, you can see that there is only one positive difference, and it has a rank of 2. Since we have agreed to always use the smaller rank sum as the test statistic, the observed value of the test statistic is $T = 2$. Since $T = 2$ falls in the rejection region, we reject H_0 and conclude that the two population frequency distributions of percentage monthly returns differ in respect to location. ▪

The Minitab command WTEST can be used to implement the signed rank test for a paired experiment. Data for the two samples are stored in two separate columns, say C1 and C2, and the differences, d_i, are generated using the command LET C3 = C1 − C2. The command WTEST 0 C3 specifies that the Wilcoxon signed rank test will be used to test for a zero median in column 3. The Minitab output for Example 16.3 is shown in Table 16.8. The value $T = 2$ is labeled WILCOXON STATISTIC and the associated P-VALUE is .093, which agrees with previous calculations.

T A B L E **16.8**
Minitab printout for the data in
Example 16.3

```
MTB > WTEST C3

TEST OF MEDIAN = 0.000000 VERSUS MEDIAN N.E. 0.000000

             N FOR   WILCOXON               ESTIMATED
          N  TEST    STATISTIC  P-VALUE     MEDIAN
C3        6  6          2.0       0.093      -1.100
MTB >
```

Although Table 15 in Appendix II is applicable for values of n (number of data pairs) as large as $n = 50$, it is worth noting that T^+, like the Mann-Whitney U statistic, will be approximately normally distributed when the null hypothesis is true and n is large (say 25 or more). This enables us to construct a large-sample z test, where

$$E(T^+) = \frac{n(n + 1)}{4}$$
$$V(T^+) = \frac{n(n + 1)(2n + 1)}{24}$$

Then the z statistic,

$$z = \frac{T^+ - E(T^+)}{\sigma_{T^+}} = \frac{T^+ - [n(n + 1)/4]}{\sqrt{[n(n + 1)(2n + 1)]/24}}$$

can be used as a test statistic. Thus for a two-tailed test and $\alpha = .05$, we will reject the hypothesis of "identical population distributions" when $|z| > 1.96$. The details of the large-sample test are given in the following display.

A Large-Sample Wilcoxon Signed Rank Test for a Paired Experiment, $n \geq 25$

1 Null Hypothesis: H_0: the population relative frequency distributions for A and B are identical

2 Alternative Hypothesis: H_a: the two population relative frequency distributions differ in location (a two-tailed test). Or H_a: the population relative frequency distribution for A is shifted to the right (or left) of the relative frequency distribution for population B (a one-tailed test).

3 Test Statistic: $z = \dfrac{T^+ - [n(n + 1)/4]}{\sqrt{[n(n + 1)(2n + 1)]/24}}.$

4 Rejection Region: Reject H_0 if $z > z_{\alpha/2}$ or $z < -z_{\alpha/2}$ for a two-tailed test. For a one-tailed test, place all of α in one tail of the z distribution. To detect a shift in the distribution of the A observations to the right of the distribution of the B observations, reject H_0 when $z > z_\alpha$. To detect a shift in the opposite direction, reject H_0 if $z < -z_\alpha$. Tabulated values of z are given in Table 3 of Appendix II.

Exercises

Basic Techniques

16.28 Suppose that you wish to detect a difference in location between two population distributions based on a paired-difference experiment consisting of $n = 30$ pairs.

a Give the null and alternative hypotheses for the Wilcoxon signed rank test.

b Give the test statistic.

c Give the rejection region for the test for $\alpha = .05$.

d If $T^+ = 249$, what are your conclusions? [*Note:* $T^+ + T^- = n(n + 1)/2$.]

16.29 Refer to Exercise 16.28. Suppose that you wish to detect only a shift in distribution A to the right of distribution B.

a Give the null and alternative hypotheses for the Wilcoxon signed rank test.

b Give the test statistic.

c Give the rejection region for the test for $\alpha = .05$.

d If $T^+ = 249$, what are your conclusions? [*Note:* $T^+ + T^- = n(n + 1)/2$.]

16.30 Refer to Exercise 16.28. Conduct the test using the large-sample z test. Compare your results with the nonparametric test results in Exercise 16.28(d).

16.31 Refer to Exercise 16.29. Conduct the test using the large-sample z test. Compare your results with the nonparametric test results in Exercise 16.29(d).

Applications

16.32 In Exercise 16.11, we used the sign test to determine whether the data provided sufficient evidence to indicate a shift in the distributions of property assessments for tax assessors A and B.

a Use the Wilcoxon signed rank test for a paired experiment to test the null hypothesis that there is no difference in the distributions of property assessments between tax assessors A and B. Test using a value of α near .05.

b Compare the conclusion of the test in part (a) with the conclusions derived from the t test in Exercise 8.43 and the sign test in Exercise 16.11. Explain why these test conclusions are (or are not) consistent.

16.33 The number of machine breakdowns per month was recorded for nine months on two identical machines, A and B, used to make wire rope. The data are shown in the accompanying table.

Month	Machine A	Machine B
1	3	7
2	14	12
3	7	9
4	10	15
5	9	12
6	6	6
7	13	12
8	6	5
9	7	13

a Do the data provide sufficient evidence to indicate a difference in the monthly breakdown rates for the two machines? Test using a value of α near .05.

b Can you think of a reason why the breakdown rates for the two machines might vary from month to month?

16.34 Two investment counselors were asked to rate each of ten investments on a scale of 1 to 100 based on each counselor's evaluation of the projected annual return and the intended risks of the investment. Do the data, shown in the accompanying table, provide sufficient evidence to indicate that one of the counselors tends to rate investments higher than the other? Test using $\alpha = .10$.

Investment	Counselor 1, x_1	Counselor 2, x_2
1	83	75
2	40	48
3	86	100
4	70	67
5	30	20
6	84	80
7	67	61
8	50	44
9	84	86
10	56	50

16.35 The personnel director at a manufacturing plant compared the number of accidents per month for two different production lines over a 12-month period. The data are shown in the accompanying table.

Production Line	Jan.	Feb.	Mar.	Apr.	May	June	July	Aug.	Sept.	Oct.	Nov.	Dec.
1	2	1	0	2	2	3	1	2	0	1	3	2
2	0	2	1	1	0	2	1	0	0	1	0	2

a Do the data provide sufficient evidence to indicate differences in the monthly rates of accidents between the two production lines? Use the Wilcoxon signed rank test with $\alpha = .05$.

b Which assumption required for the Student's t test do you think these data might violate? [*Note:* See Section 4.3 to learn why the data violate the assumptions required for a Student's t test.]

16.36 A CPA firm compared the auditing ability of two auditors by having each audit the same eight financial statements. The number of errors per audit found by each of the two auditors, A and B, for each of the eight statements is shown in the accompanying table.

	Auditor	
Statement	A	B
1	6	5
2	8	7
3	6	9
4	12	14
5	4	5
6	10	15
7	8	12
8	14	14

a Do the data provide sufficient evidence to indicate that the level of error detection differs for the two auditors? Test using the Wilcoxon signed rank test with $\alpha = .05$, and state the practical conclusions to be derived from the test results.

b Give the approximate p-value for the test and interpret it.

16.37 Exercise 8.44 gives the shipments (in thousands of cases) for the current year and the last year for six different exporters. The data are reproduced in the accompanying table.

	Year	
Exporter	Current	Last
1	4.81	4.27
2	5.03	5.97
3	2.38	2.61
4	4.26	3.96
5	5.14	4.86
6	3.93	3.17

a Do the data provide sufficient evidence to indicate a higher level of exports this year over the preceding year? Test using the Wilcoxon signed rank test with $\alpha = .05$.

b Find the approximate p-value for the test, and compare it with the results of the t test in Exercise 8.44. Are the test conclusions the same?

16.5 The Kruskal-Wallis H Test for Completely Randomized Designs

Just as the Mann-Whitney U test is the nonparametric alternative to the Student's t test for a comparison of population means, the **Kruskal-Wallis H test** is the nonparametric alternative to the analysis of variance F test for a completely randomized design. It is used to detect differences in location among more than two population distributions based on independent random sampling.

The procedure for conducting the Kruskal-Wallis H test is similar to that used for the Mann-Whitney U test. Suppose that we are comparing k populations based on independent random samples, n_1 from population 1, n_2 from population 2, $\ldots$, and n_k from population k, where

$$n_1 + n_2 + \cdots + n_k = n$$

The first step is to rank all n observations from the smallest (rank 1) to the largest (rank n). Tied observations are assigned a rank equal to the average of the ranks they would have received if they had been nearly equal but not tied. We then calculate the rank sums T_1, T_2, $\ldots$, T_k for the k samples and calculate the test statistic

$$H = \frac{12}{n(n+1)} \sum_{i=1}^{k} \frac{T_i^2}{n_i} - 3(n+1)$$

The greater the differences in location among the k population distributions, the larger will be the value of the H statistic. Thus we reject the null hypothesis that the k population distributions are identical for large values of H.

How large is large? It can be shown (proof omitted) that when the sample sizes are moderate to large—say, each sample size equal to 5 or larger—and when H_0 is true, the H statistic will possess approximately a chi-square distribution with $(k-1)$ degrees of freedom. Therefore, for a given value of α, we reject H_0 when the H statistic exceeds χ_α^2 (see Figure 16.5).

FIGURE 16.5

Approximate distribution of the H statistic when H_0 is true

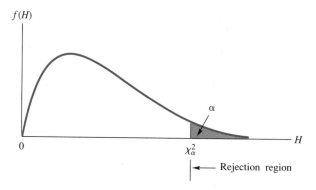

EXAMPLE 16.4 The data shown in Table 16.9 represent the achievement test scores for four different groups of sales trainees, each group taught by a different teaching technique. The objective of the experiment is to test the hypothesis of "no difference in the population distributions of achievement test scores" against the alternative that they differ in location, that is, that at least one of the distributions is shifted above the others. Conduct the test by using the Kruskal-Wallis H test with $\alpha = .05$.

Solution The first step is to rank the $n = 23$ observations from the smallest (rank 1) to the largest (rank 23). These ranks are shown in parentheses in the table. Note how the ties are handled. For example, two observations at 69 were tied for rank 5. Therefore, they were assigned the average 5.5 of the two ranks (5 and 6) that they would have occupied if they had been slightly different. The rank sums T_1, T_2, T_3, and T_4 for the four samples are shown in the bottom row of the table.

	Group 1	Group 2	Group 3	Group 4
	65 (3)	75 (9)	59 (1)	94 (23)
	87 (19)	69 (5.5)	78 (11)	89 (21)
	73 (8)	83 (17.5)	67 (4)	80 (14)
	79 (12.5)	81 (15.5)	62 (2)	88 (20)
	81 (15.5)	72 (7)	83 (17.5)	
	69 (5.5)	79 (12.5)	76 (10)	
		90 (22)		
Rank sum	$T_1 = 63.5$	$T_2 = 89$	$T_3 = 45.5$	$T_4 = 78$

TABLE 16.9
Achievement test scores for Example 16.4 (ranks in parentheses)

Substituting rank sums and sample sizes into the formula for the H statistic, we obtain

$$H = \frac{12}{n(n+1)} \sum_{i=1}^{k} \frac{T_i^2}{n_i} - 3(n+1)$$

$$= \frac{12}{23(24)} \left[\frac{(63.5)^2}{6} + \frac{(89)^2}{7} + \frac{(45.5)^2}{6} + \frac{(78)^2}{4} \right] - 3(24)$$

$$= 79.77510 - 72 = 7.77510$$

The rejection region for the H statistic for $\alpha = .05$ includes values of H satisfying $H \geq \chi_{.05}^2$, where $\chi_{.05}^2$ is based on $(k-1) = (4-1) = 3$ degrees of freedom. This value of χ^2, given in Table 5 of Appendix II, is $\chi_{.05}^2 = 7.81473$.

Since we reject H_0 for large values of H, which are those larger than $\chi_{.05}^2$ (see Figure 16.5), the observed value of the H statistic, $H = 7.77510$, does not fall in the rejection region for the test. Therefore, there is insufficient evidence to indicate differences in the distributions of achievement test scores for the four teaching techniques at the $\alpha = .05$ level of significance. ■

Table 16.10 contains the output generated using the Minitab command KRUSKAL-WALLIS for the data in Example 16.4. The data were stored in C1, and the treatment subscripts in C2.

TABLE 16.10
Minitab printout for the data in Example 16.4

```
MTB > KRUSKAL-WALLIS C1 C2

LEVEL      NOBS      MEDIAN    AVE. RANK     Z VALUE
    1         6       76.00        10.6       -0.60
    2         7       79.00        12.7        0.33
    3         6       71.50         7.6       -1.86
    4         4       88.50        19.5        2.43
OVERALL      23                    12.0

H = 7.78   d.f. = 3   p = 0.051
H = 7.79   d.f. = 3   p = 0.051 (adjusted for ties)

* NOTE  * One or more small samples
```

The details of the Kruskal-Wallis test are summarized in the following display.

The Kruskal-Wallis H Test for Comparing More Than Two Populations: Completely Randomized Design (Independent Random Samples)

1   Null Hypothesis: H_0 : the k population distributions are identical

2   Alternative Hypothesis: H_a : at least two of the k population distributions differ in location

3   Test Statistic: $H = \dfrac{12}{n(n+1)} \displaystyle\sum_{i=1}^{k} \dfrac{T_i^2}{n_i} - 3(n+1)$, where

$$n_i = \text{Sample size for population } i$$
$$T_i = \text{Rank sum for population } i$$
$$n = \text{Total number of observations}$$
$$= n_1 + n_2 + \cdots + n_k$$

4   Rejection Region: For a given α, reject H_0 when $H > \chi_\alpha^2$ with $(k-1)$ degrees of freedom.

Assumptions:

1   All sample sizes are greater than or equal to 5.

2   Ties assume the average of the ranks that they would have occupied if they had not been tied.

Exercises

Basic Techniques

16.38   Three treatments were compared by using a completely randomized design. The data are shown in the accompanying table. Do the data provide sufficient evidence to indicate a difference in location for at least two of the population distributions? Test using $\alpha = .05$.

Treatment 1	Treatment 2	Treatment 3
26	27	25
29	31	24
23	30	27
24	28	22
28	29	24
26	32	20
	30	21
	33	

16.39 Four treatments were compared by using a completely randomized design. The data are shown in the accompanying table. Do the data provide sufficient evidence to indicate a difference in location for at least two of the population distributions? Test using $\alpha = .05$.

Treatment 1	Treatment 2	Treatment 3	Treatment 4
124	147	141	117
167	121	144	128
135	136	139	102
160	114	162	119
159	129	155	128
144	117	150	123
133	109		

Applications

16.40 In Exercise 9.14, we compared the mean price of a loaf of bread (a particular brand) at four city locations using an analysis of variance. The data are reproduced in the accompanying table.

Location	Prices (dollars)			
1	1.59	1.63	1.65	1.61
2	1.58	1.61	1.64	1.63
3	1.54	1.59	1.55	1.58
4	1.69	1.70		

a Use the Kruskal-Wallis H test to determine whether the data provide sufficient evidence to indicate a difference in the level of bread prices among the four city locations. Test using $\alpha = .05$.

b Find the approximate p-value for the test.

c Consult the Minitab printout for the analysis of variance given in Exercise 9.14, and find the p-value for the analysis of variance F test used to detect differences in mean price levels among the four city locations.

d Compare the Kruskal-Wallis test results in part (a) with those of the analysis of variance F test in Exercise 9.14.

16.41 In Exercise 9.15, we used an analysis of variance F test to detect differences in the mean time to assemble a device among assemblers exposed to one of three different training programs. The data are reproduced in the accompanying table.

Training Program	Average Assembly Time (minutes)				
A	59	64	57	62	
B	52	58	54		
C	58	65	71	63	64

a Use the Kruskal-Wallis H test to determine whether the data provide sufficient evidence to indicate differences in the length of time to assemble the device among assemblers trained by the three programs. Test using $\alpha = .05$.

b Find the approximate p-value for the test.

c Use the SAS printout in Exercise 9.15 to find the p-value for the analysis of variance F test.

d Compare the results of the Kruskal-Wallis H test in part (a) with the results of the analysis of variance F test in Exercise 9.15.

16.42 A supermarket chain conducted an experiment to investigate customer response to the use of background music. Fifteen stores were selected for the experiment, and five each were assigned to one of three different types of background music: type 1, soft and slow contemporary; type 2, medium-volume, slow contemporary; and type 3, medium-volume, medium tempo. After the background music was used for one week, 100 customers per store were randomly selected and questioned to determine whether they liked the background music. The percentages that liked the music and favored its continued use are shown in the accompanying table. Do the data provide sufficient evidence to indicate differences in levels of acceptance of the three types of background music? Test using the Kruskal-Wallis H test with $\alpha = .10$.

Type 1	Type 2	Type 3
94	84	81
97	89	76
90	82	73
86	90	79
91	78	84

16.6 The Friedman F_r Test for Randomized Block Designs

The **Friedman F_r test,** proposed by Nobel prize–winning economist Milton Friedman, is a nonparametric test for comparing the distributions of measurements for k treatments laid out in b blocks, using a randomized block design. The procedure for conducting the test is very similar to that used for the Kruskal-Wallis H test. The first step in the procedure is to rank the k treatment observations within each block. Ties are treated in the usual way—that is, they receive an average of the ranks occupied by the tied observations. The rank sums $T_1, T_2, \ldots, T_k$ are then obtained, and the test statistic

$$F_r = \frac{12}{bk(k+1)} \sum_{i=1}^{k} T_i^2 - 3b(k+1)$$

is calculated.

The value of the F_r statistic will be at a minimum when the rank sums are equal, that is, $T_1 = T_2 = \cdots = T_k$, and will increase in value as the differences among the rank sums increase. When either the number k of treatments or the number b of blocks is larger than 5, the sampling distribution of F_r can be approximated by a chi-square distribution with $(k-1)$ degrees of freedom. Therefore, like the Kruskal-Wallis H test, the rejection region for the F_r test is

$$F_r > \chi_\alpha^2$$

We will illustrate the use of the test with an example.

EXAMPLE 16.5 Suppose that you want to compare consumer ratings of six different television advertisements. Each of four consumers rated each advertisement on a scale of 1 (poor) to 10 (excellent). The objective of the experiment is to determine whether differences exist in the rating levels for the six advertisements. The data are reproduced in Table

16.11 (ranks of the observations within each block are shown in parentheses). Use the Friedman F_r test to determine whether the data provide sufficient evidence to indicate differences in the ratings of the six television advertisements. Test using $\alpha = .05$.

T A B L E 16.11
Consumer ratings data for
Example 16.5

Subject	A	B	C	D	E	F
			Advertisement			
1	5 (2.5)	8 (6)	7 (5)	6 (4)	4 (1)	5 (2.5)
2	6 (3.5)	10 (6)	6 (3.5)	7 (5)	4 (1.5)	4 (1.5)
3	8 (3)	10 (6)	9 (4.5)	9 (4.5)	6 (1)	7 (2)
4	4 (2)	6 (5)	7 (6)	5 (3.5)	3 (1)	5 (3.5)
Rank sum	$T_1 = 11$	$T_2 = 23$	$T_3 = 19$	$T_4 = 17$	$T_5 = 4.5$	$T_6 = 9.5$

Solution   We wish to test

H_0 : the distributions of ratings for the six television advertisements
are identical

against the alternative hypothesis

H_a : at least two of the distributions of ratings for the six television
advertisements differ in location

The table shows the ranks (in parentheses) of the observations within each block and the rank sums for each of the six advertisements (the treatments). The value of the F_r statistic for these data is

$$F_r = \frac{12}{bk(k+1)} \sum_{i=1}^{k} T_i^2 - 3b(k+1)$$

$$= \frac{12}{(4)(6)(7)}[(11)^2 + (23)^2 + (19)^2 + \cdots + (9.5)^2] - 3(4)(7)$$

$$= 100.75 - 84 = 16.75$$

Since the number $k = 6$ of treatments exceeds 5, the sampling distribution of F_r can be approximated by a chi-square distribution with $(k-1) = (6-1) = 5$ degrees of freedom. Therefore, for $\alpha = .05$ we reject H_0 if

$$F_r > \chi_{.05}^2 \quad \text{where } \chi_{.05}^2 = 11.0705$$

This rejection region is shown in Figure 16.6.

Since the observed value of F_r, $F_r = 16.75$, exceeds $\chi_{.05}^2 = 11.0705$, it falls in the rejection region. We therefore reject H_0 and conclude that the distributions of ratings, for at least two of the advertisements, differ in location.   ∎

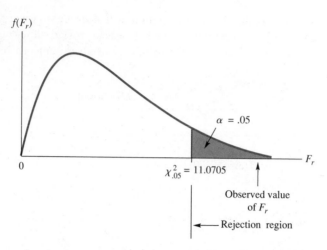

F I G U R E **16.6**
Rejection region for
Example 16.5

$f(F_r)$

$\alpha = .05$

0

$\chi^2_{.05} = 11.0705$

F_r

Observed value
of F_r

Rejection region

E X A M P L E **16.6** Find the approximate p-value for the test in Example 16.5.

Solution The values of χ^2 for $\alpha = .05, .025, .010,$ and $.005$ for 5 degrees of freedom are $\chi^2_{.05} = 11.0705$, $\chi^2_{.025} = 12.8325$, $\chi^2_{.010} = 15.0863$, and $\chi^2_{.005} = 16.7496$. The observed value of F_r, $F_r = 16.75$, is very close to $\chi^2_{.005} = 16.7496$. Therefore, the p-value for the test is approximately equal to .005. ∎

Table 16.12 contains the output generated using the Minitab command FRIED-MAN for the data in Example 16.5. The data were stored in C1, the treatment subscripts in C2, and the block subscripts in C3.

T A B L E **16.12**
Minitab printout for the data in
Example 16.5

```
MTB > FRIEDMAN C1-C3

Friedman test of C1 by C2 blocked by C3

S = 16.75   d.f. = 5   p = 0.005
S = 17.37   d.f. = 5   p = 0.004 (adjusted for ties)

                         Est.      Sum of
        C2     N       Median      RANKS
        1      4       5.625        11.0
        2      4       8.125        23.0
        3      4       7.125        19.0
        4      4       6.625        17.0
        5      4       4.125         4.5
        6      4       5.125         9.5

Grand median    =      6.125
```

The details of the Friedman F_r test are summarized in the following display.

The Friedman F_r Test for a Randomized Block Design

1 Null Hypothesis: H_0: the k population distributions are identical

2 Alternative Hypothesis: H_a: at least two of the k population distributions differ in location

3 Test Statistic: $F_r = \dfrac{12}{bk(k+1)} \sum\limits_{i=1}^{k} T_i^2 - 3b(k+1)$, where

$$b = \text{Number of blocks}$$
$$k = \text{Number of treatments}$$
$$T_i = \text{Rank sum for treatment } i, i = 1, 2, \ldots, k$$

4 Rejection Region: Reject H_0 when $F_r > \chi_\alpha^2$, where χ_α^2 is based on $(k-1)$ degrees of freedom.

Assumption: Either the number k of treatments or the number b of blocks is larger than 5.

Exercises

Basic Techniques

16.43 A randomized block design is employed to compare three treatments in six blocks. The data are shown in the accompanying table.

	Treatment		
Block	1	2	3
1	3.2	3.1	2.4
2	2.8	3.0	1.7
3	4.5	5.0	3.9
4	2.5	2.7	2.6
5	3.7	4.1	3.5
6	2.4	2.4	2.0

a Use the Friedman F_r test to detect differences in location among the three treatment distributions. Test using $\alpha = .05$.

b Find the approximate p-value for the test.

c Perform an analysis of variance, and give the ANOVA table for the analysis.

d Give the value of the F statistic for testing the equality of the three treatment means.

e Give the approximate p-value for the F statistic in part (d).

f Compare the p-values for the tests in parts (a) and (d), and explain the practical implications of the comparison.

16.44 A randomized block design is employed to compare four treatments in eight blocks. The data are shown in the accompanying table.

	Treatment			
Block	1	2	3	4
1	89	81	84	85
2	93	86	86	88
3	91	85	87	86
4	85	79	80	82
5	90	84	85	85
6	86	78	83	84
7	87	80	83	82
8	93	86	88	90

a Use the Friedman F_r test to detect differences in location among the four treatment distributions. Test using $\alpha = .05$.

b Find the approximate p-value for the test.

c Perform an analysis of variance, and give the ANOVA table for the analysis.

d Give the value of the F statistic for testing the equality of the four treatment means.

e Give the approximate p-value for the F statistic in part (d).

f Compare the p-values for the test in parts (a) and (d), and explain the practical implications of the comparison.

Applications

16.45 A management consulting firm conducted a survey to compare estimates by senior-level management, middle-level management, and a company's chief financial officer of the company's prospective percentage annual growth in earnings. Random samples of ten companies were selected for the experiment, and representatives of senior- and middle-level management were randomly selected from each company. The estimates of percentage increase (or decrease) in annual earnings are shown in the accompanying table. Do the data provide sufficient evidence to indicate differences in the levels of forecast earnings increases for the three types of forecasters? Test using the Friedman F_r test with $\alpha = .10$.

Company	Senior Management	Middle Management	Financial Officer
1	10	7	9
2	16	10	11
3	13	20	10
4	22	15	6
5	14	12	12
6	19	8	6
7	25	10	8
8	14	12	12
9	16	12	13
10	21	15	12

16.46 The price-earnings ratio (P/E) of a company's common stock is the ratio of the current price of the stock to the company's previous 12 months' earnings. For example, if a company's stock sells for $20 a share and the annual earnings are $2.00 a share, then the P/E for the stock is $20/2 = 10$. Do the common stocks of similar companies tend to sell at a common level (i.e., with nearly equal P/Es), or do some sell at a premium compared with others? Shown in the accompanying table are the average annual P/Es for three high-quality food-company stocks each year from 1989 through 1994.

	Year					
Company	**1989**	**1990**	**1991**	**1992**	**1993**	**1994**
A	7.6	6.8	8.7	9.9	11.2	14.5
B	7.4	8.2	9.8	10.3	15.2	15.5
C	6.5	6.9	8.0	9.4	10.5	13.9

a   Explain why this design is or is not a randomized block design.

b   Do the data provide sufficient evidence to indicate differences in P/E levels for the three food companies' common stock? Test using $\alpha = .05$.

16.7   Rank Correlation Coefficient

In the preceding sections, we have used ranks to indicate the relative magnitude of observations in nonparametric tests for comparisons of treatments. We now use the same technique in testing for a relation between two ranked variables. Two common **rank correlation coefficients** are the **Spearman** r_s and the Kendall τ. We present the Spearman r_s because its computation is identical to that for the sample correlation coefficient r of Chapter 11. The formula for r_s is given in the following display.

Spearman's Rank Correlation Coefficient

$$r_s = \frac{S_{xy}}{\sqrt{S_{xx}S_{yy}}}$$

where x_i and y_i represent the ranks of the ith pair of observations and

$$S_{xy} = \sum_{i=1}^{n}(x_i - \bar{x})(y_i - \bar{y}) = \sum_{i=1}^{n}x_i y_i - \frac{\left(\sum_{i=1}^{n}x_i\right)\left(\sum_{i=1}^{n}y_i\right)}{n}$$

$$S_{xx} = \sum_{i=1}^{n}(x_i - \bar{x})^2 = \sum_{i=1}^{n}x_i^2 - \frac{\left(\sum_{i=1}^{n}x_i\right)^2}{n}$$

$$S_{yy} = \sum_{i=1}^{n}(y_i - \bar{y})^2 = \sum_{i=1}^{n}y_i^2 - \frac{\left(\sum_{i=1}^{n}y_i\right)^2}{n}$$

When there are no ties in either the x observations or the y observations, the expression for r_s algebraically reduces to the simpler expression

$$r_s = 1 - \frac{6 \sum\limits_{i=1}^{n} d_i^2}{n(n^2 - 1)} \quad \text{where } d_i = x_i - y_i$$

If the number of ties is small in comparison with the number of data pairs, little error will result from using this shortcut formula.

Suppose that eight middle-level managers have been ranked by senior management on their managerial skills, and all have taken a psychological test for which the score is purported to be correlated with the potential for management. The data for the eight managers are shown in Table 16.13. Do the data suggest an agreement between the senior management's ranking of managerial skills and the examination score? We might express this question by asking whether a correlation exists between ranks and test scores.

TABLE **16.13**
Management test score data

Manager	Rank for Managerial Skills	Test Scores
1	7	44
2	4	72
3	2	69
4	6	70
5	1	93
6	3	82
7	8	67
8	5	80

The two variables of interest are rank and test score. The former is already in rank form, and the test scores may be ranked similarly, as shown in Table 16.14. The ranks for tied observations are obtained by averaging the ranks that the tied observations would occupy, as we did for the Mann-Whitney U statistic. The Spearman rank correlation coefficient r_s is calculated by using the ranks as the paired measurements on the two variables x and y in the formula for r in Chapter 11. Example 16.7 illustrates the calculation.

E X A M P L E **16.7** Calculate r_s for the management-ranking test-score data.

	Rank for Managerial	Ranked Test
Manager	**Skills, x_i**	**Score, y_i**
1	7	1
2	4	5
3	2	3
4	6	4
5	1	8
6	3	7
7	8	2
8	5	6

T A B L E **16.14**
Ranking of test scores

Manager	x_i	y_i	d_i	d_i^2
1	7	1	6	36
2	4	5	−1	1
3	2	3	−1	1
4	6	4	2	4
5	1	8	−7	49
6	3	7	−4	16
7	8	2	6	36
8	5	6	−1	1
Total				144

T A B L E **16.15**
Data calculations for
Example 16.7

Solution   The differences and squares of differences between the two rankings are as shown in Table 16.15. Substituting into the formula for r_s yields

$$
\begin{aligned}
r_s &= 1 - \frac{6\sum_{i=1}^{n} d_i^2}{n(n^2 - 1)} \\
&= 1 - \frac{6(144)}{8(64 - 1)} \\
&= -.714 \qquad \blacksquare
\end{aligned}
$$

The Spearman rank correlation coefficient can be used as a test statistic to test an hypothesis of *no association* between two populations. We assume that the n pairs of observations (x_i, y_i) have been randomly selected, and, therefore, *no association between the populations* will imply a random assignment of the n ranks within each sample. Each random assignment (for the two samples) will represent a simple event associated with the experiment, and a value of r_s can be calculated for each. Thus it is possible to calculate the probability that r_s assumes a large absolute value due solely to chance and thereby suggests an association between populations when none exists.

The rejection region for a two-tailed test is shown in Figure 16.7. If the alternative hypothesis is that the correlation between the ranks of x and y is negative, you will reject H_0 for negative values of r_s that are close to -1 (in the lower tail of Figure 16.7). Similarly, if the alternative hypothesis is that the correlation between the ranks of x and y is positive, you will reject H_0 for large positive values of r_s (in the upper tail of Figure 16.7).

FIGURE **16.7**

Rejection region for a two-tailed test of the null hypothesis of no association, using Spearman's rank correlation test

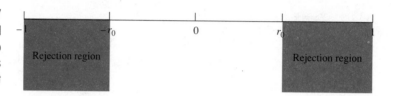

The critical values of r_s are given in Table 16 of Appendix II. An abbreviated version of Table 16 is shown here in Table 16.16. Across the top of Table 16.16 (and Table 16 in Appendix II) are recorded values of a that you may want to use for a test of the null hypothesis of *no association* between x and y. The number of rank pairs, n, appears at the left side of the table. The table entries give the critical value r_0 such that $P(r_s \geq r_0) = a$. For example, suppose that you have $n = 8$ rank pairs and the alternative hypothesis is that the correlation between the ranks is positive. Then you will want to reject the null hypothesis of "no association" only for large positive values of r_s and will use a one-tailed test. Referring to Table 16.16 and using the row corresponding to $n = 8$ and the column for $a = .05$, you read $r_0 = .643$. Therefore, you will reject H_0 for all values of r_s greater than or equal to .643.

The test is conducted in exactly the same manner if you want to test only the alternative hypothesis that the ranks are negatively correlated. The only difference is that you will reject the null hypothesis if $r_s \leq -.643$. That is, you just place a minus sign in front of the tabulated value of r_0 to get the lower-tail critical value.

To conduct a two-tailed test, reject the null hypothesis if $r_s \geq r_0$ or $r_s \leq -r_0$. The value of a for the test will be $\alpha/2$. For example, if $n = 8$ and you choose the .025 column, you will reject H_0 if $r_s \geq .738$ or $r_s \leq -.738$. The α value for the test will be $2(.025) = .05$.

EXAMPLE **16.8**   Test an hypothesis of *no association* between the populations for Example 16.7.

Solution   The critical value of r_s for a one-tailed test with $\alpha = .05$ and $n = 8$ is .643. Let us assume that a correlation between rank assigned by senior management and the middle-level managers' test scores could not possibly be positive. Lower ranks are assigned to the managers with the best skills. Therefore, a low rank should be

TABLE 16.16
An abbreviated version of Table
16 in Appendix II for Spearman's
rank correlation coefficient

n	$a = .05$	$a = .025$	$a = .01$	$a = .005$
5	.900	—	—	—
6	.829	.886	.943	—
7	.714	.786	.893	—
8	.643	.738	.833	.881
9	.600	.683	.783	.833
10	.564	.648	.745	.794
11	.523	.623	.736	.818
12	.497	.591	.703	.780
13	.475	.566	.673	.745
14	.457	.545	⋮	⋮
15	.441	.525		
16	.425	⋮		
17	.412			
18	.399			
19	.388			
20	.377			

associated with a high test score if the senior-management rankings and the tests agree in identifying managers with managerial skill. The alternative hypothesis is that the population rank coefficient ρ_s is less than zero, and we will be concerned with a one-tailed statistical test. Thus α for the test will be the tabulated value .05, and we will reject the null hypothesis if $r_s \leq -.643$.

The calculated value of the test statistic, $r_s = -.714$, is less than the critical value of $a = .05$. Thus the null hypothesis will be rejected at the $\alpha = .05$ level of significance. It appears that some agreement does exist between senior-management's ranking and the test scores. However, note that this agreement can exist when *neither* provides an adequate yardstick for measuring managerial ability. For example, the association can exist if both the senior management and those who constructed the psychological examination possessed a completely erroneous, but identical, concept of the characteristics of good managerial skills. ∎

The details of Spearman's rank correlation test are given in the following display.

Spearman's Rank Correlation Test

1 Null Hypothesis: H_0: there is no association between the rank pairs

2 Alternative Hypothesis: H_a : there is an association between the rank pairs (a two-tailed test). Or H_a : the correlation between the rank pairs is positive (or negative) (a one-tailed test).

3 Test Statistic: $r_s = \dfrac{S_{xy}}{\sqrt{S_{xx}S_{yy}}}$

where x_i and y_i represent the ranks of the ith pair of observations.

4 Rejection Region: For a two-tailed test, reject H_0 if $r_s \geq r_0$ or $r_s \leq -r_0$, where r_0 is given in Table 16 of Appendix II. Double a, the tabulated probability, to obtain the value of α for the two-tailed test. For a one-tailed test, reject H_0 if $r_s \geq r_0$ (for an upper-tailed test) or $r_s \leq -r_0$ (for a lower-tailed test). The α value for a one-tailed test is the value of a shown in Table 16 of Appendix II.

Exercises

Basic Techniques

16.47 Give the rejection region for a test to detect positive rank correlation if the number of pairs of ranks is 16 and α is as follows.

  a $\alpha = .05$ **b** $\alpha = .01$

16.48 Give the rejection region for a test to detect negative rank correlation if the number of pairs of ranks is 12 and α is as follows.

  a $\alpha = .05$ **b** $\alpha = .01$

16.49 Give the rejection region for a test to detect rank correlation if the number of pairs of ranks is 25 and α is as follows.

  a $\alpha = .05$ **b** $\alpha = .01$

16.50 The following paired observations were obtained on two variables x and y.

x	1.2	.8	2.1	3.5	2.7	1.5
y	1.0	1.3	.1	−.8	−.2	.6

  a Calculate the Spearman's rank correlation coefficient r_s.

  b Do the data provide sufficient evidence to indicate a correlation between x and y? Test using $\alpha = .05$.

Applications

 16.51 Two art critics, A and B, each ranked ten paintings by contemporary (but anonymous) artists in accordance with their estimate of the investment potential of a painting. The ratings are shown in the accompanying table. Do the critics seem to agree on their rating of contemporary art? That is, do the data provide sufficient evidence to indicate a positive correlation between the rankings of critics A and B? Test using a value of α near .05.

Painting	Critic A	Critic B	Painting	Critic A	Critic B
1	6	5	6	7	8
2	4	6	7	3	1
3	9	10	8	8	7
4	1	2	9	5	4
5	2	3	10	10	9

 16.52 A large corporation selects college graduates for employment, using both interviews and a psychological achievement test. Interviews conducted at the home office of the company were far more expensive than the tests, which could be conducted on campus. Consequently, the personnel office was interested in determining whether the test scores were correlated with interview ratings and whether tests could be substituted for interviews. The idea was not to eliminate interviews but to reduce their number. For a determination of whether correlation was present, ten prospects were ranked during interviews and tested. The paired scores are as shown in the accompanying table. Calculate the Spearman rank correlation coefficient r_s. Rank 1 is assigned to the candidate judged to be the best.

Subject	Interview Rank	Test Score
1	8	74
2	5	81
3	10	66
4	3	83
5	6	66
6	1	94
7	4	96
8	7	70
9	9	61
10	2	86

16.53 Refer to Exercise 16.52. Do the data provide sufficient evidence to indicate that the correlation between interview rankings and ranked test scores is less than zero? If this evidence does exist, can we say that tests could be used to reduce the number of interviews?

16.8 ENVELOPE PLEASE ...: A RANK CORRELATION ANALYSIS

Let us return to the case study introduced at the beginning of this chapter and address the question "Are the attributes of aroma and taste correlated?" The pertinent data from the case study are reproduced in Table 16.17.

Because we suspect that the scores on these two attributes do not satisfy the assumptions required for using Pearson's correlation coefficient, given in Chapter 11, we can ask if the ranks of the scores on these two attributes are correlated. The

TABLE **16.17**
Water ratings for aroma and
taste

Water	Aroma	Taste
Temecula	2.7 (2)	3.2 (3)
San Jacinto	3.7 (3.5)	2.7 (2)
Sparkletts[†]	4.7 (6.5)	4.2 (7)
Lake Elsinore	3.7 (3.5)	3.7 (4.5)
Norco	4.7 (6.5)	4.0 (6)
Metropolitan Water	2.5 (1)	2.2 (1)
Riverside	4.2 (5)	3.7 (4.5)

[†]Bottled water.

appropriate statistic for this question is Spearman's rank correlation coefficient, presented in Section 16.7. The ranks of the scores are shown in parentheses to the right of their respective values.

The calculated value of the Spearman rank correlation coefficient for these data is found to be

$$r_s = .908 \quad \text{with} \quad n = 7$$

In testing H_0 : there is no association between the ranks versus H_a : the correlation between the rank pairs is positive, with $\alpha = .05$, we would reject H_0 for values of $r_s \geq .714$ (see Table 16 in Appendix II). In fact, since the calculated value of .908 exceeds the critical value of .893 with $\alpha = .01$, the p-value associated with this test is

$$p\text{-value} \leq .01$$

What conclusion can we draw? The ranks of the observations for aroma and taste are positively correlated, and large scores on one attribute are associated with large scores on the other.

16.9 Summary

Nonparametric statistical procedures are particularly useful when the experimental observations are susceptible to ordering but cannot be measured on a quantitative scale. Parametric statistical procedures usually cannot be applied to this type of data; hence all inferential procedures must be based on nonparametric methods. A second application of nonparametric statistical methods is in testing hypotheses associated with populations of quantitative data when uncertainty exists concerning the satisfaction of assumptions about the form of the population distributions.

In this chapter, we presented a number of useful nonparametric methods along with illustrations of their applications. The Mann-Whitney U test can be used to compare the locations of two population frequency distributions when the observations can be ranked according to their relative magnitudes and when the samples have been randomly and independently selected from the two populations. The Kruskal-Wallis H test provides similar methodology for comparing the locations of three or more population frequency distributions. The simplest nonparametric test, the sign test, provides a rapid procedure for comparing the locations of two population distributions when the observations have been independently selected in matched pairs. If the differences between pairs can be ranked according to their relative magnitudes, you can use the Wilcoxon signed rank test for comparing the two populations. This latter test utilizes more sample information than the sign test and consequently is more likely to detect a difference in location if a difference exists. The Friedman F_r test enables us to extend this comparison to more than two population distributions when the data have been collected in matched sets—that is, according to a randomized block design. Finally, we presented a nonparametric method, Spearman's rank correlation test, for testing the correlation between two variables when the observations associated with each variable can be ranked according to their relative magnitudes.

Supplementary Exercises

16.54 A time study was conducted to compare the length of time (in seconds) to assemble a device using two different assembly methods, A and B. The data are shown in the accompanying table. So that natural person-to-person variability in the responses was removed, both assembly methods were used by each of nine workers, thus permitting an analysis of the difference between assembly times *within* each worker.

Worker	Method A	Method B
1	9.4	10.3
2	7.8	8.9
3	5.6	4.1
4	12.1	14.7
5	6.9	8.7
6	4.2	7.1
7	8.8	11.3
8	7.7	5.2
9	6.4	7.8

  a Use the sign test to determine whether sufficient evidence exists to indicate a difference in the distribution of assembly times for the two methods. Use a rejection region for which $\alpha \leq .05$.

  b Test the hypothesis of no difference in mean response using Student's t test.

16.55 Refer to Exercise 16.54. Test the hypothesis that no difference exists in the distributions of responses for the two methods, using the Wilcoxon signed rank test. Use a rejection region for which α is as near as possible to the α achieved in Exercise 16.54(a).

16.56 The coded values for a measure of brightness in paper (light reflectivity) prepared by two different processes are given in the accompanying table for samples of size nine drawn randomly from each of the two processes.

Process A	6.1	9.2	8.7	8.9	7.6	7.1	9.5	8.3	9.0
Process B	9.1	8.2	8.6	6.9	7.5	7.9	8.3	7.8	8.9

Do the data provide sufficient evidence ($\alpha = .10$) to indicate a difference in the populations of brightness measurements for the two processes?

a Use the Mann-Whitney U test.   **b** Use Student's t test.

16.57 If (as in the case of measurements produced by two well-calibrated measuring instruments) the means of two populations are equal, it is possible to use the Mann-Whitney U statistic for testing hypotheses concerning the population variances as follows:

1 Rank the combined sample.

2 Number the ranked observations "from the outside in"; that is, number the smallest observation 1; the largest, 2; the next-to-smallest, 3; the next-to-largest, 4; and so on. This final sequence of numbers induces an ordering on the symbols A (population A items) and B (population B items). If $\sigma_A^2 > \sigma_B^2$, one would expect to find a preponderance of As near the first of the sequences and, thus, a relatively small "sum of ranks" for the A observations.

a Given the following measurements produced by well-calibrated precision instruments A and B, test at or near the $\alpha = .05$ level to determine whether the more expensive instrument, B, is more precise than A. (Note that this would imply a one-tailed test.) Use the Mann-Whitney U test.

Instrument A	Instrument B
1060.21	1060.24
1060.34	1060.28
1060.27	1060.32
1060.36	1060.30
1060.40	

b Test using the F statistic of Section 8.9.

16.58 Two certified public accountants (CPAs) were asked to estimate the tax liability associated with each of six long-term investments. The data are given in the accompanying table. Suppose we suspect that CPA 1 tends to give higher estimates of tax liability than CPA 2. Do the data support this theory? Test with $\alpha = .01$.

Investment	CPA 1	CPA 2
1	89,600	86,500
2	105,000	110,000
3	75,000	72,000
4	91,000	86,000
5	63,000	60,000
6	71,000	70,000

a Use the sign test.   **b** Use the Wilcoxon signed rank test.

16.59 Does the IRS audit income tax returns in the same percentages from different regions of the country? The percentages of income tax returns audited, from five western, southern, and northeastern states, are given in the accompanying table. Do the data provide sufficient evidence of differences in the rates of auditing for the three different regions of the United States? Test using $\alpha = .10$.

South		West		Northeast	
Alabama	1.11	California	1.49	Maine	.85
Florida	1.37	Arizona	1.45	Connecticut	1.22
Georgia	1.21	Utah	1.97	Massachusetts	.82
Mississippi	1.28	Nevada	2.51	Rhode Island	1.16
Louisiana	1.31	Washington	1.42	New Hampshire	.80

16.60 Three investment advisory firms were asked to rate each of six investments on a scale of 1 (low) to 10 (high) in terms of potential appreciation for the coming year. All investments were considered to be of moderate risk. The ratings are shown in the accompanying table. Do the data provide sufficient evidence to indicate that one or more of the advisory firms tend to rate investments higher than the other firm(s)? Test using the Friedman F_r test with $\alpha = .05$.

	Firm		
Investment	1	2	3
1	7	5	9
2	7	4	7
3	5	3	3
4	8	9	10
5	6	7	8
6	7	6	9

16.61 An experiment was conducted to study the relationship between the ratings of a tobacco leaf grader and the moisture content of the corresponding tobacco leaves. Twelve leaves were rated by the grader on a scale of 1 to 10, and corresponding readings of moisture content were made. The data are as shown in the accompanying table. Calculate r_s. Do the data provide sufficient evidence to indicate an association between the grader's ratings and the moisture content of the leaves?

Leaf	Grader's Rating	Moisture Content
1	9	.22
2	6	.16
3	7	.17
4	7	.14
5	5	.12
6	8	.19
7	2	.10
8	6	.12
9	1	.05
10	10	.20
11	9	.16
12	3	.09

DESCRIPTION OF THE DATA DISK

The data stored as ASCII files on the disk are identified by the chapter and problem in which they appear. For example, the data corresponding to Exercise 10 in Chapter 2 are found in the file 2-10.DAT. Almost all data sets with more than ten entries are included on this disk. The data are stored using a column format. When the data correspond to one set or one sample, the data are found in the first column; when the data correspond to two sets or two samples, the data are found in the first two columns. For three or more sets, the data are found in the first column, and the subscripts corresponding to the sample or population number are found in the second column. In addition to the data from problem sets, five large data sets are also provided on this data disk. Information concerning these data sets follows.

Data Set A

Data set A consists of the salaries for male and female college full professors, associate professors, assistant professors, and instructors for institutions having at least five or more individuals in all of these ranks ("The Annual Report of the Status of the Profession: 1993–1994," Appendix I, *Academe*, March–April 1994). There are six columns of data, described as follows:

Column 1: college number (1 through 246)

Column 2: salaries for full professors

Column 3: salaries for associate professors

Column 4: salaries for assistant professors

Column 5: salaries for instructors

Column 6: gender (male = 0, female = 1)

Minitab output for data set A

```
MTB > Describe 'FULL'-'INSTR';
SUBC>   By 'GENDER'.
```

	GENDER	N	MEAN	MEDIAN	TRMEAN	STDEV	SEMEAN
FULL	0	246	56.484	54.150	55.750	11.037	0.704
	1	246	52.674	50.550	52.089	9.551	0.609
ASSOC	0	246	44.984	44.150	44.691	6.302	0.402
	1	246	42.554	41.550	42.269	5.705	0.364
ASST	0	246	38.343	38.200	38.093	4.777	0.305
	1	246	36.157	35.700	35.971	4.271	0.272
INSTR	0	246	29.922	28.950	29.669	4.683	0.299
	1	246	28.468	27.500	28.334	3.945	0.252

	GENDER	MIN	MAX	Q1	Q3
FULL	0	34.700	97.100	48.275	62.300
	1	34.600	89.600	45.575	58.200
ASSOC	0	32.800	72.900	40.525	48.300
	1	31.400	62.200	38.350	45.625
ASST	0	26.900	60.200	34.775	41.300
	1	26.900	52.700	33.000	38.700
INSTR	0	20.800	49.800	26.975	32.225
	1	19.000	42.400	25.800	30.875

```
MTB > Describe 'FULL'-'INSTR'.
```

	N	MEAN	MEDIAN	TRMEAN	STDEV	SEMEAN
FULL	492	54.579	52.300	53.915	10.485	0.473
ASSOC	492	43.769	42.900	43.475	6.127	0.276
ASST	492	37.250	36.700	37.021	4.657	0.210
INSTR	492	29.195	28.350	28.978	4.386	0.198

	MIN	MAX	Q1	Q3
FULL	34.600	97.100	46.925	60.500
ASSOC	31.400	72.900	39.225	47.200
ASST	26.900	60.200	33.800	40.100
INSTR	19.000	49.800	26.300	31.600

Data Set B

Data set B consists of measurements of yield characteristics of broccoli as a function of controlled ozone levels (*data source*: Dr. Patrick McCool, Statewide Air Pollution Research Center, University of California, Riverside, CA 92521). There are 317 observations on each of the following variables:

Column 1: number of the exposure chamber in which the observation is made

Column 2: fresh weight

Column 3: market weight

Column 4: dry weight

Column 5: maximum head diameter

Column 6: 0 = Variety 1, 1 = Variety 2

Column 7: average ozone level for 6-hour period 9 A.M. to 3 P.M.

Column 8: average ozone level for 10-hour period 9 A.M. to 7 P.M.

Column 9: average ozone level for 13-hour period 8 A.M. to 9 P.M.

Minitab output for data set B and histograms for the variables fresh weight, market weight, dry weight, and the maximum diameter of the head of broccoli follow.

Minitab output for data set B

```
MTB > Describe 'FRSH-WT'-'DIAM';
SUBC>    By C6.
```

	VRTY	N	MEAN	MEDIAN	TRMEAN	STDEV	SEMEAN
FRSH-WT	0	170	517.2	500.0	513.9	139.5	10.7
	1	147	482.54	483.00	479.95	107.68	8.88
MRKT-WT	0	170	140.35	135.00	138.79	41.31	3.17
	1	147	144.62	144.00	143.83	34.61	2.85
DRY-WT	0	170	10.163	10.000	10.101	3.424	0.263
	1	147	10.698	10.600	10.653	2.801	0.231
DIAM	0	170	9.146	9.000	9.097	2.228	0.171
	1	147	8.982	9.200	9.005	1.397	0.115

	VRTY	MIN	MAX	Q1	Q3
FRSH-WT	0	132.0	883.0	429.8	600.3
	1	266.00	786.00	410.00	562.00
MRKT-WT	0	55.00	270.00	111.00	160.75
	1	63.00	232.00	120.00	168.00
DRY-WT	0	2.200	20.600	8.175	12.425
	1	3.700	18.300	8.700	12.400
DIAM	0	3.700	15.400	7.675	10.400
	1	4.100	12.100	8.000	9.700

```
MTB > Describe 'FRSH-WT'-'DIAM'.
```

	N	MEAN	MEDIAN	TRMEAN	STDEV	SEMEAN
FRSH-WT	317	501.13	488.00	497.30	126.75	7.12
MRKT-WT	317	142.33	141.00	141.22	38.35	2.15
DRY-WT	317	10.411	10.300	10.364	3.157	0.177
DIAM	317	9.070	9.100	9.040	1.888	0.106

	MIN	MAX	Q1	Q3
FRSH-WT	132.00	883.00	417.50	577.00
MRKT-WT	55.00	270.00	113.00	165.00
DRY-WT	2.200	20.600	8.250	12.400
DIAM	3.700	15.400	7.800	10.200

Histograms for broccoli variables

```
Histogram of FRSH-WT   N = 317
Each * represents 2 obs.

Midpoint    Count
     150       1   *
     200       1   *
     250       4   **
     300      13   *******
     350      34   *****************
     400      30   ***************
     450      58   *****************************
     500      49   *************************
     550      45   ***********************
     600      35   *****************
     650      15   ********
     700      15   ********
     750       6   ***
     800       6   ***
     850       4   **
     900       1   *
```

```
Histogram of MRKT-WT   N = 317
Each * represents 2 obs.

Midpoint    Count
      60       4   **
      80      18   *********
     100      41   *********************
     120      64   ********************************
     140      61   *******************************
     160      61   *******************************
     180      30   ***************
     200      19   **********
     220      13   *******
     240       4   **
     260       1   *
     280       1   *
```

```
Histogram of DRY-WT   N = 317
Each * represents 2 obs.

Midpoint    Count
       2       2   *
       4       8   ****
       6      31   ****************
       8      61   *******************************
      10      89   *********************************************
      12      68   **********************************
      14      28   **************
      16      22   ***********
      18       7   ****
      20       1   *
```

Histograms for broccoli variables (continued)

```
Histogram of MAX-DIAM   N = 317
Each * represents 2 obs.

Midpoint   Count
      4       3   **
      5       7   ****
      6      13   *******
      7      37   ******************
      8      54   ***************************
      9      75   **************************************
     10      71   ************************************
     11      30   ***************
     12      14   *******
     13       5   ***
     14       5   ***
     15       3   **
```

Data Set C

Data set C consists of three variables recorded for 604 money market funds for the period ending July 13, 1994. The quotations that follow were collected by the National Association of Securities Dealers, Inc. ("Money Market Summary," *Wall Street Journal*, July 14, 1994). Yields do not include gains or losses.

Column 1: average maturity in days

Column 2: seven-day average yields

Column 3: assets

Histograms and minitab output for data set C follow.

Minitab output for data set C

	N	MEAN	MEDIAN	TRMEAN	STDEV	SEMEAN
AVGMAT	604	38.560	38.000	38.289	14.224	0.579
7DAYYLD	604	3.7324	3.7500	3.7435	0.3337	0.0136
ASSETS	604	843	261	483	3312	135

	MIN	MAX	Q1	Q3
AVGMAT	1.000	88.000	29.000	47.000
7DAYYLD	1.9700	4.5100	3.5125	3.9500
ASSETS	1	70100	98	694

Histograms for money market funds

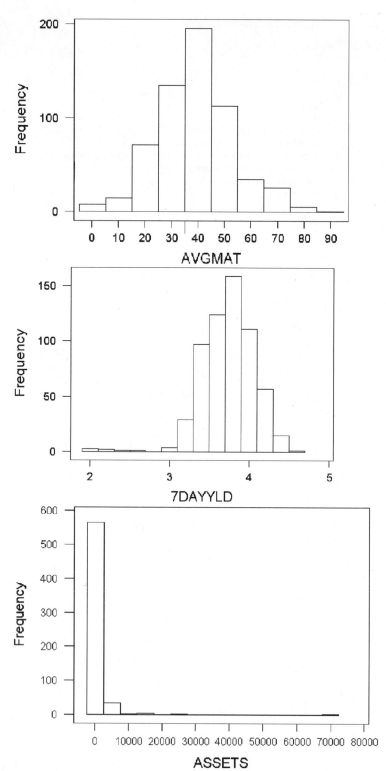

Data Set D

Data set D consists of financial data for the 100 best small companies ("The Best Small Companies," *Business Week*, May 23, 1994, pp. 101–104).

Column 1: company rank

Column 2: current sales in millions of dollars

Column 3: current earnings in millions of dollars

Column 4: three-year average return on capital

Column 5: price-earnings ratio

Column 6: market value in millions of dollars

Minitab output for data set D

	N	MEAN	MEDIAN	TRMEAN	STDEV	SEMEAN
SALES	100	59.19	52.10	57.32	37.00	3.70
EARNINGS	100	5.868	4.150	5.400	4.787	0.479
3YR-RET	100	33.39	29.55	32.45	14.52	1.45
P-E	100	25.01	21.00	23.61	14.62	1.46
MRKT-VAL	100	145.9	107.0	132.3	133.4	13.3

	MIN	MAX	Q1	Q3
SALES	11.20	146.40	26.98	87.92
EARNINGS	0.500	21.700	2.600	7.625
3YR-RET	9.10	87.50	22.80	40.67
P-E	6.00	104.00	15.25	31.00
MRKT-VAL	9.0	704.0	41.3	188.5

Data Set E

Data set E is a time series consisting of the total manufacturers' inventories, given as book value (in billions of dollars) recorded at the end of the period, unadjusted for seasonal variation (*Business Statistics 1963–93*, June 1994). The data are given monthly, beginning with January 1961 and through December 1993.

Column 1: year (1961–93)

Column 2: month (January = 1, . . . , December = 12)

Column 3: total inventories

TABLES

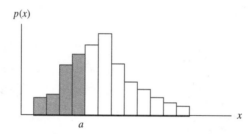

$p(x)$

a

x

T A B L E 1 Cumulative Binomial Probabilities

Tabulated values are $P(x \leq a) = \sum_{x=0}^{a} p(x)$. (Computations are rounded at the third decimal place.)

$n = 2$

| | | | | | | | p | | | | | | | |
a	0.01	0.05	0.10	0.20	0.30	0.40	0.50	0.60	0.70	0.80	0.90	0.95	0.99	a
0	.980	.902	.810	.640	.490	.360	.250	.160	.090	.040	.010	.002	.000	0
1	1.000	.998	.990	.960	.910	.840	.750	.640	.510	.360	.190	.098	.020	1
2	1.000	1.000	1.000	1.000	1.000	1.000	1.000	1.000	1.000	1.000	1.000	1.000	1.000	2

$n = 3$

| | | | | | | | p | | | | | | | |
a	0.01	0.05	0.10	0.20	0.30	0.40	0.50	0.60	0.70	0.80	0.90	0.95	0.99	a
0	.970	.857	.729	.512	.343	.216	.125	.064	.027	.008	.001	.000	.000	0
1	1.000	.993	.972	.896	.784	.648	.500	.352	.216	.104	.028	.007	.000	1
2	1.000	1.000	.999	.992	.973	.936	.875	.784	.657	.488	.271	.143	.030	2
3	1.000	1.000	1.000	1.000	1.000	1.000	1.000	1.000	1.000	1.000	1.000	1.000	1.000	3

T A B L E   **1**   (Continued)

n = 4

							p							
a	**0.01**	**0.05**	**0.10**	**0.20**	**0.30**	**0.40**	**0.50**	**0.60**	**0.70**	**0.80**	**0.90**	**0.95**	**0.99**	*a*
0	.961	.815	.656	.410	.240	.130	.062	.026	.008	.002	.000	.000	.000	0
1	.999	.986	.948	.819	.652	.475	.312	.179	.084	.027	.004	.000	.000	1
2	1.000	1.000	.996	.973	.916	.821	.688	.525	.348	.181	.052	.014	.001	2
3	1.000	1.000	1.000	.998	.992	.974	.938	.870	.760	.590	.344	.185	.039	3
4	1.000	1.000	1.000	1.000	1.000	1.000	1.000	1.000	1.000	1.000	1.000	1.000	1.000	4

n = 5

							p							
a	**0.01**	**0.05**	**0.10**	**0.20**	**0.30**	**0.40**	**0.50**	**0.60**	**0.70**	**0.80**	**0.90**	**0.95**	**0.99**	*a*
0	.951	.774	.590	.328	.168	.078	.031	.010	.002	.000	.000	.000	.000	0
1	.999	.977	.919	.737	.528	.337	.188	.087	.031	.007	.000	.000	.000	1
2	1.000	.999	.991	.942	.837	.683	.500	.317	.163	.058	.009	.001	.000	2
3	1.000	1.000	1.000	.993	.969	.913	.812	.663	.472	.263	.081	.023	.001	3
4	1.000	1.000	1.000	1.000	.998	.990	.969	.922	.832	.672	.410	.226	.049	4
5	1.000	1.000	1.000	1.000	1.000	1.000	1.000	1.000	1.000	1.000	1.000	1.000	1.000	5

n = 6

							p							
a	**0.01**	**0.05**	**0.10**	**0.20**	**0.30**	**0.40**	**0.50**	**0.60**	**0.70**	**0.80**	**0.90**	**0.95**	**0.99**	*a*
0	.941	.735	.531	.262	.118	.047	.016	.004	.001	.000	.000	.000	.000	0
1	.999	.967	.886	.655	.420	.233	.109	.041	.011	.002	.000	.000	.000	1
2	1.000	.998	.984	.901	.744	.544	.344	.179	.070	.017	.001	.000	.000	2
3	1.000	1.000	.999	.983	.930	.821	.656	.456	.256	.099	.016	.002	.000	3
4	1.000	1.000	1.000	.998	.989	.959	.891	.767	.580	.345	.114	.033	.001	4
5	1.000	1.000	1.000	1.000	.999	.996	.984	.953	.882	.738	.469	.265	.059	5
6	1.000	1.000	1.000	1.000	1.000	1.000	1.000	1.000	1.000	1.000	1.000	1.000	1.000	6

T A B L E   **1**   (Continued)

$n = 7$

							p							
a	0.01	0.05	0.10	0.20	0.30	0.40	0.50	0.60	0.70	0.80	0.90	0.95	0.99	a
0	.932	.698	.478	.210	.082	.028	.008	.002	.000	.000	.000	.000	.000	0
1	.998	.956	.850	.577	.329	.159	.062	.019	.004	.000	.000	.000	.000	1
2	1.000	.996	.974	.852	.647	.420	.227	.096	.029	.005	.000	.000	.000	2
3	1.000	1.000	.997	.967	.874	.710	.500	.290	.126	.033	.003	.000	.000	3
4	1.000	1.000	1.000	.995	.971	.904	.773	.580	.353	.148	.026	.004	.000	4
5	1.000	1.000	1.000	1.000	.996	.981	.938	.841	.671	.423	.150	.044	.002	5
6	1.000	1.000	1.000	1.000	1.000	.998	.992	.972	.918	.790	.522	.302	.068	6
7	1.000	1.000	1.000	1.000	1.000	1.000	1.000	1.000	1.000	1.000	1.000	1.000	1.000	7

$n = 8$

							p							
a	0.01	0.05	0.10	0.20	0.30	0.40	0.50	0.60	0.70	0.80	0.90	0.95	0.99	a
0	.923	.663	.430	.168	.058	.017	.004	.001	.000	.000	.000	.000	.000	0
1	.997	.943	.813	.503	.255	.106	.035	.009	.001	.000	.000	.000	.000	1
2	1.000	.994	.962	.797	.552	.315	.145	.050	.011	.001	.000	.000	.000	2
3	1.000	1.000	.995	.944	.806	.594	.363	.174	.058	.010	.000	.000	.000	3
4	1.000	1.000	1.000	.990	.942	.826	.637	.406	.194	.056	.005	.000	.000	4
5	1.000	1.000	1.000	.999	.989	.950	.855	.685	.448	.203	.038	.006	.000	5
6	1.000	1.000	1.000	1.000	.999	.991	.965	.894	.745	.497	.187	.057	.003	6
7	1.000	1.000	1.000	1.000	1.000	.999	.996	.983	.942	.832	.570	.337	.077	7
8	1.000	1.000	1.000	1.000	1.000	1.000	1.000	1.000	1.000	1.000	1.000	1.000	1.000	8

$n = 9$

							p							
a	0.01	0.05	0.10	0.20	0.30	0.40	0.50	0.60	0.70	0.80	0.90	0.95	0.99	a
0	.914	.630	.387	.134	.040	.010	.002	.000	.000	.000	.000	.000	.000	0
1	.997	.929	.775	.436	.196	.071	.020	.004	.000	.000	.000	.000	.000	1
2	1.000	.992	.947	.738	.463	.232	.090	.025	.004	.000	.000	.000	.000	2
3	1.000	.999	.992	.914	.730	.483	.254	.099	.025	.003	.000	.000	.000	3
4	1.000	1.000	.999	.980	.901	.733	.500	.267	.099	.020	.001	.000	.000	4
5	1.000	1.000	1.000	.997	.975	.901	.746	.517	.270	.086	.008	.001	.000	5
6	1.000	1.000	1.000	1.000	.996	.975	.910	.768	.537	.262	.053	.008	.000	6
7	1.000	1.000	1.000	1.000	1.000	.996	.980	.929	.804	.564	.225	.071	.003	7
8	1.000	1.000	1.000	1.000	1.000	1.000	.998	.990	.960	.866	.613	.370	.086	8
9	1.000	1.000	1.000	1.000	1.000	1.000	1.000	1.000	1.000	1.000	1.000	1.000	1.000	9

T A B L E   1   (Continued)

n = 10

							p							
a	0.01	0.05	0.10	0.20	0.30	0.40	0.50	0.60	0.70	0.80	0.90	0.95	0.99	*a*
0	.904	.599	.349	.107	.028	.006	.001	.000	.000	.000	.000	.000	.000	0
1	.996	.914	.736	.376	.149	.046	.011	.002	.000	.000	.000	.000	.000	1
2	1.000	.988	.930	.678	.383	.167	.055	.012	.002	.000	.000	.000	.000	2
3	1.000	.999	.987	.879	.650	.382	.172	.055	.011	.001	.000	.000	.000	3
4	1.000	1.000	.998	.967	.850	.633	.377	.166	.047	.006	.000	.000	.000	4
5	1.000	1.000	1.000	.994	.953	.834	.623	.367	.150	.033	.002	.000	.000	5
6	1.000	1.000	1.000	.999	.989	.945	.828	.618	.350	.121	.013	.001	.000	6
7	1.000	1.000	1.000	1.000	.998	.988	.945	.833	.617	.322	.070	.012	.000	7
8	1.000	1.000	1.000	1.000	1.000	.998	.989	.954	.851	.624	.264	.086	.004	8
9	1.000	1.000	1.000	1.000	1.000	1.000	.999	.994	.972	.893	.651	.401	.096	9
10	1.000	1.000	1.000	1.000	1.000	1.000	1.000	1.000	1.000	1.000	1.000	1.000	1.000	10

n = 11

							p							
a	0.01	0.05	0.10	0.20	0.30	0.40	0.50	0.60	0.70	0.80	0.90	0.95	0.99	*a*
0	.895	.569	.314	.086	.020	.004	.000	.000	.000	.000	.000	.000	.000	0
1	.995	.898	.697	.322	.113	.030	.006	.001	.000	.000	.000	.000	.000	1
2	1.000	.985	.910	.617	.313	.119	.033	.006	.001	.000	.000	.000	.000	2
3	1.000	.998	.981	.839	.570	.296	.113	.029	.004	.000	.000	.000	.000	3
4	1.000	1.000	.997	.950	.790	.533	.274	.099	.022	.002	.000	.000	.000	4
5	1.000	1.000	1.000	.988	.922	.754	.500	.246	.078	.012	.000	.000	.000	5
6	1.000	1.000	1.000	.998	.978	.901	.726	.467	.210	.050	.003	.000	.000	6
7	1.000	1.000	1.000	1.000	.996	.971	.887	.704	.430	.161	.019	.002	.000	7
8	1.000	1.000	1.000	1.000	.999	.994	.967	.881	.687	.383	.090	.015	.000	8
9	1.000	1.000	1.000	1.000	1.000	.999	.994	.970	.887	.678	.303	.102	.005	9
10	1.000	1.000	1.000	1.000	1.000	1.000	1.000	.996	.980	.914	.686	.431	.105	10
11	1.000	1.000	1.000	1.000	1.000	1.000	1.000	1.000	1.000	1.000	1.000	1.000	1.000	11

T A B L E   **1**   (Continued)

$n = 12$

a	\multicolumn{13}{c}{p}	a
	0.01	**0.05**	**0.10**	**0.20**	**0.30**	**0.40**	**0.50**	**0.60**	**0.70**	**0.80**	**0.90**	**0.95**	**0.99**	
0	.886	.540	.282	.069	.014	.002	.000	.000	.000	.000	.000	.000	.000	0
1	.994	.882	.659	.275	.085	.020	.003	.000	.000	.000	.000	.000	.000	1
2	1.000	.980	.889	.558	.253	.083	.019	.003	.000	.000	.000	.000	.000	2
3	1.000	.998	.974	.795	.493	.225	.073	.015	.002	.000	.000	.000	.000	3
4	1.000	1.000	.996	.927	.724	.438	.194	.057	.009	.001	.000	.000	.000	4
5	1.000	1.000	.999	.981	.882	.665	.387	.158	.039	.004	.000	.000	.000	5
6	1.000	1.000	1.000	.996	.961	.842	.613	.335	.118	.019	.001	.000	.000	6
7	1.000	1.000	1.000	.999	.991	.943	.806	.562	.276	.073	.004	.000	.000	7
8	1.000	1.000	1.000	1.000	.998	.985	.927	.775	.507	.205	.026	.002	.000	8
9	1.000	1.000	1.000	1.000	1.000	.997	.981	.917	.747	.442	.111	.020	.000	9
10	1.000	1.000	1.000	1.000	1.000	1.000	.997	.980	.915	.725	.341	.118	.006	10
11	1.000	1.000	1.000	1.000	1.000	1.000	1.000	.998	.986	.931	.718	.460	.114	11
12	1.000	1.000	1.000	1.000	1.000	1.000	1.000	1.000	1.000	1.000	1.000	1.000	1.000	12

$n = 15$

a	\multicolumn{13}{c}{p}	a
	0.01	**0.05**	**0.10**	**0.20**	**0.30**	**0.40**	**0.50**	**0.60**	**0.70**	**0.80**	**0.90**	**0.95**	**0.99**	
0	.860	.463	.206	.035	.005	.000	.000	.000	.000	.000	.000	.000	.000	0
1	.990	.829	.549	.167	.035	.005	.000	.000	.000	.000	.000	.000	.000	1
2	1.000	.964	.816	.398	.127	.027	.004	.000	.000	.000	.000	.000	.000	2
3	1.000	.995	.944	.648	.297	.091	.018	.002	.000	.000	.000	.000	.000	3
4	1.000	.999	.987	.836	.515	.217	.059	.009	.001	.000	.000	.000	.000	4
5	1.000	1.000	.998	.939	.722	.403	.151	.034	.004	.000	.000	.000	.000	5
6	1.000	1.000	1.000	.982	.869	.610	.304	.095	.015	.001	.000	.000	.000	6
7	1.000	1.000	1.000	.996	.950	.787	.500	.213	.050	.004	.000	.000	.000	7
8	1.000	1.000	1.000	.999	.985	.905	.696	.390	.131	.018	.000	.000	.000	8
9	1.000	1.000	1.000	1.000	.996	.966	.849	.597	.278	.061	.002	.000	.000	9
10	1.000	1.000	1.000	1.000	.999	.991	.941	.783	.485	.164	.013	.001	.000	10
11	1.000	1.000	1.000	1.000	1.000	.998	.982	.909	.703	.352	.056	.005	.000	11
12	1.000	1.000	1.000	1.000	1.000	1.000	.996	.973	.873	.602	.184	.036	.000	12
13	1.000	1.000	1.000	1.000	1.000	1.000	1.000	.995	.965	.833	.451	.171	.010	13
14	1.000	1.000	1.000	1.000	1.000	1.000	1.000	1.000	.995	.965	.794	.537	.140	14
15	1.000	1.000	1.000	1.000	1.000	1.000	1.000	1.000	1.000	1.000	1.000	1.000	1.000	15

T A B L E 1 (Continued)

$n = 20$

							p							
a	0.01	0.05	0.10	0.20	0.30	0.40	0.50	0.60	0.70	0.80	0.90	0.95	0.99	a
0	.818	.358	.122	.012	.001	.000	.000	.000	.000	.000	.000	.000	.000	0
1	.983	.736	.392	.069	.008	.001	.000	.000	.000	.000	.000	.000	.000	1
2	.999	.925	.677	.206	.035	.004	.000	.000	.000	.000	.000	.000	.000	2
3	1.000	.984	.867	.411	.107	.016	.001	.000	.000	.000	.000	.000	.000	3
4	1.000	.997	.957	.630	.238	.051	.006	.000	.000	.000	.000	.000	.000	4
5	1.000	1.000	.989	.804	.416	.126	.021	.002	.000	.000	.000	.000	.000	5
6	1.000	1.000	.998	.913	.608	.250	.058	.006	.000	.000	.000	.000	.000	6
7	1.000	1.000	1.000	.968	.772	.416	.132	.021	.001	.000	.000	.000	.000	7
8	1.000	1.000	1.000	.990	.887	.596	.252	.057	.005	.000	.000	.000	.000	8
9	1.000	1.000	1.000	.997	.952	.755	.412	.128	.017	.001	.000	.000	.000	9
10	1.000	1.000	1.000	.999	.983	.872	.588	.245	.048	.003	.000	.000	.000	10
11	1.000	1.000	1.000	1.000	.995	.943	.748	.404	.113	.010	.000	.000	.000	11
12	1.000	1.000	1.000	1.000	.999	.979	.868	.584	.228	.032	.000	.000	.000	12
13	1.000	1.000	1.000	1.000	1.000	.994	.942	.750	.392	.087	.002	.000	.000	13
14	1.000	1.000	1.000	1.000	1.000	.998	.979	.874	.584	.196	.011	.000	.000	14
15	1.000	1.000	1.000	1.000	1.000	1.000	.994	.949	.762	.370	.043	.003	.000	15
16	1.000	1.000	1.000	1.000	1.000	1.000	.999	.984	.893	.589	.133	.016	.000	16
17	1.000	1.000	1.000	1.000	1.000	1.000	1.000	.996	.965	.794	.323	.075	.001	17
18	1.000	1.000	1.000	1.000	1.000	1.000	1.000	.999	.992	.931	.608	.264	.017	18
19	1.000	1.000	1.000	1.000	1.000	1.000	1.000	1.000	.999	.988	.878	.642	.182	19
20	1.000	1.000	1.000	1.000	1.000	1.000	1.000	1.000	1.000	1.000	1.000	1.000	1.000	20

T A B L E **1** (Continued)

$n = 25$

							p							
a	0.01	0.05	0.10	0.20	0.30	0.40	0.50	0.60	0.70	0.80	0.90	0.95	0.99	a
0	.778	.277	.072	.004	.000	.000	.000	.000	.000	.000	.000	.000	.000	0
1	.974	.642	.271	.027	.002	.000	.000	.000	.000	.000	.000	.000	.000	1
2	.998	.873	.537	.098	.009	.000	.000	.000	.000	.000	.000	.000	.000	2
3	1.000	.966	.764	.234	.033	.002	.000	.000	.000	.000	.000	.000	.000	3
4	1.000	.993	.902	.421	.090	.009	.000	.000	.000	.000	.000	.000	.000	4
5	1.000	.999	.967	.617	.193	.029	.002	.000	.000	.000	.000	.000	.000	5
6	1.000	1.000	.991	.780	.341	.074	.007	.000	.000	.000	.000	.000	.000	6
7	1.000	1.000	.998	.891	.512	.154	.022	.001	.000	.000	.000	.000	.000	7
8	1.000	1.000	1.000	.953	.677	.274	.054	.004	.000	.000	.000	.000	.000	8
9	1.000	1.000	1.000	.983	.811	.425	.115	.013	.000	.000	.000	.000	.000	9
10	1.000	1.000	1.000	.994	.902	.586	.212	.034	.002	.000	.000	.000	.000	10
11	1.000	1.000	1.000	.998	.956	.732	.345	.078	.006	.000	.000	.000	.000	11
12	1.000	1.000	1.000	1.000	.983	.846	.500	.154	.017	.000	.000	.000	.000	12
13	1.000	1.000	1.000	1.000	.994	.922	.655	.268	.044	.002	.000	.000	.000	13
14	1.000	1.000	1.000	1.000	.998	.966	.788	.414	.098	.006	.000	.000	.000	14
15	1.000	1.000	1.000	1.000	1.000	.987	.885	.575	.189	.017	.000	.000	.000	15
16	1.000	1.000	1.000	1.000	1.000	.996	.946	.726	.323	.047	.000	.000	.000	16
17	1.000	1.000	1.000	1.000	1.000	.999	.978	.846	.488	.109	.002	.000	.000	17
18	1.000	1.000	1.000	1.000	1.000	1.000	.993	.926	.659	.220	.009	.000	.000	18
19	1.000	1.000	1.000	1.000	1.000	1.000	.998	.971	.807	.383	.033	.001	.000	19
20	1.000	1.000	1.000	1.000	1.000	1.000	1.000	.991	.910	.579	.098	.007	.000	20
21	1.000	1.000	1.000	1.000	1.000	1.000	1.000	.998	.967	.766	.236	.034	.000	21
22	1.000	1.000	1.000	1.000	1.000	1.000	1.000	1.000	.991	.902	.463	.127	.002	22
23	1.000	1.000	1.000	1.000	1.000	1.000	1.000	1.000	.998	.973	.729	.358	.026	23
24	1.000	1.000	1.000	1.000	1.000	1.000	1.000	1.000	1.000	.996	.928	.723	.222	24
25	1.000	1.000	1.000	1.000	1.000	1.000	1.000	1.000	1.000	1.000	1.000	1.000	1.000	25

T A B L E **2**   Cumulative Probabilities of the Poisson Distribution

Tabulated values are $P(x \le a) = \sum_{x=0}^{a} p(x)$. (Computations are rounded at the third decimal place.)

Mean

a	0.250	0.500	0.750	1.000	1.250	1.500	1.750	2.000	2.250	2.500
0	0.779	0.607	0.472	0.368	0.287	0.223	0.174	0.135	0.105	0.082
1	0.974	0.910	0.827	0.736	0.645	0.558	0.478	0.406	0.343	0.287
2	0.998	0.986	0.959	0.920	0.868	0.809	0.744	0.677	0.609	0.544
3	1.000	0.998	0.993	0.981	0.962	0.934	0.899	0.857	0.809	0.758
4	1.000	1.000	0.999	0.996	0.991	0.981	0.967	0.947	0.922	0.891
5	1.000	1.000	1.000	0.999	0.998	0.996	0.991	0.983	0.973	0.958
6	1.000	1.000	1.000	1.000	1.000	0.999	0.998	0.995	0.992	0.986
7	1.000	1.000	1.000	1.000	1.000	1.000	1.000	0.999	0.998	0.996
8	1.000	1.000	1.000	1.000	1.000	1.000	1.000	1.000	0.999	0.999
9	1.000	1.000	1.000	1.000	1.000	1.000	1.000	1.000	1.000	1.000
10	1.000	1.000	1.000	1.000	1.000	1.000	1.000	1.000	1.000	1.000
11	1.000	1.000	1.000	1.000	1.000	1.000	1.000	1.000	1.000	1.000
12	1.000	1.000	1.000	1.000	1.000	1.000	1.000	1.000	1.000	1.000
13	1.000	1.000	1.000	1.000	1.000	1.000	1.000	1.000	1.000	1.000
14	1.000	1.000	1.000	1.000	1.000	1.000	1.000	1.000	1.000	1.000

Mean

a	2.750	3.000	3.250	3.500	3.750	4.000	4.250	4.500	4.750	5.000
0	0.064	0.050	0.039	0.030	0.024	0.018	0.014	0.011	0.009	0.007
1	0.240	0.199	0.165	0.136	0.112	0.092	0.075	0.061	0.050	0.040
2	0.481	0.423	0.370	0.321	0.277	0.238	0.204	0.174	0.147	0.125
3	0.703	0.647	0.591	0.537	0.484	0.433	0.386	0.342	0.302	0.265
4	0.855	0.815	0.772	0.725	0.678	0.629	0.580	0.532	0.485	0.440
5	0.939	0.916	0.889	0.858	0.823	0.785	0.745	0.703	0.660	0.616
6	0.978	0.966	0.952	0.935	0.914	0.889	0.862	0.831	0.798	0.762
7	0.993	0.988	0.982	0.973	0.962	0.949	0.933	0.913	0.891	0.867
8	0.998	0.996	0.994	0.990	0.985	0.979	0.970	0.960	0.947	0.932
9	0.999	0.999	0.998	0.997	0.995	0.992	0.988	0.983	0.976	0.968
10	1.000	1.000	0.999	0.999	0.998	0.997	0.996	0.993	0.990	0.986
11	1.000	1.000	1.000	1.000	0.999	0.999	0.998	0.998	0.996	0.995
12	1.000	1.000	1.000	1.000	1.000	1.000	1.000	0.999	0.999	0.998
13	1.000	1.000	1.000	1.000	1.000	1.000	1.000	1.000	1.000	0.999
14	1.000	1.000	1.000	1.000	1.000	1.000	1.000	1.000	1.000	1.000

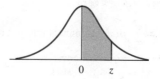

T A B L E 3 Normal Curve Areas	z	.00	.01	.02	.03	.04	.05	.06	.07	.08	.09
	0.0	.0000	.0040	.0080	.0120	.0160	.0199	.0239	.0279	.0319	.0359
	0.1	.0398	.0438	.0478	.0517	.0557	.0596	.0636	.0675	.0714	.0753
	0.2	.0793	.0832	.0871	.0910	.0948	.0987	.1026	.1064	.1103	.1141
	0.3	.1179	.1217	.1255	.1293	.1331	.1368	.1406	.1443	.1480	.1517
	0.4	.1554	.1591	.1628	.1664	.1700	.1736	.1772	.1808	.1844	.1879
	0.5	.1915	.1950	.1985	.2019	.2054	.2088	.2123	.2157	.2190	.2224
	0.6	.2257	.2291	.2324	.2357	.2389	.2422	.2454	.2486	.2517	.2549
	0.7	.2580	.2611	.2642	.2673	.2704	.2734	.2764	.2794	.2823	.2852
	0.8	.2881	.2910	.2939	.2967	.2995	.3023	.3051	.3078	.3106	.3133
	0.9	.3159	.3186	.3212	.3238	.3264	.3289	.3315	.3340	.3365	.3389
	1.0	.3413	.3438	.3461	.3485	.3508	.3531	.3554	.3577	.3599	.3621
	1.1	.3643	.3665	.3686	.3708	.3729	.3749	.3770	.3790	.3810	.3830
	1.2	.3849	.3869	.3888	.3907	.3925	.3944	.3962	.3980	.3997	.4015
	1.3	.4032	.4049	.4066	.4082	.4099	.4115	.4131	.4147	.4162	.4177
	1.4	.4192	.4207	.4222	.4236	.4251	.4265	.4279	.4292	.4306	.4319
	1.5	.4332	.4345	.4357	.4370	.4382	.4394	.4406	.4418	.4429	.4441
	1.6	.4452	.4463	.4474	.4484	.4495	.4505	.4515	.4525	.4535	.4545
	1.7	.4554	.4564	.4573	.4582	.4591	.4599	.4608	.4616	.4625	.4633
	1.8	.4641	.4649	.4656	.4664	.4671	.4678	.4686	.4693	.4699	.4706
	1.9	.4713	.4719	.4726	.4732	.4738	.4744	.4750	.4756	.4761	.4767
	2.0	.4772	.4778	.4783	.4788	.4793	.4798	.4803	.4808	.4812	.4817
	2.1	.4821	.4826	.4830	.4834	.4838	.4842	.4846	.4850	.4854	.4857
	2.2	.4861	.4864	.4868	.4871	.4875	.4878	.4881	.4884	.4887	.4890
	2.3	.4893	.4896	.4898	.4901	.4904	.4906	.4909	.4911	.4913	.4916
	2.4	.4918	.4920	.4922	.4925	.4927	.4929	.4931	.4932	.4934	.4936
	2.5	.4938	.4940	.4941	.4943	.4945	.4946	.4948	.4949	.4951	.4952
	2.6	.4953	.4955	.4956	.4957	.4959	.4960	.4961	.4962	.4963	.4964
	2.7	.4965	.4966	.4967	.4968	.4969	.4970	.4971	.4972	.4973	.4974
	2.8	.4974	.4975	.4976	.4977	.4977	.4978	.4979	.4979	.4980	.4981
	2.9	.4981	.4982	.4982	.4983	.4984	.4984	.4985	.4985	.4986	.4986
	3.0	.4987	.4987	.4987	.4988	.4988	.4989	.4989	.4989	.4990	.4990

Source: This table is abridged from Table 1 of *Statistical Tables and Formulas,* by A. Hald (New York: Wiley, 1952). Reproduced by permission of A. Hald and the publisher, John Wiley & Sons, Inc.

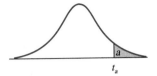

TABLE **4**	d.f.	$t_{.100}$	$t_{.050}$	$t_{.025}$	$t_{.010}$	$t_{.005}$	d.f.
Critical Values of t	1	3.078	6.314	12.706	31.821	63.657	1
	2	1.886	2.920	4.303	6.965	9.925	2
	3	1.638	2.353	3.182	4.541	5.841	3
	4	1.533	2.132	2.776	3.747	4.604	4
	5	1.476	2.015	2.571	3.365	4.032	5
	6	1.440	1.943	2.447	3.143	3.707	6
	7	1.415	1.895	2.365	2.998	3.499	7
	8	1.397	1.860	2.306	2.896	3.355	8
	9	1.383	1.833	2.262	2.821	3.250	9
	10	1.372	1.812	2.228	2.764	3.169	10
	11	1.363	1.796	2.201	2.718	3.106	11
	12	1.356	1.782	2.179	2.681	3.055	12
	13	1.350	1.771	2.160	2.650	3.012	13
	14	1.345	1.761	2.145	2.624	2.977	14
	15	1.341	1.753	2.131	2.602	2.947	15
	16	1.337	1.746	2.120	2.583	2.921	16
	17	1.333	1.740	2.110	2.567	2.898	17
	18	1.330	1.734	2.101	2.552	2.878	18
	19	1.328	1.729	2.093	2.539	2.861	19
	20	1.325	1.725	2.086	2.528	2.845	20
	21	1.323	1.721	2.080	2.518	2.831	21
	22	1.321	1.717	2.074	2.508	2.819	22
	23	1.319	1.714	2.069	2.500	2.807	23
	24	1.318	1.711	2.064	2.492	2.797	24
	25	1.316	1.708	2.060	2.485	2.787	25
	26	1.315	1.706	2.056	2.479	2.779	26
	27	1.314	1.703	2.052	2.473	2.771	27
	28	1.313	1.701	2.048	2.467	2.763	28
	29	1.311	1.699	2.045	2.462	2.756	29
	inf.	1.282	1.645	1.960	2.326	2.576	inf.

Source: From "Table of Percentage Points of the *t*-Distribution," *Biometrika* 32 (1941) 300. Reproduced by permission of the *Biometrika* Trustees.

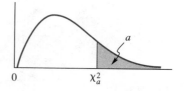

Source: From "Tables of the Percentage Points of the χ²-Distribution," *Biometrika Tables for Statisticians* 1, 3d ed. (1966). Reproduced by permission of the *Biometrika* Trustees.

TABLE 5
Critical Values of Chi-Square

d.f.	$\chi^2_{0.995}$	$\chi^2_{0.990}$	$\chi^2_{0.975}$	$\chi^2_{0.950}$	$\chi^2_{0.900}$
1	0.0000393	0.0001571	0.0009821	0.0039321	0.0157908
2	0.0100251	0.0201007	0.0506356	0.102587	0.210720
3	0.0717212	0.114832	0.215795	0.351846	0.584375
4	0.206990	0.297110	0.484419	0.710721	1.063623
5	0.411740	0.554300	0.831211	1.145476	1.61031
6	0.675727	0.872085	1.237347	0.63539	2.20413
7	0.989265	1.239043	1.68987	2.16735	2.83311
8	1.344419	1.646482	2.17973	2.73264	3.48954
9	1.734926	2.087912	2.70039	3.32511	4.16816
10	2.15585	2.55821	3.24697	3.94030	4.86518
11	2.60321	3.05347	3.81575	4.57481	5.57779
12	3.07382	3.57056	4.40379	5.22603	6.30380
13	3.56503	4.10691	5.00874	5.89186	7.04150
14	4.07468	4.66043	5.62872	6.57063	7.78953
15	4.60094	5.22935	6.26214	7.26094	8.54675
16	5.14224	5.81221	6.90766	7.96164	9.31223
17	5.69724	6.40776	7.56418	8.67176	10.0852
18	6.26481	7.01491	8.23075	9.39046	10.8649
19	6.84398	7.63273	8.90655	10.1170	11.6509
20	7.43386	8.26040	9.59083	10.8508	12.4426
21	8.03366	8.89720	10.28293	11.5913	13.2396
22	8.64272	9.54249	10.9823	12.3380	14.0415
23	9.26042	10.19567	11.6885	13.0905	14.8479
24	9.88623	10.8564	12.4011	13.8484	15.6587
25	10.5197	11.5240	13.1197	14.6114	16.4734
26	11.1603	12.1981	13.8439	15.3791	17.2919
27	11.8076	12.8786	14.5733	16.1513	18.1138
28	12.4613	13.5648	15.3079	16.9279	18.9392
29	13.1211	14.2565	16.0471	17.7083	19.7677
30	13.7867	14.9535	16.7908	18.4926	20.5992
40	20.7065	22.1643	24.4331	26.5093	29.0505
50	27.9907	29.7067	32.3574	34.7642	37.6886
60	35.5346	37.4848	40.4817	43.1879	46.4589
70	43.2752	45.4418	48.7576	51.7393	55.3290
80	51.1720	53.5400	57.1532	60.3915	64.2778
90	59.1963	61.7541	65.6466	69.1260	73.2912
100	67.3276	70.0648	74.2219	77.9295	82.3581

	$\chi^2_{0.100}$	$\chi^2_{0.050}$	$\chi^2_{0.025}$	$\chi^2_{0.010}$	$\chi^2_{0.005}$	d.f.
TABLE **5** (Continued)	2.70554	3.84146	5.02389	6.63490	7.87944	1
	4.60517	5.99147	7.37776	9.21034	10.5966	2
	6.25139	7.81473	9.34840	11.3449	12.8381	3
	7.77944	9.48773	11.1433	13.2767	14.8602	4
	9.23635	11.0705	12.8325	15.0863	16.7496	5
	10.6446	12.5916	14.4494	16.8119	18.5476	6
	12.0170	14.0671	16.0128	18.4753	20.2777	7
	13.3616	15.5073	17.5346	20.0902	21.9550	8
	14.6837	16.9190	19.0228	21.6660	23.5893	9
	15.9871	18.3070	20.4831	23.2093	25.1882	10
	17.2750	19.6751	21.9200	24.7250	26.7569	11
	18.5494	21.0261	23.3367	26.2170	28.2995	12
	19.8119	22.3621	24.7356	27.6883	29.8194	13
	21.0642	23.6848	26.1190	29.1413	31.3193	14
	22.3072	24.9958	27.4884	30.5779	32.8013	15
	23.5418	26.2962	28.8485	31.9999	34.2672	16
	24.7690	27.8571	30.1910	33.4087	35.7185	17
	25.9894	28.8693	31.5264	34.8053	37.1564	18
	27.2036	30.1435	32.8523	36.1908	38.5822	19
	28.4120	31.4104	34.1696	37.5662	39.9968	20
	29.6151	32.6705	35.4789	38.9321	41.4010	21
	30.8133	33.9244	36.7807	40.2894	42.7956	22
	32.0069	35.1725	38.0757	41.6384	44.1813	23
	33.1963	36.4151	39.3641	42.9798	45.5585	24
	34.3816	37.6525	40.6465	44.3141	46.9278	25
	35.5631	38.8852	41.9232	45.6417	48.2899	26
	36.7412	40.1133	43.1944	46.9630	49.6449	27
	37.9159	41.3372	44.4607	48.2782	50.9933	28
	39.0875	42.5569	45.7222	49.5879	52.3356	29
	40.2560	43.7729	46.9792	50.8922	53.6720	30
	51.8050	55.7585	59.3417	63.6907	66.7659	40
	63.1671	67.5048	71.4202	76.1539	79.4900	50
	74.3970	79.0819	83.2976	88.3794	91.9517	60
	85.5271	90.5312	95.0231	100.425	104.215	70
	96.5782	101.879	106.629	112.329	116.321	80
	107.565	113.145	118.136	124.116	128.299	90
	118.498	124.342	129.561	135.807	140.169	100

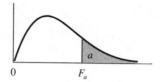

T A B L E **6**   Percentage Points of the *F* Distribution

		ν_1								
ν_2	*a*	1	2	3	4	5	6	7	8	9
1	.100	39.86	49.50	53.59	55.83	57.24	58.20	58.91	59.44	59.86
	.050	161.4	199.5	215.7	224.6	230.2	234.0	236.8	238.9	240.5
	.025	647.8	799.5	864.2	899.6	921.8	937.1	948.2	956.7	963.3
	.010	4052	4999.5	5403	5625	5764	5859	5928	5982	6022
	.005	16211	20000	21615	22500	23056	23437	23715	23925	24091
2	.100	8.53	9.00	9.16	9.24	9.29	9.33	9.35	9.37	9.38
	.050	18.51	19.00	19.16	19.25	19.30	19.33	19.35	19.37	19.38
	.025	38.51	39.00	39.17	39.25	39.30	39.33	39.36	39.37	39.39
	.010	98.50	99.00	99.17	99.25	99.30	99.33	99.36	99.37	99.39
	.005	198.5	199.0	199.2	199.2	199.3	199.3	199.4	199.4	199.4
3	.100	5.54	5.46	5.39	5.34	5.31	5.28	5.27	5.25	5.24
	.050	10.13	9.55	9.28	9.12	9.01	8.94	8.89	8.85	8.81
	.025	17.44	16.04	15.44	15.10	14.88	14.73	14.62	14.54	14.47
	.010	34.12	30.82	29.46	28.71	28.24	27.91	27.64	27.49	27.35
	.005	55.55	49.80	47.47	46.19	45.39	44.84	44.43	44.13	43.88
4	.100	4.54	4.32	4.19	4.11	4.05	4.01	3.98	3.95	3.94
	.050	7.71	6.94	6.59	6.39	6.26	6.16	6.09	6.04	6.00
	.025	12.22	10.65	9.98	9.60	9.36	9.20	9.07	8.98	8.90
	.010	21.20	18.00	16.69	15.98	15.52	15.21	14.98	14.80	14.66
	.005	31.33	26.28	24.26	23.15	22.46	21.97	21.62	21.35	21.14
5	.100	4.06	3.78	3.62	3.52	3.45	3.40	3.37	3.34	3.32
	.050	6.61	5.79	5.41	5.19	5.05	4.95	4.88	4.82	4.77
	.025	10.01	8.43	7.76	7.39	7.15	6.98	6.85	6.76	6.68
	.010	16.26	13.27	12.06	11.39	10.97	10.67	10.46	10.29	10.16
	.005	22.78	18.31	16.53	15.56	14.94	14.51	14.20	13.96	13.77
6	.100	3.78	3.46	3.29	3.18	3.11	3.05	3.01	2.98	2.96
	.050	5.99	5.14	4.76	4.53	4.39	4.28	4.21	4.15	4.10
	.025	8.81	7.26	6.60	6.23	5.99	5.82	5.70	5.60	5.52
	.010	13.75	10.92	9.78	9.15	8.75	8.47	8.26	8.10	7.98
	.005	18.63	14.54	12.92	12.03	11.46	11.07	10.79	10.57	10.39
7	.100	3.59	3.26	3.07	2.96	2.88	2.83	2.78	2.75	2.72
	.050	5.59	4.74	4.35	4.12	3.97	3.87	3.79	3.73	3.68
	.025	8.07	6.54	5.89	5.52	5.29	5.12	4.99	4.90	4.82
	.010	12.25	9.55	8.45	7.85	7.46	7.19	6.99	6.84	6.72
	.005	16.24	12.40	10.88	10.05	9.52	9.16	8.89	8.68	8.51
8	.100	3.46	3.11	2.92	2.81	2.73	2.67	2.62	2.59	2.56
	.050	5.32	4.46	4.07	3.84	3.69	3.58	3.50	3.44	3.39
	.025	7.57	6.06	5.42	5.05	4.82	4.65	4.53	4.43	4.36
	.010	11.26	8.65	7.59	7.01	6.63	6.37	6.18	6.03	5.91
	.005	14.69	11.04	9.60	8.81	8.30	7.95	7.69	7.50	7.34
9	.100	3.36	3.01	2.81	2.69	2.61	2.55	2.51	2.47	2.44
	.050	5.12	4.26	3.86	3.63	3.48	3.37	3.29	3.23	3.18
	.025	7.21	5.71	5.08	4.72	4.48	4.32	4.20	4.10	4.03
	.010	10.56	8.02	6.99	6.42	6.06	5.80	5.61	5.47	5.35
	.005	13.61	10.11	8.72	7.96	7.47	7.13	6.88	6.69	6.54

Source: A portion of "Tables of percentage points of the inverted beta (*E*) distribution" *Biometrika,* vol. 33 (1943) by M. Merrington and C. M. Thompson and from Table 18 of *Biometrika Tables for Statisticians,* vol. 1, Cambridge University Press, 1954, edited by E. S. Pearson and H. O. Hartley. Reproduced with permission of the authors, editors, and *Biometrika* trustees.

T A B L E **6**   (Continued)

				ν_1								
10	**12**	**15**	**20**	**24**	**30**	**40**	**60**	**120**	**∞**	**a**	**ν_2**
60.19	60.71	60.22	61.74	62.00	62.26	62.53	62.79	63.06	63.33	.100	1
241.9	243.9	245.9	248.0	249.1	250.1	251.2	252.2	253.3	254.3	.050	
968.6	976.7	984.9	993.1	997.2	1001	1006	1010	1014	1018	.025	
6056	6106	6157	6209	6235	6261	6287	6313	6339	6366	.010	
24224	24426	24630	24836	24940	25044	25148	25253	25359	25465	.005	
9.39	9.41	9.42	9.44	9.45	9.46	9.47	9.47	9.48	9.49	.100	2
19.40	19.41	19.43	19.45	19.45	19.46	19.47	19.48	19.49	19.50	.050	
39.40	39.41	39.43	39.45	39.46	39.46	39.47	39.48	39.49	39.50	.025	
99.40	99.42	99.43	99.45	99.46	99.47	99.47	99.48	99.49	99.50	.010	
199.4	199.4	199.4	199.4	199.5	199.5	199.5	199.5	199.5	199.5	.005	
5.23	5.22	5.20	5.18	5.18	5.17	5.16	5.15	5.14	5.13	.100	3
8.79	8.74	8.70	8.66	8.64	8.62	8.59	8.57	8.55	8.53	.050	
14.42	14.34	14.25	14.17	14.12	14.08	14.04	13.99	13.95	13.90	.025	
27.23	27.05	26.87	26.69	26.60	26.50	26.41	26.32	26.22	26.13	.010	
43.69	43.39	43.08	42.78	42.62	42.47	42.31	42.15	41.99	41.83	.005	
3.92	3.90	3.87	3.84	3.83	3.82	3.80	3.79	3.78	3.76	.100	4
5.96	5.91	5.86	5.80	5.77	5.75	5.72	5.69	5.66	5.63	.050	
8.84	8.75	8.66	8.56	8.51	8.46	8.41	8.36	8.31	8.26	.025	
14.55	14.37	14.20	14.02	13.93	13.84	13.75	13.65	13.56	13.46	.010	
20.97	20.70	20.44	20.17	20.03	19.89	19.75	19.61	19.47	19.32	.005	
3.30	3.27	3.24	3.21	3.19	3.17	3.16	3.14	3.12	3.10	.100	5
4.74	4.68	4.62	4.56	4.53	4.50	4.46	4.43	4.40	4.36	.050	
6.62	6.52	6.43	6.33	6.28	6.23	6.18	6.12	6.07	6.02	.025	
10.05	9.89	9.72	9.55	9.47	9.38	9.29	9.20	9.11	9.02	.010	
13.62	13.38	13.15	12.90	12.78	12.66	12.53	12.40	12.27	12.14	.005	
2.94	2.90	2.87	2.84	2.82	2.80	2.78	2.76	2.74	2.72	.100	6
4.06	4.00	3.94	3.87	3.84	3.81	3.77	3.74	3.70	3.67	.050	
5.46	5.37	5.27	5.17	5.12	5.07	5.01	4.96	4.90	4.85	.025	
7.87	7.72	7.56	7.40	7.31	7.23	7.14	7.06	6.97	6.88	.010	
10.25	10.03	9.81	9.59	9.47	9.36	9.24	9.12	9.00	8.88	.005	
2.70	2.67	2.63	2.59	2.58	2.56	2.54	2.51	2.49	2.47	.100	7
3.64	3.57	3.51	<u>3.44</u>	<u>3.41</u>	<u>3.38</u>	3.34	3.30	3.27	3.23	.050	
4.76	4.67	4.57	4.47	4.42	4.36	4.31	4.25	4.20	4.14	.025	
6.62	6.47	6.31	6.16	6.07	5.99	5.91	5.82	5.74	5.65	.010	
8.38	8.18	7.97	7.75	7.65	7.53	7.42	7.31	7.19	7.08	.005	
2.54	2.50	2.46	2.42	2.40	2.38	2.36	2.34	2.32	2.29	.100	8
3.35	3.28	3.22	<u>3.15</u>	<u>3.12</u>	<u>3.08</u>	3.04	3.01	2.97	2.93	.050	
4.30	4.20	4.10	4.00	3.95	3.89	3.84	3.78	3.73	3.67	.025	
5.81	5.67	5.52	5.36	5.28	5.20	5.12	5.03	4.95	4.86	.010	
7.21	7.01	6.81	6.61	6.50	6.40	6.29	6.18	6.06	5.95	.005	
2.42	2.38	2.34	2.30	2.28	2.25	2.23	2.21	2.18	2.16	.100	9
3.14	3.07	3.01	2.94	2.90	2.86	2.83	2.79	2.75	2.71	.050	
3.96	3.87	3.77	3.67	3.61	3.56	3.51	3.45	3.39	3.33	.025	
5.26	5.11	4.96	4.81	4.73	4.65	4.57	4.48	4.40	4.31	.010	
6.42	6.23	6.03	5.83	5.73	5.62	5.52	5.41	5.30	5.19	.005	

T A B L E 6   (Continued)

ν_2	a	1	2	3	4	5	6	7	8	9
					ν_1					
10	.100	3.29	2.92	2.73	2.61	2.52	2.46	2.41	2.38	2.35
	.050	4.96	4.10	3.71	3.48	3.33	3.22	3.14	3.07	3.02
	.025	6.94	5.46	4.83	4.47	4.24	4.07	3.95	3.85	3.78
	.010	10.04	7.56	6.55	5.99	5.64	5.39	5.20	5.06	4.94
	.005	12.83	9.43	8.08	7.34	6.87	6.54	6.30	6.12	5.97
11	.100	3.23	2.86	2.66	2.54	2.45	2.39	2.34	2.30	2.27
	.050	4.84	3.98	3.59	3.36	3.20	3.09	3.01	2.95	2.90
	.025	6.72	5.26	4.63	4.28	4.04	3.88	3.76	3.66	3.59
	.010	9.65	7.21	6.22	5.67	5.32	5.07	4.89	4.74	4.63
	.005	12.23	8.91	7.60	6.88	6.42	6.10	5.86	5.68	5.54
12	.100	3.18	2.81	2.61	2.48	2.39	2.33	2.28	2.24	2.21
	.050	4.75	3.89	3.49	3.26	3.11	3.00	2.91	2.85	2.80
	.025	6.55	5.10	4.47	4.12	3.89	3.73	3.61	3.51	3.44
	.010	9.33	6.93	5.95	5.41	5.06	4.82	4.64	4.50	4.39
	.005	11.75	8.51	7.23	6.52	6.07	5.76	5.52	5.35	5.20
13	.100	3.14	2.76	2.56	2.43	2.35	2.28	2.23	2.20	2.16
	.050	4.67	3.81	3.41	3.18	3.03	2.92	2.83	2.77	2.71
	.025	6.41	4.97	4.35	4.00	3.77	3.60	3.48	3.39	3.31
	.010	9.07	6.70	5.74	5.21	4.86	4.62	4.44	4.30	4.19
	.005	11.37	8.19	6.93	6.23	5.79	5.48	5.25	5.08	4.94
14	.100	3.10	2.73	2.52	2.39	2.31	2.24	2.19	2.15	2.12
	.050	4.60	3.74	3.34	3.11	2.96	2.85	2.76	2.70	2.65
	.025	6.30	4.86	4.24	3.89	3.66	3.50	3.38	3.29	3.21
	.010	8.86	6.51	5.56	5.04	4.69	4.46	4.28	4.14	4.03
	.005	11.06	7.92	6.68	6.00	5.56	5.26	5.03	4.86	4.72
15	.100	3.07	2.70	2.49	2.36	2.27	2.21	2.16	2.12	2.09
	.050	4.54	3.68	3.29	3.06	2.90	2.79	2.71	2.64	2.59
	.025	6.20	4.77	4.15	3.80	3.58	3.41	3.29	3.20	3.12
	.010	8.68	6.36	5.42	4.89	4.56	4.32	4.14	4.00	3.89
	.005	10.80	7.70	6.48	5.80	5.37	5.07	4.85	4.67	4.54
16	.100	3.05	2.67	2.46	2.33	2.24	2.18	2.13	2.09	2.06
	.050	4.49	3.63	3.24	3.01	2.85	2.74	2.66	2.59	2.54
	.025	6.12	4.69	4.08	3.73	3.50	3.34	3.22	3.12	3.05
	.010	8.53	6.23	5.29	4.77	4.44	4.20	4.03	3.89	3.78
	.005	10.58	7.51	6.30	5.64	5.21	4.91	4.69	4.52	4.38
17	.100	3.03	2.64	2.44	2.31	2.22	2.15	2.10	2.06	2.03
	.050	4.45	3.59	3.20	2.96	2.81	2.70	2.61	2.55	2.49
	.025	6.04	4.62	4.01	3.66	3.44	3.28	3.16	3.06	2.98
	.010	8.40	6.11	5.18	4.67	4.34	4.10	3.93	3.79	3.68
	.005	10.38	7.35	6.16	5.50	5.07	4.78	4.56	4.39	4.25
18	.100	3.01	2.62	2.42	2.29	2.20	2.13	2.08	2.04	2.00
	.050	4.41	3.55	3.16	2.93	2.77	2.66	2.58	2.51	2.46
	.025	5.98	4.56	3.95	3.61	3.38	3.22	3.10	3.01	2.93
	.010	8.29	6.01	5.09	4.58	4.25	4.01	3.84	3.71	3.60
	.005	10.22	7.21	6.03	5.37	4.96	4.66	4.44	4.28	4.14
19	.100	2.99	2.61	2.40	2.27	2.18	2.11	2.06	2.02	1.98
	.050	4.38	3.52	3.13	2.90	2.74	2.63	2.54	2.48	2.42
	.025	5.92	4.51	3.90	3.56	3.33	3.17	3.05	2.96	2.88
	.010	8.18	5.93	5.01	4.50	4.17	3.94	3.77	3.63	3.52
	.005	10.07	7.09	5.92	5.27	4.85	4.56	4.34	4.18	4.04
20	.100	2.97	2.59	2.38	2.25	2.16	2.09	2.04	2.00	1.96
	.050	4.35	3.49	3.10	2.87	2.71	2.60	2.51	2.45	2.39
	.025	5.87	4.46	3.86	3.51	3.29	3.13	3.01	2.91	2.84
	.010	8.10	5.85	4.94	4.43	4.10	3.87	3.70	3.56	3.46
	.005	9.94	6.99	5.82	5.17	4.76	4.47	4.26	4.09	3.96

TABLE 6   (Continued)

					ν_1						
10	**12**	**15**	**20**	**24**	**30**	**40**	**60**	**120**	**∞**	**a**	**ν_2**
2.32	2.28	2.24	2.20	2.18	2.16	2.13	2.11	2.08	2.06	.100	10
2.98	2.91	2.85	2.77	2.74	2.70	2.66	2.62	2.58	2.54	.050	
3.72	3.62	3.52	3.42	3.37	3.31	3.26	3.20	3.14	3.08	.025	
4.85	4.71	4.56	4.41	4.33	4.25	4.17	4.08	4.00	3.91	.010	
5.85	5.66	5.47	5.27	5.17	5.07	4.97	4.86	4.75	4.64	.005	
2.25	2.21	2.17	2.12	2.10	2.08	2.05	2.03	2.00	1.97	.100	11
2.85	2.79	2.72	2.65	2.61	2.57	2.53	2.49	2.45	2.40	.050	
3.53	3.43	3.33	3.23	3.17	3.12	3.06	3.00	2.94	2.88	.025	
4.54	4.40	4.25	4.10	4.02	3.94	3.86	3.78	3.69	3.60	.010	
5.42	5.24	5.05	4.86	4.76	4.65	4.55	4.44	4.34	4.23	.005	
2.19	2.15	2.10	2.06	2.04	2.01	1.99	1.96	1.93	1.90	.100	12
2.75	2.69	2.62	2.54	2.51	2.47	2.43	2.38	2.34	2.30	.050	
3.37	3.28	3.18	3.07	3.02	2.96	2.91	2.85	2.79	2.72	.025	
4.30	4.16	4.01	3.86	3.78	3.70	3.62	3.54	3.45	3.36	.010	
5.09	4.91	4.72	4.53	4.43	4.33	4.23	4.12	4.01	3.90	.005	
2.14	2.10	2.05	2.01	1.98	1.96	1.93	1.90	1.88	1.85	.100	13
2.67	2.60	2.53	2.46	2.42	2.38	2.34	2.30	2.25	2.21	.050	
3.25	3.15	3.05	2.95	2.89	2.84	2.78	2.72	2.66	2.60	.025	
4.10	3.96	3.82	3.66	3.59	3.51	3.43	3.34	3.25	3.17	.010	
4.82	4.64	4.46	4.27	4.17	4.07	3.97	3.87	3.76	3.65	.005	
2.10	2.05	2.01	1.96	1.94	1.91	1.89	1.86	1.83	1.80	.100	14
2.60	2.53	2.46	2.39	2.35	2.31	2.27	2.22	2.18	2.13	.050	
3.15	3.05	2.95	2.84	2.79	2.73	2.67	2.61	2.55	2.49	.025	
3.94	3.80	3.66	3.51	3.43	3.35	3.27	3.18	3.09	3.00	.010	
4.60	4.43	4.25	4.06	3.96	3.86	3.76	3.66	3.55	3.44	.005	
2.06	2.02	1.97	1.92	1.90	1.87	1.85	1.82	1.79	1.76	.100	15
2.54	2.48	2.40	2.33	2.29	2.25	2.20	2.16	2.11	2.07	.050	
3.06	2.96	2.86	2.76	2.70	2.64	2.59	2.52	2.46	2.40	.025	
3.80	3.67	3.52	3.37	3.29	3.21	3.13	3.05	2.96	2.87	.010	
4.42	4.25	4.07	3.88	3.79	3.69	3.58	3.48	3.37	3.26	.005	
2.03	1.99	1.94	1.89	1.87	1.84	1.81	1.78	1.75	1.72	.100	16
2.49	2.42	2.35	2.28	2.24	2.19	2.15	2.11	2.06	2.01	.050	
2.99	2.89	2.79	2.68	2.63	2.57	2.51	2.45	2.38	2.32	.025	
3.69	3.55	3.41	3.26	3.18	3.10	3.02	2.93	2.84	2.75	.010	
4.27	4.10	3.92	3.73	3.64	3.54	3.44	3.33	3.22	3.11	.005	
2.00	1.96	1.91	1.86	1.84	1.81	1.78	1.75	1.72	1.69	.100	17
2.45	2.38	2.31	2.23	2.19	2.15	2.10	2.06	2.01	1.96	.050	
2.92	2.82	2.72	2.62	2.56	2.50	2.44	2.38	2.32	2.25	.025	
3.59	3.46	3.31	3.16	3.08	3.00	2.92	2.83	2.75	2.65	.010	
4.14	3.97	3.79	3.61	3.51	3.41	3.31	3.21	3.10	2.98	.005	
1.98	1.93	1.89	1.84	1.81	1.78	1.75	1.72	1.69	1.66	.100	18
2.41	2.34	2.27	2.19	2.15	2.11	2.06	2.02	1.97	1.92	.050	
2.87	2.77	2.67	2.56	2.50	2.44	2.38	2.32	2.26	2.19	.025	
3.51	3.37	3.23	3.08	3.00	2.92	2.84	2.75	2.66	2.57	.010	
4.03	3.86	3.68	3.50	3.40	3.30	3.20	3.10	2.99	2.87	.005	
1.96	1.91	1.86	1.81	1.79	1.76	1.73	1.70	1.67	1.63	.100	19
2.38	2.31	2.23	2.16	2.11	2.07	2.03	1.98	1.93	1.88	.050	
2.82	2.72	2.62	2.51	2.45	2.39	2.33	2.27	2.20	2.13	.025	
3.43	3.30	3.15	3.00	2.92	2.84	2.76	2.67	2.58	2.49	.010	
3.93	3.76	3.59	3.40	3.31	3.21	3.11	3.00	2.89	2.78	.005	
1.94	1.89	1.84	1.79	1.77	1.74	1.71	1.68	1.64	1.61	.100	20
2.35	2.28	2.20	2.12	2.08	2.04	1.99	1.95	1.90	1.84	.050	
2.77	2.68	2.57	2.46	2.41	2.35	2.29	2.22	2.16	2.09	.025	
3.37	3.23	3.09	2.94	2.86	2.78	2.69	2.61	2.52	2.42	.010	
3.85	3.68	3.50	3.32	3.22	3.12	3.02	2.92	2.81	2.69	.005	

T A B L E 6 (Continued)

ν_2	α	ν_1 1	2	3	4	5	6	7	8	9
21	.100	2.96	2.57	2.36	2.23	2.14	2.08	2.02	1.98	1.95
	.050	4.32	3.47	3.07	2.84	2.68	2.57	2.49	2.42	2.37
	.025	5.83	4.42	3.82	3.48	3.25	3.09	2.97	2.87	2.80
	.010	8.02	5.78	4.87	4.37	4.04	3.81	3.64	3.51	3.40
	.005	9.83	6.89	5.73	5.09	4.68	4.39	4.18	4.01	3.88
22	.100	2.95	2.56	2.35	2.22	2.13	2.06	2.01	1.97	1.93
	.050	4.30	3.44	3.05	2.82	2.66	2.55	2.46	2.40	2.34
	.025	5.79	4.38	3.78	3.44	3.22	3.05	2.93	2.84	2.76
	.010	7.95	5.72	4.82	4.31	3.99	3.76	3.59	3.45	3.35
	.005	9.73	6.81	5.65	5.02	4.61	4.32	4.11	3.94	3.81
23	.100	2.94	2.55	2.34	2.21	2.11	2.05	1.99	1.95	1.92
	.050	4.28	3.42	3.03	2.80	2.64	2.53	2.44	2.37	2.32
	.025	5.75	4.35	3.75	3.41	3.18	3.02	2.90	2.81	2.73
	.010	7.88	5.66	4.76	4.26	3.94	3.71	3.54	3.41	3.30
	.005	9.63	6.73	5.58	4.95	4.54	4.26	4.05	3.88	3.75
24	.100	2.93	2.54	2.33	2.19	2.10	2.04	1.98	1.94	1.91
	.050	4.26	3.40	3.01	2.78	2.62	2.51	2.42	2.36	2.30
	.025	5.72	4.32	3.72	3.38	3.15	2.99	2.87	2.78	2.70
	.010	7.82	5.61	4.72	4.22	3.90	3.67	3.50	3.36	3.26
	.005	9.55	6.66	5.52	4.89	4.49	4.20	3.99	3.83	3.69
25	.100	2.92	2.53	2.32	2.18	2.09	2.02	1.97	1.93	1.89
	.050	4.24	3.39	2.99	2.76	2.60	2.49	2.40	2.34	2.28
	.025	5.69	4.29	3.69	3.35	3.13	2.97	2.85	2.75	2.68
	.010	7.77	5.57	4.68	4.18	3.85	3.63	3.46	3.32	3.22
	.005	9.48	6.60	5.46	4.84	4.43	4.15	3.94	3.78	3.64
26	.100	2.91	2.52	2.31	2.17	2.08	2.01	1.96	1.92	1.88
	.050	4.23	3.37	2.98	2.74	2.59	2.47	2.39	2.32	2.27
	.025	5.66	4.27	3.67	3.33	3.10	2.94	2.82	2.73	2.65
	.010	7.72	5.53	4.64	4.14	3.82	3.59	3.42	3.29	3.18
	.005	9.41	6.54	5.41	4.79	4.38	4.10	3.89	3.73	3.60
27	.100	2.90	2.51	2.30	2.17	2.07	2.00	1.95	1.91	1.87
	.050	4.21	3.35	2.96	2.73	2.57	2.46	2.37	2.31	2.25
	.025	5.63	4.24	3.65	3.31	3.08	2.92	2.80	2.71	2.63
	.010	7.68	5.49	4.60	4.11	3.78	3.56	3.39	3.26	3.15
	.005	9.34	6.49	5.36	4.74	4.34	4.06	3.85	3.69	3.56
28	.100	2.89	2.50	2.29	2.16	2.06	2.00	1.94	1.90	1.87
	.050	4.20	3.34	2.95	2.71	2.56	2.45	2.36	2.29	2.24
	.025	5.61	4.22	3.63	3.29	3.06	2.90	2.78	2.69	2.61
	.010	7.64	5.45	4.57	4.07	3.75	3.53	3.36	3.23	3.12
	.005	9.28	6.44	5.32	4.70	4.30	4.02	3.81	3.65	3.52
29	.100	2.89	2.50	2.28	2.15	2.06	1.99	1.93	1.89	1.86
	.050	4.18	3.33	2.93	2.70	2.55	2.43	2.35	2.28	2.22
	.025	5.59	4.20	3.61	3.27	3.04	2.88	2.76	2.67	2.59
	.010	7.60	5.42	4.54	4.04	3.73	3.50	3.33	3.20	3.09
	.005	9.23	6.40	5.28	4.66	4.26	3.98	3.77	3.61	3.48
30	.100	2.88	2.49	2.28	2.14	2.05	1.98	1.93	1.88	1.85
	.050	4.17	3.32	2.92	2.69	2.53	2.42	2.33	2.27	2.21
	.025	5.57	4.18	3.59	3.25	3.03	2.87	2.75	2.65	2.57
	.010	7.56	5.39	4.51	4.02	3.70	3.47	3.30	3.17	2.57
	.005	9.18	6.35	5.24	4.62	4.23	3.95	3.74	3.58	3.45

T A B L E **6**   (Continued)

					ν_1						
10	**12**	**15**	**20**	**24**	**30**	**40**	**60**	**120**	**∞**	**a**	**ν_2**
1.92	1.87	1.83	1.78	1.75	1.72	1.69	1.66	1.62	1.59	.100	21
2.32	2.25	2.18	2.10	2.05	2.01	1.96	1.92	1.87	1.81	.050	
2.73	2.64	2.53	2.42	2.37	2.31	2.25	2.18	2.11	2.04	.025	
3.31	3.17	3.03	2.88	2.80	2.72	2.64	2.55	2.46	2.36	.010	
3.77	3.60	3.43	3.24	3.15	3.05	2.95	2.84	2.73	2.61	.005	
1.90	1.86	1.81	1.76	1.73	1.70	1.67	1.64	1.60	1.57	.100	22
2.30	2.23	2.15	2.07	2.03	1.98	1.94	1.89	1.84	1.78	.050	
2.70	2.60	2.50	2.39	2.33	2.27	2.21	2.14	2.08	2.00	.025	
3.26	3.12	2.98	2.83	2.75	2.67	2.58	2.50	2.40	2.31	.010	
3.70	3.54	3.36	3.18	3.08	2.98	2.88	2.77	2.66	2.55	.005	
1.89	1.84	1.80	1.74	1.72	1.69	1.66	1.62	1.59	1.55	.100	23
2.27	2.20	2.13	2.05	2.01	1.96	1.91	1.86	1.81	1.76	.050	
2.67	2.57	2.47	2.36	2.30	2.24	2.18	2.11	2.04	1.97	.025	
3.21	3.07	2.93	2.78	2.70	2.62	2.54	2.45	2.35	2.26	.010	
3.64	3.47	3.30	3.12	3.02	2.92	2.82	2.71	2.60	2.48	.005	
1.88	1.83	1.78	1.73	1.70	1.67	1.64	1.61	1.57	1.53	.100	24
2.25	2.18	2.11	2.03	1.98	1.94	1.89	1.84	1.79	1.73	.050	
2.64	2.54	2.44	2.33	2.27	2.21	2.15	2.08	2.01	1.94	.025	
3.17	3.03	2.89	2.74	2.66	2.58	2.49	2.40	2.31	2.21	.010	
3.59	3.42	3.25	3.06	2.97	2.87	2.77	2.66	2.55	2.43	.005	
1.87	1.82	1.77	1.72	1.69	1.66	1.63	1.59	1.56	1.52	.100	25
2.24	2.16	2.09	2.01	1.96	1.92	1.87	1.82	1.77	1.71	.050	
2.61	2.51	2.41	2.30	2.24	2.18	2.12	2.05	1.98	1.91	.025	
3.13	2.99	2.85	2.70	2.62	2.54	2.45	2.36	2.27	2.17	.010	
3.54	3.37	3.20	3.01	2.92	2.82	2.72	2.61	2.50	2.38	.005	
1.86	1.81	1.76	1.71	1.68	1.65	1.61	1.58	1.54	1.50	.100	26
2.22	2.15	2.07	1.99	1.95	1.90	1.85	1.80	1.75	1.69	.050	
2.59	2.49	2.39	2.28	2.22	2.16	2.09	2.03	1.95	1.88	.025	
3.09	2.96	2.81	2.66	2.58	2.50	2.42	2.33	2.23	2.13	.010	
3.49	3.33	3.15	2.97	2.87	2.77	2.67	2.56	2.45	2.33	.005	
1.85	1.80	1.75	1.70	1.67	1.64	1.60	1.57	1.53	1.49	.100	27
2.20	2.13	2.06	1.97	1.93	1.88	1.84	1.79	1.73	1.67	.050	
2.57	2.47	2.36	2.25	2.19	2.13	2.07	2.00	1.93	1.85	.025	
3.06	2.93	2.78	2.63	2.55	2.47	2.38	2.29	2.20	2.10	.010	
3.45	3.28	3.11	2.93	2.83	2.73	2.63	2.52	2.41	2.29	.005	
1.84	1.79	1.74	1.69	1.66	1.63	1.59	1.56	1.52	1.48	.100	28
2.19	2.12	2.04	1.96	1.91	1.87	1.82	1.77	1.71	1.65	.050	
2.55	2.45	2.34	2.23	2.17	2.11	2.05	1.98	1.91	1.83	.025	
3.03	2.90	2.75	2.60	2.52	2.44	2.35	2.26	2.17	2.06	.010	
3.41	3.25	3.07	2.89	2.79	2.69	2.59	2.48	2.37	2.25	.005	
1.83	1.78	1.73	1.68	1.65	1.62	1.58	1.55	1.51	1.47	.100	29
2.18	2.10	2.03	1.94	1.90	1.85	1.81	1.75	1.70	1.64	.050	
2.53	2.43	2.32	2.21	2.15	2.09	2.03	1.96	1.89	1.81	.025	
3.00	2.87	2.73	2.57	2.49	2.41	2.33	2.23	2.14	2.03	.010	
3.38	3.21	3.04	2.86	2.76	2.66	2.56	2.45	2.33	2.21	.005	
1.82	1.77	1.72	1.67	1.64	1.61	1.57	1.54	1.50	1.46	.100	30
2.16	2.09	2.01	1.93	1.89	1.84	1.79	1.74	1.68	1.62	.050	
2.51	2.41	2.31	2.20	2.14	2.07	2.01	1.94	1.87	1.79	.025	
2.98	2.84	2.70	2.55	2.47	2.39	2.30	2.21	2.11	2.01	.010	
3.34	3.18	3.01	2.82	2.73	2.63	2.52	2.42	2.30	2.18	.005	

T A B L E 6   (Continued)

ν_2	a	1	2	3	4	5	6	7	8	9
40	.100	2.84	2.44	2.23	2.09	2.00	1.93	1.87	1.83	1.79
	.050	4.08	3.23	2.84	2.61	2.45	2.34	2.25	2.18	2.12
	.025	5.42	4.05	3.46	3.13	2.90	2.74	2.62	2.53	2.45
	.010	7.31	5.18	4.31	3.83	3.51	3.29	3.12	2.99	2.89
	.005	8.83	6.07	4.98	4.37	3.99	3.71	3.51	3.35	3.22
60	.100	2.79	2.39	2.18	2.04	1.95	1.87	1.82	1.77	1.74
	.050	4.00	3.15	2.76	2.53	2.37	2.25	2.17	2.10	2.04
	.025	5.29	3.93	3.34	3.01	2.79	2.63	2.51	2.41	2.33
	.010	7.08	4.98	4.13	3.65	3.34	3.12	2.95	2.82	2.72
	.005	8.49	5.79	4.73	4.14	3.76	3.49	3.29	3.13	3.01
120	.100	2.75	2.35	2.13	1.99	1.90	1.82	1.77	1.72	1.68
	.050	3.92	3.07	2.68	2.45	2.29	2.17	2.09	2.02	1.96
	.025	5.15	3.80	3.23	2.89	2.67	2.52	2.39	2.30	2.22
	.100	6.85	4.79	3.95	3.48	3.17	2.96	2.79	2.66	2.56
	.005	8.18	5.54	4.50	3.92	3.55	3.28	3.09	2.93	2.81
∞	.100	2.71	2.30	2.08	1.94	1.85	1.77	1.72	1.67	1.63
	.050	3.84	3.00	2.60	2.37	2.21	2.10	2.01	1.94	1.63
	.025	5.02	3.69	3.12	2.79	2.57	2.41	2.29	2.19	2.11
	.010	6.63	4.61	3.78	3.32	3.02	2.80	2.64	2.51	2.41
	.005	7.88	5.30	4.28	3.72	3.35	3.09	2.90	2.74	2.62

T A B L E **6** (Continued)

					ν_1						
10	**12**	**15**	**20**	**24**	**30**	**40**	**60**	**120**	**∞**	**a**	ν_2
1.76	1.71	1.66	1.61	1.57	1.54	1.51	1.47	1.42	1.38	.100	40
2.08	2.00	1.92	1.84	1.79	1.74	1.69	1.64	1.58	1.51	.050	
2.39	2.29	2.18	2.07	2.01	1.94	1.88	1.80	1.72	1.64	.025	
2.80	2.66	2.52	2.37	2.29	2.20	2.11	2.02	1.92	1.80	.010	
3.12	2.95	2.78	2.60	2.50	2.40	2.30	2.18	2.06	1.93	.005	
1.71	1.66	1.60	1.54	1.51	1.48	1.44	1.40	1.35	1.29	.100	60
1.99	1.92	1.84	1.75	1.70	1.65	1.59	1.53	1.47	1.39	.050	
2.27	2.17	2.06	1.94	1.88	1.82	1.74	1.67	1.58	1.48	.025	
2.63	2.50	2.35	2.20	2.12	2.03	1.94	1.84	1.73	1.60	.010	
2.90	2.74	2.57	2.39	2.29	2.19	2.08	1.96	1.83	1.69	.005	
1.65	1.60	1.55	1.48	1.45	1.41	1.37	1.32	1.26	1.19	.100	120
1.91	1.83	1.75	1.66	1.61	1.55	1.50	1.43	1.35	1.25	.050	
2.16	2.05	1.94	1.82	1.76	1.69	1.61	1.53	1.43	1.31	.025	
2.47	2.34	2.19	2.03	1.95	1.86	1.76	1.66	1.53	1.38	.010	
2.71	2.54	2.37	2.19	2.09	1.98	1.87	1.75	1.61	1.43	.005	
1.60	1.55	1.49	1.42	1.38	1.34	1.30	1.24	1.17	1.00	.100	∞
1.83	1.75	1.67	1.57	1.52	1.46	1.39	1.32	1.22	1.00	.050	
2.05	1.94	1.83	1.71	1.64	1.57	1.48	1.39	1.27	1.00	.025	
2.32	2.18	2.04	1.88	1.79	1.70	1.59	1.47	1.32	1.00	.010	
2.52	2.36	2.19	2.00	1.90	1.79	1.67	1.53	1.36	1.00	.005	

TABLE 7
Percentage Points of the
Studentized Range, $q(k, \nu)$;
Upper 5% Points

ν	2	3	4	5	6	7	8	9	10	11
1	17.97	26.98	32.82	37.08	40.41	43.12	45.40	47.36	49.07	50.59
2	6.08	8.33	9.80	10.88	11.74	12.44	13.03	13.54	13.99	14.39
3	4.50	5.91	6.82	7.50	8.04	8.48	8.85	9.18	9.46	9.72
4	3.93	5.04	5.76	6.29	6.71	7.05	7.35	7.60	7.83	8.03
5	3.64	4.60	5.22	5.67	6.03	6.33	6.58	6.80	6.99	7.17
6	3.46	4.34	4.90	5.30	5.63	5.90	6.12	6.32	6.49	6.65
7	3.34	4.16	4.68	5.06	5.36	5.61	5.82	6.00	6.16	6.30
8	3.26	4.04	4.53	4.89	5.17	5.40	5.60	5.77	5.92	6.05
9	3.20	3.95	4.41	4.76	5.02	5.24	5.43	5.59	5.74	5.87
10	3.15	3.88	4.33	4.65	4.91	5.12	5.30	5.46	5.60	5.72
11	3.11	3.82	4.26	4.57	4.82	5.03	5.20	5.35	5.49	5.61
12	3.08	3.77	4.20	4.51	4.75	4.95	5.12	5.27	5.39	5.51
13	3.06	3.73	4.15	4.45	4.69	4.88	5.05	5.19	5.32	5.43
14	3.03	3.70	4.11	4.41	4.64	4.83	4.99	5.13	5.25	5.36
15	3.01	3.67	4.08	4.37	4.60	4.78	4.94	5.08	5.20	5.31
16	3.00	3.65	4.05	4.33	4.56	4.74	4.90	5.03	5.15	5.26
17	2.98	3.63	4.02	4.30	4.52	4.70	4.86	4.99	5.11	5.21
18	2.97	3.61	4.00	4.28	4.49	4.67	4.82	4.96	5.07	5.17
19	2.96	3.59	3.98	4.25	4.47	4.65	4.79	4.92	5.04	5.14
20	2.95	3.58	3.96	4.23	4.45	4.62	4.77	4.90	5.01	5.11
24	2.92	3.53	3.90	4.17	4.37	4.54	4.68	4.81	4.92	5.01
30	2.89	3.49	3.85	4.10	4.30	4.46	4.60	4.72	4.82	4.92
40	2.86	3.44	3.79	4.04	4.23	4.39	4.52	4.63	4.73	4.82
60	2.83	3.40	3.74	3.98	4.16	4.31	4.44	4.55	4.65	4.73
120	2.80	3.36	3.68	3.92	4.10	4.24	4.36	4.47	4.56	4.64
∞	2.77	3.31	3.63	3.86	4.03	4.17	4.29	4.39	4.47	4.55

TABLE **7**
(Continued)

				k					
12	**13**	**14**	**15**	**16**	**17**	**18**	**19**	**20**	ν
51.96	53.20	54.33	55.36	56.32	57.22	58.04	58.83	59.56	1
14.75	15.08	15.38	15.65	15.91	16.14	16.37	16.57	16.77	2
9.95	10.15	10.35	10.52	10.69	10.84	10.98	11.11	11.24	3
8.21	8.37	8.52	8.66	8.79	8.91	9.03	9.13	9.23	4
7.32	7.47	7.60	7.72	7.83	7.93	8.03	8.12	8.21	5
6.79	6.92	7.03	7.14	7.24	7.34	7.43	7.51	7.59	6
6.43	6.55	6.66	6.76	6.85	6.94	7.02	7.10	7.17	7
6.18	6.29	6.39	6.48	6.57	6.65	6.73	6.80	6.87	8
5.98	6.09	6.19	6.28	6.36	6.44	6.51	6.58	6.64	9
5.83	5.93	6.03	6.11	6.19	6.27	6.34	6.40	6.47	10
5.71	5.81	5.90	5.98	6.06	6.13	6.20	6.27	6.33	11
5.61	5.71	5.80	5.88	5.95	6.02	6.09	6.15	6.21	12
5.53	5.63	5.71	5.79	5.86	5.93	5.99	6.05	6.11	13
5.46	5.55	5.64	5.71	5.79	5.85	5.91	5.97	6.03	14
5.40	5.49	5.57	5.65	5.72	5.78	5.85	5.90	5.96	15
5.35	5.44	5.52	5.59	5.66	5.73	5.79	5.84	5.90	16
5.31	5.39	5.47	5.54	5.61	5.67	5.73	5.79	5.84	17
5.27	5.35	5.43	5.50	5.57	5.63	5.69	5.74	5.79	18
5.23	5.31	5.39	5.46	5.53	5.59	5.65	5.70	5.75	19
5.20	5.28	5.36	5.43	5.49	5.55	5.61	5.66	5.71	20
5.10	5.18	5.25	5.32	5.38	5.44	5.49	5.55	5.59	24
5.00	5.08	5.15	5.21	5.27	5.33	5.38	5.43	5.47	30
4.90	4.98	5.04	5.11	5.16	5.22	5.27	5.31	5.36	40
4.81	4.88	4.94	5.00	5.06	5.11	5.15	5.20	5.24	60
4.71	4.78	4.84	4.90	4.95	5.00	5.04	5.09	5.13	120
4.62	4.68	4.74	4.80	4.85	4.89	4.93	4.97	5.01	∞

T A B L E **8**   Percentage Points of the Studentized Range, $q(k, \nu)$; Upper 1% Points

ν	2	3	4	5	6	7	8	9	10	11
1	90.03	135.0	164.3	185.6	202.2	215.8	227.2	237.0	245.6	253.2
2	14.04	19.02	22.29	24.72	26.63	28.20	29.53	30.68	31.69	32.59
3	8.26	10.62	12.17	13.33	14.24	15.00	15.64	16.20	16.69	17.13
4	6.51	8.12	9.17	9.96	10.58	11.10	11.55	11.93	12.27	12.57
5	5.70	6.98	7.80	8.42	8.91	9.32	9.67	9.97	10.24	10.48
6	5.24	6.33	7.03	7.56	7.97	8.32	8.61	8.87	9.10	9.30
7	4.95	5.92	6.54	7.01	7.37	7.68	7.94	8.17	8.37	8.55
8	4.75	5.64	6.20	6.62	6.96	7.24	7.47	7.68	7.86	8.03
9	4.60	5.43	5.96	6.35	6.66	6.91	7.13	7.33	7.49	7.65
10	4.48	5.27	5.77	6.14	6.43	6.67	6.87	7.05	7.21	7.36
11	4.39	5.15	5.62	5.97	6.25	6.48	6.67	6.84	6.99	7.13
12	4.32	5.05	5.50	5.84	6.10	6.32	6.51	6.67	6.81	6.94
13	4.26	4.96	5.40	5.73	5.98	6.19	6.37	6.53	6.67	6.79
14	4.21	4.89	5.32	5.63	5.88	6.08	6.26	6.41	6.54	6.66
15	4.17	4.84	5.25	5.56	5.80	5.99	6.16	6.31	6.44	6.55
16	4.13	4.79	5.19	5.49	5.72	5.92	6.08	6.22	6.35	6.46
17	4.10	4.74	5.14	5.43	5.66	5.85	6.01	6.15	6.27	6.38
18	4.07	4.70	5.09	5.38	5.60	5.79	5.94	6.08	6.20	6.31
19	4.05	4.67	5.05	5.33	5.55	5.73	5.89	6.02	6.14	6.25
20	4.02	4.64	5.02	5.29	5.51	5.69	5.84	5.97	6.09	6.19
24	3.96	4.55	4.91	5.17	5.37	5.54	5.69	5.81	5.92	6.02
30	3.89	4.45	4.80	5.05	5.24	5.40	5.54	5.65	5.76	5.85
40	3.82	4.37	4.70	4.93	5.11	5.26	5.39	5.50	5.60	5.69
60	3.76	4.28	4.59	4.82	4.99	5.13	5.25	5.36	5.45	5.53
120	3.70	4.20	4.50	4.71	4.87	5.01	5.12	5.21	5.30	5.37
∞	3.64	4.12	4.40	4.60	4.76	4.88	4.99	5.08	5.16	5.23

T A B L E **8**   (Continued)

					k					
12	**13**	**14**	**15**	**16**	**17**	**18**	**19**	**20**	**ν**
260.0	266.2	271.8	277.0	281.8	286.3	290.0	294.3	298.0	1
33.40	34.13	34.81	35.43	36.00	36.53	37.03	37.50	37.95	2
17.53	17.89	18.22	18.52	18.81	19.07	19.32	19.55	19.77	3
12.84	13.09	13.32	13.53	13.73	13.91	14.08	14.24	14.40	4
10.70	10.89	11.08	11.24	11.40	11.55	11.68	11.81	11.93	5
9.48	9.65	9.81	9.95	10.08	10.21	10.32	10.43	10.54	6
8.71	8.86	9.00	9.12	9.24	9.35	9.46	9.55	9.65	7
8.18	8.31	8.44	8.55	8.66	8.76	8.85	8.94	9.03	8
7.78	7.91	8.03	8.13	8.23	8.33	8.41	8.49	8.57	9
7.49	7.60	7.71	7.81	7.91	7.99	8.08	8.15	8.23	10
7.25	7.36	7.46	7.56	7.65	7.73	7.81	7.88	7.95	11
7.06	7.17	7.26	7.36	7.44	7.52	7.59	7.66	7.73	12
6.90	7.01	7.10	7.19	7.27	7.35	7.42	7.48	7.55	13
6.77	6.87	6.96	7.05	7.13	7.20	7.27	7.33	7.39	14
6.66	6.76	6.84	6.93	7.00	7.07	7.14	7.20	7.26	15
6.56	6.66	6.74	6.82	6.90	6.97	7.03	7.09	7.15	16
6.48	6.57	6.66	6.73	6.81	6.87	6.94	7.00	7.05	17
6.41	6.50	6.58	6.65	6.72	6.79	6.85	6.91	6.97	18
6.34	6.43	6.51	6.58	6.65	6.72	6.78	6.84	6.89	19
6.28	6.37	6.45	6.52	6.59	6.65	6.71	6.77	6.82	20
6.11	6.19	6.26	6.33	6.39	6.45	6.51	6.56	6.61	24
5.93	6.01	6.08	6.14	6.20	6.26	6.31	6.36	6.41	30
5.76	5.83	5.90	5.96	6.02	6.07	6.12	6.16	6.21	40
5.60	5.67	5.73	5.78	5.84	5.89	5.93	5.97	6.01	60
5.44	5.50	5.56	5.61	5.66	5.71	5.75	5.79	5.83	120
5.29	5.35	5.40	5.45	5.49	5.54	5.57	5.61	5.65	∞

T A B L E 9   Factors Used When Constructing Control Charts

Number of Observations in Sample,	Chart for Averages — Factors for Control Limits			Chart for Standard Deviations — Factors for Central Line		Chart for Standard Deviations — Factors for Control Limits				Chart for Ranges — Factors for Central Line			Chart for Ranges — Factors for Control Limits			
n	A	A_1	A_2	c_2	$1/c_2$	B_1	B_2	B_3	B_4	d_2	$1/d_2$	d_3	D_1	D_2	D_3	D_4
2	2.121	3.760	1.880	.5642	1.7725	0	1.843	0	3.267	1.128	.8865	.853	0	3.686	0	3.276
3	1.732	2.394	1.023	.7236	1.3820	0	1.858	0	2.568	1.693	.5907	.888	0	4.358	0	2.575
4	1.501	1.880	.729	.7979	1.2533	0	1.808	0	2.266	2.059	.4857	.880	0	4.698	0	2.282
5	1.342	1.596	.577	.8407	1.1894	0	1.756	0	2.089	2.326	.4299	.864	0	4.918	0	2.115
6	1.225	1.410	.483	.8686	1.1512	.026	1.711	.030	1.970	2.534	.3946	.848	0	5.078	0	2.004
7	1.134	1.277	.419	.8882	1.1259	.105	1.672	.118	1.882	2.704	.3698	.833	.205	5.203	.076	1.924
8	1.061	1.175	.373	.9027	1.1078	.167	1.638	.185	1.815	2.847	.3512	.820	.387	5.307	.136	1.864
9	1.000	1.094	.337	.9139	1.0942	.219	1.609	.239	1.761	2.970	.3367	.808	.546	5.394	.184	1.816
10	.949	1.028	.308	.9227	1.0837	.262	1.584	.284	1.716	3.078	.3249	.797	.687	5.469	.223	1.777
11	.905	.973	.285	.9300	1.0753	.299	1.561	.321	1.679	3.173	.3152	.787	.812	5.534	.256	1.744
12	.866	.925	.266	.9359	1.0684	.331	1.541	.354	1.646	3.258	.3069	.778	.924	5.592	.284	1.719
13	.832	.884	.249	.9410	1.0627	.359	1.523	.382	1.618	3.336	.2998	.770	1.026	5.646	.308	1.692
14	.802	.848	.235	.9453	1.0579	.384	1.507	.406	1.594	3.407	.2935	.762	1.121	5.693	.329	1.671
15	.775	.816	.223	.9490	1.0537	.406	1.492	.428	1.572	3.472	.2880	.755	1.207	5.737	.348	1.652
16	.750	.788	.212	.9523	1.0501	.427	1.478	.448	1.552	3.532	.2831	.749	1.285	5.779	.364	1.636
17	.728	.762	.203	.9551	1.0470	.445	1.465	.466	1.534	3.588	.2787	.743	1.359	5.817	.379	1.621
18	.707	.738	.194	.9576	1.0442	.461	1.454	.482	1.518	3.640	.2747	.738	1.426	5.854	.392	1.608
19	.688	.717	.187	.9599	1.0418	.477	1.443	.497	1.503	3.689	.2711	.733	1.490	5.888	.404	1.596
20	.671	.697	.180	.9619	1.0396	.491	1.433	.510	1.490	3.735	.2677	.729	1.548	5.922	.414	1.586
21	.655	.679	.173	.9638	1.0376	.504	1.424	.523	1.477	3.778	.2647	.724	1.606	5.950	.425	1.575
22	.640	.662	.167	.9655	1.0358	.516	1.415	.534	1.466	3.819	.2618	.720	1.659	5.979	.434	1.566
23	.626	.647	.162	.9670	1.0342	.527	1.407	.545	1.455	3.858	.2592	.716	1.710	6.006	.443	1.557
24	.612	.632	.157	.9684	1.0327	.538	1.399	.555	1.445	3.895	.2567	.712	1.759	6.031	.452	1.548
25	.600	.619	.153	.9696	1.0313	.548	1.392	.565	1.435	3.931	.2544	.709	1.804	6.058	.459	1.541
Over 25	$\dfrac{3}{\sqrt{n}}$	$\dfrac{3}{\sqrt{n}}$	—	—	—	*	†	*	†	—	—	—	—	—	—	—

Source: Reproduced by permission from *ASTM Manual on Quality Control of Materials*, American Society for Testing Materials, Philadelphia, PA, 1951.

* $1 - 3/\sqrt{2n}$

† $1 + 3/\sqrt{2n}$

T A B L E **10**   Distribution of the Total Number of Runs r in Samples of Size (n_1, n_2): $P(r \leq r_0)$

					r_0					
(n_1, n_2)	2	3	4	5	6	7	8	9	10
(2,3)	.200	.500	.900	1.000					
(2,4)	.133	.400	.800	1.000					
(2,5)	.095	.333	.714	1.000					
(2,6)	.071	.286	.643	1.000					
(2,7)	.056	.250	.583	1.000					
(2,8)	.044	.222	.533	1.000					
(2,9)	.036	.200	.491	1.000					
(2,10)	.030	.182	.455	1.000					
(3,3)	.100	.300	.700	.900	1.000				
(3,4)	.057	.200	.543	.800	.971	1.000			
(3,5)	.036	.143	.429	.714	.929	1.000			
(3,6)	.024	.107	.345	.643	.881	1.000			
(3,7)	.017	.083	.283	.583	.833	1.000			
(3,8)	.012	.067	.236	.533	.788	1.000			
(3,9)	.009	.055	.200	.491	.745	1.000			
(3,10)	.007	.045	.171	.455	.706	1.000			
(4,4)	.029	.114	.371	.629	.886	.971	1.000		
(4,5)	.016	.071	.262	.500	.786	.929	.992	1.000	
(4,6)	.010	.048	.190	.405	.690	.881	.976	1.000	
(4,7)	.006	.033	.142	.333	.606	.833	.954	1.000	
(4,8)	.004	.024	.109	.279	.533	.788	.929	1.000	
(4,9)	.003	.018	.085	.236	.471	.745	.902	1.000	
(4,10)	.002	.014	.068	.203	.419	.706	.874	1.000	
(5,5)	.008	.040	.167	.357	.643	.833	.960	.992	1.000
(5,6)	.004	.024	.110	.262	.522	.738	.911	.976	.998
(5,7)	.003	.015	.076	.197	.424	.652	.854	.955	.992
(5,8)	.002	.010	.054	.152	.347	.576	.793	.929	.984
(5,9)	.001	.007	.039	.119	.287	.510	.734	.902	.972
(5,10)	.001	.005	.029	.095	.239	.455	.678	.874	.958
(6,6)	.002	.013	.067	.175	.392	.608	.825	.933	.987
(6,7)	.001	.008	.043	.121	.296	.500	.733	.879	.966
(6,8)	.001	.005	.028	.086	.226	.413	.646	.821	.937
(6,9)	.000	.003	.019	.063	.175	.343	.566	.762	.902
(6,10)	.000	.002	.013	.047	.137	.288	.497	.706	.864
(7,7)	.001	.004	.025	.078	.209	.383	.617	.791	.922
(7,8)	.000	.002	.015	.051	.149	.296	.514	.704	.867
(7,9)	.000	.001	.010	.035	.108	.231	.427	.622	.806
(7,10)	.000	.001	.006	.024	.080	.182	.355	.549	.743
(8,8)	.000	.001	.009	.032	.100	.214	.405	.595	.786
(8,9)	.000	.001	.005	.020	.069	.157	.319	.500	.702
(8,10)	.000	.000	.003	.013	.048	.117	.251	.419	.621
(9,9)	.000	.000	.003	.012	.044	.109	.238	.399	.601
(9,10)	.000	.000	.002	.008	.029	.077	.179	.319	.510
(10,10)	.000	.000	.001	.004	.019	.051	.128	.242	.414

Source: From "Tables for Testing Randomness of Grouping in a Sequence of Alternatives," C. Eisenhart and F. Swed, *Annals of Mathematical Statistics*, Vol. 14 (1943). Reproduced with the kind permission of the editor, *Annals of Mathematical Statistics*.

T A B L E **10**   (Continued)

(n_1, n_2)	11	12	13	14	15	16	17	18	19	20
					r_0					
(2,3)										
(2,4)										
(2,5)										
(2,6)										
(2,7)										
(2,8)										
(2,9)										
(2,10)										
(3,3)										
(3,4)										
(3,5)										
(3,6)										
(3,7)										
(3,8)										
(3,9)										
(3,10)										
(4,4)										
(4,5)										
(4,6)										
(4,7)										
(4,8)										
(4,9)										
(4,10)										
(5,5)										
(5,6)	1.000									
(5,7)	1.000									
(5,8)	1.000									
(5,9)	1.000									
(5,10)	1.000									
(6,6)	.998	1.000								
(6,7)	.992	.999	1.000							
(6,8)	.984	.998	1.000							
(6,9)	.972	.994	1.000							
(6,10)	.958	.990	1.000							
(7,7)	.975	.996	.999	1.000						
(7,8)	.949	.988	.998	1.000	1.000					
(7,9)	.916	.975	.994	.999	1.000					
(7,10)	.879	.957	.990	.998	1.000					
(8,8)	.900	.968	.991	.999	1.000	1.000				
(8,9)	.843	.939	.980	.996	.999	1.000	1.000			
(8,10)	.782	.903	.964	.990	.998	1.000	1.000			
(9,9)	.762	.891	.956	.988	.997	1.000	1.000	1.000		
(9,10)	.681	.834	.923	.974	.992	.999	1.000	1.000	1.000	
(10,10)	.586	.758	.872	.949	.981	.996	.999	1.000	1.000	1.000

	Special Inspection Levels				General Inspection Levels		
Lot or Batch Size	**S–1**	**S–2**	**S–3**	**S–4**	**I**	**II**	**III**
2–8	A	A	A	A	A	A	B
9–15	A	A	A	A	A	B	C
16–25	A	A	B	B	B	C	D
26–50	A	B	B	C	C	D	E
51–90	B	B	C	C	C	E	F
91–150	B	B	C	D	C	F	G
151–280	B	C	D	E	E	G	H
281–500	B	C	D	E	F	H	J
501–1200	C	C	E	F	G	J	K
1201–3200	C	D	E	G	H	K	L
3201–10,000	C	D	F	G	J	L	M
10,001–35,000	C	D	F	H	K	M	N
35,001–150,000	D	E	G	J	L	N	P
150,001–500,000	D	E	G	J	M	P	Q
500,001 and over	D	E	H	K	N	Q	R

T A B L E **11**
Sample Size Code Letters:
MIL-STD-105D

T A B L E **12**   A Portion of the Master Table for Normal Inspection (Single Sampling):MIL-STD-105D

Sample Size Code Letter	Sample Size	Acceptable Quality Levels (normal inspection) (percent)																																						
		.010		.015		.025		.040		.065		.10		.15		.25		.40		.65		1.0		1.5		2.5		4.0		6.5		10		15		25		40		65	
		Ac	Re	Ac	Re	Ac	Re	Ac	Re	Ac	Re	Ac	Re	Ac	Re	Ac	Re	Ac	Re	Ac	Re	Ac	Re	Ac	Re	Ac	Re	Ac	Re	Ac	Re	Ac	Re	Ac	Re	Ac	Re	Ac	Re	Ac	Re
A	2	↓		↓		↓		↓		↓		↓		↓		↓		↓		↓		↓		↓		↓		↓		↓		↓		0	1	1	2	2	3	3	4
B	3	↓		↓		↓		↓		↓		↓		↓		↓		↓		↓		↓		↓		↓		↓		↓		0	1	1	2	2	3	3	4	5	6
C	5	↓		↓		↓		↓		↓		↓		↓		↓		↓		↓		↓		↓		↓		↓		0	1	1	2	2	3	3	4	5	6	7	8
D	8	↓		↓		↓		↓		↓		↓		↓		↓		↓		↓		↓		↓		↓		0	1	1	2	2	3	3	4	5	6	7	8	10	11
E	13	↓		↓		↓		↓		↓		↓		↓		↓		↓		↓		↓		↓		0	1	1	2	2	3	3	4	5	6	7	8	10	11	14	15
F	20	↓		↓		↓		↓		↓		↓		↓		↓		↓		↓		↓		0	1	1	2	2	3	3	4	5	6	7	8	10	11	14	15	21	22
G	32	↓		↓		↓		↓		↓		↓		↓		↓		↓		↓		0	1	1	2	2	3	3	4	5	6	7	8	10	11	14	15	21	22	↑	
H	50	↓		↓		↓		↓		↓		↓		↓		↓		↓		0	1	1	2	2	3	3	4	5	6	7	8	10	11	14	15	21	22	↑		↑	
J	80	↓		↓		↓		↓		↓		↓		↓		↓		0	1	1	2	2	3	3	4	5	6	7	8	10	11	14	15	21	22	↑		↑		↑	
K	125	↓		↓		↓		↓		↓		↓		↓		0	1	1	2	2	3	3	4	5	6	7	8	10	11	14	15	21	22	↑		↑		↑		↑	
L	200	↓		↓		↓		↓		↓		↓		0	1	1	2	2	3	3	4	5	6	7	8	10	11	14	15	21	22	↑		↑		↑		↑		↑	
M	315	↓		↓		↓		↓		↓		0	1	1	2	2	3	3	4	5	6	7	8	10	11	14	15	21	22	↑		↑		↑		↑		↑		↑	
N	500	↓		↓		↓		↓		0	1	1	2	2	3	3	4	5	6	7	8	10	11	14	15	21	22	↑		↑		↑		↑		↑		↑		↑	
P	800	↓		↓		↓		0	1	1	2	2	3	3	4	5	6	7	8	10	11	14	15	21	22	↑		↑		↑		↑		↑		↑		↑		↑	
Q	1,250	↓		↓		0	1	1	2	2	3	3	4	5	6	7	8	10	11	14	15	21	22	↑		↑		↑		↑		↑		↑		↑		↑		↑	
R	2,000	↓		0	1	1	2	2	3	3	4	5	6	7	8	10	11	14	15	21	22	↑		↑		↑		↑		↑		↑		↑		↑		↑		↑	

Note: ↓ = Use first sampling plan below arrow. If sample size equals or exceeds lot or batch size, do 100% inspection.
↑ = Use first sampling plan above arrow.
Ac = Acceptance number; Re = Rejection number.

TABLE **13**   Random Numbers

Line	1	2	3	4	5	6	7	8	9	10	11	12	13	14
							Column							
1	10480	15011	01536	02011	81647	91646	69179	14194	62590	36207	20969	99570	91291	90700
2	22368	46573	25595	85393	30995	89198	27982	53402	93965	34095	52666	19174	39615	99505
3	24130	48360	22527	97265	76393	64809	15179	24830	49340	32081	30680	19655	63348	58629
4	42167	93093	06243	61680	07856	16376	39440	53537	71341	57004	00849	74917	97758	16379
5	37570	39975	81837	16656	06121	91782	60468	81305	49684	60672	14110	06927	01263	54613
6	77921	06907	11008	42751	27756	53498	18602	70659	90655	15053	21916	81825	44394	42880
7	99562	72905	56420	69994	98872	31016	71194	18738	44013	48840	63213	21069	10634	12952
8	96301	91977	05463	07972	18876	20922	94595	56869	69014	60045	18425	84903	42508	32307
9	89579	14342	63661	10281	17453	18103	57740	84378	25331	12566	58678	44947	05585	56941
10	84575	36857	53342	53988	53060	59533	38867	62300	08158	17983	16439	11458	18593	64952
11	28918	69578	88231	33276	70997	79936	56865	05859	90106	31595	01547	85590	91610	78188
12	63553	40961	48235	03427	49626	69445	18663	72695	52180	20847	12234	90511	33703	90322
13	09429	93969	52636	92737	88974	33488	36320	17617	30015	08272	84115	27156	30613	74952
14	10365	61129	87529	85689	48237	52267	67689	93394	01511	26358	85104	20285	29975	89868
15	07119	97336	71048	08178	77233	13916	47564	81056	97735	85977	29372	74461	28551	90707
16	51085	12765	51821	51259	77452	16308	60756	92144	49442	53900	70960	63990	75601	40719
17	02368	21382	52404	60268	89368	19885	55322	44819	01188	65255	64835	44919	05944	55157
18	01011	54092	33362	94904	31273	04146	18594	29852	71585	85030	51132	01915	92747	64951
19	52162	53916	46369	58586	23216	14513	83149	98736	23495	64350	94738	17752	35156	35749
20	07056	97628	33787	09998	42698	06691	76988	13602	51851	46104	88916	19509	25625	58104
21	48663	91245	85828	14346	09172	30168	90229	04734	59193	22178	30421	61666	99904	32812
22	54164	58492	22421	74103	47070	25306	76468	26384	58151	06646	21524	15227	96909	44592
23	32639	32363	05597	24200	13363	38005	94342	28728	35806	06912	17012	64161	18296	22851
24	29334	27001	87637	87308	58731	00256	45834	15398	46557	41135	10367	07684	36188	18510
25	02488	33062	28834	07351	19731	92420	60952	61280	50001	67658	32586	86679	50720	94953
26	81525	72295	04839	96423	24878	82651	66566	14778	76797	14780	13300	87074	79666	95725
27	29676	20591	68086	26432	46901	20849	89768	81536	86645	12659	92259	57102	80428	25280
28	00742	57392	39064	66432	84673	40027	32832	61362	98947	96067	64760	64585	96096	98253
29	05366	04213	25669	26422	44407	44048	37937	63904	45766	66134	75470	66520	34693	90449
30	91921	26418	64117	94305	26766	25940	39972	22209	71500	64568	91402	42416	07844	69618
31	00582	04711	87917	77341	42206	35126	74087	99547	81817	42607	43808	76655	62028	76630
32	00725	69884	62797	56170	86324	88072	76222	36086	84637	93161	76038	65855	77919	88006
33	69011	65795	95876	55293	18988	27354	26575	08625	40801	59920	29841	80150	12777	48501
34	25976	57948	29888	88604	67917	48708	18912	82271	65424	69774	33611	54262	85963	03547
35	09763	83473	73577	12908	30883	18317	28290	35797	05998	41688	34952	37888	38917	88050
36	91567	42595	27958	30134	04024	86385	29880	99730	55536	84855	29080	09250	79656	73211
37	17955	56349	90999	49127	20044	59931	06115	20542	18059	02008	73708	83517	36103	42791
38	46503	18584	18845	49618	02304	51038	20655	58727	28168	15475	56942	53389	20562	87338
39	92157	89634	94824	78171	84610	82834	09922	25417	44137	48413	25555	21246	35509	20468
40	14577	62765	35605	81263	39667	47358	56873	56307	61607	49518	89656	20103	77490	18062
41	98427	07523	33362	64270	01638	92477	66969	98420	04880	45585	46565	04102	46880	45709
42	34914	63976	88720	82765	34476	17032	87589	40836	32427	70002	70663	88863	77775	69348
43	70060	28277	39475	46473	23219	53416	94970	25832	69975	94884	19661	72828	00102	66794
44	53976	54914	06990	67245	68350	82948	11398	42878	80287	88267	47363	46634	06541	97809
45	76072	29515	40980	07391	58745	25774	22987	80059	39911	96189	41151	14222	60697	59583
46	90725	52210	83974	29992	65831	38857	50490	83765	55657	14361	31720	57375	56228	41546
47	64364	67412	33339	31926	14883	24413	59744	92351	97473	89286	35931	04110	23726	51900
48	08962	00358	31662	25388	61642	34072	81249	35648	56891	69352	48373	45578	78547	81788
49	95012	68379	93526	70765	10592	04542	76463	54328	02349	17247	28865	14777	62730	92277
50	15664	10493	20492	38391	91132	21999	59516	81652	27195	48223	46751	22923	32261	85653

Source: Abridged from *Handbook of Tables for Probability and Statistics,* 2d ed. Edited by William H. Beyer (Cleveland: The Chemical Rubber Company, 1968). Reproduced by permission of CRC Press, Inc.

TABLE 13 (Continued)

						Column							
Line	1	2	3	4	5	6	7	8	9	10	11	12	13	14
51	16408	81899	04153	53381	79401	21438	83035	92350	36693	31238	59649	91754	72772	02338
52	18629	81953	05520	91962	04739	13092	97662	24822	94730	06496	35090	04822	86774	98289
53	73115	35101	47498	87637	99016	71060	88824	71013	18735	20286	23153	72924	35165	43040
54	57491	16703	23167	49323	45021	33132	12544	41035	80780	45393	44812	12515	98931	91202
55	30405	83946	23792	14422	15059	45799	22716	19792	09983	74353	68668	30429	70735	25499
56	16631	35006	85900	98275	32388	52390	16815	69298	82732	38480	73817	32523	41961	44437
57	96773	20206	42559	78985	05300	22164	24369	54224	35033	19687	11052	91491	60383	19746
58	38935	64202	14349	82674	66523	44133	00697	35552	35970	19124	63318	29686	03387	59846
59	31624	76384	17403	53363	44167	64486	64758	75366	76554	31601	12614	33072	60332	92325
60	78919	19474	23632	27889	47914	02584	37680	20801	72152	39339	34806	08930	85001	87820
61	03931	33309	57047	74211	63445	17361	62825	39908	05607	91284	68833	25570	38818	46920
62	74426	33278	43972	10119	89917	15665	52872	73823	73144	88662	88970	74492	51805	99378
63	09066	00903	20795	95452	92648	45454	09552	88815	16553	51125	79375	97596	16296	66092
64	42238	12426	87025	14267	20979	04508	64535	31355	86064	29472	47689	05974	52468	16834
65	16153	08002	26504	41744	81959	65642	74240	56302	00033	67107	77510	70625	28725	34191
66	21457	40742	29820	96783	29400	21840	15035	34537	33310	06116	95240	15957	16572	06004
67	21581	57802	02050	89728	17937	37621	47075	42080	97403	48626	68995	43805	33386	21597
68	55612	78095	83197	33732	05810	24813	86902	60397	16489	03264	88525	42786	05269	92532
69	44657	66999	99324	51281	84463	60563	79312	93454	68876	25471	93911	25650	12682	73572
70	91340	84979	46949	81973	37949	61023	43997	15263	80644	43942	89203	71795	99533	50501
71	91227	21199	31935	27022	84067	05462	35216	14486	29891	68607	41867	14951	91696	85065
72	50001	38140	66321	19924	72163	09538	12151	06878	91903	18749	34405	56087	82790	70925
73	65390	05224	72958	28609	81406	39147	25549	48542	42627	45233	57202	94617	23772	07896
74	27504	96131	83944	41575	10573	08619	64482	73923	36152	05184	94142	25299	84387	34925
75	37169	94851	39117	89632	00959	16487	65536	49071	39782	17095	02330	74301	00275	48280
76	11508	70225	51111	38351	19444	66499	71945	05422	13442	78675	84081	66938	93654	59894
77	37449	30362	06694	54690	04052	53115	62757	95348	78662	11163	81651	50245	34971	52924
78	46515	70331	85922	38329	57015	15765	97161	17869	45349	61796	66345	81073	49106	79860
79	30986	81223	42416	58353	21532	30502	32305	86482	05174	07901	54339	58861	74818	46942
80	63798	64995	46583	09785	44160	78128	83991	42865	92520	83531	80377	35909	81250	54238
81	82486	84846	99254	67632	43218	50076	21361	64816	51202	88124	41870	52689	51275	83556
82	21885	32906	92431	09060	64297	51674	64126	62570	26123	05155	59194	52799	28225	85762
83	60336	98782	07408	53458	13564	59089	26445	29789	85205	41001	12535	12133	14645	23541
84	43937	46891	24010	25560	86355	33941	25786	54990	71899	15475	95434	98227	21824	19585
85	97656	63175	89303	16275	07100	92063	21942	18611	47348	20203	18534	03862	78095	50136
86	03299	01221	05418	38982	55758	92237	26759	86367	21216	98442	08303	56613	91511	75928
87	79626	06486	03574	17668	07785	76020	79924	25651	83325	88428	85076	72811	22717	50585
88	85636	68335	47539	03129	65651	11977	02510	26113	99447	68645	34327	15152	55230	93448
89	18039	14367	61337	06177	12143	46609	32989	74014	64708	00533	35398	58408	13261	47908
90	08362	15656	60627	36478	65648	16764	53412	09013	07832	41574	17639	82163	60859	75567
91	79556	29068	04142	16268	15387	12856	66227	38358	22478	73373	88732	09443	82558	05250
92	92608	82674	27072	32534	17075	27698	98204	63863	11951	34648	88022	56148	34925	57031
93	23982	25835	40055	67006	12293	02753	14827	23235	35071	99704	37543	11601	35503	85171
94	09915	96306	05908	97901	28395	14186	00821	80703	70426	75647	76310	88717	37890	40129
95	59037	33300	26695	62247	69927	76123	50842	43834	86654	70959	79725	93872	28117	19233
96	42488	78077	69882	61657	34136	79180	97526	43092	04098	73571	80799	76536	71255	64239
97	46764	86273	63003	93017	31204	36692	40202	35275	57306	55543	53203	18098	47625	88684
98	03237	45430	55417	63282	90816	17349	88298	90183	36600	78406	06216	95787	42579	90730
99	86591	81482	52667	61582	14972	90053	89534	76036	49199	43716	97548	04379	46370	28672
100	38534	01715	94964	87288	65680	43772	39560	12918	86737	62738	19636	51132	25739	56947

TABLE 14
Distribution Function of U,
$P(U \leq U_0)$; U_0 is the
Argument; $n_1 \leq n_2$;
$3 \leq n_2 \leq 10$

$n_2 = 3$

U_0	n_1 1	2	3
0	.25	.10	.05
1	.50	.20	.10
2		.40	.20
3		.60	.35
4			.50

$n_2 = 4$

U_0	n_1 1	2	3	4
0	.2000	.0667	.0286	.0143
1	.4000	.1333	.0571	.0286
2	.6000	.2667	.1143	.0571
3		.4000	.2000	.1000
4		.6000	.3143	.1714
5			.4286	.2429
6			.5714	.3429
7				.4429
8				.5571

$n_2 = 5$

U_0	n_1 1	2	3	4	5
0	.1667	.0476	.0179	.0079	.0040
1	.3333	.0952	.0357	.0159	.0079
2	.5000	.1905	.0714	.0317	.0159
3		.2857	.1250	.0556	.0278
4		.4286	.1964	.0952	.0476
5		.5714	.2857	.1429	.0754
6			.3929	.2063	.1111
7			.5000	2778	.1548
8				.3651	.2103
9				.4524	.2738
10				.5476	.3452
11					.4206
12					.5000

Note: Computed by M. Pagano, Department of Statistics, University of Florida.

$n_2 = 6$

			n_1			
U_0	1	2	3	4	5	6
0	.1429	.0357	.0119	.0048	.0022	.0011
1	.2857	.0714	.0238	.0095	.0043	.0022
2	.4286	.1429	.0476	.0190	.0087	.0043
3	.5714	.2143	.0833	.0333	.0152	.0076
4		.3214	.1310	.0571	.0260	.0130
5		.4286	.1905	.0857	.0411	.0206
6		.5714	.2738	.1286	.0628	.0325
7			.3571	.1762	.0887	.0465
8			.4524	.2381	.1234	.0660
9			.5476	.3048	.1645	.0898
10				.3810	.2143	.1201
11				.4571	.2684	.1548
12				.5429	.3312	.1970
13					.3961	.2424
14					.4654	.2944
15					.5346	.3496
16						.4091
17						.4686
18						.5314

$n_2 = 7$

				n_1			
U_0	1	2	3	4	5	6	7
0	.1250	.0278	.0083	.0030	.0013	.0006	.0003
1	.2500	.0556	.0167	.0061	.0025	.0012	.0006
2	.3750	.1111	.0333	.0121	.0051	.0023	.0012
3	.5000	.1667	.0583	.0212	.0088	.0041	.0020
4		.2500	.0917	.0364	.0152	.0070	.0035
5		.3333	.1333	.0545	.0240	.0111	.0055
6		.4444	.1917	.0818	.0366	.0175	.0087
7		.5556	.2583	.1152	.0530	.0256	.0131
8			.3333	.1576	.0745	.0367	.0189
9			.4167	.2061	.1010	.0507	.0265
10			.5000	.2636	.1338	.0688	.0364
11				.3242	.1717	.0903	.0487
12				.3939	.2159	.1171	.0641
13				.4636	.2652	.1474	.0825
14				.5364	.3194	.1830	.1043
15					.3775	.2226	.1297
16					.4381	.2669	.1588
17					.5000	.3141	.1914
18						.3654	.2279
19						.4178	.2675
20						.4726	.3100
21						.5274	.3552
22							.4024
23							.4508
24							.5000

TABLE **14**
(Continued) $n_2 = 8$

U_0	n_1							
	1	2	3	4	5	6	7	8
0	.1111	.0222	.0061	.0020	.0008	.0003	.0002	.0001
1	.2222	.0444	.0121	.0040	.0016	.0007	.0003	.0002
2	.3333	.0889	.0242	.0081	.0031	.0013	.0006	.0003
3	.4444	.1333	.0424	.0141	.0054	.0023	.0011	.0005
4	.5556	.2000	.0667	.0242	.0093	.0040	.0019	.0009
5		.2667	.0970	.0364	.0148	.0063	.0030	.0015
6		.3556	.1394	.0545	.0225	.0100	.0047	.0023
7		.4444	.1879	.0768	.0326	.0147	.0070	.0035
8		.5556	.2485	.1071	.0466	.0213	.0103	.0052
9			.3152	.1414	.0637	.0296	.0145	.0074
10			.3879	.1838	.0855	.0406	.0200	.0103
11			.4606	.2303	.1111	.0539	.0270	.0141
12			.5394	.2848	.1422	.0709	.0361	.0190
13				.3414	.1772	.0906	.0469	.0249
14				.4040	.2176	.1142	.0603	.0325
15				.4667	.2618	.1412	.0760	.0415
16				.5333	.3108	.1725	.0946	.0524
17					.3621	.2068	.1159	.0652
18					.4165	.2454	.1405	.0803
19					.4716	.2864	.1678	.0974
20					.5284	.3310	.1984	.1172
21						.3773	.2317	.1393
22						.4259	.2679	.1641
23						.4749	.3063	.1911
24						.5251	.3472	.2209
25							.3894	.2527
26							.4333	.2869
27							.4775	.3227
28							.5225	.3605
29								.3992
30								.4392
31								.4796
32								.5204

TABLE **14**
(Continued)

$n_2 = 9$

U_0					n_1				
	1	2	3	4	5	6	7	8	9
0	.1000	.0182	.0045	.0014	.0005	.0002	.0001	.0000	.0000
1	.2000	.0364	.0091	.0028	.0010	.0004	.0002	.0001	.0000
2	.3000	.0727	.0182	.0056	.0020	.0008	.0003	.0002	.0001
3	.4000	.1091	.0318	.0098	.0035	.0014	.0006	.0003	.0001
4	.5000	.1636	.0500	.0168	.0060	.0024	.0010	.0005	.0002
5		.2182	.0727	.0252	.0095	.0038	.0017	.0008	.0004
6		.2909	.1045	.0378	.0145	.0060	.0026	.0012	.0006
7		.3636	.1409	.0531	.0210	.0088	.0039	.0019	.0009
8		.4545	.1864	.0741	.0300	.0128	.0058	.0028	.0014
9		.5455	.2409	.0993	.0415	.0180	.0082	.0039	.0020
10			.3000	.1301	.0559	.0248	.0115	.0056	.0028
11			.3636	.1650	.0734	.0332	.0156	.0076	.0039
12			.4318	.2070	.0949	.0440	.0209	.0103	.0053
13			.5000	.2517	.1199	.0567	.0274	.0137	.0071
14				.3021	.1489	.0723	.0356	.0180	.0094
15				.3552	.1818	.0905	.0454	.0232	.0122
16				.4126	.2188	.1119	.0571	.0296	.0157
17				.4699	.2592	.1361	.0708	.0372	.0200
18				.5301	.3032	.1638	.0869	.0464	.0252
19					.3497	.1942	.1052	.0570	.0313
20					.3986	.2280	.1261	.0694	.0385
21					.4491	.2643	.1496	.0836	.0470
22					.5000	.3035	.1755	.0998	.0567
23						.3445	.2039	.1179	.0680
24						.3878	.2349	.1383	.0807
25						.4320	.2680	.1606	.0951
26						.4773	.3032	.1852	.1112
27						.5227	.3403	.2117	.1290
28							.3788	.2404	.1487
29							.4185	.2707	.1701
30							.4591	.3029	.1933
31							.5000	.3365	.2181
32								.3715	.2447
33								.4074	.2729
34								.4442	.3024
35								.4813	.3332
36								.5187	3652
37									3981
38									.4317
39									.4657
40									.5000

$n_2 = 10$

U_0	1	2	3	4	5	6	7	8	9	10
0	.0909	.0152	.0035	.0010	.0003	.0001	.0001	.0000	.0000	.0000
1	.1818	.0303	.0070	.0020	.0007	.0002	.0001	.0000	.0000	.0000
2	.2727	.0606	.0140	.0040	.0013	.0005	.0002	.0001	.0000	.0000
3	.3636	.0909	.0245	.0070	.0023	.0009	.0004	.0002	.0001	.0000
4	.4545	.1364	.0385	.0120	.0040	.0015	.0006	.0003	.0001	.0001
5	.5455	.1818	.0559	.0180	.0063	.0024	.0010	.0004	.0002	.0001
6		.2424	.0804	.0270	.0097	.0037	.0015	.0007	.0003	.0002
7		.3030	.1084	.0380	.0140	.0055	.0023	.0010	.0005	.0002
8		.3788	.1434	.0529	.0200	.0080	.0034	.0015	.0007	.0004
9		.4545	.1853	.0709	.0276	.0112	.0048	.0022	.0011	.0005
10		.5455	.2343	.0939	.0376	.0156	.0068	.0031	.0015	.0008
11			.2867	.1199	.0496	.0210	.0093	.0043	.0021	.0010
12			.3462	.1518	.0646	.0280	.0125	.0058	.0028	.0014
13			.4056	.1868	.0823	.0363	.0165	.0078	.0038	.0019
14			.4685	.2268	.1032	.0467	.0215	.0103	.0051	.0026
15			.5315	.2697	.1272	.0589	.0277	.0133	.0066	.0034
16				.3177	.1548	.0736	.0351	.0171	.0086	.0045
17				.3666	.1855	.0903	.0439	.0217	.0110	.0057
18				.4196	.2198	.1099	.0544	.0273	.0140	.0073
19				.4725	.2567	.1317	.0665	.0338	.0175	.0093
20				.5275	.2970	.1566	.0806	.0416	.0217	.0116
21					.3393	.1838	.0966	.0506	.0267	.0144
22					.3839	.2139	.1148	.0610	.0326	.0177
23					.4296	.2461	.1349	.0729	.0394	.0216
24					.4765	.2811	.1574	.0864	.0474	.0262
25					.5235	.3177	.1819	.1015	.0564	.0315
26						.3564	.2087	.1185	.0667	.0376
27						.3962	.2374	.1371	.0782	.0446
28						.4374	.2681	.1577	.0912	.0526
29						.4789	.3004	.1800	.1055	.0615
30						.5211	.3345	.2041	.1214	.0716
31							.3698	.2299	.1388	.0827
32							.4063	.2574	.1577	.0952
33							.4434	.2863	.1781	.1088
34							.4811	.3167	.2001	.1237
35							.5189	.3482	.2235	.1399
36								.3809	.2483	.1575
37								.4143	.2745	.1763
38								.4484	.3019	.1965
39								.4827	.3304	.2179
40								.5173	.3598	.2406
41									.3901	.2644
42									.4211	.2894
43									.4524	.3153
44									.4841	.3421
45									.5159	.3697
46										.3980
47										.4267
48										.4559
49										.4853
50										.5147

One-Sided	Two-Sided	$n = 5$	$n = 6$	$n = 7$	$n = 8$	$n = 9$	$n = 10$
$\alpha = .05$	$\alpha = .10$	1	2	4	6	8	11
$\alpha = .025$	$\alpha = .05$		1	2	4	6	8
$\alpha = .01$	$\alpha = .02$			0	2	3	5
$\alpha = .005$	$\alpha = .01$				0	2	3

One-Sided	Two-Sided	$n = 11$	$n = 12$	$n = 13$	$n = 14$	$n = 15$	$n = 16$
$\alpha = .05$	$\alpha = .10$	14	17	21	26	30	36
$\alpha = .025$	$\alpha = .05$	11	14	17	21	25	30
$\alpha = .01$	$\alpha = .02$	7	10	13	16	20	24
$\alpha = .005$	$\alpha = .01$	5	7	10	13	16	19

One-Sided	Two-Sided	$n = 17$	$n = 18$	$n = 19$	$n = 20$	$n = 21$	$n = 22$
$\alpha = .05$	$\alpha = .10$	41	47	54	60	68	75
$\alpha = .025$	$\alpha = .05$	35	40	46	52	59	66
$\alpha = .01$	$\alpha = .02$	28	33	38	43	49	56
$\alpha = .005$	$\alpha = .01$	23	28	32	37	43	49

One-Sided	Two-Sided	$n = 23$	$n = 24$	$n = 25$	$n = 26$	$n = 27$	$n = 28$
$\alpha = .05$	$\alpha = .10$	83	92	101	110	120	130
$\alpha = .025$	$\alpha = .05$	73	81	90	98	107	117
$\alpha = .01$	$\alpha = .02$	62	69	77	85	93	102
$\alpha = .005$	$\alpha = .01$	55	68	68	76	84	92

One-Sided	Two-Sided	$n = 29$	$n = 30$	$n = 31$	$n = 32$	$n = 33$	$n = 34$
$\alpha = .05$	$\alpha = .10$	141	152	163	175	188	201
$\alpha = .025$	$\alpha = .05$	127	137	148	159	171	183
$\alpha = .01$	$\alpha = .02$	111	120	130	141	151	162
$\alpha = .005$	$\alpha = .01$	100	109	118	128	138	149

One-Sided	Two-Sided	$n = 35$	$n = 36$	$n = 37$	$n = 38$	$n = 39$	
$\alpha = .05$	$\alpha = .10$	214	228	242	256	271	
$\alpha = .025$	$\alpha = .05$	195	208	222	235	250	
$\alpha = .01$	$\alpha = .02$	174	186	198	211	224	
$\alpha = .005$	$\alpha = .01$	160	171	183	195	208	

One-Sided	Two-Sided	$n = 40$	$n = 41$	$n = 42$	$n = 43$	$n = 44$	$n = 45$
$\alpha = .05$	$\alpha = .10$	287	303	319	336	353	371
$\alpha = .025$	$\alpha = .05$	264	279	295	311	327	344
$\alpha = .01$	$\alpha = .02$	238	252	267	281	297	313
$\alpha = .005$	$\alpha = .01$	221	234	248	262	277	292

One-Sided	Two-Sided	$n = 46$	$n = 47$	$n = 48$	$n = 49$	$n = 50$	
$\alpha = .05$	$\alpha = .10$	389	408	427	446	466	
$\alpha = .025$	$\alpha = .05$	361	379	397	415	434	
$\alpha = .01$	$\alpha = .02$	329	345	362	380	398	
$\alpha = .005$	$\alpha = .01$	307	323	339	356	373	

Source: From "Some Rapid Approximate Statistical Procedures" (1964) 28, by F. Wilcoxon and R. A. Wilcox. Reproduced with the kind permission of Lederle Laboratories, a division of American Cyanamid Company.

n	$a = .05$	$a = .025$	$a = .01$	$a = .005$
5	0.900	—	—	—
6	0.829	0.886	0.943	—
7	0.714	0.786	0.893	—
8	0.643	0.738	0.833	0.881
9	0.600	0.683	0.783	0.833
10	0.564	0.648	0.745	0.794
11	0.523	0.623	0.736	0.787
12	0.497	0.591	0.703	0.780
13	0.475	0.566	0.673	0.745
14	0.457	0.545	0.646	0.716
15	0.441	0.525	0.623	0.689
16	0.425	0.507	0.601	0.666
17	0.412	0.490	0.582	0.645
18	0.399	0.476	0.564	0.625
19	0.388	0.462	0.549	0.608
20	0.377	0.450	0.534	0.591
21	0.368	0.438	.0521	0.576
22	0.359	0.428	0.508	0.562
23	0.351	0.418	0.496	0.549
24	0.343	0.409	.0485	0.537
25	0.336	0.400	0.475	0.526
26	0.329	0.392	0.465	0.515
27	0.323	0.385	0.456	0.505
28	0.317	0.377	0.448	0.496
29	0.311	0.370	0.440	0.487
30	0.305	0.364	0.432	0.478

T A B L E **16**
Critical Values of Spearman's Rank Correlation Coefficient; Tabled value is r_0 where $P(r_s \geq r_0) = a$

Source: From "Distribution of Sums of Squares of Rank Differences for Small Samples," by E. G. Olds, *Annals of Mathematical Statistics* 9 (1938). Reproduced with the permission of the editor, *Annals of Mathematical Statistics.*

CASE STUDIES

This appendix contains additional case studies representing statistical topics covered throughout the text. Coverage of the chapters in the text is a prerequisite for solving every case study in this appendix. Hence, these case studies should be assigned by your instructor, who will know whether you have covered the appropriate material in your class.

CASE STUDY

1 Screening Tests

Screening tests, which used to be associated primarily with medical diagnostic tests, are now finding application in a variety of fields. For example, camera technology and automatic test equipment (ATE) are now routinely used for inspecting parts in high-volume production processes. Other familiar applications include corporate drug testing of employees, home pregnancy tests, inexpensive tests for salmonella contamination in chickens, tests for lead poisoning of water supplies, and AIDS tests.

Very few screening tests are perfect. There is always the risk that the test will not catch all defective parts, diseased persons, or contaminated products (i.e., a "false negative" result). At the same time, there is the risk that good parts, healthy people, or safe foods could be classified as being defective, sick, or unsafe (a "false positive"). The consequences of a false positive in the case of a corporate drug test or AIDS test can be devastating to the individuals involved. A false negative would have equally important, but different, consequences.

To evaluate the effectiveness and consequences of using a screening test, we have to estimate the probabilities of getting false negatives and false positives. The following event notation will simplify this process. [*Note*: The terminology refers to disease screening, but the concepts apply to any screening test.]

T: the test indicates that the person has the disease (positive result)

$\overline{T}$: the test indicates that the person does not have the disease (negative result)

D: the person actually does have the disease

$\overline{D}$: the person actually does not have the disease.

Then

$$P(T|D) = \text{"Sensitivity" of the test}$$
$$= P(\text{positive test} \mid \text{person actually has disease})$$
$$P(\overline{T}|\overline{D}) = \text{"Specificity" of the test}$$
$$= P(\text{negative test} \mid \text{person is free of the disease})$$
$$P(\overline{D}|T) = P(\text{false positive})$$
$$P(D|\overline{T}) = P(\text{false negative})$$

In 1987, the sensitivity and specificity of the HIV test for AIDS were reported to be 98% and 99%, respectively ("Lax Labs," segment of the "MacNeil/Lehrer Newshour," August 12, 1987). The number of known AIDS cases in the United States in 1987 was 45,000 ("AIDS Diary," *Discover*, January 1988, page 38).

1   Using 242,200,000 as an estimate of the 1987 U.S. population, calculate the false positive and false negative rates for this screening test. What are the implications of the magnitude of the false positive rate?

2   Low-incidence diseases, such as AIDS, result in very high false positive rates. Can you suggest a way to improve (reduce) the false positive rate of a screening test?

3 Rephrase all of the probabilities and events above in terms of manufacturing inspection equipment, where one is testing for the presence of defective parts (instead of diseased persons).

4 In the case of inspection equipment, describe how you would determine the sensitivity and specificity of the inspection device.

CASE STUDY

2 Capital Budgeting Criteria

Estimated benefits from capital projects tend to be optimistic, and net present values of completed projects are often below those originally estimated. Over 80% of surveyed financial officers of *Fortune* 500 companies felt that revenue forecasts are typically overestimated. Therefore, an increasingly important task for the business strategist is to understand these biases and make decisions when basic estimates are known to be in error. It has been proposed that economic theory concerning expected return from new investments has practical implications for capital budgeting and that the budgeting process should be viewed as a Bayesian decision problem.

For example, suppose that a project consists of finding a store location for which the present value of operating profits exceeds the cost of establishing the store, and suppose that the cost of opening a new store is $90,000. Furthermore, suppose that only 4% of the proposed store sites will be good ones, 48% will be medium ones, and another 48% will be poor ones. The actual present value of the operating profits of a good project is $200,000, that of a medium project is $100,000, and that of a poor project is zero.

1 If 25% of all projects are overestimated, 50% correctly evaluated, and 25% underestimated, use the table that follows to find the expected value of the estimated present value of the operating profits; find the expected value of the actual present value of operating profits.

2 What is the difference between these two expected values? What are the implications for the decision maker in this situation?

3 If a project is estimated to have a present value of operating profits equal to $100,000, what is the probability that the project is, in fact, a good one? What is the probability that the project is, in fact, a medium one? a poor one? Use these probabilities to find the expected present value of a project whose estimated present value is $100,000. On the average, will a project whose estimated present value is $100,000 cover the $90,000 costs of opening a new store?

Present-value data

		Project	
Present Value	**Good (.04)**	**Medium (.48)**	**Poor (.48)**
Actual (.50)	$200,000	$100,000	0
Overestimate (.25)	300,000	200,000	$100,000
Underestimate (.25)	100,000	0	−100,000

CASE STUDY

3 Western Energy Services

Section 17 of the Mineral Leasing Act of 1920, as amended, provides that public lands "not within any known geological structure of a producing oil or gas field" are awarded bimonthly on a lottery basis by the Bureau of Land Management (BLM). Under provisions of the law, each individual or company may file only one application for each available parcel (accompanied by a nonrefundable $10 filing fee), thereby giving each applicant an equal chance of winning a lease.

In the late 1970s and early 1980s, Western Energy Services offered geological and managerial services to those interested in participating in the BLM lottery program for a fee of $25 per application in addition to the $10 BLM filing fee (information obtained from "Western Energy Services Program Guide," Western Energy Services, Las Vegas, 1980). Thus, for a $14,000 investment, Western Energy Services placed 400 applications for individuals in 400 BLM lotteries on parcels that Western Energy had found to be promising in geological testing.

As of June 1980, Western Energy reported that it had submitted a total of 64,103 filings on leases for subscribing clients and had acquired 178 leases. Thus the chance of a Western Energy client winning any given lottery was 1/360. In contrast, the success ratio for all applicants participating in BLM lotteries during 1979 was 1/552. In answering the following questions, assume that these probabilities remain the same.

1 If an individual participates with Western Energy and files 400 separate lease applications, what is his or her probability of winning at least one lease? What is his or her expected number of lease acquisitions in 400 filings through Western Energy?

2 If the individual invests $14,000 privately (without the aid of an investment service) in 1400 separate BLM lotteries, what is his or her probability of winning at least one lease?

3 Because of the cost and technicality of drilling for oil and gas, leases owned by individuals are almost always sold to a commercial organization. Western Energy reported that the average sales price (in 1980 dollars) of leases sold by its clients was $23,715. Find the expected monetary payoff to an individual who pays Western Energy $14,000 for 400 BLM lease filings.

4 Suppose the current market value of all leases sold by individuals to commercial organizations is $12,000. What is the expected monetary payoff to an individual who privately invests $14,000 in 1400 separate BLM lease filings?

CASE STUDY

4 West Coast Container Corporation

The marketing concept is generally defined as a means of organizing the plans and actions of a firm to satisfy the consumer at a profit. Some authors suggest that adoption of the marketing concept results in increased reliance on consumer responses— as opposed to research and development—for new-product ideas. Furthermore, the adoption of this concept implies greater reliance on (costly) marketing research.

For example, consider the case of the West Coast Container Corporation (WCCC), a manufacturer of paperboard containers, with primary service to the food-service industry. In response to requests from customers, WCCC has prepared a new container to protect bulk shipments of taco shells in transit to various food-service operations around the country.

The new container is more costly than competitive containers for transporting taco shells, and the additional cost must be offset by a reduction in product breakage. Thus sales personnel representing WCCC products must be able to provide convincing evidence from independent use tests to encourage adoption of the new container.

To obtain such evidence, the marketing director of WCCC contacted a private research testing firm with a request to provide an accurate estimate of the breakage rate when using the new 500-shell-capacity container. His request was for an estimate of the average number of broken taco shells per container, accurate to within one taco shell, with a probability of .99.

A preliminary report from the testing service stated that an estimate that satisfied the marketing director's requirements would cost $100,000. The testing procedure was fairly complex, involving the shipment of taco shells through the distribution channels customarily used by the industry and then a report of the number of broken shells per container at the point of destination. Using results of preliminary studies, the research firm estimated that testing costs would average about $200 per container.

In its preliminary studies, the testing firm examined the breakage in 20 containers; the results are listed in the following table.

Container	Broken Shells	Container	Broken Shells	Container	Broken Shells
1	29	8	23	15	38
2	25	9	40	16	35
3	18	10	20	17	20
4	28	11	12	18	11
5	23	12	38	19	15
6	14	13	33	20	27
7	27	14	24		

1 Using these preliminary data, would you say that the cost estimate provided by the testing service appears to be reasonable?

2 Suppose the marketing director of WCCC establishes a limit of $20,000 for testing and research for the new containers. How might he compromise his requirements of confidence and accuracy (size of the allowable bound) to accommodate this budget?

CASE STUDY

5 Noodle Soup Brand Confusion

Consumer brand confusion refers to the confusion experienced by customers regarding products available in the marketplace. Sometimes confusion results from intentional imitation strategies adopted by consumer goods companies. Imitation strategies may involve any of the marketing mix elements such as pricing, packaging, promotion, or distribution and are intended to position a product in such a manner that it may be "confused" with a better-known brand.

The courts regularly examine whether an imitation strategy has caused confusion that is damaging to the public or to the product that is the target of the imitation strategy. The Lanham Act of 1946 protects firms from imitators, and numerous court cases over time have upheld the provisions of this act to ensure fair trade in American commerce.

E. Foxman, D. Muehling, and P. Berger ("An Investigation of Factors Contributing to Consumer Brand Confusion," *Journal of Consumer Affairs*, Summer 1990, pages 170–185) examined this issue by surveying shoppers faced with the choice between two brands of noodle soup—a well-known national brand and an "imitator" that is easily confused with the well-known brand. The shoppers were divided into two groups of 30 shoppers each based on whether or not they were confused regarding the identity of the two brands. Each shopper was asked to respond to six issues on a seven-point scale, with higher numbers indicating greater familiarity, usage, experience, involvement, and judgment certainty or indicating a more positive attitude toward advertising. The results of the Foxman, Muehling, and Berger study were as shown in the table.

	Means (Standard Deviations)	
Variable Name	**Confused**	**Not Confused**
Brand familiarity	5.88 (1.49)	6.31 (1.27)
Brand usage	3.65 (1.88)	4.43 (1.93)
Experience with product class	2.53 (.91)	2.85 (.92)
Involvement with product class	3.07 (1.28)	3.50 (1.44)
Certainty of judgment	5.14 (1.65)	6.55 (.84)
Attitude toward advertising	5.35 (1.27)	5.42 (1.16)

1   Which variables provide evidence that a significant difference exists between the shoppers who were confused and those who were not confused regarding the presence of an imitation product?

2   Other than suggesting the filing of a lawsuit, how would you advise the marketing director of the well-known soup brand to protect its brand from infringement by the imitator?

CASE STUDY

6 Custom Precast

Consumer information systems are indispensable elements of consumer policy for the protection and education of consumers about product capabilities. Over the past decade, the emphasis of consumer information programs has been on consumer protection, that is, "public trusteeship." Now the aim of such programs is to "foster a self-reliant, self-actualizing consumer who can make the most of decisions and play an equal role with sellers in the market place." The key here is better information. This case study explores the way one consumer, Custom Precast, dealt with the consumer information it was given.

Custom Precast manufacturers precast concrete products principally for the construction industry. Its product line includes a variety of precast beams for bridges and buildings. In the past, the products of Custom Precast have met acceptable quality standards but have not proved to be superior in strength to those of the competition. In an attempt to gain a competitive advantage, Custom Precast has long sought a way of improving the strength of its precast concrete products. The marketing director for a chemical company claimed that an additive available from his firm will substantially improve the strength of Custom Precast's large beams. To support his claim, the marketing director noted the results of several breakage tests conducted by his firm. If his information is correct, it would prove advantageous to Custom Precast to sign an exclusive licensing agreement with the chemical company for its additive.

To test the claim, Custom Precast obtained sufficient additive from the chemical company to produce 12 test beams. The firm realized that results would be more conclusive with a larger sample size, but costs prohibited experimentation with a larger number of beams. Custom Precast then prepared 12 separate batches of concrete, each large enough to pour two beams. Each batch was then split in half, and the additive was mixed into one of the halves. The beams were then poured, allowed to set, and cured. The process was repeated 12 times, producing 12 pairs of beams. The results of the breakage tests are given in the following table.

Batch	Breaking Strength of Concrete Beams (pounds)		Batch	Breaking Strength of Concrete Beams (pounds)	
	Without Additive	With Additive		Without Additive	With Additive
1	4550	4600	7	5150	5400
2	4950	4900	8	5800	5850
3	6250	6650	9	4900	4850
4	5700	5950	10	6050	6450
5	5350	5700	11	5550	5850
6	5300	5400	12	5750	5600

Comment on the efficiency of the experimental design used by Custom Precast by comparing the two sample variances, s^2 (without pairing) and s_d^2 (with pairing). On the basis of these experimental results, what course of action should Custom Precast take?

C A S E   S T U D Y

7 Exxon's Response to the Regulators

As a result of several major studies conducted by the Environmental Protection Agency, the federal government has issued regulations mandating a gradual phaseout of tetraethyl lead (TEL) for gasoline sold in the United States. Separate legislation developed by the State of California provide even stiffer limits for the TEL content of gasoline.

For years refiners had used TEL as an addition to gasoline as a cheap and convenient way to improve the octane rating of the gasoline, thus reducing the potential for harm to motor vehicle engines. In the absence of TEL, a refinery must reprocess some of the low-octane components of gasoline to increase the octane ratings. Reprocessing can be accomplished either by breaking apart the hydrocarbon chains, through processes known in the trade as "cat cracking" or "hydrocracking," or by rearranging the bonding in the chains, through processes called re-forming or alkylation. All four processes are very costly and efficient but provide more variability in results than the simple addition of TEL to improve the octane rating of a blend.

Faced with the dual impact of federal regulations on TEL and the more stringent California limits, officials at Exxon's Benicia, California, refinery initiated a crash program to determine the most efficient means of addressing the new requirements. Experiments were undertaken using each of the four known methods of increasing octane ratings without exceeding mandated limits on the use of TEL. In their study, gasoline from Exxon's Benicia refinery was reprocessed in such a way that costs were equalized by using each experimental procedure. The data in the table show the octane ratings resulting from the application of the four reprocessing procedures to gasoline derived from each of eight storage tanks.

Storage Tank	Octane Rating When Reprocessing by			
	Cat Cracking	Hydrocracking	Re-forming	Alkylation
1	89.4	88.6	95.5	89.6
2	88.3	91.3	94.0	90.2
3	87.2	88.2	86.7	90.0
4	89.8	89.0	87.5	88.9
5	90.1	88.9	90.7	90.2
6	87.7	88.8	94.8	92.5
7	84.6	86.0	87.3	87.1
8	88.3	89.1	91.5	92.0

Source: L. Golovin, "Product Blending: A Simulation Case Study in Double-Time," *TIMS Interfaces*, November 1979.

1   Do the data suggest that a difference exists in the ability of the four reprocessing procedures to increase octane ratings?

2 Does a difference exist between the octane efficiency of processes that break apart the hydrocarbon chains (cat cracking and hydrocracking) and processes that rearrange the bonding in chains (re-forming and alkylation)?

3 If you were advising Exxon, which reprocessing procedure would you recommend for increasing octane ratings? [*Hint:* Consider the variability of octane rating when using each reprocessing procedure.]

8 Does Statistical Process Control Really Work?

The Ford Motor Company, which produces the various parts for its vehicles in many different locations and brings these parts to central assembly locations, must ensure that these parts are within specification limits in order that the parts be assembled into a working, nondefective entity with a minimum of defects. As part of its ongoing statistical process control, one of Ford's problem-solving teams selected a process used to harden the fuel pump eccentric of a 3.8-liter, V6 engine camshaft. The process produced inconsistent case hardness depth, causing 12% rework and 9% scrap. Excessive drill bit breakage occurred at the next operation, in which an oil hole was to be drilled near the hardened fuel pump eccentric.

The study group decided initially to sample parts from the hardening production process and construct $\overline{x}$ and R charts of the case hardness depth at the fuel pump eccentric lobe "nose." The process was automated; however, the electric coil used in the hardening process (coil A) could be adjusted. Thirty consecutive samples of $n = 5$ pieces were recorded, as well as any changes or repairs to the process during this time. The $\overline{x}$ and R charts, together with the trial control limits, are given in Figure 1.

At point A, the power on the coil was increased from 8.2 to 9.2; at point B, the team discovered and straightened a bent coil; at point C, power on the coil was reduced to 8.8. At point D, the coil shorted out and needed to be straightened; at this time the team devised a gauge to check the coil spacing to the camshaft. At point E, the team decreased the spacing between the camshaft and the coil; and at point F the first coil (A) was replaced with a second coil (B) of the same type.

1 How would you interpret the results of the actions taken at points A, B, C, D, E, and F?

2 After coil B was installed, the $\overline{x}$ and R charts in Figure 2 resulted. Do these charts indicate that the process has been stabilized? If specification limits are 3.5 to 10.5 mm, is the process capable of producing *individual* parts that will be within the specification limits?

3 Figure 3 was plotted after a third redesigned coil was installed. Is the process in control? Is the process capable of producing parts within the specification limits? Has statistical process control been effective in this study?

FIGURE **1**
Initial sampling for coil A: fuel pump eccentric nose (case study)

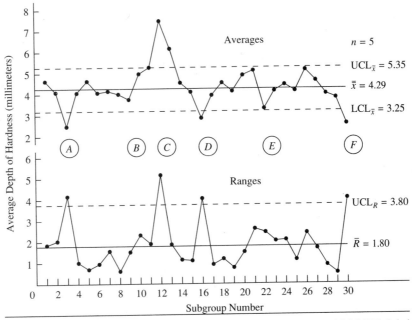

Source: Redrawn with permission from James C. Seigel, "Managing with Statistical Models," SAE Technical Paper No. 820520, Society for Automotive Engineers, Inc. (Warrendale, PA), 1982.

FIGURE **2**
Sampling for coil B: fuel pump eccentric nose (case study)

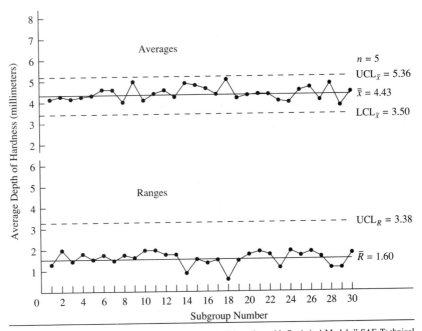

Source: Redrawn with permission from James C. Seigel, "Managing with Statistical Models," SAE Technical Paper No. 820520, Society for Automotive Engineers, Inc. (Warrendale, PA), 1982.

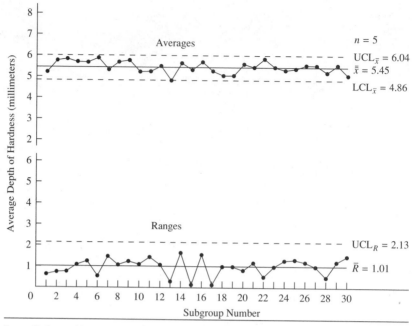

FIGURE **3**
Sampling for coil C: fuel pump
eccentric nose (case study)

Source: Redrawn with permission from James C. Seigel, "Managing with Statistical Models," SAE Technical
Paper No. 820520, Society for Automotive Engineers, Inc. (Warrendale, PA), 1982.

CASE STUDY

9 The Wood Stove Industry

Wood stoves have become increasingly popular for home heating since the advent of
higher energy prices. In many urban areas where wood stoves have become popular,
they have contributed to local air pollution, thus drawing considerable attention
from governmental agencies charged with monitoring air quality. Proposals have
been offered to legislative bodies stating that wood stoves, like automobiles, should
require pollution control devices or standards.

Some members of the wood stove industry have become concerned that imposed
standards might necessitate difficult and expensive modifications that would make
their products less competitive in the market for home heating equipment. The claim
is made that a substantial reduction in the air pollution produced by a wood stove can
be made by a slight modification of most stoves and by informing wood stove users
of better methods for operating their stoves.

As a result of concerns over the impact of such regulation, one wood stove manufacturer has undertaken research to determine what product or operational modifications would minimize the air pollution produced by its brand of wood stoves. The
research involved the measurement of particulate matter, as observed by a photoelec-
tric cell, shown by the gases venting out of a wood stove chimney. This method, used
in conjunction with a measure of air flow out of the stack, provided a relative measure of particulate matter in terms of the percentage of light blocked by (or passing
through) the venting gases.

Since the manufacturer's product could be used with two sizes of flue pipe, the experiment involved varying the air intake setting (1/4, 1/2, 3/4, and fully open) and the flue size, and measuring the temperature and the relative level of particulate matter exiting the stack. The experiment was conducted by first building a fire in the stove and then closing the stove and adjusting the air intake valve to the appropriate level. After half an hour (to allow the fire to stabilize at the set air flow level), measurements of temperature and particulate matter were recorded, as shown in the following table.

Observation Number	Relative Particulate Matter Concentration	Air Intake Setting	Flue Size	Flue Temperature
1	44	1/4	S	106
2	28	1/2	S	248
3	26	3/4	S	385
4	31	Open	S	534
5	42	1/4	L	124
6	26	1/2	L	211
7	29	3/4	L	374
8	34	Open	L	487
9	42	1/4	S	131
10	28	1/2	S	230
11	27	3/4	S	353
12	36	Open	S	515
13	40	1/4	L	144
14	26	1/2	L	255
15	27	3/4	L	286
16	33	Open	L	517
17	42	1/4	S	117
18	27	1/2	S	235
19	25	3/4	S	378
20	34	Open	S	510
21	39	1/4	L	139
22	30	1/2	L	248
23	29	3/4	L	302
24	34	Open	L	521

Analyze these data by the development of an appropriate multiple regression model relating relative particulate matter concentration to the air intake setting, flue size, and flue temperature. Interpret your results.

CASE STUDY

10 Household Energy Conservation and Expenditures

Since the oil embargo was imposed in the early 1970s, American consumers have become more sensitive to the importance of energy conservation and to the identification of energy sources for which the supply and price are reasonably stable. The following table lists the energy consumption and expenditures by American households for the years 1980, 1984, 1987, and 1990.

Application/Fuel Source	Consumption (quadrillion Btu)				Expenditures ($ billion)			
	1980	1984	1987	1990	1980	1984	1987	1990
Space heating								
Natural gas	3.32	3.51	3.38	3.37	$12.80	$20.66	$18.05	$18.59
Electricity	.28	.30	.28	.30	3.71	5.71	5.53	6.16
Distillate fuel oil and kerosene	1.32	1.10	1.05	.94	10.59	8.51	6.25	7.42
Liquefied petroleum gases	.25	.21	.22	.19	1.90	2.00	1.85	2.01
Total	5.17	5.13	4.93	4.80	29.00	36.85	31.68	34.18
Air conditioning								
Electricity	.32	.36	.44	.48	5.07	7.51	9.77	11.19
Water heating								
Natural gas	1.24	1.10	1.10	1.16	4.79	6.63	6.02	6.59
Electricity	.31	.32	.31	.34	4.54	6.44	6.45	7.21
Distillate fuel oil and kerosene	.24	.15	.17	.11	1.89	1.09	.94	.83
Liquefied petroleum gases	.07	.06	.06	.06	.59	.58	.50	.64
Total	1.86	1.62	1.64	1.67	11.80	14.76	13.91	15.27
Appliances								
Natural gas	.38	.35	.34	.33	1.71	2.31	2.02	2.03
Electricity	1.55	1.53	1.72	1.91	26.82	34.95	39.83	46.95
Liquefied petroleum gases	.04	.04	.04	.03	.41	.54	.46	.48
Total	1.97	1.92	2.10	2.27	28.94	37.81	42.33	49.46
Total	9.32	9.04	9.11	9.22	74.81	97.00	97.75	110.10
Natural gas	4.94	4.98	4.83	4.86	19.30	29.80	26.15	27.21
Electricity	2.46	2.48	2.76	3.03	40.14	54.50	61.58	71.51
Distillate fuel oil and kerosene	1.55	1.26	1.22	1.04	12.48	9.60	7.21	8.25
Liquefied petroleum gases	.36	.31	.32	.28	2.89	3.10	2.81	3.13

Source: U.S. Department of Energy, Energy Information Administration, *Annual Energy Review*, 1992. (Released June 1993.)

1 Suppose that it is of interest to compare household energy costs over the four reference years included in the table. Which index would you recommend using?

2 Using 1980 as the base year, compute the Laspeyres indexes for space heating expenditures for 1984, 1987, and 1990.

3 Using 1980 as the base year, compute the Laspeyres indexes for total household energy expenditures for 1980, 1984, 1987, and 1990.

4 What limitation exists when using the Laspeyres indexes as energy expenditure indexes with this study?

CASE STUDY

11 Clark County Port District

Clark County is a large western county with a population of approximately 300,000. The economic base of the county is provided by agriculture, wood products, commercial fishing, tourism, and some light industry. The largest community in Clark

County is its county seat, which has a population of approximately 125,000. Most of the remaining population of the county is located in smaller communities, which vary distinctly in character. The communities in the central-valley area serve as agricultural distribution centers; wood products are the dominant industry in those communities that lie in the mountainous area to the east. Most of the small communities on the coast rely on tourism and the commercial fishing industry. The economy of the county seat has elements of all these industries, with additional diversification offered by the state university, the county government offices, and most of the county's nonwood products manufacturing.

County officials are now considering the formation of a Clark County Port District to encourage the location of new industry, which might serve to further diversify the county's economic base. However, before a port district can be formed, the issue requires approval from the county's registered voters. And many residents have expressed concerns that economic diversification might result in uncontrolled population growth and danger to the county's aesthetic qualities. In addition, attitudes appear to be dividing along political party lines, with the local Republican organization favoring economic diversification, and the Democrats opposing it. Thus approval of the port district proposal is uncertain.

The county manager feels that a survey should be taken to explore voter attitudes and concerns regarding economic diversification before proceeding with a full-scale promotional campaign in support of a port district. A list of all registered voters in the county is available from the office of the county elections director. A brief description of each of the seven state legislative districts represented in Clark County is given in the following table.

District	Number of Registered Voters	Predominant Economic Base	Predominant Political Affiliation
17	11,650	Fishing, tourism	Republican
18	13,509	Agriculture	Republican
19	9684	Forest products	Democratic
21	15,235	Light manufacturing	Independent
22	14,476	Forest products, light manufacturing	Republican
27	13,695	Forest products	Democratic
32	19,128	Government, light manufacturing	Democratic

Develop a sampling plan to assist the county manager. In your discussion, consider the following aspects of your sampling design:

1   the definition of the population of interest

2   the definition of the sampling elements, units, and frame

3   monetary and time constraints

4   the reason for your particular choice of a sampling design

5   the limitations of the sampling design you have chosen

6   the method to be used in determining an appropriate sample size

7 potential biases and their effect on interpreting the results of the survey

8 sampling errors that may result

9 assumptions you have made in the selection of your chosen sampling design

C A S E   S T U D Y

12 The Gray Flannel Suit Syndrome

Does a woman's attire affect her climb up the corporate ladder? As late as a decade ago, women in executive positions were attired in somber business suits and blouses with bows at the neck. As women achieved more security in executive positions through their performance within the corporate structure, media reports indicated that successful female executives could now look more feminine. However, interviews with many male and female executives indicated that even with relaxed dress codes in the business workplace, it is still not acceptable for women to appear too feminine in the workplace.

According to the *Wall Street Journal*, this result does not mean that a dark suit is the only acceptable professional dress; tailored dresses and skirts worn with a jacket are also acceptable business attire for women.[†] Apparently, a woman whose dress is "extremely feminine gives the message that she needs to be taken care of" or that she is not as serious as someone dressing more conservatively. This dual standard concerning dress codes seems to have little to do with actual performance. Nonetheless, "women whose clothes were described as conservative were *twice* as likely to receive promotions as those whose dress was labeled as frilly, frivolous, or sexy."

As part of a survey conducted by John T. Malloy, 298 women in corporate sales positions were classified according to their perceived job performance and their type of dress. The survey results are given in the following table.

Job Performance	Attire			
	Very Professional, Very Conservative	**Appropriate**	**Sexy, Frilly, Fashionable**	**Poorly Dressed**
Top performers	12.5%	12.6%	11.5%	1.9%
Consistently above average	60.9	41.1	31.0	30.8
Average or below	21.9	44.2	54.0	50.0
Failing	4.7	2.1	3.4	17.3

Source: John T. Malloy, *Wall Street Journal*, September 1, 1987. Reprinted by permission of the *Wall Street Journal.* ⓒ Dow Jones & Company, Inc., 1987. All rights reserved.

1 Suppose that, of the 298 women surveyed, 92 were classified as very professionally dressed, 95 dressed appropriately, 59 dressed in a sexy, frilly, or fashionable

[†]Kathleen A. Hughes, "Business Women's Broader Latitude in Dress Codes Goes Just So Far," *Wall Street Journal*, September 1, 1987. Reprinted by permission of the *Wall Street Journal.* ⓒ Dow Jones & Company, Inc., 1987. All rights reserved.

manner, and 52 were poorly dressed. Does the information provided in the table appear to support the contention that job performance evaluations for women in corporate positions depend upon their choice of attire?

2   What is the statistical interpretation of the comment that women whose clothes were described as conservative were twice as likely to receive promotions as women whose dress was labeled as frilly, frivolous, or sexy?

CASE STUDY

13 Market Proxies: How Well Do They Perform?

In the capital asset pricing model (CAPM) for monitoring investment performance, an asset's measured risk factor depends upon the proxy (an observed index or measure of CAPM) selected as well as other security-specific characteristics. One viewpoint is that the choice of different proxies can lead to opposite rankings of portfolios; others contend that errors in the approximating index are unimportant in the long run.

In investigating whether a reasonable estimate of CAPM is likely to lead to invalid conclusions, Keith C. Brown and Gregory D. Brown evaluated the historical performance of publicly traded investment portfolios relative to six different market proxies (or indexes). In formulating these market proxies, Brown and Brown obtained the historical rates of return and the market values of five different classes of market assets covering the 32-year period from 1947 to 1978. These five classes were (1) common stock, (2) fixed-income corporate issues, (3) real estate, (4) U.S. government issues, and (5) municipal bonds. Brown and Brown created five market proxies (indexes), summarized as follows:

Index 1: common stock

Index 2: index 1 plus fixed-income corporate issues

Index 3: index 2 plus real estate

Index 4: index 3 plus U.S. government issues

Index 5: index 4 plus municipal bonds

Of those mutual funds in operation during this time period, the final selection of 32 funds shown in Table 1 included an equal representation of Weisenberger's (Weisenberger Financial Services, *Investment Companies*, 1987) four investment classifications: growth, income, growth and income, and balanced investments.

Using each of the five indexes, the authors ranked these 32 funds according to measures that reflect the excess return on investments when compared with the risk-free returns on investments. The rankings for each fund by each of the five indexes and a ranking of each fund according to its average rate of return are given in Table 2.

Evaluate Spearman's rank correlation coefficient for all pairs of the five indexes, as well as for each index, and the mean return for each of the 32 mutual funds. What can you conclude about the value of these indexes as they relate to one another and to the mean return of these 32 mutual funds?

TABLE 1
Mutual funds data

Fund Number	Code*	Name	Annual Return (%)	
			Mean	**Standard Deviation**
1	GI	Affiliated Fund, Inc.	10.65	14.74
2	I	Axe-Houghton Income Fund, Inc.	9.52	15.90
3	G	Axe-Houghton Stock Fund, Inc.	9.94	19.74
4	I	Century Shares Trust	10.37	20.11
5	G	Chemical Fund, Inc.	10.13	20.54
6	GI	The Colonial Fund, Inc.	9.77	16.25
7	B	Composite Bond & Stock Fund, Inc.	7.72	11.52
8	GI	Delaware Fund, Inc.	7.98	17.53
9	B	Dodge & Cox Balanced Fund	8.04	12.16
10	B	Eaton & Howard Balanced Fund	7.53	10.74
11	GI	Fidelity Fund, Inc.	11.46	17.21
12	G	Growth Industry Shares, Inc.	10.68	18.98
13	GI	The Investment Company of America	12.05	17.05
14	GI	Investment Trust of America	10.69	18.05
15	B	Investors Mutual, Inc.	7.30	11.63
16	I	Investors Selective Fund, Inc.	5.25	5.78
17	G	Johnston Mutual Fund, Inc.	9.68	15.22
18	B	Loomis-Sayles Mutual Fund, Inc.	7.37	11.12
19	G	Massachusetts Investors Growth Stock Fund	11.21	19.08
20	I	Mutual Investing Foundation—MIF Fund	9.72	15.82
21	G	National Investors Corporation	11.92	18.58
22	G	National Growth Fund	10.46	20.74
23	B	Nation-Wide Securities Company, Inc.	8.07	10.94
24	I	Puritan Fund, Inc.	11.19	15.79
25	B	The George Putnam Fund of Boston	9.07	13.42
26	G	Scudder Common Stock Fund, Inc.	9.72	17.85
27	I	Scudder Income Fund, Inc.	6.76	12.08
28	GI	Selected American Shares, Inc.	8.89	16.31
29	GI	State Street Investment Corporation	10.80	15.73
30	I	United Income Fund	10.17	16.50
31	B	Wellington Fund	7.13	11.18
32	I	Wisconsin Income Fund, Inc.	8.36	15.04

Source: Keith C. Brown and Gregory D. Brown, "Does the Composition of the Market Portfolio *Really* Matter?" *Journal of Portfolio Management*, Winter 1987, p. 28. Reprinted with permission.
* Weisenberger classifications: G, growth; I, income; GI, growth and income; and B, balanced.

TABLE **2**
Performance rankings data

Fund Number*	Index 1	Index 2	Index 3	Index 4	Index 5	Index Average	Mean Return
1	3	4	6	10	3	5.2	9
2	8	9	9	4	6	7.2	19
3	11	13	5	6	10	9.0	14
4	7	8	4	8	12	7.8	11
5	12	14	3	5	7	8.2	13
6	16	15	10	12	8	12.2	15
7	15	20	11	16	19	16.2	26
8	32	30	32	32	32	31.6	25
9	25	23	19	21	22	22.0	24
10	21	22	13	15	20	18.2	27
11	5	5	14	11	5	8.0	3
12	20	19	28	26	16	21.8	8
13	1	1	7	2	2	2.6	1
14	18	11	23	13	11	15.2	7
15	27	31	22	25	26	26.2	29
16	14	21	1	9	13	11.6	32
17	10	10	16	19	14	13.8	18
18	22	24	17	20	25	21.6	28
19	13	7	25	7	9	12.2	4
20	17	16	21	22	15	18.2	17
21	4	3	8	3	4	4.4	2
22	23	25	31	31	31	28.2	10
23	9	12	12	14	17	12.8	23
24	2	2	2	1	1	1.6	5
25	19	17	15	18	21	18.0	20
26	28	26	30	30	30	28.8	16
27	31	32	26	27	29	29.0	31
28	30	28	29	29	28	28.8	21
29	6	6	18	17	18	13.0	6
30	24	18	24	24	23	22.6	12
31	26	27	20	23	24	24.0	30
32	29	29	27	28	27	28.0	22

Source: Keith C. Brown and Gregory D. Brown, "Does the Composition of the Market Portfolio *Really* Matter?" *Journal of Portfolio Management*, Winter 1987, p. 30. Reprinted with permission.
* Fund numbers are as shown in Table 1.

PRIMARY DATA SOURCES

"American Voices," *American Demographics*, December 1990, p. 14.

Amirrezvani, Anita. "Help Is on the Way... But When?" *PC World*, Vol. 12, No. 6, June 1994, pp. 133–138.

"Antilock Brakes Have Not Reduced Car Crashes," *Consumers' Research*, Vol. 77, No. 3, March 1994, p. 14.

Arthur, Caroline. "Paper or Plastic?" *American Demographics*, December 1990, p. 14.

Automotive News: 1994 Market Data Book, p. 22.

Barnum, H. F., and Simutis, Z. M., "Computer-Based Graphics for Terrain Visualization Training," *Human Factors*, Vol. 26, 1984. Copyright 1984 by the Human Factors Society, Inc. Reproduced by permission.

Batsell, Jake. "Definitely a Major Decision," *U.: The National College Magazine*, 1994, p. 19.

Braus, Patricia. "HMOs Get High Marks from Their Members," *American Demographics*, February 1994, p. 17.

"Byte Windows Application and Low-level Benchmarks for Windows, OS/2, and NT," *Byte*, November 1993, p. 88.

Chappell, Lindsay. "Toyota Spreads Gospel of Efficiency," *Automotive News*, January 24, 1994, p. 16.

"Consumer Confidence in Economy Slips," *Press-Enterprise*, Riverside, California, July 27, 1994, p. C-6.

Craig, David. "Texaco Joins in Oil-field Dieting," *USA Today*, July 6, 1994, p. B1.

CRB Commodity Yearbook 1994, Knight-Ridder Financial/Commodity Research Bureau, New York: John Wiley & Sons, Inc., 1994.

"Crunching Numbers: New Data on How We Eat," *Consumer Reports*, July 1994, p. 431.

Darnay, Arsen J., ed. *Statistical Record of Older Americans*. Detroit, Washington, D.C., London: Gale Research, Inc., 1994.

Dholakia, Ruby Roy. "Even PC Men Won't Shop by Computer," *American Demographics*, May 1994, p. 11.

Dortch, Shannon. "What's Good for the Goose May Gag the Gander," *American Demographics*, May 1994, p. 15.

Duncan, Joseph W., and Handler, Douglas P. "The Misunderstood Role of Small Business," *Business Economics*, Vol. XXIX, No. 3, July 1994, p. 7.

Energy Facts 1992, U.S. Department of Energy, Energy Information Administration, 1992, p. 18.

Engelhardt, Gary V., and Mayer, Christopher J. "Gifts for Home Purchase and Housing Market Behavior," *New England Economic Review*, May/June 1994, p. 51.

EPA Journal, July/August 1992.

Eskey, Kenneth. "Sexes Near Parity in Medical School Entry," *Press-Enterprise*, Riverside, California, November 4, 1992, p. A13.

Exter, Thomas. "Kid's Stuff," *American Demographics*, October 1990, p. 8.

Felix, W. L. Jr., and Roussey, R. S. "Statistical Inference and the IRS," *Journal of Accountancy*, June 1985.

Filimon, V.; Maggass, R.; Frazier, D.; and Klingel, A. "Some Applications of Quality Control Techniques in Motor Oil Can Filling," *Industrial Quality Control*, Vol. 12, No. 2, 1955. Copyright 1955, American Society for Quality Control. Reprinted by permission.

Fox, Richard J., and Stephenson, Frederick J. "How to Control Corporate Air Travel Costs," *Business*, July-September 1990, pp. 3–9.

"Freshman Statistics," *U.: The National College Magazine*, 1994, p. 11.

Galifianakis, Nick, and Baumann, Marty. "How USA Feels About Term Limits," *USA Today*, June 28, 1994, p. 1A.

Gallup, George, Jr. *The Gallup Poll: Public Opinion 1993*, Wilmington, Delaware: Scholarly Resources, Inc., 1994.

"Gallup Short Subjects," *The Gallup Poll Monthly*, January 1994, p. 34.

"Getting Things Fixed," *Consumer Reports*, Vol. 59, No. 1, January 1994, pp. 34–42.

Gross, Neil, and Brandt, Richard. "Sony Has Some Very Scary Monsters in the Works," *Business Week*, May 23, 1994, p. 116.

Hamilton, Leslie. "Performance of Area Savings and Loans," *Press-Enterprise*, Riverside, California, October 12, 1990.

Holmes, Steven A. "Census to Survey More, Count Less for Year 2000," *Press-Enterprise*, Riverside, California, May 16, 1994, p. A-8.

Horovitz, Bruce. "A Frankly Tough Challenge," *Los Angeles Times*, June 28, 1994, p. D1.

"Housing Opportunity: Where Can You Afford?" *Consumers' Research*, May 1994, p. 30.

"How the Japanese See Themselves…and Us," *The American Enterprise*, Vol. 1, No. 6, November/December 1990, p. 82.

"Is College Worth It?" *U.: The National College Magazine*, 1994, p. 18.

"Japan Carmakers Lose Edge," *USA Today*, July 6, 1994, p. 2B.

Kaplan, Rachel. "The Gender Gap at the PC Keyboard," *American Demographics*, January 1994, p. 18.

Kerschner, Edward M., and Ryan, Michael P. "Paine Webber Viewpoint," June 8, 1994, p. 3.

Lynn, Frances M., and Kartez, Jack D. "Environment Democracy in Action: The Toxics Release Inventory," *Environmental Management*, Vol. 18, No. 4, 1994, p. 511.

Mayer, Christopher J., and Simons, Katerina V. "A New Look at Reverse Mortgages: Potential Market and Institutional Constraints," *New England Economic Review*, March/April 1994, p. 15.

"Midwest, South Most Affordable, Group Says," *Press-Enterprise*, Riverside, California, July 22, 1994, p. F-1.

Miko, Chris John, and Weilant, Edward, eds. *Opinions '90 Cumulation*. Detroit and London: Gale Research, 1991.

"Money Market Summary," *Wall Street Journal*, July 14, 1994, p. C27.

Moore, David W., and Gallup, Alec. "Are Women More Sexist Than Men?" *The Gallup Poll Monthly*, September 1993, No. 336, p. 20.

"Most Restaurant Meals Are Bought on Impulse," *American Demographics*, February 1994, p. 16.

Newport, Frank, and McAneny, Leslie. "The Right to Drive?" *The Gallup Monthly Poll*, October 1993, p. 34.

"Ozone Damage Gets Worse in Yosemite," *Press-Enterprise*, Riverside, California, November 24, 1990.

Palmer, Jay. "Attractive Ratios: Companies with Below-Average Multiples of Earnings and Book," *Barron's*, November 5, 1990, p. 18.

"Parental Leave," *Journal of Accountancy*, November 1990, p. 23.

Parrish, Michael. "Thinking Cheaper on Electric Cars," *Los Angeles Times*, June 28, 1994, p. D-1.

"Passenger Van Bumpers Allow Costly Damage," *Consumers' Research*, April 1994, pp. 22–23.

Phelps, David; McEnroe, Paul; and Byrne, Carol. "Pilots Confused About Location Before Collision, Officials Say," *Press-Enterprise*, Riverside, California, December 5, 1990, p. A-7.

Press-Enterprise, Riverside, California, February 11, 1993 (E-8); May 26, 1994 (F-3); June 23, 1994 (F-7).

Prodis, Julia. "Honda to Increase U.S. Production," *Press-Enterprise*, Riverside, California, July 20, 1994, p. D1.

"Profile of Smokers in the U.S.," *Press-Enterprise*, Riverside, California, November 14, 1990.

"Ratings: Garbage Bags," *Consumer Reports*, Vol. 59, No. 2, February 1994, p. 110.

"Ratings: Interior Latex Paints," *Consumer Reports*, Vol. 59, No. 2, February 1994, p. 130.

"Ratings: Refrigerators," *Consumer Reports*, Vol. 59, No. 2, February 1994, p. 84.

"Ratings: 27-inch TV Sets," *Consumer Reports*, Vol. 59, No. 3, March 1994, p. 160.

"Ratings: VCRs," *Consumer Reports*, Vol. 59, No. 3, March 1994, p. 164.

Rezaee, Zabihollah, and Elmore, Robert C. "The Need for Enhanced Defense Contractor Cost Management Systems," *Public Budgeting and Finance*, Vol. 13, No. 4, Winter 1993, p. 45.

Rickard, Leah. "Minorities Show Brand Loyalty," *Advertising Age*, May 9, 1994, p. 29.

"The ROB/D&B Poll," *Report on Business Magazine*, September 1993, p. 16.

"Saturn Workers Get Bonuses," *Automotive News*, January 24, 1994, p. 2.

Savage, David G. "High Court OKs Congress' Right to Regulate Cable TV," *Los Angeles Times*, June 28, 1994, p. A1.

Schwartz, Joe. "Will Baby Boomers Dump Department Stores?" *American Demographics*, December 1990, p. 42.

Seligman, Daniel. "The Road to Monte Carlo," *Fortune*, April 15, 1985, p. 157.

Smeltzer, L. R., and Watson, K. W. "A Test of Instructional Strategies for Listening Improvement in a Simulated Business Setting," *Journal of Business Communication*, Vol. 22, No. 2, 1985.

Stang, Patti, and Galifianakis, Nick. "Where's the Beef (or Chicken)?" *USA Today*, June 28, 1994, p. 1D.

Stokley, Sandra. "Envelope Please: The Winner Is Norco's Water," *Press-Enterprise*, Riverside, California, July 17, 1994, p. B-4.

Sullivan, Christopher. "Main Street USA Takes on Wal-Mart," *Press-Enterprise*, Riverside, California, August 16, 1993, p. C1.

"Survey Said. . . ," *Business Marketing*, June 1994, p. 42.

Tannenwald, Robert. "Massachusetts' Tax Competitiveness," *New England Economic Review*, January-February 1994, p. 45.

"The Times Poll: LAPD Approval Rating," *Los Angeles Times*, June 28, 1994, p. A19.

"Top 10 Windows Accelerators: Diamond Steals No. 1 Spot," *PC World*, Vol. 12, No. 6, June 1994, p. 175.

"Upward Mobility," *American Demographics*, April 1989, p. 13.

"Used Computer Prices," *PC Source*, June 1992, p. 96.

"Using Your College Planning Report," *The College Board*, Princeton, New Jersey, 1994, p. 6.

U.S. Department of Commerce, Bureau of the Census. In *The World Almanac and Book of Facts, 1994*. New York: Funk & Wagnalls Corporation, 1994.

U.S. Department of Commerce, Bureau of the Census. *Statistical Abstract of the United States, 1993,* 113th ed. Washington, D.C.: Government Printing Office, 1994.

U.S. Department of Commerce, Bureau of Economic Analysis. *Business Statistics, 1963–1991*. Washington, D.C.: 1992.

U.S. Department of Commerce, Bureau of Economic Analysis. *Survey of Current Business* (various issues). Washington, D.C.: 1991–1994.

Value Line Investment Survey, editions 2 and 8, Vol. XLIX, Nos. 34 and 41, 1994.

Waldrop, Judith. "Leaves of Gold," *American Demographics*, October 1990, p. 4.

"What Should Top 1994 Agenda?" *The Gallup Poll Monthly*, January 1994, p. 14.

"Who's Tops in Service?" *PC World*, November 1992, p. 198.

"You Aren't Paranoid If You Think Someone Eyes Your Every Move," *Wall Street Journal*, March 19, 1985.

Answers to Selected Exercises

Exercises

Chapter 2

2.2	**a**	discrete	**b**	continuous	**c**	discrete
	d	discrete	**e**	continuous		

2.4 less profitable

2.8 **a** 8–10 classes   **c** 43/50   **d** 33/50

2.12 **b** 36/50

2.14 **b** 2.5

2.16 **a** digits at or to the left of the tens digit
c roughly symmetric, not mound-shaped; 682; 646

2.22 **b** $\bar{x} = 2$; $m = 1$; mode = 1   **c** skewed

2.24 **a** 5.8   **b** 5.5   **c** 5 and 6

2.26 **a** skewed   **b** $\bar{x} = .5705$; $m = .49$; mode = .29

2.30 **a** $\bar{x} = 9.1$ **b** $m = 9.5$

2.32 **a** $\bar{x} = 2.4$ **b** $s^2 = 2.8$; $s = 1.673$ **c** $s \approx 1.6$

2.34 **a** 2   **b** 1.581   **c** 79.1

2.36 \$286 to \$442

2.38 **a** at least 3/4 in the interval 15.84 to 16.00; at least 8/9 in the interval 15.80 to 16.04
b approximately 68% in the interval 15.88 to 15.96; approximately 95% in the interval 15.84 to 16.00; almost all in the interval 15.80 to 16.04
c no

2.40 **a** $s \approx 3.5$ **b** $\bar{x} = 6.22$; $s = 3.497$ **c** 96%

2.42 **a** $s^2 = .078095$; $s = .279$ **b** MAD = .22;   **c** Tchebysheff's Theorem

2.44 **a** $s \approx 30.9$ **b** $\bar{x} = 26.23$; $s = 26.22$

c

k	$\bar{x} \pm ks$	Proportion in Interval
1	.01 to 52.45	.88
2	−26.21 to 78.67	.94
3	−52.43 to 104.89	.98

yes; no

2.46 **b** as x increases, y increases
2.54 z-score $= 1.55$
2.56 **a** $\bar{x} = 1842.48$; $s = 1506.0607$
2.58 z-score $= 1.43$; no
2.60 inner fences: -1 and 15; outer fences: -7 and 21; 22 is an extreme outlier
2.62 **a** inner fences: -19.2 and 60.0; outer fences: -48.9 and 89.7
          b 69.0 and 79.0 are mild outliers
2.64 **a** qualitative   **b** quantitative   **c** qualitative   **d** quantitative
2.66 **a** skewed   **b** symmetric   **c** symmetric   **d** skewed
2.70 **a** $\bar{x} = 3$; $m = 2.5$; no mode   **b** $R/4 = 1.75$; $R/2.5 = 2.8$
          c MAD $= 2$; $s^2 = 6.8$; $s = 2.60768$ **d** CV $= 86.92$

2.74

k	$\bar{x} \pm ks$	Proportion in Interval
1	4.23 to 41.55	.87
2	−14.43 to 60.21	.93
3	−33.09 to 78.87	.98

2.76 **a** $\bar{x} = 2.652$; $s = .502$

b–c

k	$\bar{x} \pm ks$	Proportion in Interval
1	2.150 to 3.154	.68
2	1.648 to 3.656	.96
3	1.146 to 4.158	1.00

yes; yes

2.78 **a** at least 75%   **b** approximately .16
2.80

k	$\bar{x} \pm ks$	Empirical Rule
1	3.279 to 6.441	$\approx .68$
2	1.698 to 8.022	$\approx .95$
3	.117 to 9.603	≈ 1.00

2.82 **b** $\bar{x} = .66$; $s = 1.387$ **c** .95; .96; yes; yes
2.84 **a** at least 3/4 of the CDs between $53 and $85
          b at most 1/4 of the CDs exceeding $85
2.86 **a** quantitative, multivariate data

Chapter 3

3.2 assignment is valid

3.4 $P(E_4) = .2, P(E_5) = .1$

3.6 **b** $P(E_i) = 1/38$ **c** $P(A) = 1/19$ **d** 9/19

3.8 **a** .84   **b** .12   **c** .21

3.10 **b** $E_1 : (1, 2, 3), E_2 : (1, 3, 2), E_3 : (2, 1, 3), E_4 : (2, 3, 1), E_5 : (3, 2, 1), E_6 : (3, 1, 2)$
    c 1/3, 1/3

3.12 **a** 1   **b** 2/5   **c** 4/5   **d** 2/5
    e 3/5   **f** 1/5   **g** 1/5   **h** 1/5
    i 1/4   **j** 1   **k** 3/5   **l** 1/2

3.14 **a** .35   **b** .70   **c** .25   **d** .80
    e no   **f** .3571   **g** .7143   **h** no

3.16 **a** .55   **b** .39   **c** .545

3.18 **a** .09   **b** .0675

3.20 **a** .43   **b** yes   **c** .0795   **d** .4191

3.22 **a** 2/35   **b** 34/35   **c** 12/35   **d** .3529

3.24 .1905; .7143; .0952

3.26 .25

3.28 .6667

3.30 **a** .4; .6   **b** .9   **c** 1.85

3.32 **a** 3.45; 2.0475; 1.4309   **c** $P(.59 < x < 6.31) = .95$ **d** yes

3.34 **a** 7.9   **b** 2.1749   **c** .96

3.36 $2050

3.38 **b** 1/2   **c** 2/3   **d** 2/3

3.42 **a** .0005   **b** .0002   **c** .3685

3.44 **a** discrete   **b** continuous   **c** discrete
    d discrete   **e** discrete

3.46 **a** .0081   **b** .4116   **c** .2401

3.48 **a** $p(3) = .2$

3.50 **a** no; yes; yes   **b** 1/2   **c** 1; 0   **d** no; no

3.52 13,800.3882

3.54 $.39

3.56 **a** .5997   **b** .0002   **c** .0061

Chapter 4

4.2 binomial; $n = 2, p = 3/5$

4.4

		$p(x)$	
x	$p = .2$	$p = .5$	$p = .8$
0	.32768	.03125	.00032
1	.40960	.15625	.00640
2	.20480	.31250	.05120
3	.05120	.31250	.20480
4	.00640	.15625	.40960
5	.00032	.03125	.32768

4.6   **a**   .995   **b**   .562   **c**   .788

4.8   .312

4.10   **a**   .748   **b**   .610   **c**   .367

       d   .966   **e**   .656

4.12   **a** $\mu = 300, \sigma = 14.49$ **b** $\mu = 4, \sigma = 1.99$ **c** $\mu = 250, \sigma = 11.18$

       d $\mu = 1280, \sigma = 16$

4.14 $p = .5$

4.16   binomial experiment

4.18   **a**   0   **b**   .885   **c**   .097

4.20   **a** $p(x) = C_x^{50}(.71)^x(.29)^{50-x}$ **b** $p(x) = C_x^{50}(.42)^x(.58)^{50-x}$

       c $\mu = 21, \sigma = 3.490$

       d   no; $x = 5$ lies more than 4 standard deviations below the mean

4.22   **a**   .125; .031; .001   **b**   machine is not working as expected

       c   when the run becomes long, stop and check the machine

4.24   **a**   .301194   **b**   .3614   **c**   .8132   **d**   .3374

4.28   **a**   .423   **b**   .632   **c**   .271

4.30   .3679; .3679; .1839; .6321

4.32   **a**   .073; .092

       b   no; $x = 9$ lies 2.5 standard deviations above the mean

4.34   **a**   .085; .125   **b**   no; $P(x > 10) = .014$

4.36   **a** $p(0) = .5965, p(1) = .3579, p(2) = .0447, p(3) = .0009$

4.38   .0163

4.40   **a**   7/210   **b**   35/210   **c**   175/210

4.42   **a** $p(0) = .125, p(1) = .375, p(2) = .375, p(3) = .125$

       c   1.5; .866

       d

k	$\mu \pm k\sigma$	Probability
1	.63 to 2.37	.75
2	−.23 to 3.23	1.00

4.44   **a**   .251   **b**   .092   **c**   .205

4.46   **a**   .737   **b**   .263   **c**   .328

       d   .409   **e**   .269

4.48   .03125; .18750

4.50 **a** $\mu = 11.25$ **b** $\sigma = 1.933$
     c yes; $x = 9$ lies 1.16 standard deviations below the mean
4.52 **a** 3,333,333.33   **b** 1490.71
     c no; $x = 3,300,000$ lies more than 22 standard deviations below the mean
4.54 .0186
4.56 **a** .3874   **b** .6513   **c** .2638
4.58 **a** 805   **b** 22.875   **c** 759.251 to 850.749
     d no; 249 lies more than 24 standard deviations below the mean
4.60 **a** .983   **b** .736   **c** .392   **d** .069
4.62 .021
4.64 **a** .001   **b** no
4.66 **a** .387   **b** .651   **c** .736
4.68 no; $x = 4$ lies more than 2 standard deviations above the mean

Chapter 5

5.2 **a** .3159   **b** .3159
5.4 **a** .8384   **b** .9544   **c** .9974
5.6 1.96
5.8 $-.53$
5.10 .63
5.12 1.645
5.14 **a** .0401   **b** .1841   **c** .2358
5.16 $\mu = 9.2$
5.18 $\mu = 8, \sigma = 2$
5.20 .1562; .0012
5.22 **a** .1335   **b** .8665   **c** .1335
5.24 **a** .0808   **b** $1820 and $3780
5.26 **a** .390   **b** .4049
5.28 .1711
5.30 .3520
5.32 **a** .245   **b** .2483
5.34 **a** .0409   **b** .3121   **d** yes
5.36 **a** ≈ 0 **b** 250 to 286
5.38 **a** .0239   **b** .9693
5.40 **a** .3849   **b** .3159
5.42 **a** .4279   **b** .1628
5.44 **a** .3227   **b** .1586
5.46 **a** .0730   **b** .8623
5.48 .9115
5.50 .36
5.52 1.65
5.54 -1.30
5.56 1.10

5.58   .1596

5.60   .8612

5.62   **a**   .1056   **b**   .8944   **c**   .1056

5.64   .16

5.66   .0344

5.68   **a**   .421   **b**   .4013

5.70   .8980

5.72   **a**   .0668; .0107   **b**   48% is too large

5.74   **a** $\sigma = 357.14$ **b**   .3594   **c**   .1492

5.76   **a**   1/2   **b**   1/4

5.78   **a**   .3   **b**   14,990 gallons

5.80   .9929

5.82 $\mu = 7.301$

5.84   **a**   .0778   **b**   .0274

5.86 $\mu = 1940.119$

Chapter 6

6.2   **a**   normal
          b   approximately normal for parts (a) and (b)

6.4   **c** $\mu = 3.5; \sigma_{\bar{x}} = .765$

6.6   the standard error decreases as the sample size increases

6.8   **a** $\mu = 1; \sigma_{\bar{x}} = .161$ **b**   .0314   **c** ≈ 0 **d**   .0132

6.12 $\mu = 1500; \sigma_x = 35.355$

6.16   **a**   normal with mean $\mu, \sigma_{\bar{x}} = 2/\sqrt{10}$
           b   .0571   **c**   21.948

6.18   **a** $\mu_{\hat{p}} = .3; \sigma_{\hat{p}} = .0458$ **b** $\mu_{\hat{p}} = .1; \sigma_{\hat{p}} = .015$ **c** $\mu_{\hat{p}} = .6; \sigma_{\hat{p}} = .0310$

6.20   **b**   .9198

6.22   **a**   .0099   **b**   .03   **c**   .0458   **d**   .05
           e   .0458   **f**   .03   **g**   .0099

6.24   **a**   .0668   **b**   .9544

6.26   **a**   approximately normal with $p =$ proportion of people whose home cost more than $112,000 and $\sigma_{\hat{p}} \approx .03162$
           b ≈ 0
           c   either the sample is not random or the national median home price may have changed

6.28   **a**   UCL $= 21.57$; LCL $= 19.91$

6.30   **a**   UCL $= 12,905.3$; LCL $= 8598.7$

6.32   UCL $= .0357$; LCL $= .0155$

6.36 $p(\bar{x}) = 1/4$ for $\bar{x} = 2, 3, 3.333, 3.667$

6.38   **a**   40   **b**   .4   **c**   .0062

6.40   .0217

6.42   **a** $\mu = 31,256; \sigma_{\bar{x}} = 310$
           b   yes, because of the Central Limit Theorem
           c   .0082; ≈ 0 **d** $30,636 to $31,876

6.44 **a** $p = .86; \sigma_{\hat{p}} = .00975$ **b** yes; yes **c** .3030

Chapter 7

7.2 **a** .620 **b** .186 **c** .960
7.4 **a** (12.496, 13.704) **b** (2.651, 2.809) **c** (28.280, 28.920)
7.6 **a** (32.550, 35.450) **b** (1047.543, 1050.457) **c** (65.973, 66.627)
7.8 **a** decreases by $\dfrac{1}{\sqrt{2}}$ **b** decreases by $\dfrac{1}{2}$
7.10 7.2% with margin of error $= .776\%$
7.12 **a** ($32,404, 32,796) **b** $24,100; 156.8 **c** ($22,722.60, 23,095.40)
7.14 (36.213, 47.987)
7.16 **a** 2.015 **b** 2.306 **c** 1.330 **d** ≈ 1.960
7.18 **a** $\bar{x} = 6.333; s = .568038$ **b** (5.866, 6.800) **c** (5.737, 6.929)
7.20 **a** (8.367, 8.633) **b** no; yes
7.22 **a** $\bar{x} = 516.857; s = 106.550$ **b** (450.707, 583.007)
     c (446.329, 639.035) **d** yes
7.24 .690
7.26 **a** 22 **b** 20 **c** 16
7.28 $s^2 = 12.4571$
7.30 (0.020, 4.780)
7.32 **a** (6.449, 7.951) **b** (1.762, 3.238)
7.34 $(-7415.45, -1784.55)$
7.36 (.690, .766)
7.38 (.241, .319)
7.40 **a** (.786, .814) **b** (.562, .598)
     c $\hat{p} = .56$ with margin of error .018
7.42 **a** (.222, .258) **b** .020
7.44 **a** .031 **b** .018; $\hat{p}$ is different **c** (.340, .400)
7.46 **a** $(-.203, -.117)$ **b** random and independent samples
7.48 **a** $(-.235, -.063)$ **b** random and independent samples
7.50 **a** (.009, .031) **b** random and independent samples
7.52 **a** $(-.121, -.059)$
     b $\hat{p}_1 - \hat{p}_2 = .02$ with margin of error .018
7.54 505
7.56 $n_1 = n_2 = 1086$
7.58 $n_1 = n_2 = 347$
7.60 1201
7.62 **a** 13,979 **b** smaller
7.68 (28.298, 29.902)
7.70 (3.867, 4.533)
7.72 **a** $.48 \pm .044$ **b** (.443, .517)
7.74 $(-.0157, .2907)$
7.76 (132.8445, 311.7805)

7.78 $(-9.441, -.359)$

7.80 a $.028; .027$

  b $\hat{p}_1 - \hat{p}_2 = -.21$ with margin of error $.039$

7.82 $(14,700.109, 14,899.891)$

7.84 44

7.86 $(.608, .772)$

7.88 $\hat{p}_1 - \hat{p}_2 = .11$ with margin of error $.0879$

7.90 6147

7.92 a $(-7.588, -1.496)$

  b yes, since $\mu_1 - \mu_2 = 0$ is not in the interval

7.94 a $\bar{x} = 4172.857; s = 1841.5831$ b $(2469.615, 5876.099)$

7.96 $(36.254, 41.946)$

7.98 92

7.100 the 44.7 driving range is probably incorrect

Chapter 8

8.2 $H_0: \mu = 2.9; H_a: \mu < 2.9$; one-tailed

8.4 a $H_a: \mu < 84$ b $H_0: \mu = 84$ c $.05$ d   no; $z = -.436$

8.6 a $H_a: \mu > 10$ b $H_0: \mu = 10$

  c reject H_0 if $t > 1.638$ d   do not reject H_0; $t = 1.602$

8.8 do not reject H_0; $t = 1.361$

8.10 $H_a: \mu > 5; H_0: \mu = 5$; test statistic: $t = .80$; rejection region: $t > 2.764$; do not reject H_0

8.12 a $H_a: \mu > 4.8$ c $H_0: \mu = 4.8$

  e reject H_0 if $z > 2.33$ f   do not reject H_0; $z = .16$

8.14 no; do not reject H_0; $t = -.647$

8.16 a $.0268$ b   reject H_0

8.18 a p-value $> .20$ d   do not reject H_0

8.20 reject H_0 if p-value $< .05$

8.22 $.4364$

8.24 a reject H_0; $t = -2.750$ b $.01 < p$-value $< .025$

8.26 a $H_0: \mu_1 - \mu_2 = 0; H_a: \mu_1 - \mu_2 > 0$ b   one-tailed

  c $z > 1.28$ e   reject H_0; $z = 2.087$

  c reject H_0; $z = 2.087$

8.28 b $H_0: \mu_1 - \mu_2 = 0; H_a: \mu_1 - \mu_2 \neq 0$

8.30 a $H_a: \mu_1 - \mu_2 > 0$ b $H_0: \mu_1 - \mu_2 = 0$ c $t > 1.440$

  d reject H_0; $t = 2.043$ e $.025 < p$-value $< .05$

8.32 a yes; reject H_0; $z = -3.807$ b   conclusion is the same

8.34 do not reject H_0; $t = 1.22$

8.36 $t = 2.372$; reject H_0; $.02 < p$-value $< .05$

8.38 62

8.40 $(.114, .146)$

8.42 a no; $t = 3.041$, do not reject H_0 b $.02 < p$-value $< .05$

  c $(.061, .567)$

8.44 a no; $t = .471$ b p-value $> .10$ c $(-.527, .763)$

8.46 **a** yes; $t = -1.90$; reject H_0 **b** $(-1233, -37)$

8.48 **a** $H_a: p > .6, H_0: p = .6$ **b** one-tailed   **c** yes; $z = 1.734$; reject H_0

8.50 p-value $= .0143$; reject H_0

8.52 **a** $H_0: p = .05, H_a: p < .05$ **b** one-tailed

      d $z = -1.538$; do not reject H_0

8.54 **a** $H_0: p = .6, H_a: p < .6$ **b** $z = -1.732$; reject H_0

      c $(.471, .609)$ **d** 9220

8.56 **a** $H_0: p_1 - p_2 = 0; H_a: p_1 - p_2 < 0$ **b** one-tailed

      c do not reject H_0; $z = -.84$

8.58 **a** $H_0: p_1 - p_2 = 0; H_a: p_1 - p_2 > 0$ **b** do not reject H_0; $z = 1.210$

8.60 **a** yes; reject H_0; $z = 3.49$ **b** yes; p-value $< .002$

8.62 **a** no; $z = .88$, do not reject H_0 **b** p-value $= .3788$

8.64 $(.190, .685)$

8.66 $(9.089, 25.617)$

8.68 **a** no; $F = 2.316$ **b** $.05 < p$-value $< .10$

8.70 $(.227, 2.194)$

8.72 $(1.408, 31.264)$

8.74 **a** $\sigma_1^2 = \sigma_2^2$ **b** no; $F = 5.897$

8.82 **a** $H_a: \mu < 60, H_0: \mu = 60$ **b** one-tailed

      c yes; $z = -1.992$; reject H_0

8.84 $\hat{p} \leq .1342$

8.86 yes; $t = 2.108$; $.025 < p$-value $< .05$

8.88 yes; $F = 3.268$

8.90 **a** no; $z = -.915$; do not reject H_0 **b** no

      c $(1.259, 1.481)$ **d** $(-.196, .056)$

8.92 **a** $H_0: p = .2, H_a: p > .2$ **b** $\alpha = .0749$

8.94 $(-10.246, -2.354)$

8.96 no; paired analysis is not appropriate

8.98 **a** $.02 < p$-value $< .05$ **b** reject H_0; $t = 2.497$

8.100 yes; reject H_0; $t = 2.096$

8.102 $\chi^2 = 12.6$; do not reject H_0

8.106 no; $\chi^2 = 7.008$

8.108 no; $t = -1.8$

8.110 **a** $H_0: \mu = 1100; H_a: \mu < 1100$ **b** $z < -1.645$

      c yes; $z = -1.897$

8.112 yes; reject H_0; $z = 1.768$

8.114 **a** Poisson

8.116 yes; reject H_0; $z = -16.0$

8.118 **a** yes; reject H_0; $t = -2.42$ **b** $.02 < p$-value $< .05$

8.120 **a** no; do not reject H_0; $t = -1.95$ **b** $(1.584, 3.282)$

8.122 no

8.124 no; do not reject H_0; $t = 1.862$

8.126 **a** yes; reject H_0; $t = -2.347$ **b** $.01 < p$-value $< .02$

c $(-2.753, -.247)$

8.128   **a** $\sigma = 4.167$

b   yes; reject H_0 ; $\chi^2 = 37.374$

Chapter 9

9.2   **a**

Source	d.f.	SS	MS	F
Treatments	5	5.2	1.04	3.467
Error	54	16.2	.30	
Total	59	21.4		

b $\nu_1 = 5, \nu_2 = 54$ **c** $F > 2.37$ **d**   yes; $F = 3.467$

9.4

Source	d.f.
Treatments	3
Error	20
Total	23

9.6   **a** $(86.181, 89.819)$ **b** $(1.528, 6.672)$

9.8   **a** $CM = 103.142857$; Total SS $= 26.8571$ **b**   SST $= 14.5071$; MST $= 7.2536$

c   SSE $= 12.3500$; MSE $= 1.1227$ **f** $F > 3.98$

g $F = 6.46$; reject H_0

9.10   **a** $(1.95, 3.65)$ **b** $(.27, 2.83)$

9.12   no; $t = -1.66$

9.14   **a**   yes; $F = 13.0$ **b** $(-.122, -.028)$

9.16   **b**   yes; $F = 8.11$ **c**   yes; $t = 3.025$

9.20

Source	d.f.	SS	MS	F
Treatments	2	11.4	5.70	4.01
Blocks	5	17.1	3.42	2.41
Error	10	14.2	1.42	
Total	17	42.7		

9.22 $(-3.833, -.767)$

9.24

Source	d.f.
Treatments	5
Blocks	3
Error	15
Total	23

9.26   yes; $F = 4.69$

9.28 $F = 2.82$ with $.05 < p$-value $< .10$; reject H_0 at $\alpha = .10$ but not at $\alpha = .05$

9.30   **a** $CM = 1064.0833$; Total SS $= 148.9167$ **b**   SST $= 25.5833$; MST $= 8.5278$

c   SSB $= 120.6667$; MSB $= 60.3333$ **d**   SSE $= 2.6667$; MSE $= .4444$

f   yes; $F = 19.19$ **g**   yes; $F = 135.75$ **h**   yes

9.32 **a** CM = 159.414; Total SS = 6.496   **b** SST = 4.476; MST = 2.238

    c SSB = 1.796; MSB = .449   **d** SSE = .224; MSE = .028

    f yes; $F = 79.93$ **g** yes; $F = 16.04$ **h** yes

9.34 **b** yes; $F = 22.03$ **c** yes; $t = 6.551$

9.36 **b** no   **c** (14.2335, 16.0415)

9.40 **a** 20   **b** 60

    c

Source	d.f.
A	3
B	4
AB	12
Error	40
Total	59

9.42 **a**

Source	d.f.	SS	MS	F
A	2	5.3	2.6500	1.30
B	3	9.1	3.0333	1.49
AB	6	4.8	.8000	.39
Error	12	24.5	2.0417	
Total	23	43.7		

    b no; $F = .39$ **c** no; $F_A = 1.30$; $F_B = 1.49$

9.44 **a** no; $F = 1.4$ **c** yes; $F = 6.51$ **d** yes; $F = 7.37$

9.46 **a**

Source	d.f.	SS	MS
A	2	3.1111	1.5556
B	2	81.4444	40.7222
AB	4	62.2222	15.5556
Error	9	21.0000	2.3333
Total	17	167.7778	

    c yes; $F = 6.67$ **d** $.005 < p\text{-value} < .01$

    e (10.057, 14.943)   **f** (4.045, 10.955)

9.48 **b** yes; $F = 8.48$ **c** .0001   **d** (−.438, 3.504)

9.50 **b** yes

    c since the interaction is significant, attention should be focused on means for the individual factor-level combinations

9.54 **a** 5.06   **b** 3.88   **c** 6.20   **d** 9.32

9.56 **a** $\omega = 6.78$ **b** $\overline{x}_4 \ \overline{x}_2 \ \overline{x}_1 \ \overline{x}_5 \ \overline{x}_3 \ \overline{x}_6$

9.58 $\omega = .631$; $\overline{x}_{11} \ \overline{x}_{22} \ \overline{x}_{12} \ \overline{x}_{21}$

9.60 $A_1 B_2 \ A_3 B_2 \ A_2 B_2 \ A_4 B_2 \ (\omega = 2.63)$

9.62 $A_3 B_1 \ A_3 B_2 \ A_2 B_1 \ A_2 B_2 \ A_1 B_1 \ A_1 B_2 \ (\omega = 23.61)$

9.64   **a**   yes; $F = 5.20$ **b**   no; $t = .88$ **c** $(-.579, -.117)$
9.66   **b**   no; $F = 1.53$ **c** $(21.456, 25.744)$ **d** $(-3.814, 2.964)$
9.68   **a**   yes; $F = 19.44$ ($F_{.05} = 4.76$)   **b**   yes; $F = 40.21$ ($F_{.05} = 5.14$)
      c $(-1.997, -.803)$
9.70   **a**   randomized block   **c**   yes; $F = 5.917$ ($F_{.05} = 4.10$)
9.72   **a**   yes; $F = 6.46$ ($F_{.05} = 5.14$)   **b**   no; $F = 3.75$ ($F_{.05} = 4.76$)
      c $(-1.85, -.85)$
9.74   **a**

Source	d.f.	SS	MS	F
Treatments	2	.01084	.00542	1.86
Error	12	.03496	.00291	
Total	14	.04580		

      b   no; $F = 1.86$ ($F_{.05} = 3.89$)   **c**   no

Chapter 10

10.2   **a** $\hat{\sigma} = 5.353$ **c**   LCL = 63.196; UCL = 77.564
10.4   UCL = 26.332; LCL = 0
10.6   UCL = 13,588.875; LCL = 0
10.8   UCL = .0270; LCL = .0242; $\bar{\bar{x}} = .0256$
10.10   **a**   UCL = 7.516; LCL = 6.964   **b**   UCL = .69525; LCL = 0
10.12   **a**   LCL = 0; UCL = .090
      b   centerline at .035; LCL and UCL as in part (a)
10.14   **a**   LCL = 0; UCL = 3.210   **b**   centerline at $\bar{c} = .7$
10.16   centerline at $\bar{\hat{p}} = .021$; UCL = .043; LCL = 0
10.18   centerline at $\bar{c} = 4.9$; UCL = 11.54; LCL = 0
10.20   **a**   .076   **b**   .043   **c**   .117   **d**   .004
10.22   **a**   5   **b**   3   **c**   6   **d**   8
10.24 $\mu_r = 28.27$; $\sigma_r = 3.643$
10.26   do not reject H_0; $z = 0$; no evidence of nonrandomness
10.28   do not reject H_0; $r = 5$ with $\alpha = .143$
10.30   **a**   .590   **b**   .168   **c**   .031   **d**   1   **e**   0
10.32   **a**   .349   **b**   .028   **c**   .001   **d**   1   **e**   0
10.34   for a given value of n and p, P(accept) increases as a increases; for a given value of a and p, P(accept) decreases as n increases
10.36   .004
10.38 $n = 80$; $a = 0$
10.44   **a**   .7495   **b**   2505
10.46 $\approx .78$, using the Poisson approximation
10.48   **a**   LCL = 0; UCL = .081   **b**   .1894
10.50   **a** $(25, 5)$ **b** $(25, 5)$
10.52 $a = 1$
10.54 $n = 200$; $a = 0$

Chapter 11

11.2 y intercept $= 1$; slope $= -2$

11.4 same line, but with negative slope; lines are perpendicular

11.6 $y = x - 3$

11.8 **a** $\hat{y} = 3.0 + 1.2x$

11.10 **a** $\hat{y} = 201.433 - 30x$ **b** $\hat{y} = 3.433$

11.12 **b** the point $(518, 68)$ may be an outlier   **c** $\hat{y} = -17.896 + 2.455x$

   d $\hat{y} = 473.1$

11.14 $(.657, 1.743)$

11.16 $(-.653, -.461)$

11.18 yes; reject H_0; $t = 3.76$

11.20 **a** yes; reject H_0; $t = 2.69$ **b** see Section 11.2; yes

11.22 **a** $\hat{y} = 63.86 + .4736x$; no, do not reject H_0; $t = 0.75$

   b $\hat{y} = 116.15 - .3773x$; yes, reject H_0; $t = -3.05$

   c median home price

11.24 $(4.667, 5.105)$

11.26 $(16.51, 23.62)$

11.28 **a** $\hat{y} = 1071.42 - 28.765x$ **c** yes; reject H_0; $t = -9.54$ **d** p-value $< .01$

   e $(586.045, 636.321)$ **f** $(518.09, 704.276)$

11.30 **a** $\hat{y} = 648 + 2049.5x$ **b** yes; reject H_0; $t = 18.29$ **c** yes

   d $(35,650, 39,428)$ **e** $(34,163, 40,916)$

11.34 **a** positive   **b** $r = .9487$; $r^2 = .9000$

11.36 **a** positive   **b** $r = .9822$ **c** 96.47%

11.38 **a** $r = .8376$ **b** $r^2 = .7016$

11.40 **a** positive   **b** yes; reject H_0; $t = 19.92$

11.42 **a** y intercept $= -2$; slope $= -2/3$

11.44 **a** y intercept $= 0$; slope $= 2$

11.46 **a** $\hat{y} = 2.643 + .554x$ **c** yes; reject H_0; $t = 3.618$

   d $(.161, .947)$ **e** $(1.209, 2.969)$

   f $(1.900, 5.600)$ **g** $r = .851$ **h** 72.36%

11.48 $\hat{y} = .0667 + .51667x$

11.50 **a** $\hat{y} = 57,885.4 - 27.9744x$ **c** no; do not reject H_0; $t = -1.04$

   d $(-77.0864, 21.1376)$ **f** $(918.326, 3346.76)$

11.52 **b** $\hat{y} = -4.58706 + 1.31176x$ **c** yes; $t = 4.18$

   d $r^2 = .6357$ **e** $(91.494, 109.213)$

   f $(82.285, 144.658)$

11.54 **a** yes; reject H_0; $t = 5.48$ **b** $(33.57, 42.83)$

11.56 $r^2 = .789$; the fit is reasonably good

Chapter 12

12.2 **b** parallel lines

12.4 **b** yes; $F = 57.44$

12.6 **a** $\hat{y} = 1.04 + 1.29x_1 + 2.72x_2 + .41x_3$ **b** parallel lines

12.8 **a** quadratic **b** $R^2 = .762$ **c** yes; $F = 27.214$

12.10 **a** $E(y) = 1.21$ **b** yes; $t = 1.952$ **c** $(.131, 2.289)$

12.12 **b** yes; reject H_0: $\beta_2 = 0$ since $t = -2.848$

12.14 **b** $R^2 = .9955$ **c** $R^2_{adj} = .9925$

12.16 **a** $E(y) = \beta_0 + \beta_1 x_1 + \beta_2 x_2$ **b** $\hat{y} = 81.6 + .356 x_1 + .591 x_2$

  c $R^2 = .992$ **d** yes; $F = 426.37$ **e** p-value $= .000$

  f models used in Exercise 11.58 were almost as effective

12.18 **a** quantitative **b** quantitative **c** qualitative

  d quantitative **e** qualitative

12.20 **a** $\hat{y} = -5125 + 1763.9 x_1 + 9533.3 x_2$ **b** $F = 170.74$ with p-value $= .000$

  c all

  d $\hat{y} = 26,625.2$; $\hat{y} = 36,158.5$; rounding error

12.22 **a** yes; $F = 23.375$

  b K/L, AGR, and SEV are significant predictor variables

  c PY

12.24 **a** SSE $= 152.17748$; $s^2 = 8.4543$ **b** 18

  c yes; $F = 31.85$ **d** p-value $< .005$

  e $S_{yy} = 1498.625$; $R^2 = .898$ **f** $\hat{y} = 41.262$

12.30 **b** 9 **d** $R^2 = .98397$

  f $\hat{y} = 4.10 + 1.04 x_1 + 3.53 x_2 + 4.76 x_3 - .43 x_1 x_2 - .08 x_1 x_3$

  g yes; $t = -2.613$ **h** $(-.802, -.058)$ **i** no; $t = -.486$

12.32 $E(y) = \beta_0 + \beta_1 x + \beta_2 x^2$

12.34 $E(y) = \beta_0 + \beta_1 x_1 + \beta_2 x_2$ where

$$x_1 = \begin{cases} 1, & \text{if level 2} \\ 0, & \text{otherwise} \end{cases} \qquad x_2 = \begin{cases} 1, & \text{if level 3} \\ 0, & \text{otherwise} \end{cases}$$

12.36 $E(y) = \beta_0 + \beta_1 x_1 + \beta_2 x_2 + \beta_3 x_3 + \beta_4 x_1^2 + \beta_5 x_2^2 + \beta_6 x_3^2 + \beta_7 x_1 x_2 + \beta_8 x_1 x_3 + \beta_9 x_2 x_3$

12.38 $E(y) = \beta_0 + \beta_1 x_1 + \beta_2 x_2 + \beta_3 x_3 + \beta_4 x_1 x_3 + \beta_5 x_2 x_3$ where

$$x_1 = \begin{cases} 1, & \text{if level 2, factor A} \\ 0, & \text{otherwise} \end{cases} \qquad x_2 = \begin{cases} 1, & \text{if level 3, factor A} \\ 0, & \text{otherwise} \end{cases}$$

$$x_3 = \begin{cases} 1, & \text{if level 2, factor B} \\ 0, & \text{otherwise} \end{cases}$$

Chapter 13

13.2 $I_{80} = 102.71$; $I_{81} = 146.90$; $I_{82} = 217.88$; $I_{83} = 275.84$; $I_{84} = 455.55$; $I_{85} = 500.80$; $I_{86} = 538.85$; $I_{87} = 536.95$; $I_{88} = 499.47$; $I_{89} = 440.53$; $I_{90} = 528.62$

13.4 $I_2 = 214.86$; $I_3 = 196.01$; $I_4 = 189.86$

13.6 $I_2 = 226.01$; $I_3 = 201.21$; $I_4 = 194.82$

13.8 **a** $I_{90} = 214.74$; $I_{94} = 289.92$ **b** $I_{90} = 206.90$; $I_{94} = 281.45$

  c $I_{90} = 207.13$; $I_{94} = 279.95$

13.20 **a** $\hat{T}_t = 836.6 + 87.702t$; $R^2 = 83.9\%$ **b** cyclic effect is present

  c $\hat{y}_{27} = 3525.01$ (using $\hat{C}_{27} \approx 1.1$)

13.26 **b** $E(y_t) = \beta_0 + \beta_1 t + \beta_2 x_1 + \beta_3 x_2 + \beta_4 x_3$ with $x_i = 1$ if quarter i, 0 if not, for $i = 1, 2, 3$

  c from part (b), $y_t = E(y_t) + z_t$, where $z_t = \phi z_{t-1} + \epsilon_t$

d $\hat{y}_t = 70.30000 + 1.7045t - 14.6078x_1 - 6.9717x_2 - 7.4449x_3 + .5586\hat{z}_{t-1}$

13.28 $\hat{y}_{18} = 6.8354$

13.30 **c** $\hat{y}_t = 11.1427 - .2037t + .00182t^2 + .33766\hat{z}_{t-1}$

13.32 $I_{80} = 69.03; I_{81} = 82.19; I_{82} = 100.00; I_{83} = 113.27; I_{84} = 114.38; I_{85} = 111.39;$
$I_{86} = 100.22; I_{87} = 93.81; I_{88} = 93.14; I_{89} = 95.13; I_{90} = 96.57; I_{91} = 94.14;$
$I_{92} = 96.57$

13.34 $I_{88} = 98.11; I_{89} = 101.45; I_{90} = 103.42; I_{91} = 102.98; I_{92} = 105.17$

13.36 **b** $E(y_t) = \beta_0 + \beta_1 t$

  c from part (b), $y_t = E(y_t) + z_t$, where $z_t = \phi z_{t-1} + \epsilon_t$

  d $\hat{y}_t = 2.051739 + 0.180217t - 0.14021658\hat{z}_{t-1}$

13.38 **a** $\hat{y}_t = 11.02788 - .19534t + .001703t^2 + .26093\hat{z}_{t-1} + .22725\hat{z}_{t-2}$

  c MSE = .31838; slightly better

13.40 **b** $\hat{y}_{47} = 6.541$

Chapter 14

14.12 8410 ± 216.879

14.16 $525,408 \pm 19,170.662$

14.18 6751 ± 400.415

14.20 $6.633 \pm .523$

14.22 408.143 ± 3.743

14.24 $.298 \pm .0443$

14.26 $13,208.633 \pm 549.2741$

14.28 $.4073 \pm .0627$

14.30 $421,543 \pm 44,997.546$

14.32 $2.472 \pm .02148$

14.34 $6.129 \pm .5472$

14.36 $.5048 \pm .0599$

14.38 $45,097,986.75 \pm 7,944,903.77$

14.42 $12.376 \pm .189$

14.44 $.4589 \pm .0554$

14.46 $11,914,614 \pm 344,056.0727$

14.48 $2.0167 \pm .2213$

14.50 $.303 \pm .0513$

14.52 $.4 \pm .1142$

Chapter 15

15.2 **a** 7.81473 **b** 20.0902 **c** 22.3072 **d** 17.2750

15.4 **a** 4 **b** $X^2 > 9.4877$

  c $H_a: p_i \neq p_j$ for some pair $i, j \ (i \neq j)$ **d** $X^2 = 8.00$; do not reject H_0

  e $.05 < p\text{-value} < .10$

15.6 yes; $X^2 = 24.48 \ (\chi^2_{.05} = 7.81)$

15.8 yes; $X^2 = 173.64 \ (\chi^2_{.05} = 12.59)$

15.10 no; do not reject H_0; $X^2 = 2.718 \ (\chi^2_{.01} = 11.3449)$

15.12   8

15.14   **a**   yes; $X^2 = 6.49$ ($\chi^2_{.05} = 5.99$)   **b** $(-.195, .013)$

15.16   **a**   no; do not reject H_0; $X^2 = 3.809$ ($\chi^2_{.05} = 3.84$)

     b $.05 < p\text{-value} < .10$; yes

15.18   reject H_0; $X^2 = 17.597$ ($\chi^2_{.05} = 15.51$)

15.20   yes; reject H_0; $X^2 = 6.274$

15.22   **a** $X^2 = 5.322$ **b** $X^2 > 5.99$

     c   do not reject H_0 **d** $.05 < p\text{-value} < .10$

15.24   yes; $X^2 = 7.193$ ($\chi^2_{.05} = 5.99$)

15.26   **a**   no; $X^2 = 2.917$ **b** $p\text{-value} > .10$

     c   estimated expected cell counts < 5

15.28   no; $X^2 = 10.4$ ($\chi^2_{.05} = 11.07$)

15.30   **a** $H_0: p_1 = .2, p_2 = .05, p_3 = .03, p_4 = .72$

     b H_a: the probabilities are different from those in H_0

     c $X^2 = 6.588$ **d** $X^2 > 6.251$ **e**   reject H_0

15.32   yes; $X^2 = 27.17$ ($\chi^2_{.10} = 6.25$)

15.34   reject H_0; $X^2 = 680.397$ ($\chi^2_{.05} = 9.49$)

15.36   no; $X^2 = 5.826$ ($\chi^2_{.01} = 20.09$)

15.38   no; $X^2 = 12.915$ ($\chi^2_{.05} = 15.5$)

Chapter 16

16.2   **b** $\alpha = .004, .014, .044, .108$

16.4   **b** $\alpha = .008, .036, .118$

16.6   **a**   no   **b** $p\text{-value} = .054$

16.8   **a** $H_0: p = .5$ **b** $H_a: p \neq .5$ **c**   reject H_0; $x = 8$ **d** $p\text{-value} = .0390625$

16.10   **a** $H_0: p = .5$; $H_a: p > .5$

     b $z = (x - .5n)/\sqrt{.25n} = .169$; reject H_0 if $z > 1.645$

     c   do not reject H_0 **d** $p\text{-value} = .4325$

16.12   no; do not reject H_0; $x = 8$

16.14   yes; reject H_0; $x = 8$

16.16   yes; $z = 4.04$

16.18   **a**   the smaller of U_A or U_B **b** $U \leq 10$ **c** $\alpha = .0812$

16.20   **a**   the smaller of U_A or U_B **b** $U \leq 2$ **c** $\alpha = .0634$

16.22   do not reject H_0; $z = -.77$

16.24 $U_B = 12.5$, reject H_0; rejection region: $\{0, 1, \ldots, 21\}$ for $\alpha = .1012$

16.26   **b** $U_B = 29.5$, do not reject H_0; rejection region: $U \leq 24$ with $\alpha = .0524$

16.28   **a** H_0: the two population distributions are identical; H_a: the two population distributions differ in location

     b $\min(T^+, T^-)$ **c** $T \leq 137$ **d** $T^- = 216$; do not reject H_0

16.30 $z = -.34$; do not reject H_0

16.32   **a** $t = 3$; rejection region: $\{T \leq 4\}$; reject H_0

     b   consistent with results of Exercise 8.43

16.34   no; do not reject H_0; $T = 18.5$

16.36   **a**   no; do not reject H_0; $T = 4$ **b** p-value $\approx .10$

16.38   yes; $H = 13.39$ ($\chi^2_{.05} = 5.99$)

16.40   **a**   reject H_0; $H = 9.064$ ($\chi^2_{.05} = 7.81$)   **b** $.025 < p$-value $< .05$

          c p-value $= .0009$ **d**   results are identical

16.42   reject H_0; $H = 8.405$ ($\chi^2_{.10} = 4.605$)

16.44   **a**   reject H_0; $F_r = 21.19$ ($\chi^2_{.05} = 7.81$)   **b** p-value $< .005$

          d $F = 75.43$ **e** p-value $< .005$

16.46   **a**   randomized block   **b**   yes; $F_r = 8.33$ ($\chi^2_{.05} = 5.99$)

16.48   **a** $r_s \leq -.497$ **b** $r_s \leq -.703$

16.50   **a** $r_s = -1$ **b**   yes; reject H_0 for $|r_s| \geq .886$

16.52 $-.845$

16.54   **a**   do not reject H_0; $x = 2$; rejection region: {0, 1, 8, 9} for $\alpha = .040$

          b   do not reject H_0 $t = -1.65$; rejection region: $|t| \geq 2.306$ for $\alpha = .05$

16.56   **a**   no; $U = 32$; rejection region: $U \leq 21$ for $\alpha = .094$

          b   no; $t = .30$

16.58   **a**   no; do not reject H_0; $x = 5$ **b**   no; do not reject H_0; $T = 5.5$

16.60   no; do not reject H_0; $F_r = 4.75$ ($\chi^2_{.05} = 5.99$)